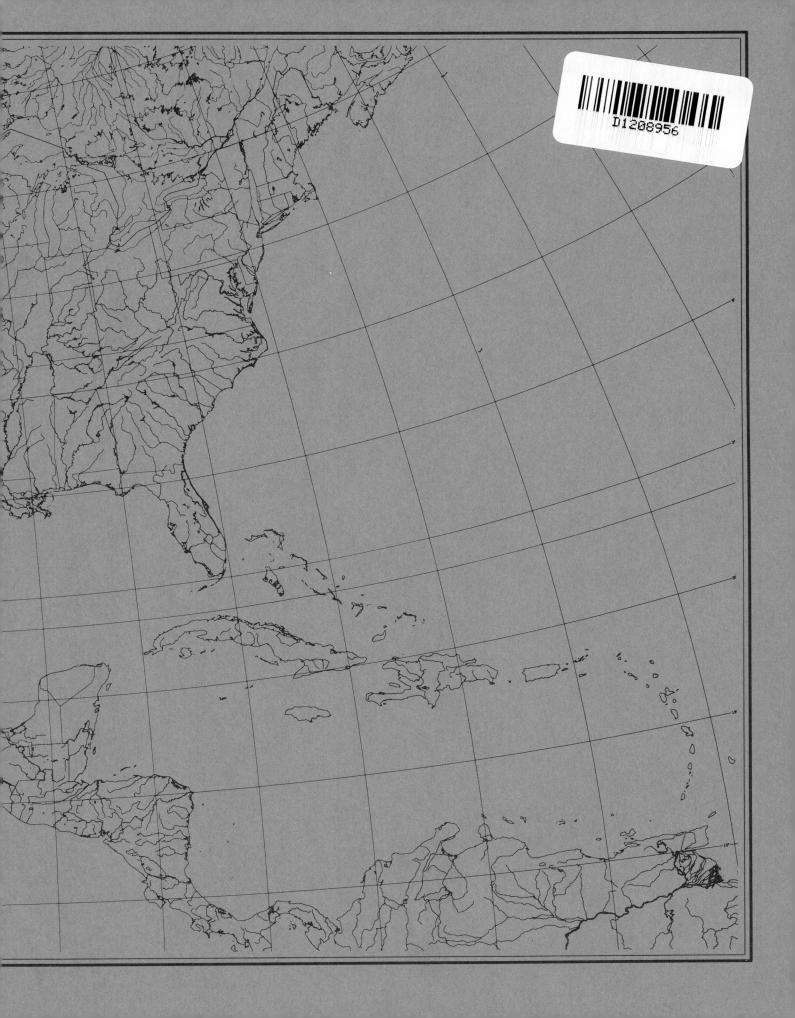

Centennial Special Volume 1

Geologists and Ideas:
A History of North American Geology

Edited by

Ellen T. Drake
School of Oceanography
Oregon State University
Corvallis, Oregon 97331

and

William M. Jordan
Department of Earth Sciences
Millersville University
Millersville, Pennsylvania 17551

1985

Acknowledgment

Publication of this volume, one of the Special Volumes of *The Decade of North American Geology Project* series, has been made possible by members and friends of the Geological Society of America, corporations, and government agencies through contributions to the Decade of North American Geology fund of the Geological Society of America Foundation.

Following is a list of individuals, corporations, and government agencies giving and/or pledging more than $50,000 in support of the DNAG Project:

ARCO Exploration Company
Chevron Corporation
Cities Service Company
Conoco, Inc.
Diamond Shamrock Exploration
   Corporation
Exxon Production Research Company
Getty Oil Company
Gulf Oil Exploration and Production
   Company
Paul V. Hoovler
Kennecott Minerals Company
Kerr McGee Corporation
Marathon Oil Company
McMoRan Oil and Gas Company
Mobil Oil Corporation
Pennzoil Exploration and Production
   Company

Phillips Petroleum Company
Shell Oil Company
Caswell Silver
Sohio Petroleum Corporation
Standard Oil Company of Indiana
Sun Exploration and Production Company
Superior Oil Company
Tenneco Oil Company
Texaco, Inc.
Union Oil Company of California
Union Pacific Corporation and its
   operating companies:
      Champlin Petroleum Company
      Missouri Pacific Railroad Companies
      Rocky Mountain Energy Company
      Union Pacific Railroad Companies
      Upland Industries Corporation
U.S. Department of Energy

Published by the Geological Society of America, Inc.
3300 Penrose Place, P.O. Box 9140, Boulder, Colorado 80301

Printed in U.S.A.

Library of Congress Cataloging-in-Publication Data
Main entry under title:

Geologists and ideas.

(Centennial special volume ; 1)
Includes index.
1. Geology—United States—History. 2. Geology—
Canada—History. I. Drake, Ellen, T. II. Jordan,
William M. III. Series.
QE13.U6G466   1985   550'.973   85-17631
ISBN 0-8137-5301-5

Contents

Contents

CONTRIBUTIONS OF INDIVIDUALS

CONTRIBUTIONS OF ORGANIZED GROUPS

APPLICATION OF SIGNIFICANT IDEAS

Preface

The Decade of North American Geology Project is part of the celebration of the Centennial of the Geological Society of America.

In addition to four topical special volumes, of which this is one, the publications of the Project will include 28 volumes of *The Geology of North America,* synthesizing the state of knowledge of the geology and geophysics of North America and adjacent oceanic regions in the 1980s. Nineteen are being produced by the Geological Society of America, and nine by the Geological Survey of Canada.

The project also includes six volumes of *Centenial Field Guides,* each describing 100 of the best geologic sites within the area of one of the six regional sections of the Society; twenty-four packets of continent-ocean transects depicting the geology across the continental margins at 23 sites around North America, plus a separate accompanying volume; and seven maps of North America at a scale of 1:5,000,000 synthesizing different aspects of geology or geophysics of the continent and adjacent oceans. A separate volume is associated with the neotectonics map.

This volume is the Centennial contribution from the History of Geology Division of the Society.

Introduction

The History of Geology Division, in celebration of the 100th anniversary of the Geological Society of America, proudly presents this collection of 33 scholarly papers.

The formation of the Geological Society of America nearly a century ago was tangible proof of the coming of age of geology in North America. All through its history the Society has been instrumental in various and creative ways in promoting the progress of our science. Many of the central figures in these articles were, and some still are, active members of the Society. The purpose of our volume, however, is not to document the influence of our Society on the development of North American geology but rather to spotlight events, ideas, and people, and thus shed light on the history of North American geology as a whole. The volume, therefore, represents our celebration of this most fascinating development, and in doing so, we also pay tribute to the Society's 100th birthday.

In spotlighting, we are bound to meet criticism that we have missed this or that important event or idea or scientist. Calls for papers appeared at various times in several publications. In addition, we, the editors, each sent numerous personal letters urging participation in the venture. The response was most gratifying when one considers the extreme high value that a scholar places on his or her time. A many-membered committee of the G.S.A. History of Geology Division reviewed the proposals submitted for approval in response to our call. Any volume that depends on the generosity of contributors, therefore, is not intended to be comprehensive. To fill what might be conceived as gaps, readers are urged to consult the ever-growing body of historical literature consisting of official and unofficial histories of government agencies, academic departments, biographies, and monographs on particular topics.

The papers in this volume are written both by geologists, themselves intimately involved in the shaping of North American geology, and by historians of the earth sciences, impartial observers of the intricate interactions of the forces in the geological community. We believe that the two groups are not mutually exclusive; each has much to contribute to the other and both provide insight in the evolution of a science.

The papers fit into four major categories: 1. The Evolution of Significant Ideas, 2. Contributions of Individuals, 3. Contributions of Organized Groups, and 4. Application of Significant Ideas. These topics were conceived at the start as essential sections of such a volume. Although the history of ideas was selected as the central theme of this collection, no major restriction on content was placed on contributors other than appropriateness to the theme of geology in North America within the approximate time-frame of the Geological Society of America. All papers were reviewed by at least two readers from the History of Geology Division Centennial Committee or from the divisional membership-at-large. Some papers have been presented orally, at either the 1982 Annual Meeting symposium, "The Development of North American Geology: 1881 to the Present," organized by Ellen T. Drake or at the 1983 Annual Meeting symposium, "Some Significant Geologic Ideas Originating from the Study of Cratons," organized by Ursula B. Marvin; some papers have been presented in technical sessions and section meeting symposia; others appear here for the first time.

The papers range in subject matter from geosynclines to plate tectonics, from the origin of life to the origins of caves and craters, from James Hall to George Gaylord Simpson, from the Arctic to Mexico, and from deep-sea drilling to orbital remote sensing. On behalf of the authors and the History of Geology Division, we wish you enjoyment in reading this volume.

ACKNOWLEDGMENTS

Many of the following members of the History of Geology Division Centennial Committee kindly evaluated manuscripts for us, some having been imposed upon more than once: Nancy S. Alexander (chairman 1978–80), William M. Jordan (chairman 1980-present), Allen P. Agnew, Claude C. Albritton, Jr., Michele L. Aldrich, Kennard B. Bork, Jules Braunstein, William R. Brice,

James X. Corgan, George H. Crowl, Ellen T. Drake, J. Thomas Dutro, Jr., Robert N. Ginsburg, Joseph T. Gregory, Ursula B. Marvin, Authur Mirsky, Clifford M. Nelson, Roland A. Pettit, Jean C. Prior, Cecil J. Schneer, Hubert C. Skinner, James S. Street, Kenneth L. Taylor, and Richard S. Williams, Jr.

Additional reviewers whose insightful critiques helped us to maintain a high quality were: Arthur L. Bloom, James E. Brooks, Preston Cloud, John C. Crowell, John G. Dennis, Clifford M. Dodge, Robert H. Dott, Jr., Harold E. Enlows, Carol Faul, Donald W. Fisher, Alan R. Geyer, Sarah E. Hoffman, Donald M. Hoskins, Rachel C. Laudan, Alan E. Leviton, Eldridge M. Moores, Paul H. Nichols, Neil D. Opdyke, Allison R. Palmer, Stephen J. Pyne, Charles K. Scharnberger, Eugene M. Shoemaker, Arthur A. Socolow, John W. Wells, and Ellis L. Yochelson. The help of all these individuals is greatly appreciated.

Ellen T. Drake
College of Oceanography
Oregon State University
Corvallis, Oregon 97331

William M. Jordan
Department of Earth Sciences
Millersville University
Millersville, Pennsylvania 17551

July 1985

Geological Society of America
Centennial Special Volume 1
1985

Mountain-building theory: The nineteenth-century origins of isostasy and the geosyncline

Dwight E. Mayo
Northern Arizona University
Flagstaff, Arizona 86011

ABSTRACT

Isostasy and the geosyncline are two major geological concepts that had their origins in the nineteenth century. Each is uniquely American in its origin, although a number of suggestions by Europeans triggered ideas in the minds of Americans. James Hall first published the basics of the geosyncline in 1859, although the concept was without its present name until that was provided later by James Dwight Dana. Isostasy-like generalizations began with J.F.W. Herschel in 1836, but it was not until 1889 that Clarence E. Dutton formally named the concept of isostasy. Isostasy and the geosyncline are closely related concepts in modern geological theory, and it is no surprise to find that they developed more or less concurrently and that certain individuals were involved in the development of each. Among geologists who either contributed to or critiqued the ideas, we find such familiar names as Hall, Dutton, Dana, T. Sterry Hunt, Joseph LeConte, G. K. Gilbert, William J. McGee and others. The development of isostasy and the geosyncline follows a process that is familiar in the history of science. Each was developed against the background of major philosophical arguments. Each was postulated minus much of the evidence required for a credible idea. Each attracted a number of proponents and detractors, and each had staying power sufficient to achieve general acceptance by the scientific community. At the present, against the backdrop of a major new theory system—plate tectonics—isostasy remains a useful and noncontroversial idea. The geosyncline, however, has a less solid future. Some geologists assert that it is no longer useful and should be junked, while others, by redefining the idea, continue to find it a major unifying concept.

INTRODUCTION

When compared with the achievements of European scientists during the nineteenth century, the record of their American counterparts is not terribly distinguished. However, in geological theory building during that century, two major conceptual developments are primarily the creation of Americans. The geosyncline and isostasy are the concepts, and as one writer has pronounced, they were "made in America" (Dott, 1979, p. 239). These two ideas often are treated separately in historical presentations, but their origins are contemporary, the individuals who created them or contributed to their creation are often the same people, and the places that each fills in mountain-building theory developments of that century are so complementary as to allow for their historical treatment in a single essay.

BACKGROUND

Isostasy and the geosyncline have played an important role in twentieth-century developments in geological theory, yet each came into being and matured against a backdrop of theory that has long since been largely discarded. Catastrophism, vulcanism, contractionalism, uniformitarianism, and creationist thinking all form a part of this backdrop in one way or another. Catastrophism and the contractional hypothesis were the most important of these general intellectual structures which guided the thinking of nineteenth-century geologists that are missing from the scene today. Uniformitarianism alone has survived as a reasonably viable philosophy in geological study, although, as happens to most major ideas, it has been significantly modified from the nineteenth-century Lyellian form.

How mountains come into being provides one of the great intellectual mysteries all down through recorded history. As modern science emerged in the seventeenth, eighteenth, and nineteenth centuries, theories of mountains became more sophisticated just as did theoretical structures in the other sciences. Mountain-building theory, however, like evolutionary thinking, was a rather slow starter when compared, for instance, to Newtonian physics. More often than not, mountain-building thought of that era conformed to a general catastrophist orientation which required:

that the forces of nature were once more stupendous in their operation than they now are, and that they from time to time devastated the earth's surface; extirpating the races of plants and animals, and preparing the ground for new creations of organized life (Geikie, 1911, p. 643).

Thomas Burnet and William Whiston, both writing in the late seventeenth century, each made extensive use of the Noachian deluge to explain the present configuration of the globe. Johann Gottlob Lehman also postulated the Biblical flood, while Abraham Gottlob Werner, as reported by one of his students, believed that a mighty inundation played a major role in deposition of maritime sediments. Peter Simon Pallas, a French geologist, while in the employ of the Russian Imperial Academy of Sciences in the 1770s, postulated a different maritime disaster—a great cataclysmic, general uprising of the volcanic islands of the Indies from Africa to Japan which thrust a huge wall of water against the lands of the entire world, transporting and depositing on the slopes of the earlier mountains all the materials which now make up the tertiary mountains. This catastrophe, he told his readers, corresponded quite well in time to the Noachian event, and was, he said, a more reasonable explanation for it than older accounts. Other geologists expounded similar theories. For example, Leopold von Buch added the idea that volcanic mountains come into being through sudden, catastrophic uplift of extensive areas of the terrain and are the remains of giant displays of internal forces. And others, the Italians especially, invoked volcanoes—and cited evidence of volcanoes that appeared from the sea (Mayo, 1968).

Meanwhile, the broad philosophy of uniformitarianism had undergone its primary developmental process. The work of James Hutton, John Playfair, and Sir James Hall culminated in the classic nineteenth-century statement by Charles Lyell that geological phenomena can be fully explained by processes now observable. Uniformitarianism, interestingly, did not grow out of a reaction to the doctrine of successive catastrophes as some writers have suggested (Mayo, 1968). Lyell concluded that natural philosophers as long ago as antiquity had espoused doctrines that had uniformitarian overtones but that in more recent decades a surge of anti-uniformitarian writings had retarded the study of geology (Lyell, 1837, v. 1, p. 30).[1] However, by the 1830s and 1840s, certain geologists had developed sufficient faith in uniformitarianism so that it provided a relatively firm background for the development of isostasy and the geosyncline.

Figure 1. James Hall, president of the American Association for the Advancement of Science, 1856. From John M. Clarke (1923).

The contractional hypothesis was another popular nineteenth-century idea. "The cause," said Archibald Geikie,

to which most geologists are now disposed to refer the corrugations of the earth's surface is secular cooling and consequent contraction. If our planet has been steadily losing heat by radiation into space, it must have progressively diminished in volume. The cooling implies contraction (Geikie, 1903, p. 394–395).[2]

James Dwight Dana, noted Yale geologist and long-time editor of the *American Journal of Science,* who often used his editorship of that publication to expound his own ideas, was America's greatest advocate of the contractional hypothesis.

HALL AND THE GEOSYNCLINE

A number of individuals stand out in the nineteenth-century study of geology in the state of New York, but James Hall, founding president of the Geological Society of America, stands head and shoulders above all others. He first went to work for New York in the initial survey of the state in 1836 and continued his employment there most of the time until his death in 1898. He served for the majority of that time as state geologist and paleontologist (Clark, 1923). Seldom is there any dissent with the idea that he originated the concept of the geosyncline. Less well known is his contribution to the development of the principle of isostasy.

Hall took a giant intellectual leap when he postulated a close relationship between large, subsiding areas of sedimentation and

Figure 2. James Hall's "office" in which he and his staff worked from the 1850s on. From John M. Clarke (1923).

the ultimate emergence of sedimentary mountain ranges. "I hold," he stated,

... that it is impossible to have any subsidence along a certain line of the earth's crust, from the accumulation of sediments, without producing the phenomena which are observed in the Appalachian and other mountain ranges (Hall, 1859, p. 69–70).

This connection had not been made before Hall first publicly postulated it to AAAS members in 1857. This initial statement of the geosynclinal concept is relatively simple compared with modern development of the idea. It, like so many other significant concepts in science, grew in strength and applicability as it was related to more and more situations.

In only a few items of Hall's long bibliography do we find extensive discussions of the idea for which he is best remembered. In 1857 he delivered the presidential address at the AAAS convention held that year in Montreal (Hall, 1857). On that occasion he laid out the basics of the concept which he described as a theory of subsidence, but that address was not published in the *AAAS Proceedings* until 26 years later (Hall, 1883). In 1858 he presented further supporting data as a result of his work on the geological survey of Iowa (Hall, 1858), and in 1859 an extensive version of the concept and much supporting data appeared as the introduction to Volume III of his monumental series, *Natural History of New York*. In 1860 a summary of the concept appeared in the *Canadian Journal* following a speech he made at Albany in that year. Perhaps the most widely distributed summary of the concept was that written by T. Sterry Hunt of the Canadian Geological Survey for the *American Journal of Science*

(1861), although therein he inserted Hall's ideas into the contractional hypothesis which Hall had earlier rejected as "not always philosophical" (Hall Papers). In none of these publications did Hall use the term geosyncline, and only years later do we find Dana providing that name for Hall's subsidence concept.

HALL'S SOURCES

The mental process by which Hall arrived at his conclusions concerning the new concept we can only guess at, as he does not reveal this in any of his writings. However, in the publications listed immediately above and in notes attached to the manuscript of his AAAS presidential address (Hall Papers) is found a wealth of information that reveals many of his sources, relating both to his general philosophy of geology and to the development of the geosyncline. Other bits of information can be filtered out of others of his publications, especially those which appeared prior to 1857.

Finding the background to Hall's geological philosophy is not difficult. In an annual report to the New York legislature, he rejected any notion of catastrophism. Nature, he stated, is always the same, and her laws do not change. Consequently, many types of geological phenomena can be explained that were previously inexplicable "except through a suspension of the natural laws, or a miraculous interposition of creative power" (Hall, 1843, p. 10–11). To Hall, only during a comparatively recent period of time had this interpretation become accepted and the stupendous conclusion reached that "nature has been operating through incalculable periods of time, with the same harmony and unity of design as we behold in her present creations." He also sharply rejected the notion that the Biblical deluge was of any importance to geological theory. Reference to that idea, he rather scornfully said, had led to a general belief that all superficial deposits are due to a single period and to one agency. "Geological phenomena," he stated emphatically,

are now studied without reference to preconceived opinions or interpretations, and by adopting more natural and rational explanations than otherwise could be done, we escape advocating numerous absurdities, without conflicting with religious opinions (Hall, 1843, p. 339).

The source of Hall's uniformitarian orientation is easily traceable. In one of his earlier annual New York survey reports, he revealed familiarity with Lyell's *Principles of Geology*, picturing its publication as "an era" in the science of geology (Hall, 1840, p. 394). Nearly 20 years later, he again listed Lyell as one of his chief sources of this philosophy, saying he had

necessarily incorporated the general philosophic views so long ago clearly set forth by Babbage, Herschel, Lyell and others; since these had early been fixed in my mind as a part of the elements and principles of geological science (Hall, 1859, p. 81).

Lyell and Hall were personally acquainted, and when Lyell visited the United States in 1841–42 and again in 1845–46, Hall escorted his British colleague to various points of geological interest in the state of New York, including Niagara Falls (Lyell, 1845, 1849).

Hall postulated that areas of continental proportions subside while being filled with enormous sedimentary deposits as a part of the mountain-building process. Two questions here had to be answered: the source of the sediments and how those sediments are transported to the place where they are deposited. In dealing with these questions, Hall asserted that ". . . we must conceive of the existence of continents where no vestige of them now remains." The process, reasoned Hall, was the same then as that which occurs in the present. Using New York's fourth geological district[3] as an example, he concluded that:

the high hills and deep valleys indicate the absence of an immense quantity of matter, which . . . was transported in the direction of the great outlets into the present ocean, there to lay the foundation of future continents in strata like those occupying our district, filled with the organic remains of successive ages, and exhibiting throughout their extent all the varying characters that we now find in the rocky strata of our continent (Hall, 1843, p. 16).

A third question—what causes the subsidence where the sediments are deposited—caused Hall all kinds of problems. He argued that it had long since been shown that "the first effect of this great augmentation of matter would be to produce a yielding of the earth's crust beneath, and a gradual subsidence would be the consequence" (Hall, 1859, p. 69). Quite early on, Dana (1866, p. 209) scornfully suggested that if the crust is so sensitive that it subsides under the added weight of a few inches or feet of

sediments, would not the enormous weight of mountain ranges also cause the crust to subside? Hall never satisfactorily answered this question.

The continental source of the sediments for the Appalachians lay to the east of the present continent, Hall speculated, noting that the Chemung group is a mass of sediments which had its origin in that direction. It includes well-preserved fragments of land plants that could have come only from land on the eastern margin of the ocean. He concluded that the chief source of materials making up the present Appalachian strata must have been from the east or southeast (Hall, 1843). A few years later, however, Hall, this time speculating on the source of sediments for the Hudson river group, stated that these probably had come from the east and northeast with the coarser materials being deposited first and the finer materials being carried further into the ocean, resulting in a very widespread formation (Hall, 1851). Again Hall did not provide a very satisfactory answer for a major question.

Hall's 1859 discussions of the means of transport of the sediments are extensive. Deposition of sediments on a scale leading to large area subsidence and deposition can take place only in ocean areas, he stated, and furthermore, the only means whereby those materials can be transported to areas of deposition are ocean currents. For this idea, Hall refers primarily to Lyell and to a lesser extent to Henry T. de la Beche and William W. Mather. Lyell wrote that ocean currents have the capability of carrying the finer particles of sediments over distances of hundreds and even thousands of miles under ideal conditions. To the casual observer, he noted, large rivers appear to be the most important of the transporting agencies, but by comparison with ocean currents, "the deltas of rivers must shrink into insignificance." Examples he cited include distribution of sediments from the Nile, Amazon, Orinoco, Rio Grande, and Mississippi river systems by currents over vast areas (Lyell, 1837, v. 2, p. 286).

Hall had been gathering evidence to support his ideas on the role of ocean currents for two decades prior to publicizing the subsidence concept. In a report on the results of his work with the New York survey, he called attention to the fossilized remains of an early brachiopod, *Lingula,* which he found in great quantities throughout several layers of sandstone in western New York. Many of these fossils were aligned in a single direction, generally northwest by north to southeast by south, with small ridges of stone extending from the beaks of the fossils to the southeast. "It is impossible to avoid the conclusion," related Hall, "that the surface of each of these layers was once the original surface of the sandy bottom of an ocean, covered with living shells, over which a gentle current flowed." The direction of the current of this ancient sea could be determined, he believed, simply by reference to the ridges of stone built up by each of these fossils. Further evidence of an ocean current appeared in scratches on the surfaces of rocks, similar to those which are impressed by flowing water on soft and yielding beds of sand or clay (Hall, 1838, p. 296).

Several years later Hall further amplified his postulate on the transporting role of ocean currents. He repeated and expanded his description of the stony formations appearing with the *Lingula* fossils which indicated to him the existence of gentle ocean currents at the time of their deposition. He also described ripple marks in Medina sandstone that were "beautifully preserved," and these, along with the diagonal lamination of this formation, indicates the existence of currents in the ancient ocean which covered the area. Wave lines, ridges of sand, slight scratches, and deeper furrows in the mud of the Portage group likewise are "beautifully illustrative of the effects of oceanic currents upon the bottom." In his conclusions to this report, Hall noted that in some cases oceanic currents may diffuse slowly, precipitating sediments over wide areas, while other types of sediments are carried only short distances before settling to the bottom. In either case, oceanic currents are essential to the process (Hall, 1843).

When Hall compared the Old Red[4] sandstone with lower strata, he became convinced that an ocean and its currents are necessary to the growth of a continent. The older groups, he speculated, are distributed over wide areas, often appearing consistently throughout an area that extends at least a thousand miles to the west of New York, but the later Old Red is not found beyond the boundaries of the state in a westerly direction. He attributed this change in deposition to a diminution of the transporting power of the ocean currents which had in earlier times carried materials over a much wider area (Hall, 1843).

HALL ON ELEVATION AND SUBSIDENCE

Nineteenth-century geologists puzzled over the phenomena of elevation and subsidence of the earth's crust. Hall was generally reluctant to derive an explanation for this problem, but he did include some discussion of it in his writings. According to him, at some locations in the Appalachian chain, there is an aggregate thickness of at least 40,000 feet of sediments, although at no place are the remaining mountains anywhere near that elevation. Since the deposits indicate a shallow-sea environment at the time of their deposition, he assumed that during the period of deposition, the region must have been steadily subsiding. Another obvious assumption is that the region which had subsided had been elevated into the existing chain of mountains at some later time. Geologists contemporary with Hall were understandably interested in trying to explain the causes of this subsidence and later elevation. Hall was no exception, and in his "Introduction" of 1859, he developed an explanation for these phenomena to the greatest extent found in any of his published works. In this discussion, he depended most heavily on ideas developed by J.F.W. Herschel, the British astronomer; Charles Babbage, a British mathematician; and Lyell (Hall, 1859).

Herschel noted Lyell's conjecture that the largest transfer of material to the bottom of the ocean is produced along a coastline by ocean currents. He then speculated that variations in local pressure due to this transfer of material may cause changes in the elevation of the affected areas. Where the materials are deposited, the surface will tend to subside, forcing yielding material from that location underneath the continent from which the materials

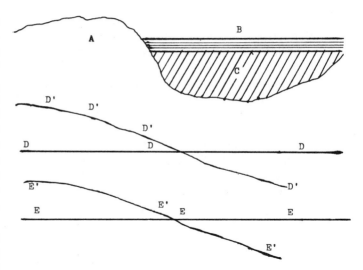

Figure 3. This diagram plus the text below is cited by both Hall and Dutton. It is from a letter written by J.F.W. Herschel to Charles Lyell in 1836. From Charles Babbage (1837). Herschel writes:

According to the general tenor of your book [Lyell's *Principles of Geology*], we may conclude, that the greatest transfer of material to the bottom of the ocean, is produced at the coast line by the action of the sea; and that the quantity carried down by rivers from the surface of continents, is comparatively trifling. While, therefore, the greatest local accumulation of pressure is in the central area of deep seas, the *greatest local relief* takes place along the abraded coast lines. Here, then, in this view, should occur the chief volcanic vents. If the view I have taken of the motionless state of the interior of the earth be correct, there appears no reason why any such influx of heat should take place under an existing continent (say Scandinavia) as to heat incumbent rocks (whose bases retain their level) 5 or 600 Fahr. for many miles in thickness. (Princ. of Geol. vol. ii. p. 384. 4th Ed.) LaPlace's idea of the elevation of surface due to columnar expansion (which you attribute, in a note, to Babbage,) is in this view inadequate to explain the rise of Scandinavia, or of the Andes, &c. But, in the variation of local pressure due to the transfer of matter by the sea, on the bed of an ocean imperfectly and unequally supported, it seems to me an adequate cause may be found. Let A be Scandinavia, B the adjacent ocean (the North Sea), C a vast deposit, newly laid on the original bed D of the ocean; EEE a semi-fluid, or mixed mass, on which DDD reposes. What will be the effect of the enormous weight thus added to the bed DDD (rock being heavier than sea)? Of course, to depress D under it, and to force it down into the yielding mass E, a portion of which will be driven laterally under the continent A, and upheave it. . . .

had come, causing a continental area to be elevated (Herschel, 1836a).

Much previous geological speculation had included an assumption that changes in the earth's crust must be attributed to internal forces. Herschel disagreed, arguing that the interior of the earth is essentially inert and that there are no internal forces present that will act without some sort of outside stimulus; that the activating principle is the ordinary process of abrasion of sediments and their mass transfer and deposition at other areas (Hershel, 1836b). Hall accepted Hershel's contentions and incorporated them into his own work. Lyell, also, said Hall, postulated that the "ordinary repose of the surface of our planet argues a wonderful inertness in the interior." Thus, from these two sources, Hall argued that geologists "must look for external influences to provoke the interior manifestations," and these external forces consist in the main of abrasion, removal, and redeposition of sedimentary materials (Hall, 1859, p. 87, 96).

Another speculative idea used by Hall is that of Babbage, who visited the Temple of Jupiter Serapis at Pozzuoli near Naples, Italy, in 1828. This temple, located on the shoreline of the Mediterranean Sea, had, according to Babbage, undergone significant changes in elevation since its construction about the end of the second century A.D. Although these changes probably could not be related to large area changes in elevation, he rather extravagantly extrapolated from the evidence he found there to a broad generalization. He concluded that:

from changes continually going on, by the destruction of forests, the filling up of seas, and the wearing down of elevated lands, the heat radiated from the earth's surface varies considerably at different periods. In consequence of this variation, and also in consequence of the covering up of the bottoms of seas, by the detritus of the land, the *surfaces of equal temperature* within the earth are continually changing their form, and exposing thick beds near the exterior to alterations of temperature. The expansion and contraction of these strata and, in some cases, their becoming fluid, may form rents and veins, produce earthquakes, determine volcanic eruptions, elevate continents, and possibly raise mountain chains (Babbage, 1847, p. 212).[5]

To the late twentieth-century reader, the speculations of Hall, Babbage, Herschel, and Lyell seem rather unsatisfactory. But because Hall had rejected any catastrophic hypothesis and since he had criticized the contractional theory as "not always philosophical for want of a basis in facts, and . . . always unsatisfactory as giving a very inadequate solution of the problem," he fell back upon the only reasonable hypothesis left to him (Hall, 1883, p. 69).[6]

HALL ON METAMORPHISM

The origin of metamorphic rocks was another topic to which Hall paid close attention. Lyell had warned his readers that "it was once a favourite doctrine, and is still maintained by many, that these rocks owe their crystalline texture, their want of all signs of a mechanical origin, or of fossil contents, to a peculiar and nascent condition of the planet at the period of their formation." He went on to say that metamorphism is not necessarily concerned with plutonic action, nor is it necessary that a contiguous mass of granite be the altering power (Lyell, 1855, p. 598, 603). In Hall's earlier writings, he had had little to say about metamorphic rocks, merely listing them as constituent parts of the strata he happened to be studying at the moment. In his "Introduction" of 1859, however, he devoted several pages to the subject and its relationship to his subsidence concept. Many of the strata in the Appalachian chain, Hall said, show striking evidence of metamorphic action to one degree or another. When the chain is approached from the west, this evidence becomes more and more noticeable, and as one nears the mountain range, the shales

become more broken, are changed in color, and contain more and more particles of a talc-like substance until this becomes a predominating factor. Limestones lose their dark color, and veins of calcareous spar can be observed traversing the mass. Fossils become less distinct in their forms, often appearing distorted. Finally, the rock appears as a homogeneous, crystalline mass, but often with fossils remaining to show the original sedimentary condition of the strata. Like Lyell, Hall denied that this metamorphism had proceeded from contact with or proximity to granitic or other rocks of plutonic origin. It was clearly evident to him that the phenomenon of metamorphism appeared over wide areas where no eruptive or intrusive granitic masses were present (Hall, 1859).

Significantly, Hall observed that only when one approaches a zone of great accumulation of sediments can evidence of metamorphism be found on a grand scale. Again using the Appalachians as an example, he observed that:

in this mountain range, and I believe also in others, the line of metamorphic action is parallel to the mountain chain, and parallel to the minor elevations or subordinate axes of the great mass; parallel, indeed, to the great line of original accumulation of the sediments constituting the mountain mass (Hall, 1859, p. 78).

The process of metamorphism, like that of mountain formation, said Hall, simply does not occur without the large accumulation of sedimentary matter that is present in mountain regions. In areas such as the Mississippi Valley, therefore, metamorphic rocks are largely absent because of the lack of significant accumulations of sediments (Hall, 1860). He also denied that folding, plication, and other alterations of the strata are requisite to the metamorphic processes. Mountains in which the strata are essentially horizontal contain rocks that show exactly the same metamorphic character as those in nearby formations which have been severely folded and plicated. Accumulation and subsequent metamorphism seem to be, speculated Hall, a more significant relationship than is relating the alteration of the position of the strata with metamorphism. He cited as an example some of the beds near the base of the Catskills where the rocks are extremely hard and dense although in beds which are essentially horizontal, indeed exhibiting metamorphic characteristics even more than those of the severely disarranged beds on the western flanks of the Green mountains (Hall, 1859).

These notions on metamorphism seem to have come from the ideas of two of Hall's closest associates in geology, Lyell and T. Sterry Hunt of the Canadian Geological Survey. To the former can be attributed a rather basic philosophy on the general role of metamorphism in geological processes. This philosophy appeared in publications that Hall had at hand and from which he quoted in 1859. The metamorphic process as applied to geology, according to Lyell, had no place in catastrophist-oriented systems. Indeed, Lyell seems to have been the individual who first used the term metamorphic to describe a whole series of classes of rocks that had formerly been called primary, although the idea that "the strata called *primitive* were mere altered sedimentary rocks" probably originated with Hutton in the preceding century (Lyell, 1837, vol. 2, p. 499, 506).

Hunt was a more direct influence on Hall's views on metamorphic rocks. Hunt, whom Hall called a "versatile and brilliant genius," maintained a close professional and personal relationship with Hall for many years. In one short passage in his 1857 AAAS address which concerned metamorphism, Hall gave this acknowledgment to Hunt:

> . . . .Laying aside all assumption of causes not known to exist, and ignoring the supposed effects of that imaginary nucleus, he [Hunt] has proceeded from known and demonstrated facts derived from existing conditions, to explain most satisfactorily certain processes in metamorphism, and instead of appealing to an unknown source for the ingredients of certain metamorphic strata has demonstrated the existence of the same elements in the unaltered beds which are known to be of the same age and prolongations of the same strata (Hall Papers, MSS of Montreal Address, p. 64).

HALL ON VOLCANOES

Hall also attempted, though less successfully than in the case of metamorphism, to establish a similar relationship between large-scale deposition and volcanic activity. In accounting for the presence of igneous matter, he speculated that whenever sediments accumulate over large areas at a slow rate, then depression of the area would proceed slowly. As a result, comparatively few extensive rents or fractures would be produced. Those formed would fill with trappean matter, but rarely would any material overflow onto the surface. But when very rapid accumulation occurs over relatively small areas, then "the crust below might give way, from the overload, and the whole be plunged into the semi-fluid mass beneath, causing it to overflow. Whether this reasoning be correct or otherwise," noted Hall, "I believe that the overflows of trappean matter are always coincident with the rapid accumulation of sedimentary materials." He cited evidence of certain accumulations in Nova Scotia and in the Connecticut and Hudson river valleys that he believed supported his conclusion, although he did admit his evidence might be scanty. In any case, he went on, "I believe the law will hold true, that all great outbursts of igneous matter are accompanied by formations of rapid accumulation" (Hall, 1859, p. 79–80).

Volcanoes proper and their products, Hall generalized, are invariably connected with tertiary or more modern geological formations. These phenomena are found in areas where sizable deposits have accumulated rapidly, and furthermore, they "can never occur except as the result of such conditions. The igneous outflows," he extravagantly stated, "therefore, I regard as produced by and dependent upon other agencies, and are but the manifestitations of rapid accumulations of sedimentary matters" (Hall, 1859, p. 80).

HALL'S FIELD EVIDENCE

By definition, a geosyncline covers an extensive geographical area. Hall was struck by the extent of the sedimentary deposits in the eastern half of the nation, how these formations extended westward nearly to the Rocky Mountains, and how, throughout the Mississippi valley, the strata tend to be thin, largely horizontal, and relatively undisturbed. To the east, as the Appalachian range is approached, the formations begin to thicken, ultimately reaching more than 40,000 feet in total thickness. Furthermore, he could not escape noticing their severely folded and disarranged character in the areas of great thickness.

As an example of the extent of the formations, he noted the Niagara group which reaches from just west of the Hudson river to a vast area in the Mississippi river valley (Hall, 1843). Hall later verified these data when he conducted a geological survey for the state of Iowa from 1855 to 1857 (Hall, 1858). Another example he cited to demonstrate the wide-ranging extent of Appalachian formations is the upper portion of the Hudson river group which he identified as reaching from the Atlantic coast in Canada to Iowa and Missouri on the west. The thickest portion of this group lies in eastern Canada (7,000 ft.) and Pennsylvania (6,000 ft.), while the thinnest is found in Iowa where it is less than 50 feet thick (Hall, 1859, p. 19–21).[7] Hall cited numerous other examples to illustrate the extent of the deposits associated with the Appalachian chain.

Following this extensive presentation of evidence, Hall (1859, p. 59–60) generalized that:

> thus we see everywhere the operation of the same law, viz. a greater accumulation, and a coarser character of sediments along the line of the Appalachian chain, with a gradual thinning to the westward, and a deposition of the finer or far transported matter in that direction. In all the great periods of sedimentary deposits which we have considered, or the transportation of shore-derived materials, this law has held true, and has governed the distribution, or the cause that originated these conditions, long before the distribution and deposition of the material commenced, giving form and contour to the eastern part of our continent, to its mountain ranges, and their elevation.

ASSIMILATION OF HALL'S IDEAS

Hall's new subsidence concept was first published the same year as another important theory in the structure of science—Darwinian evolution. And like evolution, the geosyncline did not gain immediate acceptance, although unlike evolution, it did not excite a storm of popular protest. By contrast, it rather languished, and only slowly did it gain stature in the scientific community. Its initial introduction at Montreal in 1857 was certainly anything but auspicious. Joseph LeConte, who attended that meeting, later recalled that:

> in 1857 the A.A.A.S. met at Montreal, and Hall as retiring President gave his memorable address on the formation of mountains by sedimentation. I can never forget the impression produced. The idea was so new, so utterly opposed to prevailing views, that it was wholly incomprehensible even to the foremost geologists. There was no place in the geological mind where it could find lodgment. It was curious to observe the look of perplexity and bewilderment on the faces of the audience. Guyot was sitting immediately behind me. He leaned forward and whispered in my

ear: 'Do you understand anything he is saying?' I whispered back, 'not a word' (LeConte, 1896, p. 699).

Furthermore, there was a rather decided lack of mention of it for a number of years. Part of this may be attributed to Hall's refusal to permit publication of his presidential address until 1883, and he seemed hesitant to publicize the concept. This hesitancy may be attributed, at least in part, to friendly advice from Joseph Henry, then secretary of the Smithsonian Institution. Henry noted that, had not the remarks made in the Montreal address come from Hall, he would have supposed that there was nothing to them. Henry wrote, "They may be considered at variance with what have been long regarded as established principles." He advised Hall to test his views against the widest possible collection of evidence before giving them to the world. Henry's skepticism is not surprising, given that he was not a geologist and that he firmly believed in the contractional hypothesis in which he doubted there was room for Hall's subsidence theory (Henry, 1858, letter to Hall quoted in Clarke, 1923, p. 327–328).

Hunt was Hall's staunchest advocate in publicizing the subsidence concept. During the decade of the 1860s, Hunt published a number of articles that clearly stated his agreement with Hall's views. Furthermore, Hunt attended a meeting of the Société Géologique de Paris at which he read several papers, giving some European publicity to the idea (Hunt, 1867, letter to Hall, Hall Papers). He also was one of the first investigators who attempted to stuff the concept into the contractional hypothesis, although he made contradictory statements on that theory. On the one hand, he believed that Hall's vague reference to continental elevation as the cause of mountain formation was deficient and that contractional folding provided the necessary forces of elevation, but on the other, he indicated that to view mountains as arising from local uplift, such as contractional folding, was in error (Hunt, 1861).

The necessity indicated by both Henry and Hunt to include the subsidence concept in the contractional hypothesis seemed to be a very common tendency in the last decades of the nineteenth century. Hall, exhibiting one of the interesting qualities of successful scientists down through the centuries—the ability to ignore one part of a problem, in this case the forces of elevation, and to concentrate on the solvable—found no such necessity. Dana also included the geosyncline in contractional theory—after he finally accepted and named the idea, but many years passed before he came to understand and accept the hypothesis fully. Nevertheless, Dana's role as a leading geologist and editor was crucial. Robert H. Dott, Jr., (1979, p. 329) asserts that "the geosynclinal theory—American style—was named and fully nurtured by James D. Dana," yet it was many years before Dana really began to accept much of Hall's version of the theory. As late as 1873, Dana criticized it as "seriously deficient and defective" (Dana, 1873a, p. 350). Mott Greene, in his excellent volume on nineteenth-century geological theory, effectively describes the intellectual infighting between Dana, Hall, and others concerning subsidence theory (Greene, 1982).

Figure 4. James Dwight Dana. From D. C. Gilman (1899).

Initially Dana was highly critical, indeed almost scathing in his criticism of the subsidence concept. In the first edition of his *Manual of Geology,* he disagreed with virtually every element of the idea (Dana, 1863). Three years later he repeated the same criticisms in the *American Journal,* added a comment that was often repeated in nineteenth-century geological literature: "Mr. Hall's hypothesis has its cause for subsidence," chided Dana, "but none for the lifting of the thickened sunken crust into mountains. *It is a theory for the origin of mountains, with the origin of mountains left out*" (Dana, 1866, p. 208–210. Present author's italics).

In the early 1870s, Hall's subsidence concept attracted more attention and increasingly came in for serious discussion, indicating that it had strength and ultimate credibility. Josiah D. Whitney, although he had worked with Hall in prior years, in 1871 wrote a rather severe critique of the idea. After discussing the various components of the concept, in general rather critically, he repeated Dana's frivolous commentary about the origin of mountains and then went on to add his own harsh comments. He argued that "the theory, as set forth by its author, is left in such vague form that it seems impossible to bring it to any crucial test, and one has to be content with finding in it nothing which will bear examination." Whitney also revealed a portion of his own philosophy of geology when he declared that uniformitarians "are trying to pull the cornerstone from under the fabric of the science" (Whitney, 1871, p. 249, 267).

A year later LeConte displayed a mixed reaction to Hall's idea, but he did incorporate a number of the elements of Hall's theory into his own scheme of mountain building. At the same time he, like a number of his contemporaries, fell back upon the contractional hypothesis to provide the activating mechanism. LeConte first noted that he had now rejected the hypothesis of a liquid earth, and that he had become convinced that "the whole theory of igneous agencies—which is little less than the *whole foundation of theoretic geology—must be reconstructed on the basis of a solid earth."* In reference to Hall's work, LeConte gave high praise to Hall for drawing attention to the relationship between immense masses of sediments and mountain chains, noting that "his views on this subject form, I believe, an era in the history of geological science." But, continued LeConte, the elevation of mountain chains and ranges is "beyond question, produced by horizontal thrust crushing together the whole rock mass, and swelling it up vertically; the horizontal thrust being the necessary result of secular contraction of the interior of the earth." LeConte's explanation of the mountain-building process, excluding incorporation of the contractional hypothesis, is remarkably similar to that of Hall (LeConte, 1872, p. 461–463).

Hunt wrote an extensive reply to LeConte's article, again stoutly defending Hall, again reiterating that neither he nor Hall had proposed any theory to explain the process of elevation of mountains, that Hall had attempted no explanation for continental uplift, the real cause of mountain formation (Hunt, 1873, p. 266). This is not completely correct as Hall had alluded to explanations of elevation advanced by Herschel and Babbage in his notes to his "Introduction" of 1859, but at no time did Hall develop a comprehensive theory of elevation. Later that same year, LeConte returned to the fray once again, this time in rather biting language, asserting that he thought Hunt's perspective was too narrow for him to understand what the important features of the mountain-building process are. Hall's views, on the other hand, are important, or so LeConte implied, and the association of sedimentation and mountain ranges is valid. But he insisted that such a concept is not a theory of mountain chains and that was what he himself was trying to provide (LeConte, 1873).

Dana now published another major criticism of Hall's subsidence concept, and therein he provided the name "geosynclinal" for Hall's subsiding troughs (Dana, 1873a, p. 430). He, like Whitney, felt that it was unreasonable to believe that a small mass of sediments could depress the surface while the huge masses of nearby mountains stand thousands of feet above the surface of the sea. He argued that although the combination of subsidence and sedimentation will lead to massive accumulations like those of the Appalachians, "a slow subsidence of a continental region has often been the ocassion for thick accumulations of sediments," not the other way around as Hall had asserted. Hall, he alleged, had made no provision to explain the process of continental elevation and thus his theory was "seriously deficient and defective" (Dana, 1873a, p. 348–350). Dana's critical attitude, however, did not prevent his making effective use of certain elements of the geosynclinal concept in his own theory of mountains published that same year (Dana, 1873b).

By 1875, Dana had slightly moderated his view of the relationship between the weight of surface materials and the stability of the crust of the earth, admitting that perhaps the effect of the weight of a huge ice cap might tend to cause subsidence in some small degree, but that the preponderant cause of surface oscillations remained the lateral pressure within the crust (Dana, 1875). Five years later the third edition of Dana's *Manual of Geology* appeared, and again some further modification in his views can be noted. He granted a somewhat larger role to his own version of the geosynclinal concept than he had earlier, explaining that a geosynclinal is necessary as a prerequisite to the formation of all mountain ranges since they owe their very origins to the process of geosynclinals. He also now distinguished between geosynclinal depressions and contractional depressions; that geosynclinals exist on the continents and along their borders, while contractional depressions are primarily an oceanic phenomenon (Dana, 1880, p. 824–825).

At long last Hall's presidential address of 1857 appeared (in the *AAAS Proceedings* of 1883), so that now a far larger reading audience had the basics of the concept available, although by this time it had been discussed thoroughly in other publications. He now appended a rejoinder to Dana's criticism that he had proposed a system of mountains with the mountains left out, saying that he did not:

pretend to offer any new theory of elevation, nor to propound any principle as involved beyond what had been suggested by Babbage and Herschel. I did not propose to discuss the theory of the contraction of the globe from cooling, or of the crumpling of the earth's crust from the gradual cooling and shrinking of the interior mass, because such arguments are not always philosophical for want of a basis in facts, and are always unsatisfactory as giving a very inadequate solution of the problem (Hall, 1883, p. 68–69).

The term *geosyncline* finally appeared in print for the first time in 1891 when Dana used it in another paper on the mountain-making process. He made no mention therein of his favorite contractional hypothesis, but he did make liberal reference to lateral pressure as the force which produces surface disturbances (Dana, 1891, p. 442–444). Two years later LeConte did likewise although by now he had come to doubt the validity of the contractional hypothesis. Lateral pressure, he stoutly asserted, can satisfactorily explain all the major phenomena of mountains, but as to what causes the lateral pressure, he could not provide a satisfactory answer. The most obvious answer thus far, LeConte believed, is the interior contraction of the earth, but to it, he said, "objections have recently come thick and fast from many directions. Some of these I believe can be removed; but others perhaps cannot in the present condition of science, and may indeed prove fatal" (LeConte, 1893, p. 561–564). How right he was.

In the last year of his life, Dana at long last fully acknowledged the usefulness of the geosynclinal concept. The fourth edition of his *Manual of Geology* carries a publication date of 1896, but Dana died the previous year. In that book he stoutly defended the contractional hypothesis, but he finally acknowledged Hall as the originator of the idea of the geosyncline, saying:

Figure 5. James Hall late in life. From John M. Clarke (1923).

The knowledge of the Appalachian facts led Professor James Hall to suggest in 1859 that a similar trough of deposition preceded the upturning in all cases of mountain-making. It was the first statement of this grand principle in orography (Dana, 1896, p. 357).

Thus, by 1896 the geosyncline had come of age and had found a firm place within the body of geological theory that it was to retain. Hall, then 85 years of age (Fig. 5), was given a hearty acclamation at the AAAS meeting that year, and no voices were raised in dissent to Hall's great contribution to geological theory, the concept of the geosyncline (Clarke, 1923).

THE CONCEPT OF ISOSTASY

Meanwhile, the geosyncline's companion concept, isostasy, was going through a similar process of development. On April 27, 1889, Clarence E. Dutton of the U.S. Army Corps of Engineers, working at times with the U.S. Geological Survey, read a paper to the Philosophical Society of Washington that played much the same role as Hall's AAAS address had done more than 30 years earlier. The paper, according to Dutton,

was written hastily to occupy a vacant half hour of a meeting of the Philosophical Society without thought of publication. . . .It contains a rough outline of some thoughts which have worked in my mind for the last fifteen years and which, from time to time, I have discussed at length in unpublished manuscripts and in familiar conversation with my esteemed colleagues.

It was on this occasion that Dutton publicly provided the name

isostasy for a major concept that had been germinating for many years,[8] for he had already published several times on the general principle of isostatic adjustment (Dutton, 1889, p. 64).

DUTTON'S SOURCES

The idea of compensatory elevation and subsidence can be traced back at least as far as the work of Herschel and Babbage, sources used by Hall and now by Dutton. Babbage, in his paper on compensatory subsidence and elevation, noted the reports by Charles Darwin generated during the famous voyage of the *H.M.S. Beagle* through the Pacific Ocean (Babbage, 1847). Darwin speculated that the appearance of barrier reefs, atolls, and coral islands indicates that the topographic features to which coral formations become attached are gradually subsiding into the crust of the earth beneath the ocean. It was perhaps a part of a continental-sized downwarping taking place gradually in that part of the world, while there appeared to be an area in the same general region that is being uplifted, thus the suggestion that some sort of compensatory relationship exists (Darwin, 1837).

In 1872, Dana took up a discussion of compensatory movements of elevation and subsidence of the coral islands of the Pacific Ocean area and extended it considerably beyond the rudimentary notions of Darwin. Such movements Dana described as "one of the great secular movements of the earth's crust," noting that the subsidence of islands in both the Pacific and the Atlantic associated with coral growth probably took place at the same time that certain other areas of the earth were being elevated, particularly the area of the North American continent (Dana,

Figure 6. Clarence E. Dutton. From George P. Merrill, *Contributions to the History of American Geology* (1904).

1872, p. 32–36). Dana applied the principle of crustal equilibrium in a somewhat different way the following year. He asked:

> When the material of the under-Appalachian sea was pushed aside by the subsiding Paleozoic deposits of the Appalachian region, what became of it? Some of it may have moved off southward. The chief part would have passed either to the *west*, or to the *east*. That it did not go *west* is evident from the ascertained fact that the oscillations in that direction during Paleozoic time were small; for the region was then, the larger part of the time, a mediterranean salt-water basin or sea, nurturing crinoids, corals and mollusks, and making limestones. If not westward, then it passed *eastward*: and if driven eastward, a geanticlinal elevation of a sea-border region parallel with the area of subsidence must have been in progress from lateral pressure (Dana, 1873b, p. 8).

Thus, it seems that Dana accepted an isostatic relationship by 1873, although his statements then were very general and sometimes ambiguous.

Years before, Hall had likewise speculated on the topic of compensatory subsidence and elevation that had been developed earlier by Herschel. Depression of the yielding mass and uplift at another location, Hall said, could explain a considerable number of geological phenomena. He called attention to the depression of accumulated matter along the synclinal axis of the Appalachian chain which, by displacing the yielding mass beneath, caused an uplift or bulging of the ocean to the west of that region, a bulge he had previously identified as the Cincinnati axis (Hall, 1859, p. 58). This uplift, taking place at a distance of a hundred miles or so to the west, may have prevented the accumulation of more sediments in that area. Meanwhile, the gradually subsiding sea floor allowed the formation of strata having their thickened edges toward the east, while they gradually thin out to the west, just as they now can be observed (Hall, 1859).

While he was preparing his 1857 AAAS presidential address for its 1883 publication, Hall invited comments on the essay from a few of his associates. These comments appear on the page proofs of the address (Hall Papers), and in some cases were incorporated into the final printed version. One such comment was made by John J. Stevenson, a former Hall assistant. It carried a hint of orientation to isostasy. Stevenson's note read: "Subsidence at one locality means elevation somewhere else—so while the ocean was subsiding, might not the Appalachians have risen?" Hall (1883, p. 69) apparently accepted the suggestion, elaborated on it, and had it printed as follows:

> During the long palaeozoic time the area of subsidence was in the Appalachian region, though clearly enough, during some portion of that time great uplifting occurred on the northeast, to be succeeded by subsidence which may have been equal to the elevation. Why could not the area of subsidence be changed from the Appalachian region to the ocean on the east? Subsidence in one locality means a corresponding, but not necessarily equal, elevation elsewhere; so while the ocean bed was subsiding may not the Appalachians have risen?

One could assume, surely, that Hall had specific locations in mind to support a generalization as broad as this one, but he does not include such information in this publication. Nevertheless, it appears that by 1883, Hall had joined the isostatic bandwagon.

Some years before in far-off India, John Henry Pratt, an English mathematician—theologian, then archdeacon of Calcutta, supplied an element of the doctrine of isostasy. As he was doing a trigonometric survey of the country, he was puzzled by the amount of deflection of his plumb-line when working in the vicinity of the large mountain mass in northern India. He concluded that there must be a deficiency of mass in the matter that lay below the mountains, an idea that he noted had first been suggested to him by George B. Airy, the astronomer-royal of Great Britain. Pratt speculated that when the earth had just entered a state no longer quite liquid, it was a perfect spheroid, but as the crust gradually solidified, contraction and expansion took place at various places causing corresponding depressions and elevations in the surface. If these surface movements were chiefly vertical in direction, then at any time a line is projected downward from any point on the surface to a sufficient depth, it must pass through a mass of matter which will be equal to that at any other location. Mountains were created by expansion which forced the surface upward, and the mass thus forced up must have a corresponding attenuation below (Pratt, 1859, p. 747–749). Pratt's assumption that the materials which compose mountains and their underlying structure are somewhat less dense than the materials found beneath an ocean basin appears in both the modern definition of isostasy and that stated by Dutton in 1889.

The part that glacial ice sheets can play in subsidence and elevation became a subsidiary factor in the newly emerging views on crustal equilibrium during the 1870s. Nathaniel S. Shaler (1875, p. 291) noted that accumulations of ice might depress regions when the mass became a mile or more thick. "We should expect to find," he asserted, "that such depression of one part of a continent would be attended by an uplift of another region. . . ." William J. McGee (1881) self-educated ethnologist-geologist, likewise wrote that these effects may be caused either by the deposition of sedimentary materials or by the formation of an extensive ice cap. Sedimentary accumulations, he believed, should cause a subsidence about equal to its own thickness, but ice sheets, because their density is only about one-third of the average rock stratum, would cause a lesser subsidence. In addition, because of the relatively short span of geological time during which an ice sheet is normally present and because of the extremely slow reaction of the terrestrial crust to the added weight, the actual depression would likely be much less. After the ice disappears at the end of any glacial period, the hydrostatic principles involved demand that the crust return to its original form. Shaler and McGee are but two of a number of individuals that speculated on the isostatic effects of ice sheets.

DUTTON'S THOUGHTS

Dutton did not apply the term *isostasy* to the principle of crustal equilibrium publicly until 1889, but he had enunciated many of its basic assumptions several years earlier. A summary of his thinking appeared in 1876 in a paper in which he analyzed a number of contemporary theories dealing with orogenic geology. "It has been indicated," Dutton (1876, p. 424) wrote,

that plications occur where strata have rapidly accumulated in great volume and in elongated narrow belts; that the axes of plication are parallel to the axes of maximum deposit; and that the movements immediately followed the deposition. All of these facts are covered by the cause here suggested. Wherever the load of sediments becomes heaviest, there they sink deepest, protruding the colloid magma beneath them to the adjoining areas which are less heavily weighted, forming at once both synclinals and anticlinals.

The language in this passage is strikingly similar to that used by Hall, and it suggests that he had been strongly influenced by the New Yorker, although he did not directly cite Hall in this article. Later in that same publication, Dutton (1876, p. 430–431) gave indications that he was near to the perfection of the isostatic principle when he stated that:

the transfer of great bodies of sediment from one portion of the earth's surface to others, is tantamount to a disturbance of the earth's equilibrium of figure, which the force of terrestrial gravitation constantly tends to restore, and which it inevitably will restore wholly or in part, if the materials of which it is composed are sufficiently plastic.

In this article Dutton displayed an excellent understanding of the growth of geological theory, specifically denying that his conclusions were intended to provide a comprehensive theory for the origin of the earth's physical features. Such a theory, he said, is accomplished only by the work of a generation of great men. Instead, he offered these assumptions as indispensable factors for the final theorem, and they were advanced in an effort to break the deadlock which, he said, "has hitherto beset all inquiry into this magnificent and mysterious province of scientific research, and has apparently driven a large body of geologists into a premature acceptance of the contractional hypothesis" (Dutton, 1876, p. 430).

Dutton's next major published article appeared in *Nature* in 1879. Therein he reported his observations and speculations derived from several years' participation in J. W. Powell's U.S. Survey of the Rocky Mountain Region as Powell's assistant. The primary topic of the article is the geological history of the Grand Canyon of the Colorado. He noted the vast erosion of the region, especially of the canyon itself, and evidence that convinced him that there must also have been compensatory elevation of various parts of the area. His conclusions in his own words read as follows (Dutton, 1879, p. 251):

It will be seen that these local uplifts are important in determining the subdivisions of the area and the distribution of the maxima and minima of degradation. We may see here a correspondence which is worthy of close attention. Those areas which have been uplifted most have been most denuded. I have asked myself a hundred times whether we might not turn this statement round, and say that those regions which have suffered the greatest amount of denudation have been elevated most, thereby assuming the removal of the strata as a cause and the uplifting as the effect; whether the removal of such a mighty load as ten

thousand feet of strata from an area of ten thousand square miles may not have disturbed the earth's equilibrium of figure, and that the earth, behaving as a quasi-plastic body, has reasserted its equilibrium by making good a great part of the loss by drawing upon its whole mass beneath.

When one reads through the entire text of this paper, it seems, indeed, that Dutton's whole approach to the study of the Colorado Plateau and the Grand Canyon was based upon the assumption of crustal equilibrium. It is perhaps the most revealing of any in the Dutton literature of the thought processes through which he came to the principle of isostasy.

OTHER STATEMENTS ON CRUSTAL EQUILIBRIUM

The ideas under development in America concerning crustal equilibrium did not go unnoticed across the Atlantic, especially in Great Britain. Charles Ricketts (1883), in an important article in the *Geological Magazine,* summarized the work of a number of individuals during the previous half-century including Herschel, Babbage, Darwin, Hall, Dutton, Hermite, Lyell, Fisher, and others. At one point he complained that the whole subject had hardly been taken under consideration on his side of the Atlantic, although he did contradict himself by acknowledging the thinking done by a number of Europeans on the subject of crustal equilibrium. He cited Dutton's study of the Grand Canyon, and he then applied those same ideas to various geological phenomena in Great Britain.

In Osmond Fisher's important book of 1881, *Physics of the Earth's Crust,* we find an extensive discussion of the characteristics and behavior of the crust of the earth. Dutton's review (1882) of this book reveals his concurrence with Fisher's ideas, and in a footnote Dutton noted that he had long since been convinced that Fisher's belief in a plastic substratum on which rests a solid crust "must form an important part of any true theory of the earth's evolution" (Dutton, 1882, p. 289). In a second edition of this book, Fisher revised his text to include the notion of hydrostatic equilibrium in the earth's crust; that is, using the Himalayan region as an example, sediments are removed from the mountains and deposited on the plains below, after a time the mountains will become relatively lighter and will rise, while the plains will sink from the overload of sediments in order to reestablish the equilibrium (Fisher, 1889, p. 134).

G. K. Gilbert (1886), in an oft-quoted address to the American Society of Naturalists, presented an effective argument in favor of the hypothesis of crustal equilibrium, although he admitted in his presentation that he was being highly speculative and that his example, the extinct Lake Bonneville in Utah, was limited. He first described the general region occupied by the ancient lake and how today this same general area appears to have been deformed so that the former lake bottom presents the appearance of a low, broad dome with its crest near the center of the old lake. He began his discussion with the assumption "well known to geologists" that large deposition of sediments will cause the sea floor to sink locally as rapidly as the sediments are added. At the same time, the adjacent continent, which provides the sediments, rises at a rate equivalent to the amount of surface degradation. "It is a favorite theory—at least with that large division of geologists who consider the interior of the earth as mobile—," he continued, "that the sea-bottom sinks in such cases because of the load of sediment that is added and that the land is forced up hydrostatically because it is unloaded by erosion." Similarly, draining away the water of Lake Bonneville, which had been approximately 1,000 feet in depth, would give the supposed liquid interior an irresistible force, which in turn created the dome-like structure. By a mathematical analysis of this hypothesis, he found that he could retrodict an elevation of some 200 feet which agreed quite well with the approximate elevation he ascribed to the dome (Gilbert, 1886, p. 290–295).

ISOSTASY IS NAMED

Finally, after having been in the wings for many years, the new and important concept called isostasy was finally presented in rather unauspicious circumstances. The April 27, 1889, meeting of the Washington Philosophical Society was attended by only 36 persons. Dutton's paper had been hastily prepared. Yet in terms of its importance to the science of geology, it turned out to be a landmark presentation. Not only did Dutton lay out a fully developed version of his isostasy concept, he also mounted a comprehensive attack on the contractional hypothesis as a causal factor for elevation and subsidence. "The contractional theory," he asserted,

gives us a force having neither direction nor determinate mode of action, nor definite epoch of action. It gives us a force acting with a far greater intensity than we require, but with far less quantity. To provide a place for its action it must have recourse to an arbitrary postulate assuming for no independent reason the existence of areas of weakness in a supposed crust which would have no *raison d'être* except that they are necessary for the salvation of the hypothesis.

In this presentation Dutton sidestepped what he viewed as one of the great contemporary problems of physical geology—"the cause of general elevations and subsidences"—examples of which he cited being the Himalayan plateau and the whole North American continent. He denied that isostasy could explain those phenomena, saying that:

it is sufficiently obvious that the theory of isostasy offers no explanation of these permanent changes of level. On the contrary, the very idea of isostasy means the conservation of profiles against lowering by denudation on the land and by deposition on the sea bottom provided no other cause intervenes to change those levels.

He insisted that geologists must look for some independent principle of causation that would suffice to explain the gradual and permanent change in the profiles of the land and sea bottom. "And I hold this cause to be an independent one," he argued, going on to assert that:

it has been much the habit for geologists to attempt to explain the

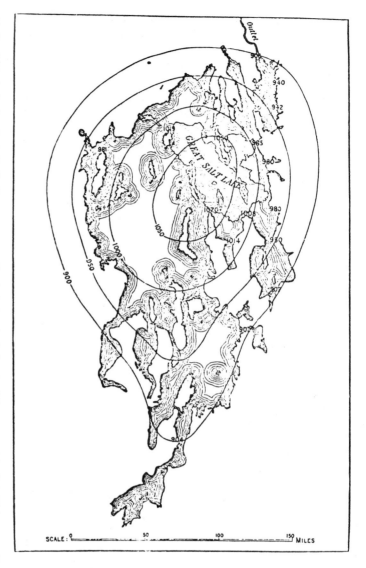

Figure 7. Map of the extinct Lake Bonneville, with hypothetic contours to illustrate the subsequent deformation of the earth's crust, by G. K. Gilbert. The horizontal figures mark determinations of the height of the old shoreline above Great Salt Lake. From Gilbert (1886).

progressive elevation of plateaus and mountain platforms, and also the foldings of the strata by one and the same process. I hold the two processes to be distinct and have no necessary relation to each other.

In contrast to a number of contemporary American geologists who relied completely on the contractional hypothesis to explain causal factors in crustal movements, Dutton freely acknowledged that he did not know how to answer these questions, saying that "what the real nature of the uplifting force may be is, to my mind, an entire mystery" (Dutton, 1889, p. 62–64).

REACTION TO THE CONCEPT OF ISOSTASY

In sharp contrast to the years of quiet that followed Hall's presentation of the basics of the geosyncline in 1859, the reaction to Dutton's presentation was vigorous and immediate. At the meeting where Dutton presented his paper on isostasy, Robert S. Woodward, of the U.S. Geological Survey, vigorously contended that secular contraction does indeed play an important part in the crumpling of the earth's surface, Dutton's remarks notwithstanding, and that secular contraction is "an essential basis for the hypothesis of isostasy" (Woodward and Gilbert, 1889, p. 536–537). Later Woodward recognized that there were some severe difficulties with the contractional hypothesis, but he asserted that it did provide a reasonable basis from which isostasy could become a viable theory; that since isostasy tends toward a condition of equilibrium at a relatively rapid rate, it would simply run down at an early geological age. By contrast, he said, the process of contraction goes ahead at a very slow rate, and all the while it tends to oppose the equilibrium toward which isostasy leads (Woodward, 1891, p. 196).

Soon thereafter McGee likewise concluded that isostasy alone could not account for all the phenomena of elevation and subsidence, although he seemed to believe that isostasy is one of the essential principles involved. He briefly reviewed the historical development of the idea of isostasy in this paper, recalling Hall's subsidence concept, the determination of the density of various regions of the earth first by Pratt and later by Fisher and others, and finally the introduction of the concept by isostasy itself by Dutton. Then, comparing the areas of degradation and sedimentation of the Mississippi river system with the known and inferred amounts of elevation and subsidence of that area, he arrived at what he thought to be some significant generalizations about the area. "The data relating to the condition of the earth's crust derived from the modern Gulf of Mexico indicate that throughout the vast geologic province of southeastern North America, isostasy is probably perfect," he reported, "*i.e.,* that land and sea bottom are here in a state of hydrostatic equilibrium so delicately adjusted that any transfer of load produces a quantitatively equivalent deformation." However well he thought the Gulf of Mexico illustrated Dutton's concept, McGee denied it could satisfactorily explain all movements of elevation and subsidence (McGee, 1892, p. 177, 185–189). Dutton (1889, p. 63) had already denied that isostasy could.

In his presidential address to the Geological Society of America in 1892, Gilbert discussed continental problems as he viewed them. At one point he contrasted a doctrine of rigidity of the earth's crust with that of isostasy. What are the conditions, he queried, under which the earth's surface is differentiated into oceanic and continental plateaus and how are the continents supported? In attempting to answer these questions, he (Gilbert, 1892, p. 179–180) stated that:

the entire weight of the continental plateau, pressing on the tract beneath it, tends to produce a transfer of material in the direction from left to right, resulting in the lowering of the higher plateau and the raising of the lower. To the question, how this tendency is counteracted, two general answers have been made: first, that the earth, being solid, by its rigidity maintains its form; second, that the materials of which consist the conti-

nental plateau and the underlying portions of the lithosphere are, on the whole, lighter than the materials underlying the ocean floor and that the difference in density is the complement of the difference in volume, so that at some level horizon far below the surface the weights of the superincumbent columns of matter are equal. The first answer regards the horizontal variations of density in the earth's crust as unimportant; the second regards them as important. The first may be called the doctrine of terrestrial rigidity; the second has been called the doctrine of isostasy. At the present time the weight of opinion and, in my judgment, the weight of evidence lie with the doctrine of isostasy.

Gilbert, never one to let his thinking get rigid, effectively reflects the unsettled state of orogenic theory of the late nineteenth century, although others such as Dana did not hesitate to come to firm conclusions as to how things really are.

In 1893 two major discussions of Dutton's isostatic concept appeared. Bailey Willis of the U.S. Geological Survey provided another instance in which the contractional hypothesis and isostasy were believed to complement each other. He first discussed several conflicting hypotheses concerning the Appalachian range, then concluding that Dutton had used faulty assumptions when he asserted that secular loss of heat could never have provided the degree of contraction seen in the phenomena. Fisher's computations, on which Dutton had based his conclusions, Willis alleged, were questionable. He recalled Dutton's conclusion that elevation of strata, on the one hand, and folding and plication, on the other, can be and often are entirely separate processes. Furthermore, examples can easily be located where elevated areas are very little plicated. Dutton, said Willis, "is shown by his own assumptions and by the opinions of his eminent supporters to have confused the lesser problem of zonal compression with the far greater one of deformation of the spheroid." Thus confused, Dutton's objection to the contractional hypothesis on quantitative grounds seemed invalid (Willis, 1893, p. 279–280).

According to Willis, Dutton had objected to the contractional hypothesis on qualitative grounds in two ways. First, Dutton had contended that any force produced by contraction would act equally in all directions, and second, there must be a zone of weakness in the strata, which, said Willis, Dutton had described as "required for the salvation of the hypothesis." Willis answered the latter objection by reference to experimental studies he himself had conducted by which he determined that conditions which ascertain places of folding and plication of the strata are inherent in the attitude of the strata, not in the thickness thereof nor in the forces involved. Concerning Dutton's first objection to the contractional thesis, Willis noted that Dutton himself had already answered that in his presentation on movements of isostatic adjustment. According to Willis, "contraction gives to isostasy a needed force; isostasy directs contraction; the two effect a result which neither alone could bring about," a relationship Willis attributed to Dutton himself. Willis thus concluded that the concepts of contraction and isostasy were not contradictory, as Dutton had asserted, but instead complemented each other (Willis, 1893, p. 279–280). One might note that Willis seems to have not understood Dutton's statements of 1889.

Willis then described how this process worked in the case of the Appalachians. Degradation of a preexisting continent progressed, and the bulk of the sediments were deposited in a narrow belt along the shore of the adjacent ocean. Subsidence of this belt produced synclines of deposition, and in their depths the temperature rose as the strata sank, producing expansion and the beginnings of complex folding. Because the condition of isostasy prevails in the earth's mass, compensation must be made to the continental area for the load taken from it, and materials at a great depth flowed landward in a quantity sufficient to restore the elevation of the continent. Meanwhile, as sedimentation continued and isostatic adjustment began, the nucleus steadily contracted, resulting in a compression strain with no determinate direction or effect. Here, said Willis, are the three conditions needed for the mountain-building process to take place: sedimentation, isostatic adjustment, and contraction. "There came a time when isostasy gave direction," he speculated, "and contraction gave a force to a movement of the submarine earth's crust toward the land, a movement extending seaward far beyond the zone of maximum sedimentary deposits, now folded and including great extent of strata, now as then flat" (Willis, 1893, p. 279–280). Willis can properly be described as eclectic in his geological speculations, using Hall's geosyncline, the contractional hypothesis of Dana and others, and Dutton's isostasy. Go along and get along, perhaps?

That same year LeConte (1983) composed yet another paper on the origin of mountain ranges, a subject which seemed dear to his heart. He now rejected Dana's view that a geosyncline must precede sedimentation, indicating that "the place of mountains while in preparation, in embryo, before birth, *was* gradually subsiding, as if borne down by the weight of accumulating sediments," a statement Hall could certainly agree with.[9] Nevertheless, he contended that this was the same sort of a trough that Dana called a geosyncline, although careful reading of Dana does not indicate this. LeConte then reviewed the isostatic principle, noting that if the earth's crust yields under an increasing load, then it ought to rise by unloading. He recalled that in the Colorado Plateau region more than 10,000 feet have been eroded away and yet 8,000 feet remain, while in the Uinta range at least 30,000 feet of material have been carried away, yet 10,000 feet remain. Evidently, he concluded, *pari passu* elevation must accompany lightening by erosion. From this he agreed with Dutton that the earth in its general form, as well as in its larger inequalities, is in a state of gravitational equilibrium (LeConte, 1893, p. 552–553).

In LeConte's view the principle of isostasy need not succumb to the supposition of a solid earth, which he believed to follow from its cosmic behavior. He rejected the liquid-interior hypothesis, concluding that the earth, although exceedingly rigid to a rapidly acting force, yields viscously to heavy pressures that act over a large area for extremely long periods of time. He also supposed that the earth is not necessarily or absolutely homogeneous either in density or in the conductivity of heat. In the process of secular cooling and contracting, denser and more conductive areas, because they cooled and contracted more quickly,

subsided and became the ocean bottoms while the light and more slowly cooling areas remained as the prominent land surfaces. "And thus to-day the ocean basins are in gravitative equilibrium with the continental areas," he speculated, "because in proportion as oceanic radii are shorter are the materials also *denser,* and in proportion as the continental radii are *longer,* are the materials also specifically lighter." Thus appears the condition of gravitative equilibrium Dutton called isostasy (LeConte, 1893, p. 552–553).

In 1895 a new series of gravitational measurements led Gilbert to modify somewhat his position on the concept of isostasy. Noting that the new data had been taken at a number of places extending from the East Coast to Salt Lake City and in California, he discussed the general postulate that continents and ocean beds are in isostatic equilibrium. He believed that the Great Plains area between the Rockies and the Appalachians had apparently been exempt from orogenic disturbances for a number of geologic periods, a time during which there seemed to have been ample opportunity for gradual relief through the various agencies of viscous flow, degradation, sedimentation, and strains brought on by gravity in connection with discrepancies in density. Nevertheless, he found the values of gravity at all the plains stations to be notably accordant. But when he compared those computations with others made at various stations in the Rocky Mountains, he estimated that if those mountains were converted to a high plateau, it would have an elevation of between 2,000 and 2,500 feet higher than the adjacent plain to the east. "The conclusion is thus reached," said Gilbert (1895), "that the whole mountain mass above the level of its base is in excess of the requirement for isostatic adjustment; or, in other words, is sustained by the rigidity of the earth." He (Gilbert, 1895, p. 331–333) then generalized that:

these results tend to show that the earth is able to bear on its surface greater loads than American geologists, myself included, have been disposed to admit. They indicate that unloading and loading through degradation and deposition cannot be the cause of the continued rising of mountain ridges with reference to adjacent valleys, but that, on the contrary, the rising of mountain ridges, or orogenic corrugation, is directly opposed by gravity and is accomplished by independent forces in spite of gravitational resistance.

What those independent forces were, Gilbert did not say. One notes that in this article, Gilbert tended to reverse his earlier position.

CONCLUSIONS

At the 1896 annual meeting of the AAAS, a session of the geology section was devoted to the celebration of Hall's sixtieth anniversary of public service. McGee addressed the group, paying tribute to Hall and his role in creation of the concept of the geosyncline. Furthermore, although Dutton and others had been most instrumental in creating the formal doctrine of isostasy, in Hall's "Introduction" of 1859 is found "one of the most important contributions ever made to the doctrine of isostasy" (McGee, 1896, p. 702–703). The general tenor of McGee's remarks, in spite of Gilbert's recently expressed reservations about isostasy, seemed to indicate that both the geosyncline and isostasy had been relatively well accepted as integral parts of theoretical orogeny. The contractional hypothesis likewise was a basic component of the science of geology in spite of the reservations of Dutton and others. Isostasy, originally conceived by Dutton as a specific replacement for the contractional theory, was frequently considered complementary to it rather than contradictory, especially among those geologists who were either thoroughly committed to the contractional view, as in the case of Dana, or, like Woodward, did not consider isostasy as a satisfactory answer to all dynamic problems of geology. These three concepts did not answer all questions concerning mountain building, but their combination satisfied many geologists of that era. Many of the arguments of that time concerned the mechanism by which mountains are elevated. Isostasy was not completely satisfactory nor was the geosyncline, for neither really provided that mechanism. Contraction did, no matter how unsatisfactory we find that theory today.

Yet it is the contractional hypothesis which has been relegated to the theoretical "junk heap" of science, while isostasy and perhaps the geosyncline have survived as viable concepts today. The latter is at present subject to considerable controversy. One critic of this paper asserts that "the geosyncline has *not* survived, and also belongs on the *historical* 'junk heap'. Plate tectonics has so completely modified our thinking about the role of thick strata in the orogenic belts, that the old (pre-plate tectonics) geosynclinal concept has no value any more *except* as an important historical artifact." In contrast to this view is one expressed by Frederick Schwab (1982, p. 5) who states that:

because our models of what ancient geosynclines represent and where their modern analogues now exist have changed, it is also necessary to modify the terminology and classification schemes applied to geosynclines. This volume demonstrates that geosynclinal theory repeatedly has been flexible enough to be adopted to new analytical models of the earth. The changes made imperative by the universal adoption of plate tectonics mechanisms are neither cosmetic or apocalyptic, although they are probably more fundamental in nature than any of the earlier modifications. The basic aspects of geosynclinal theory—the single most unifying concept in geology—can certainly be accommodated within the premises of sea-floor spreading, continental drift, and plate tectonics. For the future, the concept will probably continue to be reshaped to conform to changing modern perceptions.

In between these two extremes is Dott (1974, p. v), who is rather equivocal in asking:

is the geosyncline extinct, after all, or has the leopard merely changed its spots? Fantastic variability of sediments called 'geosynclinal,' as well as the jumbled tectonic elements in orogens, is inescapable. The difficulty of generalizing, together with the misleading deterministic overtones and semantical confusion associated with 'geosyncline,' may be sufficient argument for abandoning most existing terminology. It may be best to clean the slate and develop new language.

Again in a later publication, Dott (1979, p. 257) asserts that the geosyncline "commonly has been proclaimed as one of geology's most important unifying concepts, but lately has been scoffed at as either dead or so profoundly altered by plate tectonics as to be an anachronism." In spite of any apparent equivocation, Dott and Batten (1981, p. 115–129) seem to have little trouble making use of a classical presentation of the geosynclinal concept in their popular historical geology text.

Whatever the future of the geosynclinal concept may be, we find that the study of its origins and of its sister concept, isostasy, provides us today with an excellent example of the birth and maturation of ideas of science, how such ideas can survive despite the fact that there is often insufficient evidence to support them, and how they survive in spite of strong arguments to the contrary. Faith in one's ideas continues to be a major factor in scientific endeavor.

EXPLANATORY FOOTNOTES

[1]The first American edition of Lyell's *Principles of Geology* is cited because this is the edition that Hall had available when he was formulating the geosynclinal concept.

[2]Geikie was behind times in 1903. Although plate tectonic theory was still several decades in the future, many geologists including Lyell, Hall, Dutton, and others already found the contractional hypothesis unsatisfactory.

[3]The fourth geological district was comprised of the western 16 counties of the state plus part of a seventeenth.

[4]Although Hall ordinarily preferred North American names for North American formations, in this case he used English terminology.

[5]This quotation is from Babbage (1847). Hall referred to an abbreviated version of this article published by Babbage in 1837, but since at other times he referred to other items in the *Quarterly Journal of the Geological Society of London,* it seems reasonable to believe that he had access to Babbage's 1847 article in that journal. This article is more detailed than the 1837 version.

[6]This is the published version of Hall's 1857 address to the AAAS. The manuscript of that address is substantially the same as the published version.

[7]Interestingly, at various places in his 'Introduction" of 1859, Hall compared various formations in North America with formations in Europe, particularly in Great Britain, by examining the fossil content, similarity of position, direction of accumulation, etc. How exciting this would have been for him had he known of continental drift theory!

[8]Dutton, in his review of Fisher's *Physics of the Earth's Crust* in 1882, stated that in an earlier unpublished paper he had used the terms "isostasy" and "isostatic," but he did not make these terms public until 1889 (Dutton, 1882, p. 289).

[9]A reader of this essay asked: "The quotation sounds like an *endorsement* of the geosyncline???" Indeed, it is, although in this article LeConte was ambiguous. The quotation is pure Hall, but LeConte immediately contradicted himself by asserting that this was Dana's view. Dana himself believed that the subsidence preceded deposition and was thus not concurrent with it.

REFERENCES CITED

Babbage, C., 1837, The Ninth Bridgewater Treatise: A fragment: London, John Murray.

—— 1847, Observations on the Temple of Serapis, at Pozzuoli, near Naples; with remarks on certain causes which may produce geological cycles of great extent: The Quarterly Journal of the Geological Society of London, v. 3, p. 186–217.

Clarke, J. M., 1923, James Hall of Albany, geologist and palaeontologist, 1811–1898: Albany, n. p.

Dana, J. D., 1863, Manual of geology: treating of the principles of the science with special reference to American geological history . . .: Philadelphia Theodore Bliss & Co., 798 p.

—— 1866, Observations on the origin of some of the earth's features: American Journal of Science, series 2, v. 42, p. 205–211.

—— 1872, On the oceanic coral subsidence: American Journal of Science, series 3, v. 4, p. 31–36.

—— 1873a, On the origin of mountains: American Journal of Science, series 3, v. 5, p. 347–350.

—— 1873b, On some results of the earth's contraction from cooling: American Journal of Science, series 3, v. 5, p. 423–443; v. 6, p. 6–14, 104–115, 161–172.

—— 1875, Recent changes of level on the coast of Maine, with reference to their origin and relation to other similar changes, by Shaler, N. S., American Journal of Science, series 3, v. 9, p. 316–318.

—— 1880, Manual of geology, treating of the principles of the science, with special reference to American geological history: 3rd ed., New York, Ivison, Blakeman and Co., 911 p.

—— 1891, On Percival's map of the Jura-Trias trap-belts of central Connecticut, with observations on the up-turning, or mountain-making disturbance of the formation: American Journal of Science, series 3, v. 42, p. 439–447.

—— 1896, Manual of geology; treating of the principles of the science with special reference to American geological history: 4th ed., New York, American Book Co., 1087 p.

Darwin, C., 1837, On certain areas of elevation and subsidence in the Pacific and Indian oceans, as deduced from the study of coral formations: Proceedings of the Geological Society of London, v. 2, p. 552–554.

—— 1842, The structure and distribution of coral reefs. Being the first part of the geology of the voyage of the Beagle under the command of Capt. Fitzroy, R.N., during the years 1832 to 1836: London, Smith, Elder, and Co., 344 p.

Dott, R. H., Jr., 1979, The geosyncline—first major geological concept "Made in America," in Schneer, C., ed., Two Hundred Years of Geology in America: Hanover, N.H., University Press of New England, p. 238–267.

——and Batten, R. L., 1981, Evolution of the Earth: 3rd ed., New York, McGraw-Hill, 573 p.

Dott, R. H., Jr., and Shaver, R. H., 1974, Modern and ancient geosynclinal sedimentation: Society of Economic Paleontologists and Mineralogists, Special Publication No. 19, in honor of Marshall Kay, Tulsa, Society of Economic Paleontologists and Mineralogists.

Dutton, C. E., 1876, Critical observations on theories of the earth's physical evolution: The Penn Monthly, v. 7, p. 364–378, 417–431.

—— 1879, The geological history of the Colorado river and plateaus: Nature, v. 19, p. 247–252, 272–275.

—— 1882, Physics of the Earth's Crust: by the Rev. Osmond Fisher, M.A., F.G.S.: American Journal of Science, series 3, v. 11, p. 283–290.

—— 1889, On some of the greater problems of physical geology: Bulletin of the Philosophical Society of Washington, v. 11, p. 51–64.

Fisher, O., 1889, Physics of the earth's crust: 2nd ed., London, Macmillan and Co.

Geikie, A., 1903, Text-book of geology: 4th ed., 2 vols., London, Macmillan and Co.

—— 1911, Geology: Encyclopaedia Britannica, 11th ed., vol. 11, p. 890–896.

Gilbert, G. K., 1886, The inculcation of scientific method by example, with an illustration from the Quaternary geology of Utah: American Journal of Science, series 3, v. 31, p. 284–299.

—— 1892, Continental Problems; annual address by the president, G. K. Gilbert: Bulletin of the Geological Society of America, v. 4, p. 179–190.

—— 1895, New light on isostasy: Journal of Geology, v. 3, p. 331–334.

Gilman, D. C., 1899, The Life of James Dwight Dana, scientific explorer, mineralogist, zoologist, professor in Yale University: New York, Harper and Brothers, 409 p.

Greene, M., 1982, Geology in the nineteenth century; changing views of a changing world: Ithaca, N.Y., Cornell University Press, 324 p.

Hall, J., 1838, Second annual report of the fourth geological district of New York, State of New York, communication from the governor, relative to the geological survey of the state: Second annual report of the geological survey, Assembly No. 200, February 20, 1838.

—— 1840, Fourth annual report of the survey of the fourth geological district, State of New York, communication from the governor, transmitting several reports relative to the geological survey: Fourth annual report of the geological survey, Assembly No. 50, January 24, 1840.

—— 1843, Geology of New York, Part IV, comprising the survey of the fourth geological district: Albany, State of New York.

—— 1851, Lower Silurian system: in Foster, J. W., and Whitney, J. D., Report on the geology of the Lake Superior land district; Part II, the iron region, together with general geology: U.S. Senate, Executive No. 4, 32nd Congress, Special Session, March, 1851.

—— 1857, Contributions to the geological history of the American continent: MSS of presidential address to the AAAS, 1857, located in the Hall Papers, fol. 15, New York State Library, Albany, New York.

—— 1859, Introduction: in Vol. 3, Palaeontology of New York, 8 vols. in 13, Albany, State of New York, p. 1–96.

—— 1860, On the formation of mountain ranges: Canadian Journal, new series, v. 5, p. 542–544.

—— 1883, Contributions to the geological history of the American continent: Proceedings of the American Association for the Advancement of Science, v. 31, p. 29–69.

——and Whitney, J. D., 1858, Report on the geological survey of the state of Iowa; embracing the results of investigations made during portions of the years 1855, 56, & 57: Des Moines, State of Iowa.

Henry, J., 1858, Letter to Hall, in Clarke, James Hall of Albany, geologist and palaeontologist, 1811–1898: 1923, p. 327–328.

Herschel, J.F.W., 1836a, Letter to Lyell, Fredhausen, Cape of Good Hope, Feb. 20, 1836, in Babbage, C., 1837, The ninth Bridgewater treatise; a fragment: London, John Murray.

—— 1836b, Letter to Murchison: in Babbage, 1837, The ninth Bridgewater treatise; a fragment: London, John Murray.

Hunt, T. S., 1861, On some points in American geology: American Journal of Science, series 2, v. 31, p. 392–414.

—— 1867, Letter to Hall: Hall Papers.

—— 1873, On some points in dynamical geology: American Journal of Science, series 3, v. 5, p. 264–270.

LeConte, J., 1872, A theory of the formation of the great features of the earth's surface: American Journal of Science, series 3, v. 4, p. 345–355, p. 460–472.

—— 1873, On the formation of features of the earth-surface. Reply to the criticisms of T. Sterry Hunt: American Journal of Science, series 3, v. 5, p. 448–453.

—— 1893, Theories of the origin of mountain ranges: Journal of Geology, v. 1, p. 543–573.

——et al., 1896, Honors to James Hall at Buffalo: Science, new series, v. 4, p. 697–717.

Lyell, C., 1837, Principles of geology . . .: 1st American ed., 2 vols., Philadelphia, James Kay, Jun. & Brother, 1099 p.

—— 1845, Travels in North America in the years 1841–42; with geological observations in the United States, Canada, and Nova Scotia: 2 vols. in 1, New York, Wiley and Putnam.

—— 1849, A second visit to the United States of North America: 2 vols., London,

John Murray.

——1855, A manual of elementary geology; or, the ancient changes of the earth and its inhabitants as illustrated by geological monuments: 5th ed., London, John Murray, 647 p.

Mayo, D. E., 1968, The development of the idea of the geosyncline [unpublished Ph.D. dissertation]: University of Oklahoma, Norman, Oklahoma.

McGee, W. J., 1881, On the local subsidence produced by an ice-sheet: American Journal of Science, series 3, v. 22, p. 368–369.

——1892, The Gulf of Mexico as a measure of isostasy: American Journal of Science, series 3, v. 44, p. 177–192.

——1896, James Hall, founder of American stratigraphy: Science, new series, v. 4, p. 700–706.

Pratt, J. H., 1859, On the deflection of the plumb line in India, caused by the attraction of the Himalaya mountains and the elevated regions beyond; and its modification by the compensating effect of matter below the mountain mass: Philosophical Transactions of the Royal Society of London, v. 149, p. 745–778.

Ricketts, C., 1883, On accumulation and denudation, and their influence in causing oscillation of the earth's crust: Geological Magazine, v. 20, p. 302–306, 348–356.

Schwab, F. L., ed., 1982, Geosynclines; concept and place within plate tectonics: Benchmark papers in geology, 164, Stroudsburg, Pa., Hutchinson Ross Publishing Co., 411 p.

Shaler, N. S., 1875, Notes on some of the phenomena of elevation and subsidence of the continents: Proceedings of the Boston Society of Natural History, v. 17, p. 288–292.

Willis, B., 1893, The mechanics of the Appalachian structure, *in* Geology, part II, thirteenth annual report of the United States Geological Survey to the Secretary of the Interior, 1891–92, ed. by Powell, J. W., Washington, Government Printing Office.

Woodward, R. S., 1981, The mathematical theories of the earth: in the annual report of the board of regents of the Smithsonian Institution showing the operations, expenditures, and condition of the Institution to July, 1890, p. 190–200, Washington, Government Printing Office.

Woodward, R. S., and Gilbert, G. K., 1889, An abstract of remarks on "Some of the greater problems of physical geology" by Dutton, C. E., Bulletin of the Philosophical Society of Washington, v. 11, p. 536–537.

Printed in U.S.A.

Geological Society of America
Centennial Special Volume 1
1985

Evolving tectonic concepts of the central and southern Appalachians

Rodger T. Faill
Pennsylvania Geological Survey
P.O. Box 2357
Harrisburg, Pennsylvania 17120

ABSTRACT

The Appalachian mountain system of eastern North America comprises a complex of deformed rocks which has stimulated geologic thinking since the area was first settled in the 17th century. Little attention was paid to it in the early days of European settlement, but in the early nineteenth century interest grew rapidly, leading to the creation of numerous state surveys and research that has continued unabated to this day. The earliest tectonic concepts were understandably rather primitive and mostly concerned the simpler foreland portion of the orogen. Upheavals, explosions, and other vertical motions were the main ingredients of the theory of the Rogers brothers and the geosyncline of Hall and Dana. Awareness of the importance of horizontal movements increased as continued detailed mapping revealed the presence of large thrust faults, culminating in the middle of this century with the elucidation of a décollement tectonics that pervades the entire Appalachian foreland.

The Appalachian crystallines were originally considered to be part of the crust, and they formed a large part of Appalachia, Dana's geanticline that supplied sediment to the adjoining geosyncline. Studies early in the 1900s began to reveal the complexities of this terrane and by the middle of this century, the crystallines had been divided into a number of distinctive belts. With the application of the plate tectonics scheme in the 1970s, geologists realized that these belts probably are micro-continents, either broken off from the Laurentian craton, or having come from other plates. With the accession of deep seismic data, the crystallines are no longer thought of as being rooted in the crust, but are believed to have undergone an allochthony at least as extensive as the foreland's.

INTRODUCTION

One can still marvel at the recent change of paradigms for the Appalachian mountain system, the "Eastern Overthrust Belt," the Piedmont, the Carolina slate belt and other crystalline terranes (Fig. 1) were once thought (and not that long ago) to be firmly rooted in the continental crust, an integral part of the Laurentian craton. The probability, or even possibility, that these terranes have traveled 250 km or more on horizontal faults would have seemed fanciful, even as little as 25 years ago. That the Charlotte, the Kiokee, or the Raleigh belts are micro-continents that may have traveled hundreds or even thousands of kilometers to "dock" against the Laurentian craton would have been scoffed at as another variation of the then largely discredited theory of continental drift. But these concepts have rapidly become the cornerstones of the evolving interpretation of the Appalachian orogen. So well do the various parts merge together within this new framework that other, older viewpoints have almost completely disappeared from the recent literature.

The plate tectonics paradigm developed quickly, during the late 1960s and the 1970s, and constitutes one of the most rapid and far-reaching revolutions of geologic thinking. It grew largely out of studies of the ocean basins, but it has since been applied to continents and their mountain systems. One basic idea of plate tectonics is that portions of the Earth's crust have travelled large distances horizontally. Prior to the development of plate tecton-

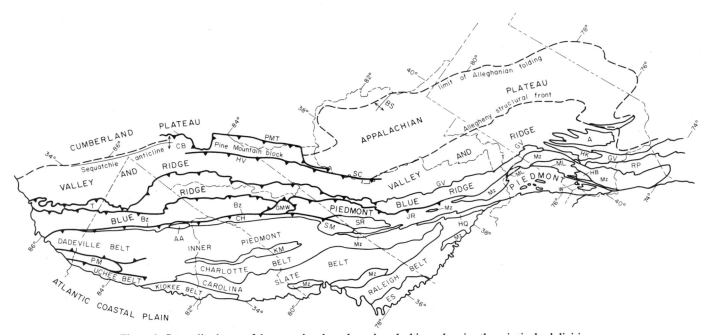

Figure 1. Generalized map of the central and southern Appalachians, showing the principal subdivisions and belts. Appalachian foreland consists of the Valley and Ridge, Appalachian Plateau and Cumberland Plateau provinces. The crystallines include the Blue Ridge and all the belts southeast to the Atlantic Coastal Plain, with the exception of the Mesozoic basins. Geologic terranes (belts) and other features referred to in the text: A — Anthracite basins, AA — Alto allochton, AQ — Arvonia and Quantico synclines, BS — Burning Springs anticline, BZ — Brevard zone, CB — Cumberland block, CH — Chauga belt, ES — Eastern slate belt, GMW — Grandfather Mountain window, GV — Great Valley, HB — Honey Brook upland, HK — Hamburg klippe, HV — Hunter Valley fault, JR — James River synclinorium, KM — Kings Mountain belt, ML — Martic line, MZ — Mesozoic basins, PM — Pine Mountain belt, PMT — Pine Mountain thrust, RP — Reading Prong, SC — St. Clair thrust, SM — Sauratown Mountains anticlinorium, SR — Smith River allochthon, T — Talladega belt, and W — Woodville structure.

ics, the extent of large horizontal transport within the Appalachians had been only gradually deduced by the many geologists who worked here (especially in the foreland portion) over the past 100 or so years. The earliest workers in the Appalachians viewed orogenesis largely in terms of a vertical tectonics. Mountanous terranes were seen as being intruded, fissured, folded, and faulted, but they moved little from their original place. Since then, tectonic thinking in the central and southern Appalachians changed from a simplistic, stationary diastrophism to one in which horizontal transport was the primary element in the orogenic processes.

BACKGROUND

The earliest tectonic models that were developed and applied to the Appalachians seem quite primitive when examined in the light of today's geologic knowledge. But in the 18th and early 19th centuries of the United States, American geology was in its infancy, just beginning to grapple with the sparsely populated, unknown, and largely unstudied Appalachian terrane. The few geological accounts in the early and mid-18th century were mostly concerned with mineral occurrences. Guettard (1752)

produced a map of eastern North America, which mostly showed mineralogical locations. The more systematic reports of this era were usually by Europeans who traveled through the colonies (or later the young country) and presented descriptions and hypotheses of what they had encountered (e.g., Evans 1755; Kalm 1771; and especially Schöpf 1787). These accounts were rather sketchy and, given the inaccessibility of the terrane, their summaries of the Appalachians were understandably lean. Sufficient observations had been made by the end of the century to enable the division of the Appalachians into five distinct regions: (1) granite (the crystallines); (2) sandstone or grit (the folded Paleozoics); (3) limestone (the folded Paleozoics in the Great Valley); (4) loose sea sand (the Coastal Plain); and (5) river alluvium (see Mitchell 1798; Volnay 1803; and Mease 1807). Maclure (1809) portrayed the areal extents of these divisions of the mountain system on the first (Guettard's mineralogical map notwithstanding) geologic map of North America. Within this limited geologic knowledge, it is understandable that comprehension of Appalachian tectonics was essentially non-existent. In fact, diatrophism and tectonics anywhere in the world was only dimly perceived as an aspect of geologic activity.

We now know that the internal heat of the earth is the

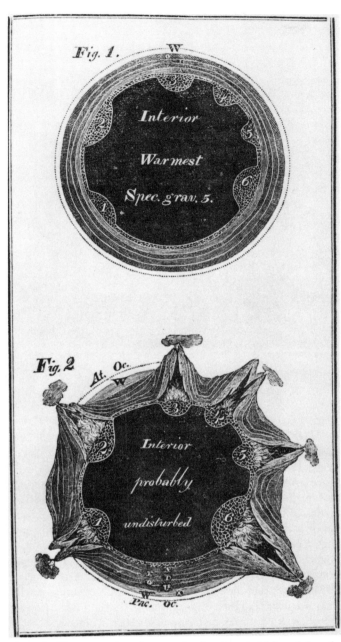

Figure 2a. The outpouring of gases and volcanic material from below the crust is a geologic concept that goes back to antiquity. Both the recognition that crystalline cores of mountain systems are related to molten effusions of volcanos, and the upending of nearby sedimentary beds, underlay the idea that mountain systems were formed by explosions of hot, interior materials. This view is wonderfully portrayed in Eaton's (1830) textbook.

primal driving force for all tectonic activity. Our present understanding of the internal structure of the Earth is based primarily on recent investigations, but the idea of a hot interior, possibly filled with burning subterranean gases, dates back to antiquity. By the mid-eighteenth century a fire in the center of the Earth was believed to be the source of volcanic activity and the principal

cause of all mountain building (Moro 1740). This concept was perpetuated into the nineteenth century (von Buch 1809; Eaton 1830), despite the problems with the empty subterranean caverns that were left behind (Fig. 2). At the end of the 18th century, Laplace (1796) proposed the nebular theory for the origin of the solar system. The theory hypothesized that a large, hot cloud of gas gradually cooled and condensed, causing a number of rings to form around a central star (the Sun). Each ring gathered itself into a globe, each of which cooled to form one of the planets. This hypothesis gave a physical rationale to the concept of an Earth with a solid shell surrounding a hot, possibly molten interior, a model that was independently suggested by others (e.g., Franklin 1793). The idea of a hot interior was to become central to the global tectonic concepts of the next century. In fact, it persists until today (though in greatly modified form), even though the nebular theory and the succeeding planetesimal theory (Chamberlin 1916) are both inadequate explanations for the origin of the Earth, and have been discarded. However, the idea of a liquid interior played an important role in the first tectonic theory developed for the Appalachians.

Philosophers from antiquity realized that the Earth's surface is not in its original state, but that it has changed under the influence of surface agents (erosion by wind and water) and subterranean processes (exploding or escaping air and vapors). Although they understood that mountains had formed since the creation of the Earth, they gave no explanations as to how they were formed (Adams 1938, p. 330). The idea of superposition, that stratified sedimentary rocks were formed as continuous horizontal layers of sediment deposited consecutively above older sediments, was first stated by Steno (1669). This concept, so fundamental to stratigraphy and sedimentology, is crucial to tectonics because any divergence from horizontal is an indication of tectonic movements. Steno attributed dipping (disturbed) beds to an outburst of air or burning gases (a common belief at that time) and to the collapse of underground caverns.

Up to the mid-eighteenth century, no pattern was seen in the locations or contents of mountains. As geological studies moved more into actual field investigations, Moro's (1740) distinction between a primary class of mountains (unstratified) and a secondary class (stratified) was elaborated. Mountain systems were soon perceived to be linear chains comprising a central primitive (unstratified) core flanked on both sides by bedded calcareous rocks and farther away by sandstones and shales (e.g., in Germany, Lehmann, 1756 and Füchsel, 1762; in Italy, Arduino, 1759; and in Russia, Pallas, 1777). The absence of fossils in the core rocks suggested that these are the oldest, and that the fossiliferous sands and clays in the far flanks are the youngest. In addition to this crude relative age assignment, it was found that layers of specific lithologies could be traced over long distances.

It was upon this foundation that Werner erected his doctrine that all the world's rocks were deposited in a specific order from a primordial ocean (Werner, 1787). The inaccessible and thus unknown (but cold) nucleus of the Earth possessed a great relief of high mountains and deep valleys, upon which the unfossiliferous

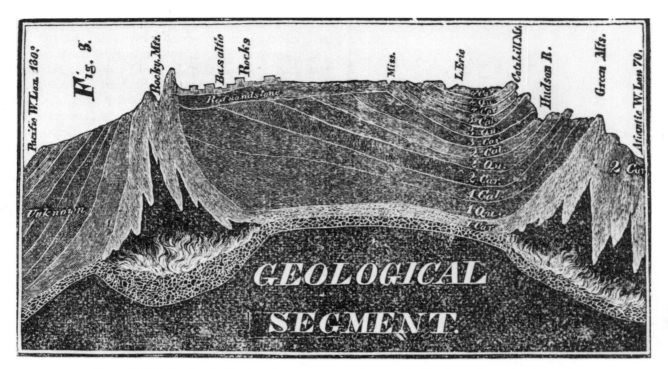

Figure 2b. Both the Appalachians and the western mountain system had hot fires underneath to heat and drive the molten effusions in each mountain chain. Sedimentary rocks lie between the mountain ranges in a broad basin that stretches across the central part of the continent. Also from Eaton's (1830) textbook.

granites, gneisses, and other crystallines were deposited. Upon these were deposited or precipitated the sequence of other lithologies, many of them fossiliferous. This 'Neptunist' theory (as it came to be called) was a nondiastrophic explanation of the origin of rocks; dipping beds were attributed to precipitation on the sides of mountains, to compaction of the sediments, or to local slumping. Because of Werner's own personal persuasiveness and the enthusiasm of his students and followers (see e.g., Geikie, 1905, p. 207–209), the Wernerian scheme dominated geologic thinking throughout Europe and North America into the first two decades of the nineteenth century.

Fossils had long been recognized in sedimentary rocks although their significance was not always understood. Some thought they were mineral deposits, or that they grew or were placed in the rock after its formation (Adams, 1938, p. 250 ff). By the eighteenth century fossils were widely understood to be the remains of organisms, but only at the end of that century was their utility appreciated, when William Smith noticed that different layers contained different fossil assemblages, and that the relative position of a layer within a stratigraphic succession could be ascertained by the enclosed assemblage (Smith, 1799, 1815). The significance of this for tectonics was enormous. Stratigraphic sequences could now be correlated over large distances with much greater certainty than by lithology alone, thereby enabling a clear delineation of intervening geologic structures. And it could now be ascertained whether an exposed sequence was overturned or right-side-up, a critical point in tectonic interpretations.

The long-standing explanation of dipping beds that was common throughout the eighteenth century was an extension of Steno's views: the beds at the edges of mountain systems were turned up because of the explosions which produced the mountains (e.g., Moro, 1740; Michell, 1761). As the concept of a fiery interior of the Earth gave way in the late eighteenth century to one of a molten interior, igneous intrusions into the mountain cores replaced the explosions as the driving force in mountain building. Recognition of angular unconformities led to the realization that deposition and diatrophism were not a single sequence of events, but that they may be cyclic within a given mountain chain. Repetition of the cycle—intrusion, tilting of beds, erosion, and renewed deposition on the upturned ends of older beds—was first presented by the Scot, James Hutton (1788, 1795). This 'Plutonist' paradigm, decidedly diastrophic, was in distinct contrast to the non-diatrophic 'Neptunist' dogma, and a vigorous conflict between the two views lasted throughout the decades spanning the turn of the century.

It was also at this time that some geologists began to look upon folds not as disrupted beds in the aftermath of explosions, but rather as a consequence of some deep-seated process in the Earth (e.g., de Saussure, 1796). It must have been about this time that it was suspected that lateral pressure applied parallel to strata could produce folds, because by 1812 Sir James Hall (not the American James Hall) had produced folds in layered clay and cloth by exerting pressure on the ends of the layers in an experimental device (Hall, 1815).

Figure 3. In 1840, in the fifth year of the first geologic survey of Pennsylvania, the artist George Lehman was the draftsman employed to sketch outcrops and other scenes in various parts of the state, some of which were eventually published in the final report of 1858. Lehman spent a month with the geologic party that was surveying the bituminous coal basins in Somerset and Fayette Counties in the southwestern part of the state. Among his other sketches, Lehman sketched the field camp itself. Ten years later, the German-American artist William T. van Starkenborch painted from Lehman's sketch this figure, entitled "The first geological survey in Pennsylvania." Henry D. Rogers apparently did not visit this camp that summer. The geologists that were stationed there, probably represented by the top-hatted figures, are J. Peter Lesley and James T. Hodge. The other figures are a geologic assistant and/or "working hands" and a cook. Painting is owned by the William Penn Memorial Museum.

Geology had come a long way by the end of the eighteenth century in its emergence from the strict orthodoxy of religious dogma. Yet the religious influence was still strong, particularly with respect to the well-known Noachian flood, the biblical account of the flooding of the Earth. This deluge had been utilized in some earlier schemes to explain the accumulations of sediments, although by the turn of the century the flood and its effects were considered quite separate from the Wernerian theories regarding an original ocean and sedimentation. It was believed then that the water from that original ocean had retreated into subterranean cavities, and then reemerged to produce the Noachian flood (see Silliman, 1837). With time, though, the possibility that subterranean caverns large enough to hold such huge volumes of water became increasingly untenable, and by the middle of the century attempts to include the deluge in geological interpretations had been largely abandoned.

The early years of the nineteenth century were the real beginning of geological investigations in North America. Geologic surveys of nearly all the states in the young country were begun in the 1830s (Fig. 3). The Wernerian doctrine held sway (Cleaveland, 1816; Dana and Dana, 1818) until the 1820s, when its influence on geologists began to fade, in part because of the new data that was accumulating in this country as well as in Europe. But it was the geology, primarily in the Appalachians, that was to give rise to the mountain building concepts that would evolve in America. Three important concepts that had been developed in earlier years were to be principal ingredients in the tectonic concepts soon to be developed for the Appalachians: (1) that the Earth possesses a hot, molten interior; (2) that many of the various lithologies were originally formed in extensive horizontal layers that could be distinguished by fossil content; and (3) that folds reflect some form of deep-seated diastrophism, and were not merely disrupted debris left over from explosions.

THE EARLIEST TECTONIC CONCEPTS

Upheaval may have been the word of choice when it came to discussing mountain building in the early years of the 19th century. For example, erosion was understood to be a continuing process; so, to have high mountains today requires that they must have been raised in some way fairly recently. Furthermore, steeply dipping and overturned beds require some rotation after deposition. It was understood that forces deep within the Earth were probably responsible for the effects, but the movements were considered to be primarily up and down—vertical tectonics. A few geologists surmised that horizontal stresses could produce upright folds, but for the most part horizontal displacements were not in the thinking of early 19th century geologists.

Upturned beds and folds had been observed throughout the Appalachians, but their occurrence could not be readily explained by the early 19th century geologists (Merrill, 1924, pp. 78, 105, 150–151, 184). For the most part they accepted the belief that a greater dip means a greater age, a belief that dated back to Pallas (1777) and other workers of that time, and which fit well within the then still widely accepted Wernerian scheme. This idea, however, was discarded after development of the technique of distinguishing and dating beds by their fossil content (Smith, 1815). Vanuxem (1829) pointed out that it was unlikely that the disturbing forces which tilted the horizontal beds acted uniformly at one time over large areas to produce a constant dip. In other words, attitude cannot be used to ascertain age—it is far better to use fossil content and mineralogical characteristics to establish relative ages and correlations. Clearing away this misconception permitted other causes of bed inclination to be considered.

The initiation of numerous state geological surveys in the 1820's and 1830's brought some of the best geological minds of the time into the first comprehensive studies of the Appalachians. Two prominent geologists in the central Appalachians, the brothers Henry Darwin and William Barton Rogers, were the first state geologists of the Pennsylvania and Virginia Geological Surveys, respectively. They accumulated much geological experience in the course of their work and greatly advanced the understanding of Appalachian geology.

Recognition of the continuity of strata across the foreland belt was one of the Rogers brothers' most important contributions (Lesley, 1876, pp. 53–55). The numerous and variable changes in bedding dip was confusing to those who did not realize the beds continued underground (or had at one time continued overhead through what is now air) to a similar-looking sequence nearby. The Rogers brothers used fossil content to demonstrate that the Valley and Ridge province is not a jumble of disconnected blocks, but that it comprises a single sequence of beds that persists across the entire province, folded into anticlines and synclines which they envisioned as waves in the rocky layers (Rogers and Rogers, 1843; Rogers, 1858, p. 885–895).

But how did these folds form? The Rogers brothers realized that the folded layers occupied a shorter horizontal distance than they did originally. They also understood that horizontal forces parallel to the layers could produce a horizontal shortening. What they did not believe was that such forces could cause the remarkable repeating pattern of long uniform folds that exists in the Valley and Ridge province. They thought that horizontal forces alone would produce a much more irregular fold pattern similar to those produced in Hall's experiments (1815). Some other agent, therefore, must have initiated the periodicity and uniformity they saw in the folds. To explain them, they put together a curious pastiche of Huttonian and explosion diastrophism. Working from the concept of the Earth's liquid interior which was commonly accepted then, they envisioned that an expansion of molten matter and gaseous vapors developed under the Appalachians, which was released through fissures in the core of the mountain system. These exhalations produced uniform pulsations on the surface of the Earth's liquid interior under the flanking foreland belt, much as a pebble dropped in a pond produces uniform ripples. As each anticline was formed in the overlying layers, molten material intruded the core of the fold from below and solidified. This solidified rock thereafter supported the arch, preventing it from collapsing.

This tectonic model is essentially one of vertical movements, yet the Rogers brothers did include a horizontal component. One of their other discoveries in the folded Appalachians was that in most of the anticlines the bedding in the northwest limbs was steeper than in the southeast limbs (the folds verge to the northwest). They realized that a simple vertical oscillation would not produce the observed vergence away from the crystalline core, so they supposed that concomitant with the oscillations, a tangential force shoved or floated the foreland northwestward, causing the observed northwestward vergence of the folds. They thought that the Appalachians are symmetric, with folds on both sides verging away from a granitic core, a pattern they also saw in the Alps (Fig. 4).

This tectonic construct by the Rogers brothers was the first coherent view of Appalachian mountain building, but it was not widely accepted, primarily because it seemed to be a rather peculiar mechanism for producing folds, and because exploding matter and escaping gases as a cause of mountain building was gradually being abandoned (despite Eaton's textbooks, 1830; Fig. 2). Furthermore, the folds they described showed no evidence of having intrusions in their cores. Other geologists were not averse to accepting folding by horizontal forces.

At the same time that the Rogers brothers were working in Virginia and Pennsylvania, the French geologist Léonce Élie de Beaumont (1828, 1852) developed a new explanation of the ultimate cause of mountain building. It followed on Laplace's

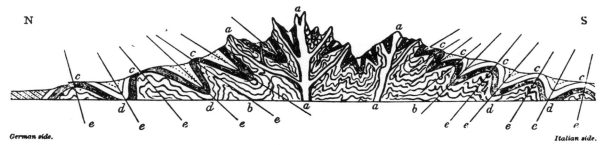

N · a · S

714.—Generalised Section of the Alps, displaying the Dipping of the Folds of the Strata on both Sides in towards the Igneous Axis.

Figure 4. Henry D. Rogers' cross-section (1858, p. 902) of the Alps, a mountain system he considered to be similar to the Appalachians. As was commonly thought then, the folds decrease in amplitude away from the mountain core, and become more upright.

idea of the Earth (and the solar system) having formed by condensation and a gravitational contraction of a hot, gaseous nebula. The globe that was to become the Earth cooled by radiating its heat into space. The surface thus cooled most rapidly, and the Earth soon possessed a solid outer shell enclosing a hot, probably molten interior. Experiments had shown that molten rock shrank upon solidification, and Élie de Beaumont utilized this effect to explain mountain building. Once the solid shell (crust) formed, continued cooling of the Earth resulted in radial shrinkage towards the core which produced tangential compressive forces in the crust. Élie de Beaumont maintained that these forces caused breaks (fissures) in the crust, with the crust on one side elevating to form a mountain system. Compressional forces in the other, depressed side produced folds which decreased in intensity away from the fissure. In addition, molten rock was forced upwards into the fissures to form the granitic core of the mountain ranges. This explanation for mountain building (radial shrinkage producing compressive tangential forces) encountered stiff resistance towards the end of the 19th century and was eventually abandoned, although its beguiling simplicity attracted the attention of some geologists well into the 20th century (e.g., Bucher, 1933). Even though it was not the first theory to propose horizontal forces for mountain building, it was the first based on demonstrable physical experimentation and on a coherent (though erroneous) theory for the origin of the Earth and the solar system. More significantly, it formed the basis for the dominant theories of Appalachian mountain building throughout the middle and late 1800s.

GEOSYNCLINES AND MOUNTAIN SYSTEMS

The Appalachian orogen was the cradle for numerous concepts, some of which rank foremost in understanding mountain systems worldwide. One of these concepts, the geosyncline, had its birth in mid-19th century studies of the Appalachians and it has been fundamental to theories of orogeny ever since.

The geosyncline was initially defined as an area for accumulation of sediments, which in and of itself is not strictly a tectonic feature. In fact, Marshall Kay (1951) maintained that the contents and location of a geosyncline were the essential elements,

and that whether or not it was subsequently deformed is not part of the definition. Considerable merit lies in this viewpoint, but on the other hand, it is a crustal movement, a tectonic activity, that forms the basin in which the sediment collected. These basins frequently form adjacent to, or even within, tectonically active zones. Not infrequently, the geosyncline becomes involved to a lesser or greater extent in the deformation originating in those zones. And it is usually the sediments from the active areas that fill the basin, providing important information about the nature of the adjacent orogenic activity.

James Hall had his earliest geological experience with the New York survey in the late 1830s. While working subsequently in the Iowa Survey in the 1850s (see Dott, this volume), Hall made note of the fact, as Vanuxem had done previously, that although the Paleozoic formations persist across much of the United States, the total section is much thicker in the Appalachians (12,000 m) than in the central part of the country (1,200 m). In addition, he recognized that the great thickness of sediments was not isolated or localized, but extended in a large, long syncline parallel to and including the Appalachians, that is the foreland part. Hall demonstrated that this syncline did not begin as a 12,000-m-deep ocean trough but had subsided concomitantly with sediment accumulation, as shown by the clearly shallow water deposits that occur throughout the entire section (Hall, 1859, 1883).

The geosyncline concept as developed by Hall was a vertical tectonic concept, focusing primarily on the origin of the sedimentary contents. When pressed for an explanation for the folds and faults, Hall was somewhat vague. He imagined that with continued sinking of the syncline the beds at the base of the syncline were extended, and that compression in the upper beds near the surface produced folds. These horizontal basal extensions and higher contractions, however, were only secondary effects of the fundamental vertical movements. This explanation did not accord well with actual observations, a situation that led Dana to deem Hall's proposal "a system of mountain formation with the mountains left out." Yet these two aspects, that deep sedimentary troughs are an integral part of the mountain systems and that the deepening of these troughs kept pace with the sediment filling, were two important features that are fundamental to understanding orogenic zones.

James D. Dana (see Mayo, this volume) was the dominant voice in Appalachian geology in the mid- to late 19th century. He possessed a breadth of knowledge of the mountain system that was probably unequalled in his day. He presented his views on the Earth's structure in general and on Appalachian tectonics in particular in three series of papers (1847; 1856; 1873a, b). Dana adopted Élie de Beaumont's views that the Earth was contracting because of cooling from a hot or molten state. He also made a clear distinction between the continents and ocean basins. The continents solidified from the molten state first; the Earth then continued shrinking with molten ocean basins. After some time, the oceanic crust finally solidified as well, but it stood at a lower elevation because the Earth's radius had decreased during the cooling period between the two solidifications. As in Élie de Beaumont's theory, the continuing radial shrinkage generated the tangential crustal forces which caused the mountain building. Dana considered the contrasting elevations between the continents and ocean basins to be weak zones in the solid shell, and it was here that the built-up horizontal forces were relieved by the diastrophic processes producing the mountains. With this boundary being a zone of weakness, it is natural that a deep, depositional syncline, called geosynclinal by Dana (1873a, p. 430) (later changed, to geosyncline) would develop along the boundary. It also explained why the geosyncline was long and narrow.

The development of a thick geosynclinal sequence along the edge of the continental crust required a descent of the geosynclinal floor by as much as 10 to 12 km. This is feasible if the Earth's interior is liquid, but in the latter part of the 19th century increasingly forceful arguments were made that the Earth's interior may be largely or entirely solid. Not that the interior was now thought to be cool, but rather it was realized by Hopkins as early as 1839 that pressure can cause a hot liquid to solidify, and that the enormous pressures inside the Earth were sufficient to solidify even hot core material. Dana accepted this constraint but he did insist that for the geosyncline to descend 10 or more kilometers, an equal volume of material below it needed to be displaced. Thus, if the Earth is largely solid, even if it is completely solid under the central parts of the continental crust, under the continental edges a large pool (fire-sea) must exist that could make room for the descending geosynclines. As the geosyncline deepened, more and more of this molten fire-sea material migrated eastward underneath the adjacent edge of the oceanic crust (the present crystallines and continental shelf). Accumulating there, it produced a large geanticline in the Archean crust, which Dana (1856) called Appalachia. This geanticline was not merely a convenient fabrication; evidence for a barrier between the geosyncline and the open sea (Atlantic Ocean) was present in the geosynclinal sediments. In contrast to the profuse marine fossil assemblages in many of the Lower and even Middle Paleozoic rocks in the geosyncline, the Upper Paleozoic formations are largely barren, and the Early Mesozoic strata are entirely devoid of such fossils. Clearly, if a geanticline were growing in tandem

with the geosyncline, it would progressively isolate the geosyncline from the Atlantic to the extent of a complete separation by the end of the Paleozoic. This geanticline persisted as a barrier throughout the Mesozoic until the Early Cretaceous, when it seems to have subsided, been eroded, or simply "foundered," allowing ocean-related sediments to reappear on the continent, as indicated by the rich marine faunas in the Coastal Plain deposits.

Dana's tectonic views, termed "Archean protaxis" (1890), were based on the concept of a paired geosyncline-geanticline positioned along the continent-oceanic boundary, the natural zone of weakness, with the geosyncline on the continent's edge and the geanticline on the edge of the ocean crust (as it was then understood). With the deepening of the geosyncline, the isotherms rose through the underlying crust, causing melting and a general weakening of the lower portions. Simultaneously, the rising geanticline provided a large eastern source for the voluminous and coarse Middle and Upper Paleozoic sediments that filled the geosyncline. Both structures, driven by the shrinkage-induced tangential forces, grew gradually *pari passu* throughout the Paleozoic to a climax near the end of the Paleozoic, at which time the crustal forces were relieved by the deformation (folding and faulting) of the geosyncline and collapse (foundering) and disappearance of the geanticline.

Both the geanticline and the geosyncline remained dominant in Appalachian tectonics during the ensuing decades. The primary strength of the "Appalachia" geanticline concept lay in its providing a southeastern source of sediment for the geosyncline. This aspect of geanticlines led to the extension of the concept to other orogenic zones around North America and the development of the "borderlands" idea. By the early part of this century, Charles Schuchert (1923) had proposed 11 geanticlines (borderlands) in various locations around the North American continent, positive source areas such as Llanoria, Cascadia, and Pearya which supplied sediment to the adjacent Ouachita, Cordilleran, and Franklinian geosynclines (Fig. 5). But by the mid-20th century, the geanticline idea itself had foundered. Geologic mapping in the crystalline terranes, particularly in New England, provided the main snags. It became increasingly evident that the areas underlying the supposed geanticlines were not simply rising, Precambrian crustal material, but were in fact complex Paleozoic terranes that had internally subsided and rose, shedding and receiving sediments while undergoing several orogenic events, all this during the time they were supposed to be passively rising geanticlines. In addition, the early seismic studies of the North Atlantic continental shelf revealed a continent-ocean transition that in no way could accommodate large crustal blocks that simply subside and vanish as some lost continent "Atlantis."

The geosyncline concept fared somewhat better. In the Appalachians, it was the dominating tectonic idea until the 1970s, when it was absorbed and refashioned within the newly evolving plate tectonics paradigm. In Europe, the geosyncline was introduced at the turn of the last century and adopted by a number of (but by no means all) geologists. The concept was incorporated into formal classifications during the ensuing decades, forming a

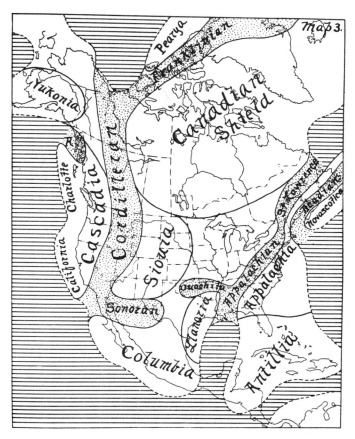

Figure 5. The North American borderlands, Schuchert's (1923) extrapolation of Dana's "archean protaxis." The borderlands (unpatterned) during the Paleozoic were topographically positive areas which shed detrital material into the adjacent geosynclines (dotted). The ruled areas were unspecified oceans.

"fixist" school of thought in which vertical crustal movements dominated and horizontal displacements were minimal. This was in distinct contrast to the more mobile school that developed somewhat earlier, an outlook epitomized by the horizontal tectonics of Albert Heim (1878) and the nappes of Marcel Bertrand (1884). One of the proponents of the "fixist" school of thought, Leopold Kober, subdivided the continents into stable regions, termed Karatogens (later called kratons, or cratons), which were surrounded by tectonic zones, the orogens, in which the geosynclines developed (Kober, 1921). The geosyncline was also central to the thinking of Hans Stille, who advocated recurrent orogenic episodes, driven by the Earth's contraction, with repeated uplifts, subsidences, and unconformities (Stille 1918), a view remarkably reminiscent of Élie de Beaumont. Stille later classified the various mobile belts, dividing the orthogeosynclines (geosynclines) into a non-volcanic miogeosyncline portion, and a volcanic-rich eugeosyncline (Stille, 1940).

Back in the Appalachians, Marshall Kay elaborated on Stille's classification by introducing terms for intracratonic basins, and further distinguishing the non-volcanic portion of orthogeosynclines on the basis of sediment source (Kay, 1951). Kay believed that the contents, location and sediment source were the essential elements that defined geosynclines. This was in marked contrast to Stille, who had subdivided orthogeosynclines into Alpino- and Germano-type, depending on the tectonic style of the mountains formed in them. Kay's disagreement lay in his returning to the original definition of geosynclines, emphasizing that it is the sedimentary aspects that specify the geosyncline, whether or not it is subsequently deformed. Thus the Michigan basin is a geosyncline (autogeosyncline) even though it is still undeformed. In Kay's view, the Appalachian geosyncline began as a true miogeosyncline with arenite and carbonate deposition throughout the Late Precambrian, Cambrian, and the Early and Middle Ordovician, all derived from the craton. The coeval eugeosyncline was oceanward, where volcanic deposits were common (Fig. 6). The miogeosyncline changed into an exogeosyncline at the beginning of the Taconian orogeny which was developing in the eugeosyncline, because Taconian flysch spread over the basin from the rising Taconian highlands. The geosyncline continued as a Silurian, Devonian, Mississippian, and Pennsylvanian exogeosyncline because the primary source of sediment remained in the repeatedly active, oceanward eugeosyncline.

The eugeosyncline also underwent modification at Kay's hand. By the 1940's, the geology of New England became sufficiently well known to understand that from the Cambrian into the Middle Devonian, central and western New England was a highly active region consisting of several basins and uplifts, and was strongly affected by two orogenic episodes, the Taconian and the Acadian. Supplanting the passive "Appalachia" borderland, Kay proposed the Magog belt, an active zone along the length of the Appalachians dominated by extensive volcanic activity, with local sources providing detritus for shale and graywacke deposition in the surrounding basins. He also demonstrated the close similarity of the Magog belt with present day island arcs, a tectonic feature that is a fundamental element in the plate tectonic scheme.

The geosyncline has evolved into several types of basins or depositional environments within the plate tectonics milieu, depending on its location relative to the continental edge and the type of underlying basement. Thus the Early Paleozoic transition from carbonates of the folded Appalachians to the more argillaceous carbonates (the Conestoga Formation in Pennsylvania) to the basinal clastics of the Wissahickon Formation (and farther south, the Evington Group and other equivalents) reflects the change in depositional environment from continental shelf, to slope, to oceanic basin. All of these depositional sites, initially farther apart than at present, have been extensively telescoped into one another by subsequent orogenic thrusting. These deposits, the non-volcanic sequences, were formed on the open eastern edge of the North American continent that now comprises the Appalachian miogeocline.

The spread of the Taconian flysch (Reedsville and Martinsburg Formations) across the miogeocline during the Late Ordovician signaled a fundamental change in the structure of this depositional area—it was no longer on the open edge of the

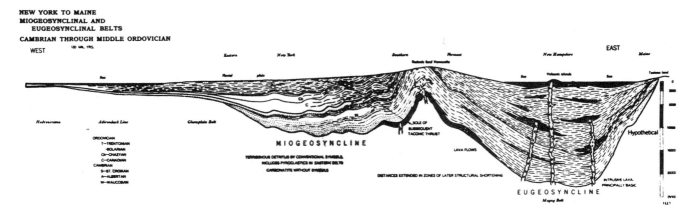

Figure 6. Marshall Kay's (1951) concept of the Appalachian orthogeosyncline across northern New England at the close of the Middle Ordovician. The two principal parts, the western non-volcanic miogeosyncline and the eastern volcanic-rich eugeosyncline, were separated by a positive area, Vermontia, which would later be the locus of the Green Mountain anticlinorium. To the west of the miogeosyncline is the Laurentian craton; to the east of the eugeosyncline is an unspecified positive source area in what is now eastern Maine.

continent. The flysch and overlying molasse from the southeast indicated that an active tectonic zone was rising, one that would separate this former shelf and slope from the ocean for the remainder of the Paleozoic. The similarity between the current interpretation and the geosynclinal classification of Kay is striking — in both, a fundamental change in the orogenic belt occurs in the Late Ordovician, which is reflected in a permanent shift in sediment source.

The geosyncline is definitely a "fixist" concept, as was noted in the discussion of early 20th century European views. A geosyncline develops in a given area and, although the edges and depositional centers may shift through time, the geosyncline itself never moves relative to the crust; it deepens, and is tectonized, but it does not change its location. Even Kay's highly mobile Magog belt, as active as it was, experienced only vertical movements as the volcanoes and other source areas rose and subsided. The juxtaposition of tectonic slices such as the shales and graywackes of the Taconic Mountains onto the adjoining carbonate shelf was viewed as a relatively local and isolated transport within an otherwise autochthonous province. This "fixist" attitude dominated the thinking of Appalachian tectonics from the mid-1800s into the 1960s. The idea of large horizontal movements within orogenic zones that were developed in Europe late in the 19th century by Bertrand, Heim and others, and certainly the early 20th century continental drift of Wegener among others, made little impression on most Appalachian geologists until recently. Even when the weight of accumulated field evidence from the Appalachian foreland itself demanded allochthonous tectonics, the controversy between "thin-skinned" and "thick-skinned" geologists continued (see, e.g. Lowry, 1964; Fisher and others, 1970, specifically articles by John Rodgers and Byron N. Cooper). Only with the force of the plate tectonics revolution and the acquisition of deep seismic profiles has the Appalachian geologic community as a whole accepted the allochthonous nature of the

central and southern Appalachians, from the outer reaches of the foreland to the innermost parts of the crystallines.

HORIZONTAL SHORTENING

The development of the geosynclinal/depositional basin concepts was an important, indeed integral aspect of the evolution of tectonic concepts of the Appalachian orogenic belt. But it is the deformation of an orogenic belt, the folding, faulting, the metamorphism, intrusions, and other changes to the rocks and their fabrics that are the core of tectonic activity. Starting from horizontal, unaltered, and undeformed bedded sequences of the basins and adjacent areas, the conversion of these rocks into simple large arches, into telescoped slices, and into contorted layers, with the development of complex fabrics and modification of the mineral constituencies, has been long the source of interest and fascination for the tectonic geologist. In the early nineteenth century, American geologists understood little of these processes and their consequences in the Appalachians. Persistent observation and studies by hundreds of workers over the ensuing decades has brought our comprehension of Appalachian tectonism to the sophistication it now possesses.

This activity began in earnest during the 1830s when numerous surveys were initiated by the individual states (see Merrill, 1924, for an extended, detailed review; also Aldrich, 1979). Several advances developed out of this work: correlation of the eastern American Paleozoic sections by fossil content, establishing the continuity of formations over long distances, recognition and delineation of folds, and the concept of the geosyncline. These advances laid the groundwork for understanding the tectonics of the Appalachian orogen.

Geologists came to understand that the tilted and upended beds were not a consequence of initial deposition or of explosions, but formed parts of folds in which originally horizontal layers were forced by some means into arches and troughs (first called anticlinals and synclinals by Rogers and Rogers, 1843). They also realized that a fold occupies a lesser horizontal or tangential shortening. To what depth this shortening extends was not considered because no distinction was made between surface structures and the rest of the crust. The folding at the surface was assumed to equal the shortening of the crust at depth. The idea that shortening at the surface might be independent of shortening at depth apparently did not occur to the geologists of the time because there was no need or evidence for such an idea. They recognized and understood faults, but they often considered faults ancillary to folds (Rogers, 1858, p. 895–899) (Fig. 7), occurring, at least in the foreland, as breaks in the beds where folds had been pushed beyond their flexible limit.

The Rogers brothers and other investigators recognized the presence of very large thrust faults that extended for tens of kilometers along the steeper northwest flanks of anticlines. They also realized that the relative proportion of folds versus faults differed between the central and southern Appalachians. In the central Appalachians, folds were the predominant structures, with faults being quite subordinate. In southern Appalachians, faults were the dominant structures. As prevalent as these faults were in the southern Appalachians though, no one attributed any great crustal shortening to them. That these faults represented tens of kilometers of horizontal transport was an idea that was to develop only gradually over the next 100 years.

Safford (1869), in one of the first geologic studies of Tennessee, recognized that the rocks in eastern Tennessee had been considerably disturbed. He described the contrast in fold style between the open, low amplitude folds of the Plateau and the much tighter, and in places overturned, folds in the Valley and Ridge province. Many of the folds in the Valley and Ridge province had been faulted, which he attributed (as did Rogers) to the limited flexibility of the layers. The forces responsible for the deformation were thought to have originated from the southeast. A horizontal force was the apparent cause, but Safford did not assign much horizontal movement to the deformation. The belief then was that these thrust faults were quite steep and reflected breaking of the folds when the movements had become too great for continued ductile bending of the beds.

The first realization that these faults may not be steep thrusts came in the late 1880s and early 1900s as a result of work in the southern Appalachians by several U.S. Geological Survey geologists. Hayes (1891) demonstrated that these faults extend for great distances, up to 600 km in a northeast-southwest direction parallel to the structural trends of the Appalachian Mountains. Workers also discovered that the dips of the faults were low to the southeast, as little as 5 and rarely more than 25 degrees. Displacements from 1 to 15 kilometers on these faults were determined on the basis of the apparent reduced width of the metamorphic gradients in the crystalline Appalachians to the southeast.

Among the several long, strike parallel faults in the southern Appalachians, the Pine Mountain fault in easternmost Kentucky

FIG. 708. — Upward displacement of uninverted side of fractured anticlinal.

FIG. 709.—Upward displacement of inverted side of fractured synclinal.

Figure 7. Faulted folds as illustrated by Rogers (1858). These two diagrams portray upward displacements (thrust, reverse, and normal faults were not yet in the tectonic vocabulary) in an anticline, and a syncline. Four other accompanying figures on the same page show other fold/fault relations.

is of particular interest, perhaps because it is most northwestern of these faults. The Pine Mountain and the Hunter Valley (the next major fault to the southeast) faults delineate the northwest and southeast edges of a single coherent block in the eastern part of the Cumberland Plateau. The block is particularly interesting because structures within it do not continue beyond the block. A bounding fault on the southwest side was subsequently described early in the 20th century but it was not until Wentworth (1921) examined the northeastern part of the block, locating the Russell Fork strike-slip fault, that the Pine Mountain block was completely delineated. The significance of this block was not only that it was bounded on all four sides by faults, but that it had moved and was deformed independently of the surrounding rocks. Within the next few years, Butts (1927) found four fensters within the middle of the Pine Mountain block which showed that the older rocks at the surface had been thrust northwestward over younger rocks. This pointed to an entirely different tectonic style: it was not the simple shortening by the type of folding and subordinate faulting common farther north in West Virginia and Pennsylvania. Rather, a huge slab of the Paleozoic rock sequence, the Pine Mountain block, had been thrust over and was now lying on top of a similar Paleozoic sequence. This was a telescoping of rocks, a form of shortening not previously recognized in the Appalachians. Butts estimated that this overriding movement of the Pine Mountain block must have covered a minimum of 11 km, a figure similar to Wentworth's. In addition, Butts could find no source for this block—there was no depression to the southeast from which the Pine Mountain block must have come. One could only conclude that in addition to the Cumberland block, all the rocks to the southeast in the Valley and Ridge had also moved. It followed by implication that the crystalline rocks of the Appalachians even farther southeast had similarly moved to the northwest because no gap or depression existed there. Butts attributed this horizontal displacement to the continued shrinkage of the Earth's interior. Even as late as the 1920s, the contraction theory seems to have been the tectonic explanation of choice as it

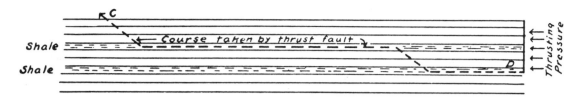

FIG. 4.—Diagram representing course of incipient thrust plane in series of sedimentary rocks. Plane follows beds of easy gliding, such as shales, and breaks diagonally across more brittle beds from one shale to another, and, finally, up to surface.

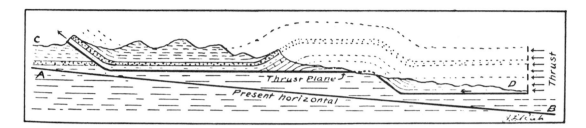

FIG. 5.—Diagram showing result of movement on thrust plane such as that shown in Figure 4. Tilting so that horizontal is represented by line *AB* and erosion down to line *CD* gives structure and topography to be compared with present cross section of Cumberland block (Fig. 3).

Figure 8. The development of the Pine Mountain block required two elements: a bed-parallel thrust plane (décollement) under the block, and ramps at its eastern and western edges, as illustrated by these diagrams of Rich (1934). Movement up the western ramp produced the eastward-dipping beds there; movement of the thicker section over the eastern ramp created the Powell Valley anticline in the east.

had been for the preceding 80 years. The décollement concept developed many years before in the Alps by Lugeon (1902), Buxtorf (1907; 1916), and Haug 1907) appears to have had little impact in the Appalachians at this time, but knowledge accumulating here was to change this attitude gradually over the next four decades (e.g., see Rodgers, 1949).

The mechanics of the Pine Mountain block development was advanced a few years later when Rich (1934) extended Butts' idea by proposing that the block had moved on a single basal fault parallel to bedding which had sheared upward across the lower beds, ramping from an older to a higher shale unit (Fig. 8). In a manner not previously realized, the rocks of the Pine Mountain block originally to the southeast, were telescoped over the underlying strata. The Powell Valley anticline, the flat-topped anticline in the southeastern part of the block, did not form independently but was a direct consequence of the ramping. Rich also concluded that the basal fault extended to the southeast under the Blue Ridge and Piedmont, progressing through a number of ramps, bringing the buried crystalline Piedmont rocks to their present surface exposure. Rich's 1934 paper did much to focus attention on large horizontal thrusting, and to crystallize the emerging décollement tectonics that has since been used to explain much of the foreland structure of the Appalachians (see also Rodgers, 1970, pp. 20–28).

Once the idea was introduced, décollement tectonics was

soon used to explain many of the features that had seemed perplexing, structures such as the Sequatchie anticline (Rodgers, 1950) and the Cumberland block (Wilson and Stearns, 1958). The great width and low amplitude of most Plateau folds in the central Appalachians had suggested to many workers that these folds persist downward into the basement (e.g., Sherrill, 1934). A major turning point occurred with the discussion of the Sandhill basement test well during a symposium in 1957 (Woodward, 1959). This well was spudded on the Burning Springs anticline, a rather peculiar conjugate (box) fold in northwestern West Virginia. This anticline is isolated from other foreland folds, and it possesses a north-south trend at variance with other Plateau folds: it was commonly believed that the fold must persist downward to basement. But it was found through the drilling that the ½-km shortening in the Devonian and younger formations was not present in the Lower Silurian and older rocks. Clearly, a break or detachment must exist in the Upper Silurian formations, at which the fold terminates. Deep drilling throughout other parts of the Plateau, primarily in West Virginia and western Pennsylvania, revealed that many other Plateau folds extend downward only to the detachment, or décollement, in the Upper Silurian Salina Formation (Gwinn, 1964). Following the suggestions of workers earlier in this century, later geologists extended the décollement concept southeastward under the folds of the Valley and Ridge province. This extension was supported in part by the discovery from seismic data that the top of basement there is quite smooth and does not exhibit the fold structure present in the rocks at the surface (Gwinn, 1970). The presence of Lower Paleozoic rocks as old as Middle Cambrian in some anticline cores indicates that the décollement is at a much deeper level in the Valley and Ridge than in the Plateau, probably in a Middle or Lower Cambrian formation.

More recently, deep seismic data have been obtained from the Valley and Ridge, Blue Ridge, and the various crystalline terranes from different parts of the southern Appalachians. Each seismic line, from Tennessee to Georgia (Cook and others, 1979), in eastern Tennessee (Milici and others, 1979), from Tennessee to North Carolina (Harris and others, 1981), and across central Virginia (Harris and others, 1982), shows strong reflecting surfaces extending continuously underneath the various provinces and terranes of the Appalachians. The subhorizontal reflectors have been interpreted as décollement faults, on which the overlying Precambrian and Paleozoic rocks of virtually the entire southern Appalachian orogen have moved, perhaps as much as two hundred or more kilometers to the northwest, onto the North American craton. Geologic relations in the central Appalachians, where comparable seismic data are currently not available, suggest a similar, completely allochthonous nature.

The widespread acceptance of décollement tectonics for the Appalachian foreland has not brought an end to basic investigations in this province—questions still remain. The nature of the décollement ramps and the process by which they form and deform the overlying rocks, is currently an active area of investigation (e.g., Wiltschko and Eastman, 1983). The sequence of folding, whether from northwest to southeast (Milici, 1975) or outward from the southeast (Perry, 1978), is also not settled. The nature of the driving force behind the décollement tectonics and how this force is transmitted throughout the décollement block is currently being examined as well. Simple gravity sliding (Gwinn, 1964; Milici, 1975) requires a northwestward dipping décollement, which would require a subsequent regional rotation of the décollement (and the foreland) to its present southeast-dipping attitude. In contrast, other theories, using "glacial analogy" models, require a northwest erosion surface during the deformation. These models would drive the décollement sheet up the inclined slope either by the gravitational potential created by the cratonward thinning wedge (e.g., Elliott, 1976) or by combining the gravitational energy with a regional compression (Chapple, 1978).

A complete reorientation of thinking about Appalachian tectonics has occurred since the early years of the nineteenth century (Rodgers, 1949; 1972). Continued geologic mapping, especially in the southern Appalachians, has demonstrated that the tectonics of the Appalachians reflects essentially large horizontal movements on subhorizontal décollements rather than vertical movements with only ancillary horizontal displacements. Those vertical movements that have occurred are largely secondary consequences of ramping and overthrusts inherent in décollement tectonics.

FOLDS AND OTHER CONTINUOUS DEFORMATIONS

Much of the early work on faulting and especially in décollement tectonics was accomplished in the southern Appalachian foreland, largely because it is here that the faults are exposed. Décollements are undoubtedly present in the central Appalachians but at such great depth that they do not appear at the surface. Splays from the basal décollement probably occupy the cores of every major, first-order fold, but only a very few breach the present erosion surface (e.g., at Birmingham on the Nittany anticline, Gwinn, 1970). The contrast between the southern and central Appalachian forelands may be attributed to one or more of several reasons: the thicker Paleozoic section in the central Appalachians may have induced a more ductile behavior, restricting the faulting to the lower, more brittle carbonate portion; a larger amount of northwestward transport in the southern part may have led to a greater amount of late faulting of earlier fold structures; or the shallower basement in the southern Appalachians may indicate that the lower, more extensively faulted part of the section is exposed and that the higher, more folded part has been eroded. Whatever the reason, the folds of the central Appalachians attracted and focused attention on the more ductile aspects of Appalachian tectonism.

The folds of the central Appalachian Valley and Ridge province are impressive. At first glance, they appear as simple arches, clearly delineated in the topographic expression of the Silurian Tuscarora and Mississippian Pocono Formations. They are generally upright (though verging northwestward), and continue along their lengths unbroken in the sweep around the arcuation in Pennsylvania, a feature that was particularly impressive to the Rogers brothers. Over the years, these folds came to be considered the archtype of foreland folding, the classic form of a simply folded terrane.

The Rogers brothers (Rogers and Rogers, 1843; and later Rogers, 1858, p. 886 ff.) were probably the first to recognize and describe these folds. They saw their length and parallelism as the primary characteristics of the folds, but they also saw variations among the folds, which H. D. Rogers enumerated in his final report (1858). One of the variations he noted was the size of the folds, with the largest, the first order ones, possessing wavelengths in excess of several kilometers and lengths of more than 200 km. Second and third order folds are progressively smaller. This was a concept that was expanded more than a century later in the 5-fold classification of Nickelsen (1963), by including yet smaller folds, those too small to show on a map. Although a complete continuum of fold sizes may actually exist, especially among smaller folds, the concept has utility for distinguishing and classifying Appalachian folds by size.

It must have been the approximately sinusoidal profiles of the foreland folds that made the Rogers brothers view the folds as waves in the rocks, a central feature of their tectonic scheme. The idea that horizontal compression could produce folds had been around for some time, and the brothers were aware of the experiments of Hall (1815). But the folds Hall produced were quite irregular, horseshoe-shaped in profile, and of short length. Clearly, to the Rogers brothers, horizontal compression could not produce the long uniform folds of the Appalachians. It was probably the association of waves at fluid surfaces that brought to their mind the flexing of the crust on an oscillating liquid interior, the core of their tectonic thesis.

Rogers used his fold-size classification to point out the correlation of fold size and fold trend in each of the Anthracite basins. The first-order synclines plunge to the east-northeast whereas all the small second- and third-order folds trend in a more easterly direction. This feature has been described more recently by Darton (1940), among others, and was attributed by Rogers to appression of the uppermost beds (the Coal measures) in the centers of the more fundamental first-order synclines. This divergence of trends has been recently interpreted as evidence for two noncoaxial phases of the Alleghanian orogeny (Geiser and Engelder, 1983).

Rogers also distinguished folds by their profiles, based both on tightness (degree of limb appression) and by the attitude of the axial surface. His "symmetric folds" are open, of low amplitude, and have vertical axial surfaces; they are characteristic of the Appalachian Plateau. Folds with steeper dips are common to the Valley and Ridge province—these are the "asymmetric folds," verging to the northwest with the northwest limb generally steeper and the axial surfaces, bisecting the angle between the limbs, usually dipping steeply to the southeast. The more highly deformed rocks of southeastern Pennsylvania exhibit the "folded flexures." In these, the northwest limb is overturned, even to the degree of the fold being isoclinal and recumbent.

Some authors have interpreted this change of fold profile as a continuum from the highly folded terrane in the southeast of Pennsylvania to the scarcely folded layers at the outer edges of Alleghanian folding in the northwest, a view that was prevalent in the nineteenth century. Perhaps some things, but not many, are further from the truth. A structural front is a boundary between two provinces of contrasting tectonic style. The first one defined in the Appalachians was proposed for the profound difference between Plateau and Valley and Ridge folds (Price, 1931). This Appalachian structural front (Allegheny structural front, Rodgers, 1964), occurs above the ramp of the décollement from Cambrian shales to Silurian evaporites (Gwinn, 1970). The marked difference in tectonic style between the Valley and Ridge folds, and the overturned to recumbent folds and thrust faults in the Great Valley to the southeast led Rodgers (1964) to define the Blue Mountain structural front. The Allegheny structural front extends from eastern Pennsylvania to southwestern Virginia, where it is replaced by the St. Clair fault (Fig. 1). On the other hand, the Blue Mountain structural front is more problematic. From the Susquehanna River eastward, it separates the Taconian (and Alleghanian) terrane of the Great Valley from the Alleghanian folds and faults of the Valley and Ridge. Southwest of the Susquehanna River, Taconian structures are not present—the transition from Blue Ridge and Great Valley to the Valley and Ridge is a continuum of Alleghanian structures (Root, 1970). But from the Pennsylvania-Maryland border, the Blue Mountain structural front is marked by the Little North Mountain fault (Rodgers, 1970).

One aspect of the folds that was not considered by geologists throughout the nineteenth century was what happens to the folds at depth. This was of no particular concern then because the folding was accepted as the surface expression of the shortening of the entire crust (Fig. 9). By the turn of the century, though, large amounts of crustal shortening were being deduced from field evidence, not only within the faulted terranes of the southern Appalachians, but also within the folded terranes in the central Appalachians (e.g., 65 km of shortening estimated by Lesley, as noted by Claypole, 1885). These estimates were larger than one would expect from simple cooling and contraction of the Earth, unless some way was devised to concentrate the shortening within the narrow orogenic zones. Chamberlin (1910) proposed that a shear zone separated the folded upper part of the crust from the lower part, which deformed in a more ductile fashion. His rather simple formula for calculating the depth of this shear zone, when applied to data he had collected from the Valley and Ridge in central Pennsylvania, yielded inconsistent depths ranging from 9 to 53 km. Although the theory was quite deficient, the concept of the folded rock sequence being separated from the underlying rocks by a detachment surface or zone was important in presaging the thin-skinned tectonics to be demonstrated some 50 years later (Gwinn, 1970). It was also remarkably similar to the relations observed in the Swiss Jura Mountains (Buxtorf, 1907), although Chamberlin may not have been aware of this work. Yet many of the geologic texts, in the first part of this century, continued to extend the folds into the Precambrian basement (Fig. 10).

Van Hise (1896) described and contrasted parallel and similar folds based primarily on change of bed thickness around the fold. Parallel (concentric) folds, which possess constant bed-normal thickness, are thought to be characteristic of the less deformed forelands of orogenic belts. Chamberlin used this geometry in his cross sections (Fig. 11), and the idea that Valley and Ridge folds are concentric has persisted for many years. Nickelsen (1963) pointed out that many Valley and Ridge folds have rather straight limbs and sharp hinges, and are not really concentric. Faill (1973) subsequently demonstrated that this geometry is characteristic of kink folding, the process which produced the vast majority of the folds, from the largest first-order ones to the smallest. In this paper, Faill described the deformation of the Valley and Ridge province as consisting primarily of a flexural-slip process with subordinate faulting.

Other processes such as flow and solution were found to be significant contributors to the total strain as well. Among the more common evidence are deformed fossils, cleavage and stylolites, distorted mudcracks, and flattened reduction spots, all of which reflect an early bulk flow in the rocks. This flow is particularly well displayed in the west limb of the Blue Ridge anticlinorium in Maryland, where strains of 100 percent have been

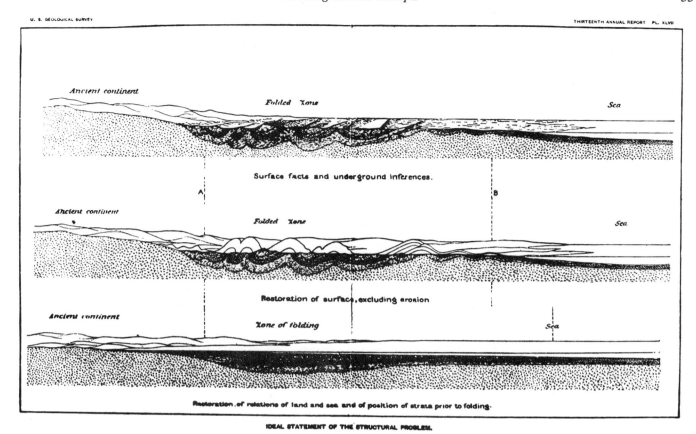

Figure 9. Bailey Willis' (1893) block diagrams reflect the view of the Appalachians widely held throughout the latter half of the nineteenth century. These diagrams, drawn unconventionally facing south, portray only the foreland portion of the Appalachians. All the terranes of the crystallines have been relegated to the "ancient continent," a prejudice widely-held at that time. The Paleozoic cover rocks and the Precambrian basement participate equally in the crustal shortening—the top of basement mirrors the surface structure. The disharmonic folding under the thrust fault in the western (right) part of the closely folded terrane is the only occurrence of a contrasting style of deformation.

measured (Cloos, 1947). It is present throughout the Valley and Ridge province (Faill and Nickelsen, 1973) and even in the Plateau (Nickelsen, 1966). These structures, including cleavage, fall within the general term "layer-parallel-shortening" (LPS), and they are generally early, prefolding structures. In other words, non-flexural-slip deformation, and it was partitioned among several mechanisms (Geiser and Engelder, 1983). Many of these aspects of Valley and Ridge deformation appear in the Bear Valley strip mine in the Anthracite region in eastern Pennsylvania. The rocks in this mine display a definite sequence of structures from early jointing and cleavage formation through folding and faulting to a late stage jointing, a sequence that is probably representative of the Alleghanian deformation throughout the Valley and Ridge and Plateau provinces (Nickelsen, 1979).

The relationship between folds and faults was another area of early investigation. The Rogers brothers had noted from the early stages of their work that the central Appalachians in Pennsylvania and West Virginia were predominantly folds, whereas faults dominated the southern Appalachians. Probably because of their more "northern" experience in the central Appalachians, they relegated faults to a subordinate position in their thinking vis-à-vis folds. Faults appeared in tightly appressed folds that could develop no further by the normal folding process, a view seemingly supported by the frequent appearance of faults in the overturned limbs on the northwest side of many folds (Fig. 6). Whereas they felt that the faulting was a late stage in the development of folds, later workers realized that faults sometimes preceded the folds or developed independently of folds.

Bailey Willis grappled with the relative role of folding and faulting at the end of the nineteenth century. He (1893) attempted to demonstrate the relationship between folds and faults by means of deforming layered materials in an experimental squeeze box. Experiments prior to this time (e.g., Hall, 1815; Favre, 1878) demonstrated that horizontal forces could produce fold shapes (see Willis, 1893, pages 230 to 235), but those models were quite crude. Willis was the first to attack the problem by attempting to

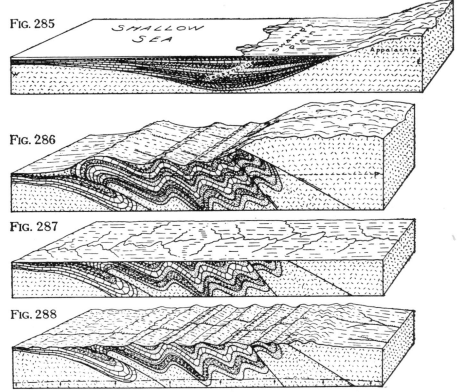

Figs. 285–288. — Four stages in the evolution of the present Appalachians.   Views looking north.

Figure 10. As late as the 1930s, the Appalachians were viewed as a geosyncline-geanticline (Appalachia) pair as shown in the diagrams of Longwell, Knopf, and Flint (1932). No detachment exists between the Paleozoic cover and the underlying Precambrian rocks. The basement fully participates in the folding, and the thrusts rise steeply from deep within the basement.

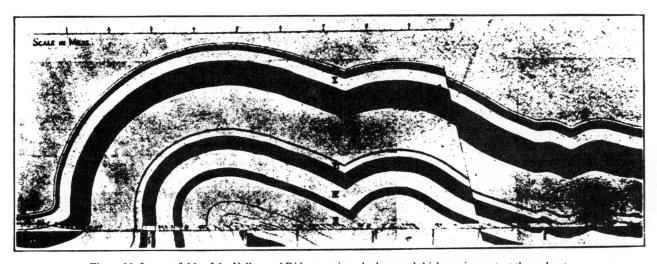

Figure 11. In most folds of the Valley and Ridge province, bed-normal thickness is constant throughout the limbs and hinges. This is characteristic of concentric (parallel) folds, so it is not surprising that most cross-sections, such as this one drawn across the province in central Pennsylvania by Chamberlin (1910), portray concentric folds. The Valley and Ridge province was the classic example of folded mountain chains; thus concentric folds became associated with simply folded terranes.

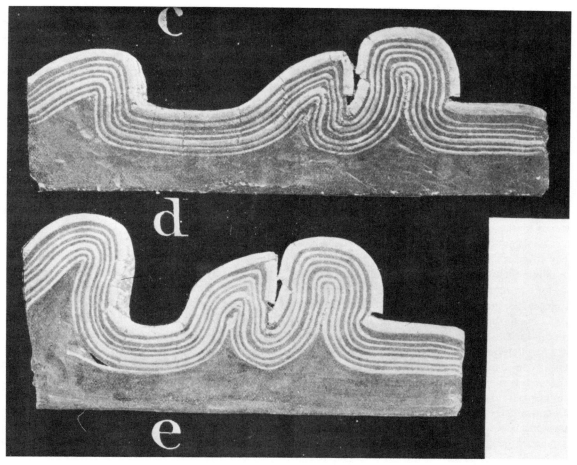

Figure 12. Bailey Willis (1893) performed a large number of experiments in a squeeze box with layered wax, plaster, and turpentine in order to reproduce Appalachian foreland folds. He varied the combinations, thicknesses, and sequences of these materials in order to ascertain how material (ductility) contrasts, bed thickness, and layer sequencing affected the fold and fault geometry. In this experiment, his Plate LXXVII, both concentric and kink fold geometries were obtained, as well as, on the left side, a hybrid form.

scale the strengths, rigidities, and viscosities of the materials down to laboratory values that would reflect the actual rock properties. The properties of various mixtures of wax, plaster, and turpentine were incorrect by several orders of magnitude, but he did produce credible results. These materials were layered in boxes in different combinations, thicknesses, and sequences, and then compressed horizontally. He created a wide variety of structures, many of which resembled the foreland structures observed in the field (Fig. 12). Although the scaling was not very accurate, the experiments did demonstrate the importance of ductility and strength, and the contrast in both strength and ductility among the various layers. The variations in fold geometry produced by using different combinations of materials, thicknesses, and sequences emphasized the control that materials had on the ultimate geometry. This, in turn, led Willis to develop the concept of competency and incompetency, an idea that has since been widely used in explaining structural relations in rocks. The concept of a "structural lithic unit," a layered sequence of rocks possessing similar mechanical attributes (Currie and others, 1962), evolved directly from this understanding of contrasting behavior.

Willis covered his layers with shot in order to keep the layers from breaking chaotically. He found that by simply varying the amount of shot, the behavior (ductility) of the layers changed. In other words, the types and forms of structures developed in a sequence could be altered merely by changing the overburden (confining) pressure. This effect of confining pressure in actual rocks was confirmed by laboratory experiments during the following few decades (e.g., Adams, 1910; Griggs, 1936; Griggs and Handin, 1960). The experiments showed that brittle rocks which deform primarily by faulting at the surface become quite ductile and deform by flow (under high confining pressures) at great depths. The fact that some rocks are inherently more ductile than others led to the conclusion that stress can be conveyed a greater distance in the stiffer or less ductile rocks than in more ductile rocks. During the past three decades, a large amount of increasingly sophisticated experimental rock deformation has added considerably to our understanding of the behavior of different lithologies under widely different conditions. This work basically substantiates the fundamental relations deduced from the early studies of the Appalachian fold and fault mechanics at the turn of this century.

CRYSTALLINE CORE OF THE APPALACHIANS

The metamorphic and igneous rocks to the southeast of the Apalachian foreland province constitute the crystalline terrane of the Appalachians. The presently exposed belt is 75 km wide in eastern Pennsylvania and reaches a width of 275 km in the southern Appalachians, along a total length of more than 1300 km. It probably extends southeastward under the coastal plain deposits and may be as much as 400 km wide along its whole length, an area more than twice as large as the foreland belt.

The crystalline rocks were considered the oldest part of the Appalachians for two reasons. The first was a legacy from Wernerian thinking that the metamorphic rocks were among the first to be deposited from the primordial sea. Secondly, fossils had not been found in this terrane, which suggested that these rocks were formed before fossils were preserved. Compounding the matter was the absence of any easily recognizable stratigraphic sequence. Large tracts of the crystalline terrane consist of slates, schists, and gneisses of various grain sizes and mineralogy which could not be ordered into a consistent sequence. These lithic vagaries produced by the metamorphism which has obliterated most of the original bedding has plagued geologists for years. In fact, some of the problems that perplexed the earliest workers are lacking resolution even today. Much relatively recent work in both the central and southern Appalachians has revealed structures and relationships unknown in the first half of this century and has brought our understanding from a rather bleak ignorance to an appreciation of a marvelously complex orogenic activity. The most impressive aspect exposed by this work is the degree to which allochthony pervades the entire crystalline terrane.

The earliest detailed studies in the crystallines were done around Philadelphia and to the southwest (Matthews, 1905; Bascom, 1905). Considering the complexity of this terrane, it is not surprising that controversy developed early, particularly concerning the suite of metamorphic rocks in the Piedmont of southeastern Pennsylvania and eastern Maryland known as the Glenarm Series (Knopf and Jonas, 1923). Initially investigators agreed that the Precambrian Baltimore gneiss was overlain by the Lower Paleozoic Glenarm Series, which contained the Wissahickon Formation. But then the water turned murky. Bascom (1909) concluded that the Wissahickon near Philadelphia was Precambrian in age, largely because those rocks exhibit a higher metamorphic grade, and because they contain plutonic intrusions that are absent to the west. At first, the lower grade schists to the west, renamed Octoraro, were still considered to be Lower Paleozoic in age, but subsequent mapping to the west (Knopf and Jonas, 1929), suggested that the Octoraro was not a separate formation;

rather it was a facies of the Wissahickon schist and thus Precambrian as well. But this created a problem in Chester Valley where the Octoraro "conformably" overlay Ordovician rocks. To resolve this, a large thrust fault (the Martic thrust) was proposed on which Precambrian Glenarm rocks were transported over the younger Lower Paleozoic rocks (Mackin, 1935; also see Swartz, 1948, for a more extended discussion of this controversy up to that time). A few years later Jonas (1929, 1932) extended the overthrust concept in general, and the Martic overthrust in particular, southward from the central into the southern Appalachians. In view of the presently known complexities of the southern Appalachians, her extension of the Glenarm series and Martic thrust was clearly unwarranted and simplistic. On the other hand, her recognition of the large amount of northwestward fault movement in this terrane was ahead of its time.

The Martic thrust was a tenuous proposal: it was created out of uncertain age assignments, and direct evidence for a thrust fault was lacking. A definitive study (Cloos and Hietanen, 1941) concluded that the evidence for an overthrust was lacking and that the contact was probably conformable. Subsequent work has tended to confirm a stratigraphic interpretation of the Martic line (e.g., Cloos and Wise, 1960; see also Higgins, 1972, for a recent review of the problem). Despite the uncertainties of the Martic thrust, the concept of an overthrust in the Piedmont of the Appalachians gained a foothold. Mapping throughout southeastern Pennsylvania, particularly west of the Susquehanna River, demonstrated the presence of numerous low angle overthrusts which dip gently to the southeast and upon which older rocks have been transported northwestwardly over younger rocks (e.g., Stose and Stose, 1944).

The Woodville dome is an odd, hook-shaped area of Baltimore gneiss, west of Philadelphia, that stimulated the introduction of the Alpine folded-nappe (Argand 1911) to this part of the Appalachians (Fig. 13). The dome is surrounded on three sides by the Setters, Cockeysville, and Wissahickon Formations, the Glenarm sequence. As the Woodville structure was originally mapped (Bliss and Jonas, 1916), the Wissahickon mica gneiss had been thrust over the dome of Baltimore gneiss, Chickies (Setters) quartzite, and Shanandoah (Cockeysville) limestone on a low-angle, undulatory thrust fault. Some twenty years later, Bailey and Mackin (1937; also Mackin, 1962) offered a quite different interpretation for the Woodville structure. Citing the persistent shallow, southwestward plunge of minor structures across the entire area, and the southwestward dipping of the Setters and Cockeysville beds under the gneiss on the northeast side, they suggested that the gneiss was not rooted in the crust, but is instead the core of a recumbent, Alpine-type nappe. In their view, a sizeable block of the Precambrian Baltimore gneiss and the overlying Lower Paleozoic (?) cover had been folded into a recumbent fold and thrust northwestward. Subsequent upright folding throughout the region refolded the recumbent nappe, imparting the present hook-shape. The significance is not so much the Woodville structure itself, but that this was the first application of the fold nappe concept to any rocks of the central and

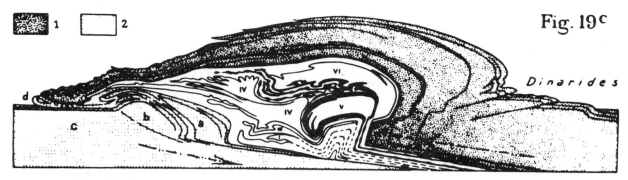

Figure 13. Some nappes are simple sheets thrust over adjacent terranes. Others resemble giant recumbent folds, with upright upper limbs, overturned lower limbs, and complex folding within, as in this cross-section of the western Alps by Argand (1916). The Appalachians do not have the spectacular exposures that are present in the Alps, and thus it was considerably later when detailed mapping was able to demonstrate that nappes as complex as above are also present in the Appalachians.

southern Appalachians, and it followed by only a few years the first introduction of this idea to the northern Appalachians.

It was perceived only gradually over the following years just how common and pervasive nappe structures are in the more highly deformed part of the central Appalachians. The carbonate rocks of Lebanon Valley in Pennsylvania are overturned and appear to comprise the lower limb of an enormous nappe (Gray, 1959), the Lebanon Valley nappe (MacLachlan, 1967). A similar carbonate sequence in the Conestoga Valley to the south is right-side-up and appears to belong to the upper limb of the same nappe.

Much more complex are clastic rocks overlying the carbonate rocks of the Lebanon Valley nappe. These disparate sequences of Late Cambrian to Middle Ordovician age were long thought to be part of the Upper Ordovician Martinsburg Formation. Rogers (1858) recognized that they were peculiar but it was Kay (1941) and Stose (1946) who proposed that these rocks were transported into the Martinsburg basin from the southeast as a single mass, the Hamburg klippe. These sequences, which do not correlate even with each other, are now thought to be individual thrust slices within the klippe, which possibly originated as gravity-driven sloughs off the top of the growing and advancing Lebanon Valley nappe (MacLachlan, 1982, personal communication). Spewing large wildflysch chunks into the basin, the klippe was itself subsequently buried by younger, 'normal' Martinsburg sediments (Root and MacLachlan, 1978; see also Lash and others, 1984).

The westernmost crystalline rocks of the central and southern Appalachians are the Precambrian and Lower Paleozoic rocks of the Blue Ridge province, which extends from Pennsylvania southward to Georgia. The superficially similar belt of rocks of the Hudson Highland terrane, or Reading Prong as its southern end is known, extends from eastern Pennsylvania into southern New York. Both of these terranes were long thought to be autochthonous, rooted in the crust, because of their Grenville affinity.

A number of large faults in the Reading Prong were recognized during the early decades of this century, but the significance of the faults was not known. Stose and Jonas (1935) proposed that the Precambrian crystallines and Cambrian quartzites comprise a large overthrust sheet, presently overlying the carbonates exposed in windows, a view that was widely disputed (e.g. Miller, 1935; also Drake, 1969, p. 99–100 for further details). In the 1950s and 1960s renewed field work and aeromagnetic studies revealed that the Precambrian masses are not rooted but are underlain by low angle, southeastward-dipping faults. The entire Reading Prong consists of a number of stacked thrust sheets composed of both crystalline and Lower Paleozoic rocks in an Alpine-type nappe mega-system (Drake, 1969; 1978). This basic structural pattern appears to extend northeastward into New York, suggesting that the entire Reading Prong-Hudson Highlands terrane is allochthonous.

The presence of Precambrian rocks of Grenville affinity in both the Reading Prong and the northern part of the Blue Ridge and their common position between the Great Valley and the Piedmont, have frequently led geologists to consider them to be separated parts of the same terrane. They decidedly are not—the two have had quite different histories. The Blue Ridge terrane was located in a Late Precambrian rift system in which a large volume of volcanics was overlain by a thick clastic sequence. The Reading Prong equivalent is a relatively thin quartzite. The tectonics of the two stand in even greater contrast. Whereas the Reading Prong was caught up in Taconian nappes and subsequently deformed and transported on Alleghanian low-angle thrusts (Drake, 1969; MacLachlan, 1979), the northern Blue Ridge has undergone only a single, décollement-style deformation, the Alleghanian (Root, 1970).

The crystalline terrane of the central Appalachians was a largely unknown area during the nineteenth century. It was part of Appalachia, the geanticline of metamorphic rocks that supplied sediment to the Appalachian geosyncline. During the ensuing decades from the beginning of this century, much has been learned about the area. Precambrian rocks once thought firmly anchored in the crust are now perceived as cores of far-traveled nappes. Metamorphism ranging from low greenschist to high pressure granulite grades occur in specific but as yet unexplained patterns (Crawford and Crawford, 1980). Not only have the rocks undergone multiple deformations (Freedman and others, 1964), but two, probably three separate orogenies have affected them. But much still needs to be explained. From where have all the various micro-terranes (Honey Brook Upland, etc.) and stratigraphic facies, such as the Lebanon Valley sequence and the Lehigh Valley sequence, come? Can any sense be made of the diverse lithologies that have been thrown into the bin called Wissahickon?

Progress in understanding the crystalline rocks of the southern Appalachians was retarded as compared to the central Appalachians, not only by the complexities of the terrane, but perhaps even more by the paucity of good, fresh exposures. The absence of fossils and the extensive metamorphism were additional hinderances. Although differences were recognized within the terrane, the area was generally considered to be a large Archean Appalachia that supplied sediment to the geosyncline to the west. Keith (1923) recognized that the Appalachian folding and deformation extended across this crystalline terrane and passed under the Coastal Plain sediments on the east. The long linear belts such as the Brevard, the Kings Mountain, and the Pine Mountain were considered to be infolds of Cambrian rocks. The complexities of the metamorphic history were not realized—in general, metamorphism was thought to increase towards the southeast. Keith also emphasized the presence of "Late Carboniferous" granites that were hardly deformed and thus must have been emplaced after the deformation.

Realization of the complex nature of the southern Appalachian crystalline terrane developed gradually over the next two decades. Certain areas had characteristics that were not seen in other areas and so it seemed a reasonable first step to divide the crystallines into various belts with anticipation of later further subdivision (Adams, 1933; Crickmay, 1952). By the mid-1950s the southern crystallines had been divided into the Carolina slate and the Charlotte belts, the Kings Mountain, Inner Piedmont and Brevard belts, and the Blue Ridge belt, among other smaller ones (King, 1955). Each of these belts contained a distinctly different lithology, deformational style, and metamorphic history. These contrasts made correlation within the crystallines virtually impossible and suggested that they had different histories and perhaps different origins. This was a fortuitous direction to take because these belts have become cornerstones in the application of the recently developed concepts of plate tectonics, suspect terranes, and docking microcontinents.

The Brevard is the most extensively studied and discussed of the various belts. It is a narrow belt, at most only a few kilometers wide and it is very linear, extending for more than 500 kilometers from Alabama to Virginia. A fault zone lies along its northwest side for much of its length, and its platform-type carbonate rocks possess an affinity to those of the Valley and Ridge province and not to any within the adjacent Inner Piedmont or Blue Ridge belts. For these reasons, the Brevard seemed to be the most important feature in the southern Appalachian crystallines. Interpretations of the Brevard's significance have been varied and imaginative. It has been postulated that it is a strike-slip fault zone, a strike-slip fault zone with a dip-slip component, a thrust zone, or a fault zone of practically any type except normal. It has been proposed that the Brevard is a root zone for nappes, that it is the front of a nappe, and that it is an isoclinal syncline. It has also been called a continental suture, and a transported continental suture (see Hatcher, 1978, for a fuller discussion and relevant references). To many workers' surprise, it was found recently that the basal décollement on which the folded and crystalline rocks moved passes completely under the Brevard without interruption or deflection (Cook and others, 1979), suggesting that the Brevard is but one of several splays, albeit a quite distinctive one.

As discussed in a preceding section, considerable horizontal transport has been inferred from the faults in the foreland of the southern Appalachians. In contrast, much less was known of the amount of transport in the crystalline portion although this situation is rapidly changing. A few large nappes have been discerned within the Inner Piedmont (Griffin, 1971), but how far the rocks in these nappes have traveled has not been determined. Two allochthonous masses, the Smith River and the Alto, have been found adjacent to the Brevard fault zone. The provenance and amount of movement of these allochthons are similarly unknown although emplacement of the Alta is at least as old as the latest movement on the Brevard (Hatcher, 1978). To the northwest, the Grandfather Mountain window in the Blue Ridge province exposes Precambrian and Cambrian rocks of a lower metamorphic grade than those in the surrounding territory, demonstrating the allochthony of the Blue Ridge (King, 1964; Bryant and Reed, 1970). From this window alone it has been estimated that the Blue Ridge terrane has moved a minimum of 70 km.

The absence of stratigraphic correlation and fossils, and the complex metamorphic effects has hindered resolution of the geology, yet this metamorphism has provided two alternative approaches to delineate the history of the southern crystallines: metamorphic grade and associated deformations, and age determinations. The paucity of age determinations before the 1970s precluded an understanding of the sequence or age of orogenic events beyond the recognition of a general decrease in age toward the southeast (Butler, 1972). The extensive accumulation of data since then has demonstrated that each of the three major Appalachian orogenies, the Taconian, Acadian, and Alleghanian, has affected the southern crystallines, but to different degrees in the various parts (Glover and others, 1983). The Taconian appears to have been the most widespread and intense dynamometamorphic event, as it was in the central Appalachian crystallines (Crawford and Crawford, 1980). The Acadian was a more localized and less intensive event, restricted largely to the Inner Piedmont and Talledega belt. It seems to have consisted more of an uplift and cooling event, although this is also the time of the most voluminous batholithic intrusion. Alleghanian metamorphism and ductile deformation was restricted to the easternmost part of the crystallines despite the fact that its brittle deformation, the décollement tectonics, extended the farthest to the northwest of the three orogenies.

SOURCE OF HORIZONTAL STRESS

When geological study of the Appalachian mountain system was just getting underway in the early nineteenth century, vertical motions were central to the tectonic thinking of that time. It became evident in the ensuing years that the Appalachian foreland folds and faults represent a substantial amount of horizontal shortening, and that this shortening required horizontal forces. The mountain building theory of the Rogers brothers did not provide a source for such forces. The geosynclinal theory of Hall was also deficient in this regard. In contrast, Élie de Beaumont's theory of the cooling of the crust and the consequent tangential contraction which provided horizontal tectonic forces was an appealing hypothesis. It was based on what seemed to be a reasonable origin of the earth (Laplace's nebular theory), it was consistent with well-known physical principles and laws, and it provided the horizontal forces necessary to form the geologic structures being delineated in the field. It supplied explanations where so many earlier ideas had fallen short.

Dana amalgamated the shrinking Earth and Hall's geosyncline, and added zones of crustal weakness along the continent-ocean boundaries and a geanticline to produce an explanation for the Appalachian orogen. At last the mountain building process had a coherent paradigm: the geosyncline filling with sediment supplied by an adjacent geanticline and a physically rational means of creating mountains from the geosynclines. Or so it seemed. At first glance, the contraction concept is attractive and simple, but closer examination by geologists and physicists in the latter part of the nineteenth century began to reveal deficiencies (e.g., Dutton, 1874; Fisher, 1881). The cooling did not produce sufficient shortening, the Earth's interior turned solid, the discovery of radioactivity reduced the efficacy of the cooling effect, and Laplace's nebular theory was replaced by a planetesimal origin of the solar system (Chamberlin, 1905). In addition, the amount of crustal shortening found in the rocks continued to grow.

First and foremost, it was becoming evident that thermal contraction of the Earth was not a large enough effect to cause more than a few kilometers of crustal shortening. In addition, this shortening would be distributed evenly around the world and would not be naturally concentrated in mountain systems. With a liquid interior, the crust could easily move over the liquid to focus the shortening on a linear mountain system. But high pressure, such as exists deep within the Earth, causes even hot liquid to condense into a solid state, which was the basis for the nineteenth century physicists to claim the Earth is solid (as early as Hopkins, 1839). And, if the Earth is solid, the distributed crustal shortening cannot be concentrated at the mountain zones. Ardent supporters of the contraction theory attempted to answer these difficulties (e.g., LeConte, 1878; Bucher, 1933). They proposed that a slip surface, perhaps even a liquid zone of detachment, exists between the crust and the solid interior that would enable the crust to deform only at the mountains. But the evidence against thermal contraction continued to accumulate. The discovery of radioactivity at the end of the century put a further crimp on the concept, for now the Earth's outward heat flow had two components: one from radioactive sources; and the second, a reduced one, the flow due to cooling. But thermal contraction was not the only way to shrink the Earth. Finding fault with Laplace's nebular theory, Chamberlin (1905, 1916) devised a planetesimal theory for the origin of the Earth and other planets. In this scheme, the Earth formed by gravitational accumulation of many small bodies, planetesimals. Once assembled into the Earth, these planetesimals continued to settle gravitationally based on density differences, resulting in a gradually contracting planet Earth. But it was soon realized that even if this process had occurred, its effect was also far too small to produce the required crustal shortening. Although the physicists pointed out the flaws in the theory of thermal and gravitational contraction as the source of horizontal shortening, ultimately it was the geological evidence of very large amounts of crustal shortening that put the required movements far beyond the limits of the Earth's contraction.

In the early part of the twentieth century, various other mechanisms were called upon to produce the horizontal movements. Notable among these in the Appalachians was Keith's (1923) resurrection of batholithic intrusion as the driving force behind mountain building. The fairly large volumes of intrusive rocks in the southern Appalachians lent credence to this concept, except when examined more closely. The batholiths were not emplaced at the end of the Paleozoic during the Alleghanian orogeny, but were mostly formed earlier, especially during the Middle Paleozoic, long before the mountain building they were supposed to initiate. In addition, batholiths of any great size are not present in the central Appalachian crystallines, unless they are presently covered by the Coastal Plain deposits.

The inability to find a suitable mechanism and explanation for crustal shortening was not the only problem. Field evidence increasingly made demands on the theories, demands that could not be met. For example, early nineteenth century estimates of horizontal shortening were on the order of kilometers. But as work progressed, it became evident that the extensive horizontal faulting in the southern Appalachians, when accumulated, amounted to many tens of kilometers. Even early estimates of shortening by folding in the central Appalachians stood at 100 or more kilometers (Claypole, 1885).

One idea that did promise to answer the large crustal shortening problem was a new one, continental drift, the movement of continents across the face of the Earth, a concept developed in the early years of the twentieth century by geologists such as Suess (1904), Taylor (1910), and in particular, Wegener (1915). It was indeed new, and quite different. To those geologists who believed the continents were fixed in the Earth's crust, continental drift must have created the same disquiet and uncertainty as the theory of relativity did to those early twentieth century physicists who believed in the immutability of time and Euclidean space.

The cornerstone of the drift theories was the large horizontal movement of continental masses, and a considerable body of supporting geologic data was gathered (e.g., du Toit, 1937). However, not all aspects could be explained and the theory encountered considerable resistance. The absence of forces or mechanisms that could drive the lighter continental cratons through or on top of the more mafic, denser oceanic crust was among the many objections. The tidal forces that Wegener invoked are far too weak to move entire continents and build mountain ranges. Even though continental drift accounted for many geological puzzles, principal among them being the geometrical shapes of the continents, and the stratigraphic, faunal, and floral similarities from one continent to another, the theory was discounted or ignored because of the apparent physical impossibility of continental movement. Enthusiastic support from a few geologists notwithstanding, Appalachian geologists such as Charles Schuchert, Arthur Keith, and Rollin Chamberlin, were definitely anti-drift (see, for example, van der Gracht and others, 1928). At about this time the idea of convection currents in the mantle was introduced to provide a physically reasonable mechanism for moving continents horizontally (Holmes, 1931). This idea was elaborated to explain mountain building (Griggs, 1939), and led directly to the concept of sea-floor spreading (Hess, 1962), the precursor of the plate tectonics scheme.

With the demise of the contraction theory, and the refusal of most Appalachian geologists to entertain continental drift, the Appalachian mountain system was left without a viable orogenic theory. Appalachian orogenic thinking fell into a stasis, and it remained there until the 1970s, when the Appalachians were incorporated into the newly-erected plate tectonics paradigm.

APPLICATION OF THE PLATE TECTONICS PARADIGM

The change in geologic thinking from a vertical to a completely horizontal tectonics for the Appalachian orogen was far advanced by the late 1960s. Geologic mapping and geophysical data had shown that décollement tectonics applies to the entire foreland belt and probably to the Blue Ridge as well. In the crystalline Appalachians, and in the Great Valley immediately to the northwest, the numerous southeast-dipping, low-angle thrust faults speak of a considerable, if unknown, amount of northwestward transport and telescoping of these rocks. The delineation of large nappes in these terranes also suggests large northwestward movements. The klippen in the Great Valley (the Hamburg allochthon) and the Inner Piedmont (the Smith River and Alto allochthons) point to the large horizontal transport of huge rock volumes into the depositional basins, reflecting an early horizontal tectonism prior to the principal orogenic activity. The juxtaposition of dissimilar stratigraphic sequences, for example around the Reading Prong (MacLachlan, 1979), is a further indication that presently adjacent rock sequences were initially much farther apart.

Two "events" in the 1970s served to amalgamate these factors into a more coherent tectonic picture: the development of the plate tectonics paradigm and the deep seismic reflection profile programs in the Appalachians. The first of these events provided the conceptual framework for bringing together all the disparate facts of Appalachian geology. The latter, the seismic data, was the first confirmational evidence that the décollements and the flat thrusts of the foreland belt extend southeastward under the crystalline Appalachians, indicating that décollement tectonics was not just a peripheral, foreland deformation, but that it applied to most, if not all, of the Appalachian mountain system.

Two aspects of plate tectonics that are particularly useful in understanding the crystalline Appalachians are the related concepts of microcontinents (Coney and others, 1980) and suspect terranes (Williams and Hatcher, 1983). With the continued development of the plate tectonic concept it became apparent that not all the plates were continental in size, i.e., thousands of kilometers across. Smaller pieces of the larger plates, tens to hundreds of kilometers in size, were found to have broken off and to have moved independently among the larger plates. As a consequence, when two larger plates converged, some of these smaller plates (microcontinents) were often caught in between, becoming incorporated within the orogenic zone. These microcontinents can be discerned by the structural, stratigraphic, and metamorphic contrasts among adjacent terranes. Such terranes do not fit well into the geologic framework of the associated miogeocline, and because their origin is uncertain, they are suspect (Williams and Hatcher, 1983). With this concept, the earlier subdivision of the southern crystalline Appalachians into separate homogeneous belts takes on a greatly enhanced significance.

A coherent scenario for the Appalachian orogen in terms of plate tectonics is developing as more of the details of the crystalline are worked out. The Appalachian orogen was active from the Late Precambrian to the Jurassic, indicating that this edge of the Laurentian craton had been converging with, diverging from, or sliding along the edges of other cratonic blocks (such as Baltica and Africa) for more than 500 million years. During the divergent phases, sedimentary basins formed. Some, such as the Late Precambrian Ocoee basin and the Mesozoic basins, are floored by extended continental crust. Others, possibly the Ashe and Tallulah Falls basins of the eastern Blue Ridge and the Wissahickon of the Piedmont, overlie newly created oceanic crust. During convergent phases, oceanic crust was subducted under or obducted over continental crust, or two continental masses collided resulting in metamorphism and/or deformation on a regional scale. The Appalachian orogen seems to have initially formed on what was originally a continuous expanse of continental crust comprising Grenville or Grenville-affinity rocks. Late in the Precambrian, perhaps 700 million years ago, a spreading center developed underneath what was to become the Appalachian orogen, producing a rift system (Wilson, 1966; Rankin, 1975). As the two continental plates moved apart, a portion of the Laurentian craton was broken off and oceanic basins (Iapetos) developed on either side of this microcontinent fragment (Fig. 14) (Hatcher and Odom, 1980). As the divergence progressed during the Late Precambrian, Cambrian, and Early Ordovician?, shelf-type carbonate sediments and more distal siliciclastic and volcani-clastic deposits were formed on the intervening shelf, slope, and basin.

The spreading reversed in the Middle Ordovician, and the continents began to converge; the resulting deformation represents the Taconian orogeny, attended by upper amphibolite-grade Barrovian metamorphism. Parts of the microcontinent were forced upwards over the edge of the Laurentian continent, causing a telescoping of depositional terranes that were originally widely separated (Fig. 14). Fragments of the microcontinent were caught up in large nappes of the Pine Mountain belt, the Sauratown Mountains anticlinorium, the Baltimore gneiss complexes and in the basinal and slope rocks on the craton margin. In addition, allochthonous klippen such as the Hamburg and Smith River were sloughed off the advancing fronts of the nappes.

Crustal conditions were relatively quiet during the Silurian, judging from the platform-type deposits in the Appalachian basin. At the beginning of the Middle Devonian, movement began again producing the Acadian orogeny, but it was transcurrent or oblique movement (Fig. 14). The central Piedmont root zone, centered on the Kings Mountain belt (Glover and others, 1983), formed at this time, separating the Avalon terrane of the eastern Piedmont, the Carolina slate belt, and the other suspect, eastern belts with Atlantic faunal affinity (Williams and Hatcher, 1983) from the western Laurentian belts (of North American affinity). The high-grade Taconian metamorphism that had affected the eastern Blue Ridge and western Piedmont (it is not present east of the King's Mountain Line) was overprinted in the Devonian by a greenschist metamorphism that pervaded all of the crystalline Appalachians (Tull, 1980, Crawford and Crawford, 1980). Numerous granite intrusions were emplaced at this time in the southern Appalachians, particularly in the eastern part of the crystalline terrane. Paleomagnetic evidence suggests that portions of southeasternmost Pennsylvania and northern Delaware had traveled from far to the south and were sutured to the North American continent at this time (Brown and Van der Voo, 1983). Little of this activity affected the miogeoclinal rocks of the foreland terrane to the west (Faill, 1985). In the central Appalachians, the wide Catskill basin progressively deepened, filling with marine and continental deposits from active orogenic zones to the east or northeast. In contrast, little or no sediment reached the foreland region in the southern Appalachians during the Middle Paleozoic. However, the arrival of coarse Carboniferous deposits in the southern foreland signaled some increase in tectonic activity in the southern crystalline terrane at that time.

Near the end of the Carboniferous or perhaps at the beginning of the Permian, a major convergence began again with the approach of the African continent against the central and southern Appalachians. The convergence produced the widespread Alleghanian deformation with metamorphism restricted to the easternmost crystalline terrane. Its main characteristic was moderately brittle, flexural-slip folding and the northwestward, low-angle thrusting of the décollement tectonics now associated with the foreland belt. The décollement extends eastward under the

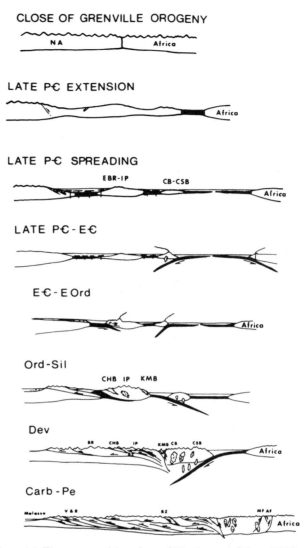

Figure 14. These sequential sections of Hatcher and Odom (1980) present a thumbnail sketch of the southern Appalachian history in terms of the plate tectonics paradigm. Extension of the Laurentian continental crust in the Late Precambrian led to the opening of small, oceanic-crust basins between fragments of the continents. Convergence in the Middle and Late Ordovician telescoped the basins and continental fragments on the edge of the continent during the Taconian orogeny. Additional convergence during the Middle and possibly Late Devonian sutured more microcontinents to the Laurentian craton with accompanying batholithic intrusions, all part of the Acadian orogeny. The Alleghanian orogeny resulted from the complete closure between the North American and African plates, and consisted primarily of a widespread décollement tectonics. In these sections, the abbreviations represent: AF — Augusta fault, BR — Blue Ridge, BZ — Brevard zone, CB — Charlotte belt, CHB — Chauga belt, CSB — Carolina slate belt, EBR — eastern Blue Ridge, IP — Inner Piedmont, KMB — Kings Mountain belt, MF — Modoc fault, and V&R — Valley and Ridge province. Black crust represents oceanic crust.

crystallines, and Alleghanian thrusting was common throughout the crystalline terrane, often utilizing existing faults.

SUMMARY

Concepts for the Appalachian orogen have changed from predominantly vertical tectonics of the early 19th century to overwhelmingly horizontal tectonics at the present time. The initial emphasis on vertical movements reflected the poor understanding of tectonism at the time and also the lack of geologic information about the Appalachian mountain system. In the early to mid-19th century, small amounts of crustal shortening were understood from observing the folds in the foreland of the Appalachians, but it took continued work, particularly in the southern Appalachians, to begin to understand that large, northwestward movement of the foreland terrane had indeed occurred.

The large simple folds and thrusts of the Appalachian foreland understandably attracted the attention of the earliest geologists. The classic folds and faults of mountain systems were associated with this terrane, and the enormously thick and easily measured sequence of Paleozoic sediments gave rise to the important concept of the geosyncline as the locus of mountain building. The more complex and metamorphosed crystalline terrane was originally relegated to a presumed geanticline, an "ancient continent" that supplied sediment to the geosyncline before foundering and disappearing from view. But when attention was focused on this terrane from the beginning of this century the presence and significance of the various crystalline terranes were gradually appreciated. Rather than being a homogeneous, complex terrane, the characteristics and distinguishing features of the various belts were delineated. Because of their crystalline nature, it was long presumed that the crystalline Appalachians were rooted in the crust. Even when the allochthony· of the foreland was widely discerned, the fact that the crystallines also moved great distances horizontally (some on décollements, others in transcurrent motions) was much slower in development. It required the recent evolution of the plate tectonics paradigm, and supporting seismic data, to work out the mechanisms of the horizontal movements, the specific histories of the various belts, and how the three principal orogenies affected both the crystalline terranes and the foreland.

ACKNOWLEDGMENTS

The author is indebted to William D. Sevon for his critical review, which appreciably improved the manuscript. The author also benefited materially from the numerous discussions with David B. MacLachlan.

REFERENCES CITED

Adams, F. D., 1910, An experimental investigation into the action of differential pressure on certain minerals and rocks, employing the process suggested by Professor Kick: Journal of Geology, v. 18, p. 489–525.

Adams, F. D., 1938, The birth and development of the geological sciences: New York, Dover Publications, Inc., 506 p.

Adams, G. I., 1933, General geology of the crystallines of Alabama: Journal of Geology, v. 41, p. 159–173.

Aldrich, M. L., 1979, American State Geological surveys, 1820–1845, *in* Schneer, C. J., ed., Two hundred years of geology in America: Hanover, N. H., University Press of New England, p. 133–143.

Arduino, G., 1759, Saggio fisico-mineralogico di lythogomia e orognosia: Atti dell' Accademia delle Scienze di Siena, T. V.

Argand, E., 1911, Les nappes de recouvrement des Alpes Pennines et leurs prolongements structuraux: Beit. geol. Karte Schweiz, Neue Folge, v. 31, p. 1–26.

Argand, E., 1916, Sur l'arc des Alpes occidentales: Eclogae Geologicae Helvetiae, v. 14, p. 146–204.

Bailey, E. B. and Mackin, J. H., 1937, Recumbent folding in the Pennsylvania Piedmont—preliminary statement: American Journal of Science, v. 233, p. 187–190.

Bascom, F., 1905, Piedmont district of Pennsylvania: Geological Society of America Bulletin, v. 16, p. 289–328.

Bascom, F., 1909, The pre-Cambrian gneisses of the Pennsylvania Piedmont Plateau [abstracts]: Science ns 30, p. 415.

Bertrand, M. A., 1884, Rapports de structure des Alpes de Glaris et du bassin houiller de Nord: Bulletin de la Société Géologique de France, 3eme ser., v. 12, p. 318–330.

Bliss, E. F. and Jonas, A. I., 1916, Relation of the Wissahickon mica gneiss to the Shenandoah limestone and Octoraro schist of the Doe Run and Avondale region, Chester County, Pennsylvania: U.S. Geological Survey Professional Paper 98, p. 9–34.

Brown, P. M. and Van der Voo, R., 1983, A paleomagnetic study of Piedmont metamorphic rocks in northern Delaware: Geological Society of American Bulletin, v. 94, p. 815–822.

Bryant, B. and Reed, J. C., Jr., 1970, Structural and metamorphic history of the southern Blue Ridge, *in* Fisher, G. W., Pettijohn, F. J., Reed, J. C., Jr., and Weaver, K. N., Studies of Appalachian Geology: New York, Interscience Publishers, p. 213–225.

Bucher, W. H., 1933, Deformation of the Earth's crust: Hafner Publishing Company, New York and London, 518 p.

Butler, J. R., 1972, Age of Paleozoic regional metamorphism in the Carolinas, Georgia, and Tennessee southern Appalachians: American Journal of Science, v. 272, p. 319–333.

Butts, C., 1927, Fensters in the Cumberland overthrust block in southwestern Virginia: virginia Geological Survey Bulletin 28, 12 p.

Buxtorf, A., 1907, Geologie des Weissensteintunnels, Beitrage zum Geologie Karte d. Schweiz.

Buxtorf, A., 1916, Prognosen und Befunde beim Hauenstein-basis und Grenchenberg tunnel, und die Bedeutung der letzteren für die Geologie des Juragebirges: Naturforschende Gesellschaft in Basel, Verhandlungen, v. 27.

Chamberlin, R. T., 1910, The Appalachian folds of central Pennsylvania: Journal of Geology, v. 18, p. 228–251.

Chamberlin, T. C., 1905, The planetesimal hypothesis: Carnegie Institution Yearbook, p. 195–254.

Chamberlin, T. C., 1916, The origin of the Earth: Chicago, The University of Chicago Press, 271 p.

Chapple, W. M., 1978, Mechanics of thin-skinned fold and thrust belts: Geological Society of America Bulletin, v. 89, p. 1189–1198.

Claypole, E. W., 1885, Pennsylvania before and after the elevation of the Appalachian Mountains—a study in dynamical geology: American National, v. 19, p. 257–268.

Cleaveland, P., 1816, An elementary treatise on mineralogy and geology: Boston,

668 p.; 1822, 2nd edition, 2 volumes: Boston, 818 p.

Cloos, E., 1947, Oolite deformation in the South Mountain fold, Maryland: Geological Society of America Bulletin 58, p. 843–918.

Cloos, E. and Hietanen, A., 1941, Geology of the "Martic overthrust" and the Glenarm series in Pennsylvania and Maryland: Geological Society of America Special Paper 35, 207 p.

Cloos, E. and Wise, D. U., 1960, The Martic problem and the New Providence railroad cut, *in* Bricker, O. P., Hopson, C. A., Kauffman, M. E., Lapham, D. M., McLaughlin, D. B., and Wise, D. U., eds., Some tectonic and structural problems of the Appalachian Piedmont along the Susquehanna River: Pennsylvania Geological Survey, 4th ser., Guidebook 25th Annual Field Conference of Pennsylvania Geologists, p. 39–48.

Coney, P. J., Jones, D. L., and Monger, W. H., 1980, Cordilleran suspect terranes: Nature, v. 188, p. 329–333.

Cook, F. A., Albaugh, D. S., Brown, L. D., Kaufman, S., and Oliver, J. E., 1979, Thin-skinned tectonics in the crystalline southern Appalachians; COCORP seismic-reflection profiling of the Blue Ridge and Piedmont: Geology, v. 7, p. 563–567.

Crawford, M. L. and Crawford, W. A., 1980, Metamorphic and tectonic history of the Pennsylvania Piedmont: Journal of the Geological Society, v. 137, part 3, p. 311–320.

Crickmay, G. W., 1952, Geology of the crystalline rocks of Georgia: Georgia Geological Survey Bulletin 58, 54 p.

Currie, J. B., Patnode, H. W., and Trump, R. P., 1962, Development of folds in sedimentary strata: Geological Society of America Bulletin, v. 73, p. 655–674.

Dana, J. D., 1847, On the origin of continents: American Journal of Science, 2nd ser., v. 3, p. 94–100.

Dana, J. D., 1847, The geological results of the Earth's contraction in consequence of cooling: American Journal of Science, 2nd ser., v. 3, p. 176–188.

Dana, J. D., 1847, Origin of the grand outline features of the Earth: American Journal of Science, 2nd ser., v. 3, p. 381–398.

Dana, J. D., 1847, A general review of the geological effects of the Earth's cooling from a state of igneous fusion: American Journal of Science (2nd series), v. 4, p. 88–92.

Dana, J. D., 1856, On American geological history: American Journal of Science, 2nd ser., v. 22, p. 305–334.

Dana, J. D., 1873a, On some results of the Earth's contraction from cooling, including the origin of mountains and the nature of the Earth's interior: Part I: American Journal of Science, 3rd ser., v. 5, p. 423–443.

Dana, J. D., 1873b, On some results of the Earth's contraction from cooling; Part II, the condition of the Earth's interior, and the connection of the facts with mountainmaking; III, metamorphism: American Journal of Science, 3rd ser., v. 6, p. 6–14, 104–115, 161, 172.

Dana, J. D., 1890, Areas of continental progress in North America, and the influence of the conditions of these areas on the work carried forward within them: Geological Society of America Bulletin, v. 1, p. 36–48.

Dana, J. F. and Dana, S. L., 1818, Outlines of the mineralogy and geology of Boston and vicinity, with a geologic map: Boston, 108 p.

Darton, N. H., 1940, Some structural features of the northern anthracite coal basin, Pennsylvania: U.S. Geological Survey Professional Paper 193-D, p. 69–81.

de Saussure, H. B., 1796, Voyages dans les Alpes: Neuchatel, 4 volumes.

Drake, A. A., Jr., 1969, Precambrian and Lower Paleozoic geology of the Delaware Valley, New Jersey-Pennsylvania, *in* Subitzky, S., ed., Geology of selected areas in New Jersey and eastern Pennsylvania and guidebook of excursions: New Brunswick, New Jersey, Rutgers University Press, p. 57–131.

Drake, A. A., Jr., 1978, The Lyon Station-Paulins Kill nappe—the frontal structure of the Musconetcong nappe system in eastern Pennsylvania and New Jersey: U.S. Geological Survey Professional Paper 1023, 20 p.

du Toit, A., 1937, Our wandering continents: Edinburgh, Oliver and Boyd.

Dutton, C. S., 1874, A criticism upon the contractional hypothesis: American Journal of Science, 3rd ser., v. 8, p. 113–123.

Eaton, A., 1830, Geological Textbook: Albany, 63 p.; 1832, Geological Textbook, 2nd edition: Albany, 134 p.

Élie de Beaumont, L., 1828, Observations géologiques sur les différentes formations qui, dans le système des Vosses, separent la formation houillère de celle du lias: Paris, Mme Huzard, 199 p.

Élie de Beaumont, L., 1852, Notice sur les systèmes de montagnes: Paris, P. Bertrand, 3 volumes.

Elliott, D., 1976, The energy balance and deformation mechanisms of thrust sheets: Proceedings of the Royal Society of London, Ser. A., v. 283, p. 289–312.

Evans, L., 1755, Analysis of a general map of the middle British colonies in America: Philadelphia, B. Franklin and D. Hall.

Faill, R. T., 1973, Kink-band folding, Valley and Ridge privince, Pennsylvania: Geological Society of America Bulletin, v. 84, p. 1289–1314.

Faill, R. T., 1985, The Acadian orogeny and the Catskill Delta, *in* Woodrow, D. L. and Sevon, W. D., eds., The Catskill Delta: Geological Society of America Special Paper 201, p. 15–38.

Faill, R. T. and Nickelsen R. P., 1973, Structural geology, *in* Faill, R. T., Wells, R. B., Nickelsen, R. P., and Hoskins, D. M., Structure and Silurian-Devonian stratigraphy of the Valley and Ridge province, central Pennsylvania: Pennsylvania Geological Survey, 4th ser., Guidebook, 38th Annual Field Conference of Pennsylvania Geologists, p. 9–38.

Favre, M. A., 1878, Expériences sur les essets des refoulements ou ecrasements lateraux en géologie: Archives des Sciences Physiques et Naturelles, 2nd ser., v. 62, no. 246, p. 193–211.

Fisher, G. W., Pettijohn, F. J., Reed J. C., Jr. and Weaver, K. N., eds., 1970, Studies of Appalachian Geology: central and southern: New York, Interscience Publishers, 460 p.

Fisher, O., 1881, Physics of the Earth's Crust: London, Macmillan, 272 p.

Franklin, B., 1793, Conjectures concerning the formation of the Earth, etc. in a letter to the Abbe Soulavie: Transactions of the American Philosophical Society, v. 3, p. 1–5.

Freedman, J., Wise, D. K., and Bentley, R. D., 1964, Pattern of folded folds in the Appalachian Piedmont along Susquehanna River: Geological Society of America Bulletin, v. 75, p. 621–638.

Füchsel, J. C., 1762, Historia terrae et maris ex historia Thuringiae per montium descriptionen erecta: Acta, Academie Elect. moguntinae zu Erfuhrt, v. 2, p. 44–209.

Geikie, A., 1905, Founders of Geology: The Macmillan Company, London, 488 p.

Geiser, P. and Engelder, T., 1983, The distribution of layer parallel shortening fabrics in the Appalachian foreland of New York and Pennsylvania: evidence for two non-coaxial phases of the Alleghenian orogeny, *in* Hatcher, R. D., Jr., Williams, H., and Zietz, I., eds, Contributions to the tectonics and geophysics of mountain chains: Geological Society of America Memoir 158, p. 161–175.

Glover, L., III, Speer, J. A., Russell, G. S., and Farrar, S. S., 1983, Ages of regional metamorphism and ductile deformation in the central and southern Appalachians: Lithos, v. 16, p. 223–245.

Gray, C., 1959, Nappe structures in Pennsylvania [abst.]: Geological Society of America Bulletin, v. 70, p. 1611.

Griffin, V. S., Jr., 1971, The Inner Piedmont belt of the southern crystalline Appalachians: Geological Society of America Bulletin, v. 82, p. 1885–1898.

Griggs, D. T., 1936, Deformation of rocks under high confining pressures: Journal of Geology, v. 44, p. 541–577.

Griggs, D. T., 1939, A theory of mountain building: American Journal of Science, v. 237, p. 611–650.

Griggs, D. T. and Handin, J., eds., 1960, Rock Deformation: Geological Society of America Memoir 79, 382 p.

Guettard, J.-E., 1752, Mémoire dans lequel on compare le Canada à la Suisse, par rapport à ses mineraux: Histoire de l'Académie Royale des Sciences, B, p. 189–220; C, p. 323–361. Also Sur la comparaison du Canada avec la Suisse, par rapport à ses mineraux: A, p. 12–16; and Addition au Mémoire dans lequel on compare le Canada à la Suisse, par rapport à ses mineraux: D,

p. 524–538.

Gwinn, V. E., 1964, Thin-skinned tectonics in the Plateau and northwestern Valley and Ridge provinces of the central Appalachians: Geological Society of America Bulletin, v. 69, p. 863–900.

Gwinn, V. E., 1970, Kinematic patterns and estimates of lateral shortening, Valley and Ridge and Great Valley provinces, central Appalachians southcentral Pennsylvania, *in* Fisher, G. W., Pettijohn F. J., Reed, J. C., Jr., and Weaver, K. N., eds., Studies in Appalachian Geology, central and southern: New York, Interscience Publishers, p. 127–146.

Hall, J., 1815, On the vertical position and convolution of certain strata and their relation with granite: Royal Society of Edinburgh Transactions, v. 7, p. 79–108.

Hall, J., 1859, Description and figures of the organic remains of the Lower Helderberg Group and the Oriskany Sandstone: Natural History of New York, Part VI, Paleontology: v. 3, p. 1–96.

Hall, J., 1883, Contributions to the geological history of the North American continent: Proceedings of the American Association for the Advancement of Science, v. 31, p. 29–69.

Harris, L. D., deWitt, W., Jr., and Bayer, K. C., 1982, Interpretive seismic profile along Interstate I-64 from the Valley and Ridge to the Coastal Plain in central Virginia: U.S. Geological Survey Oil and Gas Investigations Chart OC-123.

Harris, L. D., Harris, A. G., deWitt, W., Jr., and Bayer, K. C., 1981, Evaluation of southern eastern overthrust belt beneath Blue Ridge-Piedmont thrust: American Association of Petroleum Geologist Bulletin, v. 65, p. 2497–2505.

Hatcher, R. D., Jr., 1978, Tectonics of the western Piedmont and Blue Ridge, southern Appalachians: review and speculation: American Journal of Science, v. 278, p. 276–304.

Hatcher, R. D., Jr. and Odom, A. L., 1980, Timing of thrusting in the southern Appalachians, U.S.A.: model for orogen?: Journal of Geological Society, London, v. 137, p. 321–327.

Haug, E., 1907, Tráté de géologie: Paris, A. Colin, v. 1, 536 p.

Hayes, C. W., 1891, The overthrust faults of the southern Appalachians: Geological Society of America Bulletin, v. 2, p. 141–152.

Heim, A., 1878, Untersuchungen über den Mechanismus der Gebirgsbildung im Anchluss an die geologische Monographie der Todi—Windgallen Gruppe: Basel, Switzerland, Benno Schwabe, 2 volumes.

Hess, H. H., 1962, History of Ocean basins, *in* Engle, A.E.J., James, H. L., and Leonard, B. F., eds., Petrologic studies: A volume in honor of A. F. Buddington: Geological Society of America, p. 599–620.

Higgins, M. W., 1972, Age, origin, regional relations, and nomenclature of the Glenarm Series, central Appalachian Piedmont: a reinterpretation: Geological Society of America Bulletin, v. 83, p. 989–1026.

Holmes, A., 1931, Radioactivity and earth movements: Geological Society of Glasgow Transactions, v. 18, p. 559–606.

Hopkins, W., 1839, Transactions of the Royal Society for 1839, p. 381; for 1840, p. 192; for 1842, p. 43; for 1851, Pt. II, p. 495.

Hutton, J., 1788, Theory of the Earth: Royal Society of Edinburgh Transactions, v. 1, p. 216–304.

Hutton, J., 1795, Theory of the Earth with proofs and illustrations: Edinburgh, 2 volumes.

Jonas, A. I., 1929, Structure of the metamorphic belt of the central Appalachians: Geological Society of America Bulletin, v. 40, p. 503–514.

Jonas, A. I., 1932, Structure of the metamorphic belt of the southern Appalachians: Geological Society of America Bulletin, v. 43, p. 228–243.

Kalm, P., 1771, Travels into North America (translated by J. R. Forster), 2nd ed.: London, T. Lowndes. 2 vols., 414, 423 p.

Kay, M., 1941, Taconic allochthone and the Martic thrust: Science, v. 94, no. 2429, p. 73.

Kay, M., 1951, North American geosynclines: Geological Society of America Memoir 48, 143 p.

Keith, A., 1923, Outlines of Appalachian structure: Geological Society of America Bulletin, v. 34, p. 309–380.

King, P. B., 1955, A geological section across the southern Appalachians: an outline of the geology in the segment in Tennessee, North Carolina, and South Carolina, *in* Russell, R. J., ed., Guides to southeastern geology: Geological Society of America Guidebook, p. 332–373.

King, P. B., 1964, Geology of the central Great Smoky Mountains, Tennessee: U.S. Geological Survey Professional Paper 349-C, 148 p.

Knopf, E. B. and Jonas, A. I., 1923, Stratigraphy of the crystalline schists of Pennsylvania and Maryland: American Journal of Science, 5th ser., v. 5, p. 40–62.

Knopf, E. B. and Jonas, A. I., 1929, Geology of the McCalls Ferry-Quarryville district, Pennsylvania: U.S. Geological Survey Bulletin 799, 156 p.

Kober, 1921, Der Bau der Erde. Gebruder Borntrager, Berlin.

Laplace, P. S., 1796, Exposition du système du monde; 1830, The system of the world [translation].

Lash, G. G., Lyttle, P. T., and Epstein, J. B., 1984, Geology of an accreted terrane: the eastern Hamburg Klippe and surrounding rocks, eastern Pennsylvania: Guidebook, 49th Annual Field Conference of Pennsylvania Geologists, Pennsylvania Geological Survey, 151 p.

Le Conte, J., 1878, On the structure and origin of mountains, with special reference to recent objections to the "contractional theory": American Journal of Science, 3rd ser., v. 16, p. 95–112.

Lehmann, J. G., 1756, Versuch einer Geschichte von Flötzgeburgen: Berlin.

Lesley, J. P., 1876, Historical sketch of geological explorations in Pennsylvania and other states: Pennsylvania Geological Survey, 2nd ser., Report of Progress A, 200 p.

Longwell, C. R., Knopf, A., and Flint, R. F., 1932, A textbook of geology: part 1—physical geology: New York, John Wiley & Sons, 514 p.

Lowry, W. D., ed., 1964, Tectonics of the southern Appalachians: Virginia Polytechnic Institute, Department of Geological Sciences Memoir 1, 114 p.

Lugeon, M., 1902, Les grandes nappes de recouvrement des Alpes du Chablais et de la Suisse: Bulletin de la Société Géologique de France, ser. 4, v. 1, p. 723–823.

Mackin, J. H., 1935, The problem of the Martic overthrust and the age of the Glenarm series in southeastern Pennsylvania: Journal of Geology, v. 43, p. 356–380.

Mackin, J. H., 1962, Structure of the Glenarm Series in Chester County, Pennsylvania: Geological Society of America Bulletin, v. 73, p. 403–410.

MacLachlan, D. B., 1967, Structure and stratigraphy of the limestones and dolomites of Dauphin County, Pennsylvania: Pennsylvania Geological Survey, 4th ser., General Geology Report 44, 168 p.

MacLachlan, D. B., 1979, Geology and mineral resources of the Temple and Fleetwood quadrangles, Berks County, Pennsylvania: Pennsylvania Geological Survey, 4th ser., Geological Atlas A187ab, 71 p.

Maclure, W., 1809, Observations on the geology of the United States, explanatory of a geological map: Transactions of the American Philosophical Society, v. 6, p. 441–428.

Mathews, E. B., 1905, Correlation of Maryland and Pennsylvania Piedmont formations: Geological Society of America Bulletin, v. 16, p. 329–346.

Mease, J., 1807, A geological account of the United States, comprehending a short description of their animal, vegetable, and mineral production, antiquities, and curiosities: Birch and Small, Philadelphia, 510 p.

Merrill, G. P., 1924, The first one hundred years of American geology: New Haven, Yale University Press, 773 p.

Michell, J., 1761, Conjectures concerning the cause and observations upon the phenomena of earthquakes; particularly of that great earthquake of the first of November 1755, which proved so fatal to the city of Lisbon and whose effects were felt as far as Africa and more or less throughout all Europe: Philosophical Transactions of the Royal Society of London, v. 11, Part 2, p. 582.

Milici, R. C., 1975, Structural patterns in the southern Appalachians—evidence for a gravity glide mechanism for Alleghenian deformation: Geological Society of America Bulletin, v. 86, p. 1316–1320.

Milici, R. C., Harris, L. D., and Statler, A. T., 1979, An interpretation of seismic cross sections in the Valley and Ridge of eastern Tennessee: Tennessee Division of Geology Oil and Gas Seismic Investigations Series 1.

Miller, B. L., 1935, Age of the schists of the South Valley Hills, Pennsylvania: Geological Society of America Bulletin, v. 46, p. 715–756.

Mitchill, S. L., 1798, A sketch of the mineralogical history of the State of New York: Medical Repository, New York, v. 1, p. 293–314, 445–452.

Moro, A. L., 1740, De' Crostacei e degli altri marini Corpi che si truovano su' Monti: Venice.

Nickelsen, R. P., 1963, Fold patterns and continuous deformation mechanisms of the central Pennsylvania Folded Appalachians, *in* Tectonics and Cambrian-Ordovician stratigraphy, central Appalachians of Pennsylvania: Pittsburgh Geological Society Guidebook, p. 13–29.

Nickelsen, R. P., 1966, Fossil distortion and penetrative rock deformation in the Appalachian Plateau: Journal of Geology, v. 78, p. 609–630.

Nickelsen, R. P., 1979, Sequence of structural stages of the Allegheny orogen, at the Bear Valley strip mine, Shamokin, Pennsylvania: American Journal of Science, v. 279, p. 225–271.

Pallas, P. S., 1777, Observations sur la formation des montagnes et les changements arrives au globe, particulièrement de l'Empire Russe: St. Petersburg.

Perry, W. J., Jr., 1978, Sequential deformation in the central Appalachians: American Journal of Science, v. 278, p. 518–542.

Price, P. H., 1931, The Appalachian structural front: Journal of Geology, v. 39, p. 24–44.

Rankin, D. W., 1975, The continental margin of eastern North America in the southern Appalachians: the opening and closing of the proto-Atlantic ocean: American Journal of Science, v. 275-A, p. 298–336.

Rich, J. L., 1934, Mechanics of low-angle overthrust faulting as illustrated by Cumberland thrust block, Virginia, Kentucky, and Tennessee: American Association of Petroleum Geologists Bulletin, v. 18, p. 1584–1596.

Rodgers, J., 1949, Evolution of thought on structure of middle and southern Appalachians: American Association of Petroleum Geologists Bulletin, v. 33, p. 1643–1654.

Rodgers, J., 1950, Mechanics of Appalachian folding as illustrated by Sequatchie anticline, Tennessee and Alabama: American Association of Petroleum Geologist Bulletin, v. 34, p. 672–681.

Rodgers, J., 1964, Basement and no-basement hypotheses in the Jura and the Appalachian Valley and Ridge, *in* Lowry, W. D., ed., Tectonics of the southern Appalachians: Blacksburg, Va., Virginia Polytechnic Institute, Department of Geological Science Memoir 1, p. 71–80.

Rodgers, J., 1970, The tectonics of the Appalachians: New York, Wiley-Interscience, 271 p.

Rodgers, J., 1972, Evolution of thought on structure of middle and southern Appalachians: second paper, *in* Kanes, W. H., ed., Appalachian structures, origin, evolution, and possible potential for new exploration frontiers: a seminar: Morgantown, West Virginia University and West Virginia Geological Survey, p. 1–15.

Rogers, H. D., 1858, On the laws of structure of the more disturbed zones of the Earth's crust, *in* The Geology of Pennsylvania, vol. 2: Philadelphia, p. 885–916.

Rogers, W. B. and Rogers, H. D., 1843, On the physical structure of the Appalachian chain as exemplifying the laws which have regulated the elevation of great mountain chains generally: Transactions of American Geologists and Naturalists for 1840–1842, v. 1, p. 474–531.

Root, S. I., 1970, Structure of the northern terminus of the Blue Ridge in Pennsylvania: Geological Society of America Bulletin, v. 81, p. 815–830.

Root, S. I. and MacLachlan, D. B., 1978, Western limit of Taconic allochthons in Pennsylvania: Geological Society of America Bulletin, v. 89, p. 1515–1528.

Safford, J. M., 1869, Geology of Tennessee: Nashville, 550 p.

Schöpf, J. D., 1787, Beytrage zur mineralogischen Kenntniss des östlichen Theils von Nord-Amerika und seiner Geburge: Erlangen, Germany, John Jacob Palm, 194 p.

Schuchert, C., 1923, Sites and nature of the North American geosynclines: Geological Society of America bulletin, v. 34, p. 151–230.

Sherrill, R. E., 1934, Symmetry of northern Appalachian foreland folds: Journal of Geology, v. 42, p. 225–247.

Silliman, B., 1837, Consistency of the discoveries of modern geology with the sacred history of the creation and deluge: London, 148 p.

Smith, W., 1799, Tabular view of the order of strata in the vicinity of Bath with their respective organic remains: manuscript.

Smith, W., 1815, A geologic map of England and Wales, with part of Scotland; exhibiting the collieries, mines, and canals, the marshes and fenlands originally overflowed by the sea; and the varieties of soil, according to the variations of the substrata; illustrated by the most descriptive name of places, and of local districts; showing also the rivers, sites of parks, and principal seats of the nobility and gentry; and the opposite coast of France: London, 15 sheets.

Steno, N., 1669, De solido intra solidum naturalitum contento: Florence.

Stille, H., 1918, Uber Hauptformen der Orogenese und ihre Verknupfung, Nachr, K. Gesell, Wiss. Gottingen, math.-phys. Klasse, pp. 1–32.

Stille, H., 1940, Einfuhrung in den Bau Amerikas. Gebruder Borntrager, Berlin.

Stose, A. J. and Stose, G. W., 1944, Geology of the Hanover-York district, Pennsylvania: U.S. Geological Survey Professional Paper 204, 84 p.

Stose, G. W., 1946, The Taconic sequence in Pennsylvania: American Journal of Science, v. 244, p. 665–696.

Stose, G. W. and Jonas, A. I., 1935, Highland near Reading, Pa.: an erosional remnant of a great overthrust sheet: Geological Society of America Bulletin, v. 46, p. 757–780.

Suess, E., 1904, The Face of the Earth: Oxford, Clarendon, 5 volumes; translated from German, 1885–1909, Dzs Antlitz der Erde, Wien, Tempsky, 4 volumes.

Swartz, F. M., 1948, Trenton and sub-Trenton of outcrop areas in New York, Pennsylvania, and Maryland: American Association of Petroleum Geologists Bulletin, v. 32, p. 1493–1595.

Taylor, F. B., 1910, Bearing of the Tertiary mountain belt on the origin of the Earth's plan: Geological Society of America Bulletin, v. 21, p. 179–226.

Tull, J. F., 1980, Overview of the sequence and timing of deformational events in the southern Appalachians: evidence from the crystalline rocks, North Carolina to Alabama, *in* Wones, D. R., ed., The Caledonides in the U.S.A.: Blacksburg, Virginia Polytechnic Institute and State University Memoir no. 2, p. 167–177.

van der Gracht, W.A.J.M. van Waterschoot, Willis, B., Chamberlin, R. T., Joly, J., Molengraaff, G.A.F., Gregory, J.W., Wegener, A., Schuchert, C., Longwell, C. R., Taylor, F. B., Bowie, W., White, D., Singewald, J. T., Jr., and Berry, E. W., 1928, Theory of Continental Drift: A symposium on the origin and movement of land masses both inter-continental and intra-continental, as proposed by Alfred Wegener: American Association of Petroleum Geologists, 240 p.

Van Hise, C. R., 1896, Principles of North American Pre-Cambrian geology: U.S. Geological Survey Annual Report 16 (1894–1895), pt. 1, p. 571–843.

Vanuxem, L., 1829, Remarks on the characters and classification of certain American rock formations: American Journal of Science, 1st ser., v. 16, p. 254–257.

Volney, C. F., 1803, Tableau du Climat et du Sol des Etats Unis d'Amérique: Paris, 2 vols., 532 p.

von Buch, L., 1809, Geognostische Beobachtungen auf Reisien durch Deutschland und Italian: Berlin.

Wegener, A., 1915, Die Entstehung der Kontinente und Ozeane: Braunschweig, Vieweg.

Wentworth, C. K., 1921, Russell Fork fault of southwest Virginia: Virginia Geological Survey Bulletin 21, p. 53–66.

Werner, A. G., 1787, Kurze Klassification und Beschreibung der verschiedenen Gebirgsarten: Dresden.

Williams, H., compiler, 1978, Tectonic lithofacies map of the Appalachian orogen: Memorial University of Newfoundland, Map. No. 1.

Williams, H. and Hatcher, R. D., Jr., 1983, Appalachian suspect terranes, *in* Hatcher, R. D. Jr., Williams, H., and Zietz, I., eds., Contributions to the tectonics and geophysics of mountain chains: Geological Society of America Memoir 158, p. 33–53.

Willis, B., 1893, The mechanics of Appalachian structure: U.S. Geological Survey 13th Annual Report, Part 2, p. 211–281.

Wilson, C. W., Jr. and Stearns, R. G., 1958, Structure of the Cumberland Plateau, Tennessee: Geological Society of American Bulletin, v. 69, p. 1283–1296.

Wilson, J. T., 1966, Did the Atlantic close and then re-open?: Nature, v. 211, no. 5050, p. 676–681.

Wiltschko, D. and Eastman, D., 1983, Role of basement warps and faults in localizing thrust fault ramps: Geological Society of America Memoir 158, p. 177–190.

Woodward, H. P., compiler, 1959, A symposium on the Sandhill deep well, Wood County, West Virginia [Tucker County, West Virginia, 1957]: West Virginia Geological and Economic Survey Report of Investigations 18, 182 p.

Printed in U.S.A.

Geological Society of America
Centennial Special Volume 1
1985

A basement of melanges:
A personal account of the circumstances leading to the
breakthrough in Franciscan research

Kenneth J. Hsü
Geological Institute
Swiss Federal Institute of Technology
Zurich, Switzerland

ABSTRACT

The Franciscan of the California Coast Ranges was considered a lithostratigraphic unit (a formation or group of formations) for more than a century. Normal stratigraphical methodology was practised, and few questioned the applicability of the laws of superposition, lateral continuity, and paleontological dating. Despite evidence to the contrary, researchers portrayed deformation of coherent strata. Metamorphic conditions were deduced on the basis of presumed field relations, and the geosynclinal theory of mountain building was the paradigm. A subjective choice of data allowed researchers to fit the Franciscan into a straight-jacket, and for many years it was described as a Late Jurassic Tithonian formation, deposited in a eugeosyncline on the western margin of the North American continent after the Nevadan Orogeny.

The classical model failed to take into account a number of key observations that challenged accepted beliefs. Micropaleontological evidence for the Cretaceous age of some Franciscan rocks was known but not taken seriously. Finally, however, the discovery of an ammonite of undoubted Cretaceous age in a Franciscan sandstone at the type locality led to a crisis. It became impossible to ignore mounting evidence that the Franciscan contains rocks coeval to the Great Valley Sequence, which ranges in age from Tithonian to Late Cretaceous, despite the fact that rocks of typical Franciscan lithology (e.g., ophiolites) underlie the Knoxville and should therefore be older than the Great Valley Sequence.

The application of the mélange concept and the assumption of underthrusting resolved the Franciscan-Knoxville Paradox, and resulted in a revolutionary change in our thinking on Franciscan geology. At about the same time, a new paradigm in geology, the new plate-tectonic theory, was introduced. With plate tectonics, crises could be overcome, contradictions eliminated, and a period of "normal science" could prevail.

Franciscan research went through five stages: 1) random observations; 2) synthesis; 3) crisis; 4) revolution, or overthrow of the old paradigm and establishment of the new; and 5) mopping up. A history of the process illustrates the theory of scientific revolution advocated by Thomas Kuhn.

INTRODUCTION

I went to study at UCLA in 1950 after finishing a library research thesis on mountain-building theories. I took a course on the geology of California from Cordell Durrell, who told us to forget all about the rules and theories we had learned; California geology is unique. I was not impressed at first, but soon learned that Durrell knew what he was talking about: he grew up in the Bay Area, studied at Berkeley with Lawson and Taliaferro and, most important, he had seen the Franciscan.

On a field trip to Point Sal, I soon learned about the "uniqueness" of California geology. We walked across a coastal strip underlain by coarse-grained peridotite and gabbro. They seemed to be intrusive bodies, yet at the next outcrop we saw sandstone and shale, and there was no sign of contact metamorphism anywhere. Farther down the beach were pillow basalt and chert, and we finally encountered a porphyritic rock, which was spilitic. There appeared to be no meaningful order in the succession; we could not fit the outcrop pattern to any known geologic structure, and we could conceive of no tectonic process which could bring all those strange bedfellows together. Durrell told us that the rocks were Jurassic in age (more exactly, Tithonian) but Cretaceous fossils had been found here and there in the Franciscan terranes.

I did not choose to study the Franciscan for my dissertation research. None of us did then; it was Berkeley's terrane, and we at L.A. were not particularly anxious to tackle the problem. Eventually, I went to work for an oil company and would probably have had nothing to do with the Franciscan, but for a chance visit by Emil Kündig to the United States in 1956. Kündig was a chief geologist for Royal Dutch Shell and he wanted to see the Franciscan before he retired. I was designated to lead a trip for him through the Coast Ranges, and I was given a month to prepare for my assignment.

My friends at the U.S. Geological Survey, Menlo Park, referred me to George Schlocker, Ed Bailey, Porter Irwin, and Dave Jones. Schlocker, an engineering geologist working on Bay Area problems, knew where the outcrops were. Kündig wanted a piece of glaucophane schist for a museum, and since occurrences of the metamorphics are rarely shown on maps, I needed Schlocker to tell me where to find them. From him I learned for the first time the expression "knockers," a term used by bulldozer drivers to designate exotics in the Franciscan terranes. The most accessible glaucophane schist, as it turned out, was located among the "knockers" on the parking lot of a shopping center. The "knockers" were a big nuisance for the Bay Area builders; they dented the blades of bulldozers and some had to be dynamited before they could be removed. Schlocker then took me to see Ed Bailey, who was a specialist on mercury deposits and very knowledgable about the Franciscan. I also met Irwin who was carrying out a reconnaissance of the Northern Coast Ranges, and kept finding fossils in the Franciscan. Jones, a paleontologist, identified one of the fossil finds as Cretaceous, the same fauna as

that found in the Cretaceous formations on the west side of the Great Valley.

The Kündig excursion started from Los Angeles. We drove into the Franciscan country shortly after we passed San Luis Obispo. It was a wonderful spring day and the Coast Ranges were still green. Sticking out of the meadows were big exotic blocks. Kündig had spent a number of years working near Ankara and he was much impressed by the similarity in climate, in landscape, and in geology between California and Turkey. The comparison did not mean much to me, except I had read an article by Sir E. B. Bailey on the Ankara Melange.

The trip was to have a lasting impact on me. I managed to persuade the Shell Oil Company to start a project in 1963 to investigate Franciscan geology. The timing was ideal: I entered the scene on the eve of a small rebellion in science, which was to become part of the Earth Science Revolution of the sixties.

Two decades later, the image of the Franciscan has been completely altered. Instead of a unique puzzle, the Franciscan is now held up as the most thoroughly investigated reference model of tectonic melanges in convergent plate-margins or suture zones of continental collision. The breakthrough was the result of the efforts of many people. I suggested, therefore, to the editors of this volume that the history of Franciscan research be written by a colleague who had followed the progress from the sidelines. Ellen Drake wrote me that she would welcome an article on the Franciscan, but that I had to do it myself. I was reluctant, lest I should be accused of blowing my own horn, but eventually consented to take up the chore, when I was persuaded that I did not have to brag. After all, politicians and generals write their memoirs as supplementary histories of wars, and many revolutionaries record their experiences in successful or failed ventures. A participant has information not available to others, and also, perhaps, a deeper insight into the thought-processes which led the establishment of a new paradigm.

In writing this article, I cannot free myself from personal judgements, but I shall try my best to be fair. If there are too many "I"s in the text, I hope to be excused on the grounds that this effort is fundamentally a personal account.

I would like to take the opportunity to offer peace and good will to those with whom I disagrees and argued, and I would like to thank the numerous persons who have inspired me, given me support, or understood my efforts. Finally, I am grateful to Ellen Drake for the gift of this unorthodox assignment.

This work is dedicated to the memory of my late wife Ruth; I left my geochemistry laboratory in Houston to start field work in Morro Bay because of her love of the outdoors.

RANDOM OBSERVATIONS AND A CLASSICAL PARADIGM, 1850–1950

Geological investigations of the Franciscan rocks started shortly after the gold rush to California. In 1855 W. P. Blake made the first reconnaissance of the San Francisco area in connection with a survey for railroad routes between the Mississippi and the Pacific. The distribution of the sandstone and of the serpentine were fairly accurately delineated in his map, published with his Geological Report to the 33rd Congress of the United States (Blake, 1856). Later Blake reported his geological reconnaissance in California to the War Department in a more comprehensive report (Blake, 1858). Blake put all the rocks of the San Francisco Peninsula in one basket and suggested calling the unit the San Francisco Sandstone or California Sandstone. He noted the wide extent of this formation, stating that "it forms the greater part of the hills and mountains around the bay, and, so far as explored, a considerable part of the mass of the Coast Mountains." The sandstone unit was considered Tertiary because of a Tertiary fossil in a sandstone block washed up on the beach.

In 1860 the Geological Survey of California was established. J. D. Whitney (1865) recognized, in addition to sandstone and serpentine, jasper, diabase, and schist in the Bay Area. He used the expression "*metamorphosed* Cretaceous" to designate the Franciscan. Whitney puzzled over the different degrees of metamorphism of the various rock types, but he convinced himself of the Cretaceous age of the "metamorphics" on the basis of an *Inoceramus* found in a barge load of rocks from Alcatraz Island.

After the state geological survey was abolished, not much attention was paid to the Franciscan terranes until George Becker was engaged to investigate the quicksilver deposits of the Pacific Slope. Becker followed Whitney's lead, designating the Franciscan rocks as "metamorphosed Lower Cretaceous." Serpentine, diabase, and basalt were mistaken as metamorphosed sandstones, and age assignment was made on the grounds that they were apparently older than the Upper Cretaceous Chico Formation on the east side of Sacramento Valley (Becker, 1885). Becker also adopted the name "Knoxville Series," recently introduced by C. A. White (1885), not only for the fossiliferous Cretaceous sediments on the west side of the Great Valley, but also for serpentine, diabase and chert—rocks that are now considered Franciscan (Becker, 1888).

In a review paper on the geology of the California Coast Ranges read before the Geological Society of America in 1894, H. W. Fairbanks objected very strongly to Whitney's use of the name "Cretaceous metamorphics," pointing out that they were neither Cretaceous nor were they much metamorphosed. Fairbanks also corrected the mistaken notions of his time about the origin of jaspers. J. S. Newberry (1857) had considered the jasper around San Francisco "a metamorphosed form of the associated rocks;" Whitney referred to those rocks as silicified shales; Becker (1888, p. 106) spoke of the jaspers as "shales silicified to chertlike masses of green, brown, red or black colors, intersected by innumerable veins of silica." Becker did notice, however, that "under the microscope the most highly indurated specimens are found to contain fossils." Those fossils had a circular or elliptical outline up to a millimeter across, and had been certified by "the authority of a Professor Leidy" (Joseph Leidy of *Hadrosaurus* fame?) as *bona fide* organic remains. European geologists of Fairbanks' time were familiar with the finding of radiolarian oozes by *H.M.S. Challenger,* and the Alpine jaspers had been recognized "to be clearly ancient representatives of the modern 'Radiolarian ooze' of the deep sea" (Nicholson, 1890). Fairbanks had to cite a passage, however, from an elementary textbook, "*The Manual of Paleontology* by Nicholson and Lydekker, to support his arguments that the Franciscan jaspers are Jurassic "silicified deep-sea Radiolarian ooze." His sedimentologic conclusion was confirmed immediately by Hinde (1894) who found radiolaria in a jasper from Angel Island in San Francisco Bay. Fairbanks was apparently unaware of the fact that radiolarian faunas had been found in rocks as old as the Ordovician (Hinde, 1890) or as young as the Tertiary (Nicholson, 1890). He was impressed by the elementary textbook's statement that "many of the Jurassic Radiolaria occur in jasper."

Fairbanks (1894, p. 102) described the Franciscan as "a series of uncrystalline rocks in the Coast ranges of greater age than Cretaceous. This series underlies the Cretaceous unconformably, and rests on the worn surface of the crystalline basement complex." The Cretaceous was by then known to have a wide distribution in the Coast Ranges, occurring "on the eastern slope nearly the whole length of the San Joaquin and Sacramento valleys. Within the Coast ranges it occurs more extensively south than north of San Francisco" (Fairbanks, 1894, p. 95). The Cretaceous thus included the rocks which were later called the Great Valley Sequence by Bailey and others (1964). Those rocks are well-stratified, rarely disturbed by shearing parallel or subparallel to bedding, and they are fossiliferous, including *Buchia* (formerly known as *Aucella*). The oldest unit of "this formation, consisting almost wholly of black shale, ...extending along the eastern side of the (San Luis) range for several miles, crosses it on the head of Morro Creek, and, on Toro Creek appears enclosed between two high serpentine ridges which form the crest of the range. It was found again on Pine mountain on the summit of the range back of San Simeon." (Fairbanks, 894, p. 95–96). Many years later, I was to work in the San Luis quadrangle mapped by Fairbanks (1904), and I had to agree with the old master. Although the contact is modified by minor faulting at many localities, a basal conglomerate is present here and there, indicating an unconformable depositional contact between the serpentine and overlying black shale, which has yielded a late Jurassic *Buchia* fauna since Fairbanks' time.

In 1895 Fairbanks proposed the name Golden Gate Series for the Franciscan; later when his San Luis quadrangle was published, he was advised by the U.S. Geological Survey to adopt local names. None of Fairbanks' terms found favour with posterity, and the name which eventually worked its way into the literature is Franciscan, proposed by Andrew Lawson in 1895.

It has been generally thought that Lawson was motivated to study the Franciscan because of the Great Earthquake of 1906. In fact, he published his first contributions more than a decade before the event (Lawson, 1895, 1897). The earthquake may have given added impetus to the effort, however, and the San Francisco quadrangle of the Geological Atlas Series of the U.S. Geological Survey was finally published in 1914. Lawson divided the "Franciscan Series" or "Franciscan Group," according to current practice, into five formations. They are, in descending order, Bonita Sandstone, Ingelside Chert, Marin Sandstone, Sausalito Chert and Cahil Sandstone. The last included a pelagic limestone member, the Calera Limestone, recognized by Lawson previously in 1902.

Lawson assumed a normal coherent sequence of five formations, and the interbedding implied alternate epochs of sandstone and chert sedimentation. We now know that this assumption is not valid, so it is easy to understand why the Franciscan was represented nowhere else by a correlative alternation of three sandstone and two chert formations. It was more difficult for Lawson. He was puzzled by the stratigraphic relationship between the Franciscan and various other units of the Coast Ranges. He found an unconformity between the Knoxville and the Franciscan, and concluded that the Franciscan should be pre-Cretaceous. However, he included in the Franciscan some sandstone beds above the Coast Range Granite, west of the Pilacitos Fault, which were then considered Cretaceous in age. Such a stratigraphic interpretation would date the Franciscan as Late Cretaceous or Tertiary. Lawson was not able to resolve the paradox and speculated that perhaps the Franciscan and the Coast Range Granite were deposited and emplaced during a "lost" time-interval represented by the unconformity between the Jurassic and Cretaceous elsewhere in the world.

Lawson's student E. F. Davis published two monographs in 1918 on the sedimentology of Franciscan rocks. He concluded that the largely non-fossiliferous coarse clastics of the Franciscan were arkosic sediments deposited in arid alluvial environments. Assuming an interbedding of the sandstone and chert layers, Davis found it difficult to accept the interpretation of Fairbanks, who found in radiolarian ooze the modern analogue of Franciscan chert. Davis denied the deep-sea origin for the chert, stating (Davis, 1918a, p. 362):

In the neighbourhood of San Francisco there are two formations of radiolarian chert which divide the sandstone into three formations. If we regard the cherts as abyssal radiolarian oozes, like those of the present day, we must believe that the region sank from sea level to a depth of from 12,000 to 15,000 feet, and that this tremendous displacement occurred twice within Franciscan time, with two reversals of movement.

Davis did recognize the presence of scattered radiolarian tests in Franciscan chert, but he preferred a chemical origin for the matrix silica and postulated its deposition by "siliceous springs" in shallow sea. Davis also commented on the presence of pillow lavas in the Franciscan. Citing Lawson (1914) and F. L. Ransome (1894), Davis thought that those rocks could be both intrusive and extrusive. When those intrusives "were magma reservoirs beneath the surface, slowly cooling, they would give rise to siliceous springs." The common association of the radiolarian cherts with submarine basalt was thus explained.

Ralph Reed (1933), in his *Geology of California* regarded the "Franciscan Series" as one of the three major groups of rocks in California, the other two being the "Granitic Basement" and "the post-Jurassic Sedimentary Blanket." He added that the "ser-

ies may with equal propriety be considered 'blanket' or 'basement,' " because its base "has never been seen; the contact between the Franciscan and the granite basement is invariably a fault contact" (p. 27). Reed further emphasized the great contrast in the geologic history of the terrane underlain by the "Franciscan Basement" and that underlain by the "Granitic Basement." Contrary to Davis, Reed thought the Franciscan sandstones were largely marine. He noted (1933, p. 90) that "no identifiable vertebrate remains have been described from Franciscan rocks. On the theory that these rocks, or at least the widespread sandstones are non-marine, this absence of vertebrate materials has been felt as a difficulty." Later Reed and Hollister (1936, p. 67) contrasted the Franciscan and "those formations of other regions that are classed as belonging to the geosynclinal facies." They speculated that the Franciscan may include both Triassic and Jurassic deposits and that the geosynclinal facies had its age equivalent in the "phyllites and slates of the Santa Monica and Santa Ana Mountains."

N. L. Taliaferro was the last grandmaster of the pre-War generation to write a monograph on the Franciscan, and he made a valiant effort to fit conflicting evidence into a new synthesis. Where Reed had considered the Franciscan a basement and the Knoxville a "post-Jurassic Sedimentary Blanket," Taliaferro (1943) failed to see an unconformity between the two. He thought instead that the Knoxville strata are stratigraphically intercalated between rocks of Franciscan lithology. Consequently the Franciscan and the Knoxville were made into one single stratigraphic unit, deposited during the late Jurassic Tithonian Stage. Using the paradigm of layer-cake stratigraphy to cut the geological Gordian knot, Taliaferro (1943, p. 110) made a vivid portrait of late Jurassic paleogeography:

A geosyncline came into existence as a result of the Nevadan orogeny (Portlandian) that strongly folded the earlier Mesozoic sediments of the present Sierran region. As the ancestral Sierra Nevada emerged, a long and broad trough was formed and the site of the present Coast Ranges was flooded for the first time in the Mesozoic. At the same time a high and rugged land mass came into beeing . . . west of the present coast line. . . .In (this) great trough, . . . the Franciscan-Knoxville rocks accumulated . . . Local disturbance took place on the margins of the geosyncline, . . . but at no time was the entire trough affected. . . .There is no record of widespread diastrophism until the beginning of the Lower Cretaceous.

Taliaferro thought that the Franciscan suffered many periods of folding and faulting during the Cretaceous and Tertiary Periods, but that the Franciscan sediments remained unmetamorphosed except where volcanic or intrusive rocks are present; the glaucophane and related schist were thus local alterations of "pneumatolytic contact action," induced by serpentine intrusives.

Taliaferro made Lawson's "lost interval of time between the Jurassic and Cretaceous" more credible by replacing it with the more concrete "Tithonian Stage," but in doing so he actually amplified the problem. Now all the action—namely, the accumulation of a sedimentary sequence more than 8 km thick, the extrusion and intrusion of basic igneous rocks, the deformation and the metamorphism—had to happen within a timespan of 5 or 6 million years. Taliaferro's strong personality overcame logical objections to this telescoped time span. His students from Berkeley accepted the Taliaferro model in their investigations of the Franciscan terranes, and his model of the Franciscan world was the one presented by Cordell Durrell to us students in his class on the geology of California. The Franciscan was no longer a catch-all of uncertain age and of no stratigraphic significance. It was now pigeon-holed and, together with the Knoxville, became a fossiliferous formation of well-defined age like many others.

The patterns of progress in Franciscan research followed the general outline suggested by T. S. Kuhn in his analysis of *The Structure of Scientific Revolution* (1962). It started with random observations and ended with a grand synthesis in harmony with the paradigm of classical geology. This eagerness to conform, to fit all facts into a conventional scheme, resulted in anomalies, contradictions, and led inevitably to crisis and revolution.

The crisis was first of all an identity crisis; the very meaning of the name "Franciscan" was never clearly defined. Lawson invented the term in 1914 for a supposedly lithostratigraphic unit, long before formal nomenclatural rules in stratigraphy were defined. The Franciscan includes a wide assortment of rock types; sedimentary (sandstones, shales, conglomerates, breccias, cherts, limestones); igneous (periodotites, gabbros, serpentinites, pillow basalts, diabases, albite porphyries) and metamorphic (glaucophane schist, lawsonite schist, eclogite, amphibolite, etc.). Yet there is an apparent "uniformity" in this heterogeneity, emphasized by Fairbanks in 1894, and again by E. F. Davis in 1918: the same suite of rocks has been found throughout the California Coast Ranges, extending from the Santa Ynez Mountains northward to Humboldt County. Terranes underlain by a similar assortment are present in Southern California, Baja California, and in the Pacific Northwest, where different formational names were assigned. Yet the association of rock types and the style of their deformation are so distinctive everywhere that one almost never has any difficulty in identifying the Franciscan. Serpentines are Franciscan; glaucophane schist and other metamorphic rocks are Franciscan; basic igneous rocks (gabbros, basalts, porphyries) that are Franciscan are easily distinguished from Tertiary gabbros, diabases, and basalts; radiolarian cherts (red, green or yellow) are Franciscan. The Calera Limestone has been defined as Franciscan. We always knew what was Franciscan, but we were sometimes not so sure what was not Franciscan. Fossiliferous sandstones and shales were especially liable to initiate a dispute. Two geologists could stand before an outcrop and engage in a hot debate as to whether that particular rock does or does not belong to the kaleidoscopic mixture of rocks that have been known collectively as the Franciscan. This peculiar difficulty was, in fact, the root of all the troubles for the first hundred years of the Franciscan research.

We all know that mistakes have been made. I remember my own excitement in 1963 when I found an oyster reef between a sandstone and a serpentine outcrop in a Franciscan terrane; I thought I had found fossils that could be studied to date the Franciscan. When I took my collection back for my paleontologist friends to identify, they laughed at me. It was an Early Miocene fauna, typically found in the Vaqueros Formation of California. The small patch-reef was not a part of the Franciscan, even though I had no way to show separately on my map this unit,—unmappable on an inch-to-the-mile scale. W. P. Blake had the same problem in 1855. He could not separate the Pliocene Merced Formation from the Franciscan during his reconnaissance mapping of the Bay Area; the fossils from the Merced gave a Tertiary age for his San Francisco Sandstone. Today none would place the Merced strata into the Franciscan. It is taken for granted that the Cenozoic, or at least Neogene, rocks of the Coast Ranges constitute separate formations; they cannot be Franciscan. Lawson (1897) made the same kind of mistake when he failed to differentiate the sediments above the Coast Range Granite from the Franciscan. We now know that those sediments near Half Moon Bay are Paleocene (Martinez); they are the sedimentary cover of the Coast Range Batholith and are, therefore, not Franciscan. Lawson's inclusion of those rocks in the Franciscan led him to the mistaken notion that the Franciscan was younger than the Coast Range Granite.

Eventually we realized that the Tertiary rocks are not Franciscan. But what should we do with sedimentary formations which yielded Cretaceous fossils? The common practice was to leave them out. Problems arose, however, when paleontologists *changed age assignments* of their faunas. The fossil-bearing Knoxville strata of the Southern Coast Ranges, for example, are clearly post-Franciscan. Fairbanks included them as the basal unit of his "post-Jurassic Sedimentary Blanket," and his interpretation was accepted by Jennings when he compiled the State Geologic Map in 1958. The Knoxville, or *Buchia piochii* beds, constituted the basal unit of the Great Valley Sequence of Bailey and others (1964), separate and distinct from the Franciscan. My investigations of the Morro Bay-San Simeon area supported Fairbanks' conclusion that the Knoxville unconformably overlies an ultramafic body of the Franciscan and is therefore younger than that part of the Franciscan (Hsü, 1968). Although many of us agreed, there was one problem—the Knoxville had been made a part of the Franciscan by Taliaferro in 1943, after *B. piochii* was assigned a late Jurassic Tithonian age!

Taliaferro was probably led to his conclusion by the fact that the age of *B. piochii* seemed to be identical to that of an ichthyosaur fossil from a Franciscan chert boulder, which was compared to the Tithonian *I. posthumus* of the Solenhofen Limestone by Camp (1942). Also, the distribution of the Knoxville is about the same as that of the Franciscan and finally, typical Knoxville fossils are found in the shales and sandstones both above and below Franciscan pillow basalts in various parts of the Coast Ranges. Assuming normal stratigraphical superposition, the field relation was taken as proof that the Franciscan-Knoxville constituted one and the same rock-stratigraphic unit of Tithonian age.

While the Knoxville became Taliaferro's Franciscan when supposedly-Cretaceous fossils were assigned a Jurassic age, another sedimentary sequence was excluded from the Franciscan when supposedly-Jurassic fossils were given a Cretaceous age. Fairbanks (1894) found poorly-preserved *Inoceramus* specimens in a sequence of sandstone and shale beds near Slate Springs south of Big Sur. T. W. Stanton (in Fairbanks, 1896) gave a faunal age not older than the Jurassic and thought it might be Cretaceous. J. C. Merriam (1899) stated that the *Inoceramus* was of Cretaceous, rather than Jurassic, aspect. Yet C. H. Davis (1913) and C. H. Crickmay (1931) both assigned a Jurassic age to the Slate Springs fauna because it was found in the Franciscan, which had to be older than Cretaceous. Nomland and Schenk (1932) collected new fossils and examined the old ones from the Slate Springs locality, and they were certain that the fauna was Late Cretaceous, and not Jurassic in age. Nevertheless they failed to persuade Taliaferro that the Franciscan included sediments of Cretaceous age; the Franciscan could remain Jurassic when the Slate Springs Sequence was excluded from the Franciscan, like the Merced of the San Francisco Peninsula or the Martinez of Half Moon Bay.

There were fossiliferous Cretaceous rocks, however, which could not be so easily dismissed from the Franciscan. The pelagic limestone of the Calera Valley is characterized by a *Globotruncana* fauna, and Hans Thalmann knew already in 1942 that the limestone beds were definitely Cretaceous and stated so in 1943. One cannot simply claim that the Calera Limestone, like the Slate Springs Sequence, is not Franciscan, because this limestone is a member of the original Cahil Sandstone formation of the type Franciscan (Lawson, 1914). Furthermore, the limestone is interbedded in submarine basalt flows at a number of localities in the Bay area, and occurs as xenoliths in the basalt. I was able to convince myself of the stratigraphic relationship after George Thompson showed me the outcrop of the Palo Alto Quadrangle described by Bailey and others (1964, p. 71). The Calera Limestone has to be Franciscan. Reed (1933, p. 89) did not deny this interpretation; he merely stated that the foraminifera were "not specifically identifiable" and had no stratigraphic significance. He did not bother to change this conclusion when his 1933 opus was reprinted in 1950. Taliaferro (1943, 1944, 1951) took little notice of the *Globotruncana* fauna; he thought the micropaleontological dating technique unreliable (Cordell Durrell, personal communication). The Calera Limestone thus remained Franciscan and was listed as an Upper Jurassic unit on the Geologic Map of the San Francisco Bay region (Bowen and Crippen, 1951, p. 163).

To belong or not to belong depends upon an understanding of mutual relations of the strange lithologic bedfellows which make up the Franciscan. These associations were familiar to geologists of the Old World. J.J.H. Teall noted the association of pillow lavas and radiolarian cherts in 1895 and this relation was cited by Dewey and Flett (1911) as evidence of submarine volcanism in the deep sea, where radiolarian oozes were normally sedimented. Davis (1918a), however, was adamant that the Franciscan radiolarites were not abyssal sediments and his proposal of deposition by siliceous springs in a shallow-water geosyncline was a manifestation of the spirit of his time: *geology became mythified as geologists preferred to speculate on hypo-*

thetical, non-existent processes, while modern analogues were ignored.

Teall (1899) was also impressed by the common presence of gabbros, peridodites, and serpentines near pillow lavas, an intimate association which led Steinmann (1927) to propose the term "ophiolite suite" to designate those extrusive and intrusive rocks. Steinmann had no doubt that the Alpine radiolarites were deepsea oozes, although he erred in assuming an intrusive relation between the cherts and the ophiolites. Steinmann's misconception of intrusive ophiolites was shared by his American colleagues. Taliaferro repeatedly portrayed field relations of serpentinites intruding into sediments and volcanics; glaucophane schists were considered the product of contact metamorphism, even though contact aureoles were nowhere to be seen. Blocks of schists were found in the vicinity of ophiolites, but blocks of unaltered sediments were also there. Taliaferro (1943, p. 165) thought, however, he had found "non-selective and selective types of alteration."

Looking back, we see that the first century of Franciscan research ended up in a dead end. Provincialism and strict adherence to dogmas led our pioneers down a primrose path of errors. The inviolate faith in the *law* of superposition gave rise to mistaken interpretations of stratigraphic relationship. The naive assumption of lateral stratacontinuity led to postulates of rapid facies changes wherever lenticular blocks of sedimentary rocks were observed. The conventional wisdom of paleontological dating (based on the fossil content of a "formation") resulted in conflict and controversy concerning the identity of the "Franciscan Formation." The ignorant belief that sands could only be deposited in continental or nearshore realms was responsible for the complete misunderstanding of the Franciscan depositional environment. The textbook approach of attributing all coarse-grained igneous rocks an intrusive origin caused a failure to appreciate the tectonic emplacement of the ophiolites. Finally, the simplistic view that local occurrences of highly metamorphosed rocks near unaltered sediments implied contact metamorphism culminated in a misconception about the origin of blue schists.

Lawson, Fairbanks, Reed and Taliaferro were great geologists. Their work brought Franciscan research far beyond the stage of random observations, and their positive contributions cannot be overlooked. But they were also victims of the *Weltanschauung* of their time. The first half of the 20th century was a time of "normal science," of complacency and stagnancy in geology, while the world was devastated by two wars. A paradigm in the realm of Franciscan geology was established, but the suppression of discordant data paved the way for a rebellion, for a revolution in earth science that was to occur in the 1960s.

THE MAKING OF A REVOLUTION, 1950–1970

A systematic investigation of the California Coast Ranges was carried out by the students of the University of California at Berkeley after the Second World War under the supervision of N. L. Taliaferro. The mapping was done by students of summer field classes, and by candidates for Master's and Ph.D. degrees. Much of the work was not published, but the maps served as a basis for the Olaf P. Jenkins edition of the Geological Map of California on a scale of 1:250,000. Some of the Berkeley theses did get published as publications of the California Division of Mines (e.g., Jenkins, 1951; Manning and Ogle, 1950; Gealey, 1951; Crittenden, 1951; Travis, 1952; Ogle, 1953; Briggs, 1953; Darrow, 1963). The Berkeley products were standardized according to the Taliaferro model: the Franciscan-Knoxville was designated Upper Jurassic, a group or two formations, and the Franciscan strata were portrayed as forming simple structures of anticlines and synclines despite the fact that the measured dip and strike of deformed strata bore no correspondence to structures depicted (Figure 1).

Several other graduate students from Berkeley and Stanford chose to study aspects of the Franciscan for their Ph.D. dissertations. They discovered regionally metamorphosed jadeite graywackes (Maddock, 1955; Bloxam, 1956; McKee, 1962; Soliman, 1958; Coleman, 1961), blue schists (Brothers, 1954; Kilmer, 1961; Ghent, 1963), or even eclogites (Switzer, 1945; Borg, 1956). The widespread occurrence of rocks typical of high-pressure metamorphism, did not harmonize with the postulate of "pneumatolitic alteration by ultramafic magmas." A first assault on the establishment had begun.

The frontal attack was launched by the paleontologists. In 1948 micropaleontologists Joseph Cushman and Ruth Todd described a foraminiferal fauna from a Calera-like limestone in the Franciscan terrane near New Almaden; they concluded that the limestone could not be as old as Jurassic. Martin Glaessner (1949) analysed their results and dated the limestone as middle Cretaceous (Albian). The pioneering micropaleontologists did not have a strong influence within the community of geologists, and the micropaleontological dating of the Franciscan continued to be ignored until an ammonite was found by two high-school students.

On June 1, 1950, George Topham and Richard Sessions climbed down the cliff on the north shore of the City of San Francisco. They found an ammonite in a sandstone that had been mapped as the Marin Sandstone of the Franciscan Group. The specimen was taken to Julius ("George") Schlocker, who was then preparing a geologic map of the city. Schlocker and his colleagues, M. G. Bonilla and Edgar Bailey, showed the fossil to Sy Muller of Stanford, who identified the ammonite as *Douvilleceras* sp. of Albian age (Schlocker and others, 1954). The die was cast: a sandstone formation of the Franciscan in the type area was definitely Cretaceous. We could no longer discard evidence that we didn't like as if we were playing some geological card game.

The dam broke and out flooded information that had been repressed for years. Schlocker and others recalled that fossils had been found on Alcatraz Island and identified as Cretaceous by W. M. Gabb as early as 1869. Later F. M. Anderson (1938) had revisited the location, made additional collections, studied the fauna, and he too had given an Early Cretaceous age to the Cahil

Sandstone of the Franciscan Group. Micropaleontological evidence also became respectable, and Glaessner's determination of the Albian age of the New Almaden limestone five years earlier was accepted.

Schlocker and others now had to accept the Cretaceous age for the Franciscan, but the new interpretation created a paradox on the Franciscan-Knoxville relation.

"In the Berkeley Hills," they wrote (Schlocker and others, 1954, p. 2380):

a range about 10 miles east of San Francisco, graywacke-type sandstone and other rocks typical of the Franciscan group are overlain by the Knoxville (Lawson 1915, p. 60, field edition; 1914, p. 8, library edition; Anderson, 1938, p. 53; Taliaferro, 1943, p. 205). The Knoxville formation in the Berkeley Hills is predominantly shale in which *Aucella piochii* and other Upper Jurassic fossils are found in several localities in a belt more than 10 miles long. Evidently the Franciscan group is Late Jurassic or older in this belt. Because of the lack of criteria for separating the Franciscan of Late Jurassic age in this belt from that of Early Cretaceous age in the Golden Gate cliffs area and elsewhere, it is proposed that the Franciscan group be assigned to the Jurassic (?) and Cretaceous periods until more precise age designations can be made for subdivisions within the group.

The Knoxville Formation was named after exposures at Knoxville in Napa County, California. The formation is the basal unit of a conformable sequence of largely Cretaceous sediments, cropping out on the west side of the Great Valley and at numerous localities in the Coast Ranges. The Knoxville had been considered Early Cretaceous (White, 1885; Diller and Stanton, 1894), before it was assigned a Tithonian age on account of its *Buchia* (=*Aucella*) *piochii* fauna (Anderson, 1933; 1938; Taliaferro, 1943; Bailey and others, 1964). Now we had an anomaly: A normally superposed sequence containing fossils ranging from latest Jurassic (Tithonian) to Late Cretaceous in age is superposed upon another "formation" or "group" of rocks which also yielded fossils of latest Jurassic (Tithonian) to Late Cretaceous age. What could be the nature of the superposed relation?

The first discovery of the ammonite inspired others to search. Soon another hiker found another ammonite, also in the Marin Sandstone but on the opposite shore of the Golden Gate Strait, just north of Bird Rock at the north end of the Golden Gate Bridge. Hertlein (1956), after having consulted with F. M. Anderson and Ralph Imlay reported that the specimen "might be referable to the genus *Mantelliceras* Hyatt," and this second ammonite confirmed the middle Cretaceous age-assignment to the Marin Sandstone. Meanwhile the dating of the Calera Limestone was re-evaluated. Thalmann (cited in Irwin, 1957) now assigned a late Albian to early Cenomanian age to the microfauna in the Calera Limestone at the quarries of the Permanente Cement Company. Church (1952) described a foraminiferal assemblage from the Calera Limestone at Rockaway Beach, San Mateo County, and the new find was considered late Cenomanian by Küpper (1955). The microfossils in the limestone near New Almaden were now also dated Cenomanian. Küpper compared the fauna of the Calera Limestone at Rockaway Beach to that of the

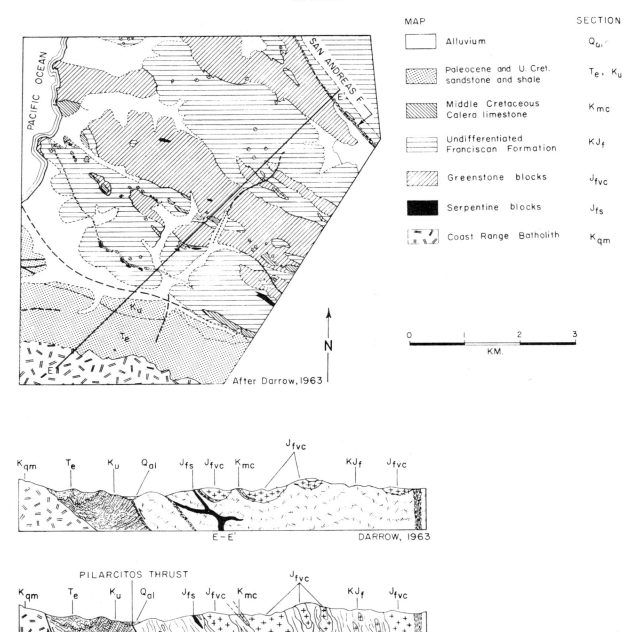

Figure 1. Two structural interpretations of Franciscan geology. The geology of the Montara Mountain area, south of San Francisco, was mapped by Darrow (1963). Two structural sections are identically located (line E-E′ on map). The upper represents the interpretation by Darrow according to Taliaferro's model: the layered rocks were assumed to form gentle anticlines and synclines and the serpentinite was considered an igneous intrusive. The lower section represents an interpretation according to the mélange model (from Hsü, 1967).

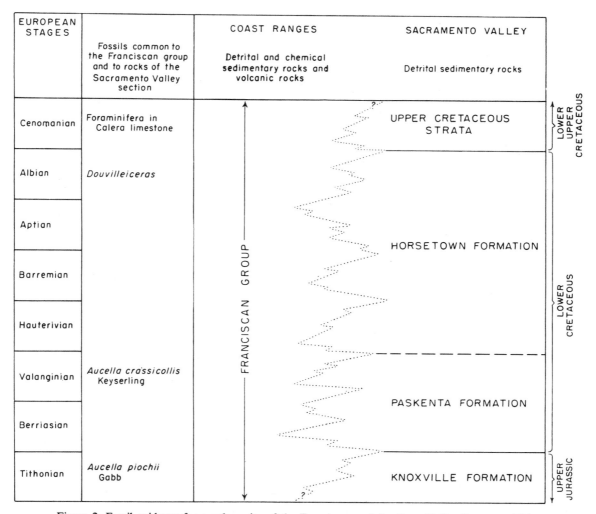

EUROPEAN STAGES	Fossils common to the Franciscan group and to rocks of the Sacramento Valley section	COAST RANGES Detrital and chemical sedimentary rocks and volcanic rocks	SACRAMENTO VALLEY Detrital sedimentary rocks
Cenomanian	Foraminifera in Calera limestone		UPPER CRETACEOUS STRATA
Albian	*Douvilleiceras*		
Aptian			HORSETOWN FORMATION
Barremian			
Hauterivian			
Valanginian	*Aucella crassicollis* Keyserling		PASKENTA FORMATION
Berriasian			
Tithonian	*Aucella piochii* Gabb		KNOXVILLE FORMATION

Figure 2. Fossil evidence for synchroneity of the Franciscan and the Great Valley Sequence. This diagram was prepared by Irwin in 1957 to summarize the paleontological evidence correlating the Franciscan to the Great Valley Sequence. Not shown is the fact that many Franciscan rocks are not fossiliferous. Radiometric dating and radiolarian stratigraphy have since indicated a pre-Tithonian age for at least some of the ophiolites and radiolarian cherts.

"Antelope Shale" of the Great Valley, and believed that those two lithostratigraphic units were exactly synchronous.

While middle Cretaceous fossils were being described from the type Franciscan in the Bay Area, earlier Cretaceous and Jurassic fossils were collected from the Franciscan terranes of the Northern and Southern Coast Ranges. Students and staff of the summer field classes of the University of Southern California, working in the Shell Peak area of the Branch Mountain Quadrangle, found fossils at four Franciscan localities and an ammonite in rocks mapped as the Knoxville (Easton and Imlay, 1955). Fossils were also discovered at three localities in the north by H. D. McGinitie, and by Porter Irwin with D. B. Tatlock, and with Edgar Bailey (cited in Irwin, 1957). The Shell Peak localities yielded a *Buchia piochii* assemblage of late Jurassic age, whereas the northern localities yielded both the Jurassic *B. piochii* and the Early Cretaceous (Valanginian) *B. crassicolis* faunas. The pres-

ence of still older Jurassic strata in the Franciscan was suggested by the discovery of a plesiosaur in a limestone block enclosed in the Franciscan shale of Stanley Mountain area, San Luis Obispo County. This vertebrate fossil bears close resemblance to certain reptiles of Kimmeridgian and Oxfordian ages found elsewhere in the world (Wells, 1953).

Summarizing paleontological evidence, Irwin concluded in 1957 (p. 1294) that the Franciscan group and the strata of the Sacramento Valley (later called the Great Valley Sequence) "reflect two different depositional environments of a single stratigraphic section, judged by their contemporaneity of deposition and their distribution." The oldest Franciscan fossils (*B. piochii* fauna) are also found in the Knoxville, the oldest unit of the Great Valley Sequence. Furthermore, the same *Globotruncana* fauna occurred in the pelagic Calera Limestone of the Franciscan Group and in the hemipelagic "Antelope Shale" of the Great

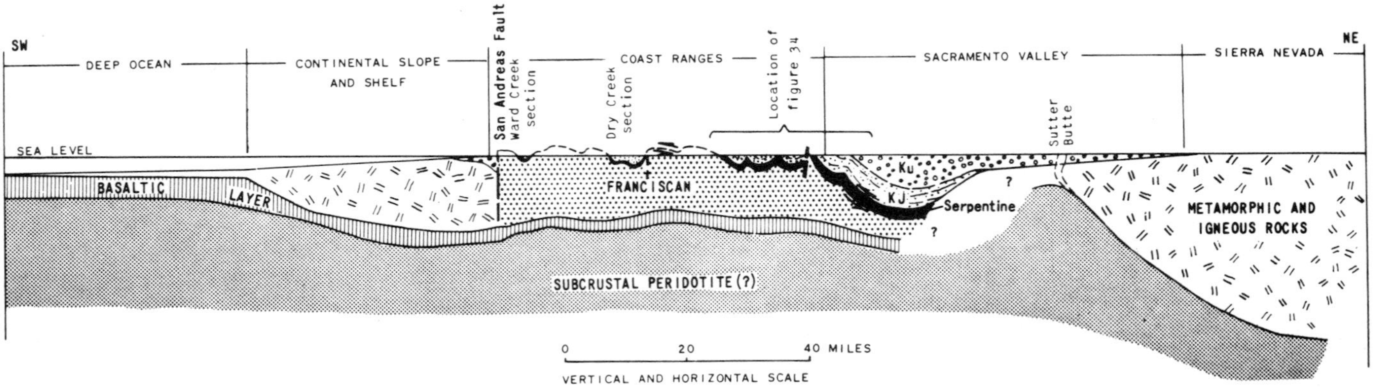

Figure 3. The Great Valley Overthrust. The interpretation that two largely coeval sequences, ranging in age from Tithonian to Late Cretaceous, are present in California, formed the basis of an overthrust interpretation. Bailey, Irwin, and Jones drew this idealized vertical section across western California and offshore area, illustrating the hypothesis that the Upper Cretaceous (Ku) and the Lower Cretaceous and Jurassic (KJ) formations of Great Valley Sequence were thrust over the Franciscan, with a serpentine layer along the thrust zone (from Bailey and others, 1964).

Valley Sequence (Figure 2). The fossil discoveries of the 1950s disproved Taliaferro's postulate of one "Franciscan-Knoxville" unit of Tithonian age.

The first shot of the revolution was fired. If the Knoxville was deposited at the base of another sequence in another realm of sedimentation, the superposition between the Franciscan and the Knoxville cannot be a transitional sedimentary contact as Taliaferro asserted. Irwin (1957, p. 2295) stated, "The impressively abrupt transition between the Franciscan group and Sacramento Valley sequence along much of the boundary between the two provinces may (thus) be one of juxtaposition chiefly by faulting."

The main attack followed quickly. In 1964 Edgar Bailey, Porter Irwin, and David Jones published their monograph on Franciscan and related rocks and their significance in the geology of western California. The appearance of this important work heralded a new era of Franciscan research.

Bailey, Irwin and Jones reviewed the fossil evidence first; there was little doubt that the fossil-bearing Franciscan strata were largely coeval with the sedimentary strata of the Great Valley Sequence. Following the Taliaferro tradition, the three authors assumed the start of Franciscan sedimentation after the Nevadan Orogeny, but they liberated the Franciscan from a Tithonian formation straight jacket. They wrote in 1964 (p. 145):

If we assume that the Knoxville is younger than the earliest phase of the Nevadan orogeny, during which the Galice Formation and older rocks of the Klamath-Sierra Nevada were folded, faulted, intruded, and eroded, the oldest part of the Franciscan probably was formed of debris eroded during this period. Thus, the oldest part of the Franciscan probably is younger than the Galice Formation (late Oxfordian to middle Kimmeridgian) and older than the Knoxville Formation (Tithonian). The geographic extent of these Franciscan rocks west of and parallel to the Klamath Mountains and Sierra Nevada for hundreds of miles, is a suitable position for collection of erosion products: and the composition of the older Franciscan graywacke, with its general lack of K-feldspar is comparable with derivation from a nearby rising metamorphic terrane.

Bailey and others thus envisioned two parallel belts of sedimentation: a miogeosynclinal Great Valley Sequence on the east, close to the newly uplifted Sierra Nevada, and a eugeosynclinal Franciscan on the west, farther out on the continental margin. The two coeval sequences are found in close proximity at a number of places in the Coast Ranges. This fact implies "telescoping" (superposition) by tectonic deformation (Figure 3). The piercement structures in the Coast Ranges clearly indicate that the Franciscan underlies the coeval Great Valley Sequence. Bailey and others (1964, p. 163ff.) were further impressed by the "common occurrence of serpentine as an intervening sheet beneath the Great Valley Sequence but above the Franciscan." They concluded, therefore, that

the gross contemporaneity of the Franciscan and the Great Valley sequence, and the general separation of the two facies by serpentine over broad areas, make an attractive working hypothesis of low-angle dislocation of great magnitude. Perhaps some of the enormously thick Upper Jurassic and Lower Cretaceous part of the Great Valley sequences glided westward off an ancient continental shelf and slope now occupied by the eastern part of the Great Valley and western part of the Sierra Nevada as an essentially cohesive and little deformed mass. The ease of movement across the bordering Franciscan may have been enhanced by synchronous extrusion of serpentine along the edge of the continental mass. Or, alternatively, the Franciscan rocks may have been carried eastward beneath the Great Valley sequence by eastward-moving subcrustal currents that plunge under the continental crust.

My sedimentological research program for Shell during the 1950s afforded me opportunities to make reconnaissance trips to the Coast Ranges. Graded bedding and other turbidite-structures are present in both the Franciscan sandstones and those of the Great Valley Sequence. I found that these two largely coeval sequences are not exactly comparable to the classical mio- and eugeosynclines, which underlie as a rule, carbonate-shelf and hemipelagic sediments. I was also troubled by the order of tectonic superposition and by the direction of tectonic transport postulated by the hypothesis of Bailey and others: In the Alps, for example, the eugeosynclinal sequence is thrust over, not under, the miogeosynclinal, and the thrusting is directed toward the craton, not away from the craton. Aside from considerations of comparative tectonics, the main argument against an overthrust hypothesis was field observations on the contact relationships between the Franciscan and the Knoxville. The contact is sedimentary, although it may have been modified locally by Cenozoic faulting. The Knoxville definitely was deposited on an ophiolitic basement. Fairbanks saw this field evidence, Taliaferro saw it, Schlocker and others saw it, and I saw the same after I started working on my Franciscan project in 1963. If we define the Knoxville as the basal formation of the Great Valley, and if we define the ophiolite as a part of the Franciscan, we have to conclude: 1) that the Great Valley Sequence depositionally overlies a Franciscan basement, and 2) the Franciscan has a pre-Knoxville, or pre-Tithonian history. An alternative was to depart from the century-old tradition and to exclude the pre-Knoxville ophiolites from the Franciscan. Bailey, Blake, and Jones chose this alternative in 1970 and referred to the ophiolites on the west side of the Great Valley as "on land oceanic crust in California Coast Ranges." However, excluding ophiolites from the Franciscan opens Pandora's box again, and accentuates the identity crisis of the Franciscan. Furthermore the fact remains that pre-Knoxville rocks are present in the Coast Ranges whether or not one chooses to call them Franciscan, or basement of the Great Valley Sequence.

I believed that we did not have to omit from the Franciscan those rocks which did not fit into our scheme of things. We could accept the fact that some of the Franciscan graywackes and shales are coeval to the Great Valley sediments without denying the possibility that the radiolarites and ophiolites (constituting the so-called Steinmann's Trinity) were largely pre-Tithonian. The stratigraphic relations between the Knoxville and the ophiolites indicate a pre-Knoxville age for the ophiolites, or at least for some of the ophiolites, and the one marine reptile fossil reported by Wells (1953) suggested pelagic deposition in Oxfordian or Kimmeridgian time, when marine sediments were being laid down in the Klamath and the Sierra Nevada regions. Like Fairbanks, I was convinced that the Coast Ranges had a pre-Knoxville history, and the first years of my effort were devoted to finding a place for the pre-Tithonian Franciscan in the tectonic framework of California.

Assuming an Oxfordian-Kimmeridigian age for the Franciscan cherts and an even older age for the ophiolite, an assumption which was later confirmed by radiolarian biostratigraphy (Kocher, 1981) and by radiometric dating (Hopson and others, 1975), I envisioned an oceanic realm west of the Jurassic continental margin of North America. The pre-Tithonian Franciscan sediments were deposited upon an oceanic crust, the Franciscan ophiolite, while hemipelagic and volcanogenic sediments, constituting the pre-batholithic formations of the Sierra Nevada, were deposited on the continental margin of North America. According to this scheme, the pre-Tithonian deformation of the older Franciscan was a part of the late Jurassic Nevadan Orogeny, *sensu lato,* induced by the collision of Salinia and the North American continent (Hsü, 1965b).

My overemphasis on the pre-Tithonian history of the Franciscan may have led to a mistaken belief that I ignored the paleontological evidence. In fact, I always took into consideration the presence of Cretaceous fossils in the Franciscan. In my interpretative scheme the Franciscan rocks coeval to the Great Valley Sequence (=San Mateo Unit) were deposited on a deformed pre-Tithonian Franciscan (=San Francisco Unit) before the old and the young Franciscan rocks were mixed together tectonically (Hsü and Ohrbom, 1969).

I did not learn the concept of tectonic mixing in school. I stumbled onto the idea in April, 1963, when I took a day off and went to the beach near San Simeon with my family, after weeks of frustrating attempts to map the grassy knolls north of Morro Bay. I noted for the first time that the Franciscan

graywacke layers commonly have a 'pinch-and-swell' appearance. This structure results from the stretching of brittle, originally even-bedded graywacke layers. Angular boudins, commonly lenticular and bounded in part by shear fractures, were produced when graywacke beds were disrupted by extension. Shear fractures, indicating extention parallel to bedding, are common in the graywacke boudins, but are rare or absent in the surrounding shaly matrix. The shales were deformed ductilely by flowage, which induced further stretching or bodily rotation of the boudins. On a medium-scale, particularly at seashore outcrops several hundred meters wide, low-angle shear planes, sub-parallel to bedding, are commonly observed. The strata between the shear planes form slabs a few decimeters to many meters thick. Some such slabs have been stretched and sheared on a minute scale as described above. Others show little internal strain. Minor recumbent folding of the slabs has been observed, but is rare and was probably produced as drag folds by movements along low-angle shears.

Those observations, cited from my field notebook, were published in my 1967 paper (p. 287–288), and they led me to appreciate the role of tectonic fragmentation and mixing in the deformation of the Franciscan rocks (Figure 4).

In June 1963, I read Ben Page's description of the Apennine *argille scagliose,* which seemed to be analogous to Franciscan. The next summer I joined the Summer Field Institute of the American Geological Institute (supervised by John Maxwell), and found considerable similarity in the deformational style between the *argille scagliose* and the Franciscan (1965b). I moved from New York to California in late 1964 to be close to my field area, and I invited a number of my friends to see the "structural wonders" of the Coast Ranges. One spring day in 1966 John Crowell, Cliff Hopson, Bob Garrison, and a guest professor at UCLA, Heli Badoux, visited me in the San Simeon area. Tectonic mixing was a new idea to them, too. Crowell, my mentor at UCLA, suggested that a new concept should be given a new name, and Badoux, a Welsh Swiss, supplied the exotic French word *mélange*. He called to our attention expressions such as Ankara Mélange, Ligurian Mélange, etc., in the writings of Bailey and McCallien (1950; 1963), who had adopted the term of

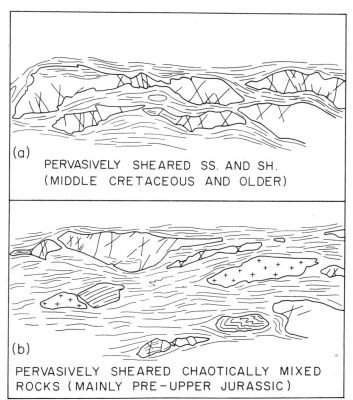

(a)

PERVASIVELY SHEARED SS. AND SH.
(MIDDLE CRETACEOUS AND OLDER)

(b)

PERVASIVELY SHEARED CHAOTICALLY MIXED
ROCKS (MAINLY PRE–UPPER JURASSIC)

Figure 4. Franciscan style of deformation. Pervasive shearing and chaotic mixing are observable on a small-scale at individual outcrops. Note the presence of extensional shear fractures in graywacke boudins. The upper sketch is a broken formation near the San Simeon coast, probably of middle Cretaceous age. The lower sketch shows a typical mélange, composed mainly of pre-Tithonian blocks (from Hsü, 1967).

Greenly (1919). I finally found a copy of the classical memoir and realized that I was not discovering anything new; I was only rediscovering something Greenly told us half a century ago.

The fragmented nature of the Franciscan had always been well known. Schlocker (personal communication, 1956) introduced the informal use of "knockers" to designate exotic blocks. Brown (1964) spoke of a Franciscan "friction carpet" beneath the Great Valley Overthrust. Dickinson (1966, p. 469) noted "the brecciated state of the Franciscan rocks," and considered "the whole terrane a giant shear zone, or a complex of successive and coalescing shear zones." Just the same, the implications of fragmentation and mixing were not always made clear. I recommended, therefore, in 1966 that we should use the term Franciscan *Mélange* in place of Franciscan *Formation* or *Group*. The adoption of a new term would symbolize the need for a new approach to the old problem.

Some felt that words are just words, that scientists should not get involved in a semantic quibble. But words are required to convey scientific concepts or to manipulate logical operations. Words are more than just words; they often imply far more than their dictionary meaning. For example, a formation is a lithostratigraphic unit. The word conveys an impression that we are

speaking of a sequence of conformable strata in normal superposition, and we therefore tend to take for granted the lateral continuity of layered rocks in a formation. Furthermore, we date a formation by the fossils found, because the unfossiliferous rocks of the same unit should not be much different in age. By using the word mélange, on the other hand, we clearly state that we are not talking of a lithostratigraphic unit. Mélange (sensu stricto) is more a tectonic unit, a giant tectonic breccia composed of fragments of all sizes, disintegrated from pre-existing formations. Fragments are separated by shear planes, and we cannot assume for a pile of broken slabs a normal stratigraphic superposition, nor much of a lateral continuity.

We all know, of course, that the fossils in a mélange fragment indicate the age of the fragment, neither less nor more, but many of us seem to forget that fossils cannot stratigraphically date a mélange. A mélange, like a mylonite in a fault zone, has an age of deformation only; it is not a stratigraphic unit and does not have a stratigraphic age, nor a paleogeographic realm of deposition. Finally, we should remember that a lithostratigraphic unit is defined by a specific interval of strata in a specific section or area, bounded, as a rule, by depositional surfaces at its base and its top (Hedberg, 1976). Tectonic units, on the other hand, are separated by dislocation surfaces, which define overthrust sheets, nappes, or schuppen zones. A main source of confusion in dealing with the Coast Range geology can be traced to the fact that mélange units are distinguished by the lithology of their components as well as their deformational style. Figure 5 illustrates the current model of paleogeographical reconstruction and tectonic interpretation, and shows why the Franciscan Mélange is separated from the Great Valley Sequence by a dislocation surface in the Coast Ranges and a depositional surface on the margin of the Great Valley. We have thus an explanation for the fact that Jurassic sediments unconformably overlie a mélange unit which includes Cretaceous components (Hsü, 1968).

On the eve of the Earth Science Revolution the paradigm of permanence of continents and oceans reigned supreme. Tectonic theories emphasized vertical movements as the primary moving force, lateral displacements in orogenic belts were deemed secondary, having resulted from sliding under gravity. A quantitative analysis by Hubbert and Rubey (1959) on the role of pore pressure in overthrusting provided support to proponents of gravity-tectonics. An erroneous assumption of zero cohesive-strength in shear zones was responsible for the postulated theoretical facility in gravity-sliding (Hsü, 1969), but the *Zeitgeist* of the time was to cite Hubbert and Rubey and to emphasize gravity. Bailey, Irwin and Jones (1964, p. 164) thought that the Great Valley Thrust originated when the "Great Valley Sequence glided westward off an ancient continental shelf and slope." I, too, proposed land-slipping movement as an analogue to interpret the Late Cretaceous deformation of the Franciscan, while I continued to envision pre-Tithonian deformation of the older Franciscan by collision-tectonics (Hsü and Ohrbom, 1969). Intuitively I did not like the idea, because the postulated distance of sliding was far too short to account for the magnitude of shearing observed in the mélange,

Coast Ranges   Great Valley

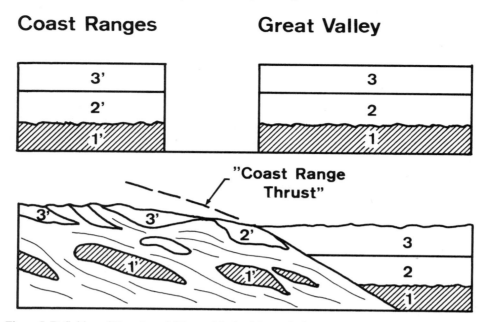

Figure 5. Definition of Franciscan Mélange. Much of the confusion concerning Franciscan geology can be traced to the fact that Franciscan components have been identified by (1) their lithology, and (2) their deformational style and tectonic history. The upper sketch shows the current paleogeographical reconstruction. We now know that the mélange of the Coast Ranges includes slabs derived from pre-Tithonian formations (1'), which include ophiolites, radiolarian cherts and other rocks, as well as sediments coeval with the Great Valley Sequence (2', 3'). The latter overlies a pre-Tithonian ophiolite (1) along the eastern margin of the Coast Ranges, and this ophiolite has been considered Franciscan, because the ophiolitic components of the Coast Ranges are, by definition, Franciscan. The lower sketch is the current tectonic interpretation. All the Coast Range rocks (1', 2', and 3') are considered Franciscan, mainly because of their deformational style. The pre-Tithonian ophiolite is considered by Bailey and others the base Great Valley Sequence, but I consider it Franciscan, because of its lithology and its correlation to other pre-Tithonian ophiolites in the Coast Ranges. If we accept this definition, as is the traditional practise, we have the unusual situation that the Franciscan Mélange (including components 1', 2', 3' and 1) is separated from the Great Valley Sequence (2, 3) by a dislocation surface at many places in the Coast Ranges, but by a depositional contact along the margin of the ranges.

but I was all too human to omit a discussion of this fatal vulnerability of my own hypothesis. Besides most of us are conformists at heart.

The current concept of relating Franciscan tectonics to subduction along active margins had its root in the theory of sea floor spreading. Dietz (1963, p. 949) considered Alpine serpentinites broken fragments of "oceanic rinds," and "eugeosynclinal sequences" accumulations on continental rises. He postulated Franciscan deformation by underthrusting (Figure 6) as a consequence of "the mechanism of sea-floor spreading, which relates this thrusting to mantle thermal convection." Bailey, Irwin, and Jones (1964, p. 165) also mentioned the possibility of an "eastward-moving subcrustal current that plunges under the continent crust" as an alternative hypothesis to explain the Coast Range Thrust. The underthrusting of the Franciscan was portrayed diagrammatically by Ben Page in 1966. Nevertheless the critical evidence leading to the final acceptance of this model, I believe, has been provided by investigations on metamorphic petrology.

Students of the 1950s had, as I mentioned previously, identi-

fied high-pressure minerals in the Franciscan rocks. A tectonic superposition of the Great Valley Sequence on top of the Franciscan was indicated by an "upside down metamorphic zonation" in the Northern Coast Ranges (Kilmer, 1961; Blake and others, 1967). A detailed study by Coleman and Lee (1963) in the Cazadero area revealed four different types of Franciscan metamorphic rocks, indicative of different grades of severity. Ernst (1965) found similar types in the Panoche Pass district, and he noted the utterly chaotic distribution of the "mélange of rock types" (p. 880). The ages of metamorphism were determined to range from Late Jurassic to Cretaceous (Lee and others, 1964; Ghent, 1963; Suppe, 1969). Ernst postulated in 1965 that the high-pressure, low-temperature metamorphism was caused by rapid deposition and burial of sediments in an oceanic trench setting. He later related Franciscan deformation and metamorphism to subduction down the Benioff Zone, and held thrust-imbrication processes "responsible for the observed intimate association of feebly and more thoroughly metamorphosed mélange rocks as well as complex tectonic contacts with roughly coeval miogeosynclinal strata" (Ernst, 1970, p. 897). At about

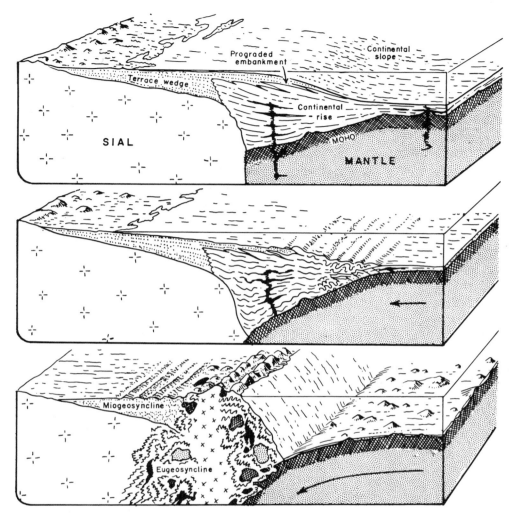

Figure 6. Original model of underthrusting. This diagram was prepared by Dietz in 1963 to illustrate his "view on the origin of spilite-keratophyre suite, serpentines, and ultramafic rocks in alpine-folded eugeosynclinal prisms" (p. 949). The sediments were deposited on a continental rise. The serpentines were "oceanic rinds." Dietz explained the evolution of the Franciscan as follows: "Upon thrusting of the ocean floor toward the continents (by sea floor spreading?) the continental rise is converted into a eugeosyncline and the terrace wedge into a miogeosyncline. Bodies of spilite, serpentine, and ultramafic rock become incorporated as fragments of the sea floor." We have made progress since then, and have abandoned the terminology and concepts of geosynclines, but the essence of his idea has remained (from Dietz, 1963).

the same time Hamilton (1969, p. 2415) also postulated that the "Coast Range Overthrust of Bailey and others was "the Benioff zone along which the oceanic crust and mantle rode relatively eastward beneath the continent and its fringing sedimentary shelf."

When I went to California in 1969 to lead a field trip prior to the Penrose Conference on New Global Tectonics, the concept of underthrusting had become generally established. A geological traverse across the Southern Coast Ranges provided, however, an opportunity to put across the idea that the Coast Range Overthrust is not an overthrust in the usual sense. The "thrust" is in fact a boundary separating the Great Valley Sequence and its

underlying ophiolite basement from the zone of penetrative deformation that produced the Franciscan Mélange. The mélange zone, but not the thrust itself, is the Benioff Zone, marking the successive positions of a migrating plate-junction (Hsü, 1971).

It was also clear to many of us then that the concept of the eugeosyncline could be replaced by a plate-tectonic model. One might recall that Gustav Steinmann recognized as early as 1892 during a field trip with Lawson to the Bay area, the similarity between the Franciscan igneous rocks and the Alpine ophiolites (Steinmann, 1905). His misconception of an ophiolitic intrusive into a geosyncline was, however, so deeply rooted that it was retained in the first formulations of the plate-tectonic theory (e.g.,

Dewey, 1969). Some American geologists, like F. C. Wells (1949), may have envisioned a correlation of the Franciscan ophiolites to the ocean floor, when the term "ensimatic geosyncline" was invented to designate the Franciscan, but a convincing case was not made until the Alpine ophiolites were studied in the light of sea floor spreading theory (e.g., Moores, 1969). In the early sixties Alexis Moiseyev identified the serpentinites at the base of the Knoxville as an ophiolite, while he was still a graduate student at Stanford. For this "heresy" he was almost kicked out of school (E. Moores, personal communication). By the end of the decade, however, no more obstacles stood in the way of comparing ophiolites and ocean lithosphere. With this new insight, we were finally liberated from the erroneous notion of Franciscan sedimentation in an elongate trough with contemporary volcanism. I could then write (Hsü, 1971, p. 1162):

The Franciscan mélange includes ophiolites emplaced at a mid-ocean ridge, radiolarites deposited on an abyssal (hill), and flysch turbidites poured into a deep-sea trench . . . (This) segment of the Northeast Pacific, which can be called the Mesozoic Neo-Franciscan plate . . . was plunged under the North American continent, the only vestiges of this once extensive plate are the blocks of the various Franciscan rocks now intimately mixed in the mélange.

The breakthrough was complete. With the gradual acceptance of plate-tectonic theory, most of the former conflicts and contradictions were resolved, and the model of a subduction mélange has become the guiding principle in interpreting the geology of California (and Circum-Pacific) Coast Ranges.

THE NEW REIGN SINCE 1970

According to Thomas Kuhn's theory (1962) a scientific revolution is followed by a period of normal science, when the new paradigm is elaborated. Since the purpose of this paper is to give an account of the circumstances leading to the breakthrough, I shall leave to future historians the task of recording the new period of progress, followed by new conflicts, new crises, new revolution, and another new paradigm. I shall make no citations of the new work lest I omit some contributors in my brief account. I would like to mention a few facts, however, to convey my appreciation of the recent discoveries.

Field mapping and laboratory investigation have made possible tectonic and metamorphic zonations of the Franciscan Mélange. The tectonic setting of the Great Valley Sequence has been placed in a plate-tectonic framework. The role of sediment-gravity flows in the genesis of mélange has been clarified. The physics of fragmentation and of mixing has been analysed by computer modelling. Last, but not least, "allochthonous terranes" in the Coast Ranges have been identified, and an analysis of the kinematics of the Pacific sea floor spreading confirmed earlier estimates that displacement as large as 10,000 km since the Jurassic may have been involved in the subduction that produced the Franciscan Mélange. From the reprints I receive, I see that many of the "old revolutionaries," people like Ben Page, Gary Ernst,

Bob Coleman, David Jones, Clarke Blake, Bill Dickinson, John Maxwell, Cliff Hopson, and others are still active in the new reign. The young generation, many of them pupils of the "old guards," including John Suppe, Darrel Cowan, Mark Cloos, Steve Bachman, J. O. Berkland, K. R. Aalto, J. M. Mattinson, E. A. Pessagno, Walter Alvarez and others have also left their "footprints" in the march of progress. The list is incomplete, reflecting the imperfection of my library collection; my oversights, I hope, may be forgiven, as I have been too distant and stayed too long away from the "scene of battle."

CONCLUSION

When one compares the history of geological research in the California Coast Ranges to that in the Alps, one sees a great contrast in rate of progress. The heroic days of the Alpine geology came during the two decades just before and after the turn of the century; the breakthrough was made possible when Marcel Bertrande (1884) proposed the theory of tectonic superposition by overthrusting (see Bailey, 1935). Complicated as the Alpine structures are, the layered rocks there form coherent sequences; ophiolitic mélanges constitute only a minor fraction of the Alpine rocks. Consequently the work of the pioneers, based on the conventional methodology of their century withstood the great Earth Science Revolution of the 1960s. The understanding of the Coast Range geology was delayed, however, because a radically different methodology had to be devised. The Laws of Superposition, of Lateral Continuity, of Paleontological Dating, applied so very successfully in investigating coherent rock-sequences are not readily applicable to the mapping of mélange terranes like the Franciscan.

The Franciscan was considered at various times a Jurassic, a Cretaceous, a Late Jurassic Tithonian, or a Tithonian to Late Cretaceous formation, deposited before or after the Nevadan Orogeny. Paradoxically, the Franciscan *seems* to be all of those things, but *is* none of those things. The mélange *includes* Jurassic and Cretaceous formations and, as I was told, probably also Tertiary formations), some of which were laid down before, and others after, the Nevadan Orogeny, but *is* itself not a formation nor a group of formations. The mélange is a giant tectonic breccia formed by subduction down the Benioff Zone at a convergent plate-margin.

John Verhoogen (1983, p. 8) musing over time and chance pondered, "what could have happened to the science of geology if William Smith had first surveyed the chaotic Franciscan formation of California instead of the well-ordered geological strata of England." He need not have "shuddered at the thought of the chaos that might otherwise have ensued," because there would have been none. William Smith would have remained an obscure surveyor if he had to work in the Coast Ranges, while someone else, working in an ordered terrace, would have come up with exactly the same set of stratigraphic principles. The progress of science has a logical sequence: Newtonian physics had to be established before a theory of relativity could be developed. The principles of stratigraphy likewise had to be formulated before we could realize that the interpretation of chaotic mélanges would require another set of rules.

The famous astronomer Cecilia Payne-Gaposchkin told us of the dilemma in her student days at Cambridge: One professor, Lord Rutherford, stressed observation, and declared that "one thorough experiment . . . is worth all the theories in the world . . . (Yet) years later Eddington uttered the dictum that he would not believe an observation unless it was supported by a good theory" (Haramundanis, 1984, p. 117). The history of Franciscan geology reveals the wisdom of Eddington.

Taliaferro was not a bad observer, and his field sketches portrayed accurately what he saw, but he did not have a good theory to help interpret his observations. One cannot see, through a grass-covered soil-mantle, if a layered rock is laterally continuous; one cannot be certain if the occurrence of a metamorphic rock near a coarse-grained igneous rock indicates contact metamorphism or tectonic mixing. Only a good theory could have

helped him interpret his observations. During the two decades after the Second World War, we all saw the fragmented blocks and the sheared matrix and we had ample evidence of the high-pressure metamorphism. There were many good observations and good experiments, but we did not have a good theory. We had many arguments, sometimes unnecessarily acrimonious, but we all came to agree finally when the new plate-tectonics model provided us with enlightenment.

Eddington might be right about the need for a good theory but Rutherford's belief in observation should not be lightly dismissed either. We would have had an empty theory convincing to nobody, if no fossils were discovered in the Franciscan, no observations were made on its deformational style, no experiments were performed on metamorphic equilibria. Good theories can help us make good observations or to perform good experiments, but usually theories are not the cause, but the consequence, of good observations and good experiments, as Thomas Kuhn advocated in his theory of scientific revolutions.

REFERENCES CITED

Anderson, F. M., 1933, Knoxville-Shasta succession in California: Geological Society America Bulletin, v. 44, p. 1237–1270.

Anderson, F. M., 1938, Lower Cretaceous deposits in California and Oregon: Geological Society America Special Paper 16, 339 p.

Bailey, E. B., 1935, Tectonic Essays, mainly Alpine: Oxford University Press, Oxford, 200 p.

Bailey, E. B., and McCallien, W. J., 1950, The Ankara mélange and the Anatolian thrust: Nature, v. 166, p. 938–940.

Bailey, E. B., and McCallien, W. J., 1963, Liguria nappe: Northern Apennines: Royal Society of Edinburgh Transactions, v. 65, p. 315–333.

Bailey, E. H., Blake, M. C., Jr. and Jones, D. L., 1970, On-land Mesozoic oceanic crust in California Coast Ranges: U.S. Geological Survey Professional Paper 700-C, p. 70C–81C.

Bailey, E. H., Irwin, W. P. and Jones, D. L., 1964, Franciscan and related rocks and their significance in the geology of western California: California Division of Mines and Geology Bulletin, v. 83, 177 p.

Becker, G. F., 1885, Notes on the stratigraphy of California: U.S. Geological Survey Bulletin 19, 28 p.

Becker, G. F., 1888, Geology of the quicksilver deposits of the Pacific slope, with atlas: U.S. Geological Survey Monograph 13, 486 p.

Bertrand, M., 1884, Rapports de la Société géologique de France, 3e série, v. 12, p. 318 ff.

Blake, M. C., Jr., Irwin, W. P., and Coleman, R. G., 1967, Upside-down metamorphic zonation, blueschist facies, along a regional thrust in California and Oregon: U.S. Geological Survey Professional Paper 575-C, p. C1–09.

Blake, W. P., 1855, Observations on the characters and probable geologic age of the sandstone of San Francisco: Proceedings of American Association for the Advancement of Science, v. 9, p. 220–222.

Blake, W. P., 1856, Report on the Geology of the route near third-second parallel (Pope's reconnaissance), U.S. 33rd Congress, 2nd Session, v. 2, 55 p. U.S. Pacific Railroad Report.

Blake, W. P., 1858, Report of a geological reconnaissance in California: U.S. War Department Explorations and Surveys for Railroads Mississippi to Pacific Ocean Reports, v. 5, part 2, 370 p.

Bloxam, T. W., 1956, Jadeite-bearing metagraywacke in California: American Mineralogist, v. 41, p. 488–496.

Borg, I. Y., 1956, Glaucophane schists and eclogites near Healdsburg, California: Geological Society America Bulletin, v. 67, p. 1563–1584.

Bowen, O. E., and Crippen, R. A., Jr., 1951, Geologic map of the San Francisco Bay region: California Division of Mines Bulletin 154, p. 161–174.

Briggs, L. I., Jr., 1953, Geology of the Ortigalita Peak quadrangle, California: California Division of Mines Bulletin 167, 61 p.

Brothers, R. N., 1954, Glaucophane schists from the North Berkeley Hills, California: American Journal of Sciences, v. 262, p. 614–626.

Brown, R. D., Jr., 1964, Thrust-fault relations in the northern Coast Ranges, California: U.S. Geological Survey Professional Paper 475-D, p. D7–D13.

Camp, C. L., 1942, *Ichthyosaur rostra* from central California: Journal of Paleontology, v. 6, p. 362–371.

Church, C. C., 1952, Cretaceous foraminifera from the Franciscan Calera limestone of California: Cushman Foundation of Foraminifera Research Contributions, v. 3, p. 68–70.

Coleman, R. G., 1961, Jadeite deposits of the Clear Creek area, New Idria district, San Benito County, California: Journal of Petrology, v. 2, p. 209–247.

Coleman, R. G., and Lee, D. E., 1962, Glaucophane-bearing metamorphic rock types of Cazadero area, California: Journal of Petrology, v. 4, p. 261–301.

Crickmay, C. H., 1931, Jurassic history of North America: its bearing on the development of continental structure: Proceedings of American Philosophical Society, v. 70, p. 15–105.

Crittenden, M. D., Jr., 1951, Geology of the San Jose-Mount Hamilton area, California: California Division of Mines Bulletin 157, 74 p.

Cushman, J. A., and Todd, M. R., 1948, A foraminiferal fauna from the New Almaden district, California: Cushman Foundation of Foraminiferal Research contributions, v. 24, p. 90–98.

Darrow, R. L., 1963, Age and structural relationships of the Franciscan Formation in the Montara Mountain quadrangle: California Division of Mines Special Report 78, 23 p.

Davis, C. H., 1913, New species from the Santa Lucia Mountains, California, with a discussion of the Jurassic age of the slates at Slate's Springs: Journal of Geology, v. 21, p. 453.

Davis, E. F., 1918a, The radiolarian cherts of the Franciscan group: California University, Department of Geology Bulletin, v. 11, p. 235–432.

Davis, E. F., 1918b, The Franciscan sandstone: California University, Department of Geology Bulletin, v. 11, p. 1–44.

Dewey, H., and Flett, J. S., 1911, On some British pillow-lavas and rocks associated with them: Geological Magazine, Decade 5, v. 8, p. 202–209, 241–248.

Dewey, J. F., 1969, Evolution of Appalachian/Caledonian orogen: Nature,

v. 222, p. 124–129.

Dickinson, W. R., 1966, Table Mountain serpentinite extrusion in California Coast Ranges: Geological Society of America Bulletin, v. 77, p. 451–472.

Dietz, R. S., 1963, Alpine serpentinites as oceanic rind fragments: Geological Society of America Bulletin, v. 74, p. 947–952.

Diller, J. S., and Stanton, T. W., 1894, The Shasta-Chico Series: Geological Society of America Bulletin, v. 5, p. 435–464.

Easton, W. H., and Imlay, R. W., 1955, Upper Jurassic Fossil localities in Franciscan and Knoxville formations in southern California: American Association of Petroleum Geologists Bulletin, v. 39, p. 2336–2340.

Ernst, W. G., 1965, Mineral parageneses in Franciscan metamorphic rocks, Panoche Pass, California: Geological Society of America Bulletin, v. 76, p. 879–914.

Ernst, W. G., 1970, Tectonic contact between the Franciscan mélange and the Great Valley sequence-crustal expression of a Late Mesozoic Benioff Zone: Journal of Geophysical Research, v. 75, p. 886–901.

Fairbanks, H. W., 1894, Review of our knowledge of the geology of the California Coast Ranges: Geological Society of America Bulletin, v. 6, p. 71–102.

Fairbanks, H. W., 1895, The stratigraphy of the California Coast Ranges: Journal of Geology, v. 3, p. 415–433.

Fairbanks, H. W., 1896, Stratigraphy at Slate's Springs with some further notes on the relation of the Golden Gate series to the Knoxville: American Geologist, v. 18, p. 350–356.

Fairbanks, H. W., 1904, Description of the San Luis quadrangle (California): U.S. Geological Survey Geological Atlas, Folio 101, 14 p.

Gabb, W. M., 1869, Cretaceous and Tertiary fossils, in Paleontology of California: California Geological Survey, Paleontology, v. 2, p. 127–254.

Gealey, W. K., 1951, Geology of the Healdsburg quadrangle, California: California Division of Mines Bulletin 161, p. 7–50.

Ghent, E., 1963, Fossil evidence for maximum age of metamorphism in part of the Franciscan Formation, northern Coast Ranges, California: California Division of Mines and Geology Special Rept. 82, p. 41.

Glaessner, M. F., 1949, Foraminifera of Franciscan (California): American Association of Petroleum Geologists Bulletin, v. 33, p. 1615–1617.

Greenly, E., 1919, The Geology of Angelsey: Great Britain Geological Survey Memoir, 980 p.

Hamilton, W., 1969, Mesozoic California and the underflow of Pacific Mantle: Geological Society of America Bulletin, v. 80, p. 2409–2430.

Haramundanis, L. (ed.), 1984, Cecilia Payne-Gaposchkin: an autobiography and other recollections: Cambridge, Cambridge University Press, 285 pp.

Hedberg, H. D. (ed.), 1976, International Stratigraphic Guide: Wiley, London, 200 p.

Hertlein, L. G., 1956, Cretaceous ammonite of Franciscan group. Marin County, California: American Association of Petroleum Geologists Bulletin, v. 40, p. 1985–1988.

Hinde, G. J., 1890, Notes on a new fossil sponge from the Utica shale formation (Ordovician) at Ottawa Canada: Geological Magazine, Decade 3, v. 8, p. 22–24.

Hinde, G. J., 1894, Note on the radiolarian chert from Angel Island and from Buri-Buri Ridge, San Mateo County, California, Appendix to Ransome, F. L., The geology of Angel Island: California University, Department of Geology Bulletin, v. 1, no. 7, p. 235–240.

Hopson, C. A., Frano, C. J., and Pessagno, E. A., Jr., 1975, Preliminary report and geologic guide to the Jurassic ophiolite near Point Sal, southern California coast: Geological Society of America, 36 p.

Hsü, K. J., 1965a, Franciscan rocks of Santa Lucia Range, California, and the argille scagliose of the Apennines, Italy: A comparison in style of deformation: Geological Society of America, Special Paper 87, p. 210–211.

Hsü, K. J., 1965b, On the klippe origin of the Franciscan rocks of the Santa Lucia Range - a working hypothesis (abs.): Geological Society of America, Special Paper 87, p. 82–83.

Hsü, K. J., 1967, Mesozoic geology of the California Coast Ranges - A new working hypothesis, in Schaer, J., ed., Etages tectoniques: Neuchâtel, à la Baconnière, p. 279–296.

Hsü, K. J., 1968, Principle of mélanges and their bearing on the Franciscan-Knoxville paradox: Geological Society of America Bulletin, v. 79, p. 1063–1074.

Hsü, K. J., 1969, Role of cohesive strength in the mechanics of overthrust faulting and of landsliding: Geological Society America Bulletin, v. 80, p. 927–952.

Hsü, K. J., 1971, Franciscan mélanges as a model for geosynclinal sedimentation and underthrusting tectonics: Journal of Geophysical Research, v. 75, p. 1162–1170.

Hsü, K. J., and Ohrbom, R., 1969, Mélanges of San Francisco Peninsula - Geologic reinterpretation of type Franciscan: American Association of Petroleum Geologist Bulletin, v. 53, p. 1348–1367.

Hubbert, M. K., and Rubey, W. W., 1959, Role of fluid pressure in mechanics of overthrusting faulting: Geological Society America Bulletin, v. 70, p. 115–166.

Irwin, W. P., 1957, Franciscan group in Coast Ranges and its equivalents in Sacramento Valley, California: American Association Petroleum Geologists Bulletin, v. 41, p. 2284–2297.

Jenkins, O. P. (ed.), 1951, Geological Guidebook of the San Francisco Bay counties: California Division of Mines Bulletin 154, 392 p.

Kilmer, F. H., 1961, Anomalous relationship between the Franciscan formation and metamorphic rocks, northern Coast Ranges, California, in Abstracts for 1961: Geological Society America Special Paper 68, p. 210.

Kocher, R. N., 1981, Biochronostratigraphische Untersuchungen oberjurassischer radiolarienführender Gesteine, insbesondere der Südalpen, Dissertation, ETH, Zürich.

Küpper, Klaus, 1955, Upper Cretaceous Foraminifera from the "Franciscan Series," New Almaden district, California: Cushman Foundation of Foraminifera Research Contributions, v. 6, p. 112–118.

Kuhn, T. S., 1962, The Structure of Scientific Revolutions: Chicago: Chicago University Press, 210 p.

Lawson, A. C., 1895, Sketch of the geology of the San Francisco Peninsula (California): U.S. Geological Survey 15th Annual Report, p. 339–476.

Lawson, A. C., 1897, The geology of San Francisco Peninsula: Journal of Geology, v. 5, p. 173–174.

Lawson, A. C., 1902, A geological section of the Middle Coast Ranges of California: Science, n.s., v. 15, p. 415–416.

Lawson, A. C., 1914, San Francisco district: U.S. Geological Survey Geological Atlas, Folio 193, 24 pp.

Lawson, A. C., 1915, San Francisco district: U.S. Geological Survey Atlas, Folio 193, (Field ed.), 180 pp.

Lee, D. E., Thomas, H. H., Marvin, R. F., and Coleman, R. G., 1964, Isotopic ages of glaucophane schists from the area of Cazadero, California: U.S. Geological Survey Professional Paper 473-D, p. 1050–1070.

McKee, B., 1962, Widespread occurrence of jadeite, lawsonite, and glaucophane in central California: American Journal of Science, v. 260, p. 596–610.

Maddock, M. E., 1955, Geology of the Boardman quadrangle: [Ph.D. thesis]: Berkeley, University of California.

Manning, G. A., and Ogle, B. A., 1950, Geology of the Blue Lake quadrangle, California: California Division of Mines Bulletin 148, 36 p.

Merriam, J. C., 1899, Letter to L. F. Ward: U.S. Geological Survey 20th Annual Report, Part 2, p. 338.

Moores, E., 1969, Petrology and structure of the Vourinos Ophiolitic Complex of northern Greece: Geological Society of America Special Paper 118, 74 p.

Newberry, J. S., 1857, Report upon the geology of the route (Williamson's Survey in California and Oregon) U.S. Pacific Railroad Reports, U.S. 33rd Congress, 2nd Session, U.S., Part 2, p. 5–68.

Nicholson, H. A., 1890, Address on recent progress in paleontology as regards invertebrate animals: Transactions of Edinburgh Geological Society, v. 6, p. 53–69.

Nomland, J. O., and Schenck, H. G., 1932, Cretaceous beds at Slate's Hot Springs, California: California University, Department of Geological Science Bulletin, v. 21, p. 37–49.

Ogle, B. A., 1953, Geology of Eel River Valley area, Humboldt County, California: California Division of Mines Bulletin 164, 128 p.

Page, B. M., 1963, Gravity tectonics near Passo della Cisa, northern Apennines, Italy: Geological Society of America Bulletin, v. 74, p. 655–672.

Page, R. M., 1966, Geology of the Coast Ranges of California: California Division of Mines and Geology, Bulletin 190, p. 255–276.

Ransome, F. L., 1894, The geology of Angel Island: California University, Department of Geology Bulletin, v. 1, p. 193–234.

Reed, R. D., 1933, Geology of California: Tulsa, Oklahoma, American Association of Petroleum Geologists, 355 p.

Reed, R. D., and Hollister, J. S., 1936, Structural evolution of southern California: American Association of Petroleum Geologists Bulletin, v. 20, p. 1529–1704.

Schlocker, J., Bonilla, M. G., and Imlay, R. W., 1954, Ammonite indicates Cretaceous age for part of Franciscan group in San Francisco Bay area, California: American Association of Petroleum Geologists Bulletin, v. 38, p. 2372–2381.

Soliman, S. M., 1958, General geology of the Isabel-Eylar area, California, and petrology of the Franciscan sandstones [Ph.D. thesis]: Stanford University, Stanford, California.

Steinmann, G., 1905, Die geologische Bedeutung der Tiefseeabsätze und die ophiolitischen Massengesteine: Berichte der naturforschenden Gesellschaft Freiburg, i.B., v. 16, p. 44–65.

Steinmann, G., 1927, Die ophiolithischen Zonen in den mediterranen Kettengebirgen: Congrès géologique international, comptes rendus de la XIVᵉ Session, en Espagne, v. 2, p. 637–667.

Suppe, J., 1969, Times of metamorphism in the Franciscan terrain of the northern Coast Ranges, California: Geological Society of America Bulletin, v. 80, p. 135–142.

Switzer, George, 1945, Eclogite from the California glaucophane schists: American Journal of Science, v. 243, p. 1–8.

Taliaferro, N. L., 1943, Franciscan-Knoxville problem: American Association of Petroleum Geologists Bulletin, v. 27, p. 109–219.

Taliaferro, N. L., 1944, Cretaceous and Paleocene of Santa Lucia Range, California: American Association of Petroleum Geologists Bulletin, v. 28, p. 449–521.

Taliaferro, N. L., 1951, Geology of the San Francisco Bay counties: California Division of Mines Bulletin 154, p. 117–150.

Tealle, J.J.H., 1895, On greenstones associated with radiolarian cherts: Transactions of Royal Geological Society of Cornwall, v. 9, p. 560.

Tealle, J.J.H., 1899, The Silurian Rocks of Great Britain, v. 1, Scotland: Memoir, Geological Survey, p. 84–87.

Thalmann, H. E., 1942, Globotruncana in the Franciscan limestone, Santa Clara County, California (abs.): Geological Society of America Bulletin, v. 53, p. 1838.

Thalmann, H. E., 1943, Upper Cretaceous age of the "Franciscan" limestone near Laytonville, Mendocino County, California (abs.): Geological Society of America Bulletin, v. 54, p. 1827.

Travis, R. B., 1952, Geology of the Sebastopol quadrangle, California: California Division of Mines Bulletin 162, 32 p.

Verhoogen, J., 1983, Personal notes and sundry comments: Annual Review of Earth and Planetary Sciences, v. 11, p. 1–9.

Wells, F. C., 1949, Ensimatic and ensialic geosynclines (abstract): Geological Society of America Bulletin, v. 60, p. 1927.

Wells, S. P., 1953, Jurassic plesiosaur vertebrae from California: Journal of Paleontology, v. 27, p. 743–744.

White, C. A., 1885, On the Mesozoic and Cenozoic paleontology of California: U.S. Geological Survey Bulletin 15, 33 p.

Whitney, J. D., 1865, Geological Survey of California, Report of progress and synopsis of the field work from 1860 to 1864; California Geological Survey, Geology, v. 1, 498 p.

Geological Society of America
Centennial Special Volume 1
1985

The Coon Butte Crater controversy

Ellen T. Drake
College of Oceanography
Oregon State University
Corvallis, Oregon 97331

ABSTRACT

G. K. Gilbert's (1891) first assessment of the crater at Coon Butte, Arizona (now called Meteor Crater), was that it was caused by a meteorite impact. His classical paper *The Moon's Face* (1893) eloquently argued for the impact hypothesis to explain lunar craters. But by 1896 he had changed his mind on the Coon Butte Crater, as he could not find the "buried star," even though several tons of meteoritic material had been found in the area. The steam explosion explanation, suggested to him by Willard D. Johnson and tentatively adopted by Gilbert, then became the accepted hypothesis for the origin of Coon Butte Crater among most geologists. In spite of a great deal of evidence for the impact idea, the steam explosion explanation, with the exception of a few notable individuals, took root in the geological community, much to the frustration of D. M. Barringer. The latter insisted on the existence of the "main mass" of the meteorite and spent a lifetime and fortune attempting to prove the impact origin of the crater by looking for the meteoritic body. He also had visions of a lucrative mining adventure. Barringer opposed G. P. Merrill, a distinguished scientist within the geological community, who supported the impact idea but explained the absence of a "main mass" of meteoritic material by its volatilization. The modern explanation for the creation of the crater is by shock waves produced from the dissipation of kinetic energy of the meteorite travelling at hypervelocity.

Several reasons might explain the long delay in acceptance of the impact hypothesis: (1) in the absence of more evidence, the geological community naturally followed the lead of the much-admired Gilbert; (2) to the geologists, Barringer and his associates were outsiders, rich and respected men in other fields and in industry, but not in geology and therefore could be ignored; (3) Barringer was "ahead of his time," and therefore could not expect much support from his contemporaries. The reasons were probably multiple and complex and may involve fundamental philosophical concepts. For example, as Eugene Shoemaker (1984) suggested, geologists were so imbued with Lyellian uniformitarianism that they could not accept a catastrophic explanation, i.e., the fall of an extra-terrestrial object causing the creation of a terrestrial feature. It is not clear, however, that any one answer explains the puzzle. The Coon Butte Crater controversy also tests two other ideas much touted by geologists, the principle of multiple working hypotheses and the idea of tolerance for unorthodoxy in science. Both concepts seemed to have failed the test.

Figure 1. Coon Butte (Meteor) Crater.

INTRODUCTION

The history of geology is marked by many controversies; the science seems actually to advance through one series of confrontations after another, as, for example, the battles between neptunists and plutonists, the struggle between catastrophists and uniformitarians, arguments over the origin of the ice ages, disputes concerning the age of the earth, and the more recent debate about continental drift (Hallam, 1983). Between such revolutionary issues that caused changes in the direction of the science and replaced paradigms is a series of debates which also may be of fundamental importance to the development of geology but whose significance has not been fully recognized by geologists. The characters involved were no less advocates and devotees of their ideas than the principals in the other issues. The Coon Butte (Meteor) Crater controversy is one such battle arena in which the drama of geology unfolded in a fascinating manner. In this article I will relate the story of the controversy and attempt to assess the reasons why geologists took decades to adopt the now accepted view of the origin of Meteor Crater

A REMARKABLE GEOLOGICAL PHENOMENON

In June 1891, alerted by reports from a prospector in Ari-

zona regarding the discovery of a "vein" of metallic iron near Cañon Diablo, A. E. Foote, a mineralogist, made a visit to the locality and reported on the "remarkable mineralogical and geological features" he observed. After a brief description of the size and shape of the depression and the surrounding circular elevation known locally as Crater Mountain, or more commonly, as Coon Butte (Fig. 1), Foote (1891) wrote:

> The rocks which form the rim of the so called 'crater' are sandstones and limestones and uplifted on all sides at an almost uniform angle of from thirty-five to forty degrees. A careful search, however, failed to reveal any lava, obsidian or other volcanic products. I am therefore unable to explain the cause of this remarkable geological phenomenon.

G. K. Gilbert, chief geologist of the United States Geological Survey (Fig. 2), asked Willard D. Johnson to investigate the site. Johnson suggested to Gilbert that the crater might have been caused by a steam explosion (Gilbert, 1896). In November 1891, Gilbert himself went to the site. With Marcus Baker, a magnetics expert, Gilbert was to map the Coon Butte region where much meteoritic iron was reported to have been found. Realizing that the crater at Coon Butte did not conform to his idea of a volcanic crater (Gilbert, 1982, p. 98), Gilbert at first suspected that it was

Figure 2. Portrait of Grove Karl Gilbert (Courtesy of the U.S. Geological Survey Photographic Library).

"the scar produced on the earth by the collision of a star" (Gilbert, 1896, p. 4). Gilbert and Baker spent 16 days investigating the hole. Baker set up stations along two lines extending at right angles to each other in the crater with one of the lines extending onto the plain for more than three miles. The results of this magnetic survey showed the magnetism to be constant in direction and intensity at all the stations (Gilbert, 1896). With the results contradicting the existence of a great buried iron mass, a telegram was sent to the *San Francisco Examiner* on Tuesday, December 1, 1891, dated November 30, 1891, reporting that

The theory is, from the appearance of the walls and from the fact that there have been found many pieces of meteoric iron around the hole, that the meteor penetrated the earth to a depth of 700 or 800 feet before it *exploded* [my italics] and this accounts for the strange phenomenon.

The entire text of this telegram, which is several paragraphs longer, is printed in the *Astronomical Society of the Pacific Publications* (1892, p. 37). Gilbert's first instinct, therefore, was brilliant and right on target, if indeed this communiqué reported accurately his first assessment of the phenomenon. The identity of the person who actually sent the telegram is unknown, but I presume it was a reporter who interviewed either Gilbert or

Baker or both, as the dateline is "Flagstaff (Arizona), Nov. 30, 1891." The first sentence of the communiqué states (*Astronomical Society of the Pacific Publications,* 1892, p. 37):

G. K. Gilbert and Marcus Baker, the former being Chief Geologist of the U.S. Geological Survey, have returned from the Cañon Diablo, where they were sent by the Government to make a map of the region where so much meteoric iron has recently been found.

In his 1893 report to John Wesley Powell, the then director of the Survey, Gilbert expostulated on both the steam explosion explanation and the impact idea:

In one case it represents a factor of volcanism not elsewhere known to be isolated, and therefore instructive in its contribution to the physical history of volcanoes; in the other, it is of important practical value as indicating the presence beneath its hollow of many thousands of tons of nickeliferous iron, and it is of scientific importance as illustrating a method of planetary aggregation by the falling together of smaller masses (Gilbert, 1893a, p. 187).

Sometime between the 1891 telegram reported in the *Astronomical Society of the Pacific Publications* and the 1892 and 1893 Annual Reports of Gilbert to the director of the U.S.G.S., the idea that the meteorite *exploded* seems to have disappeared, as by now Gilbert thought that if the impact idea was correct, then the meteoritic mass would still be buried. Some writers (e.g., William G. Hoyt, personal communication) question whether Gilbert ever thought that the meteorite exploded. If Gilbert never entertained the explosion idea, then it is curious from whom the reporter who sent the telegram got the idea. As far as I know, neither Baker nor Gilbert ever retracted this report.

The idea of an explosive origin for the crater is not new. Gilbert himself reported that some shepherds in 1886 invented a theory to explain the crater "which is admirable in its simplicity: the crater was produced by an explosion, the material of the rim being thrown out from the central cavity, and the iron was thrown out from the same cavity at the same time" (Gilbert, 1896, p. 3). Why Gilbert thought the "buried star" would be there in 1893 when the magnetic survey in 1891 denied its existence is puzzling.

By 1896, however, in his now famous presidential address to the Geological Society of Washington on the origin of hypotheses, he denied the presence of a "buried star," by invoking Baker's negative magnetic results. He stated (1896, p. 9):

So if the crater contains a mass of iron its attraction is too feeble to be detected by the instruments employed. That we might learn the precise meaning of this result, the delicacy of the instruments was afterward tested at the Washington Navy Yard, by observing their behavior when placed in certain definite relations to a group of iron cannon whose weight was known, and the following conclusions were reached: If a mass of iron equivalent to a sphere 1500 feet in diameter is buried beneath the crater it must lie at least 50 miles below the surface; if a mass 500 feet in diameter lies there its depth is not less than 10 miles. So the theory of a great iron meteor is negatived by the magnetic results, unless we may suppose either that the meteor was quite small as compared to the diameter of the crater, or that it penetrated to a very great depth.

His purpose in this speech was to demonstrate the methodology of the principle of multiple working hypotheses. At the same time, however, he provided the rationale, once and for all as far as many geologists were concerned, why the crater at Coon Butte should be considered the result of a steam explosion. He estimated that the volume of the rim material closely matched that of the crater; he reasoned that had a "star" entered the hole and partly filled it, the remaining hollow would have been less in volume than the rim material. He then proceeded to accept the steam explosion idea that had been proposed by Johnson. Gilbert agreed with the argument that since igneous intrusions rise to various degrees of nearness to the surface, one such intrusion, or at least the extremely hot surrounding rocks, could have come in contact with near-surface waters developing a volume of confined steam which then produced the gigantic explosion. The more than ten tons of meteorite fragments found by then on the rim and in the vicinity he dismissed as a coincidence.

When one considers these rather weak and tentative suggestions posed for the sake of argument in favor of the steam explosion hypothesis, it is remarkable that this explanation was accepted by so many geologists. Although there is evidence that many scientists who personally examined the site favored the impact idea, the majority of the geological community conservatively adhered to Gilbert's conclusion. It was taught in geologic textbooks, as, for example, in that of another proponent of the principle of multiple working hypotheses, T. C. Chamberlin (with R. D. Salisbury, 1904, 1909, p. 596):

Fragments of a meteorite were found on the rim and in the vicinity, but this association appears to be accidental. Computation shows that the volume of the material of the rim closely matches the size of the pit. The source of the explosion is not demonstrable, and it may be an error to connect it with an intrusion of lava below; but since intrusions rise to various degrees of nearness to the surface, and in innumerable cases reach the surface, there is every reason to entertain the conception of a class of intrusions which develop explosive phenomena by close approach to the surface, without actually reaching it. . . .There are grounds for thinking that the remarkable craters of the moon, assuming they are truly volcanic, may belong to this class, for they are very similar to the Coon Butte pit.

This dismissal of the occurrence of meteoritic material as a coincidence was also rather remarkable when one considers that Gilbert's own estimate had shown that the probability of such a coincidence was 800 times less than the probability of noncoincidence, suggesting strongly, therefore, a causal relation. As S. J. Pyne (1980, p. 191) stated, "Even his methodological theories, for all their acute insight into the practical machinery of scientific thought, had let nature deceive him into taking 800:1 odds."

In all fairness to Gilbert, however, it must be related that when geologist Edwin E. Howell suggested that some of the rocks in the crater might have been compressed by the shock of the impact of the "star" so that they occupied less space and could accommodate the volume of the extra-terrestrial body and still preserve the volume of the hole to equal that of the rim material, Gilbert immediately conceded the pertinence of the argument. He admitted that this argument in favor of his original meteoric explanation had the "ability to unsettle a conclusion that was beginning to feel itself secure" (Gilbert, 1896, p. 12). He continued to say, "In the domain of the world's knowledge there is no infallibility." Gilbert's acceptance of the steam explosion idea, therefore, can only be termed "tentative," a word he used himself with regard to "the hypotheses of Science" (Gilbert, 1896, p. 12). No mention, however, was made again of an exploded meteorite until much later and by others. The steam explosion idea seemed then to take root.

GILBERT'S ESSAY ON THE MOON

Ironically Gilbert had gone through a similar thought process in coming to the conclusion that the craters on the moon *were* mostly impact craters. He made a careful comparison between terrestrial and lunar craters. In the words of Pyne (1980, p. 156):

Perhaps his failure to satisfactorily resolve the origin of the crater at Coon Butte haunted his imagination. It is as though, since his investigations reluctantly forced him to abandon meteoric impact as a cause for the terrestrial crater, he projected it on to lunar craters.

His experience with the largest then known terrestrial craters showed that they have a mean diameter of 11 miles, while the mean for the 10 largest lunar craters was 275 miles. The individual terrestrial diameters are closely grouped near their maximum "as though constrained by a limiting condition," while the individual lunar diameters are widely scattered, the distances "of aberrant shots from the bull's eye." He predicted that any new discoveries of craters on earth would have diameters about as great as those now known, none much larger; no such prediction, he asserted, could be made about unknown craters on the other side of the moon. He then carefully reviewed the different types of volcanic craters on earth. Most terrestrial volcanoes, 19 out of 20 in his estimation, are of the Vesuvian type, i.e., like the crater of Mt. Vesuvius. Ninety-nine out of 100 lunar craters, he observed, have bottoms that lie lower than the outer plain, whereas 99 out of 100 terrestrial Vesuvian craters have bottoms higher than the outer plain—that is, "the lunar crater is sunk in the lunar plain; the Vesuvian is perched on a mountain top." If a central hill is present in the Vesuvian, it could be higher than the rim, but the central hill of the lunar crater never rises to the height of the rim and rarely to the level of the outer plain. The smooth, inner plain characteristic of so many lunar craters is either rare or unknown in craters of the Vesuvian type. Gilbert concluded, "Through the expression of every feature the lunar crater emphatically denies kinship with the ordinary volcanoes of the earth" (Gilbert, 1893b, p. 248–250).

Gilbert also rejected other theories such as the notions that the crater walls were formed about the vortices of a primeval liquid moon or that the craters are "remnants of Cyclopean bubbles that have burst." He went on to present his version of a meteoric impact hypothesis. His highly imaginative idea postulated that originally the earth was surrounded by a ring of debris similar to Saturn's rings and the moon was formed through a process of accretion of the debris. The lunar craters are the marks left by the impacts. The bombardment actually caused, by degrees, changes in the moon's axis of rotation so that the whole surface eventually became vulnerable to attack as new equators were created from these repeated jolts to the stability of the rotational axis. As was shown by Drake and Komar (1984), this idea has recently been revitalized in a new theory by Keith Runcorn (1982) who expressed ideas very similar to those of Gilbert.

After formulating this extraordinary hypothesis, Gilbert turned his thoughts back to the earth and realized that he could not have devised moonlets that bombarded the moon and even repeatedly changed its axis of rotation without having at least some of them narrowly escape collision with the moon and collide instead with the earth. He wrote (Gilbert, 1893b, p. 291):

. . .any moonlet which narrowly escaped collision with the moon was enormously perturbed acquiring an entirely different orbit about the earth. Many must have been so directed as to collide with the earth, and the traces of their collision, if ever discovered, will tie together at a new point the chronologies of satellite and planet. . . .

Does the earth exhibit impact craters? If not, then erosion and sedimentation have destroyed them, and the Cenozoic era did not witness the building of the moon. Is any horizon of stratified rocks generally or widely characterized by molten disjecta? If not, then the moon was already a finished planet in Paleozoic time.

The very fresh-looking and recent crater at Coon Butte might have helped to deter Gilbert from accepting a meteoric cause for this terrestrial hole, even though the Coon Butte crater met all the criteria he had established for the lunar craters in its shape and dimensional ratios, and the existence of both asteroids and meteorites was well known by then. In the words of W. G. Hoyt (personal communication), Gilbert's "buried star" was his *"sine qua non."* Without his "buried star," Gilbert reluctantly abandoned the impact hypothesis and supported the steam explosion origin for the Coon Butte crater.

SUPPORT FOR GILBERT

For about 15 years or so after Gilbert's speech in 1895,[1] the geological community seemed to be satisfied with Gilbert's explanation. Some, recognizing the difficulties of the explanation, tried to rationalize in various ways to make the steam explosion idea feasible. F. N. Guild (1907, p. 24–25), for example, after remarking on the "complete absence of any evidence of volcanism," wrote that "in the minds of the uninitiated" Coon Mountain could have been formed by some other agencies different from the San Francisco Mountains, which are of volcanic origin; but he explained there could be

an explosion lacking the energy necessary to bring the igneous mass or even fragments of it to the surface. . . .

It would seem then the phenomenon exhibited here can be satisfactorily explained as having been produced by an explosion followed by an entire lack of volcanic activity, as first explained by G. K. Gilbert of the United States Geological Survey. The meteorites found here probably had nothing to do with the formation of the depression.

Guild's reference to the "uninitiated" could have been directed only against one D. M. Barringer who spent almost a lifetime and a fortune trying to prove the meteorite impact origin of the crater by finding the "main mass" of the meteorite and thereby also "striking it rich."

Barringer had found large deposits of crushed and powdered silica grains that were angular in shape which he felt could be produced only by a tremendous blow like that of a meteorite strike and could not have been the product of any steam action. J. M. Davison (1910, p. 724–726) felt that he solved the "chief difficulty in the way of acceptance of the volcanic origin of Coon Butte," which was "an assumed impossibility of breaking the grains of the gray sandstone into angular fragments by hot water action." He claimed to have performed experiments to prove that geyser action had deposited the powdered rock, and "a final explosion or series of explosions . . . closed the drama." In Davison's view,

all this mass of pulverized rock has been broken by hot water action; not, of course, by solution which would give amorphous silica, nor by a single violent steam explosion, but in part by explosion of superheated water within the pores of the rock fragments and within the grains themselves, but mainly by attrition of grains and fragments of rock churned by boiling water in geyser tubes.

In contrast to Davison's complicated rationalization in support of Gilbert's hypothesis, C. R. Keyes (1911, p. 29) was succinct in his evaluation:

Contrary to the recently expressed views regarding the origin of this remarkable crater, the most critical evidences seem to indicate that this feature of the local landscape is only one of the many manifestations of the explosive type of vulcanism so prevalent throughout the region.

N. H. Darton (1910) made a reconnaissance of parts of northwestern New Mexico and northern Arizona and reported his findings in a U.S.G.S. *Bulletin*. He became a strong supporter of the steam explosion idea for most of his life. In the face of much evidence for impact, Darton stated that Gilbert's idea appealed to him "most strongly." Clearly, Gilbert's reputation was such that geologists were willing to go to great lengths to support his semivolcanic, or rather crypto-volcanic, steam explosion explanation for the creation of the Coon Butte Crater. It is safe to say that this hypothesis was the dominant one against which all other explanations must be measured. Geology students were certainly taught it if they used the Chamberlin and Salisbury textbook, the most distinguished American textbook of its day.

Eugene Shoemaker (1984) attributes the geologists' acceptance of the steam explosion idea to their innate aversion to any phenomenon that could be termed catastrophic. He maintains that geologists, having been for generations imbued with the principle of Lyellian uniformitarianism, naturally found it easier to adopt a hypothesis that involved known terrestrial processes. The idea that a geological, and therefore earthly, feature could be caused by an extra-terrestrial object is one that was fundamentally repugnant to geologists even though intellectually they knew of the existence of such things as asteroids and meteorites. Only a century earlier, educated people in the western world had denied the possibility of stones falling from the sky. So Shoemaker's assumption is a reasonable one. Certainly this philosophical consideration in conjunction with the other human tendencies of scientists in general helped to create the controversy over the origin of the Coon Butte Crater.

BARRINGER'S CHALLENGE

Daniel Moreau Barringer (Fig. 3) was born in Raleigh, North Carolina, in 1860, the son of a congressman. He graduated from Princeton University in 1879 at age 19 and continued there to receive a master's degree in 1882 as well as a law degree from the University of Pennsylvania in the same year. He practiced law with his brother from 1882 to 1889. He then discovered his love for geology. He took many courses in geology, mineralogy, and chemistry both at Harvard University and the University of Virginia. From 1890 on he worked as a consulting mining engineer and geologist. He married Margaret Bennet in 1879, fathered eight children, and became a highly respected citizen and successful businessman, becoming the president and director of several mining companies as well as trustee of educational institutions and hospitals. He was the author of two books, *The Law of Mines and Mining in the U.S.* and *Minerals of Commercial Value.*

Perhaps it is because Barringer was a wealthy and distinguished citizen in his own right, with rich and powerful friends, or that he was an outsider, not a card-carrying geologist, that he was unafraid to attack the theory of such an esteemed scientist as Grove Karl Gilbert. Barringer's own investigations convinced him that the Coon Butte Crater was unequivocally caused by a meteorite. Furthermore, the prospect of great financial return in mining a potentially rich nickel-iron field prompted him to buy this big hole in the ground.[2] Shoemaker (personal communication) believes that Barringer's obsession was not with proving impact origin but rather with proving that there was a minable body of meteorite—it was also a matter of an engineer's pride and honor. I agree with Shoemaker that it was much more than the possible financial return that was his driving force; for the sake of honor, he had to be proved right since he had invested a large sum of money, his and other people's, into the venture. But I believe that Barringer was also obsessed with proving the meteoritic origin of this crater, with great deliberation and in opposition to the most respected geologist of his day. If he were only interested in the problem as a mining venture, and upholding his honor by producing the ore, I think he would not have bothered to write and publish his many papers on the subject (1905, 1909, 1910, 1914, 1924, 1926) in distinguished scientific and technical jour-

Figure 3. Portrait of Daniel Moreau Barringer (Courtesy of the Seeley G. Mudd Manuscript Library, Princeton University).

nals. He could have adopted the following attitude: Let the geologists think what they will—the last laugh will be mine when I find the meteorite.

On the contrary, Barringer (1905) was unrelenting in his criticism of Gilbert. He claimed his own work disproved Gilbert's steam explosion idea, a theory adopted

by Mr. G. K. Gilbert on what seems to be very insufficient evidence. . . . Perhaps it would be more accurate and just to say that he has adopted this theory because of an inadequate examination of the phenomena at Coon Butte, for had he examined the surface carefully, it does not seem possible to me that any experienced geologist could have arrived at such a conclusion (Barringer, 1905, p. 863).

It was clear to Barringer that the crater could not have been volcanic or even semi-volcanic in origin. The crater is situated in an area of level beds of Permian and Triassic sandstones, limestones, and shales of the Southern Colorado Plateau; these beds extend uninterruptedly for "easily seventy miles in every direction" except for Sunset Mountain 12 miles away in a west-southwest direction, Black Mesa 20 miles away in a southeast direction, and the San Francisco Mountains 45 to 50 miles in a northwest direction. Gilbert had felt that these three localities of eruptive rocks were sufficiently close to Coon Butte to warrant postulation of a near-surface flow at the butte. Barringer disagreed.

Barringer also noted the differences between the typical volcanic crater and the Coon Butte Crater, almost the same differences that Gilbert himself had noted between Vesuvian-type terrestrial craters and the lunar craters. Coon Mountain, as it was sometimes called, had an average elevation of only about 130 feet above the surrounding plain, and only at very close range would anyone suspect the existence of such a large crater within. Besides the piling up of the fragments of sedimentary rocks on the rim and its outside slope, another reason for Coon Butte's elevation, he wrote, is that

the uppermost strata exposed in the walls of the interior crater dip quaquaversally, or generally speaking in every direction from the exact center of the crater, at an angle usually varying from ten to 40 degrees, and in one case from sixty to seventy degrees. . . .It is evident that a great, presumably wedge-shaped piece of the material of the cliffs which form the sides of the crater and the rim, has nearly been turned out bodily by the force which produced this enormous hole in the earth's surface. The effect of this would be, of course, to turn the strata nearly on edge at this place.

At this time Barringer had not noticed that small sections of the rim stratigraphy are actually overturned, lying upside down. By 1909, however, he was aware of the overturned nature of the strata. In a paper read before the National Academy of Science meeting at Princeton, he noted:

The vertical bodily uplifting of the strata, forming a portion of the southern wall for one-half mile in length, with the turning backward of the same strata on either side in a way . . . precludes the possibility of the effect having been produced by a steam explosion. . . .

The rock fragments are distributed concentrically around the crater, more occurring at the rim than beyond, although some fragments weighing 50 tons have been found more than a mile from the crater center. The sharp angularity of the fragments, Barringer reasoned, indicates how recently created they were. Meteor Crater is now estimated to be about 50,000 years old (Sutton, 1984), a very young age in geologic time.

Another piece of evidence (mentioned earlier) that Barringer found which he could not understand how Gilbert had missed is that at the bottom of the crater, beneath some lacustrine sedimentary material, is a thick layer of pulverized sandstone containing here and there angular rock fragments. Barringer's drill holes produced this material to a depth of 850 feet; it was so fine that it could be termed a powder at places, but under microscopic examination the tiny grains show very sharp edges. According to Barringer, the material was especially prevalent on the south side of the Butte where several dry washes of this snow-white silica are exposed horizontally for hundreds of feet. He was really referring to two different units, (1) the Coconino breccia beneath the lake beds in the crater and (2) the Coconino debris layer in the ejecta. The latter is exposed but the breccia is concealed except in one small place (Shoemaker, 1963, and personal communication).

Nevertheless, at this point Barringer launched another attack on Gilbert. He wrote (Barringer, 1905, p. 870):

It is difficult to understand how this exposure could escape the eye of any careful geologist making a circuit of the crater. If noticed by him it would certainly seem that he would have examined it and ascertained its nature. Having done this, it would seem that he would have been impelled to make a few shallow trenches at different places around the crater, in order to determine how much of this material there was. Having then proved it to exist on all sides of the crater in enormous quantities, it would seem to me that he could not have explained its presence in any other way than that which we have adopted; especially in view of the fact of there being so much corroborative evidence of even a more convincing character. Briefly, it seems to me impossible that this silica could be produced by volcanic action, or by a steam explosion, and I assume that it could be produced only by the pulverizing effect of an almost inconceivably great blow.

His corroborative evidence of a more convincing character was the finding of meteoritic material itself. Meteoritic iron had been found all around the Butte on the surrounding plain and on the rim. By that time Barringer and his workers had found several thousand pieces of such fragments of meteoritic iron, known as Cañon Diablo siderites. The largest piece found by his workers weighed 225 pounds although much larger pieces had been found earlier. The pieces were concentrically distributed around the crater, as far away as several miles. They were mostly lying on the surface or thinly covered by soil. Within the crater itself only four speciments at that time had been found, weighing 3 or 4 pounds each.

Dedicated as he was to his theory, Barringer began to have open cuts made through the silica and rock fragments on the outside of the rim and shallow shafts sunk through this material to look for meteoritic pieces. One of Barringer's employees in his mining company, S. J. Holsinger, camped on the crater rim for four years directing the exploration. Barringer's efforts were rewared by the finding of nearly 100 pieces of meteoritic material, some as much as 50 pounds in weight, "a number of feet beneath the surface in the silica, overlaid and underlaid in no particular order by the various kinds of rock fragments" occurring in the area.

RECEPTION OF THE IDEA

Barringer's close friends, among them Benjamin Chew Tilghman (1905), Elihu Thomson, and W. F. Magie, all expressed support for his ideas. Within the geological community some papers published during the first dozen years or so of the 20th century were cautiously in support of Barringer and his impact hypothesis. George P. Merrill (1907), an eminent professor of geology and head curator of geology at the United States National Museum (see Yochelson, this volume), at first felt that he had no really good reason to deny this theory, then later (1908) supported it. Merrill proposed that the meteorite, because of its tremendous velocity, had exploded and volatilized. William Pickering (1909) reviewed both Barringer's and Merrill's articles.

While opposing a volcanic origin, he did talk much about meteors and meteorites in general, so that by inference one could conclude he supported the impact hypothesis. He wrote (Pickering, 1909, p. 333):

> Drill holes sunk within the crater show that at the center, at a depth of only 1000 feet, the original strata are undisturbed. It is therefore clearly a surface formation, a pseudo crater, and cannot owe its existence in any way to volcanic forces.

On the grounds that "no variety of animal (either carnivore or human) ever known by the name [coon]" was found in the area, H. L. Fairchild (1906), another distinguished geologist and president of GSA in 1912, proposed the name Meteor Crater for the site. The name *crater,* he reasoned, was appropriate to cavities made either by a projectile or by a steam explosion, and under either theory of crater origin, one cannot deny the existence of the meteorite material in the area. Joseph Iddings (1914) supported the impact idea but credited Gilbert for the idea with no mention of Barringer.

The publication of these papers probably caused Chamberlin and Salisbury to soften somewhat towards the meteorite impact hypothesis in their 1909 single-volume version of their three-volume textbook on geology by adding to their discussion this statement (Chamberlin and Salisbury, 1909, p. 369), without mentioning any names:

> It has been suggested that a large meteorite fell at the site of the butte, and penetrating the earth a few hundred feet, exploded. This sequence of events would account for the pit and the rim.

Somehow the support of these men for the impact idea was not sufficient to turn the tide. On the contrary, by the early 1920s, in the words of meteorite expert H. H. Nininger writing in 1942, "opposition to Barringer's idea had developed into a sort of tradition. To oppose was orthodox; to endorse was unsound." The steam explosion idea was so entrenched through the authority of Gilbert and other "big shots" [Nininger's term] that when the large crater and several smaller ones near Odessa, Texas, were discovered in 1921, University of Texas geologist E. H. Sellards who investigated the site was "slow to be convinced of the meteoritic origin of the craters. . . .The object of much of the work at the craters was simply the recovery of fragments of the meteorite and the building of a display at the Texas Memorial Museum in Austin" (King, 1976, p. 102).

Barringer's geologist son, D. Moreau Barringer, Jr., however, did not equivocate. In 1928, he published in his father's favorite journal, the *Proceedings of the Philadelphia Academy of Natural Sciences,* and declared the large crater at Odessa a meteorite crater very similar to Barringer's crater at Coon Butte. Sellards and his associate G. Evans finally published in 1941 on the meteor craters at Odessa.

Some geologists, like Hager (1926) or Dellenbaugh (1931), in an attempt to explain Meteor Crater by some mechanism other than either impact or steam explosion, advocated that the hole was caused by dissolution of the Kaibab limestone. Chester R. Longwell (1931) of Yale, who supported the impact hypothesis, finally tried to put a stop to this notion by writing an article with the authoritative title, "Meteor Crater is not a limestone sink." Even so, Hager, as late as 1953 and 1956, when most geologists were already convinced of the impact idea, persisted with his dissolution idea. His article described the crater wall's folds and faults, the crater's domes and grabens, subterranean erosion and the dissolution of gypsum, salt beds, and limestone. When he could have cut the Gordian knot with a meteorite strike, Hager (1953) insisted that "a meteorite shower occurred in the vicinity but it bears no visible relationship to the origin of the crater. . . ." But it would be a mistake to belittle Hager's effort. Many of the structures he noted were present and real. Historically and within the geological community collectively, he also served a purpose—i.e., he was advancing *one* of the multiple working hypotheses that geologists profess to be so fond of. That it was a discarded hypothesis does not lessen its usefulness as a trial in the evolutionary process.[3]

In 1927 Barringer collaborated with his son in writing a series of three articles which were published by the *Scientific American.* The editor at that time was Orson D. Munn; it is not known whether Munn added the editor's note to the article, but it said:

> Because certain people, reluctant to believe the unprecedented, regard as sensational the theory that Meteor Crater was formed by the impact of a giant meteor which struck the earth, we have obtained the following definite statements from two well-known scientists.

The editor's note then quoted W. F. Magie of the Palmer Physical Laboratory, Princeton University, and Elihu Thomson, director of the Thomson Laboratory at the General Electric Company, who wrote statements in support of the meteoritic hypothesis. Both Magie and Thomson supported Barringer's efforts unswervingly. Thomson corresponded with Barringer for years on the subject (Abrahams, 1983). That the editors of *Scientific American* felt the necessity to publish this note is an indication of the still strong reluctance to accept the impact hypothesis.

In this series of articles, the two Barringers were still convinced that the main body, or at least a closely packed swarm of small meteorites, could be found. They described the various effects associated with the crater and made a number of astute observations that were original and still valid today. For example, in describing the metamorphic products of the white Coconino sandstone, they wrote (Barringer, 1927, p. 54):

> The explanation of the phenomenon is not easy, but it would appear that a shock-wave, of sufficient intensity to crack the sand-grains, ran through the solid rock ahead of the impacting meteorite, and ahead of the excavating effect.

They concluded (p. 246) that:

> no terrestrial explosion could cause it [the crater] without disturbing the

underlying rocks, no means other than a slanting impact could cause the symmetry of the rim, nothing but a smashing blow could pulverize the grains of the white sandstone.

They also gave clear descriptions of the objects that Barringer, Sr., had earlier found in great abundance termed "iron shale" or "shale balls." Many of these shale balls, they claimed, contain unoxidized nuclei of nickel-iron "identical in all respects with the typical Canyon Diablo meteorites," even as to the Widmanstätten lines. Barringer, Jr., also discussed the circular shape of impact craters (p. 144):

Largely by accident, my father observed one day that by firing a rifle into mud he could make an excellent replica of the crater, and, moreover, that the rifle need not be fired vertically downward, but might be held even less than 45 degrees from the horizontal. Naturally one would suppose that a shot at such an angle would make an elongated hole. But it will not. The hole will be just as round as though the shot had come straight down, although the projectile will lodge under one edge of the hole instead of in the center. A charge of shot fired from a shotgun at close range will produce the same effect.

In addition, the articles reviewed an earlier *Scientific American* article by Barringer, Sr. (1924), entitled "Volcanoes—or Cosmic Shell-Holes" in which the older Barringer related his findings at Coon Butte to the occurrence and origin of craters on the moon.

When one of Barringer's drills became permanently stuck at 1,376 feet, they were certain that the "main cluster of meteorites" was encountered. This was not an unreasonable assumption at the time, as Meteor Crater had yielded more iron meteorites than all the rest of the world combined. By publication time of the last article of the series, when no main mass had been found, they conceded that there were three possibilities: the main cluster still lay buried; it bounced out; or it vaporized and disappeared. They concluded, however, "It could not have vaporized for the vapor would have left abundant and indestructible stains. It could not have bounced out or the hole would be a trough instead of a circular basin. Therefore it must still be in the hole" (Barringer, 1927, p. 246).

MORE EVIDENCE COMES IN

As it turned out, all the other evidence provided by Barringer, and further data gathered by later investigators, made the discovery of the main mass unnecessary to designate the crater the result of meteoritic impact. In 1945, when the U.S. Board of Geographic Names finally proposed the name Meteor Crater, N. H. Darton of the U.S. Geological Survey, who was still against the impact origin, opposed the suggestion (Hoyt, 1983). In 1946, two years before Darton died, the Board officially designated the site as Meteor Crater. Prior to that time it was variously known as Meteor Crater, Diablo Cañon Crater, Coon Butte, Coon Mountain, or Crater Mound. The Meteoritical Society (Leonard, 1950) has sought to honor Barringer by naming the crater Barringer Crater, but Meteor Crater remains the official name.

The work of H. H. Nininger, an expert on meteorites and meteorite craters, and of other investigators, largely vindicated Barringer. Most geologists, however, were convinced of the impact origin before all the evidence was reported and before the definitive works of Shoemaker (1960, 1963) were published. Nininger (1951) reported finding five different forms of particles which evidently constituted the condensation products from an iron vapor cloud (now considered to be melt droplets), the product of the impact "explosion." He found, with the aid of magnets, large quantities of minute spherules and droplets of nickel-iron in an area of more than 100 square miles around the crater. Nininger estimated that there were several thousands of these particles per cubic foot in the topsoil as far away as four miles north of the crater rim. The amount and distribution of fine-grained meteoritic material have been estimated and described by Rinehart (1958). In addition, there were globules of silica glass coated with oxides "as if a droplet of silica had cooled and subsequently received a deposit of metallic condensation before reaching the soil" (Nininger, 1951, p. 80). Nininger was of the opinion that the meteoritic mass was "suddenly transformed into metallic vapor"; this sudden transformation was the blast that gave the crater its final outlines and that "propelled skyward the gigantic metallic-vapor cloud from which rained down the millions of condensed vapor droplets" discovered by him. Present knowledge indicates that only a very small fraction of the meteorite was vaporized; most of it was broken up into small pieces, melted, or occurs as microscopic drops embedded in glass in the breccia underneath the crater floor.

From examination of the different types of meteoritic specimens found in the area, Nininger reconstructed the scenario and concluded that there must have been a parent mass that fragmented and then largely disintegrated, resulting in a swarm in effect. Associated with this larger mass was a smaller one that struck a few seconds later. The angle of both strikes he calculated to be about 30 degrees (zenith angle, 60 degrees). Shoemaker's 1963 estimate of the total mass of the meteorite was 63,000 tons for an impact velocity of 15 km/sec. The apparent origin of the shock also led him to believe that the zenith angle of impact was small (i.e., high above the horizon). Based on a recent analysis of Shoemaker (1983), the mass of the meteorite is now calculated to be about 300,000 tons. Also, there is now some evidence that the zenith angle of impact may have been large (i.e., low in the horizon) (Shoemaker and Kieffer, 1974). At the magnitude of the impact energy, it makes no difference whether the target material was sand or very hard rock. This observation had been made by Alfred Wegener in 1918 and 1919 when he conducted his impact experiments (Drake and Komar, 1984). Neither does it matter how great is the angle of incidence as long as it is not so great (i.e., large zenith angle) as to approach a glancing blow; the hole the impact "explosion" creates is almost always circular or nearly so. Barringer himself had observed this fact as reported by his son in the *Scientific American* articles written with his collaboration.

Although the structure of Meteor Crater is similar to a nuclear explosion crater as exemplified by the Teapot Ess crater of

Yucca Flat, Nevada, the word "explosion" is not adequate in describing the complex mechanism characterizing an impact event that has been worked out, and its mathematical theory elegantly derived, by Eugene Shoemaker (1963). The latter showed that the shock wave that formed the crater was produced by the impact, not an explosion. But for purposes of discussing the historical and philosophical ramifications of the controversy, and because the effects of a meteoritic strike are so similar to those of a nuclear detonation, the word "explosion" will continue to be used in this paper—but advisedly.

One piece of definitive evidence is the occurrence of the mineral coesite, a dense, high-pressure polymorph of silica that was synthesized by L. Coes, Jr., in 1953. Although it had never been recognized in nature, Nininger suggested in 1956 that Meteor Crater would be a likely place where this mineral could possibly be found because of the presence of silica-rich rocks there that had been subjected to tremendous pressures by the meteorite impact. In 1960, Chao, Shoemaker and Madsen reported their discovery of the first naturally occurring coesite in Meteor Crater. Since then the mineral has been found in other impact craters. It has also been reported in other high-pressure environments as inclusions in diamonds (Smyth and Hatten, 1977), but meteorite craters remain the major places where natural coesite is found. Another feature associated with meteorite impacts is the occurrence of shatter cones, conical fracture patterns in rocks caused by shock waves (Dietz, 1960). These have not so far been found in Meteor Crater. The structure of Meteor Crater, coupled with the shock metamorphism, in Shoemaker's opinion (personal communication) constitutes the definitive evidence for impact origin. Few would now deny the impact origin of this "remarkable geological phenomenon."

FINAL ASSESSMENT

In the light of our present knowledge about crater formation, it seems strange that the crypto-volcanic idea could have had so much influence over the thinking of geologists at the turn of the century. I am puzzled by the reluctance of the geological community to accept the testimony of such distinguished geologists as Fairchild and Merrill. Was it the prestige of G. K. Gilbert that effected the dominance of the steam explosion idea over impact? Does this story demonstrate the power of widely used textbooks? Did the geologists as students learn the concepts as explained by authors like Chamberlin and Salisbury which then could not be dislodged? It is all the more ironic when one considers that Gilbert and Chamberlin were both devotees of the principle of multiple hypotheses, and Gilbert himself was not thoroughly convinced of his own conclusion. It would be a huge injustice, however, to judge Gilbert's many great achievements by this one issue. He deserves to be remembered as the honored scientist that he was. By the same token, Barringer should be given much credit for his perceptiveness and his persistence in his work on Meteor Crater.

Seddon (1970) believes that the controversy over Meteor Crater "represents much that is typical of the way that geologists operate." To him, geologists behave according to the ideal procedure in geological reasoning as expostulated by Gilbert (1896) and Chamberlin (1890) in that they must, and do, keep clearly in their minds many competing hypotheses while their aim is to eliminate all but one of them. Seddon goes on to say that it is characteristic of geologists to be ready to compromise, that they are a cautious race, having a tendency to keep all options open, and are, in fact, trained to do so. Surely this view is somewhat romantic. Several important examples come to mind that speak against such a picture. Most geologists joined in the chorus in ridiculing Alfred Wegener's hypothesis of continental drift. The 1926 AAPG Symposium on Continental Drift hardly presented a panel of speakers keeping their minds and options open. Echoes of geologists' snickering at Harlan Bretz's channeled scablands still resound. Although denial by most thesis professors would be expected, sadly, geology graduate students often instinctively, in favor of survival, suppress their creativity to grasp on to the opinions received from their professors.

Seddon makes some of his most astute statements, however, in the fine print of the footnotes to his article. For example, he refers to the term *laccolith,* the history of which he claims "would make a chapter in the history of fashion in science." In another footnote he states, in relation to the method of multiple hypotheses, "The danger with geologists is that earlier alternatives which could not rigorously be excluded on scientific and logical grounds, may nevertheless be neglected on psychological ones" (Seddon, 1970, p. 3, 5).

The picture of geologists, or any other group of scientists for that matter, jumping on the bandwagon of "fashion" and rejecting alternative hypotheses on "psychological" grounds is not very flattering. But geologists do seem to talk about the method of multiple working hypotheses more than other groups of scientists and adhere to it less. The irony in the Coon Butte story is that in strictly adhering to the principle of multiple working hypotheses, Gilbert went from the true (and now accepted) idea of the crater origin to an untenable one. The principle not only worked counterproductively for Gilbert, it also seemed to have helped delay the acceptance of the impact hypothesis. Other geologists, accepting that Gilbert had penetrated through the paths of multiple hypotheses, felt no need to repeat the same procedure. In keeping with the natural cautiousness or conservativeness of geologists, one explanation for the antipathy toward the impact theory during Barringer's days, therefore, could be perhaps the tendency of geologists to follow the lead of an eminent, proven-successful scientist. History has demonstrated such a possibility. Geologists over age 45 should remember the "accepted" reaction (i.e., laughter or sneers) one must at one time adopt or feign when the words "continental drift" were mentioned (Drake, 1976).

Another explanation for the geologists' resistance to Barringer's impact hypothesis, and one that might be more acceptable to geologists themselves, could be the animosity that they felt toward an outsider. This animosity, whether intentional or not, was certainly felt by Barringer, and it fanned the flames of his bitter and strenuous opposition to what he considered the geological establishment. The bitterness is evidenced in many statements in his letters to his friend Elihu Thomson. An example is the following quote from a 1911 letter (Abrahams, 1983, p. 21):

Some very eminent scientists do not seem to be gifted with imagination and devote their lives to describing things which others have found and proved. . . .I do confess at times to a little irritation that I should meet with so much incredulity and be generally looked upon as a harmless yet unmitigated distorter of truth.

This description of the geologists' image of him is probably close to how they really felt about him, "a harmless yet unmitigated distorter of truth," i.e., a "nut" who is best ignored.

In the three-volume geology text by Chamberlin and Salisbury (1904, 1909) used by college geology classes of that era, the authors cited Gilbert in relation to the Coon Butte Crater but never mentioned Barringer. Being treated thus as if he did not exist must have rankled him. But even when he was mentioned, as in a U.S.G.S. *Bulletin* by Darton (1910), the effect was to anger Barringer. The report of Darton's findings on a reconnaissance of northeast Mexico and northern Arizona was a well-organized and well-written article giving the geography, stratigraphy, structures, and mineral resources of the area. In a low-key style characteristic of modern scientific publications, Darton (1910, p. 73–74) stated:

After an examination of the crater and consideration of all that has been written, I believe we have no evidence adequate to explain its origin. The hypothesis that it was caused by impact of a meteor, as urged by Tilghman, Barringer, Fairchild, and Merrill, is in accordance with some of the features but does not accord with the all-important fact that no meteor is present, as has been demonstrated by many borings. The suggestion of Merrill that the meteor may have been partly volatilized by the heat of the concussion and its residue blown out of the crater by the explosion is difficult to accept.

Darton agreed that the occurrence of the few tons of meteoric iron found in the vicinity was an enigma and then proceeded to say that Gilbert's steam explosion appealed to him "most strongly, notwithstanding its purely gratuitous character." Darton became an avid opponent of the impact hypothesis for the rest of his life. Barringer's associate Holsinger reacted to this rather mildly expressed report on Coon Butte in a letter to Barringer, literally blasting it. Barringer had apparently seen a copy first and was "worked up a bit over it." In addition, he felt that Holsinger's remarks regarding the Darton report, although "rather crude" were "fully warranted." This is what Holsinger wrote on May 5, 1911 (Abrahams, 1983, p. 22–23):

Just to think of the people of the United States and the world, for that matter, being obliged or at least inveigled into depending for their information upon the *damned* [Holsinger's ital.] ignorance and criminal carelessness of this man. To think that we are paying for the printing of a book so utterly unreliable and that the Government will keep in their employ under salary a man who has so little regard for truth and states as facts things he has not investigated and knows nothing about. . . . I know

that half he says is absolutely false . . . but the Gov. with our money will continue to pica type on double sized paper and illustrate with photographs such reports. Fortunately he had no photographs of the Crater as I do not think he has ever been nearer than six miles to it. If he has been here during recent years and writes like this then he should be sent to a blind asylum. . . .He is within the sacred shrine and is immune. Neither the King nor the Geological Survey can do wrong. But what some good men think of the G.S., and justly too, would smell of brimstone if it could be made public.

By the "sacred shrine" and the "King," Holsinger obviously meant the U.S.G.S. and Gilbert, respectively. Here is clearly a case of frustrated outsiders ranting against the establishment.

An interesting aspect of this story is that even when members of the establishment such as Merrill and Fairchild, both distinguished geologists of their day, supported the impact hypothesis, Barringer failed to ally with them, preferring instead to close ranks with two other eminent scientists who were better known in fields other than geology, Elihu Thomson, a physicist and famous inventor, and William F. Magie, a physicist at Princeton.

Thomson (1912) wrote a paper on the fall of a meteorite and concluded that small meteorites might burn up before reaching the earth but large ones like the one that made Coon Butte Crater would survive both the atmosphere and the impact. Merrill, on the other hand, had repeatedly hypothesized the volatilization of the meteorite travelling at high velocity as explanation for the absence of a "main mass," which idea was supported by later investigators like Nininger. Barringer's original reason for buying the crater was, of course, the chance of making a great deal of money. If the meteorite "volatilized" as Merrill suggested, his money-making scheme would vaporize as well. This suggestion, therefore, was unacceptable to him. Merrill's explanation also contradicted Thomson's prediction that the meteorite at Coon Butte would have survived the atmosphere and impact, and the human element is unmistakable in Thomson's caveat to Barringer (Abrahams, 1983, Nov. 22, 1920, letter, p. 173):

Really, if I were you, I would pay little attention to these "prophets" and I think the prospectors [searching for the main mass] are sufficiently in earnest to use common sense instead of chimera as their guide. Unfortunately there is no way of enjoining a man from making a ninny of himself as we think is being done in this case by those who advance such unreasonable notions. . . .My own impression is that there is danger in giving too much importance to Merrill; dignifying his view with too much attention. Wouldn't it be better to ignore it altogether?

Barringer apparently took Thomson's advice. Thomson's next letter to Barringer (Nov. 26, 1920) said: "I am glad that you see the Merrill matter as I do." The friendship between Barringer and Thomson was perhaps typical of some friendships in any age in which loyalty is the great test. In addition, his stubbornness in trying to prove the existence of the main mass and his visions of its money-making potential blinded him to other possibilities, especially when his loyal friend Thomson had his own reputation to guard.

Besides loyalty towards one's friends, therefore, faith in one's conclusions was, and is, another necessary ingredient in keeping a controversy alive. For example, Thomson reassured Barringer that they were justified in having faith in their theory and in the existence of the main mass, because the facts

accord with my experience of a lifetime in forming clear notions of all sorts and ranges of actions. That has been my job for 50 years or more and success has been the result of following the leads given by facts and extending the range of vision by the imagination under control of such facts (Abrahams, 1983, p. 173).

With an ingenuousness and a modesty that probably characterizes the secret self-assessment of every scientist alive or dead, Thomson went on to write (Abrahams, 1983, p. 173):

Without wishing to indulge in self assertion or to pat one's self on the back, I am known in my profession as one whose physical conceptions are clear and crisp; and those who know, realize that my inventive work has often been so far ahead of the time, that the patents would expire just as the art was waking up to use them.

A not-altogether-wrong assessment of the whole phenomenon is possibly that these gentlemen were indeed "ahead of the time," imaginative and bold in proposing and adhering to a hitherto unheard-of explanation for the origin of a geologic feature—long before all the evidence was in. They were right about the meteorite being the ultimate cause of the creation of the hole. But they had not realized the "explosive" nature of the meteorite impact, so that in this respect they were off the target. Although Thomson's paper on the fall of a meteorite was advanced for its time, his and Barringer's view of the meteorite strike was still essentially a geometric one—i.e., a big rock struck the ground, punched a hole, and lodged itself underground. As mentioned earlier, the word *explosion* is used under advisement; as King (1976, p. 89) notes, "Hypervelocity impacts are, for all practical purposes, explosive events." On the other hand, the adherents of the steam explosion explanation recognized the appearance of an explosive event so were not completely wrong. For the present-day explanation of the phenomenon, the reader is directed to the clear exposition by Shoemaker (1960, 1963) of the very complex mechanism.

On a much more fundamental level, another plausible explanation for the early non-acceptance of impact hypothesis is one that has been suggested by Shoemaker (1984). Upon the occasion of receiving the G. K. Gilbert Award, Shoemaker stated:

Although meteorites are now accepted without question as having fallen on Earth, most geologists just don't like the idea of stones the size of hills or small mountains falling out of the sky. While they may concede, at an intellectual level, that such things might happen, at a visceral level it still seems vaguely outrageous. In part this is due, I think, to an overdose of Lyellian uniformitarianism in their geological education, and, in part, to their failure to view the Earth constantly as a member of the solar system.

Time, however, will allow geologists to adjust to accepting, not only intellectually but also philosophically, repeated impacts throughout earth's history, under a much-evolved concept of uniformitarianism from the strict Lyellian sense. In a way, therefore, the resolution of the controversy at Coon Butte could be considered the germ of another revolution in earth science. Meteor Crater represents dramatic evidence of the earth's vulnerability to all the processes that take place in the solar system. Geologists with our picks and hammers staring down on the ground will learn, as a matter of course, occasionally to look over our shoulders toward the sky.

CONCLUSION

The history of geology repeatedly demonstrates that controversies are not usually a struggle between total right and total wrong (Hallam, 1983). The major lines of thought in the Coon Butte Crater story, i.e., Barringer's impact hypothesis on the one hand and Johnson-Gilbert's crypto-volcanic idea on the other, each possessed partial truths. I have given several possible explanations for the long period of resistance against accepting the impact hypothesis. A bit of the truth probably lies in each of these suggestions. The Coon Butte Crater controversy, rather than representing how geologists operate as Seddon (1970) suggests, could better serve as another reminder to geologists to be tolerant of unorthodox ideas. Even when the use of the principle of multiple working hypotheses yielded a wrong conclusion as it did for Gilbert, the latter's tolerance for the tentatively rejected idea was admirable. I believe that during this Decade of North American Geology (DNAG) in celebration of 100 years of geological achievements since the founding of our Society, it is not inappropriate, in addition to looking back over the marvelous advances of our science, to assess also some of our mistakes and how we made them.

ACKNOWLEDGMENTS

I thank Paul D. Komar of the Oregon State Univeristy (OSU) College of Oceanography for his encouragement and many helpful discussions and Michael P. Kinch of the OSU Kerr Library for his assistance in bibliographic search. I also greatly appreciated the critical reading of the manuscript by the following individuals: Claude C. Albritton, Southern Methodist University; Charles W. Drake, chairman, OSU Department of Physics; William G. Hoyt, Lowell Observatory; and Eugene M. Shoemaker, U.S.G.S. Astrogeology Branch, all of whom offered helpful suggestions that immeasurably improved the manuscript. Charles W. Drake and Gene Shoemaker deserve special thanks for their time and patience in helping me understand the present-day concepts of a meteorite impact event.

ENDNOTES

[1]Gilbert's speech, given on 11 December 1895 as president of the Geological Society of Washington to a joint session of the Scientific Society of Washington, was published in *Science* in January 1896. "By special arrangement, through the Joint Commission of these societies," this number of *Science* was mailed to all members. It seems likely, therefore, the address had wide circulation and was read by most members of the much smaller (compared with today's) scientific community that existed then.

[2]I assumed that Barringer bought the site from this passage in his 1905 publication, p. 875: "Since we have come into possession of the property we have found several thousand pieces, in all something over a ton, of various sized fragments of meteoric iron. . . ." According to Shoemaker (personal communication), Barringer only "filed four placer claims on the mineral rights in 1903. These claims were formally patented, and the patents signed by his friend, President T. R. Roosevelt."

[3]Shoemaker showed Hager in the 1960s in Salt Lake City the detailed maps that Shoemaker constructed, and Hager was finally convinced, and he too accepted the impact origin. Shoemaker writes (personal communication), "Hager, in fact, has a very open mind about the origin of Meteor Crater. Shortly after his AAPG paper came out [1953], he learned from my friend Art Flint that we had discussed his paper at a lunch bull session in Grand Junction, Colorado. Art told Dorsey [Hager] that I had wondered aloud whether Meteor Crater might be a maar, as I was then working on the maars and diatremes of the nearby Hopi Buttes. Hager promptly wrote to me, inquiring about this idea, and I wrote back to him indicating that I really had no informed opinion (at that time) about Meteor Crater, but I was very interested in the pumice-like material which he mentioned had been recovered from shafts on the crater floor. By return mail I received a sample of the lechatelierite from Hager, which I immediately sent off for spectrographic analysis. It so happened that I was also working on the geochemistry of Colorado Plateau rocks and was soon able to satisfy myself that the lechatelierite was, indeed, melted Coconino sandstone. At that point (early 1954), I began to develop my own conviction that the Coconino must have been shock melted and that the crater was produced by impact. Hager's openness first led me into the Meteor Crater problem, and it is not at all surprising that he was later convinced by the structural details."

REFERENCES CITED

Abrahams, J., ed., 1983, Heroic Efforts at Meteor Crater. Arizona: London and Toronto, Association of University Presses (Fairleigh Dickinson University Press), 322 p.

Astronomical Society of the Pacific Publications, 1892, A meteor crater, v. 4, p. 37.

Barringer, D. M., 1905, Coon Mountain and its crater: Philadelphia Academy of Natural Sciences Proceedings, v. 57, p. 861–886.

——1909, Meteor Crater (formerly called Coon Mountain or Coon Butte) in northern central Arizona: Read before the National Academy of Science autumn meetings at Princeton November 16, 1909, published by the author, 24 p.

Barringer, D. M., Jr., 1927, The most fascinating spot on earth: Scientific American, v. 137, p. 52–54, 144–146, 244–246.

——1928, A new meteor crater: Philadelphia Academy of Natural Sciences Proceedings, v. 80, p. 307–311.

Chamberlin, T. C., and Salisbury, R. D., 1904, Geology, v. I of III, Geologic processes and their results, 1st ed., and 2nd ed. 1909: New York, Henry Holt and Co., 684 p.

——1909, A College Text-Book of Geology: New York, Henry Holt and Co., 978 p.

Chao, E.C.T., Shoemaker, E. M., and Madsen, B. M., 1960, First natural occur-

rence of coesite: Science, v. 132, p. 220.

Darton, N. H., 1910, A reconnaissance of parts of northwestern New Mexico and northern Arizona: United States Geological Survey Bulletin no. 435, 88 p.

Dellenbaugh, F. S., 1931, Meteor Butte: Science, v. 73, p. 38–39.

Dietz, R. S., 1960, Meteorite impact suggested by shatter cones in rocks: Science, v. 131, p. 1781–1784.

Drake, E. T., 1976, Alfred Wegener's reconstruction of Pangea: Geology, v. 4, p. 41–44.

Drake, E. T., and Komar, P. D., 1984, The origin of impact craters: ideas and experiments of Hooke, Gilbert and Wegener: Geology, v. 12, p. 408–411.

Fairchild, H. L., 1907, Origin of Meteor Crater (Coon Butte), Arizona: Geological Society of America Bulletin, v. 18, p. 493–504.

Foote, A. E., 1892, A new locality for meteoric iron with a preliminary notice of the discovery of diamonds in the iron: American Association of the Advancement of Science Proceedings, v. 40, p. 279–283.

Gilbert, G. K., 1892, Report of Mr. G. K. Gilbert, Chief Geologist: 13th Annual Report of the Director of the United States Geological Survey, p. 98.

——1893a, Report of Mr. G. K. Gilbert, geologist in charge: 14th Annual Report of the Director of the United States Geological Survey, p. 187.

——1893b, The Moon's Face, A Study of the Origin of its Features: Presidential address delivered December 10, 1892, Philosophical Society of Washington Bulletin, v. 12, p. 241–292.

——1896, The origin of hypotheses, illustrated by the discussion of a topographic problem: Science, v. 3, p. 1–13.

Hager, D., 1926, Meteor Crater: Engineering and Mining Journal-Press, v. 121, p. 374.

——1953, Crater Mound (Meteor Crater), Arizona, a geological feature: American Association of Petroleum Geologists Bulletin, v. 37, p. 821–857.

——1956, Additional notes on Crater Mound: American Association of Petroleum Geologists Bulletin, v. 40, p. 161–162.

Hallam, A., 1983, Great Geological Controversies: Oxford, Oxford University Press, 182 p.

Hoyt, W. G., 1983, Meteor Crater: historical note on nomenclature: Meteoritics, v. 18, p. 159–163.

Iddings, J. P., 1914, The Problem of Volcanism: New Haven, Yale University Press, 273 p.

King, A., 1976, Space Geology, An Introduction: New York, John Wiley and Sons, Inc., 349 p.

Leonard, C., 1950, The name of the Barringer meteorite crater of Arizona: Meteoritical Society Contributions, v. 4, p. 309.

Longwell, C. R., 1921, Meteor Crater is not a limestone sink: Science, v. 73, p. 234–235.

Merrill, G. P., 1908, The Meteor Crater of Canyon Diablo, Arizona; its history, origin, and associated meteoric irons: Quarterly Issue of the Smithsonian Miscellaneous Contributions, v. I, p. 461–498.

Nininger, H. H., 1942, A Comet Strikes the Earth: El Centro, California, Desert Magazine Press, 34 p.

——1951, A résumé of researches at the Arizona meteorite crater: The Scientific Monthly, v. 72, p. 75–86.

Pickering, W. H., 1909, The chance of collision with a comet, iron meteorites and Coon Butte: Popular Astronomy, v. 17, p. 329–339.

Pyne, S. J., 1980, Grove Karl Gilbert, a great engine of research: Austin, Texas, University of Texas Press, 306 p.

Rinehart, J. S., 1958, Distribution of meteoritic debris about the Arizona meteorite crater: Smithsonian Contributions in Astrophysics 2, p. 145–160.

Runcorn, K., 1982, The moon's deceptive tranquility: New Scientist, v. 96, p. 174–180.

Seddon, G., 1970, Meteor Crater: a geological debate: Geological Society of Australia Journal, v. 17, p. 1–12.

Sellard, E. H., and Evans, G., 1941, Statement of progress of investigation at Odessa meteor craters: University of Texas Bureau of Economic Geology, 12 p.

——1944, Odessa meteor craters, views in Texas Memorial Museum: Museum Notes, v. 7, p. 13.

Shoemaker, E. M., 1960, Penetration mechanics of high velocity meteorites, illustrated by Meteor Crater, Arizona: in McCall, G.J.H., ed., Meteorite Craters: Benchmark Papers in Geology 36, Stroudsburg, Pa., Dowden, Hutchinson & Ross, Inc., p. 170–186.

——1963, Impact mechanics at Meteor Crater, Arizona, in Middlehurst, B. M.,and Kuiper, G. P., eds., The Solar System, v. 4, The Moon, Meteorites and Comets: Chicago, University of Chicago Press, p. 301–336.

——1983, Asteroid and comet bombardment of the earth: Annual Review of Earth and Planetary Sciences, v. 11, p. 461–494.

——1984, Response upon the receipt of the G. K. Gilbert Award: Geological Society of America Bulletin, v. 95, p. 1001–1002.

Shoemaker, E. M., and Kieffer, S. W., 1974, Guidebook to the geology of Meteor Crater, Arizona: 37th Annual Meeting of the Meteoritical Society, 66 p.

Smyth, J. R., and Hatten, G. J., 1977, Coesite-sanidine Grosypdite from the Roberts Victor Kimberlite: Earth and Planetary Science Letters, v. 34, p. 289–290.

Sutton, S. R., 1984, Crater dated: Nature, v. 309, p. 203.

Thomson, E., 1912, Fall of a meteorite: American Academy of Arts and Sciences Proceedings, v. 47, p. 726–729.

Tilghman, B. C., 1905, Coon Butte, Arizona: Proceedings of the Academy of Natural Science of Philadelphia, v. 57, p. 887–914.

Printed in U.S.A.

Geological Society of America
Centennial Special Volume 1
1985

Development of quantitative geomorphology

Marie Morisawa
Department of Geological Sciences
State University of New York at Binghamton
Binghamton, New York 13901

ABSTRACT

Despite numerous precursors, real quantification in geomorphology did not take hold until after Horton's (1945) paper outlining a new approach to the study of drainage basins. The use of quantitative methods in landform analysis has varied for different geomorphic processes.

Quantitative studies in weathering have been done mostly in the fields of geochemistry, pedology, and soil mechanics. Little has been done to quantify weathering forms. Quantitative analyses on slopes and mass movements have included both processes and morphology, leading to geometric and mathematical models. In contrast, quantitative analyses of arid regions and aeolian forms and processes are immature and tentative. Although great strides have been taken in using physics and mathematics for understanding glacier flow, little has been done on morphometry of glacial landforms. Progress in quantitative coastal research has been most rapid since World War II, although the mathematical basis for understanding coastal processes was laid in the 19th century. As a result of Horton's (1945) provocative study, fluvial geommorphology has perhaps made the most advances in quantitative analyses in geomorphology.

The infusion of ideas from physics, chemistry, biology, engineering, pedology, soil mechanics, hydraulics, hydrology, and geography and the application of statistics and mathematics in analysis accomplished the quantitative revolution. In addition to changing techniques of analysis, quantification has altered the fundamental approach to the study of landforms and affected basic concepts. The dynamic approach (Strahler, 1952) became focused on process and mechanics rather than history and description. Landscape units are treated as open systems in equilibrium where any change in the system causes an adjustment to offset effects of the change.

Electronic instrumentation, remote sensing, and advanced methods of data processing and storage by high-tech computers will open even more vistas for exploration by quantitative geomorphologists of the future.

INTRODUCTION

Geomorphology is defined as "the science that treats the general configuration of the earth's surface; specifically the study of the classification, description, nature, origin, and development of present landforms and their relationship to underlying structures, and of the history of geologic changes as recorded by these surface features" (AGI Glossary of Geology, 1980, p. 259).

Thus, geomorphologists are concerned with the description of landforms: their geometric configuration, size, shape, height, etc., and with the relationship of these morphologic factors to the rock type and structure composing the landform. Moreover, understanding the landscape also involves an analysis of the processes which created, developed, and are still changing the surface features. Therefore, geomorphology has both an historic aspect, and a dynamic one.

The development of the understanding of landforms was traced by Fenneman (1939) through three stages. The first stage is the simple awareness that the land is being worn away. Previous to this, primitive man's thinking about natural processes was geared to catastrophic events. Landslides, floods, etc. were noticed as singular events related to personal or tribal relationships, to some being, or to some supernatural cause. However, the ancient Greeks, Romans and Arabs, as well as other peoples, realized that there were natural processes denuding the land. The Persian, Avicenna, expressed his views in a book written c. 1021.

> . . . certain parts of the ground become hollowed out while others do not, by the erosive action of winds and floods which carry away one part of the earth but not another. That part which suffers the action of the current becomes hollowed out, while that upon which the current does not flow is left as a height. The current continues to penetrate the first-formed hollow until at length it forms a deep valley, while the area from which it has turned aside is left as an eminence. This may be taken as what is definitely known about mountains and the hollows and passes between them . . .
>
> From Adams (1938, p. 334)

The next step, a giant one (Fenneman's stage two), was the perception that rivers carved their own valleys, i.e. the efficacy of running water. Although this idea had been propounded by others before him, this concept was most clearly and unequivocally pronounced by Hutton in 1795 and reaffirmed to the scientific world by Playfair (1802).

The keen eyes which made Leonardo Da Vinci (1452–1519) a great artist also focused clearly on the natural world about him:

> . . . the incessant rains have caused the rivers to increase, and by repeated washing, have stripped bare parts of the lofty summits of these mountains, leaving the site of the earth, so that the rock finds itself exposed to the air, and the earth has departed from these places . . .
>
> From Mather & Mason (1939, p. 1)

Another notable precursor of Hutton was Agricola (George Bauer, 1494–1555) who in his *De re Metallica* discussed many aspects of what we now call geology. In this treatise he clearly states that the earth's surface is being modified:

> Now we can plainly see that a great abundance of water produces mountains, for the torrents first of all wash out the soft earth, next carry away the harder earth, and then roll down the rocks, and thus in a few years they excavate the plains or slopes to a considerable depth; this may be noticed in mountainous regions even by unskilled observers.
>
> From Mather & Mason (1939, p. 10)

Agricola also realized from the slow rate of denudation that erosion had been going on for a long time.

Giovanni Targioni-Tozzetti (1712–1784) not only maintained that rivers carved their own channels but did so in refutation of the Diluvialists:

> From the Ponte a Moriano to the mouth of the Valley of Anchiano, the Serchio runs through a narrow and winding channel and by the impact of its waters is shaping, eroding, and dividing the vast mountain masses. . . .Surely the deep channel of the Serchio from the Ponte a Moriano clear to Anchiano, that of the same river to Tipafratta, those of the Arno through the Valle dell'Infierno and from the Incisa to the Ponte a Rignana, and all the other channels of rivers that I have not yet seen were not thus excavated when these regions were covered with the

waters of the sea. On the contrary, they were not formed until after the lowering of the level of the sea, when the fresh water, flowing toward the lowered sea, began to descend and acquired velocity. Then it was that the mountain streams, arranged in a network, began to erode and have continued to deepen their channels . . .

> From Mather & Mason (1939, p. 74)

Hutton's pronouncements were not readily accepted and it was up to Playfair to convince the "geological" world that Hutton was correct. Playfair's statement referring to the adjustment of tributary gradients to that of the main stream is semi-quantitative in meaning and reflects present-day equilibrium concepts:

> Every river appears to consist of a main trunk, fed from a variety of branches, each running in a valley proportioned to its size, and all of them together forming a system of vallies, communicating with one another, and having such a nice adjustment of their declivities, that none of them join the principal valley, either on too high or too low a level; a circumstance which would be infinitely improbable, if each of these vallies were not the work of the stream that flows in it.
>
> Mather and Mason (1939, p. 132)

Despite Playfair's championing of his cause, Hutton's ideas had to wait until documentation led to general agreement as to their validity. Final acceptance of the ability of the various erosional processes to create landforms led to the cognizance that surface processes do not act aimlessly but work according to a pattern (stage three of Fenneman, 1939). This last stage involved the description of a landscape in terms of its origin and historical development under the influence of geological factors such as structure and lithology.

Each of these stages of geomorphic understanding had to wait until the stage was set for the next advance. The search for pattern and order has led to a fourth stage which we may add to Fenneman's list: the application of quantitative methods to the analysis of the morphology of landforms and to the relationships among the form elements in order to more fully understand the processes going on at the earth's surface and the landforms created by these forces.

It is the purpose of this paper to trace the history of the development of quantitative methods and analysis in geomorphology. A detailed history would fill volumes, the bibliography alone would have considerable weight. Hence, I will only touch upon what I consider to be high points. If your favorite author or paper is omitted, I beg forgiveness for my shortsightedness. Further discussions and lists of papers with quantitative aspects in geomorphology can be found in Rouse and Ince, 1957; Chorley, Dunn, and Beckinsale, 1964; Chorley, 1966; and Salisbury, 1971.

VESTIGE OF A BEGINNING

Although it is generally accepted that the "quantitative revolution" in geomorphology dates from the publication of Horton's classic paper in 1945, quantitative methods have been used for a long time in landscape analyses and the study of geomorphic processes. Rouse and Ince (1957) discuss the gradualness of the change of science in general from the philosophical to the observational.

The first active step in a quantitative analysis often is making measurements. Frontinus (40-103 A.D.) indicated that the Romans measured discharge in order to regulate the distribution of water (in Rouse and Ince, 1957). Measurements of rainfall and runoff, sporadically taken earlier, began to be gathered systematically in the seventeenth century (Chorley, and others, 1964). The work of Pierre Perrault in 1674 (in Adams, 1938, p. 448) represents a landmark as the early use of quantitative data to prove a point. He compared measured yearly precipitation on the Seine River watershed to the measured annual runoff from the basin. The data showed that the rainfall was six times the discharge, thus proving to the doubting Thomases of his day that there actually was more than enough rainfall to support the year-round flow of the river.

Measurement of relief and hypsometric analyses using clinographic curves to describe large areas came into general practice in the latter part of the nineteenth century (Chorley, and others, 1964). At this same time, natural scientists (there were as yet no geologists or geomorphologists) also became intrigued by estimating rates of erosion. Early notable estimates are those of Tylor (1853) and Croll (1867) on erosion by the Mississippi and those by Geikie (1868), as follows

Comparing the measurements which have been made of the proportion of sediment in different streams we shall probably not assume too high an average if we take that of the carefully elaborated Survey of the Mississippi. This gives an annual loss over the area of drainage equal to 1/6000 of a foot (.00005 m). If then a country is lowered by 1/6000 of a foot (.00005 m) in one year, should the existing causes continue to operate undisturbed as now, it will be lowered 1 foot (0.3 m) in 6000 years, 10 feet (3 m) in 60,000 years, 100 feet (30 m) in 600,000 years, and 1000 feet (300 m) in 6,000,000 years.

Mather & Mason (1939, p. 524)

The basis for modern quantitative geomorphology was established by engineers, physicists, mathematicians, and biologists. Quantitative geomorphologists have borrowed concepts and techniques from these disciplines in their efforts to understand more fully the processes which operate on the earth's surface and the landforms these processes generate. In particular, geomorphologists have drawn substantially on the concepts developed in the field of fluid mechanics as they apply to the processes of rivers, waves, wind, and mass movement. Hence, a brief survey of devlopments in this field is appropriate.

The earliest known treatise on hydraulics is that by Ctesias, written in 250 B.C. (Graf, 1971). A volume by Vitrunus Pollo in the first century B.C., as well as their extensive aqueduct works, indicates that the Romans had a good working knowledge of hydraulics. Noteworthy individuals in the field were Gugliemini (1655-1710), the founder of the Italian school of hydraulics, and Bernoulli, whose 1738 treatise on hydrodynamics gave us the laws of fluid movement.

Da Vinci (1452-1519) not only wrote four volumes of his notebooks dealing with water, its character and its work, but used a flume as well as natural rivers for his observations. In 1755 Euler developed equations of motion for a non-viscous fluid, deriving also the "Bernoulli" equation

$$\frac{V^2}{2g} + \frac{\rho}{\gamma} + z = \text{a constant}$$

where V is velocity, g is the gravitational constant, ρ is density, γ is specific weight, and z is elevation.

Chézy, in 1775, compared flow conditions of two similar streams and gave us the essence of the equation that bears his name (although not in the form we now know it). Du Buat, founder of the French school of hydraulics, gave a comprehensive treatment of channel hydraulics in a series of "Principles" from 1779–1816. Among the topics he discussed were the importance of eddies and resistance to the flow through both pipes and natural channels. He set the stage for W. M. Davis by discussing equilibrium and grade, and established velocities required for transport of grains of given sizes.

Some hydraulicists were busy at the shore. Solitary wave motion and velocity of propagation were subjects considered by LaGrange in 1788. Gerstner (1802 in King, 1972, p. 43) treated deep water waves, determining that their profile forms were trochoidal and that the motion of fluid particles in the wave was in concentric orbits.

Thus, the background was established for the modern development of fluid mechanics in pipes and open channels, correlating theory and experiments to arrive at the present-day concepts of turbulence, drag, and boundary theory of Prandtl, Von Karman and others, and for mechanics of waves, setting up basic theory for movement of fluid in the ocean.

QUANTITATIVE GEOMORPHOLOGY AND ITS EFFECTS

When is geomorphology quantitative? Is it quantitative geomorphology to say a mountain is 4,402 m high? Is the following statement quantitative? Or is it conceptual? Thomson in 1877 stated,

It seems to me almost self-evident that wherever there is a slope, be it ever so gentle, the soil-cap must be in motion, be the motion ever so slow.

From Chorley and others (1964, p. 605)

I would define quantitative geomorphology as the application of mathematics and statistical techniques to the study of landforms, their description and the processes by which they are created and changed. It includes also the application of physical and chemical theories to the analysis of landscapes. Morisawa (1971, p. 3) wrote,

. . . The mere collection of measurements does not in itself constitute a quantitative study. Rather, the key lies in the method of approach, in the treatment and analysis of the data. These may be subjected to a rigorous statistical analysis from a simple t-test to a more complicated factor analysis. Or the treatment may be in the form of mathematical modelling. Often the complexity of the problem can be best solved by computer simulation. Nor are quantification of data and mathematical analysis ends in themselves. Quantification is just a technique, a tool, which leads us toward an understanding of a landform and the dynamics of the force or forces which created it.

Quantification of the discipline has affected geomorphology in five basic ways: (1) It has changed the general approach to the study of landforms with new techniques of analysis. (2) It has resulted in a more precise description of the landscape. (3) It has shifted the emphasis from stage to process. (4) It has changed many concepts and (5) It has allowed prediction of some geomorphic events and other practical applications.

The general approach in many present-day geomorphic studies is one in which landforms, their characteristics and the processes involved are treated as a unified system. Mackin (1948) introduced the idea of systems analysis into geomorphology. This approach was strongly advocated by Strahler (1952, p. 934–935):

Geomorphology will achieve its fullest development only when the forms and processes are related in terms of dynamic systems and the transformations of mass and energy are considered as functions of time . . .
Many of the geomorphic processes operate in clearly defined systems that can be isolated for analysis. A drainage system—whether of water or ice—within the geographical confines of a watershed represents such a dynamic system. A cross-sectional belt of unit width across a shore line or sand dune, or down a given slope from divide to stream channel, would constitute another, more limited, type of dynamic system.

The system concept was elaborated upon more fully by Chorley (1962) and by Chorley and Kennedy (1971). Chorley and Kennedy not only define various types of systems but identify and describe various morphological systems common in landscapes. Krumbein (1963) treated beach morphology and processes as parts of a system.

A set of geomorphic features with their characteristics and interrelationship may comprise a system. In general, such systems on the earth's surface are open systems, so complex that it is impossible to make sense out of the interactions unless we treat the system statistically. Thus, many geomorphologists have adopted statistical techniques in analysing surface features. In treating a system statistically we assume randomness. Now, we know that geomorphic systems do not act randomly but, because they act over long periods of time at varying rates, the *effect* is random. Also, in dealing with a large population of measurements of a feature in a complex system which varies spatially, the characteristics of the feature may be treated as stochastic. The complexities are so great that even though the events or features are not random, the net effect is *as if they were* random (Scheidegger and Langbein, 1966). Thus, we now use statistics to determine means, modes, and standard deviation of measurements of surface features; to evaluate significant differences between

landform characteristics and to calculate empirical relationships within the system. But although we treat geomorphic systems and events as if they were random, it is important to remember that they are in actuality, deterministic.

For example, "random walk" models have been used to explain network topology (Howard, 1971b; Shreve, 1966; Smart, 1969), stream gradients (Dacey, 1968), angles of junction (Howard, 1971c), deltaic networks (Morisawa, in press), and slope development (Scheidegger and Langbein, 1966). Such stochastic approaches are simplistic and are based upon assumptions that may or may not be valid so that results should be looked upon critically from a knowledge of geomorphic principles (Howard, 1971d). A random walk investigation can, if properly used, however, give insight into process and form.

Other statistical techniques which have been used are trend surface analysis (Krumbein, 1959). Application of a trend surface analysis was proposed by Svensson (in Chorley, 1972) in studying a series of upraised shorelines. The method was applied by Smith, Sissons, and Cullingford (1969) to determine if there were two or more distinctly different levels on the upraised coast of Perth and if the surface conformed to that of an elliptical dome. King (1969) reconstructed a partially dissected planation surface in the Central Pennines by a trend surface analysis. The method was used by Alexander and Eyton (1973) to distinguish between flood plain and lacustrine terrace levels. Results of a trend surface analysis used to analyze topographic data in the Basin and Range indicated sufficient topographic differences to warrant division of the province into three distinct areas (Lustig, 1969). Multiple regressions (Melton, 1957; Morisawa, 1962), analysis of variance (Chorley, 1957; Melton, 1960) and spectral analysis (Speight, 1965) are other types of statistical analyses that have been applied in geomorphic studies.

However, statistical analyses are not an end in themselves, they are only techniques to be used to gain an understanding of the subject under investigation. As Strahler (1954, p. 1–2) wrote,

. . . . Statistics is frequently completely misunderstood by geologists . . .statistical analysis is not an end in itself; it is a versatile and powerful tool for use in an intermediate stage of . . . quantitative investigations. . . .Statistical methods may be put in the same category as the Brunton compass, the petrographic microscope, the spectroscope, or the X-ray camera. In time, it is to be hoped, the geologist will accept mathematical statistics . . . as an instrument of routine scientific operations. . . .

The statistical concept of minimum variance has enabled us better to grasp the complexity of adjustments a river in equilibrium makes in response to some change in the watershed environment (Williams, 1978).

Furthermore, quantitative treatment of the morphology of landforms has led to more precise, mathematical descriptions. This aspect of quantitative geomorphology has been termed *morphometry*. Quantification of landform characteristics can facilitate understanding, standardize terminology, and make descriptions more meaningful. Quantitative analysis of the hydraulic geometry of river channels has led to a better understanding of the complex relationships and mutual adjustments of all channel characteristics. Differences in the relation of slope angle of talus to size of rock fragments allowed Caine (1969) to distinguish the processes at work. Strahler (1950) took the symmetry of the distribution of slope angles as indicative of a form-equilibrium existing in an open system of erosion and transportation. Tests of the significance of differences in slope means in different areas that Strahler examined showed that (1) lithology was not the only factor controlling slope angle and (2) no effect was evident from direction of slope exposure. These examples illustrate that morphometry is not an end in itself but should be used to understand the surface features more fully and to determine systematic relationships. Many form ratios have become standardized in the vocabulary of geomorphology. For example, drainage density defined by Horton (1945) is known by geomorphologists as the ratio of total stream length to basin area. This ratio can also be used for quantitatively describing the dissection of a region. The gradient index, or SL ratio, is a term proposed by Hack (1973). It is a quantitative way of describing stream gradient which conveys meaning as to the equilibrium of the stream.

Additionally, quantification has shifted the emphasis in geomorphic studies from stage to process. Application of mathematics and basic concepts of physics and chemistry and engineering has added to our knowledge of the mechanics of geomorphic processes. Our understanding of erosion, transportation, and deposition by flowing water has been increased by the mathematical analysis of turbulence by hydraulicists (Graf, 1971). Perception of the mechanics of the flow of glacial ice was brought about by the application of the physics of plastic flow (Nye, 1951). Hindcasting of storm waves by computer programs allowed better analysis of the dynamics of shore processes (Colonell and Goldsmith, 1971). The theory of kinematic waves has been applied to glacial flow to explain surges (Nye, 1958) and to river discharge to explain mechanics of transport (Bagnold, 1966).

Moreover, the mathematical and statistical treatment of quantitative data allow more accurate prediction of geomorphic events and more practical applications to environmental problems. Morisawa (1962) used a multiple regression to predict peak flows from geomorphic parameters. Stall (1971) predicted time-of-travel in stream systems from hydraulic geometry of their channels. Geomorphologists are identifying and delineating slide-prone areas (Colton and others, 1976) for hazard protection. It is thought that identification of threshold conditions for landsliding in terms of amount and intensity of rainfall can allow prediction of landslides in the Santa Monica Mts., California (Campbell, 1975). Practical applications of quantitative geomorphology during World War II led to advances in many fields of geomorphology, particularly in terrain analyses (Wood and Snell, 1960) and coastal processes.

Finally, quantification has led to a change in many basic concepts in geomorphology. The erosion cycle of Davis with traditional stages of youth, maturity, and old age has been challenged by those who state that landforms reach a state of equilib-

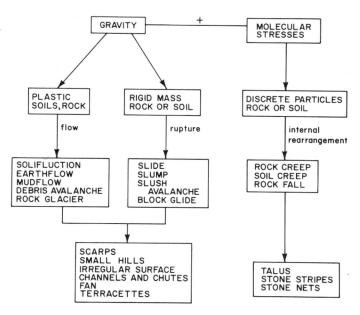

Figure 1. Landform scheme for mass movements using Strahler's (1952) classification, showing response of different materials to stress.

rium which is time-independent (Hack, 1960). It has resulted also in a re-examination of the concept of the graded stream in terms of open systems in a steady state (Morisawa, 1964). Beaches and slopes have also been examined in terms of a steady state in open systems (Tanner, 1958; Hack, 1960).

Strahler (1952) proposed a system of quantitative geomorphology with concepts based on the treatment of geomorphic processes as manifestations of shear stresses acting on some kind of material to give a strain. Stresses involved are gravitational and molecular. Rocks and soil, water, ice, and air are classified by type of material, based on their properties and behavior. Rocks and soils may be rigid, elastic solids, or plastic; water is a Newtonian fluid; ice is a plastic solid, and air is a gaseous fluid. Each type of material has definite characteristics which cause it to react in a specific manner to a given stress. The resulting strain gives rise to some kind of surface feature (Fig. 1).

Quantification has, thus, provided us with a new way of looking at the landscape around us so that we may arrive at a better understanding of its origin and dynamic evolution. In the following pages, I will examine the way in which quantification proceeded in the various geomorphic processes and landforms. By now, so many papers have appeared presenting quantitative analysis of landforms that of necessity the discussion will be limited. I will not discuss quantitative development in Karst landforms and will mention terrain analyses only in passing. These have been adequately covered in Chorley (1972).

WEATHERING

Many of the quantitative aspects of the weathering processes have been the result of investigations by engineers and scientists other than geomorphologists. For example, experiments and measurements of mechanical processes of weathering by ice were carried on by Taber (1930), who was really investigating frost heave on highways. His results aided our understanding of the disintegration of rocks by the force exerted by confined growth of ice crystals.

Our knowledge of chemical weathering has been advanced primarily by geochemical studies of the degradation of feldspar (Loughnan, 1969) and other minerals (Garrels and Mackenzie, 1971). Studies on pressures caused by hydration of anhydrite to gypsum (Winkler and Wilhelm, 1970) confirmed that such stresses can cause rock fracture.

Much quantitative data on weathering was contributed by Jenny (1941) in his classic volume on soil formation and composition. Quantitative studies of soils have been primarily in the realm of soil mechanics (Terzaghi, 1943) although geomorphologists need to know soil characteristics such as shear strength, porosity, Atterberg limits, etc. to understand the behavior of the soil mantle (Strahler, 1952) and its reaction to stress.

The question of the efficacy of heat in mechanical weathering of rocks was investigated in experiments by Blackwelder (1933) and Griggs (1936), who concluded that thermal changes would not induce breakage. However, later measurements have indicated otherwise. Greenwood (1966) inserted remote reading thermometers in boulders in Aden and Somaliland. He discovered that a high temperature gradient was induced by solar heating from the surface inward (0.45 m), a gradient high enough to cause thermal stresses. Temperature readings to depths of 0.63 m beneath the surface of boulders in Barstow, California and calculation of the difference between volume expansion and linear expansion of the surface convinced Roth (1965) that differential thermal stresses were strong enough to cause rock breakage.

As early as 1950, Peltier used the quantitative data of temperature and precipitation to classify geomorphic zones of weathering. Quantitative data on ice wedges (Black, 1954) and on patterned ground (Sharp, 1942; Hopkins and Sigafoos, 1951) added to our knowledge of weathering in cold regions. Lachenbruch (1962) presented the mechanics of ice-wedge polygons using a non-linear viscoelastic model of thermal stress which explains size, depth, and evolution of polygons.

Reiche (1943) devised a quantitative weathering-potential index and a method of graphically depicting the progress of decomposition,

$$WPI = \frac{100x \text{ moles } (Na_2O + K_2O + CaO + MgO - H_2O)}{\text{moles } (Na_2O + K_2O + CaO + MgO + SiO_2 + Al_2O_3 + Fe_2O_3)}$$

Chemical alteration during weathering is shown graphically by plotting composition as the rock decomposes.

Estimates of rates of weathering have been made by a number of researchers. Geikie (1880) calculated weathering rates of different rock types from tombstones in Edinburgh. Goodchild (1890) also used tombstones to compare weathering rates of various limestones. Emery (1960) calculated weathering rates from the amount of talus surrounding the Great Pyramid of Giza.

Methodology used by Hessink (1938 in Ollier, 1969) in determining rate of weathering in Holland was by amount of calcium carbonate leaching. More accurate methods of calculating weathering rates can now be done with the advent of radioactive dating (Ruxton, 1966).

Quantitative studies in weathering have been mostly in the fields of chemistry, pedology, and soil mechanics. Little has been done by geomorphologists quantifying weathering forms.

SLOPE STUDIES

In order to understand the development of landforms, we must be able to describe their geometry in quantitative terms. Slopes are readily quantified, for obviously an angle is involved and it can be easily measured, as can the height and length of slope. According to Young (1972) the earliest slope surveys were those of Tylor (1875). Lake (1928) mathematically compared slopes he measured both to arcs of circles and to parabolas. Slope surveys were used by Fair (1947) and Savigear (1952) in their investigations of slope evolution, and later by many others.

An important function of measurements is to supply data for statistical analyses from which geomorphological conclusions can be drawn. The impetus for the use of statistical techniques in analysis of slope data came from Strahler (1950). For slope angles the type of statistics commonly used are frequency distributions (Fig. 2), mean and maximum angles, and some measure of the variation of the angles (i.e. coefficient of variation, standard deviation, etc.).

In his study Strahler (1950) used the symmetry of the frequency distribution of valley-wall slope angles as evidence of form-equilibrium. He suggested that a high correlation between channel gradients and valley-wall slopes indicated an adjustment among the morphologic components of the terrain. He also carried out a significance test of the differences in slope means of different localities which showed that (1) characteristics other than lithology may control slope angles, (2) orientation of the slope has little effect on the inclination, and (3) slopes modified by other than fluvial processes and with no stream erosion at the base decline in angle.

Field studies by Leblanc (1842) gave maximum scree angles and measurements by Fair (1948) related particle size to angle of scree surface. Behre (1933) observed talus above timber in the Rocky Mts. and, from measurements, stated that the angle of rest (repose) depended upon the size of the rock fragments composing the slope. The relation of particle size to scree angle has been investigated by many geomorphologists (Table 1). Experiments by Van Burkalow (1945) with artificial materials provided data on relationships between angle of repose and various characteristics of rock fragments (Table 2). Rapp (1957, 1960) carried out extensive quantitative studies of scree in the Karkevagge, making continuous recordings of slope processes as well as measurements of components of slope forms.

Another quantitative technique that has been applied to the study of scree is fabric analysis (Caine, 1967 and McSaveney,

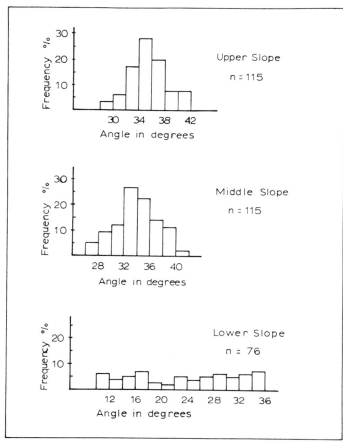

Figure 2. Frequency distribution of slope angle for middle, upper and lower talus surfaces, Buckskin Gulch, Colorado. From Morisawa (1966).

TABLE 1. ANGLE OF REPOSE AND FRAGMENT SIZE

	Higher Angle with inc. size	Lower Angle with inc. size
* Auerback		x
n* Leblanc	x	
* Linck		x
* Thoulet		x
* Van Burkalow (1945)		x
n Morisawa (1966)		x
n Behre (1933)	x	
n Caine (1967)	x	

*From Van Burkalow, 1945.
n - natural slopes; unless noted thus, the materials are artifical

TABLE 2. FACTORS INFLUENCING ANGLE
OF REPOSE*

	Change in angle of debris slope
Increasing fragment size	Lower angle
Increasing fragment density	Lower angle
Increasing height of fall	Higher angle
Increasing roundness	Lower angle
Increasingly smoother surface	Lower angle

*Van Burkalow, 1945.

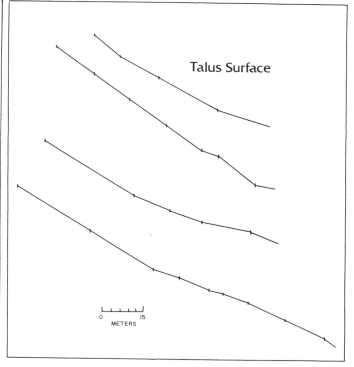

Figure 3. Representative talus profiles, McClellan Mountain, Colorado. From Morisawa (1966).

1971). They determined that rockfall debris has a random orientation, whereas a definite fabric results from mass transport.

Both Tinkler (1966) and Blong (1970) studied the frequency distribution of scree slope angles. Morisawa (1966), from examination of scree slopes in Colorado, confirmed that talus profiles are concave upward overall but are composed of short linear segments (Fig. 3). Caine (1969) used profiles of scree slopes in the New Zealand Alps to derive empirical equations relating process to rate of accumulation of scree. Accumulation rates were measured in relation to distance from the head of the talus. The processes taking place were rock fall and slush avalanching, each resulting in different scree profiles.

One of the intriguing questions in geomorphology is the change of slopes with time. Qualitative models of slope evolution were devised by many geomorphologists, including Davis (1909b) and Penck (1924). Penck's model, although qualitative, was the precursor of later quantitative geometric models of parallel cliff retreat, such as those of Fisher (1866), Lehmann (1933), and Gerber (1934), as well as those models dealing with nonparallel retreat (Bakkar and LeHeux, 1947; Van Dijk and Le Heux, 1952; and Bakkar and Strahler, 1956). Fisher (1866) provided the first mathematical model of slope development. This model of Fisher was based on the parallel retreat of a cliff with the accumulation of debris at the base. The profile of the bedrock surface, buried and protected by rock fragments, would be expressed by a parabolic equation, the form depending on the angle of repose of the talus and initial cliff height. The combined Fisher-

Lehmann model (Young, 1972) dealt with the parallel recession of a free cliff face with scree accumulation at the base, which gradually buried the free face. Under presumed conditions the slope of the rock face could be expressed as

$$\frac{dx}{dy} = \frac{h \cot \beta - (\cot \alpha - (c \cot \alpha - \cot \beta) \, y)}{h - cy}$$

where x and y are the vertical and horizontal coordinates of the rock face respectively, β is the cliff angle, α is the angle of slope of the scree, h is height of the free face and $c = 1 - c_1 c_2$, and c_1 is the increase in scree volume and c_2 is the proportion of scree remaining.

The convexity and concavity of slope profiles, long noted by geologists (Gilbert, 1909) can be calculated. White (1966) showed that slope profiles plot as straight line segments on logarithmic graph paper. These segments define three form elements: (1) an upper convex form where gradient increases with length; (2) a middle slope where gradient is constant with length; and (3) a lower concave form where gradient decreases with length. The general equation for these segments is

$$H = k \, L^n + H_o$$

where H is change in elevation, L is length and k, n, and H_o are constants for the given slope element.

Equations can also be fitted to the complete slope profile. One such is by Hack and Goodlett (1960):

$$H = c \, L^f$$

where H is height to a given point of the face, L is length to the point, and c and f are empirical constants.

Such models are based on assumptions so they are only as valid as are the assumptions. For example, Scheidegger (1960), in constructing his model of slope evolution, arbitrarily set a rate of recession and assumed that it acted vertically and normal to the face. Using the given rate, he then developed an evolutionary model for three cases: (1) where recession was uniform from top to bottom of the face, (2) where recession was related to the altitude, and (3) where recession depended on the angle of slope. Another caveat is that it is not necessarily true that if the final form of the model is similar to that of real slopes observed, then the evolution depicted in the model is correct. The geographic principle of *equifinality* states that the same geometric form can be achieved in different ways.

Whereas geometric models analyze changes in the form over time, process-response models are based on the interaction of process and form through time. In making the process-response model an initial form is assumed and the mechanics of a given process are applied. Cullings' (1963, 1965) process-response models of slope development were based on the assumptions that (1) the downslope movement of individual grains was random, (2) the main process at work on slopes was soil creep, and (3) that rates of grain movement were related to slope angle. The resulting equation for the configuration of the slope was

$$\frac{du}{dt} = a \, \frac{d^2u}{dx^2}$$

where u is elevation, x is distance downslope, t is time, and a is the coefficient of diffusion.

If one assumes a different process, a different evolutionary model is derived. Souchez (1961, discussed in Young, 1972) assumed that plastic flow was the process by which the slope was molded. In this case, the rate of flow of a plastic material would depend upon the stress applied and the viscosity of the material flowing. The stress, in turn, depends upon the specific gravity of the material and the angle of slope. Young (1963) and Ahnert (1966) computed models with slope form evolving through the multiple processes of soil creep, slope wash, and solution. Ahnert's model was improved in 1973 by a Fortran program using the processes of weathering, fluvial transport, and mass movements. The program allowed varying combinations of both processes and quantitative parameters governing the processes. The model, thus, was able to adjust its simulation to various materials, structures, climates, and lithologies, as well as process.

Techniques for trend surface analyses were developed by Troeh (1964) uniting both profile and plan form. The trend surface can be expressed by

$$Z = a + bX + cY = dX^2 + eXY + fY^2 + gX^2 + \ldots \ldots$$

where Z is elevation and X and Y are coordinates mutually perpendicular in the horizontal plane.

Slope maps were developed by Savigear (1956, 1963) and Waters (1958). Morphological units are mapped according to slope inclination. However, use of such maps is limited because studies have shown operator variance results in different maps of the same area drawn by different compilers. Speight (1968) proposed maps that were compilations of four maps, each showing different factors such as slope angle, profile curvature, plan curvature, and unit watersheds.

In arid regions both badland slopes and pediment surfaces have been the subject of quantitative studies. Tator (1952) simply recorded the gradient of pediment surfaces. The morphology of pediment surfaces has been examined by Bryan (1922) in reference to the longitudinal profile; and by Blackwelder (1931), Gilluly (1937), and Melton (1965) all of whom dealt with slope angles. Gilluly (1937) related pediment surface profiles to the gradients of streams running subparallel to them. Langford-Smith and Dury (1964) suggested the fit of the pediment profile to a logarithmic curve. Dury and others (1967) derived a best fit regression for log tan pediment slope on length. The relation between pediment-surface-slope and relief length ratio was established by Cooke (1970). According to Mammerickx (1964) there is no statistically significant relationship between mean slope of pediments and lithology in the Western United States, although Mabbutt (1966) found that pediments in Australia had a steeper slope on schists than on granite.

Based on profile studies, geometric models of parallel slope retreat of pediment surfaces have been proposed by Lawson (1915), and Ruxton and Berry (1961). Schumm (1956a) and Smith (1958) provided measurements of pediment erosion on badland slopes showing that rates of removal of material varied with slope and that the rectilinear slopes receded in a parallel fashion. Using rose diagrams, Sharp (1940) showed graphically that 40 percent of pedimentation is the result of lateral planation by feeder streams. Measured changes in the surface of pediments by Schumm (1962) showed progressive regrading by sheetwash. Doehring (1970) attempted a quantitative technique to distinguish between pediment and fan surfaces on maps using counts of contour crenulations. Recent work by Campbell (1980) has used trend surface analysis to present measurements of badland erosion. In addition, the graphed surfaces were photographed in succession at time intervals, showing actual pulses of change in the surface configuration of the slopes. Processes worked intermittently.

Quantitative analysis of slopes has come a long way!

MASS MOVEMENT

Quantitative studies of mass movements differ from qualitative ones by the emphasis on the response of different types of materials to stress. Geomorphologists have moved from qualitative descriptions of slope failure to an analysis of rate and mechanism of the failure, following Strahler's (1952) recommended approach. This change has come from the application of soil mechanics and engineering principles of failure and safety factors to mass movements, adapted from texts such as Terzaghi (1943), Terzaghi and Peck (1948), and Scott (1963).

The early quantitative analyses of mass movements are found in engineering literature. For example, Rankine (1857) developed the theory of limiting stability of materials on a slope, where stability depends on cohesion of the material and its angle of shearing resistance. Strahler (1952) suggested analysing mass movements by considering the type of stress acting on a material, which in turn responds to the stress with a deformation depending on its strength. Materials are classified according to their behavior under stress as plastic, a rigid solid, or discrete particles (Fig. 1). A plastic (viscous) material reacts to stress in various ways, depending on its viscosity which offers resistance to flow. As viscosity increases, plastic materials can resist deformation until a threshold (yield strength) is reached. Rigid solids fracture when a threshold of stress is reached. Loose individual grains move under stress, maintaining their individuality. Thus, the shear strength of a material is its resistance to deformation. So, if shear strength equals the stress applied, no deformation (movement) will take place. Hence, we can define a safety factor, F, where

$$F = s/\tau$$

and s is shear stress and τ is stress applied. If F is equal to or greater than 1, no failure occurs; if F is less than 1, deformation takes place. This little equation, the basis for quantitative analyses of downslope movement is derived from Coulomb (1776) who defined the strength of material, s, as

$$s = c + \sigma \tan \phi$$

where c is cohesion, σ is the normal stress and ϕ is the angle of internal friction.

MacDonald (1915) explained the steepness of natural slopes by relating the angle of slope to the shear strength of the rock in the walls of canyons he examined. Brice (1958), in determining the origin of loess-mantled slopes in Nebraska, related shearing resistance to the shearing force. This ratio defined the possibility of failure: if it were less than 1, failure occurred. Lohnes and Handy (1968) used a stability analysis on slab failures in loess, calculating the critical height (H_c) of the face beyond which it fails.

$$H_c = \frac{4c}{\gamma} \tan (45 + \phi /2) - Z$$

where c is a cohesive factor, γ is the specific weight of the material, ϕ is the angle of internal friction and Z is depth of tension cracks. A form of this equation had been used by Fellenius (1927) and Taylor (1937) to derive stability charts which related the steepest stable slope to the total stress on the slope. These types of charts were brought up to date by Bishop and Morgenstern (1960), who used effective stress (pore pressure) instead of total stress. Carson and Kirkby (1972) give an excellent quantitative discussion of failure analysis.

In order to understand the behavior of the earth's soil mantle and rocks, geomorphologists had to learn some concepts already familiar in soil and rock mechanics: bearing capacity, liquid limits, plasticity, Atterberg limits, etc. Properties such as these are important in determining the response of soil and rocks to stress. For example, the natural water content of earthflows is greater than the liquid limit of the soil in the flow (Bjerrum, 1954). It is for this reason that Strahler (1952) based his approach on type of material and its properties.

Jones and others (1961) carried out a statistical analysis of factors measured in connection with a large landslide in the Columbia River valley. They found they could express the stability of the slopes in terms of a regression equation

$$Y = 0.022 \log x_1 + 0.003 \log x_2 + 0.009 \log x_3 + 0.007 \log x_4$$

where x_1 is the original slope of the bank, x_2 is the percent of the bank submerged by the river, x_3 is height of the terrace surface above the water, x_4 is a ranking of the ground water level where a high water table equals 1, and a low water table equals zero. The stability condition was $y = 0.0106$, i.e. sliding occurred if $y > 0.0106$. This paper illustrates the type of empirical studies done on slope failure.

Strahler (1956) devised a method of slope analysis that ena-

TABLE 3. MORPHOMETRIC INDICES FOR MASS MOVEMENTS

Index	Definition	Use
Depth/length	D/L	Indicates process group
Flowage	$100(W_x/W_c-1)(L_m/L_c)$	Expresses relative speed of movement and fluidity
Tenuity	L_m/L_c	Indicates cohesiveness
Displacement	L_r/L_c	Low value indicates instability measures safety factor
Viscous flow	L_f/D_c	Indicates viscosity
Dilation	W_x/W_c	Describes spreading Indicates water content

where D is maximum depth of scar of the movement

L is maximum length from tip of displaced mass to top of scar

W_x is width of convex part of displaced mass

W_c is width of concave scar of the movement

L_m is length of displaced mass

L_c is length of concave scar

L_r is length from top of headscarp to head of displaced mass

L_f is length of overlap of displaced mass on the scar

D_c is depth of concave scar

bles prediction of instability. An isosinal map is constructed. This indicates the shear forces acting on the soil particles because shear depends on gravity and sin of the angle of slope, Θ

$$\tau = Fg \sin \Theta$$

Hence one can deduce areas which may be slide prone by examining the isosinal map.

The pattern of curved failure surfaces and stability analyses of slides (slumps) is usually found by the Mohr circle method (see Carson, 1971, for a discussion of this and other surfaces of failure in natural materials). The method of slices (Hough, 1957 and Wu, 1966) was used by Swanston (1970) in a stability analysis of debris avalanches in Alaska. These papers are typical of many similar ones in the field.

Except for descriptive measurements of size (length, width, area, volume) very little real quantitative work (as defined in this paper) has been done on the morphometry of mass movements. Heim (1932) described rockslides in terms of the height of the top of the scar to the horizontal distance from that point to the endpoint of the slide. Skempton (1953) used the depth/length ratio to distinguish among surface slips, deep rotational slips, and slumps. Crozier (1973) proposed six morphometric indices which he considered could be used to distinguish among various types of mass movements (i.e. slides and flows) (Table 3). The most important index was the depth/length ratio. The flowage index,

which applied to plastic movements, expresses the fluidity of the material and the relative speed of deformation. The tenuity index also indicates fluidity but, in addition, reflects the effects of the angle of slope and microrelief of the surface. The displacement index is important in that low values indicate instability and the possibility of further movement. A statistical analysis of the significance of these factors showed that the depth/length ratio was the most diagnostic process. Crozier's paper is an excellent example of quantification of the morphology of mass movements in a meaningful way. Not only do the morphometric indices tell us about the character of the material, they also give indications of the processes operating.

Blong (1973) measured 19 characteristics of 92 landslides and carried out a statistical analysis of the correspondence of these different attributes to classification (type of movement). He found that no single characteristic can be used to distinguish among debris slides, debris flows, or debris avalanches, that there is no correspondence between classification and the attributes of those in the classification, and that groupings done by the computer on the basis of similarity of properties was not similar to traditional classifications.

Rates of downslope movements have been measured for all types of mass movements in plastic materials (Table 4). Creep has been measured intensively with increasing sophistication through the years. Davison (1889) used the simple method of visible markers on the surface. More recently, electronic methods

TABLE 4. RATES OF MOVEMENT, PLASTIC FLOWAGE PHENOMENA

Type of Flow	Rate of Movement	Source
Continuous creep (Surface)	0.10-2.23 cm/yr 0.079-0.152 cm/yr 0.03-0.047 cm/yr	Kojan, 1967 Swanston and Swanston, 1977 Lewis, 1975
Rock glacier	87.9-123 cm/yr	Wahrhaftig and Cox, 1959
Earthflow	96 cm/day 0.109 cm/day 30 cm/day 0.08 cm/yr	Hadley, 1964 Crozier, 1973 Varnes, 1949 Swanston and Swanston, 1977
Mudflow	16 (max.) ms^{-1} 4.4 (max.) ms^{-1} 7-30 ms^{-1}	Curry, 1966 Sharp and Noble, 1953 Janda and others, 1981
Debris flow	10-16 ms^{-1} 60-120 ms^{-1}	Brunsden, 1979 Ericksen and others, 1970

(Everett, 1963), flexible tubes and strain gauges (Kojan, 1967) have been used. Rudberg (1962) depicted the velocity profile of creep from the surface downward. Brundsden (1979) summarized figures for rates of creep.

Solifluction rates were measured by Taber (1929), who drew up empirical curves for his data; by Williams (1957), by Rapp (1960), and by Washburn (1967). In a classic study, Rapp (1960) measured denudation rates of a number of downslope processes. Data indicated that, in his study area, rates for earth slides and mudflows were greater than for avalanches, which were greater than for rockfalls.

Mathematical and theoretical models have been developed to explain processes at work in downslope movement of earth materials. Culling (1963) based his mathematical model for creep on heat conduction equations, whereas Kirkby (1967) built a theoretical model of creep on principles of soil mechanics. Bagnold (1956) elaborated upon a mechanism proposed by Heim (1932) for downslope movement of material on the theory of inertial grain flow. That is, grains are dispersed in an interstitial fluid in order to flow. In such a situation there is little resistance to flow and the slide or flow can move great distances. Another mechanism that has been proposed is movement on a cushion of air (Shreve, 1968). Hsü (1975), after analysing some catastrophic debris streams by a morphometric approach, came to the conclusion that they moved as a mass of cohesionless grains dispersed in a fluid medium, thus supporting Bagnold's thesis. He also introduced the parameter "excessive travel distance" as a measure of mobility of the mass, at the same time pointing out the correlation of excessive travel distance (T) to size of the mass as,

$$T = L - H/\tan 32°$$

where L is horizontal movement and H is height of fall.

Thus, we see that the approach to the study of mass movements has become quantitative due primarily to input from soil and rock mechanics of which geomorphologists have recently taken advantage. Consequently, we have gained a great deal of insight into the behavior of earth materials under stress. Quantitative analysis shows promise!

EOLIAN AND ARID REGION LANDFORMS

Although quantitative aspects of windformed and arid region landscapes are restricted in the geomorphic literature per se, development of the principles of aerodynamics is important for an understanding of the mechanics of eolian processes and the landforms produced. These principles elucidated by Prandtl and others in the early 1900's applied both to flowing water and air. With the advent of the airplane and World War II, research in aerodynamics of airfoils became critical and resulted in a spurt of research. Some of this was applied to geomorphic studies (Zingg and Chepil, 1950).

Although early work on eddies in the lee of dunes was done by Cornish in 1914, this aspect of aerodynamics in natural situations is still a problem because quantitative measurements are difficult to make.

The most important geomorphic work in this field is the book by Bagnold (1941) in which he clearly and quantitatively set forth basic principles of erosion, transportation, and deposition by wind. His work has been refined by others (e.g. Chepil and Woodruff, 1963) but it still remains the main source of information. Kawamura (1951) further elaborated on the relation of sand transport rates to the wind force, which he divided into wind shear and shear from the impact of saltating grains. Ford (1957) carried out a physical-mathematical analysis of the motion of an individual sand grain in transport by wind and found that its trajectory agreed with that of grains observed in a wind tunnel. Wind tunnel experiments of Kawamura, Ford, and others have been the main source of quantitative data on eolian processes.

On the other hand, quantitative descriptions of naturally-occurring dunes (dimensions, orientation) have been recorded by many geomorphologists including Long and Sharp (1964) on dunes in the Imperial Valley, California; Hastenrath (1967) on barchans in Peru, and general geometry by Wilson (1972). Hack (1941) based a semi-quantitative classification of dunes on supply of sand, wind strength, and vegetative cover. Relations of wind direction and dune orientation have been studied by Cooper (1958), McKee and Tibbitts (1964), Inman and others (1966), and Sharp (1966).

Dune migration rates and mechanics of movement were discussed by Finkel (1959) and also by Williams (1964), Long and Sharp (1964), and Sharp (1966), for example. In a simulation model of erosion and deposition of barchan dunes using data gathered over a dune near the Salton Sea, Howard and others (1977) found the resulting patterns resembled closely that in the field. In addition, he found the shape of the dunes was a function of grain size, velocity, saturation of the flow, and variability in wind direction. In a study of dunes at White Sands National Monument, McKee and others (1971) measured migration of as much as 11.5 m per year.

Quantitative studies of arid region landforms have been morphological rather than process oriented. Key papers have been those by Bull (1964, 1968) on alluvial fans. Bull established the form of fan profiles and related fan morphometry to characteristics of the drainage basin of the fanhead feeder stream. For example

$$A_f = cA_b{}^n$$

where A_f is area of the fan, A_b is drainage area of the fanhead stream and c and n are empirical constants which vary from fan to fan, depending on other basin characteristics such as lithology and structure.

Denny (1967) in relating fan size to source area applied Hack's theory of dynamic equilibrium in open systems to the study. Hooke (1968) attributed variation in rates of deposition on fans he studied to incomplete equilibrium. The steady-state model would require that the area of a fan be proportional to the debris added per unit time,

$$V = \frac{dT}{dt} A_f$$

where V is volume of debris added per unit time, dT/dt is the rate of increase in thickness during that time and A_f is fan area. Bull also concluded that fan slope was closely related to gradient of the feeder stream. This was confirmed by studies of Hooke (1968) and Denny (1965). Fan slope has also been related to increasing fan area by Eckis (1928) and Melton (1965); to drainage basin area by Bull (1964); and to discharge and lithology by Hooke (1968). Hooke derived his conclusions from both field measurements and laboratory experiments.

Troeh (1965) showed that fan shape could be expressed by an equation

$$Z = P + SR + LR^2$$

where Z is elevation at any point, P is elevation at the apex, S is slope of the fan at the apex, R is radial distance from P to Z, and L is half the change of slope along R.

One can see the nebulous nature and immaturity of quantitative analyses in the geomorphology of eolian forms and processes in contrast to fluvial and slope systems. Most of the quantitative work has been done in the laboratory. As stated by Ritter (1978, p. 309), "... significant geomorphic work is simply not available ... because definitive, quantitative data concerning eolian processes and features are woefully few."

Much more work needs to be done in this field to develop techniques of measurement in the field and quantitative analysis of form and process.

GLACIAL LANDFORMS

One of the earliest quantitative measurements in glacial geomorphology was the experiment of Simony (1846) when he determined the amount (minimum limiting value) of erosion by cutting a 3 mm deep scratch in the rock over which the Dachstein glacier advanced in 1856 and obliterated the mark (in Embleton and King, 1968, p. 249). Other measurements of erosion by moving ice were done by boreholes (de Quervain, 1920, and Lutschg, 1926), as mentioned by Embleton and King, p. 249.

Erosion rates were estimated from measurement of the silt loads of glacial meltwater streams by Reid (1892) on the Muir Glacier, Alaska and by Thorarinsson (1939) in Iceland. Davis (1909a) calculated the deepening of valleys in Wales by measuring elevations of hanging valleys compared to that of the main valley. Matthes (1930) used the same technique to determine deepening of Yosemite Valley. Boulton (1974) presented a detailed general theory of glacial erosion processes by analysing erosion, transportation, and deposition as a continuum. His mathematical theory of abrasion beneath a glacier allowed both formation of small and large-scale streamlined bedrock forms (for example, roches moutonnees and p-forms) and production of depositional features (flow till, drumlins). By assuming an isotropic linear elastic response of bedrock to shear stresses from ice pressures, he calculated the loci of failure in the rock, creating erosive features.

Not many workers have yet used morphometry of glacial landforms in describing glacial features or in applications of form to process. Exceptions are in studies of drumlins and cross-valley profiles. Chorley (1959) compared the plan form of drumlins to airfoils using the equation of a lemniscate loop to define the streamlined shape. Reed and others (1962) also attempted to quantify drumlin shape, applying the equation of an ellipsoid. They also carried out an examination of the spacing and orientation of groups of drumlins in upstate New York and in the Boston area. Differentiation of the origin of drumlins in the Eden Valley and Solway Lowland areas was accomplished by establishment of a statistically significant variation in elongation of the two groups and from orientation data (Embleton and King, 1968).

The form of cross-profiles of glacial valleys, qualitatively defined as U-shaped, was quantitatively expressed as a parabola by Svensson (1959).

$$y = ax^b$$

where y is elevation of the surface at a point, P, and x is the distance to P from the valley center.

Till fabric analysis is a common technique used in studying glacial deposits. The fact that stones were not randomly dispersed in glacial till was noted as early as 1859 and was discussed by Miller in 1884 (Embleton and King, 1968, p. 307). Since then till fabric measurements have been treated statistically to determine ice flow and process of deposition. Holmes (1941) examined in detail the effect of till-stone shape on its orientation. Fabric analysis was used by West and Donner (1956) to differentiate till sheets. Harrison (1957) examined the fabric from moving ice. Both Hoppe (1959) in Sweden and Harrison (1957) in Minnesota used fabric analysis to determine the direction of ice flow and origin (process) of drumlins in the fields they studied. Andrews (1963) established different origins for a single moraine by analysing the orientation and dip of stones on two sides of the moraine. Three dimensional till fabrics were discussed by Andrews and Shimizu (1966).

Finally, the mechanism of flow of glaciers, despite intensive study, was not fully understood until a breakthrough by Nye (1951, 1952) who applied concepts from physics and materials science to glacial ice. He explained variations in velocity and thickness by the theory of compressive and extending flow. This was followed by theories of glacier sliding presented by Weertmann (1957, 1962) and Nye (1970). The dynamics of cirque glaciers were discussed by Lewis (1960).

Quantitative analysis in glacial landforms is in its infancy. Although great strides have been taken in using physics and mathematics in the understanding of the flow of glacial ice, except in the case of drumlins, little has been done in morphometry and its application to mechanics of formation of the given feature.

COASTAL GEOMORPHOLOGY

Progress in coastal research has been rapid since World War II because of the impetus given to the study of waves and beaches as a result of the need for amphibious landings on Pacific Islands, the development of monitoring instrumentation, and recent environmental problems relating to the loss of beaches. With the advent of quantitative methods, coastal geomorphologists have responded by looking at beaches as complex open systems with feedback in response to processes operating on the system. The use of large quantities of data gathered by more sophisticated instrumentation is now allowed by computers which can handle multivariate systems and quickly process large amounts of data (Fox and Davis, 1971a and 1971b).

The basis for understanding coastal processes was laid in the 19th century with proposals on wave theory. Airy (1845) developed equations for ideal, deep-water wave velocity. Both his work and the wave theory of Stokes (1847) dealt with waves of finite height and assume irrotational flow and mass transport. Levi-Civita (1925) and Struik (1926) were later workers who derived the mathematics of such waves.

Taking a different set of premises, Gerstner (1802) and Rankine (1863) developed the theory of the deep-water trochoidal wave with a circular orbit. The theory of the shallow-water trochoidal wave was worked out by Gaillard (1904).

Dynamics of wave generation were discussed by Jeffreys (1925), who attributed the development to differential pressure, and by Stanton and others (1932), who used drag theory. Turbulent flow generated waves according to Sverdrup and Munk (1947), Eckart (1953), and Phillips (1957). Miles (1965) combined both turbulent pressure fluctuations and shear.

Many of these studies were done in wave tanks because of the difficulty of field measurements. More recently, numerous monitoring techniques (shipborne recorder, wave gauges, flowmeters, stereophotography, remote sensing) have been used to record wave phenomena and beach characteristics. The vast amount of data collected has resulted in many advances, such as in wave forecasting, which is both empirical and theoretical. Sverdrup and Munk (1947) and Bretschneider (1952), as well as other researchers, used many wave records to derive empirical equations for forecasting based on significant wave height and period. Pierson and others (1955) used both empirical data and theory to formulate their forecasting methods on the basis of analysis of wave spectra. Darbyshire (1952, 1955) derived equations relating wave energy to wave and wind characteristics, basing his results on spectral distribution during storms. In 1963, Darbyshire and Draper published graphs for forecasting waves from wind speed, duration, and fetch; the graphs gave maximum wave height and period.

A great many studies have dealt with sediment transport on the beach and offshore. These are ably discussed in King (1972). Noteworthy quantitative contributors to the field have been Trask (1952) and Ingle (1966). However, according to Derbyshire and others (1979), most papers using tracers to measure rates of littoral drift are qualitative in technique rather than quantitative and do not take into account regimes of nearshore flow, waves, or seaward topography. Hence, the real value of tracer techniques is questionable. Mass transport has been discussed in theoretical terms by Longuet-Higgins (1953) and investigated experimentally in wave tanks by many, including Bagnold (1947) and Russell and Osorio (1958). Inman and Bagnold (1963) applied principles of physics and hydrodynamics to determine theoretically the movement of particles on a beach.

The generation, properties, and effects of longshore currents have been discussed by many researchers. Putnam and others (1949) used a mathematical analysis in a momentum flux approach to generation of longshore currents and estimate of velocities. Inman and Bagnold (1963) based their analysis on the principle of the conservation of mass. The work of these and others were evaluated by Galvin in 1967. More recent work has been done by Komar and Inman (1970) and Komar (1971).

Saville (1950) modelled the movement of sand on the shore and found that velocities were close to those derived by Putnam and others (1949). Among those who have made field measurements of longshore sand transport are Shepard and Inman (1950) off the California coast, McKenzie (1958) on Australian beaches, and Harrison and others (1968) at Virginia beach on the Atlantic coast of the United States.

Morphologically, the beach profile has been a major topic of discussion by coastal geomorphologists. In a model experiment, Bagnold (1940) determined that angle of slope of a beach was independent of wave depth or amplitude but confirmed Meyers' (1933) conclusion that it depended instead upon size of grains comprising the beach material. The relation of beach slope to sand size has since been confirmed by field studies by Bascom (1951), Wiegel (1964), Krumbein and Graybill (1965), and others. However, in analysing profiles of Cape Cod beaches, Shepard (1950) determined that steep storm waves affected the beach to give it lower gradients. And Rector (1954) found that for any given size material the beach profile depended upon wave steepness. The equilibrium beach profile was discussed by Bruun (1954) from a study on the coast of Jutland and by Zenkovich (1967). A quantitative treatment giving empirical equations relating profile changes to 14 process variables was performed as a linear multiple regression by Harrison (1969).

Mathematical curves have been developed for describing the geometry of beach forms. Hoyle and King (1958) determined that a stable equilibrium beach had a curved outline, in plan form, of a circle with an angle of 0.25 radians subtended by the radii to the bench end. Dicken (1961) in a study of Oregon beaches, proposed a C/P Index, the ratio of the chord length (C) of the beach to the length of a perpendicular (P) constructed at the midpoint of the chord. The equilibrium form would have a C/P Index of 15. McLean (1967) described beach forms in New Zealand in terms of a radius of curvature (R) so that

$$R = \frac{C^2 + 4P^2}{8P}$$

where C and P are defined in the same way as Dicken defined them. Silvester (1960) proposed that equilibrium beaches assumed a half-heart shaped form as determined in wave tank experiments. He also suggested this form could be simulated in nature to stabilize beaches which were being eroded.

Yasso (1964) simulated the form of the spit at Sandy Hook, N. J. by using the equation of a log spiral:

$$r = C^{\theta \cot \alpha}$$

where r is the radius vector of the log spiral center at an angle θ and α is the constant angle between a radius vector and the tangent at the point (Fig. 4). The work of Yasso is problematic because it may not really describe other spits and has not been of any use in understanding the process of spit formation. McCullagh and King (1970) simulated the formation of a recurved spit using a stochastic model based on various orientations of wave

approach and current generation. By varying the independent variables, they were able to construct spits with different dimensions and shapes. This type of model could prove able to allow discrimination of variables determining spits of different sizes and shapes

Size and spacing of beach cusps were measured by Johnson (1919) and related to wave properties. Evans (1938) also studied characteristics of beach cusps on shores of Lake Michigan and pertinent wave properties. Mii (1958) derived data relating cusp dimensions and wave characteristics on beaches of Japan. These various relationships have been verified by Kuenen (1948) and Otvos (1964). Longuet-Higgins and Parkin (1962) related spacing of cusps to length of swash. King (1965) found a tendency for size of cusps to be dependent upon both the size of sand on the beach and wave characteristics.

Rates of beach erosion, transportation, and deposition have been frequently and sporadically noted, but regular quantitative measurements and analyses are more recent (King, 1972). These have been made possible by data gathering of several agencies such as Scripps Institute at La Jolla, California, the Hydraulic Research Station at Wallingford, England, and the Coastal Engineering and Research Center (CERC), Fort Belvoir, Va. Their monitoring systems have promoted the advance of quantitative analysis of coastal forms and processes.

FLUVIAL GEOMORPHOLOGY

As can be seen in most geomorphology textbooks, rivers have been the focus of attention more than any other geomorphic agent. Nascent quantitative notions of the Romans, da Vinci, and others, as well as the development of fluid dynamics, have already been mentioned. Quantitative fluvial geomorphologists owe a great debt to those researchers in hydrodynamics, as well as to hydrologists (Horton, for example, was a hydrologist). Geomorphologists interested in fluvial processes and landforms must of necessity delve into the hydrologic and hydraulic literature.

Early twentieth century fluvial geomorphologists' thinking was dominated by the ideas of William Morris Davis, in particular his theory of the erosion cycle. Despite the experimental and empirical work of Gilbert (1914) and Rubey (1933, 1938), Leighley's (1932, 1934) discussion of turbulence and erosion processes in rivers, and Hjulstrom's (1935) monumental study on the River Fyris, the time was not yet ripe for quantitative analyses to become popular.

The springboard for the "quantitative revolution" in fluvial geomorphology was Horton's (1945) provocative study of drainage basins, the ideas of which stemmed from his earlier (1932) paper. In these papers he set up a methodology of ordering based on Gravelius (1914) and principles for a quantitative approach to the study of running water—overland flow as well as channelized. He covered many aspects, pointing the way for numerical analysis of networks, basin systematics, origin and development of drainage basins, and "laws" of network composition (Table 5).

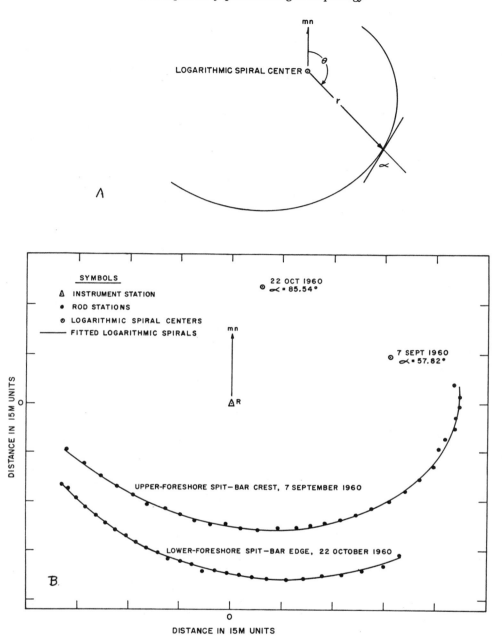

Figure 4. A. Nomenclature for a log spiral simulation. B. Log spiral
fitted to actual spit. Both from Yasso (1964).

Because the time was ripe, his 1945 paper stimulated two geo-morphologists to follow his lead: Strahler and Leopold. Strahler and his students at Columbia University began a program of drainage basin studies which verified, applied and enlarged upon the principles and techniques that Horton had proposed. Published papers in this early period included: analyses of badland erosion in terms of Horton's principles (Smith, 1958; Schumm, 1956a and b), application of dimensional analysis to fluvial land-forms (Strahler, 1958), river profile equations (Broscoe, 1959), correlation structure of morphometric parameters of watersheds

and feedback loops (Melton, 1958), and application of a Horton analysis relating basin characteristics to peak discharge in a multiple regression (Morisawa, 1962). Their work brought about a change in concepts of the development of fluvial landforms and equilibrium in drainage systems, and in their approach they demonstrated the usefulness of statistical techniques.

Meanwhile, at the U.S. Geological Survey, Leopold and his co-workers were detailing the relationship of discharge and river channel morphology (Leopold and Maddock, 1953; Leopold and Miller, 1956; Wolman, 1955). This group, by their investigation

TABLE 5. HORTON'S "LAWS"

Law	Empirical Relation
Stream numbers	$N_u = R_b^{s-u}$
Stream lengths	$T_u = T_1 R_L^{u-1}$
Basin areas	$A_u = A_1 R_a^{u-1}$
Stream gradients	$S_u = S_1 R_s^{s-u}$

N_u = number of streams of order u

R_b = bifurcation ratio

s = highest order stream in the basin

T_u = mean length of stream of order u

T_1 = mean length of first order stream

R_L = length ratio

A_u = mean area of basin of order u

A_1 = mean length of first order basin

R_a = area ratio

S_u = mean gradient of stream of order u

S_1 = mean gradient of first order stream

R_s = gradient ratio

of hydraulic geometry stimulated the dynamic analysis of river morphometry (Fig. 5). They also emphasized the probabilistic approach in examining stream patterns (Leopold and Wolman, 1957). They applied thermodynamics and the concept of entropy to river basins (Leopold and Langbein, 1962) and introduced the concept of minimum variance in determining morphologic adjustments in stream channels (Langbein and Leopold, 1964) and in variation of streamflow direction in meandering (Langbein and Leopold, 1966).

With the advent of the stochastic approach, random walk and simulation models became an accepted method of analysis (Leopold and Langbein, 1962; Schenck, 1963; Smart, Surkan, and Considine, 1967; Langbein and Leopold, 1966; Milton, 1967; Howard and others, 1970; Smart, 1970; Howard, 1971a and b; Pickup, 1977) but there were also warnings as to their pitfalls (Howard, 1971d).

Questioning of the Hortonian "laws" and concepts also occurred, primarily by Shreve (1966, 1967) and Smart (1967, 1969), who examined networks in terms of probabilities and suggested they were random in composition. Woldenberg (1966) upheld the validity of Horton's "laws" and insisted on spatial order in fluvial systems as hexagonal hierarchies (Woldenberg, 1969).

By 1970 the quantitative revolution in fluvial geomorphology was well on its way and beginning to influence geomorphic thought and methods of analysis outside the Strahler and Leopold groups. In fact, as early as 1961, Scheidegger wrote a geomorphology text that was highly statistical and mathematical and based on concepts of physics. We will now concentrate on specific aspects of fluvial landforms and dynamics and examine how they were influenced by quantification.

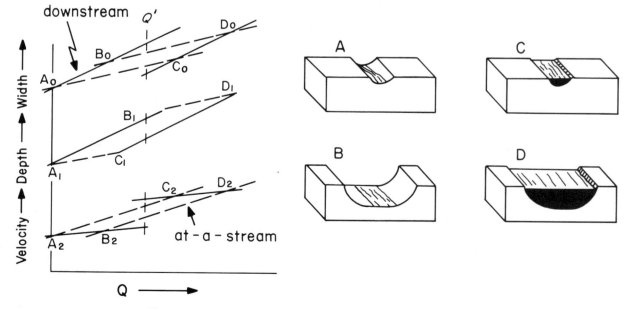

Figure 5. Generalized hydraulic geometry of a stream channel. A-B and C-D represent downstream changes; A-C and B-D represent at-a-station changes with increasing discharge. After Leopold and Maddock, 1953.

Quantitative characteristics of overland flow elucidated by Horton (1945) have been verified and amplified by Emmett (1970) and Pearce (1973), who used hydraulic geometry exponents to express changes in flow downslope. Previously, erosion rates of overland flow had been quantified in terms of change in kinetic energy and in terms of distance downslope (Smith and Wischmeier, 1962).

Systematic changes in rill or gully systems formed by overland flow with development into full-fledged drainage networks have been the subject of numerous topological studies. Both Schumm (1956b) and Morisawa (1964) traced changes in developing networks, verifying much of the qualitative framework theorized by Glock (1931). Much topological information and interpretation was supplied by Smart in a series of papers (1969, 1970, 1972, and 1978). Various terminology and methods of attack were given by Smart in these papers. Others also introduced new terms and approaches: James and Krumbein (1969) proposed the use of *cis* and *trans* links; Mock (1971) distinguished exterior links as source and tributary source links and interior links as bifurcating or tributary bifurcating. Flint (1980) proposed a series of terms which would distinguish interior branching from exterior branching of tributaries. The concepts of topologic distance (Jarvis, 1972) or path length (Werner and Smart, 1973) and network diameter allow valid comparisons of networks of the same diameter (Jarvis, 1976b).

Attempts to ascertain controls on network development and configuration include numerous papers by Abrahams (1975, 1976, 1977 among many), Woldenberg (1969, 1971) and Flint (1980). Orientation of stream patterns has been quantitatively investigated by Judson and Andrews (1955), by Morisawa (1963), by Jarvis (1976a) and by Cox and Harrison (1979), all of whom found a high correlation between direction of stream flow and rock structure.

Topology has also been studied using angles of junction, as approach suggested by Horton. Lubowe (1964) verified Horton's equation

$$\cos \Theta_j = \tan S_m / \tan S_t$$

where Θ_j is angle of junction, and S_m and S_t are gradients of the main stream and tributary, respectively. In addition Lubowe found that the angle increased as order of the receiving stream increased. Howard (1971c) studied angles of junction of ephemeral streams and determined the relation between the angles and discharge. Smart and Moruzzi (1972) set up nomenclature and procedures for analysing distributary networks of deltas. Their concepts were adapted to braided nets by Krumbein and Orme (1972). Their methods were used by Morisawa (in press) who compared natural and random model deltaic patterns.

Hydraulic geometry studies, because they were easily done, had a large bank of data already collected, and could give results from analysis, became popular. Only a few early and key papers will be mentioned here—a bibliography of hydraulic geometry studies (compiled by Williams, 1979, pers. comm.) consisted of

86 entries. Richards (1973, 1976a) raised questions concerning the influence of the shape of a cross section on the hydraulic geometry. Knighton (1974) found that real hydraulic geometry data did not form a parallel family of curves as theorized by Leopold and Maddock (1953). Investigations of hydraulic geometry of meandering and braided channels were carried out by Fahnestock (1963) on braided outwash of the White River, Washington, by Church (1972) on an Icelandic sandur, and by Knighton (1975a) in Britain. Richards (1976b) differentiated pool and riffle sequences by the hydraulic geometry.

The question of channel forming discharge is still an open one. Because this problem is ably discussed by Dury (1973) it will not be addressed here. The question of the influence of bed and bank material on channel morphology is also an open one despite the general acceptance of the statement by Schumm (1960) that the width/depth ratio of a channel depends on the percent of silt/clay in the perimeter. His mathematical equations were attacked as invalid by Melton (1961) and by Miller and Onesti (1979).

The relation of channel morphology to sediment transport was elucidated by Rubey (1952) in an equation. Leopold and Maddock (1953) related the hydraulic geometry to sediment load. Rhodes (1977), in a ternary diagram, indicated the relation of hydraulic geometry exponents to hydraulic transport variables. Andrews (1979a and b) showed the hydraulic geometry exponents to be constrained by minimum total work and uniform expenditure of energy. Attempts to determine the theoretical hydraulic geometry exponents were made by Smith (1974), Williams (1978), and Hey (1978). As yet such attempts have been indeterminate.

Early quantification of the longitudinal profile of rivers (Jones, 1924) resulted in exponential equations such as that proposed by Green (1934) in terms of distance from the mouth, x, and elevation, y, in general form

$$x = e^{-y}$$

or Shulits (1941), generalized as

$$y = e^{-ax}$$

where a is an abrasion coefficient, after the concept of Sternberg (1875) that size of grains decrease downstream. Hack (1957) formulated a semi-logarithmic equation to relate fall of a stream channel (H) to length (L),

$$H = k \log_e L + c$$

where k and c are empirical constants. He later (1973) related the constant K to SL (termed the gradient index) where S is slope of a reach and L is the distance from the head to the midpoint of the reach. He maintains that the SL value is related to rock type or size of bed material over which the river is flowing. Langbein and Leopold (1964) derived the theoretically most probable stream gradient based on assumptions of stream power and the rate of work.

The irregularity of the longitudinal profile because of the alteration of pools and riffles was pointed out by Tinkler (1970). The regularity of spacing was established by Keller (1972) and Keller and Melhorn (1978). Cherkauer (1973) studied the relation between pool/riffle spacing and power expenditure by streams.

A final morphological aspect to be considered is the plan-form of rivers. Engineering attempts to provide stable irrigation canals in India led to regime equations relating discharge to parameters of canal plan-form morphology (Lacey, 1930; Inglis, 1949; Blench, 1966). Experimental work on meandering includes studies by Friedkin (1945). Both Friedkin and Wolman and Brush (1961) dealt with the initiation of bends and their growth. The flume experiments of Schumm and Khan (1972) modelled development of meanders in resistant material. Hooke (1975) used models to examine distribution of boundary shear at bends. Leopold and Wolman (1957) and Brice (1964) defined sinuosity indices for the plan-form of a channel. Schumm (1963) related sinuosity of rivers on the Great Plains to bed and bank material. Dury (1964, 1965) gave a comprehensive treatment to the relation of discharge to morphology of river bends. Spectral analysis has been applied to meander bends by Speight (1965) and Ferguson (1977).

Field measurements on river bends and channel migration are more difficult than model studies. Sundborg's (1956) study of the Klaralven is a classic to be read by all fluvial geomorphologists. Measurements of flow velocities at bends have been made by Jackson (1975), Bridge and Jarvis (1976), and Hickin (1978). Field surveys of bend migration were done by Wolman (1959) and Leopold (1973) on Watts Branch, Maryland; by Coleman (1969); by Daniel (1972); and by Hickin (1978), to cite a few.

Theoretical quantitative studies of meanders include that of Bagnold (1960) relating flow resistance to curvature; Langbein and Leopold (1966) who proposed that meanders are sine-generated curves which express a minimum variation in change of flow direction; and Thakur and Scheidegger (1968) using the stochastic basis that meandering involves random variations in direction of flow. Yang (1971a) proposed that meandering results from an attempt by the river to equalize its time rate of energy expenditure. Despite much quantitative work, it seems, the reason a stream meanders is still a puzzle.

Adjustment of all morphological characteristics to change has been explained by Leopold and others (1964) as a result of the effort of a river to balance the tendency toward least work with equalizing the distribution of stream power. Yang (1971b) proposed in a thermodynamic analog that river morphology was an expression of the potential energy change downstream. Kirkby (1977) suggested that a river adjusted its configuration in order to carry its load efficiently. In other words, the limiting process determining morphology is sediment transport. This was supported by Yang and Song (1979) who attributed morphologic adjustment to a stream power concept different from that of

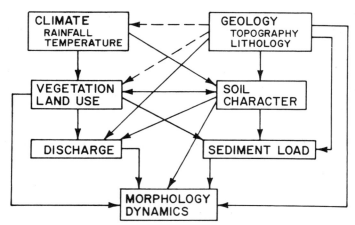

Figure 6. Constraints on river dynamics and morphology. From Morisawa and Vemuri, 1975.

Leopold and others (1964). Yang and Song maintained that river morphology expressed the relation between sediment transport and unit stream power: at equilibrium the river adapted that configuration and plan-form which would allow the minimum rate of energy dissipation while carrying its load. We are back to Mackin's definition (1948, p. 464) of a graded stream:

A graded stream is one in which, over a period of years, slope is delicately adjusted to provide, with available discharge and with prevailing channel characteristics, just the velocity required for transportation of the load supplied . . .

As a result of Horton's (1945) paper the drainage basin has become a fundamental unit of landscape study (Gregory and Walling, 1973). Morphological and dynamic process aspects of rivers and their watersheds are now routinely analysed in quantitative terms by statistical techniques. Such studies have changed many concepts and led to the formulation of new principles of landform origin and development by fluvial action. Rivers are now viewed as systems in equilibrium such that any change in the system will cause a response in the dynamics and morphology so as to offset the effects of the change. Morphometric variables are adjusted, within the constraints of rock type and structure and other environmental factors (Fig. 6), so that there is a minimum variance in the change of each variable during adjustment. The response varies depending upon the time frame (Schumm and Lichty, 1965) and any lag or relaxation time, as well as the basin environment. Threshold concepts (Schumm, 1973) have taken on new meaning in the context of catastrophic theory (Graf, 1979) and both can aid in our understanding of the complexities of changes in the fluvial system.

Measurements, statistical analyses, and empirically-derived equations have aided in detailing the magnitude of the impact of humans on rivers. Numerous studies have documented the effects of logging and deforestation on the hydrology of a watershed. Bates and Henry (1928) measured an increase of 15 percent in the mean annual discharge after deforestation in the Rio Grande

TABLE 6. REPRESENTATIVE PAPERS SHOWING EFFECTS OF URBANIZATION
ON WATERSHED HYDROLOGY

Date	Author	Area Studied	Remarks
1960	Carter	Near Washington, D.C.	Derived coefficient of impermeability
	Waananen	Syracuse, N.Y.	Hydrographs
	Savini and Kammerer	----	General effects
1964	Harris and Rantz	Permanente Cr., Calif.	RF/RO relation
	Seaburn	East Meadow Br., L.I.	Changes with sewering
1965	Crippen	Sharon Cr., Calif.	Changes in unit hydrograph
	James	Morrison Cr., Calif.	Digital computer program
1966	Espey, Morgan, and Masch	Austin, Tex.	Unit hydrograph, peak flows, multiple regression
1967	Wilson	Jackson, Miss.	Increase in peak discharge
1968	Leopold	Many	General summary
1969	Brater and Suresh	Detroit, Mich.	Peak flows
	Narayana, Riley, and Israelson	Waller Cr., Tex.	Analog computer simulation
1970	Anderson	N. Virginia	Derived coefficient of impermeability
1973	Knott	Colma Cr., Calif.	Storm runoff
1974	Stankowski	New Jersey	Estimating equation for flood peaks
1979	Knight	United Kingdom	Peak and total flows
1982	Gregory and Madew	Many	Review

National Forest; Hoyt and Troxell (1934) determined that there was a 29 percent increase in mean annual discharge after a California watershed was devastated by fire; clearcutting resulted in an augmented water yield of 18 percent in a watershed studied by Hewlett and Helvey (1970). Hoover (1944) showed that, although logging in Coweeta basin, N.C., increased discharge significantly the first year after cutting, the increase in the second year was less. This result was confirmed by long-term monitoring which showed that, after reforestation, experimental watersheds in New York gradually decreased peak discharges as well as mean annual discharges (Ayer, 1968). Hornbeck (1973) found that the initial increased streamflow after logging progressively disappeared over a 4-year period. Many such quantitative studies of this type have been made by hydrologists of the United States Department of Agriculture and United States Forest Service.

Similarly, it has been shown that urbanization of a watershed drastically affects its hydrologic regime, increasing mean annual discharge, total runoff, and peak discharges. Table 6 lists a sampling of such papers.

Although strictly speaking these studies are not geomorphological, they are an important aspect because the increased discharge increases erosion by overland flow, causes rivers to erode their channels, and increases the overall sediment yield. Increased sediment yield brought on by logging, deforestation, and changing land use has been quantitatively documented by Van Dijk and Vogelgang (1948), Shallow (1956), Guy (1965), Ursic and Dendy (1965), Douglas (1967, 1969), Wolman (1967), Fredriksen (1970), Walling and Gregory (1970), Brown and Krygier (1971), Megahan and Kidd (1972), Costa (1975), and Yorke and Herb (1978), among others. Vice and others (1969) and Reed (1980) are among those who dealt with changes in sediment yield brought on by highway construction. Adjustments in the hydraulic geometry in response to increased discharge from urbanization was affected by channel enlargement. Table 7 gives a list of some studies documenting augmentation of channel size as watersheds

TABLE 7. REPRESENTATIVE CHANNEL ENLARGEMENT STUDIES

Date	Author
1967	Wolman
	Wolman and Schick
1969	Vice, Guy, and Ferguson
1972	Hammer
	Leopold and Emmett
	Miller and Viessman
1973	Leopold
1974	Gupta and Fox
1975	Morisawa and Vemuri
1976	Fox
	Hollis and Luckett
1977	Park
1978	Yorke and Herb
1979	Knight
	Morisawa and La Flure
	Nunnally and Keller
1982	Gregory and Madew

are developed. These reports indicate that there may be a threshold of urbanization below which channel enlargement may not occur, that there is a lag time before the river responds by increasing channel size, and that the short and long term adjustments vary (Fig. 7).

Disruption of stream equilibrium by interference directly in the channel (channelization) was illustrated by Daniels (1960) and Emerson (1971). Winkley (1973) estimated the effects on morphology caused by man-made cutoffs on the Mississippi River. Keller (1975) pointed out deleterious effects of artificial alterations in a river channel as did Morisawa and Vemuri (1975) in several cases. Warner and Pickup (1978), in a detailed study of channel morphology, showed the effects of human actions on a tidal estuary in Australia.

Quantitative predictions of channel erosion below dams was made by Tinney (1962) and Komura and Simons (1967). Dams on rivers of the Piedmont-Coastal Plain region of southeastern United States have not only changed their rates of seasonal transport but have decreased the amount of sediment brought to the sea (Meade and Trimble, 1974). The hydraulic geometry of a river channel above a reservoir is different from that below (Petts, 1977), indicating an upset in river equilibrium. Measurements before and after installation of gully-control structures showed how streams adjust their longitudinal profiles after a disturbance

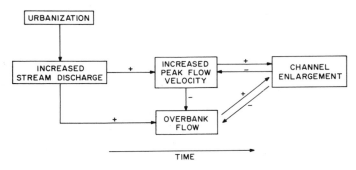

Figure 7. Short and long term adjustments of a river to urbanization. From Morisawa and LaFlure (1979).

(Woolhiser and Lenz, 1965). Recent studies of the effects of alterations in gravel-bed rivers are discussed by various authors in Hey and others (1982).

Thus monitoring, measuring, and analysis of the data have conclusively demonstrated how man affects rivers and their watersheds and the magnitude of the effect. This knowledge can be used to ameliorate or avoid deleterious results (Keller, 1975; Morisawa and Vemuri, 1975).

The revolution Horton started in fluvial geomorphology has had far-reaching effects in methodology and concepts. Use of quantitative techniques has greatly aided our understanding of "problem" streams as we alter watershed environments.

NO PROSPECT OF AN END

In tracing the development of the quantification of geomorphology in this paper I have limited myself primarily to the literature written in English. Moreover, I have omitted quantification in karst landforms, the history of which has been adequately discussed by Williams (1972) and McConnell and Horn (1972). These are among the acknowledged shortcomings of this review.

However, I have tried to demonstrate how a quantitative approach with timid beginnings has changed the techniques of analysis and altered many basic concepts in the field of geomorphology. I have indicated how indebted we geomorphologists are to the engineering disciplines such as soil mechanics, hydraulics, and hydrology, as well as to the basic sciences, physics, chemistry, and biology, in borrowing and adapting some of their principles and methodology to the study of landforms.

From small sporadic beginnings to more and more sophisticated methods of treatment, quantitative geomorphology has grown until a bibliography of quantitative papers since Horton (1945) would comprise a volume. These studies, too numerous to mention them all here, effected a change in geomorphic thinking. No longer are landscapes timebound; they have become timeless. Landform systems achieve an equilibrium which is time-independent (Hack, 1960). Strahler's (1952) proposal to treat landforms as parts of open systems and as a result of stress acting on some type of material which responds by deforming has insinuated itself into our geomorphic consciousness so that it is permeating present-day analyses.

The application of statistical techniques and mathematical formulae have become commonplace in geomorphic literature. Morphometry, used to quantitatively define and describe landforms more precisely, contributes to our understanding of the origin and evolution of surface features. New methods of data gathering and storage are available: electronic instruments, landsat imagery and other types of remote sensing, and high-tech computers. With these ever more sophisticated methods of monitoring, measuring, and handling vast amounts of data there seems to be no prospect of an end to the growth of quantitative geomorphology. It is hoped that future advances will be used to aid our understanding of geomorphic processes and the landforms they create, thus proving the worth of the quantitative revolution.

REFERENCES CITED

Abrahams, A. D., 1975, Initial bifurcation process in natural channel networks: Geology, Vol. 3, No. 6, p. 307–308.

——1976, Evolutionary changes in link lengths: further evidence for stream abstraction: Transactions of the Institute of British Geographers, v. 1, p. 225–230.

——1977, The factor of relief in the evolution of channel networks in mature drainage basins: American Journal of Science, v. 277, p. 626–646.

Adams, F. D., 1938, The birth and development of the geological sciences: New York, Dover Publication Inc., 506 p.

Ahnert, F., 1966, Zur Rolle der elektronische Rechenmaschine und des mathematischen Modells in der Geomorphologie: Zeitschrift fur Geographie, v. 54, p. 118–133.

Airy, G. B., 1845, On tide and waves, in Encyclopedia Metropolitana: v. 5, p. 241–396.

Alexander, C. S. and Eyton, R. J., 1973, Trend surface analysis of flood plain and alluvial terraces in Southern Illinois and Western Kentucky: Geological Society of America Bulletin, v. 84, p. 1069–1073.

American Geological Institute, 1980, in Bates, R. L., and Jackson, J. A., eds., Glossary of Geology: Falls Church, Virginia, American Geological Institute, 749 p.

Anderson, D. G., 1970, Effects of urban development on floods in northern Virginia, U.S. Geological Survey Water Supply Paper: 2001C, 22 p.

Andrews, E. D., 1979a, Scour and fill in a stream channel, East Fork River, Western Wyoming: U.S. Geological Survey Professional Paper 1117, 49 p.

——1979b, Hydraulic adjustment of the East Fork River, Wyoming, to the supply of sediment, in Rhodes, D. D., and Williams, G., eds., Adjustments of the fluvial system: Winchester, Mass., Allen and Unwin, p. 69–94.

Andrews, J. T., 1963, Cross-valley moraines of the Rimrock and Isotog river valleys, Baffin Island. A descriptive analysis: Geographical Bulletin, v. 19, p. 49–77.

——and Shimizu, K., 1966, Three-dimensional vector technique for analysing till fabrics: discussion and Fortran program: Geographical Bulletin, v. 8, p. 151–165.

Ayer, G. R., 1968, Reforestation with conifers—its effects on stream flow in central New York: Water Resources Bulletin, v. 4, p. 13–24.

Bagnold, R. A., 1940, Beach formation by waves: some model experiments in a wave tank: Journal of the Institute of Civil Engineers, v. 15, p. 27–52.

——1941, The physics of blown sand and desert dunes: London, Methuen & Co., 264 p.

——1947, Sand movement by waves: some small-scale experiments with sand of very low density: Journal of the Institute of Civil Engineers, Paper No. 5554, p. 447–469.

——1956, The flow of cohesionless grains in fluids: Philosophical Transactions of the Royal Society of London, Series A, v. 249, p. 235–297.

——1960, Some aspects of the shape of river meanders: U.S. Geological Survey Professional Paper No. 282E, 10 p.

——1966, An approach to the sediment transport problem from general physics: U.S. Geological Survey Professional Paper No. 422-I, 37 p.

Bakker, J. P. and LeHeux, J.W.N., 1947, Theory on central rectilinear recession of slopes I and II: Nederlandse Akademie van Wetenschoppen Proceedings, v. 50, p. 959–966 and 1154–1162.

——and Strahler, A. N., 1956, Report on quantitative treatment of slope recession problems: International Geographical Union, Commission pour l'Etude des Versants, Rapport 1, p. 30–41.

Bascom, W. N., 1951, The relationship between sand size and beach fore slope: American Geophysical Union Transactions, v. 32, p. 866–874.

Bates, C. G. and Henry, A. J., 1928, Forest and streamflow experiments at Wagon Wheel Gap, Colorado: U.S. Department of Agriculture, Monthly Weather Review Supplement No. 30.

Behre, Charles H., Jr., 1933, Talus behavior above timber in the Rocky Mountains: Journal of Geology, v. 41, p. 622–635.

Bishop, A. W. and Morgenstern, N., 1960, Stability coefficients for earth slopes: Geotechnique, v. 10, p. 129–150.

Bjerrum, L., 1954, Geotechnical properties of Norwegian marine clays: Geotechnique, v. 4, p. 49–69.

Black, R. F., 1954, Permafrost—a review: Geological Society of America Bulletin, v. 65, p. 839–855.

Blackwelder, E., 1931, Desert plains: Journal of Geology, v. 39, p 133–140.

——1933, The insolation hypothesis of rock weathering: American Journal of Science, v. 26, p. 97–113.

Blench, T., 1966, Mobile-bed fluviology—a regime theory treatment of rivers for engineers and hydrologists: Edmonton, Alberta, T. Blench and Associates Ltd., 300 p.

Blong, R. J., 1970, The development of discontinuous gullies in a pumice catchment: American Journal of Science, v. 268, p. 369–383.

——1973, Relationships between morphometric attributes of landslides: Zeitschrift fur Geomorphologie Supplementband, v. 18, p. 66–77.

Boulton, G. S., 1974, Processes and patterns of glacial erosion, in Coates, D. R., ed., Glacial geomorphology: State University of New York at Binghamton, Publications in Geomorphology, p. 41–87.

Brater, E. F. and Suresh, S., 1969, Effects of urbanization on peak flows, in Moore, W. L., and Morgan, C. W., eds., Effects of watershed changes on stream flow: Water Resources Symposium No. 2, Austin, University of Texas Press, p. 201–213.

Bretschneider, C. L., 1952, The generation and decay of wind waves in deep water: American Geophysical Union Transactions, v. 33, No. 3, p. 381–389.

Brice, James C., 1958, Origin of steps on loess-mantled slopes: U.S. Geological Survey Bulletin 1071-C, p. 69–85.

——1964, Channel patterns and terraces of the Loup Rivers in Nebraska: U.S. Geological Survey Professional Paper 422-D, p. D1–D41.

Bridge, J. S. and Jarvis, J., 1976, Flow and sedimentary processes in the meandering River South Esk, Glen Cove, Scotland: Earth Surface Processes, v. 1, p. 303–336.

Broscoe, Andy J., 1959, Quantitative analysis of longitudinal stream profiles of small watersheds: Office of Naval Research, Technical Report No. 18, Contract N6 271-30, Department of Geology, Columbia University, New York, 73 p.

Brown, G. W. and Krygier, J. T., 1971, Clear-cut logging and sediment production in the Oregon Coast Range: Water Resources Research, v. 7, p. 1189–1198.

Brunsden, D., 1979, Mass Movements, in Embleton, C., and Thornes, J., ed., Process in geomorphology: New York, John Wiley and Sons, p. 130–186.

Bruun, P., 1954, Coast erosion and the development of beach profiles: U.S. Beach Erosion Board Technical Memo. 44, p. 65–79.

Bryan, K., 1922, Erosion and sedimentation in the Papago Country, Arizona: U.S. Geological Survey Bulletin 730B, p. 19–90.

Bull, W. B., 1964, Geomorphology of segmented alluvial fans in Western Fresno County, California: U.S. Geological Survey Professional Paper 352-E, p. 89–129.

——1968, Alluvial fans: Journal of Geological Education, v. 16, p. 101–106.

Caine, N., 1967, The texture of talus in Tasmania: Journal of Sedimentary Petrology, v. 37, p. 796–803.

——1969, A model for alpine talus slope development by slush avalanching: Journal of Geology, v. 77, p. 92–100.

Campbell, I., 1980, Accelerated erosion rates—the badlands factor: Geological Society of America, Abstracts with Programs, v. 12, No. 7, p. 398.

Campbell, R. H., 1975, Soil slips, debris flows, and rain storms in the Santa Monica Mountains and vicinity, Southern California: U.S. Geological Survey Professional Paper 851, 51 p.

Carson, M. A., 1971, The mechanics of erosion: London, Pion, 174 p.

——and Kirkby, M. J., 1972, Hillslope form and process: London, Cambridge University Press, 475 p.

Carter, R. W., 1960, Magnitude and frequency of floods in suburban areas: U.S. Geological Survey Professional Paper 424-B, p. 9–11.

Chepil, W. S. and Woodruff, N. P., 1963, The physics of wind erosion and its control: Advances in Agrononomy, v. 15, p. 211–302.

Cherkauer, D. S., 1973, Minimization of power expenditure in a riffle-pool alluvial channel: Water Resources Research, v. 9, p. 1613–1628.

Chorley, R. J., 1957, Illustrating the laws of morphometry: Geological Magazine, v. 94, p. 140–150.

—— 1959, The shape of drumlins: Journal of Glaciology, v. 3, p. 339–344.

—— 1962, Geomorphology and general systems theory: U.S. Geological Survey Professional Paper 500-B, 10 p.

—— 1966, The application of statistical methods to geomorphology, in Dury, G. H., ed., Essays in Geomorphology: Amsterdam, Elsevier Publishing Co., p. 275–387.

—— 1972, Spatial analysis in geomorphology: New York, Harper and Row, 393 p.

—— Dunn, A. J. and Beckinsale, R. P., 1964, The history of the study of landforms, Volume 1, Geomorphology before Davis: London, Methuen and Company, 678 p.

—— and Kennedy, B. A., 1971, Physical geography, a systems approach: London, Prentice-Hall, 370 p.

Church, M. A., 1972, Baffin Island sandurs: A study of Arctic fluvial processes: Canada Geological Survey Bulletin 216, 208 p.

Coleman, J. M., 1969, Brahmaputra River: channel process and sedimentation: Sedimentary Geology, v. 3, p. 129–239.

Colonell, J. M., and Goldsmith, V., 1971, Computational methods for analysis of beach and wave dynamics, in Morisawa, M., ed., Quantitative geomorphology: some aspects and applications: State University of New York, Binghamton, Publications in Geommorphology, p. 199–222.

Colton, R. B., Holligan, J. A., Anderson, L. W., and Patterson, P. E., 1976, Preliminary map of landslide deposits in Colorado: U.S. Geological Survey Map I-964.

Cooke, R. U., 1970, Morphometric analysis of pediments and associated landforms in the western Mojave desert, California: American Journal of Science, v. 269, p. 26–38.

Cooper, W. S., 1958, Coastal sand dunes of Oregon and Washington: Geological Society of America Memoir 72, 169 p.

Cornish, V., 1914, Waves of sand and snow: Fisher-Unwin, London, 383 p.

Costa, J. E., 1975, Effects of agriculture on erosion and sedimentation in Piedmont Province, Maryland: Geological Society of America Bulletin, v. 86, p. 1281–1286.

Coulomb, C. A., 1776, Essai sur une application des regles des maximus et minimus a quelques problemes de statique relatifs a l'architecture: Memoires de l'Academie Royale presentees par divers savants, v. 7 (also included in J. Hayman, 1972, Coulomb's memoire on statics: an essay in the history of civil engineering: New York, Cambridge University Press, 1x + 211 p. on p. 1–40).

Cox, J. C. and Harrison, S. S., 1979, Fracture-trace influenced stream orientation in glacial drift, northwestern Pennsylvania: Canadian Journal of Earth Science, v. 16, p. 1511–1514.

Crippen, J. R., 1965, Changes in character of unit hydrographs, Sharon Creek, California, after development: U.S. Geological Survey Professional Paper 525-D, p. D196–D198.

Croll, J., 1867, On the eccentricity of the Earth's orbit and its physical relations to the Glacial Epoch: Philosophical Magazine, 4th Series, v. 33, p. 119–131.

Crozier, M. J., 1973, Techniques for the morphometric analysis of landslips: Zeitschrift fur Geomorphologie, v. 17, p. 78–101.

Culling, W.E.H., 1963, Soil creep and the development of hillside slopes: Journal of Geology, v. 71, p. 127–161.

—— 1965, Theory of erosion on soil-covered slopes: Journal of Geology, v. 73, p. 230–254.

Curry, R. R., 1966, Observation of alpine mudflows in the Tenmile Range, central Colorado: Geological Society of America Bulletin, v. 77, p. 771–776.

Dacey, M. F., 1968, The profile of a random stream: Water Resources Research, v. 4, p. 651–654.

Daniel, J. F., 1972, Channel movement of meandering Indiana streams. Physio-

graphic and hydraulic studies of rivers: U.S. Geological Survey Professional Paper 732-A, p. A1–A18.

Daniels, R. B., 1960, Entrenchment of the Willow Drainage Ditch, Harrison County, Iowa: American Journal of Science, v. 258, p. 161–176.

Darbyshire, J., 1952, The generation of waves by wind: Royal Society of London Proceedings, Series A, v. 215, p. 299–328.

—— 1955, An investigation of storm waves in the North Atlantic Ocean: Royal Society of London Proceedings, Series A, v. 230, p. 560–569.

Darbyshire, M., and Draper, L., 1963, Forecasting wind-generated sea waves: Engineering, v. 195, p. 482–484.

Davis, W. M., 1909a, Glacial erosion in North Wales: Quarterly Journal of the Geological Society of London, v. 65, p. 281–350.

—— 1909b, Geographical essays: New York, Ginn, 777 p.

Davison, C., 1889, On the creeping of the soil cap through the action of frost: Geological Magazine, v. 36, p. 255–261.

Denny, C. S., 1965, Alluvial fans in the Death Valley region, California and Nevada: U.S. Geological Survey Professional Paper 466, 62 p.

—— 1967, Fans and pediments: American Journal of Science, v. 265, p. 81–105.

Derbyshire, E., Cooper, R., and Page, L.W.F., 1979, Recent movements on the cliff at St. Mary's Bay, Brixham., Devon.: Geographical Journal, v. 145, part I, p. 86–96.

Dicken, S. S., 1961, Some recent physical changes of the Oregon coast: Office of Naval Research, Final Technical Report, Contract 2771(04), Project NR 388-062, Department of Geography, University of Oregon, Eugene, 151 p.

Doehring, D., 1970, Discrimination of pediments and alluvial fans from topographic maps: Geological Society of America Bulletin, v. 81, p. 3109–3116.

Douglas, I., 1967, Natural and man-made erosion in the humid tropics of Australia, Malaysia and Singapore: International Association of Scientific Hydrology Publication No. 75, p. 17–30.

—— 1969, Sediment yields from forested and agricultural lands, in Taylor, J. A., ed., University College of Wales, Aberystwyth Memo, 12, El-22.

Dury, G. H., 1964, Principles of underfit streams: U.S. Geological Survey Professional Paper 452-A, p. A1–A67.

—— 1965, Theoretical implications of underfit streams: U.S. Geological Survey Professional Paper 452-C, p. C1–C43.

—— 1971, Channel characteristics in a meandering tidal channel: Crooked River, Florida: Geografiska Annaler, Series A, v. 53, p. 188–197.

—— 1973, Magnitude frequency analysis and channel morphology, in Morisawa, M., ed., Fluvial geomorphology: State University of New York, Binghamton, Publications in Geomorphology, p. 91–122.

—— Ongley, E. D., and Ongley, V. A., 1967, Attributes of pediment form on the Barrier and Cobar pediplains of New South Wales: Australian Journal of Science, v. 30, p. 33–34.

Eckart, C., 1953, The generation of wind waves over a water surface: Journal of Applied Physics, v. 24, p. 1485–1494.

Eckis, R., 1928, Alluvial fans of the Cucamonga district, southern California: Journal of Geology, v. 36, p. 224–247.

Embleton, C., and King, C.A.M., 1968, Glacial and periglacial geomorphology: Edinburgh, Edward Arnold, 608 p.

Emerson, John W., 1971, Channelization: A case study: Science, v. 173, p. 325–326.

Emery, K. O., 1960, Weathering of the Great Pyramid: Journal of Sedimentary Petrology, v. 30, p. 140–143.

Emmett, W. W., 1970, Hydraulics of overland flow on hillslopes: U.S. Geological Survey Professional Paper No. 622A, p. A1–A68.

Ericksen, G. E., Pflacker, G., and Fernandez, J. V., 1970, Preliminary report on the geological events associated with the May 31, 1970 Peru earthquake: U.S. Geological Survey Circular 639, 25 p.

Espey, W. H., Morgan, C. W., and Masch, F. D., 1966, Study of some effects of urbanization on storm runoff from a small watershed: Texas Water Development Board Report 23, 109 p.

Evans, O. F., 1938, The classification and origin of beach cusps: Journal of Geology, v. 46, p. 615–627.

Everett, K. R., 1963, Slope movement, Neotoma Valley, Southern Ohio: Report of the Institute of Polar Studies, Ohio State University, No. 6, 59 p.

Fahnestock, R. K., 1963, Morphology and hydrology of a glacial stream—White River, Mount Rainier, Washington: U.S. Geological Survey Professional Paper No. 422-A, 70 p.

Fair, T. J., 1947, Slope form and development in the interior of Natal: Geological Society of South Africa Transactions, v. 50, p. 105–120.

—— 1948, Slope form and development in the coastal hinterland of Natal: Geological Society of South Africa Transactions, v. 51, p. 37–53.

Fellenius, W., 1927, Erdstatische Berechnungen: Berlin, Ernst and Sohn, 40 p.

Fenneman, N. M., 1939, The rise of physiography: Geological Society of America Bulletin, v. 50, p. 349–360.

Ferguson, R. I., 1977, Meander migration: equilibrium and change, in Gregory, K. J. (ed.), River channel changes: New York, Wiley & Sons, p. 235–248.

Finkel, H. J., 1959, The barchans of Southern Peru: Journal of Geology, v. 67, p. 614–647.

Fisher, O., 1866, On the disintegration of chalk cliff: Geological Magazine, v. 3, p. 354–356.

Flint, J. J., 1980, Tributary arrangements in fluvial systems: American Journal of Science, v. 280, p. 26–45.

Ford, E. F., 1957, The transport of sand by wind: American Geophysical Union Transactions, v. 38, p. 171–174.

Fox, H. L., 1976, The urbanizing river: a case study in the Maryland Piedmont, in Coates, D. R., ed., Geomorphology and engineering symposium: London, Allen and Unwin, p. 245–272.

Fox, W. T., and Davis, R. A., Jr., 1971a, Fourier analysis of weather and wave data from Holland, Michigan, July, 1970: Technical Report, No. 3, ONR Task No. 388-092, Williams College, Williamstown, Mass., 79 p.

—— 1971b, Computer simulation model of coastal processes in eastern Lake Michigan, Technical Report No. 5, ONR Task No. 388-092, Williams College, A Williamstown, Mass., 114 p.

Frederiksen, R. L., 1970, Erosion and sedimentation following road construction and timber harvest on unstable soils in three small western Oregon watersheds: U.S. Department of Agriculture, Forest Service Research Paper PNW 104, 15 p.

Friedkin, J. F., 1945, A laboratory study of the meandering of alluvial rivers: U.S. Army Corps of Engineers, Mississippi River Commission, U.S. Waterways Experiment Station, Vicksburg, Mississippi, 40 p.

Gaillard, D. D., 1904, Wave action in relation to engineering structures: U.S. Army Corps of Engineers Professional Paper, no. 31.

Galvin, C. J., 1967, Longshore current velocity: a review of theory and data: Reviews of Geophysics, v. 5, p. 287–304.

Garrels, R. M., and Mackenzie, F. T., 1971, Evolution of the sedimentary rocks: New York, Norton, 376 p.

Geikie, J., 1868, On denudation in Scotland since glacial times: Geological Magazine, v. 5, p. 19–25.

Geikie, A., 1880, Rock weathering as illustrated in Edinburgh churchyards: Proceeding of the Royal Society of Edinburgh, v. 10, p. 518–532.

Gerber, E., 1934, Zur Morphologische wachsender Wande: Zeitschrift fur Geomorphologie, v. 8, p. 213–223.

Gerstner, F., 1802, Theorie der Wellen: Abhandlung Konig. Bohm. Ges. Wissenschaft, p. 412–445.

Gilbert, G. K., 1909, The convexity of hilltops: Journal of Geology, v. 17, p. 344–350.

—— 1914, The transportation of debris by running water: U.S. Geological Survey Professional Paper 86, 259 p.

Gilluly, J., 1937, Physiography of the Ajo region, Arizona: Geological Society of America Bulletin, v. 48, p. 323–348.

Glock, W. S., 1931, The development of drainage systems: a synoptic view: Geographical Review, v. 21, p. 74–83.

Goodchild, J. G., 1890, Notes on some observed rates of weathering of limestones: Geological Magazine, v. 27, p. 463–466.

Graf, W. H., 1971, Hydraulics of sediment transport: New York, McGraw-Hill, 513 p.

Graf, W. L., 1979, Catastrophe theory as a model for change in fluvial systems, in

Rhodes, D. D., and Williams, G. P., eds., Adjustments of the fluvial system: Dubuque, Iowa, Kendall/Hunt Publishing Company, p. 13–22.

Gravelius, H., 1914, Flusskunde: Berlin, Goschensche Verlagshandlung, 176 p.

Green, J.F.N., 1934, The River Mole: its physiography and superficial deposits: Proceedings of the Geologists Association of London, v. 45, p. 35–59.

Greenwood, J.E.G.W., 1966, Rock weathering in relation to the interpretation of igneous and metamorphic rocks in arid regions: Proceedings of the Symposium on Photo Interpretation, Delft, Netherlands, 1966, p. 93–99.

Gregory, K. J., and Madew, J. R., 1982, Land use change, flood frequency and channel adjustments, in Hey, R. D., ed., and others, Gravel-bed rivers; fluvial processes, engineering and management: United Kingdom, John Wiley and Sons, p. 757–781.

Gregory, K. J. and Walling, D. E., 1973, Drainage basin form and process: New York, John Wiley and Sons, 456 p.

Griggs, D. T., 1936, The factor of fatigue in rock exfoliation: Journal of Geology, v. 44, p. 781–796.

Gupta, A., and Fox, H., 1974, Effects of high magnitude floods on channel form: a case study in Maryland Piedmont: Water Resources Research, v. 10, p. 499–509.

Guy, H. P., 1964, An analysis of some storm-period variables affecting stream sediment transport: U.S. Geological Survey Professional Paper 462-E, p. E1–E46.

—— 1965, Residential construction and sedimentation of Kensington, Maryland: U.S. Department of Agriculture Miscellaneous Publication 970, p. 30–37.

Hack, J. T., 1941, Dunes of the Western Navajo Country: Geographical Review, v. 31, p. 240–263.

—— 1957, Studies of longitudinal streams profiles in Virginia and Maryland: U.S. Geological Survey Professional Paper 294-B, p. 45–97.

—— 1960, Interpretation of erosional topography in humid temperate regions: American Journal of Science, Bradley volume, 258-A, p. 80–97.

—— 1973, Stream-profile analysis and stream-gradient index: U.S. Geological Survey Journal of Research, v. 1, p. 421–429.

—— and Goodlett, J. C., 1960, Geomorphology and forest ecology of a mountain region in the central Appalachians: U.S. Geological Survey Professional Paper 347, p. 1–66.

Hadley, J. B., 1964, Landslides and related phenomena accompanying the Hebgen Lake earthquake of August 17, 1959: U.S. Geological Survey Professional Paper No. 435, p. 107–138.

Hammer, T. R., 1972, Enlargement due to urbanization: Water Resources Research, v. 8, p. 1530–1540.

Harris, E. E., and Rantz, S. E., 1964, Effect of urban growth on streamflow regimen of Permanente Creek, Santa Clara County, California: U.S. Geological Survey Water Supply Paper 1591-B, 18 p.

Harrison, P. W., 1957, A clay-till fabric: its character and origin: Journal of Geology, v. 65, p. 275–308.

Harrison, W., 1969, Empirical equations for foreshore changes over a tidal cycle: Marine Geology, v. 7, p. 529–551.

—— Rayfield, E. W., Boon, J. D. III, Reynolds, G., Grant, J. B., and Tyler, D., 1968, A time series from the beach environment: E.S.S.A. Research Laboratory Technical Memo. A OL-1, 25 p.

Hastenrath, S. L., 1967, The barchans of the Arequipa region, southern Peru: Zeitschrift fur Geomorphologie, v. 11, p. 300–331.

Heim, A., 1932, Bergsturz und Menschenleben: Zurich, Fretz Wasmath Verlag, 218 p.

Hewlett, J. D., and Helvey, J. D., 1970, Effects of forest clear felling in the storm hydrograph: Water Resources Research, v. 6, p. 768–782.

Hey, R. D., 1978, Determinate hydraulic geometry of river channels: American Society of Civil Engineers Proceedings, Journal of Hydraulic Division, v. 104, No. HY6, p. 869–885.

Hey, R. D., Bathurst, J. C. and Thorne, C. R., 1982, Gravel-bed rivers: fluvial processes, engineering and management: New York, John Wiley and Sons, 895 p.

Hickin, E. J., 1978, Mean flow structure in meanders of the Squamish River,

British Columbia: Canadian Journal of Earth Science, v. 15, p. 1833–1849.

Hjulstrom, F., 1935, The morphological activity of rivers as illustrated by River Fyris: University of Upsala Geology Institute Bulletin No. 25, p. 221–527.

Hollis, G. E., and Luckett, J. K., 1976, The response of natural river channels to urbanization; two case studies from Southeast England: Journal of Hydrology, v. 30, p. 351–363.

Holmes, C. D., 1941, Till fabric: Geological Society of America Bulletin, v. 52, p. 1299–1354.

Hooke, R. L., 1968, Steady-state relationships on arid-region alluvial fans in closed basins: American Journal of Science, v. 266, p. 609–629.

—— 1975, Distribution of sediment transport and shear stress in a meander bend: Journal of Geology, v. 83, p. 543–565.

Hoover, M. D., 1944, Effects of removal of forest vegetation upon water yield: American Geophysical Union Transactions, v. 4, p. 969–977.

Hopkins, D. M., and Sigafoos, R. S., 1951, Frost action and vegetation patterns on Seward Peninsula, Alaska: U.S. Geological Survey Bulletin 974-C, p. 51–101.

Hoppe, G., 1959, Glacial morphology and the inland ice recession in northern Sweden: Geografiska Annaler, v. 41, p. 193–212.

Hornbeck, J. W., 1973, Storm flow from hardwood forested and cleared watersheds in New Hampshire: Water Resources Research, v. 9, p. 346–354.

Horton, R. E., 1932, Drainage basin characteristics: American Geophysical Union Transactions, v. 13, p. 350–361.

—— 1945, Erosional development of streams and their drainage basins: hydrophysical approach to quantitative morphology: Geological Society of America Bulletin, v. 56, p. 275–370.

Hough, B. K., 1957, Basic soils engineering: New York, Ronald Press, 513 p.

Howard, A. D., 1971a, Simulation model of stream capture: Geological Society of America Bulletin, v. 82, p. 1355–1376.

—— 1971b, Simulation of stream networks by headward growth and branching: Geografiska Annaler, v. 3, p. 29–50.

—— 1971c, Optimal angles of stream junction: geometric, stability to capture, and minimum power criteria: Water Resources Research, v. 7, p. 863–873.

—— 1971d, Problems of interpretation of simulation models of geologic processes, in Morisawa, M., ed., Quantitative geomorphology: State University of New York, Binghamton, Publications in Geomorphology, p. 61–82.

—— Keetch, M. E., and Vincent, C. L., 1970, Topological and geometrical properties of braided streams: Water Resources Research, v. 6, p. 1674–1688.

—— Morton, J. B., Gal-el-Hak, M., and Pierce, D. B., 1977, Simulation model of erosion and deposition: National Aeronautics and Space Administration Contractor Report No. NASA CR-2838, University of Virginia, Charlottesville, VA, 77 p.

Hoyle, J. W., and King, G. T., 1958, Origin and stability of beaches: Proceedings of the 6th Coastal Engineering Conference, Palm Beach, Florida, Dec. 1957, p. 281–301.

Hoyt, W. G. and Troxell, H. C., 1934, Forests and streamflow: American Society of Civil Engineers Transactions, v. 99, p. 1–30.

Hsü, K. J., 1975, Catastrophic debris streams generated by rockfalls: Geological Society of America Bulletin, v. 86, p. 129–140.

Hutton, J., 1795, Theory of the Earth: with proofs and illustrations: 2 vols., Edinburgh.

Ingle, J. C., 1966, The movement of beach sand—An analysis using fluorescent grains: New York, Elsevier Publishing Company, 221 p.

Inglis, C. C., 1949, The behavior and control of rivers and canals: Central Water-Power Irrigation and Navigation Research Station, Research Publication No. 13, Poona, India, 2 volumes.

Inman, D. L., and Bagnold, R. A., 1963, Littoral processes, in Hill, M. N., ed., The Sea III: New York, John Wiley and Sons, p. 529–553.

—— Ewing, G. C., and Corliss, J. B., 1966, Coastal sand dunes of Guerrero Negro; Baja, California, Mexico: Geological Society of American Bulletin, v. 77, p. 787–802.

Jackson, R. G., 1975, Hierarchical attributes and a unifying model of bed forms composed of cohesionless material and produced by shearing flow: Geological Society of America Bulletin, v. 86, p. 1523–1533.

James, L. D., 1965, Using a digital computer to estimate the effects of urban development on flood peaks: Water Resources Research, v. 1, p. 223–234.

James, W. R. and Krumbein, W. C., 1969, Frequency distribution of stream link lengths: Journal of Geology, v. 77, p. 544–565.

Janda, R. J., Scott, K. M., and Martinson, H. A., 1981, Lahar movement, effects and deposits, in Lipman, P. W., ed., and others, The 1980 eruptions of Mount St. Helens, Washington: U.S. Geological Survey, Professional Paper No. 1250, p. 461–478.

Jarvis, R. S., 1972, New measure of the topologic structure of dendritic drainage networks: Water Resources Research, v. 8, p. 1265–1271.

—— 1976a, Stream orientation structures in drainage networks: Journal of Geology, v. 84, p. 563–582.

—— 1976b, Link length organization and network scale dependencies in the network diameter model: Water Resources Research, v. 12, p. 1215–1225.

Jeffreys, H., 1925, On the formation of water waves by wind: Royal Society of London Proceedings, Series A, v. 107, p. 189–206.

Jenny, H., 1941, Factors of soil formation: New York, McGraw Hill, 281 p.

Johnson, D. W., 1919, Shore processes and shoreline development: New York, John Wiley and Sons, 584 p.

Jones, F. O., Embody, D. R., and Peterson, W. L., 1961, Landslides along the Columbia River Valley, Northeastern Washington: U.S. Geological Survey Professional Paper 367, 98 p.

Jones, O. T., 1924, The longitudinal profiles of the Upper Towy drainage system: Quarterly Journal of the Geological Society of London, v. 80, p. 568–609.

Judson, S., and Andrews, G. W., 1955, Pattern and form of some valleys in the driftless area, Wisconsin: Journal of Geology, v. 63, p. 328–336.

Kawamura, R., 1951, Study on sand movement by wind: Report of the Institute of Science and Technology, University of Tokyo, Tokyo, Japan, v. 5, No. 314, p. 95–112.

Keller, E. A., 1972, Development of alluvial stream channels: a five-stage model: Geological Society of America Bulletin, v. 83, p. 1531–1536.

—— 1975, Channelization: A search for a better way: Geology, v. 3, p. 246–248.

—— 1976, Environmental geology: Columbus, Charles Merrill, 488 p.

—— and Melhorn, W. N., 1978, Rhythmic spacing and origin of pools and riffles: Geological Society of America Bulletin, v. 89, p. 723–730.

King, C.A.M., 1965, Some observations on the beaches of the west coast of County Donegal: Irish Geographer, v. 5, no. 2, p. 46–50.

—— 1969, Trend-surface analysis of Central Pennine erosion surfaces: Transactions Institute of British Geographers, v. 47, p. 47–59.

—— 1972, Beaches and coasts: New York, St. Martin's Press, 570 p.

King, L. J., 1969, The analysis of spatial forms and its relation to geographic theory: Annals of the Association of American Geographers, v. 59, p. 572–595.

Kirkby, M. J., 1967, Measurement and theory of soil creep: Journal of Geology, v. 75, p. 359–378.

—— 1977, Maximum sediment efficiency as a criterion for alluvial channels, in Gregory, K. J., ed., River channel changes: Chichester, John Wiley and Sons, p. 429–442.

Knight, C., 1979, Urbanization and natural stream channel morphology: the case of two English new towns, in Hollis, G. E., ed., Man's impact on the hydrological cycle in the United Kingdom: Norwich, England, Geo Abstracts Ltd., p. 181–198.

Knighton, A. D., 1974, Variation in width-discharge relation and some implications for hydraulic geometry: Geological Society of America Bulletin, v. 85, p. 1069–1076.

—— 1975a, Variations in at-a-station hydraulic geometry: American Journal of Science, v. 275, p. 186–218.

—— 1975b, Channel gradient in relation to discharge and bed material characteristics: Catena, v. 2, p. 263–274.

Knott, J. M., 1973, Effects of urbanization on sedimentation and flood-flows in Colma Creek Basin, California: U.S. Geological Survey Open File Report, 54 p.

Kojan, E., 1967, Mechanics and rates of natural soil creep, in Proceedings of the 5th Annual Engineering Geology and Soils Engineering Symposium: Pocatello, Idaho, Idaho Department of Highways, Boise, Idaho, p. 233–253.

Komar, P. D., 1971, The mechanics of sand transport on beaches: Journal of Geophysical Research, v. 76, p. 713–721.

——and Inman, D. L., 1970, Longshore sand transport on beaches: Journal of Geophysical Research, v. 75, p. 5914–5927.

Komura, S. and Simons, D. B., 1967, Riverbed degradation below dams: American Society of Civil Engineers, Proceedings, Journal of the Hydraulics Division, v. 93, H.Y. 4, p. 1–14.

Krumbein, W. C., 1959, Trend surface analysis of contour-type maps with irregular control point spacing: Journal of Geophysical Research, v. 64, p. 823–834.

——1963, A geological process-response model for analysis of beach phenomena: U.S. Beach Erosion Board Bulletin, v. 17, p. 1–15.

——and Graybill, F. A., 1965, An introduction to statistical models in geology: New York, McGraw-Hill, 475 p.

——and Orme, A. R., 1972, Field mapping and computer simulation of braided-stream networks: Geological Society of America Bulletin, v. 83, p. 3369–3380.

Kuenen, P. H., 1948, The formation of beach cusps: Journal of Geology, v. 56, p. 34–40.

Lacey, G., 1930, Stable channels in alluvium: Institute of Civil Engineers Proceedings, v. 229, p. 259–285.

Lachenbruch, A. H., 1962, Mechanics of thermal contraction cracks and icewedge polygons in permafrost: Geological Society of America Special Paper 70, 69 p.

Lake, P., 1928, On hill slopes: Geological Magazine, v. 65, p. 108–116.

Langbein, W. B., and Leopold, L. B., 1964, Quasi-equilibrium states in channel morphology: American Journal of Science, v. 262, p. 782–794.

——1966, River meanders-theory of minimum variance: U.S. Geological Survey Professional Paper 422-H, p. 1–15.

Langford-Smith, T., and Dury, G. H., 1964, A pediment survey at Middle Pinnacle, near Broken Hill, New South Wales: Journal of the Geological Society of Australia, v. 11, p. 79–88.

Lawson, A. C., 1915, The epigene profiles of the desert: University of California, Department of Geology Bulletin, v. 9, p. 23–48.

Leblanc, F., 1842, Observations sur le maximum d'inclinasion des talus dans les montagnes: Bulletin de la Societe Geologique de France, v. 14, p. 85–98.

Lehmann, O., 1933, Morphologische Theorie der Verwitterung von Steinschlagwanden: Vierteljahrsschrift der Naturforschende Gesellschaft, v. 78, p. 83–126.

Leighley, J., 1932, Toward a theory of the morphological significance of turbulence in the flow of water in streams: University of California Publications in Geography, Berkeley, v. 6, p. 1–22.

——1934, Turbulence and the transportation of rock debris by streams: Geographical Review, v. 24, p. 453–464.

Leopold, L. B., 1968, Hydrology for urban land planning—a guidebook on the hydrologic effects of urban land use: U.S. Geological Survey Circular 554, 18 p.

——1973, River channel change with time: Geological Society of America Bulletin, v. 84, p. 1845–1860.

——and Emmett, W. W., 1972, Some rates of geomorphological processes: Geographia Polonica, p. 27–35.

——and Langbein, W. B., 1962, The concept of entropy in landscape evolution: U.S. Geological Survey Professional Paper 500-A, p. A1–A20.

——and Maddock, T. Jr., 1953, The hydraulic geometry of stream channels and some physiographic implications: U.S. Geological Survey Professional Paper 252, 57 p.

——and Miller, J. P., 1956, Ephemeral streams—hydraulic factors and their relation to the drainge net: U.S. Geological Survey Professional Paper 282-A, p. 1–37.

——and Wolman, M. G., 1957, River channel patterns: braided, meandering, straight: U.S. Geological Survey Professional Paper 282-B, p. 39–85.

——Wolman, M. G. and Miller, J. P., 1964, Fluvial processes in geomorphology: San Francisco, Freeman and Co., 522 p.

Levi-Civita, T., 1925, Determination rigoureuse des ondes d'ampleus finie: Mathematische Annalen, v. 93, p. 264–314.

Lewis, A., 1975, Slow movements of earth under tropical rainforest conditions: Geology, v. 2, p. 9–10.

Lewis, W. V., ed., 1960, Norwegian cirque glaciers: Royal Geographical Society, Research Series No. 4, 104 p.

Lohnes, R. A. and Handy, R. L., 1968, Slope angles in friable loess: Journal of Geology, v. 76, p. 247–258.

Long, J. T., and Sharp, R. P., 1964, Barchan-dune movement in Imperial Valley, California: Geological Society of America Bulletin, v. 75, p. 149–156.

Longuet-Higgins, M. S., 1953, Mass transport in water waves: Philosophical Transactions of the Royal Society of London, Series A, v. 245, p. 535–581.

——and Parkin, D. W., 1962, Sea waves and beach cusps: Geographical Journal, v. 128, p. 194–200.

Loughnan, F. C., 1969, Chemical weathering of the silicate minerals: New York, Elsevier Publishing Co., 154 p.

Lubowe, J. K., 1964, Stream junction angles in the dendritic drainage pattern: American Journal of Science, v. 262, p. 325–339.

Lustig, L. K., 1969, Trend-surface analysis of the Basin and Range Province, and some geomorphic implications: U.S. Geological Survey Professional Paper 500-D, 70 p.

Mabbutt, J. A., 1966, Mantle-controlled planation of pediments: American Journal of Science, v. 264, p. 79–81.

MacDonald, D. F., 1915, Some engineering problems of the Panama Canal in their relation to geology and topography: U.S. Bureau of Mines Bulletin, Series B, v. 86, 88 p.

Mackin, J. H., 1948, Concept of the graded river: Geological Society of America Bulletin, v. 59, p. 463–512.

Mammerickx, J., 1964, Quantitative observations on pediments in the Mojave and Sonoran deserts (southwestern United States): American Journal of Science, v. 262, p. 417–435.

Mather, K. F., and Mason, S. L., 1939, A source book in Geology: New York, McGraw-Hill, 702 p.

Matthes, F. E., 1930, Geologic history of the Yosemite Valley: U.S. Geological Survey Professional Paper 160, 137 p.

McConnell, H., and Horn, J. M., 1972, Probabilities of surface karst, *in* Chorley, R. J., ed., Spatial analysis in geomorphology, New York, Harper and Row, p. 111–133.

McCullagh, M. J., and King, C.A.M., 1970, Spitsym, a fortran IV computer program for spit simulation: Computer Contribution of the Kansas Geological Survey, 9 p.

McKee, E. D., Douglass, J. R., and Rittenhouse, S., 1971, Deformation of lee-side laminae in eolian dunes: Geological Society of America Bulletin, v. 82, p. 359–378.

——and Tibbitts, G. C. Jr., 1964, Primary structures of a seif dune and associated deposits in Libya: Journal of Sedimentary Petrology, v. 34, p. 5–17.

McKenzie, P., 1958, Rip current systems: Journal of Geology, v. 66, p. 103–113.

McLean, R., 1967, Plan shape and orientation of beaches along the east coast, South Island, New Zealand: New Zealand Geographer, v. 23, p. 16–22.

McSaveney, E. R., 1971, The surficial fabric of rockfall talus, *in* Morisawa, M., ed., Quantitative geomorphology: some aspects and applications: State University of New York, Binghamton, Publications in Geomorphology, p. 181–197.

Meade, R. H., and Trimble, S. W., 1974, Changes in sediment loads in rivers of the Atlantic drainage of the United States since 1900: International Association of Hydrology, Scientific Publication 113, p. 105–129.

Megahan, W. F., and Kidd, W. J., 1972, Effects of logging and logging roads on erosion and sediment deposition from steep terrain: Journal of Forestry, v. 70, p. 136–141.

Melton, M. A., 1957, An analysis of the relation among elements of climate, surface properties, and geomorphology: Office of Naval Research, Technical Report No. 11, Contract N6 ONR 271-30, Department of Geology, Columbia University, New York, 102 p.

——1958, List of sample parameters of quantitative properties of landforms: their use in determining the size of geomorphic experiments: Office of Naval Research Technical Report No. 16, Contract N6 ONR 271-30, Department of Geology, Columbia University, New York, 17 p.

—— 1960, Concept of state of a drainage system (abstract): Geological Society of America Bulletin, v. 71, p. 1928.

—— 1961, Discussion—The effect of sediment type on the shape and stratification of some modern fluvial deposits: American Journal of Science, v. 259, p. 231–233.

—— 1965, The geomorphic and paleoclimatic significance of alluvial deposits in southern Arizona: Journal of Geology, v. 73, p. 1–38.

Meyers, R. D., 1933, A model of wave action on beaches: M.Sc. Thesis, University of California, Berkeley.

Mii, H., 1958, Beach cusps on the Pacific coast of Japan: Scientific Report of Tohoku University, Sendai, Japan, v. 29, p. 77–107.

Miles, J. W., 1965, A note on the interaction between surface waves and wind profiles: Journal of Fluid Mechanics, v. 22, p. 823–827.

Miller, C. R., and Viessman, W., 1972, Runoff volumes from small urban watersheds: Water Resources Research, v. 8, p. 429–434.

Miller, T. K. and Onesti, L. J., 1979, The relationship between channel shape and sediment characteristics in the channel perimeter: Geological Society of American Bulletin, v. 90, p. 301–304.

Milton, L. E., 1967, The geometric irrelevance of some drainage net laws: Australian Geographical Studies, v. 4, p. 89–95.

Mock, S. J., 1971, A classification of channel links in stream networks: Water Resources Research, v. 7, p. 1558–1566.

Morisawa, M. E., 1962, Quantitative geomorphology of some watersheds in the Appalachian Plateau: Geological Society of America Bulletin, v. 73, p. 1025–1046.

—— 1963, Distribution of stream-flow direction in drainage patterns: Journal of Geology, v. 71, p. 528–529.

—— 1964, Development of drainage system on an upraised lake floor: American Journal of Science, v. 262, p. 340–354.

—— 1966, Talus slopes in high mountain areas: Department of the Army, Office of Research and Development, Contract No. DA-ARO-49-092-64-G52, Earth Sciences Department, Antioch College, Yellow Springs, Ohio, 61 p.

—— ed., 1971, Quantitative geomorphology: some aspects and applications: State University of New York, Binghamton, Publications in Geomorphology, 315 p.

—— (in press), Topologic properties of delta distributary networks, *in* Woldenberg, ed., Models in Geomorphology: London, Allen & Unwin.

—— and LaFlure, E., 1979, Hydraulic geometry, stream equilibrium and urbanization, *in* Rhodes, D. D., and Williams, G. P., eds., Adjustments of the fluvial system: Dubuque, Iowa, Kendall/Hunt Publishing Co., p. 333–350.

—— and Vemuri, R., 1975, Multiobjective planning and environmental evaluation of water resources systems: Office of Water Research and Technology Project C-6065, Final Report, State University of New York, Binghamton, 134 p.

Narayana, V.V.D., Riley, J. P. and Israelson, E. K., 1969, Analog computer simulation of the runoff characteristic of an urban watershed: Utah Water Research Laboratory Utah State, Logan, Utah, 83 p.

Nunnally, N. R., and Keller, E., 1979, Use of fluvial processes to minimize adverse effects of stream channelization: Report No. 144, Water Resources Research Institute, University of North Carolina, Charlotte, 115 p.

Nye, J. F., 1951, The flow of glaciers and ice-sheets as a problem in plasticity: Royal Society of London Proceedings, Series A, v. 207, p. 554–572.

—— 1952, The mechanics of glacier flow: Journal of Glaciology, v. 2, p. 82–93.

—— 1958, Surges in glaciers: Nature, v. 181, No. 4621, p. 1450–1451.

—— 1970, Glacier sliding without cavitation in a linear viscous approximation: Royal Society of London Proceedings, Series A, v. 315, p. 381–403.

Ollier, C. D., 1969, Weathering: New York, American Elsevier Publishing Company, 304 p.

Otvos, E. G., 1964, Observations of beach cusps and beach ridge formation on the Long Island Sound: Journal of Sedimentary Petrology, v. 34, p. 554–560.

Park, C. C., 1977, Man-induced changes in stream channel capacity, *in* Gregory,

K. J., ed., River channel changes: Chichester, United Kingdom, John Wiley and Sons, p. 121–144.

Pearce, A. J., 1973, Mass and energy flux in physical denudation; defoliated areas, Sudbury, Ontario: Technical Report No. 75-1, McGill University, Department of Geological Sciences, 235 p.

Peltier, L., 1950, The geographic cycle in periglacial regions as it is related to climatic geomorphology: Annals of the Association of American Geographers, v. 40, p. 214–236.

Penck, W., 1924, Morphological analysis of landforms: translated by H. Čzech and K. C. Boswell, 1953, MacMillan, London, 429 p.

Petts, G. E., 1977, Channel response to flow regulation: the case of the River Derwent, Derbyshire, *in* Gregory, K. J., ed., River channel changes: New York: John Wiley and Sons, p. 145–161.

Phillips, O. M., 1957, On the generation of waves by turbulent wind: Journal of Fluid Mechanics, v. 2, p. 417–445.

Pickup, G., 1977, Simulation modelling of river channel erosion, *in* Gregory, K. J., ed., River channel changes: Chichester, United Kingdom, John Wiley and Sons, p. 47–60.

Pierson, W. J., Neumann, G. and James, R. W., 1955, Practical methods for observing and forecasting ocean waves by means of wave spectra and statistics: Hydrograph Office Publications, v. 603, Hydrograph Office, U.S. Navy.

Playfair, J., 1802, Illustrations of the Huttonian Theory of the Earth: Edinburgh, William Creech, 528 p.

Putnam, J. A., Munk, W. H., and Taylor, M. A., 1949, The prediction of longshore currents: American Geophysical Union Transactions, v. 30, p. 337–345.

Rankine, W. J. H., 1857, On the stability of loose earth: Philosophical Transactions of the Royal Society of London, Series A, v. 147, p. 9–27.

—— 1863, On the exact form of waves near the surface of deep water: Philosophical Transactions of the Royal Society of London, Series A, v. 153, p. 127–138.

Rapp, A., 1957, Studien uber Schutthalden in Lappland und auf Spitzbergen: Zeitschrift fur Geomorphologie, v. 1, p. 179–200.

—— 1960, Recent development of mountain slopes in Karkevagge and surroundings, Northern Scandinavia: Geografiska Annaler, v. 42, p. 65–206.

Rector, R. L., 1954, Laboratory study of the equilibrium profiles of beaches: U.S. Beach Erosion Board Technical Memo. no. 41, 38 p.

Reed, B., Calvin, C. J., Jr., and Miller, J. P., 1962, Some aspects of drumlin geometry: American Journal of Science, v. 260, p. 200–210.

Reed, L. A., 1980, Suspended-sediment discharge, in five streams near Harrisburg, Pennsylvania, before, during, and after highway construction: U.S. Geological Survey Water Supply Paper 2072, 37 p.

Reiche, P., 1943, Graphic representation of chemical weathering: Journal of Sedimentary Petrology, v. 13, p. 58–68.

Reid, H. F., 1892, Studies of Muir Glacier, Alaska: National Geographic, v. 4, p. 19–84.

Rhodes, D. D., 1977, The b-f-m diagram: graphical representation and interpretation of at-a-station hydraulic geometry: American Journal of Science, v. 277, p. 73–96.

Richards, K. S., 1973, Hydraulic geometry and channel roughness—A non-linear system: American Journal of Science, v. 273, p. 877–896.

—— 1976a, Complex width-discharge relations in natural river sections: Geological Society of America Bulletin, v. 87, p. 199–206.

—— 1976b, Channel width and the riffle-pool sequence: Geological Society of America Bulletin, v. 87, p. 883–890.

Ritter, D. F., 1978, Process geomorphology: Dubuque, Iowa, William C. Brown Publishing Co., 603 p.

Roth, E. S., 1965, Temperature and water contents as factors in desert weathering: Journal of Geology, v. 73, p. 454–468.

Rouse, H., and Ince, S., 1957, History of hydraulics: Iowa Institute of Hydraulic Research, State University of Iowa, Ames, Iowa, 269 p.

Rubey, W. W., 1933, Settling velocities of gravel, sand and silt: American Journal of Science, v. 225, p. 325–338.

—— 1938, The force required to move particles on a stream bed: U.S. Geological

Survey Professional Paper 189-E, p. 121–141.

—— 1952, Geology and mineral resources of the Hardin and Brussels Quadrangles (Illinois): U.S. Geological Survey Professional Paper No. 218, 179 p.

Rudberg, S., 1962, A report on some field observations concerning periglacial geomorphology and mass movement on slopes in Sweden: Biuletyn Peryglacjalny, v. 11, p. 311–323.

Russell, R.C.H., and Osorio, J.D.C., 1958, An experimental investigation of drift profiles in a closed channel: Proceedings of the 6th Coastal Engineering Conference, University of California, p. 171–183.

Ruxton, B. P., 1966, The measurement of denudation rates: Institute of Australian Geographers, 5th Meeting, Sydney, Australia.

—— and Berry, L., 1961, Weathering profiles and geomorphic position on granite in two tropical regions: Revue de Geomorphologie dynamique, v. 12, p. 16–31.

Salisbury, N., 1971, Threads of inquiry in quantitative geomorphology, *in* Morisawa, M., ed., Quantitative geomorphology: some aspects and applications: State University of New York, Binghamton, Publications in Geomorphology, p. 9–60.

Savigear, R.A.G., 1952, Some observations on slope development in South Wales: Transactions of the Institute of British Geographers, v. 18, p. 31–51.

—— 1956, Technique and terminology in the investigation of slope forms: Premier Rapport, Commission pour l'Etude des Versants, International Geographical Union, p. 66–75.

—— 1963, The morphological basis of geomorphological mapping: Deltion tis Ellinikis Geogr. Etaireias, v. 4, p. 104–111.

Saville, T., 1950, Model study of sand transport along an infinitely long straight beach: American Geophysical Union Transactions, v. 31, p. 555–565.

Savini, J., and Kammerer, J. C., 1961, Urban growth and the water regimen: U.S. Geological Survey Water Supply Paper 1591-A, p. A1–A43.

Scheidegger, A. E., 1960, Analytical theory of slope development by undercutting: Journal of the Alberta Society of Petroleum Geologists, v. 8, p. 202–206.

—— 1961, Theoretical geomorphology: Englewood, Prentice-Hall, 333 p.

—— and Langbein, W. B., 1966, Probability concepts in geomorphology: U.S. Geological Survey Professional Paper No. 500-C, 14 p.

Schenck, H., 1963, Simulation of the evolution of drainage basin networks with a digital computer: Journal of Geophysical Research, v. 68, p. 5739–5745.

Schumm, S. A., 1956a, The role of creep and rainwash on the retreat of badland slopes: American Journal of Science, v. 254, p. 693–706.

—— 1956b, Evolution of drainage systems and slopes in badlands at Perth Amboy, New Jersey: Geological Society of America Bulletin, v. 67, p. 597–646.

—— 1960, The shape of alluvial channels in relation to sediment type: U.S. Geological Survey Professional Paper 352-B, p. 17–30.

—— 1962, Erosion of miniature pediments in Badlands National Monument, South Dakota: Geological Society of America Bulletin, v. 73, p. 719–724.

—— 1963, Sinuosity of alluvial rivers on the Great Plains: Geological Society of America Bulletin, v. 74, p. 1089–1100.

—— 1973, Geomorphic thresholds and complex response of drainage systems, *in* Morisawa, M., ed., Fluvial geomorphology, State University of New York, Binghamton, Publications in Geomorphology, p. 299–310.

—— and Khan, H. R., 1972, Experimental study of channel pattern: Geological Society of America Bulletin, v. 83, p. 1755–1770.

—— and Lichty, R. W., 1965, Time, space and causality in geomorphology: American Journal of Science, v. 263, p. 110–119.

Scott, R. F., 1963, Principles of soil mechanics: Reading, Addison-Wesley, 550 p.

Seaburn, C. E., 1964, Effects of urban development on direct runoff to East Meadow Brook, Nassau County, Long Island, New York: U.S. Geological Survey Professional Paper 627-B, 14 p.

Shallow, P.G.D., 1956, River flow in the Cameron Highlands: Hydro-electric Tech. Memo., no. 3 of the Central Electricity Board.

Sharp, R. P., 1940, Geomorphology of the Ruby-East Humboldt Range, Nevada: Geological Society of America Bulletin, v. 51, p. 337–371.

—— 1942, Soil structures in the St. Elias Range, Yukon Territory: Journal of Geomorphology, v. 5, p. 274–301.

—— 1966, Kelso dunes, Mohave Desert, California: Geological Society of America Bulletin, v. 77, p. 1045–1074.

—— and Noble, L. H., 1953, Mudflow of 1941 at Wrightwood, Southern California: Geological Society of America Bulletin, v. 64, p. 547–560.

Shepard, F. P., 1950, Longshore-bars and longshore-troughs: U.S. Beach Erosion Board Technical Memo, No. 15, 32 p.

—— and Inman, D. L., 1950, Nearshore circulation related to bottom topography and wave refraction: American Geophysical Union Transactions, v. 31, p. 196–212.

Shreve, R. L., 1966, Statistical law of stream numbers: Journal of Geology, v. 74, p. 17–37.

—— 1967, Infinite topologically random channel networks: Journal of Geology, v. 75, p. 179–186.

—— 1968, The Blackhawk landslide: Geological Society of America Special Paper 108, 48 p.

Shulits, S., 1941, Rational equation of river-bed profile: American Geophysical Union Transactions, 22nd Annual Meeting, Part 3, p. 622–630.

Silvester, R., 1960, Stabilization of sedimentary coastlines: Nature, v. 188, No. 4749, p. 467–469.

Skempton, A. W., 1953, Soil mechanics in relation to geology: Proceedings of the Yorks Geological Society, v. 29, p. 33–62.

Smart, J. S., 1967, A comment on Horton's law of stream numbers: Water Resources Research, v. 3, p. 773–776.

—— 1969, Topological properties of channel networks: Geological Society of America Bulletin, v. 80, p. 1757–1773.

—— 1970, Use of topologic information in processing data for channel networks: Water Resources Research, v. 6, p. 932–936.

—— 1972, Channel networks: Advances in Hydroscience, v. 8, p. 305–346.

—— 1978, The analysis of drainage network composition: Earth Surface Processes, v. 3, p 129–170.

—— and Moruzzi, V. L., 1972, Quantitative properties of delta channel networks: Zeitschrift fur Geomorphologie, v. 16, p. 268–282.

—— Surkan, A. J. and Considine, J. P., 1967, Digital simulation of channel networks: Publication of the International Association of Scientific Hydrology, v. 75, p. 87–98.

Smith, D. D., and Wischmeier, W. H., 1962, Rainfall erosion: Advances in Agronomy, v. 14, p. 109–148.

Smith, D. E., Sissons, J. B., and Cullingford, R. A., 1969, Isobases for the Main Perth raised shoreline in south-east Scotland as determined by trend surface analysis: Transactions of the Institute of British Geographers, v. 46, p. 45–52.

Smith, K. G., 1958, Erosional processes and landforms in Badlands National Monument, South Dakota: Geological Society of America Bulletin, v. 69, p. 975–1008.

Smith, T. R., 1974, A derivation of the hydraulic geometry of steady-state channels from conservation principles and sediment transport laws: Journal of Geology, v. 82, p. 98–104.

Souchez, R., 1961, Theorie d'une evolution des versants: Societe royale Belge de Geographie Bulletin, v. 85, p. 7–18.

Speight, J. G., 1965, Meander spectra of the Angabunga River: Journal of Hydrology, v. 3, p. 1–15.

—— 1968, Parametric description of land form, *in* Stewart, G. A., ed., Land evaluation: Sydney, Macmillan of Australia, p. 239–250.

Stall, J. B., 1971, Predicting time-of-travel in stream systems, *in* Morisawa, M., ed., Quantitative geomorphology: some aspects and applications: State University of New York, Binghamton, Publications in Geomorphology, p. 259–271.

Stankowski, S. J., 1974, Magnitude and frequency of floods in New Jersey with effects of urbanization: U.S. Geological Survey Special Report 38.

Stanton, T. E., Marshall, D., and Houghton, R., 1932, The growth of waves on water due to the action of wind: Royal Society of London Proceedings, Series A, v. 137, p. 283–293.

Sternberg, H., 1875, Untersuchungen uber langen- und querprofil geschiebefuhrende flusse: Zeitschrift, Bauwesen, v. 25, p. 483–506.

Stokes, G. G., 1847, On the theory of oscillatory waves: Cambridge Philosophical Society Transactions, v. 8, p. 441–455.

Strahler, A. N., 1950, Equilibrium theory of erosional slopes approached by frequency distribution analysis: American Journal of Science, v. 248, p. 673–696, and p. 800–814.

—— 1952, Dynamic basis of geomorphology: Geological Society of America Bulletin, v. 63, p. 923–938.

—— 1954, Statistical analysis in geomorphic research: Journal of Geology, v. 62, p. 1–25.

—— 1956, Quantitative slope analysis: Geological Society of America Bulletin, v. 67, p. 571–596.

—— 1958, Dimensional analysis applied to fluvially eroded landforms: Geological Society of America Bulletin, v. 69, p. 279–300.

Struik, D. J., 1926, Determination rigoureuse des ondes irrotationelles periodiques dans un canal a profondeur finie: Mathematische Annalen, v. 95, p. 595–634.

Sundborg, A., 1956, The River Klaralven, a study of fluvial processes: Geografiska Annaler, v. 38, p. 127–136.

Svensson, H., 1959, Is the cross-section of a glacial valley a parabola?: Journal of Glaciology, v. 3, p. 362–363.

Sverdrup, H. U., and Munk, W. H., 1947, Wind, sea and swell-theory of relationships in forecasting: U.S. Navy, Hydrograph Office Publication No. 601, 47 p.

Swanston, D. N., 1970, Mechanics of debris avalanching in shallow till soils of Southeast Alaska: U.S. Department of Agriculture Forest Service Research Paper PNW-103, 17 p.

Swanston, F. J., and Swanston, D. N., 1977, Complex mass-movement terrains in the western Cascade Range, Oregon: Reviews in Engineering Geology (Geological Society of America), v. 3, p. 113–124.

Taber, S., 1929, Frost heaving: Journal of Geology, v. 37, p. 428–461.

—— 1930, Mechanics of frost heaving: Journal of Geology, v. 38, p. 303–317.

Tanner, W. F., 1958, The zig-zag nature of type I and type IV curves: Journal of Sedimentary Petrology, v. 29, p. 408–411.

Tator, B. A., 1952, Pediment characteristics and terminology: Annals of the Association of American Geographers, v. 42, p. 295–317.

Taylor, D. W., 1937, Stability of earth slopes: Journal of the Boston Society of Civil Engineers, v. 24, p. 197–246.

Terzaghi, K., 1943, Theoretical soil mechanics: New York, John Wiley and Sons, 510 p.

—— and Peck, R. B., 1948, Soil mechanics in engineering practise: New York: John Wiley and Sons, 566 p.

Thakur, T. R. and Scheidegger, A. E., 1968, A test of the statistical theory of meander formation: Water Resources Research, v. 4, p. 317–329.

Thorarinsson, S., 1939, Observations on the drainage and rates of denudation in the Hoffellsjokull district: Geografiska Annaler, v. 21, p. 189–215.

Tinkler, K. J., 1966, Slope profiles and scree in the Eglwyseg Valley, North Wales: Geographical Journal, v. 132, p. 379–385.

—— 1970, Pools, riffles and meanders: Geological Society of America Bulletin, v. 81, p. 547–552.

Tinney, E. R., 1962, The process of channel degradation: Journal of Geophysical Research, v. 67, p. 1475–1480.

Trask, P. D., 1952, Sources of beach sands at Santa Barbara, California, as indicated by mineral grain studies: U.S. Beach Erosion Board Technical Memo No. 28, 24 p.

Troeh, F. R., 1964, Landform parameters correlated to soil drainage: Soil Society of America Proceedings, v. 28, p. 808–812.

—— 1965, Landform equations fitted to contour maps: American Journal of Science, v. 263, p. 616–627.

Tylor, A., 1853, On the changes of sea-level effected by existing physical causes during stated periods of time: Philosophical Magazine, 4th Series, v. 5, p. 258–281.

—— 1875, Action of denuding agencies: Geological Magazine, v. 22, p. 433–473.

Ursic, S. J. and Dendy, F. E., 1965, Sediment yields from small watersheds under various land uses and forest covers: U.S. Department of Agriculture, Miscellaneous Publications 970, p. 47–52.

Van Burkalow, A., 1945, Angle of repose and angle of sliding friction: an experimental study: Geological Society of America Bulletin, v. 56, p. 669–708.

Van Dijk, W., and LeHeux, J.W.N., 1952, Theory of parallel rectilinear slope recession I and II: Nederlandse Akademie van Weternschappen Proceedings, Series B, v. 55, p. 115–122 and p. 123–129.

Van Dijk, J. W., and Vogelgang, W.L.M., 1948, The influence of improper soil management on erosion velocity in the Tjiloetoeng Basin (Residency of Cheribon, West Java): Meded, alg. Proefstn. Landb. Buitenz., 71, 10 p.

Varnes, H. D., 1949, Landslide problems of Southwestern Colorado, U.S. Geological Survey Circular 31, 13 p.

Vice, R. B., Guy, H. P., and Ferguson, G. E., 1969, Sediment movement in an area of suburban highway construction, Scott Run Basin, Fairfax County, Virginia, 1961–64: U.S. Geological Survey Water Supply Paper 1591-E, p. E1–E41.

Waananen, A. O., 1961, Hydrologic effects of urban growth—some characteristics of urban runoff: U.S. Geological Survey Professional Paper 424-C, p. C353–C356.

Wahrhaftig, C., and Cox, A., 1959, Rock glaciers in the Alaska Range: Geological Society of America Bulletin, v. 70, p. 383–436.

Walling, D. E., and Gregory, K. J., 1970, The measurement of the effects of building construction on drainage basin dynamics: Journal of Hydrology, v. 11, p. 129–144.

Warner, R. F., and Pickup, G., 1978, Channel changes in the Georges River between 1959 and 1973/76 and their implications: Public Works Department Survey, Department of Geography, University of Sidney, Australia, 84 p.

Washburn, A. L., 1967, Instrumental observations of mass-wasting in the Mesters Vig district, N. E. Greenland: Meddelelser om Gronland, v. 166, p. 1–296.

Waters, R. S., 1958, Morphological mapping: Geography, v. 43, p. 10–17.

Weertman, J., 1957, On the sliding of glaciers: Journal of Glaciology, v. 3, p. 33–38.

—— 1962, Catastrophic glacier advances: U.S. Army Corps of Engineers Cold Regions Research and Engineering Lab., Research Report No. 102, 8 p.

Werner, C. and Smart, J. S., 1973, Some new methods of topologic classification of channel networks: Geographical Analysis, v. 5, p. 271–295.

West, R. G., and Donner, J. J., 1956, The glaciations of East Anglia and the East Midlands: a differentiation based on stone-orientation measurements of the tills: Quarterly Journal of the Geological Society of London, v. 112, p. 69–91.

White, J. F., 1966, Convex-concave landslopes: a geometrical study: Ohio Journal of Science, v. 66, p. 592–608.

Wiegel, R. L., 1964, Oceanographical engineering: London, Prentice-Hall, 532 p.

Williams, G., 1964, Some aspects of the eolian saltation load: Sedimentology, v. 3, p. 257–287.

Williams, G. P., 1978, Hydraulic geometry of river cross sections—theory of minimum variance: U.S. Geological Survey Professional Paper 1029, 47 p.

Williams, P. J., 1957, Direct recording of solifluction movements: American Journal of Science, v. 255, p. 705–715.

Williams, P. W., 1972, The analysis of spatial characteristics of karst terrain, *in* Chorley, R. J., ed., Spatial analysis in geomorphology: New York, Harper and Row, p. 135–163.

Wilson, I. G., 1972, Aeolian bedforms—their development and origins: Sedimentology, v. 19, p. 173–210.

Wilson, K. V., 1967, A preliminary study of the effect of urbanization on floods in Jackson, Mississippi: U.S. Geological Survey Professional Paper 575D, p. 259–261.

Winkler, E. M., and Wilhelm, E. J., 1970, Salt burst by hydration pressures in architectural stone in urban atmospheres: Geological Society of America Bulletin, v. 81, p. 567–572.

Winkley, B. R., 1973, Metamorphosis of a river; a comparison of the Mississippi River before and after cutoffs: Water Resources Research Institute, Mississippi State University, 42 p.

Woldenberg, M. J., 1966, Horton's laws justified in terms of allometric growth and steady state in open systems: Geological Society of America Bulletin, v. 77, p. 431–434.

——1969, Spatial order in fluvial systems: Horton's laws derived from mixed hexagonal hierarchies of drainage basin areas: Geological Society of America Bulletin, v. 80, p. 97–112.

——1971, Two dimenional spatial organization of Clear Creek and Old Man Creek, Iowa, *in* Morisawa, M., ed., Quantitative geomorphology: some aspects and applications: State University of New York, Binghamton, Publications in Geomorphology, p. 83–106.

Wolman, M. G., 1955, The natural channel of Brandywine Creek, Pennsylvania: U.S. Geological Survey Professional Paper 271, 56 p.

——1959, Factors influencing erosion of a cohesive river bank, American Journal of Science, v. 257, p. 204–216.

——1967, A cycle of sedimentation and erosion in urban river channels: Geografiska Annaler, v. 49A, p. 385–395.

——and Brush, L. M., Jr., 1961, Factors controlling the size and shape of stream channels in coarse non-cohesive sands: U.S. Geological Survey Professional Paper 282-G, p. 191–210.

——and Schick, A. P., 1967, Effects of construction on fluvial sediment, urban and suburban areas of Maryland: Water Resources Research, v. 3, p. 451–464.

Wood, W. F., and Snell, J. B., 1960, A quantitative system for classifying landforms: Quartermaster, Res. and Eng. Command, U.S. Army Technical Report EP-124, Natick, Mass.

Woolhiser, D. A., and Lenz, A. T., 1965, Channel gradients above gully-control structures: American Society of Civil Engineers, Journal of the Hydraulic Division, No. HY 3, p. 165–187.

Wu, T. H., 1966, Soil mechanics: Boston, Allyn and Bacon, 429 p.

Yang, C. T., 1971a, On river meanders: Journal of Hydrology, v. 13, p. 231–253.

——1971b, Potential energy and stream morphology: Water Resources Research, v. 7, p. 311–322.

——and Song, C.C.S., 1979, Dynamic adjustments of alluvial channels, *in* Rhodes, D. D., and Williams, G. P., eds., Adjustments of the fluvial system: Dubuque, Iowa, Kendall/Hunt Publishing Company, p. 55–67.

Yasso, W. E., 1964, Geometry and development of spit-bar shorelines at Horseshoe Cove, Sandy Hook, New Jersey: U.S. Naval Research Project NR 388-057, Contract No. 266 (08), Technical Report 5, Department of Geology, Columbia University, New York, 104 p.

Yorke, T. H., and Herb, W. J., 1978, Effects of urbanization on streamflow and sediment transport in the Rock Creek and Anacostia River basins, Montgomery County, Md, 1962–74: U.S. Geological Survey Professional Paper 1003, 71 p.

Young, A., 1963, Deductive models of slope evolution: Nachrichten der Akademie der Wissenschaften in Gottengen-Mathematisch-Physikalische Klasse #5, v. 3, p. 45–66.

——1972, Slopes: Edinburgh, Oliver and Boyd, 288 p.

Zenkovich, V. P., 1967, Processes of coastal development (Translated from the Russian by O. G. Fry): Edinburgh, Oliver and Boyd, 738 p.

Zingg, A. W., and Chepil, W. S., 1950, Aerodynamics of wind erosion: Agricultural Engineering, v. 31, p. 279–284.

Geological Society of America
Centennial Special Volume 1
1985

The history of American theories of cave origin

Richard A. Watson
Department of Philosophy
and
Department of Earth and Planetary Sciences
Washington University
St. Louis, Missouri 63130

William B. White
Department of Geosciences
and
Materials Research Laboratory
Pennsylvania State University
University Park, Pennsylvania 16802

ABSTRACT

In 1930, William Morris Davis published "Origin of Limestone Caverns" in the *GSA Bulletin,* initiating an episode of mistaken interpretation and resistance to field evidence almost unique in American geology. Davis proposed a two-cycle, deep-circulation theory of cave genesis by phreatic solution. J Harlan Bretz made heroic efforts to defend the theory, but geologic facts show it to be mostly wrong.

Davis ignored European literature, particularly the work of Alfred Grund and Jovan Cvijić whose empirical studies demonstrate that many caves are formed by solutional and mechanical action near and above the water table. Davis was misled by his lack of geochemistry and by fanciful maps of Mammoth Cave that show a labyrinth rather than the actual modified dendritic pattern of passages.

Among others, Claude A. Malott and James H. Gardner presented theories that displaced Davis' phreatic theory. In particular, Allyn C. Swinnerton developed a theory of temperate zone cave formation by flow along a seasonally fluctuating water table that is best supported by geologic evidence today. Recent work by Franz-Dieter Miotke, Arthur N. Palmer, and others relates episodes of cave formation to Pleistocene glacial periods.

Geochemical work initiated by Clifford A. Kaye and others caused a revolution in cave studies, and these data integrated with geologic, geomorphic, and hydrologic details by Derek C. Ford, Ralph O. Ewers, and others have led to sophisticated models of cave origin adaptable to all situations.

Some caves are formed by deep phreatic solution, although Carol A. Hill argues that in Carlsbad Caverns the active agent was sulfuric, not carbonic acid. Davis' prestige was so great, however, that his incorrect hypothetical model was an obstacle to the understanding of cave origin, and remains so today because of its continued uncritical incorporation in some elementary texts.

INTRODUCTION

In 1899, William Morris Davis toured the Dalmation karst with Albrecht Penck. Davis was at that time interested mainly in glacial processes, and in his written account of the tour he comments only briefly on karst features, referring the reader to Cvijić (1893). Thirty-one years later, Davis published "Origin of Limestone Caverns" (1930), an almost entirely theoretical paper in which he proposes a theory of cave origin involving solution during an initial period of deep-seated phreatic circulation, then regional uplift and lowering of the water table, and then deposition during a subsequent period of shallow vadose water flow.

This two-cycle theory was tested, utilized, and expanded in the field by J Harlan Bretz, who defended it until his death. Davis' two-cycle theory was and is contrary to most European work on the origin of caves, including Penck's and Cvijić's. It embodies a process of anastomosing deep circulation that was accepted by few field karst geologists other than Bretz. Davis' theory did, however, make its way prominently into American textbooks, where it continues to be taken more seriously than it ever should have been. It is inevitable, therefore, that this present study—primarily in the history of ideas—must have at its core an examination of the Davis theory. Nevertheless, our thesis is that numerous other American geologists are more important than Davis in developing theories of cave origin that correspond to and derive from field data, and one of our conclusions is that Davis' influence on karst studies has been generally regressive.

There are three periods of American cave research. The earliest ends in 1930. Then there is a "classic" period in the wake of Davis from 1930 to 1942. This is followed by a hiatus initiated by World War II. It is important to note, however, that this is a period during which many cave explorers and researchers were organized. The National Speleological Society was incorporated in 1941, and in 1957 a group of geologically oriented NSS members incorporated as the Cave Research Foundation. In 1957, the third "modern" period began and extends to the present day.

I. THE EARLY PERIOD

The general ideas of cave origin in America prior to 1930 are best expressed by Greene (1908) and Matson (1909). In discussing caves of the Mitchell Limestone, Greene says that rain water soaks through the soil to drain into the limestone through joint planes. He infers that downward flow is easiest where joint planes meet, and that sinkholes are most likely to form at these intersections. Horizontal flow is then initiated where joints get tighter at depth, where groundwater level or an impervious bed is encountered, or "what is probably most important, a level corresponding to that of the local base level of erosion may be reached and divert the downward moving water" (Greene, 1908, p. 177–178). Obviously, local base level can influence the downward flow of water only if it determines the water table level or if the vertically flowing water enters an underground

stream adjusted to that base level. This last is the situation Greene apparently envisions developing as water pursues the lines of least resistance vertically and horizontally along joint and bedding planes. He says that young caves are formed by solution, containing clear water flowing through passages with bare limestone walls. Later, the sinkhole entrances enlarge permitting soils and other clastic materials to enter and then cave streams gradually take on the features of surface streams through corrasion and deposition. Cave passages are cut narrow and deep if the input is high above the water outlet, low and wide if there is only a low gradient. In sum, caves are formed by the solutional action of vadose water above and along the water table, and are enlarged both by corrosive and corrasive processes.

Matson (1909) also believed that caves form by solution where water flows through limestone, most usually where an open joint intersects a prominent bedding plane. Active water flow is necessary, so caves are formed primarily in the vadose zone above the level of the surface stream into which underground water discharges. Matson says deep phreatic circulation is possible but does not cite it as a factor in cave origin. Instead, he stresses the release of phreatic water initiated by the lowering of base level as surface streams deepen their valleys by erosion. Passages formed by the flow of released phreatic water provide routes for water entering from the surface. Rapid vertically directed flow causes pits and domes, and rapid horizontally inclined flow cuts canyons. Cave streams migrate from higher to lower levels during their development until they reach grade with the surface stream into which underground water exits.

Both Greene and Matson stress the corrosive power of water as it enters the limestone from the surface. This solutional power weakens as the water becomes saturated on its way through the limestone. $CaCO_3$ dissolved early in the water's career underground may be deposited deeper in the cave system.

The basic theory of cave origin in America prior to Davis, then, is one of invasion of surface water into limestone and subsequent solutional activity along joint and bedding planes in the vadose zone followed by corrasive enlargement of cave passages as underground streams adjust to the local base level and water table determined by the surface stream into which the cave waters discharge. In the cases of Greene and Matson, this theory fits the field situations they investigated in Indiana and Kentucky.

The views of a number of other American geologists are summarized in Davis (1930, p. 447–480). None that Davis cites is as complete as the above, but they add up to the same vadose theory of cave origin.

Greene, Matson, and their contemporaries apparently assumed that water moving through the limestone mass would behave like water flowing over the land surface. It would descend vertically through open joints until it reached the water table or surface base level and then turn abruptly and flow horizontally toward some point of discharge to the surface. Such views permeated much early thinking about groundwater movement but were shown to be incorrect in light of mathematical analyses of groundwater motion. Hubbert (1940) shows that the concept of water running down the water table surface violates fundamental conservation laws. Water does indeed move vertically through the vadose zone, but when it reaches the water table it continues vertically following flow lines whose shapes are dictated by the distribution of hydraulic heads. There is no abrupt right angle turn. The success of Hubbert's groundwater flow model in all rock types other than limestone is in part responsible for the later difficulty in replacing the Davis-Bretz hypothesis in spite of overwhelming field evidence against it. It was necessary to produce a theory of groundwater flow in karstic rock with the same degree of mathematical rigor as Hubbert's theory of flow in porous media, and this has not yet been accomplished to everyone's satisfaction.

II. THE CLASSIC PERIOD

William Morris Davis takes it to be obvious from observation that "Two chief epochs in cavern history are . . . recorded; first, an earlier [longer] epoch of solutional or corrasional excavation; second a later [shorter] epoch of depositional replenishment" (1930, p. 477). Thus Davis begins with a presumption of two epochs in the history of cave development, and he takes as his problem the explanation of "the change from the greater and earlier excavational work to the smaller and later depositional work" (p. 477), a problem he finds that other geologists do not address. In fact, Greene and Matson explain both clastic deposition as the result of the enlargement of cave entrances and passages, and chemical deposition as the result of the destabilization of saturated water. They also imply that processes of excavation and deposition take place at the same time in a cave system. In Europe, among others, Cvijić (1893) and Kyrle (1923) describe excavation and deposition as integrated processes in the vadose zone.

Davis, however, immediately poses the hypothesis

that large caverns are ordinarily excavated by ground-water solution during an epoch when the body of limestone in which they occur lies below the water table of its district; and that the change from this epoch of solutional excavation to the following epoch depositional replenishment takes place when the water table sinks below the cavern level in consequence of regional uplift or other effective cause (p. 480).

What is striking about this hypothesis is that it is presented to explain two epochs of cave history which Davis merely *assumes* without explicit argument to have taken place. Presumably, he reasons that the cave opening must have been excavated before it can be filled with deposit. This is reasonable, but it is gratuitous to go on to assert that there is first an epoch of excavation and then an epoch of deposition. If one *does* assume that these successive epochs took place, then the problem *does* arise as to what caused the change from excavation to deposition. What Davis has done is take as a given two-epoch picture of cave origin that should really be part of his hypothesis. This is particularly obvious in the light of theories such as Cvijić's, Kyrle's, Greene's,

and Matson's which are based on field observations of excavation and deposition taking place in the same caves at the same time.

Why does Davis see the problem of cave origin in this two-epoch, two-cycle way? It is most probable that he is fitting a theory of cave origin to his larger theory of regional landform development according to which long epochs of stable peneplanation are interrupted by regional uplift which is followed by the erosional formation of landforms in stages from youth to old age. The long epoch of phreatic cave solution coincides with the long old age stage of peneplanation; the epoch of deposition in caves coincides with the stage of youth initiated after regional uplift. It seems likely, then, that Davis' two-cycle theory of cave origin is more the result of deduction from his general theory of geographic cycles, than the result of induction from field observations of cave features. This deduction presents him with a solution to a problem that does not exist, the problem of what causes the change from excavation processes in one epoch to deposition processes in a succeeding epoch. In fact, excavation and deposition take place coextensively and cotemporally in all caves. This is not to deny that some caves are nearly empty of deposits and that other caves are nearly full of deposits. Such caves are known. And it is also the case that deposition may be more prevalent in some parts of a cave and solution more prevalent in other parts. But what is obvious is that these process take place at the same time, in the vadose zone, in identical regional circumstances. What Davis does is take locational or spatial separations as evidence for the temporal separation of the operation of the one process from the operation of the other process.

The usual sequence of development for a cave passage is enlargement under completely pipe-full conditions, followed by a period when it may have stalactites growing on the ceiling while a stream occupies its floor, followed by a period when the flowing water is gone but deposition of travertine may continue. Certainly the processes of solution and deposition are sequential in time although there is no abrupt transition between them. In many caves in which there are interconnected sequences of passages on different levels, all processes run simultaneously with solution in completely flooded passages below base level, free surface stream action near base level, and deposition in dry, upper-level passages. Such caves record a long, complex history of both solution and deposition. In any single passage or conduit of the plumbing in such caves, solution and deposition are time-sequential. Davis thought that all of Mammoth Cave formed deep below an old peneplain more or less at one time, followed by lowering of base level and deposition of travertine, instead of seeing a complex history of both processes running in parallel, but sequentially offset in time for passages on higher to lower levels from the late Tertiary to the present.

There is plenty of observational evidence for contemporaneous excavation of caves and deposition in caves in the vadose zone. What evidence does Davis cite for deep-phreatic solution of cave passages? Davis cites Grund (1910) as anticipating the two-cycle theory by inferring from water-filled caves that they were formed below the water table, drained when the water table was lowered, and then were filled with flowstone deposits. A second observation Davis cites is the presence of very small streams in large caves, a misfit similar to that of small streams in enormous glacial valleys that they did not excavate. There are also caves with walls, floors, and ceilings covered with crystals that apparently were precipitated when the caves were under water. But the feature that Davis stresses most is the apparent network, maze, or anastomosing pattern of passages in many caves as contrasted to the branchwork or dendritic pattern of surface streams. Davis argues that the network pattern with blind passages, closed loops, dead-end rooms, and random pockets with no integrated lines of water flow indicate that these openings were dissolved by deep phreatic circulation of water to depths of thousands of feet along all possible lines of weakness, rather than by vadose flow which seeks the most direct route to a local base level. Mazes with closed loops, passages branching in a downstream direction, large passages connected by small ones, and undulating longitudinal profiles all indicate to Davis that caves cannot have been formed by underground streams in the vadose zone.

Davis (1930, p. 484) apologizes that he is too old to do any field work in caves and comments that "the facts on which the essay is based are therefore derived almost entirely from the work of others" and that his own contribution is "largely deductive." Thus, it is understandable how Davis was misled by superficial observations of others, and particularly in one crucial case by impressionistic maps of network passages in Mammoth Cave (Davis, 1930, fig. 14, p. 514; fig. 54, p. 597). Most known caves—even those with network patterns—are composed of passages forming an integrated drainage system. Because these systems are expressed in three dimensions and include many passages too small for human penetration or casual notice or that are blocked by breakdown or clastic fills, the impression that many caves consist of dead-end rooms, blind passages, and network systems is easily gained. In any event, Davis took as fact that many caves do not show evidence of being formed by water flowing in an integrated drainage system, and he took this to be a crucial datum for his inference that most caves are formed by the solutional activity of slowly circulating water in the phreatic zone. Phreatic water was thought to circulate because of the hydrostatic head between the higher parts of the water table and base level, and because of currents caused by gradients in temperature and mineral content.

Davis agreed that some linear caves must have formed under vadose conditions. Other than crystal-lined pockets—giant geodes with no entrance—are there any caves that probably did form in the phreatic zone as Davis argues? Yes, Carlsbad Caverns—of which Davis had a good map (Davis, 1930, fig. 50, p. 587)—appears to have formed at depth. We do not fault Davis for generalizing from Carlsbad Caverns and similar caves. He did assume that Carlsbad Caverns was formed by the solutional activity of regular underground water, noncarbonated or in the form of a weak carbonic acid. Recent detailed geochemical studies suggest an alternative hypothesis, however, that Carlsbad Caverns and some other deep-seated caves in similar lithologic situations

were formed by the action of a sulfuric acid solution (Hill, 1981, in press). This process is not generalizable to many caves.

Davis also points to springs discharging from outlets hundreds of feet below sea level off the coast of Florida, and the depth of the passages that bring water up in the big springs of the Ozarks. Neither of these cases supports the notion of phreatic circulation as Davis conceives it. Many submerged caves along the coasts were most likely formed when sea level was lower, and the routes of water flowing out of big springs is restricted to arterial lines by confinement of soluble beds between impervious strata. We believe that Davis—who recognized these alternate possibilities—rejected them because he was in the grip of his overall geographic model of landform development which suggested to him the two-cycle theory of cave origin.

To be sure, models in science are based on and tested by empirical data, but in this case Davis appears to be fitting data to his preconceived model rather than testing the model with the data. The line between "testing" and "fitting," however, is a thin one. In his lifetime Davis had accumulated an enormous amount of data to support his theory of geographic cycles. We are not arguing that his reasoning about the origin of caves was bad. It was just wrong. We show here why it was mistaken: because Davis was too confident of the model he took to the data (leading him to discount the idea that coastal caves were formed above the water table then submerged), because some of the data was false (the Mammoth Cave map) or lacking (the Carlsbad Caverns geochemical situation), and because he could not himself undertake the field investigations necessary to test his theory.

Davis also misgauged the solutional behavior of water. He thought that water flowing through vadose passages would not have time or enough contact with the limestone to dissolve much rock, but that extensive solutional activity can take place only in long periods of time deep in the phreatic zone (Davis, 1930, p. 505, 524–525, 558–561). He thought that the enormous quantity of dissolved mineral matter in discharging karst waters could not have come from solutional activity by surface and vadose water.

From all evidence, Davis did not know much chemistry. Cave enlargement is mainly a chemical process. By claiming that the tens of millions of years of stagnant water percolation under a peneplain is necessary for water to reach saturation, Davis was making guesses about rates of chemical reactions that are six to eight orders of magnitude too large. The pivotal role of CO_2 uptake and discharge is what makes it possible to have solution and deposition in the same cave passage. The chemical principles were well known at the time Davis wrote, but he did not apply them.

Recent studies show, however, that vadose and shallow phreatic waters are aggressive, while deep phreatic water is often saturated (Langumir, 1971; Shuster and White, 1971; Thraikill and Robl, 1981). Davis cites in particular his skepticism about the power of trickling films of water to corrade and corrode, suggesting that deposition by evaporation from films should take place instead. Again, recent studies have shown that the corrosional power of thin films of water can be great, and that vertical shafts can form actively and rapidly by solution even while chemical deposition is taking place in parts of them (Brucker, Hess, and White, 1972; Miotke, 1973b). These studies show that solution and deposition in the vadose zone almost always go together.

Davis also uses King's theory of the ideal paths of ground-water movement which shows lines of circulation far below the surface (Davis, 1930, fig. 31, p. 549). Throughout Davis' text there is conflict between his dependence on pervious rock and his recognition that joint and bedding planes in massive, nearly impermeable limestones concentrate the flow of water. These structural conditions that Davis admits lead to integrated passage formation in the vadose zone are said by him merely to provide weak zones for random solution in the phreatic zone.

Extensive solution in the phreatic zone would release clay fractions in the limestone that would sink to the floor. Thus Davis cites beds of clay in caves as further evidence of phreatic solution (Davis, 1930, p. 554).

Davis also makes much of the differences in passage profile that would result from vadose as contrasted to phreatic flow. The profile of a water-filled passage would be elliptical, whereas vadose streams would cause rectangular profiles (Davis, 1930, fig. 32, p. 555). We discuss other differences below as stressed by Bretz (1942).

Davis took passages at different levels, e.g., in Mammoth Cave (Davis, 1930, p. 598), to indicate phreatic origin because vadose flow would be adjusted to one base level. He did not consider, as have recent workers (Miotke and Palmer, 1972; Miotke, 1973a) that such multi-level caves may be very old. Levels of Mammoth Cave now have been correlated with base levels determined by the Green River during the successive epochs of Pleistocene glaciation (Palmer, 1981).

Davis took dome pits, great vertical shafts such as in Mammoth Cave, as definitive evidence of phreatic formation, because (again) he could not conceive of their formation in the vadose zone by thin films of water running down their walls. Further investigation has shown, however, that such vertical shafts are controlled stratigraphically, and that vadose solution is adequate to form them (Pohl, 1955; Brucker, Hess, and White, 1972).

Davis and Bretz cite vertical shafts up to more than 100 feet high and over 30 feet in diameter as evidence for phreatic solution. They believed that vadose water seeping down the walls of such shafts was not adequate to form them. This assumption is challenged by Pohl (1955, written in the mid-1930s) with an argument based largely on geomorphic and stratigraphic factors. He shows that vertical shafts in the Mammoth Cave region are clustered below the heads of surface valleys and under valley walls wherever the sandstone caprock of the region is truncated to expose the underlying limestone. He postulates that water flowing over the truncated edges of the impermeable sandstone is directed downward at the easiest points of access. This would probably be at the intersection of two major joints. Vertical shafts would then intersect all levels of horizontal cave development to bottom at or near present base level. Pohl argues that far from

being ancient, vertical shafts are among the youngest active cave features. Field evidence supports Pohl's theory, and geochemical work cited herein confirms his hypothesis that seeping vadose water has the solutional power to form vertical shafts.

In conclusion, Davis (1930, p. 623) stresses the network pattern to support his two-cycle theory of cave origin. His theory fails on this basis alone because most caves—when mapped in three-dimensional detail—do not exhibit network patterns but rather show integrated drainage systems that have evolved through time.

There is also the question of scale. Karst drainage basins typically have areas of 10-300 km² and the integrated drainage patterns are spread out on this scale. A small cave, say 100-1,000 meters long, does not contain enough of the drainge system to demonstrate its pattern clearly. Even in Mammoth Cave, with some 500 km of interconnected passages mapped, only a fraction of the regional drainage system is shown (Watson, 1966; Quinlan and Rowe, 1977).

We have examined Davis theory in some detail because it is so influential, because it is generally wrong, and because it is an example both of Davis' great synthesizing power and of how limited data considered under the influence of a dominant general model can give rise to a specific theory (here of deep-phreatic cave origin) coordinate with the general model, but false to much of the factual data. We now examine several geologists who used the same data that Davis had, but on its basis come to the conclusion contrary to Davis that most caves are of vadose or shallow phreatic origin.

We speculate that Davis was led to his deep phreatic theory of cave origin by his theory of geographic cycles in which regional uplift provides the answer to the supposed need of a cause to explain why there is a shift from excavation to deposition processes in caves. Piper (1932) shows, however, that a vadose theory of cave origin can be coordinated equally well with Davis' geographic cycles. Piper draws an analogy between the erosion cycle on the surface and the underground erosion cycle. Depending on the permeability of the limestone, the underground cycle may either precede or lag behind the surface one. In rocks with densely spaced, open joints, drainage may be diverted to the subsurface rapidly and a mature underground system may develop before the surface pattern has time to develop. Inversely, if joints are sparse and tight, surface erosion may dissect the system before an integrated underground drainage can develop. The stages of development fit into the traditional Davisian pattern as follows:

Youth: Small cave passages advance headward from valleys into the limestone beneath the uplands. As they advance, they split into dendritic branches. Some of the branches intersect forming looped passages without main trunk channels.

Maturity: Most surface catchments are diverted into the cave system. Trunk passages are established and these grow rapidly to become the main conduits of the cave. Underground streams approach grade to a surface base level.

Old Age: An underground "peneplain" level is established by lateral widening of the cave system as far as structure and stratigraphy will permit. Collapse sinks are common. The cave system is dissected as surface old age and underground old age features merge. Finally, the entire area is reduced to a surface peneplain.

Piper's general picture is most appropriate for flat-lying limestone. Gardner (1935) stresses such control by structure. He assumes that most large caves are formed on the up-dip side of surface valleys. The downcutting of surface streams initiates cave formation by tapping zones of static water. Then vadose water flows down-dip along the bedding planes. As downcutting continues, more and more levels develop and piracy takes place from level to level. Surface waters drain into the cave system. Integrated subsurface streams then develop in which the ratio of solution to abrasion is about what it is in ordinary surface streams.

Clyde A. Malott was a pioneer student of karst phenomena in the United States with a special interest in southern Indiana. Most of the support for his "invasion" theory of cave origin appears in a series of descriptive papers. The formal statement of his theory occurs only as an abstract (Malott, 1937) quoted below in its entirety.

Primitive, ill developed, and somewhat selective three-dimensional passageways are developed in limestone regions below the water table. Vadose waters under a flow-head develop underground flow routes from selected parts of the earlier prepared primitive systems. Fairly large streams are diverted from their surface courses into these underground routes. The diverted waters carve out and fully integrate the large caverns at or near the water table. The full developed caverns bear unmistakable evidence of streams that coursed through them, though when long vacated they tend to fill with dripstone, become encumbered with fallen rock, and lose parts of their stream-carved floors, as is well illustrated by the upper galleries of some large cavern systems.

Although Malott might appear to support the phreatic theory, he does not mean that caves are formed below the water table. All he insists is that some openings, however small, must be available for water flow. Water from the surface invades the underground through these openings and forms caves by enlargement in the vadose zone.

One aspect of vadose theories is that underground streams grade to a base level determined by the surface streams they discharge into. This implies that underground streams cannot erode below local water table level. This water table level, however, rises and falls seasonally. Thus it is suggested that major drainage passages in caves form in a shallow phreatic (seasonally vadose) zone defined by a fluctuating water table, by both corrosion and corrasion processes.

Swinnerton (1932) came to this view on the basis of the observation that most cave passages are nearly horizontal and that some are stacked in levels (tiered caves), even in folded limestones where the horizontal passages cannot be regarded as simply a result of low dip. To account for horizontal passages that truncate geologic structure, Swinnerton hypothesizes that they form in the interval between the high and low stands of the water table as it fluctuates between wet and dry seasons. The main points of Swinnerton's version of the shallow phreatic theory are:

1. Cave enlargement takes place by means of lateral flow of groundwater which begins after entrenching of surface streams has begun. This means that caves near the water table are related to the present erosion cycle.

2. The lateral enlargement of caves is similar to the headward cutting of a surface valley.

3. Piracy occurs between levels and between caves on the same level.

4. Major cave solution begins just as soon as downcutting surface streams provide sufficient hydrostatic head. Thus cave development begins in the youth stage of the erosion cycle.

5. Underground drainage becomes well-integrated only when surface topography reaches a mature stage. In maturity there is maximum relief and therefore maximum water table gradient.

6. During maturity the erosion of the cave system is faster than the erosion of the surface. Much drainage is diverted underground and surface erosion is correspondingly retarded. Thus the water table reaches old age or peneplain stage long before the surface does. In the old age stage there is little hydrostatic head and consequently little solution takes place.

7. Formation of crystal caves with passages lined with dogtooth spar takes place under the static water conditions of these later stages.

8. During old age, cave enlargement still takes place in the floodwater zone seasonally due to torrential rains raising the water table level for short times.

9. This is strictly a one-cycle theory. Different cave levels are formed merely as a result of the downcutting of surface valleys to provide new underground base levels. Underground streams are pirated to lower and lower levels. Only with the relatively static water table level of old age does breakdown and the wearing away of the overburden begin to destroy caves. Deposition of dripstone occurs in the upper levels concurrently with the removal of limestone from the lower levels.

J Harlan Bretz was a brilliant thinker who defended Davis' two-cycle theory with many observations in the field. In his "Vadose and Phreatic Features of Limestone Caverns" (1942), Bretz listed a large number of cave features that he believed could be formed only under phreatic conditions. These include sponge-work, network passages, horizontal chambers in dipping beds, anastomosing passages, pockets, tubes, and flow markings on ceilings as well as on walls and floors (Bretz, 1942, p. 699–722). He found vadose features such as entrenched channels, clastic deposits, and flowstone to be imposed on phreatic features everywhere, which indicated to him that a phreatic cycle precedes a vadose cycle. Bretz (1942, p. 773–774) found an enormous amount of unctuous clay fill in caves of the Ozarks and argued that these caves record an epoch between Davis' first and second: "the first occurring beneath the water table and being entirely solutional, the second described as one of replenishment, the epoch of dripstone and flowstone deposition after air enters the cave."

Bretz argued that the many levels, e.g., in Mammoth Cave, could form at the same time in the phreatic zone, draining only as surface stream entrenchment lowers the water table (p. 789). But as remarked above, levels of cave passages have now been correlated with surface features such as river terraces which represent periods of base-level stability.

Bretz modifies Davis' two-cycle theory by inserting an epoch of clay filling between Davis' two stages:

Stage I. Groundwater circulation forms caves by deep-seated solution beneath an old land surface.

Stage II. Caves lying beneath a peneplain are saturated with stagnant water. Fine clay and silt, filtering down from the land surface above fills the caves completely (or nearly so) with unctuous clay.

Stage III. Cave systems are drained of water. Invading vadose streams begin to remove clay fills. Sometimes this continues until the vadose streams flow on bedrock floors. In other cases most of the cave is still clogged with clay. Vadose features are superimposed on the phreatic record. Sometimes enlargement takes place through the formation of floor slots and meander niches.

Bretz later (1949, 1955) came to the same conclusion after examining many other cave systems: Most caves form below the water table and are only modified and cleared of fill by the streams that now flow through them. However, a careful reading of the 1942 paper is instructive. There is much discussion of sub-water table streams, but nowhere is there any implication of flow at vast depths below the water table. Although Bretz was a firm supporter of the Davis hypothesis, one gets the distinct impression that Bretz's deep phreatic was nowhere near as deep as Davis' deep phreatic.

Is there any evidence for cave formation by deep-phreatic solution? There does seem to be some. Moneymaker (1941) reports large solution cavities more than 100 feet below the bedrock floors of streams in the Tennessee Valley. One might conclude, however, that solution of these cavities is related to drainage along stream valleys, and is not necessarily a widespread phenomenon. At least very few if any extensive cave systems have been discovered at depth below the water table through drilling. Nevertheless, enough bedrock solution seems to occur deep below the water table that toward the end of the classic period, Rhoades and Sinacori (1941) reopened the question of groundwater flow. They demonstrate the theoretical principles of flow-net circulation (fig. 1, p. 788), and discuss variations of flow according to hydrostatic head. In particular, they show (fig. 4, p. 793) how a horizontal passage at water table level can form through being the paramount location for solutional processes in the flow net. Flow is concentrated on this line and a master passage elongates headward as the shortest distance between the hydrostatic flow-net head and the base-level exit of underground water. In conclusion, however, they (p. 794) state that:

> The postulated conditions of homogeneity [of permeability and solubility] upon which the foregoing analysis is based are not encountered in limestone terrains where flow occurs exclusively through secondary openings. However, limestones possessing intersecting joint patterns of more or less uniform distribution may approach the postulated homogeneity to the degree that similar principles govern the mechanism of underground flow.

Even so, like Bretz, Rhoades and Sinacori defend deep circulation and solution with a theory that concentrates primary solutional activity at or just below water table level. What seems best to fit the data as the classic period closes in 1942 is the shallow phreatic theory of Swinnerton according to which main trunk passages form in the zone of a seasonally fluctuating water table, so that in the course of a single year caves are subjected to both phreatic and vadose processes of excavation and deposition. As indicated in Figure 1, however, during the classic period, for each zone of the aquifer, there was at least one hypothesis to the effect that caves originated in that zone.

III. THE HIATUS (1942–1957)

Then came World War II and a virtual absence of studies of cave origin from 1942 to 1957. It was not so much that no papers about caves were being published as it was a dearth of new ideas about cave origin. Cave geologists were using the old inadequate concepts, and most descriptions of caves during this period include a gross classification of cave origin as either vadose or phreatic. Nevertheless, this period is very important to American cave studies. As remarked above, the National Speleological Society was incorporated in 1941. As the only national speleological organization, the NSS has done a remarkable job of organizing the providing a center of communication for all Americans interested in caves. The quarterly *NSS Bulletin* is the only North American scientific periodical devoted to speleology. The NSS also encourages research by offering annual grants. The NSS provides a model of cooperation between amateurs and professionals that is the envy of speleologists in most other countries of the world. This cooperation was formalized by E. Robert Pohl who recruited cave explorers to check his theory of vertical shaft

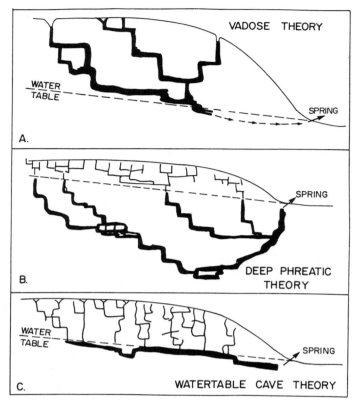

Figure 1. Zones of maximum cave development according to vadose, deep phreatic, and shallow-phreatic hypotheses. From Ford and Ewers (1978).

formation (1955) in the field. Then he, with James W. Dyer and Philip M. Smith were instrumental in incorporating the Cave Research Foundation in 1957 (Smith and Watson, 1970). Since 1957, over a thousand scientific publications have appeared under the auspices of CRF, which like the NSS, offers annual grants for research.

Among other developments, in 1953 the Cave Research Associates started *Cave Notes*, later *Caves and Karst*, a distinguished bi-monthly under the direction of Raymond deSaussure until it ceased publishing in 1973.

Since the early 1900s, there have been a few graduate thesis studies on caves, but most of these were isolated affairs. There was no "school" of cave research in North America. In the late 1950s and early 1960s, many more theses were written about caves, and the number has been increasing exponentially for two decades. Centers, formal or informal, grew up at McMaster University. The University of Kentucky, The Pennsylvania State University, and Western Kentucky University. And by 1960, enough work had been done that George W. Moore could edit a symposium titled "Origin of Limestone Caves" (1960) with nine contributors.

What we want to stress is that the explosion of cave studies that began in 1957 was a result of a great deal of ferment between 1942 and 1957. We are products of this era, and what we were

doing—like many others—was going to school. After 1945, many of us were entitled to the G. I. Bill of Rights and universities expanded. Some of us amateur cave explorers were severely discouraged by our professors from working on karst and caves. In a cave was not where *it* was at. But there were a fair number of us, we knew and encouraged one another, we were supported by organizations such as the NSS and CRF, and eventually we became professors ourselves who direct our students to work on caves where *it* is at, after all. And so that is how it was then, and how it is now.

IV. THE MODERN PERIOD (1957-PRESENT)

As renewed interest in cave research picked up momentum in the late 1950s, at least four distinct trends emerged.

First, a vastly improved data base was being accumulated. Methods and standards for cave cartography improved greatly, and maps and cave descriptions were produced in quality and profusion far beyond those available to Davis 30 years earlier. Mostly through the work of National Speleological Society cavers, systematic cave surveys were underway on a county-by-county and state-by-state basis. By attempting to describe *all* caves within a geographic area, cavers were providing data that could be used in quantitative calculations.

Second, process geomorphology also influenced the new cave studies. It was slowly being realized that the process of cave formation is a process of chemical solution coupled with an unusual sort of groundwater flow coupled with some mechanical transport. The appropriate models were built from carbonate chemical equilibria, dissolution kinetics, and the mechanics of fluid flow and mass transport. The appropriate disciplines are closer to chemical engineering than to traditional geomorphology.

Third, various statistical approaches were also introduced into geomorphology in the early 1950s. One looked for "measures" of landforms and landform processes and then relied on regression analysis, components analysis, and other methods to bring out relationships between variables connected in complex and generally unknown ways.

Fourth, work along traditional lines continues. The vadose/shallow-phreatic/deep-phreatic questions continued to be debated, but from positions supported by investigators' greatly enlarged field experience and vastly expanded data base. The milestones here are the work of William E. Davies in the late 1950s and the contemporary work of Derek C. Ford and his associates.

Swinnerton (1932) emphasized the floodwater zone at the top of the water table. Davies (1960) realized that the same arguments apply if solution takes place immediately below the water table and so proposed a shallow-phreatic hypothesis that consists of four stages:

Stage I. Random solution at depth. Primitive tubes, pockets, and other small solution openings are formed at random depths in the aquifer.

Stage II. Integration of mature development of solution openings. This occurs at water table level during a period when the water table is uniform in elevation and flow for a long period of time. Coincidence in the elevation of cave passages with the elevation of nearby river terraces suggests a causal relationship between the position of the water table and the position of local base level. This relationship also ties the evolving cave pattern to the evolving history of the drainage basin in which the cave is located.

Stage III. Deposition of fill. Cave fills of clastic materials are deposited after the integration and mature development of the cave passages. The sediments in Appalachian caves where Davies made his most important observations are stream-born sands and gravels which are evidence of fast-moving water in Stage II. Davies' sedimentation stage is quite different from Bretz's stage of infilling of unctuous clays under stagnant water conditions.

Stage IV. Uplift and erosion. With the end of stable conditions, the cave passages are drained as the water table is lowered and becomes air filled. In this stage the passages are modified by

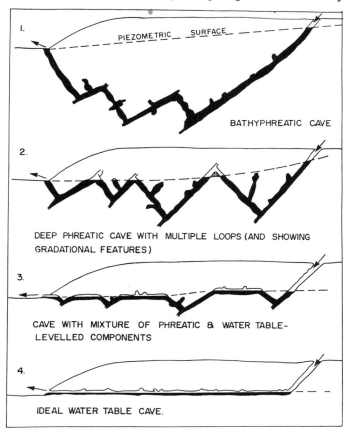

Figure 2. The four states of phreatic cave development according to the model of Ford and Ewers. From Ford and Ewers (1978).

deposition of speleothems and erosion of fill material. The ultimate development in this stage leads to total destruction of the cave by collapse and erosion.

Taken within the context of physiographic theorizing, the weight of supporting evidence is on the side of some sort of shallow-phreatic mechanism of cave origin. Solutional sculpturing as well as overall passage morphology suggest that many cave passages formed under pipe-full conditions. As more and more caves are mapped, it is apparent that many of them are nearly horizontal. The three-dimensional networks of tubes and shafts required by the most extreme interpretation of the deep-phreatic theory are seldom found. Caves in the folded Appalachians retain a horizontal aspect and terminate in both up-dip and down-dip directions as would be expected if they were controlled by some aspect of the groundwater table but not if there was extensive three-dimensional circulation (White, 1960). Davies (1960) and White and White (1974) relate caves to river terraces in the Potomac River Valley, and Miotke and Palmer (1972) and Miotke (1973a) relate Mammoth Cave passage levels to river terraces in Kentucky.

D. C. Ford and Ralph Ewers in a series of papers (Ford, 1965, 1968, 1971; Ford and Ewers, 1978; Ewers, 1982) offer what seems to be the ultimate answer to the question: Do caves form in the vadose zone above the water table, near the water table, or deep below the water table? The answer is: Yes. The Ford-Ewers model is framed in terms of hydrogeologic setting rather than according to general principles of geographic cycles, fluid mechanics, or solutional chemistry. The principal variables are the relief, the regional dip, and the concentration of joints and fractures.

Ford and Ewers distinguish several varieties of vadose caves, four "states" of phreatic cave, and two special cases, the isolated pocket cave and the artesian cave.

The basic concept in the model is that an optimum hydraulic path is selected from all possible fractures, faults, and bedding plane partings. This path need not, and indeed usually is not, a curvilinear flow line. Figure 2 shows the phreatic "states."

State 1, the bathyphreatic cave, occurs when fracture frequency is very small and as a result there is a deep loop beneath the piezometric surface. Their example is the cave system of the Sierra de El Abra, Mexico, where both descending and ascending tubes have been identified (Fish, 1977).

State 2 is cave development through an aquifer of low fracture frequency so that the optimum flow path follows a sequence of loops. With higher fissure frequency, the flow paths come up to points of zero pressure at the tops of the loops, the locus of which defines the piezometric surface. In the Ford-Ewers model, no special role is assigned to the position of the water table as a causative factor in cave genesis. Instead, the water table is regarded as a result of a particular flow path unless it is fixed by other barriers within the hydrogeologic setting. The amplitude of

State 2 loops may range from a few tens of meters to a few hundred meters. The type example is the Hölloch Cave System in Switzerland where the loop relief is several hundred meters.

State 3 is an intermediate case that appears with low to moderate fissure frequency. It consists of long reaches of water table cave with occasional phreatic loops.

State 4 caves occur at moderate to large fissure frequencies and are the true water table caves. A continuous grade between input and discharge point defines a drain of negligible flow resistance and thus a low-gradient water table. At high fissure frequency, all parts of the aquifer are in good hydraulic communication. Deep flow paths either do not form or are quickly closed by sediment.

The phreatic loops of states 2 and 3 are modified as the water table is lowered. There is entrenchment by free-surface streams that downcut across the tops of the old phreatic loops. These appear in caves as canyon passages that give way to tubes at both ends. It should be kept in mind that the drawings shown here to illustrate the flow paths have considerable vertical exaggeration. Particularly in limestones of low dip, the loops and their modifications may be very subtle with a relief of only a few meters over hundreds or thousands of meters of passageway. The lower phreatic loops act as sediment traps. If these fill as downcutting proceeds and the cave is still subject to annual fluctuations in water level, bypass tubes may develop across the tops of the loops. Once a bypass is available for the main water flow, the older loop may be completely filled with no trace remaining for the casual observer. One is suspicious of places where a spacious phreatic conduit of easy walking size suddenly gives way to a bedding plane crawl of considerably smaller dimensions.

Phreatic loops are also subject to a process described by Ford and Ewers as *paragenesis*. A cave passage at the bottom of a phreatic loop carries a substantial sediment load. As the cross section of the phreatic loop becomes larger, flow velocities decrease because the hydrostatic head across the loop has not changed and the sediment-carrying capacity of the conduit decreases. Sediment accumulated on the floor of the phreatic passage protects the floor from further solution, but solution attack on the ceiling continues. The passage thus dissolves its way upward, remaining always at the top of a growing pile of clastic sediment. A balance is developed between the sediment load and the cross-section size of the open portion of the top of the paragenetic passage. Upward solution continues until the passage reaches the piezometric surface.

Figure 3 illustrates two types of vadose caves. The first is the drawdown vadose cave. Drawdown caves develop where there is a catchment on an adjacent non-karstic borderland so that water collected in this catchment is emptied into the limestone. Initially the water table must be close to the land surface, but as the cave system develops, the water table quickly drops until it either is a piezometric surface across the top of a series of phreatic loops or is oriented along a water table cave. The rapid discharge through

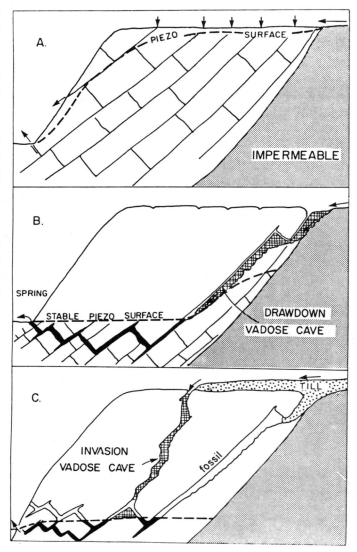

Figure 3. Two types of vadose caves according to the model of Ford and Ewers. From Ford and Ewers (1978).

the cave system produces "drawdown" in the same sense that drawdown is created by a pumped well. However, the injection of water continues at the swallow hole because there is no way for the input to readjust itself across the impermeable barrier. There may be considerable enlargement of the cave between the shallow hole and the phreatic tube that was formed primarily by free-surface streams entirely within the vadose zone. As with the phreatic caves, the relief and gradients of the drawdown vadose caves can be quite variable.

Once an open underground system has been established, further inputs of surface water can create caves, usually predominantly vertical caves through the vadose zone. Ford and Ewers call these *invasion vadose caves*. Vertical shafts in the plateau karsts of the eastern United States, and the deep vertical caves of many alpine regions are forms of vadose invasion caves.

Non-integrated caves are isolated cavities of various sizes found within the phreatic zone. They are not part of an integrated drainage system and the flow of water through them is limited by the flow of water in the bedrock that surrounds them. Cavities encountered by deep drilling operations are thought to be of this type (where they cannot be ascribed to paleo-karst). One must postulate special chemical mechanisms to account for solution at depth without connections, except through the diffuse flow field, to the conduit drainage system.

True artesian caves are formed within deep artesian aquifers. Ford and Ewers claim that the features of these caves are sufficiently different from those of bathyphreatic caves that a separate category is necessary, although this can be questioned. The type examples in the United States are the gigantic three-dimensional network caves in the Pahasapa limestone aquifer in the Black Hills of South Dakota (Howard, 1964).

The Ford-Ewers model provides a very nice integration of many of the concepts and models that have gone before. By emphasizing the geologic setting, it allows the integration of many of the European ideas, mostly based on observations of alpine karst, with ideas of the classic American theorists who were working on low-relief karsts in nearly flat-lying limestones.

Unlike most theoretical models, the Ford-Ewers model works best when applied to dipping limestones. At low fissure frequency, the distinction between dip tubes and tubes cutting across the bedding is very clear. Passages in flat-bedded limestones tend to follow bedding plane partings and although the passages usually cross-cut the bedding, the lift tubes are often very short and difficult to identify in the field.

The phreatic loops predicted for States 2 and 3 of the phreatic caves are difficult to observe in the field because later processes of vadose modification, sediment-infilling, and breakdown obscure the passage relations, particularly when the loops are of low amplitude and long length. In recent years, however, divers have provided information about the shallow phreatic environment and there seems to be no doubt that dip tubes and lift tubes exist.

The availability of comprehensive regional collections of cave descriptions, particularly the cave surveys of Pennsylvania (Stone, 1953) and West Virginia (Davies, 1949) allowed analysis of the properties of populations of caves, rather than restricting data to those from examinations of the features of individual caves to support a given hypothesis. Curl (1958) first looked at the distribution of cave entrances and found the number of entrances in a population of caves to be Poisson-distributed from which he deduced a very large population of caves with no entrances. In a following paper, Curl (1960) showed that cave lengths also are Poisson-distributed but are coupled with the

number of entrances. Curl (1965) then developed independent statistical models that couple the probability of occurrence of a cave entrance to cave length through a unique "karst constant" that appears to be a characteristic of the geologic setting.

The significance of Curl's work is that it shows a pronounced stochastic component in the processes of cave development. Caves are fragments of conduit drainage systems. The passages available to exploration are determined by a succession of random events of sediment chokes and ceiling collapses. Formation of an entrance giving explorers access to the fragment is also a stochastic event. The lesson for the process geomorphologist is clear: One should use the available cave population to reconstruct the regional drainage system in its present and paleoforms. It is not very productive to regard each cave as a unique object by itself.

Those like Davies and Pohl who continued to work on the classical theories clearly regarded the problem of cave origin as primarily geomorphological, and their arguments are cast almost entirely in terms of base levels, peneplains, water tables, and the youth through old age stages of Davis' geographic cycle. Although caves are recognized as having been formed by solution, no specific use is made of solution chemistry. And although some, particularly those advocating a phreatic origin, speak of patterns of groundwater flow, no specific calculations are offered.

Geochemical work—and the true beginning of the modern period—was initiated by Clifford Kaye (1957). Kaye (who, incidentally, is not a cave explorer turned cave scientist) is among the first to examine cave development as a kinetic phenomenon. He is also one of the first in America to apply laboratory experiments to the problem of cave origin. Kaye dissolved specimens of limestone and marble in dilute hydrochloric acid and showed that the rate of solution increases with the velocity of fluid flow. He then postulates what can be called the "principle of self-acceleration" of cave conduits. Of the various pathways through the aquifer, some will have less resistance to flow than others and thus flow velocities will be higher along these pathways. If the rate of solution increases with flow velocity, openings along these favored route will dissolve faster. As they become larger, flow velocities will increase more, thus increasing the rate of solution still more. This is a runaway process in which the favored paths enlarge at the expense of others and ultimately become the master conduits. This principle depends on the untrue assumption that the dissolution of limestone in dilute carbonic acid in the field follows the same kinetics as the dissolution of limestone in dilute hydrochloric acid in the laboratory (Watson, 1972; Plummer and Wigley, 1976).

If enclosed cavities are water-filled, water can enter and leave them only at the rate of water moving through pores in unaltered bedrock. Flow velocities remain small and are independent of the size of the cavity they flow into. On the other hand, if the cavity is part of a continuous conduit, the velocity and discharge must increase as the cavity enlarges. For each conduit size, at some velocity the flow will become turbulent and the laws of fluid flow in porous media will no longer apply. In an early effort, White and Longyear (1962) applied simple pipe flow theory, assuming the hydraulic gradients measured in real aquifers, and found that the onset of turbulence occurs when the solutionally modified fracture or bedding plane parting reaches a width of 5 or 10 mm. They postulated that the turbulent mixing would greatly enhance the rate of solution and thus produce a discontinuity between the rate of enlargement due to laminar flow in fractures and turbulent flow in proto-passages. The argument assumes a particular model for the dissolution kinetics of limestone that turns out to be false, but the general principles of laminar and turbulent flow remain.

These and other early efforts to analyze the rate of development of cave passages were frustrated by the lack of knowledge of the chemical kinetics of limestone dissolution. Most were based on Weyl's (1958) experiments in which he specifically assumed that diffusion of reactants to and from the liquid/solid interface is the rate-controlling step. Later work has shown that this mechanism is appropriate only to strongly acid solutions and not to the near-equilibrium, near-neutral solutions of carbonic acids that form caves. Further progress had to await new laboratory experiments on solution kinetics.

Thrailkill (1968) constructed one of the first comprehensive models to take both flow hydraulics and solution chemistry into account. The model has three parts.

First, the flow of fluid in a network of pipes is analyzed by using the Hagen-Poiseulle equation for laminar flow and the Darcy-Weisbach equation for turbulent flow. The pipes in the network are of fixed and equal size. Water is inserted into one corner of the network and is allowed to discharge from the opposite corner. Under these rather restricted circumstances, Thrailkill finds that the flow rates are essentially the same in all pipe segments and that the same distribution is obtained regardless of whether the flow is laminar or turbulent. This seems to be an argument for the deep phreatic theory.

Second, water entering actual cave systems is found by Thrailkill to be distinguishable into *vadose seeps* and *vadose flows* on the basis of chemistry. Vadose seeps consist of percolation water moving slowly through joints and small fractures. Seepage water is generally supersaturated with respect to calcite at the carbon dioxide pressure of the cave atmosphere and is responsible for the formation of dripstone deposits. Vadose flows are all concentrated flows of water in the vadose zone including sinking streams, water flowing down chimneys and vertical shafts, and concentrated flows entering through large fracture systems. In contrast to vadose seeps, vadose flows are usually undersaturated with respect to calcite and are thus held to be responsible for such vadose zone modifications of caves as the cutting of vadose canyons in the floors of tubular passages and the formation of vertical shafts. Thrailkill thus answers one of the questions that greatly bothered Davis. It is possible for deposition and dissolution to occur in a cave at the same time. All that is needed is a contrast in water chemistry, not a two-cycle theory of cave genesis. Thrailkill still faced the problem, however, of how undersaturated water is maintained in the phreatic zone, and thus resorted to general temperature, mixing, and flowrate effects.

Third, Thrailkill examined the flow paths expected in "open" and "capped" aquifers. The results of the pipe-network calculations suggest that the flow paths should not be dramatically different from those observed in porous media aquifers. Injection of water uniformly over the karst surface (Fig. 4) would produce deeply curving flow paths since the flow lines theoretically must be perpendicular to the water table surface. Recharge into sinkholes or shallow holes that locally penetrate an otherwise capped aquifer (Fig. 4) builds up groundwater mounds from which emerge flow lines some of which are parallel to the water table beneath the caprock. This model is offered to explain the nearly horizontal tiers of passages in Mammoth Cave and its associated cave systems. There is a curious result in the case of capped aquifers with multiple inputs: the additional hydrostatic head provided by inputs downstream in the flow field depresses the flow lines from inputs farther upstream in the flow field. If flow in open cave conduits in fact approximates Darcy flow, there is no mixing of water across flow lines except by a slow diffusion process.

The main drawback of Thrailkill's model is the use of pipes of fixed size. Cave passages enlarge with time and one of the most interesting questions is what the rate of enlargement is. Of further interest are the questions of if and when and whether the rate of enlargement is the same for all pipes, thus leading to caves with maze patterns, or whether, as Kaye earlier argued, some pipes enlarge faster than others leading to single-conduit caves.

There was an immense upsurge of interest in both the equilibrium chemistry of carbonate minerals and of the dissolution kinetics of these minerals in both fresh and sea water during the middle to late 1970s. As a result, the equilibrium chemistry of the system $CaO-MgO-CO_2-H_2O$ is one of the best known in geochemistry. Plummer and Busenberg (1982), for example, give precise equations for the solubilities of the $CaCO_3$ polymorphs calcite, aragonite, and vaterite as a function of temperature.

Giving precise descriptions of the rates and mechanisms of carbonate dissolution have proved to be continuing and difficult problems. For a long time most laboratory data on rate of calcite dissolution was for experiments in dilute hydrochloric acid. Such experiments measure dissolution rate far from equilibrium and have limited application to dissolution in the near-neutral groundwaters found in cave streams and in other limestone groundwaters. In recent years there have been many laboratory measurements on the dissolution kinetics of calcite (Berner and Morse, 1974; Plummer and Wigley, 1976; Sjöberg, 1976; Plummer, Wigley, and Parkhurst, 1978; Plummer, Parkhurst, and Wigley, 1979; Sjöberg and Rickard, 1984) and dolomite (Busenberg and Plummer, 1982; Herman and White, in press). These provide a choice of rate equations and some insight into reaction mechanisms although many questions remain unan-

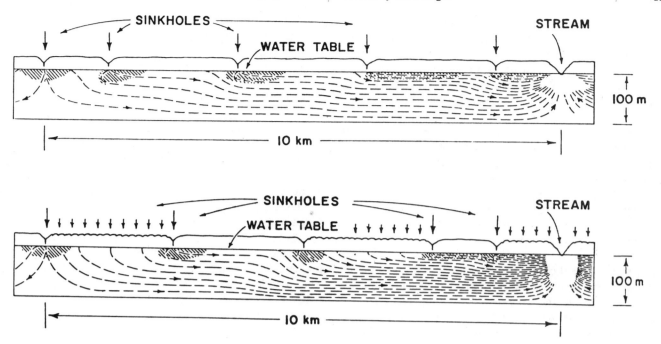

Figure 4. Flow lines in a horizontally bedded, relatively thin karst aquifer. The upper figure shows flow lines from multiple discrete sources; the lower figure shows flow lines for a spatially irregular recharge to the water table. Shaded areas indicate zones of solution. Vertical exaggeration 10X. From Thrailkill (1968).

swered. The dissolution rate of limestone is decidedly non-linear. It varies with carbon dioxide pressures and with degree of saturation. In acid regimes the rate of solution increases with increasing flow velocity of the water; in near-neutral waters solution rate is essentially independent of flow rate.

Attempts are being made to construct models of cave development that predict cave passage patterns and give a time scale for the development of caves using these various empirical and semi-theoretical rate equations. Such an elementary question as why some caves are formed as single conduits and others are complex mazes remains elusive. Palmer (1975) shows that mazes result from either hydrogeologic conditions that restrain flow through the cave system or hydrogeologic conditions where rapid in-flooding of water occurs regularly. It appears that the enlargement of a statistically optimum hydraulic path along joints and bedding planes can be explained by the non-linear behavior of the limestone dissolution rate (White, 1977; Palmer, 1984), but detailed mechanisms are lacking and other calculations (Curl, 1971) conclude that all hydraulic pathways enlarge at the same relative rate. More quantitative models have been developed by Dreybrodt (1981a, 1981b).

In contrast to the stages of cave development outlined from geomorphologic reasoning, the calculations based on flow hydraulics, mass transport, and chemical reaction kinetics support the postulation of the following stages in the development of a single reach of cave conduit.

1. *Initiation.* Precursor, mechanically opened joints and bedding plane partings with a width of perhaps 10-50 μm are gradually enlarged by percolating waters slightly undersaturated with respect to calcite. The flow regime is laminar and the aquifer behaves, in a statistical sense over a large volume of rock, rather like a porous medium aquifer. These are fracture aquifers and, although heterogeneous on a local scale, are amenable to the usual equations of fluid flow on a regional average.

2. *Thresholds.* There are three thresholds in geological settings where other factors do not interfere with the free response of groundwater movement. (i) A threshold for the onset of turbulent flow. (ii) A kinetic threshold marked by the more rapid increase in rate of solution as the enlarged proto-cave allows undersaturated water to circulate through the aquifer. (iii) A threshold for the onset of mechanical transport of clastic sediment when flow velocities become sufficiently large. All three thresholds occur for typical hydraulic gradients when the solutionally widened joint or bedding plane partings reach a width of 5 to 10 mm. Depending on available CO_2 and other details, it takes 3,000 to 10,000 years to dissolve the passage from the initial joint to the critical thresholds.

3. *Enlargement.* The growth of the selected conduit from threshold size to a full-sized cave passage, perhaps one meter in diameter, may take place under pipe-full conditions in which case the passage is an elliptical tube, or by a free-surface

stream in which case a rectangular canyon is formed. The time scale depends on which rate equations are used and on details of both chemistry and geologic setting. The enlargement stage may last from 10,000 to one million years. There is much evidence that cave passage enlargement is a rapid process compared with other geological processes.

4. *Stagnation.* If a cave conduit remains completely water-filled, it will continue to enlarge until it collapses. Mostly, however, caves are drained by lowering base levels and by piracy of water to lower and lower routes. Once a passage is drained, except for occasional breakdown and infilling of travertine by dripping water, a cave can survive thousands of years with little further modification. It is during this stage that caves (those with entrances) are explored.

5. *Decay.* A stagnant cave is fixed in space but the land surface above continues to be eroded. Eventually, the land surface is lowered through the cave and destroys it. Destruction is gradual and piecemeal as deepening valleys induce breakdown and truncation of cave passages into shorter and short-

er segments under ridges. In this model, caves form as an integral part of the landscape. Surviving fragments, like the surviving hills in which they occur, may have ages that extend back into the Tertiary.

V. CONCLUSION

Are there any big cave systems formed by deep-phreatic solution as Davis envisioned it? No. Nevertheless, the Davis model—in opposing the general vadose theories—sent cave explorers and cave scientists underground to check and see. As a result, geomorphic, hydrologic, and kinetic factors have been combined to provide an integrated model of cave origin that best fits currently known field data.

In the introduction to his classic paper, Davis (1930, p. 484) quotes the following from Swinnerton (1929, p. 82): "The whole problem of cave origin awaits systematic study and presentation." This is no longer true. There are, however, many problems remaining, and so we shall end our paper by saying that no one should assume that the last word has been said about the origin of caves.

REFERENCES CITED

Berner, R. A., and Morse, J. W., 1974, Dissolution kinetics of calcium carbonate in sea water IV, Theory of calcite dissolution: American Journal of Science, v. 274, p. 108–134.

Bretz, J H., 1942, Vadose and phreatic features of limestone caverns: Journal of geology, v. 50, p. 675–811.

——1949, Carlsbad Caverns and other caves of the Guadalupe Block, New Mexico: Journal of Geology, v. 57, p. 447–463.

——1955, Cavern-making in a part of the Mexican Plateau: Journal of Geology, v. 63, p. 364–375.

Brucker, R. W., Hess, J. W., and White, W. B., 1972, Role of vertical shafts in the movement of ground water in carbonate aquifers: Ground Water, v. 10, no. 6, p. 5–13.

Busenberg, E., and Plummer, L. N., 1982, The kinetics of dissolution of dolomite in CO_2-H_2O systems at 1.5 to 65° C and 0 to 1 atm P_{CO_2}: American Journal of Science, v. 282, p. 45–78.

Curl, R. L., 1958, A statistical theory of cave entrance evolution: National Speleological Society Bulletin, v. 20, p. 9–22.

——1960, Stochastic models of cavern development: National Speleological Society Bulletin, v. 22, p. 66–76.

——1965, Caves as a measure of karst: Journal of Geology, v. 74, p. 798–830.

——1971, Cave conduit competition—I: Power law models for short tubes: Caves and Karst, v. 13, p. 39.

Cvijić, J., 1983, Das Karstphänomen: Geographische Abhandlungen von A. Penck, v. 5, no. 3, p. 217–329.

Davies, W. E., 1949, Caverns of West Virginia: West Virginia Geological Survey, v. 19, p. 1–353.

——1960, Origin of caves in folded limestone: National Speleological Society Bulletin, v. 22, p. 5–18.

Davis, W. M., 1930, Origin of limestone caverns: Bulletin of the Geological Society of America, v. 41, p. 475–628.

Dreybrodt, W., 1981a, Kinetics of the dissolution of calcite and its application to karstification: Chemical Geology, v. 31, p. 245–269.

——1981b, Mixing corrosion in $CaCO_3$-CO_2-H_2O systems and its role in the karstification of limestone areas: Chemical Geology, v. 32, p. 221–236.

Ewers, R. O., 1982, An analysis of solution cavern development in the dimensions

of length and breadth: [Ph.D. Thesis]: Hamilton, Ontario, McMaster University.

Fish, J. E., 1977, Karst hydrogeology and geomorphology of the Sierra de el Abra and the Valles—San Luis Potosi region, Mexico [Ph.D. Thesis]: Hamilton, Ontario, McMaster University, 469 p.

Ford, D. C., 1965, The origin of limestone caverns: A model from the central Mendip Hills, England: National Speleological Society Bulletin, v. 27, p. 109–132.

——1968, Features of cavern development in central Mendip: Transactions of the Cave Research Group of Great Britain, v. 10, p. 11–25.

——1971, Geologic structure and a new explanation of limestone cavern genesis: Transactions of the Cave Research Group of Great Britain, v. 13, p. 81–94.

Ford, D. C., and Ewers, R. O., 1978, The development of limestone cave systems in the dimensions of length and depth: Canadian Journal of Earth Sciences, v. 15, p. 1783–1798.

Gardner, J. H., 1935, Origin and development of limestone caverns: Bulletin of the Geological Society of America, v. 46, p. 1255–1274.

Greene, F. C., 1908, Caves and cave formations of the Mitchell limestone: Proceedings of the Indiana Academy of Science, v. 18, p. 175–184.

Grund, A., 1910, Beiträge zur morphologie des Dinarischen Gebirges: Geographische Abhandlungen (Penck), v. 9, p. 348–570.

Herman, J. S., and White, W. B., in press, Dissolution kinetics of dolomite: Effects of lithology, temperature, and fluid flow velocity: Geochimica et Cosmochimica Acta.

Hill, C. A., 1981, Speleogenesis of Carlsbad Caverns and other caves of the Guadalupe Mountains: Proceedings of the Eighth International Congress of Speleology, v. 1, p. 143–144.

——in press, Geology of Carlsbad Caverns and other caves of the Guadalupe Mountains, Part I, Speleogenesis, Part 2, Mineralogy: Socorro Bureau of Mines Memoir.

Howard, A. D., 1964, A model for cavern development under artesian ground water flow, with special reference to the Black Hills: National Speleological Society Bulletin, v. 26, p. 7–16.

Hubbert, M. K., 1940, The theory of ground-water motion: Journal of Geology, v. 48, p. 785–944.

Kaye, C. A., 1957, The effect of solvent motion on limestone solution: Journal of Geology, v. 65, p. 35–46.

Kyrle, G., 1923, Theoretischen Speläologie: Austrian State Press, Vienna, p. 1–353.

Langmuir, D., 1971, The geochemistry of some carbonate ground waters in central Pennsylvania: Geochimica et Cosmochimica Acta, v. 35, p. 1023–1045.

Malott, C. A., 1937, The invasion theory of cavern development: Geological Society of America Proceedings, p. 323.

Matson, G. C., 1909, Water resources of the Blue Grass Region, Kentucky: U.S. Geological Survey Water Supply Paper 640, p. 42–45.

Miotke, F.-D., 1973a, Die höhlen im Mammoth Cave-gebiet (Kentucky): Böhler-Verlag, Würzburg, 133 p.

—— 1973b, Der karst im zentralen Kentucky beí Mammoth Cave: Jahrbuch 1973 der Geographischen Gesellschaft Hannover, Hannover, 360 p.

Miotke, F.-D., and Palmer, A. N., 1972, Genetic relationships between caves and landforms in the Mammoth Cave National Park area: Böhler-Verlag, Wüurzburg, 69 p.

Moneymaker, B. C., 1941, Subriver solution cavities in the Tennessee Valley: Journal of Geology, v. 49, p. 74–86.

Moore, G. W., 1960, The origin of limestone caves—a symposium with discussion: National Speleological Society Bulletin, v. 22, p. 1–84.

Palmer, A. N., 1975, The origin of maze caves: National Speleological Society Bulletin, v. 37, p. 56–76.

—— 1981, A geological guide to Mammoth Cave National Park: Zephyrus Press, Teaneck, N.J., 196 p.

—— 1984, Geomorphic interpretation of karst features: in Groundwater as a Geomorphic Agent, LaFleur, R. G., ed., Allen & Unwin, Winchester, Mass., p. 173–209.

Piper, A. M., 1932, Ground water in north-central Tennessee: U.S. Geological Survey Water Supply Paper 640, p. 69–89.

Plummer, L. N., and Wigley, T.M.L., 1976, The dissolution of calcite in CO_2-saturated solutions at 25° C and 1 atmosphere total pressure: Geochimica et Cosmochimica Acta, v. 40, p. 191–202.

Plummer, L. N., Wigley, T.M.L., and Parkhurst, D. L., 1978, The kinetics of calcite dissolution in CO_2-water systems at 5° to 60° C and 0.0 to 1.0 Atm CO_2: American Journal of Science, v. 278, p. 179–216.

Plummer, L. N., Parkhurst, D. L., and Wigley, T.M.L., 1979, Critical review of the kinetics of calcite dissolution: in Chemical Modeling in Aqueous Systems, Jenne, E. A., ed., American Chemical Society Symposium Series 93, p. 537–573.

Plummer, L. N., and Busenberg, E., 1982, The solubilities of calcite, aragonite, and vaterite in CO_2 - H_2O solutions between 0 and 90° C, and an evaluation of the aqueous model for the system $CaCO_3$-CO_2-H_2O: Geochimica et Cosmochimica Acta, v. 46, p. 1011–1040.

Pohl, E. R., 1955, Vertical shafts in limestone caves: Occasional Papers of the National Speleological Society, no. 2, 24 p.

Quinlan, J. F., and Rowe, D. R., 1977, Hydrology and water quality in the Central Kentucky Karst: Phase I: Research Report No. 101, University of Kentucky Water Resources Institute, Lexington, 91 p.

Rhoades, R., and Sinacori, M. N., 1941, Pattern of ground-water flow and solution: Journal of Geology, v. 49, p. 785–794.

Shuster, E. T., and White, W. B., 1971, Seasonal fluctuations in the chemistry of limestone springs: A possible means for characterizing carbonate aquifers: Journal of Hydrology, v. 14, p. 93–128.

Sjöberg, E. L., 1976, A fundamental equation for calcite dissolution kinetics: Geochimica et Cosmochimica Acta, v. 40, p. 441–447.

Sjöberg, E. L., and Rickard, D. T., 1984, Temperature dependence of calcite dissolution kinetics between 1 and 62° C at pH 2.7 to 8.4 in aqueous solution: Geochimica et Cosmochimica Acta, v. 48, p. 485–493.

Smith, P. M., and Watson, R. A., 1970, The development of the Cave Research Foundation: Studies in Speleology, v. 2, p. 81–92.

Stone, R. W., 1953, Descriptions of Pennsylvania's undeveloped caves: National Speleological Society Bulletin, v. 15, p. 51–139.

Swinnerton, A. C., 1929, The caves of Bermuda: Geological Magazine, v. 66, p. 79–84.

—— 1932, Origin of limestone caverns: Bulletin of the Geological Society of America, v. 43, p. 663–694.

Thrailkill, J., 1968, Chemical and hydrologic factors in the excavation of limestone caves: Geological Society of America Bulletin, v. 79, p. 19–46.

Thrailkill, J., and Robl, T. L., 1981, Carbonate geochemistry of vadose water recharging limestone aquifers: Journal of Hydrology, v. 54, p. 195–208.

Watson, R. A., 1966, Central Kentucky Karst Hydrology: Bulletin of the National Speleological Society, v. 28, p. 159–166.

—— 1972, Limitations on substituting chemical reactions in model experiments: Zeitschrift fur Geomorphologie, v. 16, p. 104–109.

Weyl, P. K., 1958, The solution kinetics of calcite: Journal of Geology, v. 66, p. 163–176.

White, W. B., 1960, Terminations of passages in Appalachian caves as evidence for a shallow phreatic origin: National Speleological Society Bulletin, v. 22, p. 43–53.

—— 1977, Role of solution kinetics in the development of karst aquifers: International Association of Hydrogeologists Memoir 12, p. 503–517.

White, W. B., and Longyear, J., 1962, Some limitations on speleogenetic speculation imposed by the hydraulics of groundwater flow in limestone: Nittany Grotto Newsletter, National Speleological Society, v. 10, p. 155–167.

White, W. B., and White, E. L., 1974, Base level control of underground drainage in the Potomac River basin: Proceedings of the Fourth Conference on Karst Geology and Hydrology, West Virginia Geological and Economic Survey, Rauch, H. W., and Werner, E., eds., p. 41–53.

Printed in U.S.A.

Geological Society of America
Centennial Special Volume 1
1985

Glacial geology and the North American craton: Significant concepts and contributions of the nineteenth century

Stanley M. Totten
Department of Geology
Hanover College
Hanover, Indiana 47243

George W. White*
Department of Geology
University of Illinois
Urbana, Illinois 61801

ABSTRACT

The craton was the focal point for the development of glacial geology in North America. Glacial deposits in the Ohio Valley were illustrated and described early in the century by Volney (1803, 1804), who, however, did not understand their origin and Drake (1815, 1825), who attributed the origin of erratics to icebergs. The iceberg hypothesis for the deposition of at least a part of the glacial drift persisted until late in the century.

The first glacial geologist in America was Charles Whittlesey who made early but significant studies in several states in the 1840s and 1850s. Whittelsey published the earliest maps and sections in America showing end moraines, kettle holes, and the glacial boundary. He was the first in America to classify drift, describe the "Forest Bed," and trace drift in the subsurface.

In the 1860s and 1870s, the state geological surveys assumed a leadership role in glacial studies. Most notable of these was the Ohio Geological Survey directed by John Newberry, who was recognized as the leading glacial geologist of his time. An understanding of glacial lobes, glacial erosion of lake basins, moraines, multiple glaciation, and climatic change developed during this period.

In the 1880s, the U.S. Geological Survey under the guidance of T. C. Chamberlin established dominance in glacial studies of a regional nature. Chamberlin's recognition that the glacial drift represented several glacial advances separated by warmer, nonglacial periods led to a formal classification for the North American Pleistocene.

By 1886, Chamberlin considered his phase of glacial studies essentially complete, and he assigned the responsibility of detailed moraine mapping to one of his assistants, Frank Leverett. Leverett proceeded to map in considerable detail the glacial drift and related geomorphic features of the Illinois and Erie lobes, covering an area extending from Iowa to New York. Leverett's detailed mapping, keen observations, and ability to synthesize large volumes of data provided a fitting climax to nineteenth century glacial studies.

---

*Editor's Note: With profound sorrow we received the news from Mildred White of the death of her husband, George W. White, on the 20th of February, 1985. George was instrumental in the founding of the GSA History of Geology Division in 1976 and was the first recipient of the Divisional Award when it was established in 1982. George will be sorely missed by his friends and colleagues, especially those of us in the History of Geology Division.

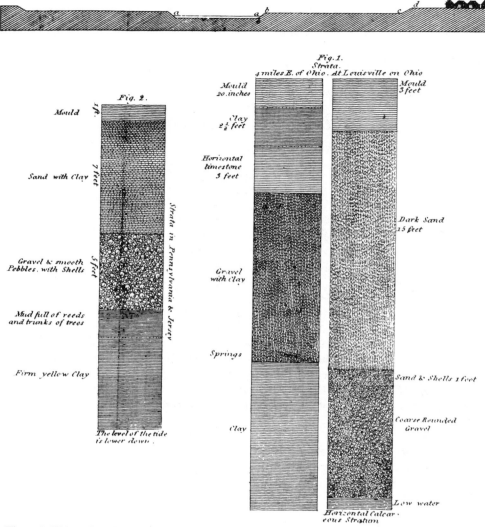

Figure 1. The earliest published sections of American Pleistocene deposits (Volney, 1804, p. 69, Pl. 1). The upper cross section is of the Ohio River terraces which Volney observed at Cincinnati in August 1796. The two columnar sections at the right are of deposits in the Ohio River Valley. The section on the left is from the Coastal Plain in the Philadelphia region.

INTRODUCTION AND EARLY STUDIES

The craton was the focal point for the birth and development of glacial geology in North America. The critical period in the development of the glacial theory, from 1840 to about 1880, occurred at a time when great emphasis was being placed on geologic investigations for economic materials across the broad midsection of the United States between the Appalachians and the Rockies. The emphasis attracted a large number of geologists to the "Midwest" where geologic surveys had been organized in several states. Many of these geologists soon focused their attention on the surficial deposits, and nearly every glaciated state on the craton had one or more highly respected geologists studying glacial deposits and landforms. It was no coincidence that many

important concepts of North American geology originated from studies of the craton.

The earliest recognition that the style of geology west of the Appalachians was "different" was by Andrew Ellicott, a surveyor and geodesist from Philadelphia. It was while surveying the western boundary of New York in 1789–1790 that Ellicott had the occasion to note the horizontal strata of Niagara Falls and Niagara Gorge (Maclay, 1927; White, 1982). Ellicott (1799, p. 227–228) wrote "that the country has never been disturbed by those terrible convulsions which a great part of this globe must have experienced at some remote period of antiquity." He was much impressed by the relationship of Lake Ontario to Lake Erie and by the Niagara River, which had eroded a deep gorge in a steep slope or escarpment during the time it had been connecting the two lakes. He continued, "The waters of this cataract formerly fell from the northern side of the slope, near the landing place; but the action of such a tremendous column of water falling from such an eminence, through a long succession of ages, has worn away the solid stone for a distance of seven miles, and formed an immense chasm which cannot be approached without horror." Maclay (1927) termed Ellicott's account of Niagara Falls "amazing" and used his data to determine the time required to cut the gorge to be 55,400 years, which is the first recorded attempt to date a Pleistocene event.

The earliest illustrations of Quaternary deposits in America were published by Volney (1803, 1804). These illustrations (Fig. 1) show the materials, now known to be glacial outwash, that occur in the Ohio River Valley at Cincinnati, Ohio, and at Louisville, Kentucky. Volney recognized that the terrace sands and gravels had been transported by the Ohio River from the northeast, but he did not suspect the glacial origin of the outwash.

The influence of ice on drift was first suggested in America by Drake (1815), who believed the erratics in the Cincinnati area of southwestern Ohio had originated in a distant ice field and had floated to their resting places. He described erratics as "foreign and adventitious debris . . . consisting of granite," and he gave the first description in America of "geest" (till) (Drake 1825, p. 133–137). In a summary statement, Drake (1825, p. 137) proposed that the origin of the erratics was from "large fields of ice in a region for beyond the lakes and floated hither by the same inundations that brought down and spread over the surface of this country the geest in which they are imbedded." Maclure (1823) also suggested that the erratics had been transported by icebergs at a time when Lake Erie was much more extensive.

Only two years after the Swiss geologist Louis Agassiz announced verification of the glacial theory (in 1837), the American paleontologist Timothy Conrad (1839) reported polished rock surfaces in western New York which he attributed to glaciers.

Hitchcock (1841) also had noticed striations which he believed could be explained by Agassiz' ice sheet theory. Many geologists on the west side of the Atlantic, however, took a dim view of the Agassiz theory which led the Association of American Geologists and Naturalists to appoint a committee, in 1842, to report on the subject of drift. The committee concluded (Jackson, 1843) that drift was the product of icebergs, and they considered the glacier theory of Agassiz to be insufficient; the opposition to Agassiz was unanimous, and it included Hitchcock who had been considered a proponent. The transactions of the discussion which ensued at the meeting (Association of American Geologists and Naturalists, 1843, p. 324–325) are enlightening:

Prof. Hitchcock remarked that so disastrous had been his experience in respect to the glacial theory of Agassiz, that he was almost afraid to say anything more on the subject. His view had been so much misunderstood on both sides of the Atlantic, that he was satisfied that the fault lay in the language which he had used on former occasions. . . .Nay, he invented a new term, viz. *glacioaqueous,* to express the final conclusions of his mind on the subject. By this term he meant to say that the

Figure 2. Charles Whittlesey (1808–1886).

phenomena of drift were the result of the joint action of ice and water, without saying which of these agents had exerted the greatest influence.

The committee action denying the validity of the Agassiz glacial theory set back glacial geology in America by almost a generation.

The first glacial geologist in North America was Charles Whittlesey (Fig. 2), who classified and described the drift of Ohio in 1848. Although he did not openly embrace the Agassiz glacial theory in his first (1848) paper, his summary report of 1866 demonstrated he had been a supporter of the glacial theory for a considerable length of time. Among Whittlesey's "firsts" are discovery of the "Forest Bed" in drift, mapping end moraines, mapping the glacial boundary, explanation of the origin of kettle holes, regional classification of drift, glacial erosion of lake basins, and the tracing of drift in the subsurface. These contributions are discussed in detail elsewhere in this paper. For good measure, he was the first to calculate that sealevel was 350 to 400 feet lower during the time of maximum ice advance (Whittlesey, 1868).

The first "permanent" convert in America to Agassiz' ice sheet origin of drift apparently was Samuel St. John, a professor of chemistry, mineralogy, and geology at Western Reserve College, then located at Hudson, Ohio (White, 1967). St. John in the

Figure 3. John Strong Newberry (1822–1892).

first geology textbook published in Ohio (1851) incorporated observations from northern Ohio, in a clear statement of the theory of continental glaciation, the first such statement published west of the Appalachians (White, 1967). It is likely that St.John was strongly influenced by Whittlesey whose family home was at Tallmadge, located only a few miles from Hudson. Significantly, one of St. John's students, John S. Newberry (Fig. 3) would soon succeed Whittlesey as the leading glacial geologist in America.

GLACIAL EROSION

The extent to which glaciers have eroded bedrock on the craton was debated with much enthusiasm in the nineteenth century, and it continues to generate interest to the present day (White, 1972).

The concept that glacial ice and its contained rock fragments had carved large lake basins including the Great Lakes was first developed by Newberry (1862, 1870a) and Whittlesey (1866). Whittlesey's geologic investigations in the Great Lakes region began about 1837, and his definitive but little known work, *Fresh-Water Glacial Drift of the Northwestern States,* published in 1866, was a summary of more than 25 years of work. Newberry's interest in Ohio geology began in 1841 (Aldrich, 1974), and his important ideas regarding glacial erosion were formulated by 1860. Neither author referred to the other in publications; thus it is assumed that each developed similar ideas regarding the origin of the Great Lakes independently and nearly concurrently.

Whittlesey was much impressed by glacial grooves and striations on bedrock—proof (to Whittlesey) that ice had eroded the

preglacial regolith and, in places, a considerable amount of bedrock as well. His section entitled "Lakes of Erosion" (1866) is a classic description of the erosive effects of ice. Whittlesey noted that ice erosion was most effective in rocks of softer lithologies where the direction of ice movement was parallel to the strike of the rocks. Many of his examples of lake basins formed or modified by ice erosion were drawn from Wisconsin, Minnesota, and the Upper Peninsula of Michigan, areas where he had a great deal of familiarity. Applying his concepts of erosion toward a larger scale, Whittlesey noted (1866, p. 30):

But we must extend our ideas of glacier excavation to larger bodies of water. The basins of the North American Lakes, constituting the valley of the St. Lawrence, have been modified by the same agent. . . .It is remarkable that most of them have the long axis nearly in the directions of the bearing of the rocks. In some cases, the strata which have least resisting power lie at the bottom of a lake; and more than this, the course of the arrows [striations] shows that the glacier moved along the outcrop of these beds, having the same general direction, thus combining all the circumstances favorable to erosive effect.

Whittlesey mentioned that the "interior lakes of Middle New York" (the Finger Lakes) also were carved by ice moving parallel to the long axis of the lakes.

In noting the erosive effect of ice, Newberry (1870a, p. 214) stated that "the track of a glacier is as unmistakable as that of a man or a bear." With reference to the Great Lakes, he concluded (p. 232), "No other agent than glacial ice, as it seems to me, is capable of excavating broad, deep, boat-shaped basins, like those which hold our lakes." As evidence for such a strong statement, he first reminded his readers that proof existed that the lake basins had been filled with "great moving glaciers" and that the great glacial grooves and furrows which parallel the axis of the west end of Lake Erie were carved out of limestone by "ice action." Newberry also was the first to propose that glaciers had eroded old river valleys to form the Great Lakes Basins. In answer to apparent, growing opposition to his theory, he wrote (1874) that there was no question about the effectiveness of glacial erosion, for he had observed glacial erosion in the Alps and in the western United States. Although relatively little opposition to Newberry appeared in print before 1890, Fairchild (1905, p. 16–17) cited an exchange between Newberry and J. P. Lesley at a meeting of AAAS in Minneapolis in 1883:

With a resounding blow on the table, Lesley said that "ice has no eroding power." There was a warm debate on a cold subject. To the writer (Fairchild) Newberry was a demi-god and Lesley was a very impious mortal, but with the passing of time the conviction has come that Lesley was essentially right.

Shortly after the first Newberry report (1862), other geologists began recognizing the erosional features of continental glaciers. Logan (1863, p. 889), in a report on the geological features of Canada, wrote:

The great basins [Great Lakes] are depressions, not of geological structure, but of denudation; and the grooves on the surfaces of the rocks, which descend under their waters, appear to point to glacial action as one of the great causes which have produced these depressions.

In a footnote to the above passage, Logan credited Professor A. C. Ramsay as being the first (in 1862) to show that many lakes in Europe were produced by glacial erosion, and he credited Dr. J. S. Newberry for showing the same for North America.

T. C. Chamberlin (Fig. 4), in his classic paper (1883a) reviewing the distribution of end moraines on the craton, argued for deep erosion by ice to form the basins of the Finger Lakes of New York. He mentioned that his views were opposed by some investigators who believed the lake basins had a preglacial canyon origin. Chamberlin's equally impressive monograph of 1888, in which he described and catalogued striations, grooves, and other glacial erosional features, is graphic testimony to the great effectiveness of glacial erosion. Like Newberry, Chamberlin (1888) was so impressed with the giant grooves on the islands of western Lake Erie that he included 13 illustrations (Fig. 5) from Kelly's Island.

The first major opposition refuting the Newberry-Whittlesey hypothesis concerning the origin of the Great Lakes was by Spencer (1881, 1890). Spencer believed that the lake basins were formed entirely by stream erosion of the ancient St. Lawrence River system, after which crustal warping caused the St. Lawrence River to be dammed in several places forming lakes. Spencer's argument for crustal warping was based in part on information from G. K. Gilbert that the raised beaches surrounding the present Great Lakes are raised more toward the northeast. Apparently, Spencer was unaware of the possibility that the tilting of beaches was a result of postglacial isostatic rebound. Spencer was convinced that ice had no erosive power, and with

Figure 4. Thomas Chrowder Chamberlin (1843–1928).

reference to the effects of glaciation in the Great Lakes region, he stated (Spencer, 1890, p. 529):

At the present stage in the investigation this subject can be quickly dismissed. The question whether glaciers can erode great lake basins is hardly pertinent, for nowhere about the lakes is the glaciation parallel to the shores or vertical escarpments which are associated with the lakes.

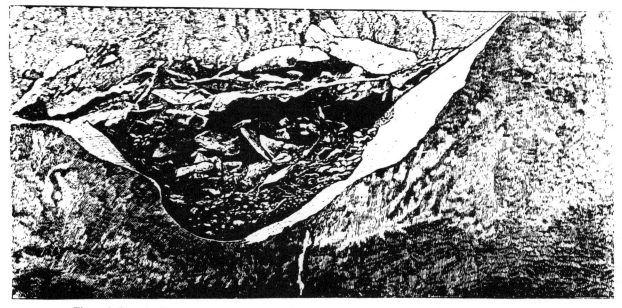

Figure 5. Chamberlin's (1888, p. 212) cross section of glacial groove, North Quarry, Kelly's Island, Lake Erie.

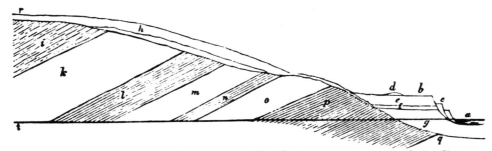

Section from Cleveland, Ohio, to the Highlands, southwest.—a, Lake Erie; b, Cleveland; c, slides; d, ridge; efg, Lacustrine; e, gravel; f, sand; g, marl; h, superficial deposits; i, coal series; k, conglomerate; l, olive shales and bands of sandstone; m, grindstone grit; n, red shale; o, fine grained sandstone; p, black shale with bands of iron and sandstone; q, cliff limestone. Height of lake level above the ocean, 564 feet; height rt, 600 feet.

Figure 6. Whittlesey's cross section showing bedrock and six units of glacial drift in northern Ohio. Note two units of drift on the upland and four units of drift, including a beach ridge, near Lake Erie. Two slump blocks are shown along the Lake Erie shore. The vertical scale is greatly exaggerated, thus causing the bedrock units to appear to be steeply dipping, which they are not (Whittlesey, 1848, p. 209).

Others of note who agreed with Spencer's river valley origin of lake basins were Davis (1882) and Wright (1891, 1892).

Hinde, in a report describing the famous glacial deposits exposed in Lake Ontario cliffs at Scarboro Heights near Toronto, took issue with a growing number of geologists who believed that tectonic movements formed the Great Lakes. Hinde (1878, p. 396) wrote:

If the earlier glacier was massive enough to overflow mountains 3000 and 4000 feet in height, as is known to have been the case in New England, it ought to have been a comparatively easy task for such a mass to plough out the hollows of our inland lakes to depths of only 300 to 400 feet below the sea level. . . . When the path of the glacier can be thus traced following the axis of the lake (Ontario) from north-east to south-west, and masses of till which have been eroded from the rocks outcropping in the area of the lake are met with, heaped up on the banks at its south-westerly end, the only conclusion which can be drawn is that the lake basin is due to the powerful eroding influence of a glacier.

The "revolt" against those favoring the erosive power of glaciers was effectively squelched by Lincoln (1892, 1894) and Tarr (1894) in their separate studies regarding the origin of the Finger Lakes of New York. Tarr included an excellent review of the pros and cons regarding the effectiveness of glacial erosion, and he recognized the extreme polarization that had recently arisen. He included Spencer, Davis, Upham, and Wright as notables who believed that the Finger Lakes were river valleys clogged with drift, whereas he mentioned Newberry, Carll, Shaler, Johnson, and Chamberlin as those who believed glacial erosion was the major factor in formation of the lakes. Tarr (1894) admitted to being a recent convert to the ice erosion hypothesis, a conversion effected by the study of Lake Cayuga. He concluded that the Cayuga valley was a preglacial valley enlarged by ice erosion, citing as proof the rock-enclosed tributaries to Cayuga valley which had their lowest parts above the present lake surface. He pointed out that similar evidence existed

for the Lake Ontario basin, likewise indicating it had been produced by ice erosion. (See also W. R. Brice on Tarr, this volume.)

So convincing were these arguments in support of glacial erosion that a lengthy, but weak paper by Culver (1895) rejecting all arguments for glacial erosion scarcely made an impact. The major works that followed, particularly those of Leverett (1899, 1901), considered the question settled and incorporated the hypothesis of ice erosion as a matter of course.

STRATIGRAPHY AND MULTIPLE GLACIATION

Stratigraphic subdivision of drift began almost as soon as geological investigation of the craton began. As mentioned previously, Volney (1803, 1804) in a very early study included an illustration (Fig. 1) showing the stratigraphy of glacial outwash sands and gravels in the Ohio River Valley. This study had little impact at the time, and it was more than 40 years later that stratigraphic studies of drift resumed. The Owen Survey (Lane, 1966), the major geologic survey of the craton in the 1830s and 1840s, was directed toward study of the economic potential of the bedrock, and the overlying drift was studied in a most incidental manner. The same was true for the state surveys, several of which were started in the 1830s.

Whittlesey (1848) was the first of a new generation of geologists to publish a stratigraphic section of drift deposits (Fig. 6). He divided the drift of Ohio into eight units based on stratigraphic position, topographic appearance, and physical properties. His descriptions of the various types of hard pan, blue clay, and outwash deposits are detailed and compare favorably with modern stratigraphic units in Ohio. He noticed regional variation in the drift deposits and his classification recognized regional differences as well as the stratigraphic sequence.

Whittlesey (1848) was one of the very first geologists to make extensive use of well records in tracing drift in the subsurface. He discovered that "buried timber" and other plant remains appeared in about 10 percent of the well records from Hamilton County, Ohio. He noted that nearly all of the wood was contained in his lowest stratigraphic unit, the blue hard pan. Whittlesey's report was unique among early papers in that it included much useful economic and environmental information. For instance, it mentioned the conditions causing severe slumping along the Lake Erie bluffs, the two very differing compositions of gravel deposits, the high perched water table conditions that developed on the impervious blue clay, and the easily tilled soil that developed in some areas of yellow clay. No doubt Whittlesey was well aware at the time of the controversy surrounding the Agassiz glacial theory, and he avoided the issue by referring to his report as "facts unencumbered by theory." One of his most important comments was a notation that if so much of the drift in Ohio was lacustrine, as it was then considered by other geologists, then it should contain fossil shells, which it didn't.

Worthen (1858) apparently was the next geologist to publish a stratigraphic section of drift deposits. He divided the drift of Lee County, Iowa, into a lower blue clay with pebbles and an upper sand and clay containing boulders. Worthen made no mention of the origin of the drift; he did suggest that the sand and clay were of economic value, and that the best water wells were completed in the drift.

Sir William Logan (1863) constructed the first stratigraphic column of drift for Canada which contained reference to the glacial origin of at least some of the deposits. Logan's (1863) column consisted of three major drift units, a lower coarse drift (till) of glacial origin, a middle variable unit containing mud and gravel, but also some clay thought to have a fresh water origin, and an upper unit consisting of peat, shell marl, and alluvium. The clay was named by Logan the "Erie Clay," a name which was soon adopted by geologists in Ohio for a widespread clay-rich layer at or near the surface south of Lake Erie.

Newberry's (1862) concept of the stratigraphy of the drift in Ohio closely resembled the units established by Whittlesey and Logan. Newberry was so impressed by the glacial pavement developed on bedrock by glacial erosion that he believed that no unconsolidated material could withstand erosion beneath a glacier's bed. Thus he did not believe the erratics that littered the surface of Ohio could have been transported by glaciers (or by running water) "as either of these must have torn up and scattered the soft clays below." It was not until 25 years later (1878) that Newberry accepted the concept of multiple glaciation.

The first paper to emphasize "buried timber" and other plant remains in detailed drift sections was by Whittlesey (1866), who recorded buried organic remains from Ohio, Wisconsin, Minnesota, and Iowa. In this summary paper, he revised his earlier stratigraphic classification to include three members: laminated sandy and marly clay resting on bedrock and buried in valleys and lake basins, a sand and gravel member, and a member consisting of sand, gravel, and boulders, sometimes known as hard pan which occurred on the uplands and comprised the moraines. Like Newberry, Whittlesey proposed that the drift orig-

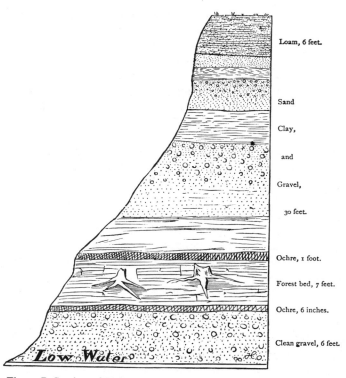

Figure 7. Section of Pleistocene deposits located 15 miles west of Cincinnati, Ohio, as measured in 1870 by Edward Orton. This is the earliest diagram to include an interglacial or interstadial deposit, the "Forest Bed" (Orton, 1873, p. 427).

inated from a single glacial episode, and he concurred with Newberry that icebergs had rafted the "hard pan" and other drift that was deposited on top of the wood and other organic matter.

Shortly thereafter, Worthen (1868) described organic deposits buried between drift deposits in Illinois. Neither Whittlesey nor Worthen speculated to any extent on the origin of the organic remains although both believed the remains would prove to be very important. As stated by Worthen (1868, p. 87), "Wherever these beds are penetrated in sinking wells or are otherwise exposed, a careful examination should be made . . . for any organic remains . . . as these would no doubt throw some light on their true origin, and the conditions under which they have been deposited."

The "Forest Bed" as the buried, organic-rich drift soon became known, was described in detail by Orton (1870) who, along with Gilbert, Winchell, and Read, was an assistant for the Ohio Geological Survey under Newberry. A short time later, Orton (1873, p. 430) first used the term "the interglacial stage" in his description of the Forest Bed he had traced across southwestern Ohio into southeastern Indiana. His illustration of a section near Cincinnati (Fig. 7) is the earliest to label a deposit specifically "interglacial." With reference to the location illustrated, Orton noted that the outwash deposits above and below the Forest Bed required two glacial stages which were separated by deposits requiring much warmer climatic conditions.

About the same time as Orton described interglacial organic remains from Ohio, Winchell (1873) described buried organic remains from Minnesota, and he suggested that ice advances both preceded and followed a period of plant growth. In 1875, Winchell (p. 364) recorded peat from Mower County, Minnesota, which he believed "seems to mark a period of interglacial conditions." He stated that exposures of much weathered drift in Fillmore County near the Driftless Area "are believed to belong to a glacial epoch that preceded the epoch that produced the great drift sheet of the Northwest" and an "interglacial epoch separated them." In 1878, Winchell (p. 44) reported discovery of wood in Goodhue County, Minnesota, in deposits occurring "between the two drift periods."

A buried soil and organic deposits were recorded from wells in Sangamon County, Illinois, by Worthen in 1873. These interglacial deposits later were designated by Leverett (1898a) as the type locality of the Sangamon Soil, which is recognized as the most famous Quaternary paleosol in North America.

Newberry stressed the stratigraphy of the drift in his report on Ohio (1874), but he professed disagreement with his staff regarding multiple glaciation. Rather surprisingly, Newberry remained for several years one of the staunchest proponents of the iceberg origin of drift, a concept which could explain all drift deposits with a single glaciation that spewed forth great numbers of icebergs during a prolonged retreat. When Winchell (1873) proposed that the Erie Clay was debris from the top of the glacier that was dropped when the ice melted, Newberry countered (1874) with the argument that no highlands existed to provide boulders to the ice surface.

An important contribution to glacial stratigraphy was made by Hinde (1878) who recorded three tills separated from each other by interglacial sand and clay in the Scarboro bluffs of Lake Ontario, Canada. He noted that the ice movement was up the slope of the land west of Lake Ontario, as proven by the distribution of erratics in the till. Hinde mentioned that a boulder pavement, developed in till and containing a striated upper surface, was similar to an example from the Miami Valley of Ohio. He concluded that the Scarboro beds of clay and sand 14 feet thick, which contained plant and animal remains and were buried beneath till, were deposited in an interglacial lake that had a climate similar to the present day. To Hinde, the presence of till overlying the clay and sand indicated a return to cold Arctic temperatures. In fact, Hinde postulated that two different interglacials were present at Scarboro, the second of which was based on the presence of unfossiliferous laminated clay and sand near the top of the section. Hinde's most skillful argument was to convince Newberry and others that the erratic boulders scattered over the ground surface of the craton were deposited directly by glaciers and not by icebergs floated from glaciers wasting in Canada. Hinde (1878) pointed out that erratic boulders were widely distributed at elevations exceeding 1200 feet, and water at these elevations would have prevented the existence of glaciers in the Laurentian Range. He also expressed the opinion (p. 405) that the amount of material on the craton supposedly rafted by icebergs would form "a very respectably-sized mountain range," and he believed that it was improbable that icebergs could have transported so much material.

In the concluding pages of his paper, Hinde (1878) discussed the relationship and significance of the Scarboro beds to the area south of the Great Lakes. He expressed the belief that the "Forest Bed" of Ohio and the organic-rich deposits of several other states represented one and possibly two interglacial periods. Taking issue specifically with Newberry's steadfast belief that glaciers must erode all soft materials in their path, Hinde (1878, p. 411) stated that "the fact is indubitable of the second glacier passing over Canada, and yet sparing some of the fossiliferous clays."

When confronted with the strong evidence by Hinde, Newberry began to alter although not abandon completely his iceberg theory of drift. He agreed (1878, p. 38) with Orgon that the Forest Bed "proved conclusively that the southern part of the state [Ohio] was covered with an interglacial forest, the first indication found on this continent of an interval of mild climate in the ice period." He also agreed, up to a point, with Hinde's suggestion that the thin (less than 10 feet thick) boulder clay at the surface of southwestern Ohio was a true glacial deposit, which meant that unconsolidated sands and clays had been overridden by ice without being distorted. Yet Newberry couldn't imagine that thin tills were common, or that undisturbed material beneath overriding ice could be widespread; thus he clung to his iceberg hypothesis for some of the drift at a time when the hypothesis had been abandoned by most of his contemporaries.

Elsewhere, McGee (1878, p. 339) also recognized the importance of the Ohio "Forest Bed" and reported that a similar bed was "observed at many points in Illinois, all through southwestern Indiana, and at many localities in Wisconsin, throughout northeastern Iowa, in Canada, and in many other places." He supported his statements with illustrative sections and noted (p. 341) that "the forest bed is overlaid by true glacial drift, and hence must be of interglacial age."

Chamberlin began glacial studies in Wisconsin at the time that the Forest Bed was gaining popularity, and he had an intimate knowledge of the glacial geology over much of the craton. His writings over nearly two decades (1877–1895) display a particularly keen understanding of Pleistocene stratigraphy (multiple drifts) and glacial landforms (moraines, drainage changes, weathering). In his first glacial paper, Chamberlin (1877, p. 214) described multiple layers of drift from eastern Wisconsin which he believed were formed by "alternating advance and retreat of the ice mass." Chamberlin (1878) utilized geomorphic evidence to distinguish among different ages of drift in the Kettle Moraine region (Fig. 8). His approach permitted detailed mapping and the subdivision of a single glacial stage (Wisconsinan of modern usage) into several substages. He mapped the Kettle Moraine (see Fig. 5, 1878) and distinguished between the drift in the Kettle Moraine and the drift outside on the basis of the degree of weathering of the drift and on the extent to which drainage had developed. Chamberlin (1878, p. 33–34) believed that the moraine represented "a definite historical datum line . . . separating the

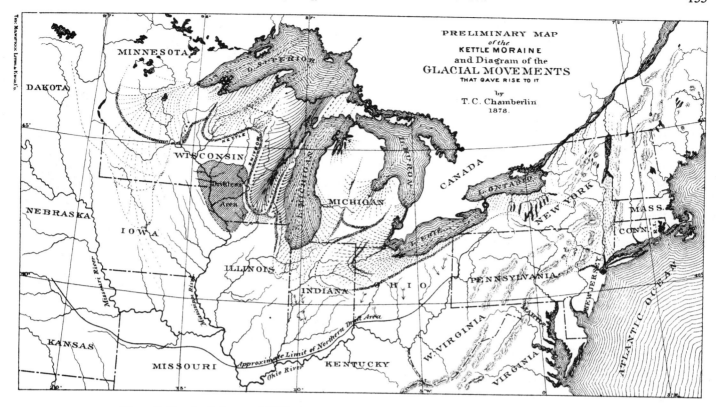

Figure 8. Chamberlin's map of 1878, showing "Approximate limit of northern drift area," the "Driftless Area," and "Kettle Moraine." This is the earliest map to show explicitly drift of more than one age because the Kettle Moraine formed the margin of drift definitely younger than between it (Kettle Moraine) and the drift limit (Chamberlin, 1878, p. 10)

formations on either hand by a chronological barrier." He continued the mapping of different ages of drift as related to moraines, and in 1882 (p. 93–94) described the "extensive range of terminal moraines extending across the craton from Pennsylvania to Dakota, beyond which was earlier drift. . . .The moraine marks a second glacial advance separated from the former by a considerable interval of time." Chamberlin summarized his concepts of glacial geology in *The terminal moraine of the second glacial epoch* (1883a). He recognized the extramorainal drift as being older based on four criteria: greater weathering, more erosion, more advanced drainage development, and thinner, less continuous drift.

Chamberlin considered his glacial studies to be finished in 1886, and he turned over most of the responsibilities of the United States Geological Survey glacial mapping to Frank Leverett (Fig. 9), who was one of his assistants. Leverett's work resulted in three monumental monographs of the United States Geological Survey: *The Illinois Glacial Lobe* (1899), *Glacial formations and drainage features of the Erie and Ohio Basins* (1901), and (with Frank Taylor) *The Pleistocene of Indiana and Michigan and the history of the Great Lakes* (1915). Like Chamberlin, Leverett depended on the surface character of the drift for most of his classification (Leverett, 1909). Leverett mapped and described more glacial landforms over a wider geographic area

than anyone had ever mapped before, and as mentioned by White (1973, p. 19), he took delight in describing in exquisite detail "the rounded cone of a single two-acre kame." Leverett constructed detailed glacial maps of the various lobes (Fig. 10) that compare favorably with modern maps (see Glacial map of the United States, east of the Rocky Mountains by Flint and others, 1959).

The recognition that the glacial deposits of the craton belonged to more than one episode of glaciation and that they could be traced at the surface and in the subsurface in the same manner as bedrock units soon led to a formal stratigraphic classification (Table 1).

TABLE 1. ORIGIN OF GLACIAL AND INTERGLACIAL STAGE NAMES IN NORTH AMERICA

Glacial	Interglacial	Reference	Date
Wisconsinan		Chamberlin	1894, 1895a
	Sangamonian	Leverett	1898a
Illinoian		Leverett	1899
	Yarmouthian	Leverett	1898b
Kansan		Chamberlin	1894, 1895a
	Aftonian	Chamberlin	1894
Nebraskan		Shimek	1909

In the last decade of the nineteenth century, all but one of the major formal names currently used for the Pleistocene of North America were first proposed by either Chamberlin or Leverett. It is remarkable that this classification could be developed so quickly, and the modern Pleistocene classification stands as a monument to their labors.

MORAINES AND GLACIAL LOBES

The lobate nature of glaciation on the craton, emphasized by Chamberlin (1883a), Leverett (1899, 1901), and numerous others, is best illustrated by the patterns made by end moraines on glacial maps.

The earliest map depicting end moraines in North America (Fig. 11) was made by Whittlesey (1866, map dated 1864) for his report summarizing 25 years of study of the craton. Whittlesey's map shows "morainic knolls and cavities" in northeastern Ohio, eastern Wisconsin, and eastern Minnesota. The knolls are shown on the map as pairs of ridges having linear trend in each of the three states. Incidentally, this map also is thought to be the earliest to show the glacial boundary in North America. The ridge-like nature of the moraines and their composition are well-illustrated on the cross sections (Fig. 12) accompanying the map. Whittlesey's description of the moraines is brief and is notable in that the stratigraphy, rather than the geomorphology, is stressed, as in the following passage (p. 2):

> Member number *one* [Till] is the seat of the Morainic hillocks and depressions that mark the summits of the country. It is always coarse and imperfectly stratified. The gravel is not derived wholly from distant and northern rocks. The strata, which underlie the drift at different points, are also represented. Where these strata are soft, the fragments, torn off by the ice movement, are more easily pulverized, and are, therefore, not transported as far as those of the hard, and especially of the tough igneous rocks.

In the same report, Whittlesey described in detail the kettles which occur in the morainic tracts (see White, 1964) and from his observations in the Alps, he attributed the kettles to masses of ice which became entangled with the loose rock debris piled into moraines by glaciers.

The next record of end moraines was made by C. A. White (1870) who described two east-west trending ridges in Iowa as being end moraines. White gave excellent descriptions of the moraines, and he interpreted the moraines as marking "periodical arrests" in the retreat phase of melting glaciers.

Early glacial maps of high quality were made by G. K. Gilbert (1871, 1873), who mapped two end moraines (Fort Wayne and Defiance of modern usage) and several beach ridges in three counties in extreme northwestern Ohio. Gilbert utilized the drainage patterns of the St. Joseph and St. Mary's Rivers in his moraine mapping, and he concluded that the resulting lobate morainic pattern was a result of lobate ice margins during ice retreat. On his map of Williams County, Ohio (Fig. 13), Gilbert (1873) identified the Fort Wayne moraine as a "buried glacial

Figure 9. Frank Leverett (1859–1943).

moraine" primarily because the "Erie Clay" occurred at the surface of the moraine. The Erie Clay, actually a fine-grained till, was interpreted by Gilbert to be a lake deposit, and he theorized that the core of the moraine would contain coarser material of glacial origin. Recent studies (Bleuer, 1974; Totten, 1969) have confirmed the occurrence of overridden and "buried" moraines in Indiana and Ohio, supporting the ideas of Gilbert who apparently never uncovered coarse till in the moraines.

Moraine mapping on the craton received a boost when T. C. Chamberlin (1883b) announced in June 1881 an extensive investigation by the U.S. Geological Survey of the "Terminal Moraine" from the Atlantic Ocean westward to the Rockies. Principal investigators, besides Chamberlin, were J. E. Todd and Professor L. C. Wooster. Reflecting the important status that the glacial mapping program had in the U.S.G.S., Chamberlin (1883a, p. 295) wrote:

> Perhaps no department of geological investigation has greater need of careful discrimination than that which deals with the complex deposits of the Quaternary age. . . .It is only by critical discrimination of their special and often quite unobtrusive features that they can be decisively referred to the several agencies that produced them.

This "critical discrimination" allowed Chamberlin to recognize 12 large morainic "loops," only one of which (Lake Champlain-Hudson River) was not on the craton. These loops, he noted, were formed at the margins of enormous ice tongues which occupied great lowlands. Chamberlin emphasized that the strike of striae was nearly always perpendicular to the strike of the moraine, and

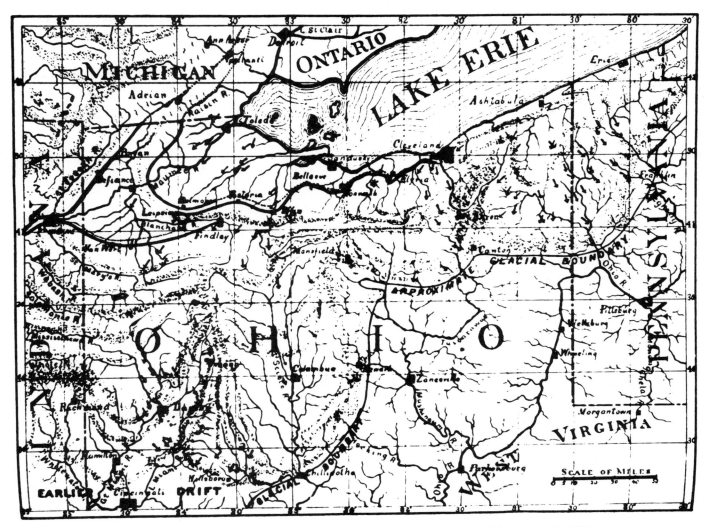

Figure 10. "Map of the Western Erie Basin and Adjacent Territory," published by Leverett in 1892.
This is one of Leverett's earliest maps and was incomplete. Moraines are shown by faint stippled pattern.
A later edition of this map (Leverett, 1901, Pls. 11 and 13) is colored, and the moraines stand out much
more clearly.

to him these two sets of data were proof of direction of ice movement. One of the important contributions of Chamberlin to glacial geology was his recognition that the moraines were the product of more than one ice advance. He noted (1883a) that younger moraines have a "fresher," less subdued surface when compared with older moraines.

Wright (1884), working independently, mapped the glacial boundary across Ohio and was particularly impressed by the extreme difference in distance of ice travel in western Ohio as compared with eastern Ohio. He not only believed that the lobate pattern was influenced by topography, as claimed by Newberry and Chamberlin, but he believed that greater snowfall in Michigan led to greater ice thickness and to significantly greater ice advance in the Scioto and Miami lobes of western Ohio.

Detailed studies of the moraines on the craton were begun in

1886 by Leverett for the U.S. Geological Survey. Originally an assistant to Chamberlin, and assigned to a mapping team that also included R. D. Salisbury and L. C. Wooster, Leverett soon inherited the entire project for himself. Leverett proved to be a tireless field geologist who literally walked on, over, and along the moraines of several states in the Midwest. Primarily a geomorphologist, Leverett, like Chamberlin before him, relied heavily on glacial landforms to reconstruct the glacial history of the various lobes. His Illinois (1899) and Erie (1901) lobe monographs, complete with colored maps, were and are outstanding comprehensive reports on the glacial geology of an area extending from Iowa to New York. As he had been directed, the moraines were particularly emphasized, both in the descriptive report and in the printing of the maps; his maps set the standard by which subsequent glacial maps were measured.

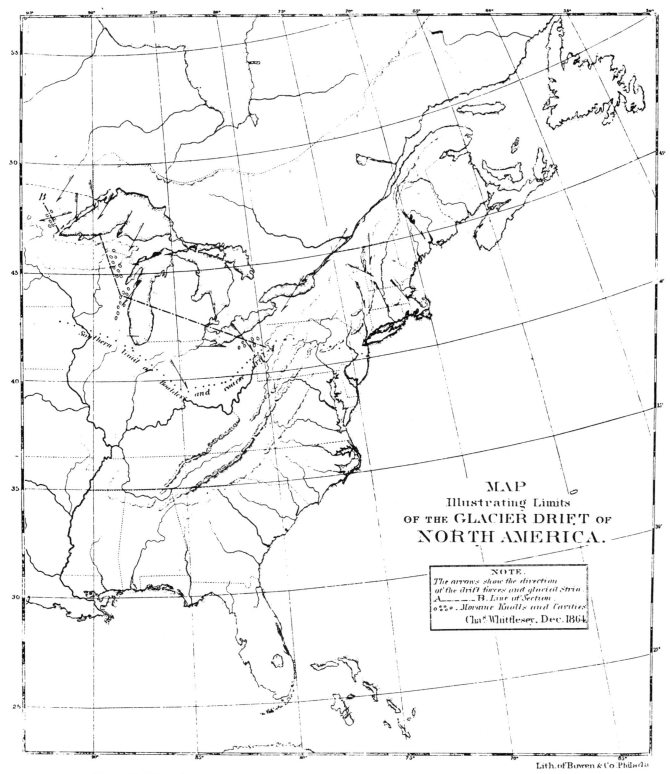

Lith. of Bowen & Co. Philada

Figure 11. The earliest map of the glacial boundary in North America constructed by Charles Whittlesey in 1864. It shows "Southern limits of boulders and coarse drift" in a line extending from New Jersey to Iowa. This is also the earliest map to show moraines. Moraines are shown by circles having linear trend in northeastern Ohio, northern and eastern Wisconsin, and northeastern Minnesota (Whittlesey, 1866).

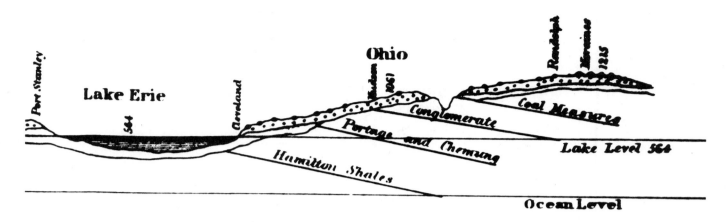

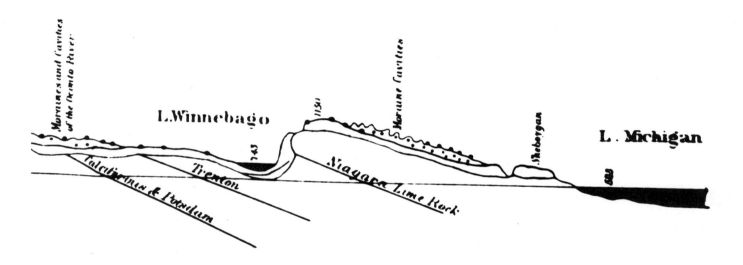

Figure 12. Cross sections of northern Ohio and Lake Erie (above) and Lake Michigan and eastern Wisconsin (below) constructed by Whittlesey in 1864. Original sections are in color, and only a portion of the sections is included in this figure. Note two drift units shown above the bedrock, morainic topography with "cavities" (kettle holes), and boulders (solid circles) littering the landscape and in the drift (Whittlesey, 1866).

LOESS

The loose, unconsolidated silt that comprises the surface material over much of the craton was noted by early investigators, and the remarkable vertical cliffs along such major valleys as the Missouri and the Mississippi must have greatly impressed early travelers. The early Owen survey of 1839 (Owen, 1840, 1844), although concerned primarily with the economic potential of the bedrock of the Mississippi Valley, recognized the agricultural potential of the soil and collected samples for analysis. Owen (1840, p. 52–53; 1844, p. 61–62) wrote:

A striking feature in the character of the Iowa and Wisconsin soils, as the table shows, is the entire absence in most of the specimens of clay, and the large proportion of silex [silica]. This silex, however, does not commonly show itself here in its usual form—that of a quartzose sand. It appears as a fine, almost impalpable, silicious powder, frequently occurring in concreted lumps that resemble clay; and, indeed, it was often reported to me incorrectly as clay—an error ultimately detected by analysis.

There is no doubt that what Owen described as silicious powder (silt) is one of the earliest descriptions of loess. Later, Owen called the silt deposit "silicious marl," and he regarded it to be the deposit of an ancient lake (White, 1870).

The silt deposits of the Mississippi Valley were given the name "Bluff formation" by Swallow (see White, 1870), state geologist of Missouri, in reference to the manner in which the loess stood in nearly vertical slopes along the valley margins. As investigations continued in several states (Newberry, 1870b;

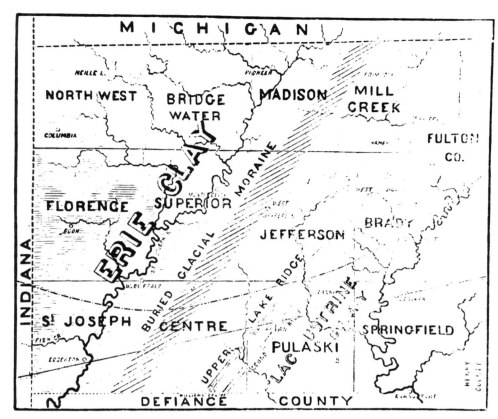

Figure 13. Map of glacial geology of Williams County, Ohio (Gilbert, 1873). The "Buried Glacial Moraine" is the Fort Wayne Moraine, and the lake ridges are strandlines of Lake Maumee. Gilbert interpreted the Erie Clay west of the Upper Lake Ridge as lacustrine rather than fine-grained till which it actually is.

White, 1870; Bradley, 1870), it was soon recognized that the fine-grained homogeneous silt compared favorably with the silt deposits called loess in the Rhine Valley of Germany. From the very beginning, both in Germany and North America, loess was considered to be a lacustrine deposit, and the lacustrine theory remained the popular theory in North America for many years, even extending into the twentieth century.

Why did it take so long to recognize the eolian origin of loess, particularly when almost all of the glacial geologists of the time were studying it? One major line of evidence that no doubt influenced Owen and others was the fine grain size and the lack of any coarse material. The fine-grained bedrock of the craton was considered to be the deposit of an inland sea, and it was not difficult for these early geologists to imagine either a return of the sea, or an extension of the Great Lakes, to provide an aqueous environment for the loess. Another major line of evidence, and one that is cited most frequently, regarded the thickness and distribution of the loess as shown on early maps. The loess was extremely thick in the river valleys such as the Mississippi and the Missouri, whereas the loess thinned rapidly and dramatically away from the valleys, exactly as would be expected if lakes had existed in the valleys, and if the silt had been transported by the rivers. As mapping continued away from the valleys and toward

the uplands where the loess was thin and had become incorporated in the soil profile, recognition problems ensued. St. John (1870) recognized that some upland areas in Iowa had a covering of silt which he believed was loess, and these observations no doubt led him to question a lacustrine origin for the loess. He observed (p. 6) that "in Guthrie County, whose western confines rest upon the great divide, the Bluff Deposit overlaps so to speak, the watershed and caps the divides on the Mississippi slope. . . ." Also the study of Warren County revealed (St. John, 1870, p. 50): "Upon some uplands a fine light colored clay deposit . . . of uncertain origin . . . resembles the Bluff deposit and differs from soils derived from clayey drift."

In the easern part of the craton, in Indiana and Ohio where the loess becomes much thinner and more difficult to recognize, alternate interpretations arose. The loess, or "white clay" of southwestern Ohio, was considered by Orton (1873) to be the fine material brought up to the surface by earthworms and other burrowing animals. From 1870 until after 1900, these upland silts in Ohio were regarded as the accumulation of fine-grained material *in situ* from soil-forming processes, and it remained for Leverett (1901) to recognize that loess, often only a foot or two thick, was widespread over the eastern craton, even extending outside the glacial boundary.

The report by Richthofen (1877) claiming that the loess of China was wind-blown silt derived from rock disintegration surfaces apparently had little impact on American geologists. Pumpelly (1879), who had studied the Chinese loess 14 years earlier and who had regarded the deposits as lacustrine, reversed his stance and supported the eolian hypothesis of Richthofen. However, Pumpelly believed that most of the American loess was derived from the semiarid regions of the Great Plains and Rocky Mountains, and thus he lost the support of T. C. Chamberlin, who by then had collected mineralogic evidence that the silt was washed from glacial debris.

Chamberlin and Salisbury's (1885) discussion of the origin of loess is an excellent review of the loess problem by the leading authorities of the time. Although biased toward a lacustrine origin, they did admit (p. 280) that "the eolian hypothesis of Baron Richthofen presents many attractions," and (p. 286) that loess "may have been formed at different times, possibly by different agencies, and not improbably under different specific conditions." Evidence for the lacustrine origin of loess cited by Chamberlin and Salisbury (1885) included topographic and areal distribution, stratification of some thicker portions, aquatic shells, a mineralogic composition similar to the drift, and a close regional association with drift. The occurrence of terrestrial gastropods in loess didn't seem to bother Chamberlin and Salisbury, but the recognition that loess apparently capped the uplands at the highest elevations certainly was a severe test of their lacustrine hypothesis. They went into great detail to explain how a lake could cover almost the entire craton, first considering and rejecting the gravitational effect of the ice sheet as proposed by R. S. Woodward of the United States Geological Survey (see Wright, 1889, p. 359) and then concluding that the crust had been depressed from the weight of ice, thereby allowing the craton to be flooded. Chamberlin and Salisbury (1885, p. 288) concluded: "But in setting aside the aeolian hypothesis, we exclude the agency which seems best fitted to free us from the topographic difficulties which the singular distribution of the loess presents." Apparently Chamberlin and Salisbury, in rejecting the eolian hypothesis, never seriously considered that loess could be wind-blown from floodplains and braided streams carrying glacial meltwaters, a source that Pumpelly (1879) originally proposed but mainly rejected. Chamberlin later moderated his views somewhat and proposed a dual origin of loess, aqueous and eolian, but he still believed that the Mississippi Valley loess was lacustrine.

The eolian hypothesis of loess was given renewed impetus by Shimek in 1896. Shimek demonstrated that the fossils, texture, and topographic relationships of loess were consistent with the entrapment of wind-blown silt or dust by vegetation and favorable topography in a strictly terrestrial environment.

Leverett (1899, 1901), who succeeded Chamberlin as the leading glacial geologist in America, mainly dodged the loess issue by restating Chamberlin's (1895b) argument of a dual origin. Leverett (1899), however, subtly pointed out that the occurrence of loess at all elevations put the flood origin to a severe test, and he also stated that the occurrence of thicker loess on the east side of valleys could best be attributed to eolian deposition by prevailing winds. These observations by Leverett were soon followed by others, and it was not long thereafter that the eolian origin of loess in the craton became firmly entrenched.

SUMMARY

Glacial drift was described in relatively few reports during the first half of the nineteenth century. Glacial geology had its beginning about 1850 with early but significant studies by Charles Whittlesey, America's first glacial geologist. Among Whittlesey's "firsts" were mapping the glacial boundary, mapping end moraines, explanation of kettle holes, classification of drift, glacial erosion of lake basins, and the tracing of drift in the subsurface.

In the 1860s and 1870s, the state geological surveys assumed a leadership role in glacial studies. Most notable of these was the Ohio Geological Survey directed by John Newberry who was recognized as the leading glacial geologist of his time. An understanding of glacial lobes, glacial erosion of lake basins, moraines, multiple glaciation, and climatic change developed during this period.

In the 1880s the U.S. Geological Survey, under the guidance of T. C. Chamberlin, established dominance in glacial studies of a regional nature. Chamberlin and his assistants mapped the moraines and related features over most of the craton and established criteria for recognizing several episodes of glaciation. Chamberlin's recognition that the glacial drift represented several glacial advances separated by warmer, nonglacial periods led to a formal classification for the North American Pleistocene. Chamberlin and most of his followers believed loess to be a lacustrine deposit, and it was not until the late 1890s that the eolian origin of loess became generally accepted by most American geologists.

By 1886, Chamberlin considered his phase of glacial studies essentially complete, and he assigned the responsibility of detailed moraine mapping to one of his assistants, Frank Leverett. Leverett proceeded to map in considerable detail the glacial drift and related geomorphic features of the Illinois and Erie lobes, covering an area extending from Iowa to New York. Leverett's detailed mapping, keen observations, and ability to synthesize large volumes of data provided a fitting climax to nineteenth century glacial studies.

ACKNOWLEDGMENTS

We gratefully acknowledge the following persons and institutions who contributed the photos used in this paper. Sturgis Bailey of the University of Wisconsin-Madison and the State Historical Society of Wisconsin provided the photos of Whittlesey and Chamberlin; William Farrand of the University of Michigan provided the Leverett photo; and Merrianne Hackathorn of the Ohio Geological Survey provided the Newberry photo.

REFERENCES CITED

Agassiz, L., 1837, Discours prononcé à l'ouverture des séances de la Société Helvétique des Sciences Naturelles, at Neuchâtel, 24 July 1837: Actes Société Helvétique des Sciences Naturelles, 22nd session, Neuchâtel, 24–26 July 1837, p. 5–32.

Aldrich, Michele, 1974, John Strong Newberry, *in* Gillispie, C. C., ed., Dictionary of Scientific Biography: v. 10, New York, N.Y., Charles Scribners Sons, p. 32–33.

Association of American Geologists and Naturalists, 1843, Abstract of the Proceedings of the Fourth Session: American Journal of Science and Arts, v. 45, Art. 10, p. 310–353.

Bleuer, N. K., 1974, Buried till ridges in the Fort Wayne, Indiana area and their regional significance: Geological Society of America Bulletin, v. 87, p. 917–920.

Bradley, F. H., 1870, Geology of Grundy County, Will County, Kankakee and Iroquois Counties, Vermillion County, Champaign, Edgar and Ford Counties, *in* Worthen, A. H., et al., Geology and Paleontology: Geological Survey of Illinois, v. 4, p. 190–275.

Chamberlin, T. C., 1877, Geology of eastern Wisconsin: Geology of Wisconsin, Survey of 1873–1877: Wisconsin Geological Survey, v. 2, p. 91–405.

——1878, On the extent and significance of the Wisconsin Kettle Moraine; n.p., n.d. (1878), 36 p: a separate publication of a paper in Wisconsin Academy of Science, Arts, and Letters Transactions, v. 4, p. 201–234.

——1882, The bearing of some recent determinations on the correlation of the eastern and western terminal moraines: American Journal of Science, v. 124, p. 93–97.

——1883a, Preliminary paper on the terminal moraine of the second glacial epoch: U.S. Geological Survey, Annual Report, v. 3, p. 291–402.

——1883b, Report of T. C. Chamberlin: U.S. Geological Survey, Annual Report, v. 3, p. 17–21.

——1888, The rock-scorings of the great ice invasions; U.S. Geological Survey, Annual Report, v. 7, p. 147–248.

——1894, Glacial phenomena of North America, *in* Geike, James, The great ice age and its relation to the antiquity of man: 3rd ed., London, Stanford, 850 p.

——1895a, The classification of American glacial deposits; Journal of Geology, v. 3, p. 270–277.

——1895b, Sedimentary hypothesis respecting the origin of the loess of the Mississippi Valley: Journal of Geology, v. 5, p. 795–802.

Chamberlin, T. C., and Salisbury, R. D., 1885, Preliminary paper on the Driftless Area of the upper Mississippi Valley: U.S. Geological Survey, Annual Report, v. 6, p. 199–322.

Conrad, T. A., 1839, Notes on American geology: American Journal of Science, v. 35, p. 237–251.

Culver, G. E., 1895, The erosive action of ice: Wisconsin Academy of Sciences, Arts, and Letters Transactions, v. 10, p. 339–366.

Davis, W. M., 1882, On the classification of lake basins: Proceedings of the Boston Society of Natural History, v. 21, p. 315–381.

Drake, Daniel, 1815, Natural and statistical view or picture of Cincinnati and the Miami country: Cincinnati, Looker & Wallace, 251 p., 2 maps.

——1825, Geological account of the Valley of the Ohio: American Philosophical Society Transactions, v. 2, p. 124–139.

Ellicott, Andrew, 1799, Miscellaneous observations relative to the western parts of Pennsylvania, particularly those in the neighborhood of Lake Erie: Transactions of the American Philosophical Society, p. 224–229.

Fairchild, H. L., 1905, Ice-erosion theory a fallacy: Geological Society of America Bulletin, v. 16, p. 13–74.

Flint, R. F., Colton, R. G., Goldthwait, R. P., and Willman, H. B., 1959, Glacial map of the United States east of the Rocky Mountains: New York, The Geological Society of America.

Frye, J. C., 1968, Development of Pleistocene stratigraphy in Illinois: *in* Bergstrom, R. E., ed., The Quaternary of Illinois: University of Illinois College of Agriculture [Urbana] Special Publication 14, p. 3–10.

Gilbert, G. K., 1871, Report on the geology of Williams, Fulton, and Lucas Counties: Ohio Geological Survey, Report of Progress, 1870, pt. 7, p. 485–499.

——1873, Geology of Williams County, Ohio Geological Survey, v. 1, pt. 1, p. 557–566.

Hinde, G. J., 1878, The glacial and interglacial strata of Scarboro Heights and other localities near Toronto, Canada: Canadian Journal, v. 15, p. 388–413.

Hitchcock, Edward, 1841, First anniversary address before the Association of American Geologists and Naturalists: American Journal of Science, v. 41, p. 232–275.

Jackson, C. T., 1843, Communication on the subject of drift, *in* Abstract of the proceedings of the Fourth Session of the Association of American Geologists and Naturalists: American Journal of Science and Arts, v. 45, Art. 10, p. 320–323.

Lane, N. G., 1966, New Harmony and pioneer geology: Geotimes, v. 11, no. 2, p. 18–22.

Leverett, Frank, 1892, On the correlation of moraines with raised beaches of Lake Erie: American Journal of Science, v. 43, p. 281–301.

——1898a, The weathered zone (Sangamon) between the Iowan loess and Illinoian till sheet: Journal of Geology, v. 6, p. 171–181.

——1898b, The weathered zone (Yarmouth) between the Illinoian and Kansan till sheets: Journal of Geology, v. 6, p. 238–243.

——1899, The Illinois glacial lobe: U.S. Geological Survey, Mon. 38, 818 p., 24 pls.

——1901, Glacial formations and drainage features of the Erie and Ohio Basins: U.S. Geological Survey, Mon. 41, 802 p., 26 pls.

——1909, Weathering and erosion as time measures: American Journal of Science, v. 177, p. 349–368.

Leverett, Frank, and Taylor, F. B., 1915, The Pleistocene of Indiana and Michigan and the history of the Great Lakes: U.S. Geological Survey, Mon. 53, 529 p., 32 pls.

Lincoln, D. F., 1892, Glaciation in the Finger-Lake region of New York: American Journal of Science, v. 44, p. 290–301.

——1894, The amount of glacial erosion in the Finger-Lake region of New York: American Journal of Science, v. 47, p. 105–113.

Logan, W. E., 1863, Superficial geology, *in* Geology of Canada: Geological Survey of Canada, p. 886–930.

Maclay, William, 1927, The journal of William Maclay, United States Senator from Pennsylvania, 1789–1791: New York, C. A. Boni, 429 p.

Maclure, Williams, 1823, Some speculative conjectures on the probable changes that may have taken place in the geology of the continent of North America: American Journal of Science, v. 6, p. 98–102.

McGee, W. J., 1878, On the relative positions of the forest bed and associated drift formations in northeastern Iowa: American Journal of Science, v. 15, p. 339–341.

Newberry, J. S., 1862, Notes on the surface geology of the basins of the Great Lakes: Boston Society of Natural History Proceedings, v. 9, p. 42–46.

——1870a, On the surface geology of the basin of the Great Lakes, and the valley of the Mississippi: Annals Lyceum Natural History New York, v. 9, p. 213–234.

——1870b, Report of progress of the Geological Survey of Ohio in 1969: Report of the Geological Survey of Ohio.

——1874, Geology of Ohio; Surface geology: Geological Survey of Ohio, Report 2, pt. 1, p. 1–80.

——1878, Review of geological structure of Ohio: Report of the Geological Survey of Ohio, v. 3, pt. 1, p. 1–51.

Orton, Edward, 1870, On the occurrence of a peat bed beneath deposits of drift in southwestern Ohio: American Journal of Science, v. 50, p. 54–57, 293.

——1873, Report on the third geological district; geology of the Cincinnati group: Hamilton, Clermont, Clarke Cos.: Ohio Geological Survey, Report 1, pt. 1, p. 365–480.

Owen, D. D., 1840, Report of a geological exploration of part of Iowa, Wisconsin, and Illinois: *in* House Executive Documents of the United States, 26th

Congress, 1st Session, Document 239, p. 10–161.

—— 1844, Report of a geological exploration of part of Iowa, Wisconsin, and Illinois: *in* Public Documents of the Senate of the United States, 28th Congress, 1st Session, v. 7, Document 407, 191 p.

Pumpelly, Raphael, 1879, The relation of secular rock-disintegration to loess, glacial drift and rock basins: American Journal of Science, v. 17, p. 133–144.

Richthofen, Ferdinand von, 1877, China: Berlin, D. Reimer, v. 1, 758 p.

Shimek, Bohumil, 1896, A theory of the loesses: Iowa Academy of Science Proceedings, v. 3, p. 82–86.

—— 1909, Aftonian sands and gravels in western Iowa: Geological Society of America Bulletin, v. 20, p. 399–408.

Spencer, J. W., 1881, Discovery of the preglacial outlet of the basin of Lake Erie into that of Lake Ontario: Proceedings of the American Philosophical Society, v. 19, p. 300–337.

—— 1890, Origin of the basins of the Great Lakes of America: Quarterly Journal of the Geological Society of London, v. 46, p. 523–531.

St. John, O. H., 1870, Geology of the middle region of western Iowa and other counties, *in* Report of the Geological Survey of the State of Iowa, v. 2, p. 1–200.

St. John, S., 1851, Elements of geology, intended for the use of students: Hudson, Ohio, Pentagon Press, 334 p.

Tarr, R. S., 1894, Lake Cayuga a rock basin: Geological Society of America Bulletin, v. 5, p. 339–356.

Totten, S. M., 1969, Overridden recessional moraines of north-central Ohio: Geological Society of America Bulletin, v. 80, p. 1931–1946.

Volney, C. F., 1803, Tableau du climat et du sol des États Unis d'Amérique. Suivi d'éclaircissements sur la Floride, sur la colonie Française au Scioto, sur quelque colonies Canadiennes et sur les sauvages. Enrichi de quatre plaunches gravées, dont deux cartes géographiques et use coupe figurée de la chute de Niagara: Paris, Courcier, Dentu, An XII-1803, 2 v., xvi, 300 and 532 p., 2 folding maps, 2 folding plates.

—— 1804, A view of the soil and climate of the United States of America: C. B. Brown, trans., Philadelphia, J. Conrad & Co., 446 p., maps (Reprinted, with introduction by G. W. White, New York, Hafner Publishing Company, Inc., 1968).

White, C. A., 1870, Surface deposits, *in* Report of the Geological Survey of the State of Iowa, v. 1, p. 82–121.

White, G. W., 1964, Early description and explanation of kettle holes, Charles Whittlesey (1808–1886): Journal of Glaciology, v. 5, no. 37, p. 119–122.

—— 1967, The first appearance in Ohio of the theory of continental glaciation: Ohio Journal of Science, v. 67, p. 210–217.

—— 1973, History of investigation and classification of Wisconsinan Drift in North-Central United States, *in* Black, R. F., Goldthwait, R. P., and Willman, H. B., eds., The Wisconsinan Stage: Geological Society of America Memoir 136, p. 3–34.

—— 1982, Andrew Ellicott's geological observations in the Mississippi Valley and Florida, 1796–1800; *in* Corgan, J. X., ed., The geological sciences in the antebellum South: University, Alabama, The University of Alabama Press, p. 9–25.

White, W. A., 1972, Deep erosion by continental ice sheets: Geological Society of America Bulletin, v. 83, p. 1037–1056.

Whittlesey, Charles, 1848, Notes upon the drift and alluvium of Ohio and the West: American Journal of Science, v. 5, p. 205–217.

—— 1866, On the fresh-water glacial drift of the northwestern states: Smithsonian Contributions to Knowledge 197 (also in v. 15, 1867), 32 p.

—— 1868, Depression of the Ocean during the ice period: American Association for the Advancement of Science Proceedings, v. 16, p. 92–97.

Winchell, N. H., 1873, The surface geology; Minnesota Geological and Natural History Survey, 1st Ann. Report, p. 61–62.

—— 1875, The geology of Mower County: Minnesota Geological and Natural History Survey, 3rd Ann. Report, p. 347–366.

—— 1878, The geology of Goodhue County: Minnesota Geological and Natural History Survey, 6th Ann. Report, p. 44–45.

Worthen, A. H., 1858, Geology of certain counties, *in* Report of the Geological Survey of the State of Iowa, v. 1, pt. 1, p. 183–258.

—— 1868, Geology and paleontology: Illinois Geological Survey, v. 5, 574 p., 20 pls.

—— 1873, Geology of Peoria County, McDonough County, Monroe County, Macoupin County, Sangamon County, Illinois: Geological Survey of Illinois, v. 5, p. 235–319.

Wright, G. F., 1884, The glacial boundary in Ohio, Indiana and Kentucky: Cleveland, Ohio, Western Reserve Historical Co., Tract No. 60, p. 199–263.

—— 1889, The ice age in North America and its bearing on the antiquity of man: New York, D. Appleton and Company, 622 p.

—— 1891, The ice age in North America and its bearing on the antiquity of man: New York, D. Appleton and Company, third edition, 648 p.

—— 1892, Man and the Glacial Period: New York, D. Appleton and Company, 385 p.

Geological Society of America
Centennial Special Volume 1
1985

Development of ideas about the Canadian Shield:
A personal account

J. Tuzo Wilson
Chancellor, York University
4700 Keele Street
Downsview, Toronto, Ontario M3J 1P3
Canada

INTRODUCTION

When the editors invited me to write they knew that I had begun field work on the Canadian Shield sixty years ago and had known many who had contributed to the great changes and developments in Precambrian geology. Unfortunately for the past seventeen years I have been fully engaged in administration and have not read the recent literature. This paper is thus a personal account and poorly documented. I'm not even sure which of the ideas advanced are accepted and which are still regarded as heretical.

In 1924, I started work in the woods, while still a schoolboy, and for much of the summers of 1926, 1927, and 1928 was field assistant to Noel E. Odell who had just returned from an expedition to Mount Everest on which he had been geologist and had made great efforts to find G. Mallory and A. Irvine, lost in their attempt to reach the summit. He was such an attractive teacher that on returning to the University of Toronto in 1927 I applied to transfer from a major in physics to one in geology. My professors were appalled. Physics was then in its heyday, but geology was held in very low esteem. Ernest Rutherford had compared it to postage-stamp collecting for it consisted of making maps by identifying and locating rocks and fossils. Instruments and methods were primitive and geology lacked general theories, which were scorned as "armchair geology." This was in striking contrast to the precise theories common in physics, but few considered that Wegener's concept of continents drifting slowly about had any merit, and no one, that I can recall, realized that therein lay the explanation for the lack of theories in geology. All attempts to explain the behavior of a mobile earth on the supposition that it was static were of course doomed to failure.

Although I did not appreciate this at the time, I nevertheless preferred the excitement of the trackless woods to being immured in physics laboratories, making endless measurements with instruments still primitive before the advent of electronics or computers.

The upshot was that I took a double major in both physics and geology and thus became one of the earliest students in geophysics at the universities of Toronto, Cambridge, and Princeton (Wilson, 1982). Although a few pioneers had been studying earthquakes, gravity, and geomagnetism for more than a century, geophysics only emerged as a separate subject about that time due to the discovery of radioactivity and to the application of concepts gained in artillery sound-ranging during World War I to prospecting for petroleum.

It is difficult now to appreciate how little was then known of Precambrian shields. Only a few classic localities had been well studied by P. Peach and J. Horne in Scotland, by J. J. Sederholm in Finland, and by miners in South Africa.

In North America the best known regions were the Adirondack Mountains and the mining areas of Minnesota and Michigan in the United States and those to the east and north of Lake Huron in Ontario and in adjacent parts of Quebec. Vigorous efforts to make regional maps of the areas around mining camps were beginning. I worked on those near Michipicoten, Sudbury, Kirkland Lake, Noranda, Thetford Mines, and Yellowknife (Fig. 1).

Apart from these areas the Canadian Shield had only been examined by traverses along the railways and major rivers. In the whole vast stretch of the Northwest Territories, from Hudson Bay west to the Mackenzie River Valley, there had only been two traverses down major rivers, by J. B. Tyrrell in the 1890s, while in Labrador A. P. Low had discovered most of what was then known in a similar way. Around mines and settlements geological surveys had spread out along township lines surveyed and blazed through the woods and along the few, unpaved roads. Aerial photographs only became generally available during the 1930s. Most geology was done from canoes or by pace-and-compass traverses with no instruments except a hammer, a Brunton compass, a tape measure to check one's pacing, and a counter in hand to count one's steps. Magnetic dip needles were used around mines.

In the laboratory the microscope reigned supreme. Chemical analyses were made by wet chemical methods, which were slow

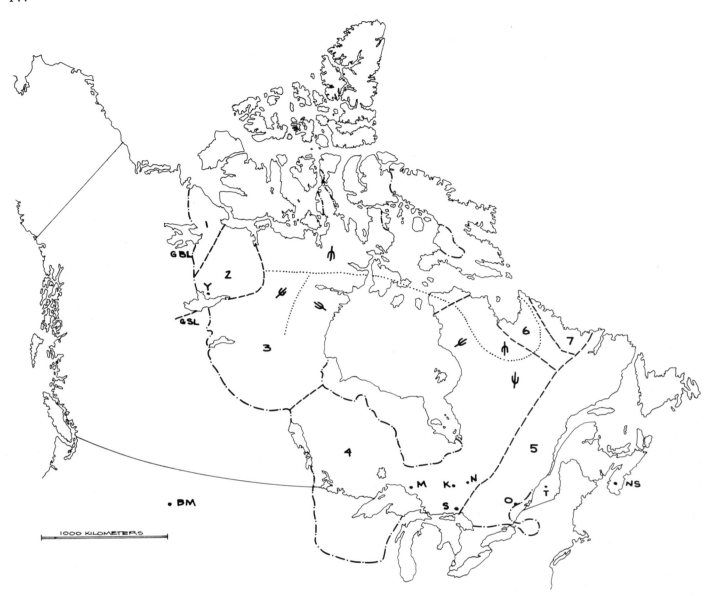

Figure 1. Index to provinces of the Canadian Shield and localities or features mentioned in the text. Legend: Boundary of Canadian Shield -·-·- ; Boundary between provinces of shield - - - -; Divide in directions of ice-sheet flow . . . . . .; Directions of late ice-sheet flow ; Key to provinces of shield: 1 = Bear; 2 = Slave; 3 = Churchill; 4 = Superior; 5 = Grenville; 6 = Labrador; 7 = Nain. Key to localities: GBL = Great Bear Lake; GSL = Great Slave Lake; Y = Yellowknife; BM = Beartooth Mountains; M = Michipicoten; K = Kirkland Lake; S = Sudbury; N = Noranda; O = Ottawa; T = Thetford Mines; NS = Nova Scotia.

and imprecise. H. V. Ellsworth of the Geological Survey of Canada was analysing uraninite crystals to determine their age by methods pioneered by B. B. Boltwood and A. Holmes (1960). (References to many of the early papers mentioned in this article can be found in Jacobs, Russell and Wilson (1974)).

THE AGE OF THE EARTH

Ideas about the division of Precambrian time were still dom-

inated in North America by the report of two committees published early in the century. Although they included the ablest American and Canadian geologists of the time, they did not agree, because one group knew the rocks west of Lake Superior and the other had worked on those to the east.

The western committee was dominated by T. C. Chamberlin and R. D. Salisbury who had written the standard textbooks of the time and whose ideas were thus widely disseminated. They considered that the earth's history had been simple. The earth had

Figure 2. Lining canoes up rapids on Plamondon River, Quebec. (Photo by J. T. Wilson, 1937, photo no. 83250 in collection of Geological Survey of Canada.)

been formed hot by the accumulation of planetesimals and it had cooled rapidly, so that the degree of metamorphism of rocks increased with their age. That theory applied nicely to the succession of rocks west of Lake Superior where relatively unaltered Keweenawan, Animiki, and Huronian sediments overlie more altered Keewatin rocks of Archean age.

On the other hand, their theory ignored a problem faced by the eastern committee, which could not place the Grenville rocks in that succession. They were more highly altered, but seemed to be younger, than the Archean rocks. Nor did that simplistic theory offer any explanation for metamorphic rocks in the Phanerozoic.

In 1907 Boltwood, acting on a suggestion by Rutherford, used assays of uraninites made much earlier by the U.S. Bureau of Mines to calculate the first radiometric ages. Arthur Holmes and members of a committee chaired by A. C. Lane soon made others. During the 1920s H. V. Ellsworth (1932) made good chemical determinations which suggested that the Grenville rocks in Ontario had last been metamorphosed about 1000 million years ago, but that the Archean of Manitoba had been undisturbed for 2500 million years. Great ages had long been advocated by a few geologists, but physicists regarded them as impossible until Eva Curie showed that radioactive decay produces heat and that the earth need not be cooling fast. Radiometric methods were soon used to suggest that the earth, meteorites, and solar system had all formed about 4600 million years ago.

In 1938 A. O. Nier at the University of Minnesota made the first isotopic age determinations and introduced precision to the subject. In 1948 A. Holmes showed the wide variation in ages of provinces in Africa. In 1956 Wilson, Russell, and Farquhar suggested that all Archean rocks might be older than 2500 million years and that they differed in composition as well as age from

Proterozoic rocks, which are younger. With H. Shillibeer they dated rocks provided by A. Holmes and proved for the first time that some Scottish rocks are also Archean (Holmes and others, 1955).

THE ORIGIN OF GNEISSES

Related to this was the question of the origin of foliated gneisses. In the 1840s W. E. Logan, the first Director of the Geological Survey of Canada, examined the Grenville rocks between Montreal and Ottawa. Observing their widespread and continuous foliation, he interpreted them as altered sediments. This was accepted, at least in Canada, until the introduction of the petrographic microscope. Then geologists, seeing interlocking crystals, took the view that the gneisses had originated from melts. F. D. Adams and A. E. Barlow (1912), members of the eastern committee, and whom I once met, made an excellent map of the gneisses around Bancroft, Ontario, showing the complex foliation which they believed had formed in a sea of molten granite. Later A. F. Buddington had similar views about the Adirondacks and in his lectures tried to explain that the foliation formed, as it does when one stirs porridge! I doubted his explanation as much as I questioned his view that many sheets of liquid granite, each a few millimeters thick, could have intruded layered gneisses to form multiple, parallel, intrusive layers in gneisses over large areas. That seemed physically impossible, but was widely advocated among geologists.

For most summers between 1926 and 1939 I worked on the Canadian Shield and everywhere I went the basement rocks were largely gneisses. Where all the soil had been scraped away by ice sheets and then washed, there were splendid exposures, especially on the shores of many of the largest lakes, and I was puzzled by

Figure 3. Supply plane arriving at camp on Splendid Island, Harricanaw River, Quebec. (Photo by J. T. Wilson, 1937, photo no. 83257 in collection of Geological Survey of Canada.)

the common references in textbooks to the granites of shields. Two explanations seemed possible. One is that in the United States outcrops are poor because those regions had also been the site of much deposition by ice-sheets. In small outcrops the foliation does not show well. Another possible explanation lies in the distance of departments of geology in most Midwestern universities from the Canadian Shield. The closest outcrops of Precambrian rocks to many of them are a few knobs which protrude through younger cover in Missouri.

G. Amstutz once took me to Iron Knob where a quarry master said that he shipped tombstones to places as far away as Montreal. This was obviously because the rock he provided was a uniform red granite, with no trace of foliation, which customers preferred to the basement rocks near Montreal that were all gneisses. The outcrops in Missouri are all knobs which had weathered out as hills and become islands in Paleozoic seas.

The situation reminded me of two isolated high hills rising out of a plain of gneisses between Great Slave and Great Bear Lakes. My investigation showed that each hill was an intrusion of unfoliated, porphoyritic granite which had resisted erosion better than the gneisses. If the origin of the Missouri knobs had been similar it seemed possible that generations of student geologists had been shown unusual granite intrusives as typical Precambrian basement and had thus failed to realize the dominance of gneisses in shields.

MAJOR STRUCTURES IN SHIELDS

That was the confused state of Precambrian geology in the late 1920s when I was an undergraduate, but between 1929 and 1934 I was introduced in the field to two developments in structural geology which were destined to improve understanding.

These were methods of deciphering complex folding and the recognition of great faults.

In 1929 I had the good fortune to be a field assistant to H. C. Cooke (1948) who was beginning to elucidate the folded structures in the Archean rocks near Noranda, Quebec. He introduced me to graded bedding, a structure seen in some sedimentary rocks in which large grains had settled to the base of each bed, enabling the top and bottom of the succession to be identified. I believe he had been introduced to this idea by R. M. Field who had seen it in the Allegheny region. Field was a professor at Princeton. He was eccentric and his colleagues regarded him with amusement because of his wild ideas, because he wrote such elementary but popular books, and because he did not do any detailed research. Nevertheless, he had a far greater impact on geology than most of his more orthodox colleagues. Instead of doing research he specialized in three other pursuits. He had a great imagination and foresaw long before others the future importance of some matters. In particular he promoted work in geophysics, in the structure of shields, and in the geology of ocean basins. He attracted brilliant students to work on these problems, among them Maurice Ewing, Harry Hess, and G. P. Woollard. He was also good at popularizing geology. A great entrepreneur, he raised money to employ students and colleagues to work on his ideas.

R. M. Field, for reasons I don't know, had earlier got money to take Cooke and other North American geologists to meet E. B. Bailey and to examine his work on Precambrian rocks in the Northwest Highlands of Scotland. Bailey was then Director of the Geological Survey of Scotland. Using graded bedding, the party had proved that a section of apparently flat-lying, late Precambrian sedimentary rocks, which extended for 10 kilometers along the side of a valley, was inverted and formed the bottom side of a huge fold. Bailey accepted this view with grace.

Figure 4. Pulling canoes through swamp at end of portage near Sucker Creek, Reindeer Lake, Saskatchewan. (Photo by F. J. Alcock, 1937, photo no. 82359 in collection of Geological Survey of Canada.)

After the trip to Scotland, Cooke worked on structure at Noranda. There he found graded beds in the Archean and also pillow lavas interbedded with them. He realized that the shape of the tops and bottoms of pillows differed and could be used to determine the orientation of folded flows. The diagnostic shape of pillows was particularly well shown in a small cliff behind the Amulet Mine. He took a photograph of this but the result was not clear. He then told me to get a ladder and paint the chilled margins of all the pillows white. His next photographs were striking and he gave one of them to M. E. Wilson (1939 and 1941), who used it in his article on the Canadian Precambrian in the Fiftieth Anniversary volume of this Society.

Cooke was a character and typically didn't tell Morley Wilson about the paint job or Morley would not have published it. In the bush Cooke also chewed tobacco all day so as not to set any fires. In his youth he had found two books he liked and took them with him every summer. On wet days in our tent, if he read Carlyle's *French Revolution,* I read Buckle's *History of Civilization in England,* and vice versa. It rained a lot and I read both twice.

At about that time, I met C. H. Stockwell (1924), who first mapped the system of huge and conspicuous faults which bound the south side of Great Slave Lake. His maps were clear and impressed me, especially since the rocks on the two sides of the principal fault bore no resemblance to one another and because the important lead-zinc deposits at Pine Point lay directly above the fault trace.

In succeeding years A. W. Jolliffe (1952) mapped other faults east of Great Bear Lake and through Yellowknife. J. F. Henderson and I added extensions, although the conservative editor of the Geological Survey refused to publish all of them. Large faults had not been a feature of older maps of the Shield, nor indeed of any others parts of the world, except in the Alps.

In 1934 and 1935 I had the good fortune to do the field work for my Ph.D. thesis along a large fault which crossed the Beartooth Mountains and Stillwater Complex in Montana. The geology was varied but simple, which was fortunate, because access was difficult and I had to solve most of the problems alone. D. C. Skeels in an adjacent area and Professor W. T. Thom provided encouragement, the latter because he had the view, extraordinary at the time, that many large faults had broken up the basement of the United States, although they were now hidden from view by later strata.

THE DIVISION OF SHIELD INTO PROVINCES OF DIFFERENT AGES

Returning from World War II, I was reminded of these great faults in the Canadian Shield by examining air photographs and by field work south of Sudbury where a large fault along the northern boundary of the Grenville Province is plainly exposed in some places.

The largest faults cut right across the shield or could be imagined to do so. Assuming that they separated provinces of widely differing ages and using the few available age determinations to assign approximate ages to the provinces, in 1949 I produced a map of the provinces of the Canadian shield. J. E. Gill (1949) of McGill University came to much the same conclusion and published a similar map at the same time and quite independently. Later D. R. Derry led a committee which added details and produced the first Tectonic map of Canada (Derry and others, 1950).

When the Tectonic map was ready, Derry and I took the manuscript to J. B. Tyrrell, who had been government geologist in the Klondike gold rush in 1898, and who had earlier crossed the Barren lands of the District of Keewatin. He was in his mid-nineties but was so interested that he got down on his hands and knees to look at it.

Also, about 1946, H. W. Wellman (1955) in New Zealand and W. Q. Kennedy (1946) in Scotland found convincing evidence that huge faults crossed those islands and that the two sides had been displaced by scores of miles. In 1948 A. Holmes divided Southern Africa into provinces, while in India C. S. Pichamuthu (1968) was perhaps the first to recognize different provinces there by noting that the rocks of Mysore resemble Archean rocks elsewhere while the surrounding provinces are gneisses. A. H. Saha measured many old dates.

In Australia mining geologists early recognized that the gneisses around Broken Hill and in the extreme south west of Australia resemble the Grenville, whereas the gold-bearing lavas around Kalgoorlie resemble the Archean in North America. In 1950 I recall discussing with de C. Clark and R. Prider the possibility that the Darling Scarp in West Australia and an east and west fault south of it separated different provinces, an idea which Prider published.

About 1954 W. T. Thom asked me to accompany him to a conference called by the California (now Chevron) Oil Company in California at which M. L. Hill and T. W. Dibblee (1953) presented their case for hundreds of miles of displacement along the San Andreas fault. Few agreed with them and one geologist considered that the 20 feet measured in 1906 was all the displacement for which he could see any evidence. Since no one there considered continental drift, I don't think we realized how crucial the debate was. Large horizontal displacements are not possible in a static earth, but are easily understood if the crust is regarded as mobile.

It was not until paleomagnetic observations became accurate and abundant some years later that the concept developed that shields might be mosaics of provinces which had coalesced.

PENEPLAINS ON SHIELDS

As I write I am sitting in a cottage set on a splendid peneplain eroded on bare Grenville rocks on the east shore of Georgian Bay, Lake Huron. This peneplain extends for a hundred miles or more in every direction, although to the west and south it dips gently beneath the waters of the Great Lakes and the Paleozoic succession which forms the Niagara Escarpment and the Bruce Peninsula.

The peneplain was thus formed in pre-Ordovician time, was buried, and has since been uncovered. Nevertheless, much of the older surface remains, and apart from a slight dip, is extraordinarily flat, so that in some places one could attempt to play tennis on it.

It is still scarred by glacial striae and grooves made more than 10,000 years ago—evidence that the present rate of erosion is very slow. Indeed it is only conceivable that these hard rocks, which once formed the roots of mountains, could have been eroded to such a level surface if the climate had once been tropical, permitting more active erosion. The fossils in the overlying

Ordovician appear to be tropical, giving credence to that view. Most shields, and indeed most continents, have been peneplained, suggesting that they have all wandered across the equator.

Another good example of a peneplain underlies the coastal plain of Eastern North America and emerges in New England and Nova Scotia. It also is tilted and the higher parts inland may form the "accordant summit levels" of the Appalachians. After studying geology in Canada and England I was surprised to see that these mountains had flat tops. Although at Princeton the concept was regarded as commonplace, I had thought that all peaks were pointed!

In the Western United States I was particularly struck by more of these flat tops, especially when I scrambled to the top of Mount Hague in the Beartooth Mountains of Montana. At the time I was excited, because I believed that I had just made the first ascent. I was alone, evening was approaching, and I was able to walk around on a flat surface, smaller than a football field, but level and covered with loose pieces of rock not one of which had been disturbed. Nearby I could see the larger flat summits of Granite Peak and Mount Wood. All clearly formed parts of a widespread summit level at a little over 12,000 feet.

I believe that H. W. Menard (1960) suggested the origin of this extensive surface when he identified the vast East Pacific Rise and proposed that it had uplifted the western third of the United States and still underlies that huge region.

Physical oceanographers later seemed to deny this, for they suggested that the Rise ended abruptly in the San Andreas fault. A compromise is possible. The northward motion of the Pacific plates does indeed terminate by shearing through California, but upwelling can still continue beneath the western part of the North American plate. D. I. Gough (1984) has produced evidence for this. The low passes through the western mountains in British Columbia (3500 feet at Yellowhead pass) and adjacent part of Washington and again in the extreme southwestern United States are in contrast to the high passes in mid-latitudes of the western states (which exceed 12,000 feet). This appears to be evidence for regional uplift due to an extension of the East Pacific Rise.

Presumably other continents have sometimes overridden upwellings. This provides a mechanism by which uplift of shields and rapid erosion can have occurred without accompanying compression and folding.

LOPOLITHS AND THE ORIGIN OF THE SUDBURY IRRUPTIVE

Shields are the home of both lopoliths and meteoric craters and many are preserved on the Canadian Shield and its borders. Much as I doubted F. F. Grout's ideas on petrology of gneisses, I was always intrigued by his recognition and naming of lopoliths (1918), those funnel-shaped basic intrusions of which the Bushveldt, Stillwater, and Duluth bodies are the largest examples and which seem to grade through Sudbury and Vredefort to dozens of smaller meteoric impact craters. Before that relationship was supported by knowledge of lunar craters and shatter cones, I had occasion in 1933 to work on the Sudbury rocks collected by W. H. Collins (1934).

He believed, although no one else did, that at Sudbury the lower norite intrusive and the upper micropegmatite intrusive had both separated from a silicate melt by liquid immiscibility. He had me measure the specific gravity of many specimens and it puzzled me greatly why the densest rocks formed his transition zone at the base of the norite in the middle of the intrusive, if the two intrusives had separated by gravity.

About 15 years later I began a reconnaissance study of air photographs in the vicinity of Sudbury, or rather I trained two undergraduates in physics, who had studied no geology, to identify features readily seen on aerial photographs, including outcrops, foliation, bedding, dikes, occasionally faults and contacts, and many glacial phenomena. When Mary James Downie and Anita Evans had finished their careful work and I compared their results with geological maps, I was astonished to discover that the conspicuous bedding which they had mapped in the Whitewater sediments within the Sudbury Basin continued for some distance into the micropegmatite, where it faded out before reaching the norite.

On comparing chemical analyses which A. G. Burroughs and H. C. Rickaby (1930) had made of the lowest sedimentary rocks in the basin with those which Collins had had made of the micropegmatite, I found that they matched closely. The average of the norite of course is very similar to that of the Nipissing diabases and almost identical to one analysis in Frechette township.

R. S. Dietz (1964) had not then advanced his idea of meteoric impacts, but I concluded that after some catastrophic event, then supposed to be volcanic, norite had risen and been intruded beneath the sediments that filled the basin. Its heat had metamorphosed the base of the sediments to micropegmatite. That view found support from an unexpected source. In an early paper N. L. Bowen (1910), the great exponent of differentiation, had suggested that the acidic, aplite fringes to Nipissing diabases are due to alteration of the intruded country rocks, a situation like that at Sudbury. Unfortunately I mislaid the paper, and it was never published, although later Gillian McTaggart-Cowan Elliot confirmed the earlier work. This observation supports a similar conclusion published later (F. W. Beales and G. P. Lozej, 1974).

GLACIAL GEOLOGY OF THE CANADIAN SHIELD

I began the examination of early air photographs in the hope that I could by that means map large structures, but apart from limited areas devoid of cover most photographs showed far more of the glacial geology than of the bedrock. This, however, was interesting, for the maps of the glacial geology were quite different from those made of the United States. The reason is obvious. The Shield was a collecting ground and transportation area, whereas the United States and parts of southern Canada were regions of melting of ice sheets and deposition.

In Nova Scotia J. W. Goldthwait had mapped many drum-lins (1924). In 1936, using air photographs, I continued his work and mapped hundreds in the southern half of the province.

The axes of all the drumlins were parallel and lay in a northwest-southeast direction, as though the ice sheet had crossed the Bay of Fundy. In contrast, the eskers which road building had not then removed, formed a radial pattern in the southern part of the province, as though the rivers of melt water which had formed them had poured off a residual cap of ice left after the ice in the Bay of Fundy had melted.

Most of the soil in Nova Scotia had been accumulated into drumlins. Between them outcrops were abundant and the aerial photographs showed much of the gently folded structure of the late Precambrian sediments which formed the bedrock. The drumlins were fairly uniform in size and shape and their axes were at right angles to the general trend of these sedimentary beds, which were of sandstone and slate in bands a few miles in width.

Drumlins were numerous on the slate, but died out within a few miles on the sandstone. It seemed evident that drumlins were streamlined shapes like snow drifts or ripple marks which could form in the clayey soil over slate, but could not form where the bedrock was sandy. The ice had eroded the upstream ends and deposited clay on the lee ends, so that the drumlins had migrated downstream in the direction of ice flow and hence could be found on the slate and for two or three miles onto the sandstone beds, but not farther (J. T. Wilson 1938a).

The next summer, in 1937, I was working in the basin of the Harricanaw River north of Noranda in the clay belt which formed the bottom of Lake Barlow-Ojibway. This vast lake formed when the ice sheets still occupied Hudson Bay, but had melted away in an area lying between Hudson Bay and the height of land south of it. The melt waters off the ice had filled a great basin between the ice-front and the height of land. There were raised beaches on some of the hill sides, and clay covered most of the Precambrian bedrock in the low ground (Wilson, 1938b).

The best walking was along so-called sand plains. On the air photos one could trace these and show that they were a series of belts of sand, each a few hundred yards wide, but elongated in a roughly north-south direction. In an adjacent area G. H. Norman had already suggested that when a stream off the ice reached Lake Barlow-Ojibway it formed a sandy delta. As the ice front receded the delta moved north, and the annual deposits overlapped to form a sand plain.

In a large region there was only one trace of a moraine. It dawned on me that this glacial geology was not like that which R. F. Flint and R. P. Goldthwait, the son of the Nova Scotian geologist, had mapped in the northern United States. the glacial geology of the gathering grounds and transmission routes in Canada is quite different from the geology of the dumping grounds in the United States.

In 1938 and 1939 I saw many huge eskers running for many miles over the terrain north and south of Great Slave Lake. The eskers were roughly parallel with the glacial striae and some eskers bifurcated, like rivers (Wilson, 1945).

After World War II, I was able to get grants to support

students to map large glacial features from air photographs. In a few years they covered over a million square miles and plotted the results on the maps which had been made by that time from the same photographs.

For the International Geophysical Year in 1957 it seemed appropriate to form a committee to combine all the glacial geology done by many geologists, including W. H. Mathews in British Columbia, J. R. Mackay in the Northwest Territories, J. A. Elson and others in the Prairies, and L. J. Chapman, D. R. Putnam and V. K. Prest in Ontario.

The Geological Society of America and the newly formed Geological Association of Canada helped to publish the first map of the glacial geology of Canada (Falconer and others, 1958).

Whereas J. B. Tyrrell (1896) had suggested that the ice flowed away from point centres, the new map showed that the direction of ice flow, represented by drumlins, and the flow of melt-water, represented by eskers, both flowed away from linear divides, as rivers do. Except as already mentioned in southern Nova Scotia, both flows are generally parallel, their direction being determined by the top surface of the ice.

There are scarcely any moraines on the whole Shield. Where the ice had flowed down steep grades its carrying power had apparently increased, and the slopes leading down to the margins of the shield, especially around glint lakes, are relatively bare. On the other hand, the central gathering grounds are heavily drift-covered.

It is surprising to note that the principal divide and greatest height of the ice surface evidently crossed from the District of Keewatin via Southampton Island to form a U-shaped divide in Northern Quebec and Labrador. As a result, the ice from Hudson Bay flowed south over the height of land into Ontario.

My own field work in the Canadian Shield came to an end in 1939. After that I only made brief trips, organized studies of air photographs and wrote a few comments. My main interests moved elsewhere. In consequence this paper says nothing about the tremendous advances in the study of shields made in the past thirty or forty years, but that was not its intention.

REFERENCES CITED

Adams, F. D., and Barlow, A. E., 1912, Geology of the Haliburton and Bancroft areas: Geological Survey of Canada, Memoir 6.

Beales, F. W., and Lozej, G. P., 1974, Sudbury Basin sediments and the meteoritic impact theory of origin for the Sudbury structure: Canadian Journal of Earth Sciences, v. 12, p. 629–635.

Bowen, N. L., 1910, Diabase and granophyre of the Gowganda Lake district, Ontario: Journal of Geology, v. 18, p. 658–674.

Burroughs, A. G., and Rickaby, H. C., 1930, Sudbury Basin area: Ontario Department of Mines, 38th Annual Report, v. 38, pt. 3,55 p.

Collins, W. H., 1934, Life History of the Sudbury nickel irruptive (1): Transactions, Royal Society of Canada, ser. 3, v. 28, pt. 4, p. 123–172.

Cooke, H. C., 1948, Back to Logan: Transactions, Royal Society of Canada, ser.

3, v. 42, pt. 4, p. 20–40.

Derry, D. R., Link, T. A., Stevenson, R. S., and Wilson, J. T., 1950, Tectonic map of Canada: Geological Association of Canada and Geological Society of America, 1 sheet.

Dietz, R. S., 1964, Sudbury structure as an Astrobleme: Journal of Geology, v. 72, p. 412–434.

Ellsworth, H. V., 1932, Rare-element minerals of Canada: Canada, Geological Survey, Economic Geology series, v. 11, 272 p.

Falconer, G., Mathews, W. H., Prest, V. K., and Wilson, J. T., 1958, Glacial map of Canada: Geological Association of Canada, 1 sheet.

Gill, J. E., 1949, Natural divisions of the Canadian Shield: Transaction of the Royal Society of Canada, ser. 3, v. 43, pt. 4, p. 61–69.

Goldthwait, J. W., 1924, Physiography of Nova Scotia: Canada, Geological Survey, Memoir 140, 179 p.

Gough, D. I., 1984, Mantle upflow under North America and plate dynamics: Nature, v. 311, p. 428–432.

Grout, F. F., 1918, The Lopolith: an igneous form exemplified by the Duluth gabbro: American Journal of Science, v. 46, p. 518–522.

Hill, M. L., and Dibblee, T. W., Jr., 1953, San Andreas, Garlock and Big Pine faults, California: Geological Society of America, Bulletin, v. 64, p. 443–458.

Holmes, A., 1948, The sequence of Precambrian orogenic belts in south and central Africa: 18th International Geological Congress, Great Britain, pt. 14, p. 254–269.

Holmes, A., Shillibeer, H. A., and Wilson, J. T., 1955, Potassium-argon ages for some Lewisian and Fennoscandian pegmatites: Nature, v. 176, p. 390–391.

Jacobs, J. A., Russell, R. D., and Wilson, J. T., 1974, Physics and Geology: McGraw-Hill, 2nd ed., 622 p.

Jolliffe, A. W., 1952, The northwestern part of the Canadian shield: 18th International Congress, Great Britain, pt. 13, p. 141–149.

Kennedy, W. Q., 1946, The Great Glen fault: Quarterly Journal Geological Society of London, v. 102, p. 41–76.

Menard, H. W., 1960, The East Pacific Rise: Science, vol. 132, p. 1737–1746.

Nier, A. O., 1938, Variations in the relative abundances of lead from various sources: Journal of the American Chemical Society, v. 60, p. 1571–1576.

Pichamuthu, C. S., 1968, The Precambrian of India, in Rankama, K. (ed), The Precambrian, v. 3, New York, Wiley-Interscience, 3271.

Stockwell, C. H., 1924, Map of eastern portion of Great Slave Lake: Canada, Geological Survey, Maps 377A ana and 378A.

Tyrrell, J. B., 1896, Explorations west of Hudson Bay: Canada, Geological Survey, vol. 9, pt. F.

Wellman, H. W., 1955, New Zealand Quaternary tectonics: Geologische Rundschau, Bd. 43, H.1., p. 248–257.

Wilson, J. T., 1938a, Drumlins of southwest Nova Scotia: Transactions Royal Society of Canada, ser. 3, v. 32, sect. 4, p. 41–47.

Wilson, J. T., 1938b, Glacial geology of part of northwestern Quebec: Transactions Royal Society of Canada, ser. 3, v. 32, sect. 4, p. 49–59.

Wilson, J. T., 1945, Further eskers north of Great Slave Lake: Transactions Royal Society of Canada, ser. 3, v. 39, sect. 4, p. 151–153.

Wilson, J. T., 1949, The origin of continents and Precambrian history: Transactions Royal Society of Canada, ser. 3, v. 43, sect. 4, p. 157–184.

Wilson, J. T., 1982, Early days in university geophysics: Annual Reviews of Earth and Planetary Science, v. 10, p. 1–14.

Wilson, J. T., Russell, R. D., and Farquhar, R. M., 1956, Radioactivity and the age of minerals: Handbuch der Physik, ed. S. Flugge, v. 47, chap. 11, pp. 288–363.

Wilson, M. E., 1939, The Canadian Shield, Geology of North America: Geologie der Erde, (Krenkel, E., ed.), Berlin, P. 232–311.

Wilson, M. E., 1941, Precambrian: Geology Society of North America, 50th Anniv. Vol. p. 271–305.

Geological Society of America
Centennial Special Volume 1
1985

Vestiges of a beginning

Preston Cloud
Department of Geological Sciences
University of California
Santa Barbara, California 93106

ABSTRACT

James Hutton's proclamation that he saw "no vestige of a beginning" was an irresistible challenge. Geologists soon began searching downward through the "interminable greywackes" of Murchison for just such vestiges, especially of life. After Logan's demonstration in the mid-1800s of thick sedimentary successions beneath fossiliferous Cambrian, Dawson described the generally discredited but perhaps in part biosedimentary *Eozoon* from rocks then thought to be among Canada's most ancient. Stromatolites eventually became widely accepted as evidence of pre-Phanerozoic life. They are now recorded in carbonate rocks from perhaps 3.5 billion years (Gyr) ago onward. Many early reports also claimed the occurrence of Metazoa in rocks below the Cambrian. All reports turned out to be mistaken or dubious, however, except for those that now define the Ediacarian System which is basal to the Paleozoic Era and, in my view, a part of that era.

We see at last why searches for older Metazoa failed. The records of pre-Ediacarian life are paleomicrobiological, sedimentological, and biogeochemical. Microbial evidence is demonstrably present back to ~2 Gyr ago and still convincing back to ~ 2.8 Gyr ago. Stromatolites are presumptive evidence (not proof) for still older life. Geochemical interpretation is consistent with the presence of life back to the oldest known sediments. Metazoa began with the Ediacarian. They were the last main event of biological evolution. Thereafter, it involved primarily elaboration on the multicellular theme. The significant older records are cratonal, edited by the caprices of preservation, erosion, discovery, and subduction.

INTRODUCTION

Were it not for the camel droves at the temple of Jupiter Ammon in Egypt, geology might have been denied the seminal thoughts of James Hutton, prime architect of its central principle of naturalistic causes. Hutton's partnership in the manufacture of sal ammoniac (NH_4Cl), initially derived only from camel dung at this locality, was the source of the wealth that enabled him to retire to Edinburgh at age 43, thereafter to devote himself exclusively to geology and his circle of scholarly friends (Geikie, 1897). As we learn from mathematician John Playfair (1802), Hutton observed widely and reflected deeply on the meaning of his observations, but he was in no rush to publish. Only his involvement in the founding of the Royal Society of Edinburgh in 1783, when he was nearly 60, moved him to present and publish his first scientific paper (Geikie, 1897). In this succinct paper,

read before the Royal Society in 1785 and containing the essence of his 1788 *Theory of the Earth,* Hutton made his oft-quoted remark, "We find no vestige of a beginning—no prospect of an end" (White, 1956).

We must now emend this statement for reasons that would have rejoiced Hutton, a man of multidisciplinary talents committed to the view that the ultimate court of appeal is nature itself. Vestiges of a beginning we now know aplenty, thanks particularly to advances of recent decades. We see them in a progressive decrease backward through time of biotic complexity and diversity. We observe them in meteoritic evidence for the nearly simultaneous origin of the Sun and planets. We detect them in isotopic evidence for the presolar origin of elements heavier than helium. We even hear them, faintly, in the radio static that records the waning heat from the Big Bang origin of the universe itself, some 13 to 20 billion years (Gyr) ago. As for prospect of an end, astrophysical studies of the main-sequence stars show that we are halfway there. It will be a firey denouement as the sun expands to become a red giant star 4 or 5 Gyr from now, reclaiming the stuff it left behind as the initial solar nebula condensed.

HUTTON TO LOGAN AND DAWSON

Hutton's judgment, nevertheless, was both profound and irrefutable in his time. How could he have foreseen these findings of modern biogeology, isotope geochemistry, meteoritics, and astronomy? Widely traveled and studied as he was, his knowledge of the oldest rocks was limited. A concept of the craton and its growth did not then exist. How could he have foreseen that geologists would one day have a system of age determination, expressible in radiometric years, that would order 3.76 Gyr of Earth history in the correct sequence and show the planet to have condensed from cosmic dust and gas 800 million years (Myr) earlier? How could he have foreseen that a record of cellular microbial life would be extended 2.8 Gyr into the past with indirect suggestions of life processes as far back as the oldest terrestrial rocks we know? He could not have, of course, but his statement posed an irresistible challenge to all natural scientists.

Vestiges of a beginning are now being reported from rocks of, or marginal to, the cratons in growing abundance, and although still plagued with uncertainty, error, and confusion, perhaps a third or a quarter of the reports might satisfy Hutton's demanding criteria for acceptable evidence.

Despite the existence of many household names for local rock units in Hutton's day—lovely descriptive names like Old Red, Millstone Grit, and the Great Oölite—no systems, eras, or eons in the formal sense had yet been designated or accepted; Tertiary and Quaternary are the only terms now a part of the principal time scale that were then already in use. R. I. Murchison, who named the Silurian System in 1835, included in it rocks then called Cambrian by Adam Sedgwick, as well as all those rocks later called Ordovician. He dismissed all underlying rocks as the "interminable greywackes." It is in rocks associated with these "interminable greywackes," in, on, or marginal to the cratons, that we find the _biological_ vestiges of a beginning here to be discussed.

An important step in the search for beginnings came with the appointment of William Logan, later Sir William, as the first director of the Geological Survey of Canada in 1842, 27 yr before the founding of the United States Geological Survey. Logan, born in Montreal in 1798, the year following Hutton's death, was then 44 yr old, best known for his mapping of Welsh coal seams (Merrill, 1906). Realizing that the location of mineral deposits was his principal justification for public employment, he turned his attention to the already long-known but little-studied ancient gneisses and schists of Quebec and Ontario, thus inaugurating the detailed study of sub-Cambrian rocks that has become one of the glories of Canadian geology. He quickly confirmed that the oldest

visibly fossiliferous strata in North America lay unconformably on crystalline rocks which he recognized as probably altered sediments and termed metamorphic. Camping, living off the land, and making his own base maps, Logan, together with Alexander Murray and Sterry Hunt, soon showed the vast extent of these rocks and worked out a rough stratigraphy of three main sequences in eastern Canada. The oldest and most metamorphosed he called Laurentian. A sequence of less metamorphosed rocks, including conglomerates that contained fragments of the old gneisses, clearly younger, he called Huronian. Unconformably above the Huronian were the little-altered and only slightly deformed copper-bearing clastic rocks and flood basalts of the Lake Superior region, now called Keweenawan. Although no fossils were then known from any of these rocks, Logan realized as early as the 1850s that he would need paleontological help. He thus offered a position to James Hall, 13 years his junior, but state geologist of New York and a power in his own right. Hall declined, and Elkanah Billings, lawyer and editor turned paleontologist, got the job (Winder, 1978), much to Hall's later discomfort.

Meanwhile Charles Darwin, in the first edition of _The Origin of Species_ in 1859, further stimulated interest in the search for beginnings with his recognition that the diversity of the Cambrian biota called for prior ancestry. Canada, possessing a vast expanse of primordial rocks, many only lightly metamorphosed, was the best place then under active study in which to search for such ancestors. A useful summary of Canadian activities in this search, with reference to detailed published discussions by others, is given by Hofmann (1971).

Then, as now, the critical edge of reasoning was sometimes dulled by over-eagerness for results. During the 1850s Logan received samples of alternately banded metasedimentary rocks from Ontario and Quebec, which he exhibited as fossils at the 1859 meetings of the American Association for the Advancement of Science at Springfield, Massachusetts. They were interpreted as animal remains (later considered a giant foraminifer) and named by J. W. Dawson (1865) as _Eozoon canadense_. This move ignited a published discussion that exceeded in volume "many times that of all other described Canadian fossils and pseudofossils combined" (Hofman, 1971). Dawson himself, principal of McGill University, president of the Geological Society of America in 1893, and later knighted, was the main contributor to this discussion. He became obsessed with his interpretation, defending it through the rest of his professional life. Elkanah Billings, on the other hand, although the official paleontologist of the Geological Survey of Canada, wisely detached himself from the controversy, maintaining an eloquent silence on the subject throughout his life. In 1888 the International Geological Congress Subcommittee on the Archean in America, thinking to resolve the dispute democratically, voted nine to four that _Eozoon canadense_ was not of organic origin (Frazer, 1888)—overruling Dawson, C. D. Walcott, Sterry Hunt, and A. Winchell.

As one may suspect, that did not settle matters any more than committee rulings or majority opinion settle them nowadays or ever should in science. Although it seems clear from published discussion and illustrations, particularly from the field studies of Bonney (1895) and a summary by Merrill (1906, p. 635–646), that most specimens called *Eozoon* are of metamorphic origin, a primary biogenic structure might be reflected in some. Deformed but recognizably columnar stromatolites of probable microbial construction are known from Grenville rocks at a number of localities, and some structures referred to *Eozoon canadense* could be stromatolitic or cryptalgal. The issue is now of interest only as a historical lesson in the futility of arguments based on superficial resemblance. Where the subject is geological or biogeological, it must be understood in terms of the whole rock— including its mineralogy, sedimentology, geochemistry, and diagenetic and metamorphic history. Where purely physical processes are capable of producing structures observed, one is not justified in interpreting them as biogenic in the absence of genetically associated, uniquely biological features. Where not clearly syngenetic with enclosing sediments, organisms or their traces may not safely be regarded as of the same age or even as fossils (Cloud and others, 1980). If we have learned nothing else about the beginnings of life, we should have learned these things by now.

CHARLES DOOLITTLE WALCOTT

Hofmann (1971, Fig. 3) provided a convenient tabulation of Canadian sub-Cambrian fossils, pseudofossils, and dubiofossils described up to 1969, beginning with Dawson's *Eozoon canadense*. The branching columnar stromatolite *Archaeozoon acadiense,* however, described by Matthew (1890) was the first structure of reasonably convincing biological origin to be named from the ancient rocks of Canada. Meanwhile, south of the border, in his very first paper on the upper Proterozoic strata of the Grand Canyon, Charles Doolittle Walcott (1883) reported the first genuine body fossil ever to be described from pre-Phanerozoic rocks. He referred to it as "a small Discinoid shell," named 16 years later (Walcott, 1899) as *Chuaria circularis,* a flat compression of a macroscopic spheroidal alga, not a discinoid brachiopod and not Cambrian as he then supposed. These rocks (Kwagunt Formation, Chuar Group) are now considered to be about 900 Myr old. Also in his 1883 paper Walcott recorded "an obscure *Stromatopora*-like form," later compared with *Cryptozoon,* a member of a group of synoptic sedimentary buildups of characteristically (but not invariably) biogenic origin, known since 1908 as stromatolites (Kalkowsky, 1908). Years later, Walcott (1914, 1915) described and illustrated from rocks of the Belt Supergroup chains of coccoid cells that appear to be the first recognizably microbial fossils recorded from below the Cambrian rocks.

Indeed, Walcott fairly deserves his name as the father of pre-Phanerozoic paleontology. The field was new and full of pitfalls. Like all pioneers, he made mistakes, a goodly number in fact, but he also provided impetus and he was right about the central issue. From 1882 to 1916, despite his ongoing Cambrian studies and heavy administrative duties as director of the United States Geological Survey, secretary of the Smithsonian Institution, and president of both the American Philosophical Society and the National Academy of Sciences, this self-taught "country boy" remained the most productive student of primordial life. Contrary to much then-current belief, he established, nearly a century ago, that there is, indeed, material evidence for a pre-Phanerozoic biosphere of long duration.

Neither Walcott nor any of his contemporaries or successors established, on or off the craton, a single convincing record of multicellular animal life in strata older than about 670 Myr. Although more than 200 reports of such early animal life have since been published, all structures reported are now known to be misdated, not syngenetic, non-metazoans, pseudofossils, or, at best, dubiofossils (e.g., Cloud, 1968, 1973; Cloud and others, 1980). To be sure, the Ediacarian assemblage of soft-bodied metazoan imprints and tracks found in South Australia and related faunas found worldwide precede the currently agreed-upon conventional Cambrian and are therefore pre-Cambrian in an informal sense. But they also belong properly to a basal Paleozoic and Phanerozoic system and period that began about 670 Myr ago with the first appearance of manifest animal life. For this new post-Proterozoic period, the name Ediacarian has been proposed (Cloud and Glaessner, 1982).

THE RECENT DECADES— SOME MISINTERPRETATIONS

What do we now know about pre-Phanerozoic life? Obviously we know abundant and diverse indirect implications of biogenic $CaCO_3$ precipitation and sediment-binding in the form of stromatolites, abundant in Proterozoic and less common in Archean rocks. Unfortunately, the biological information content of such structures is low. More informative biologically are the Proterozoic microbial assemblages of which we currently know several score. I will not here discuss, tabulate, or illustrate these assemblages or assign confidence levels. I shall try to show by reference to a few key examples how the sequence points toward an early beginning of life on Earth some 600 to 800 Myr after the initial aggregation of the planet ~4.57 Gyr ago.

First, some cautionary comments. Because I have observed that human nature can be wondrously receptive to claims reported from fields beyond personal experience, I repeat here my warning that critical faculties may suffer from lapses of wariness. Regrettably, this observation has been, and continues to be, true of the search for the oldest records of life, where results do not come easily and may be uncertain. Where ambiguous, the level of confidence should be specified. The subject is a mine field of traps. I hardly know of anyone in it, myself included, who has not at one time or another miscalculated the facts or reached untenable conclusions. Horodyski (1981) has shown, quite conclusively I think, that the chains and clusters of objects from the Baraga Group of Michigan which I thought (*in* Cloud and Morrison, 1980) "may be the remains of . . . microbes," consist, instead, of apatite crystals. Two spurious records have even made it to the cover of *Science*: Electron micrographs of almost certainly contaminant bacteria (Schopf and others, 1965) were misinterpreted as 2-Gyr-old bacteria and the spindle-shaped fillings of abridged but distinctively triradiate contraction cracks in ripple troughs (Hofmann, 1967) were thought to be probable Metazoa of early Proterozoic age. Indeed, a high percentage of the ostensibly paleomicrobiological papers published over the recent decades of accelerating interest as well as all of those claiming pre-670-Myr-old Metazoa are to be seriously questioned, either as to the biogenicity of their material, its syngenicity, or particular microbial affinities. Caution is especially called for regarding works authored by otherwise competent geologists who may not combine in themselves or in tandem with others the needed blending of biological and geological expertise and experience with pseudofossils and contaminants to deal critically with biogeological problems. Few of us could not profit from being more critical in our own work. Some elaboration of this difficulty is contained in papers by Cloud (1976), Cloud and Morrison (1979), Schopf (1976), Schopf and Walter (1980), Horodyski (1981), and Wang and Luo (1982). I will not here recite the long list of papers that had better not been published and are now best forgotten.

It is important for a clear appreciation of the progress that has been made in elaborating beginnings to keep such dubious or misinformed reporting to a minimum. Spurious reports dilute the value of the sparse but precious list of demonstrably biogenic and surely syngenetic microbial records. Consider only some of the known kinds of traps. They include: bubbles, crystallites, circular or spheroidal clusters of fine particles confined by surface tension, artifacts of preparation, and other pseudofossils (Cloud, 1976, Pl. 1); fungal and bacterial spores, pollen, and other airborne contaminants common in all laboratories; and endolithic crack-dwelling or sediment-penetrating organisms that are likely to be indigenous without being syngenetic (Cloud and Morrison, 1979), as well as other postdepositional intrusions. Some of the most extraordinary micropseudofossils are the "filamentous" tracks of ambient pyrite in chert (Tyler and Barghoorn, 1963; Knoll and Barghoorn, 1974), the minute hornlike composite tubes ("anellotubulates") that result from the processing of pyritic samples with hydrogen peroxide (Pickett and Schreibnerová, 1974), and modern stalked bacteria whose spores survive distillation and filtration to grow in the presence of carbonaceous residues (Cloud and Hagen, 1965). The best safety measures are: (1) know contaminants by periodic examination of laboratory dust; (2) never believe anything that cannot be verified in thin section and shown to be both biogenic and syngenetic; and (3) beware of unique specimens, unusual crystals and spheroids, specimens that are too abundant or well preserved for the type or age of associated rock, and preparational artifacts.

THE RECENT DECADES—SUCCESSES

What then do I consider to be well-documented and demonstrably syngenetic paleomicrobiotas? They may seem pitifully few when ranged against a time scale of billions of years, but collectively they make a fair number. Classic among them is

the Gunflint microbiota, best known from S. A. Tyler's 1953 discovery of 2-Gyr-old stromatolitic and nannofossiliferous cherts of the Gunflint Iron Formation along the north shore of Lake Superior. Actually, elements of that microbial assemblage were first reported by Moore (1918) from the approximately contemporaneous Kipalu Formation of the Belcher Islands in eastern Hudson Bay. The geological world, however, was not then ready to accept preservation of microfossils in rocks so old. What finally brought this discovery to notice was Tyler's find of an even more diverse microbiota in the cherts west of Schreiber, Ontario (Tyler and Barghoorn, 1954), followed by convincing illustrations (Barghoorn and Tyler, 1965). As I can testify from my own observations of its paragenetic relations to the enclosing stromatolitic cherts (Cloud, 1965) this Gunflint microbiota is unimpeachably syngenetic. It is essentially limited to the stromatolitic columns and excluded from interspaces between them. It is concentrated in the darker zones of the stacked columnar laminae. Its filamentous species cross grain boundaries (although this is not conclusive evidence) and are offset by microfractures. The chert probably crystallized from a primary silica gel which hermetically sealed this microbiota against later degradation. The fossils are demonstrably biogenic, shown not only by their general similarity in morphology and dimensions to living procaryotic species but also by details of cellular differentiation.

Some specimens of *Gunflintia* Barghoorn display a cellular differentiation so similar in detail to that of modern nostocacean algae (heterocysts, akinites, hormogonia) as to imply familial relationship and ambient oxygen levels high enough to require shielding of nitrogenases in the special microenvironment of the heterocystic cell, or >0.1% of present atmospheric level—still very low but growing. Another species is so similar in some 14 different growth stages to the living, Cretaceous, and younger Proterozoic, budding bacterium *Metallogenium* that no grounds can be found for giving it a different name other than separation in time; it has been named *Eoastrion.*

The Gunflint microbiota, therefore, is an important benchmark along the dim trail into the past. It appears to be wholly procaryotic—a biota of small non-nucleate cells such as are universal among the cyanophytes and more conventional bacteria. Although it teemed in variety and numbers in the silica gels that crystallized to make the Gunflint chert and is widely known from other localities in Canada, the north-central United States, and western and northern Australia, it reveals no convincing evidence of the nucleate or eucaryotic cell that is the building block of all higher forms of life—true algae, fungi, plants, and animals. The advance from procaryotic to eucaryotic forms of life, therefore, occurred at some time during the 1.3 Gyr or so that separate the Gunflint microbes from the Ediacarian Metazoa.

Working back in time from the Ediacarian Metazoa, the oldest known, we learn from the work of Schopf (1968) on the microflora from cherts of the Bitter Springs Formation of central Australia that a variety of probable eucaryotes was well established by about 800 to 900 Myr ago. The Bitter Springs microbiota, another benchmark, excels even that of the Gunflint in variety and quality of preservation, although not in numbers or evidence of syngeneity. The only truly conclusive evidence for a eucaryotic presence is either a metazoan or metaphytic level of organization, or electron-microscopical evidence of membrane-bound nuclei and other cell organelles, but the Bitter Springs microbiota provides other kinds of evidence. The consistently large size of some cells, the similarity in structure of some filaments to those of eucaryotic algae, the presence of a few structures that strongly resemble eucaryotic cell tetrads, and ultra-microscopical evidence that permits the interpretation of some intracellular bodies as nuclei (Schopf and Oehler, 1976) converge to provide strongly persuasive support for a eucaryotic component in the Bitter Springs biota.

Proterozoic strata younger than about 1 to 1.2 Gyr are richer than older rocks in their content of microbial fossils, particularly the mainly spheroidal bodies called acritarchs (meaning "of unknown origin"). They also show a general increase in size, as the Soviet paleomicrobiologist Timofeev (1969, 1974) was the first to recognize. Skipping over these to what may be the oldest eucaryotes, we now consider a more generalized line of evidence entering on the generally larger size of eucaryotic cells and the excessive rarity and peculiarity of true branching among filamentous procaryotes. Procaryotic cells are generally under about 5 μm in diameter and rarely exceed 40 μm. The eucaryotic cell generally exceeds 20 μm and may be millimeters across. In addition, true branching is almost limited to eucaryotes. Three studies involving these criteria imply that eucaryotes were already present 1.3 to 1.4 Gyr ago. First was the recognition (Cloud and others, 1969; Licari, 1978) that cells having diameters up to 62 μm and similarly large-diameter, branching, pauciseptate filaments from cherty portions of the Beck Spring Dolomite of eastern California were probably eucaryotic; these rocks are considered on indirect evidence to be about 1.3 Gyr old, although an age as young as 0.9 to 1 Gyr is possible. Second is the presence of Walcott's *Helminthoidichnites,* reinterpreted as a megascopic alga, in rocks of similar age in the Belt Supergroup (Walter and others, 1976). Third is a tabulation by Schopf (1977) that showed a general increase in mean cell diameter of known microbiotas younger than about 1.4 Gyr, implying that to be the approximate age of the incoming of a eucaryotic component. In addition, Licari (1978) found convincing affinities among the Beck Spring microbial fossils with green and yellow-green eucaryotic algae. Another line of reasoning involving the onset of a permanently oxygenous atmosphere beginning about 2 Gyr ago suggests that eucaryotes could have made their debut that early, although we as yet see no convincing evidence of it. What one can say with a high degree of confidence is that the eucaryotic cell probably first appeared at some time between about 2 and 1.4 Gyr ago, following the wholly procaryotic biotas of earlier times.

THE TIME OF THE BEGINNING

How early were those earlier times? Before the Gunflint microbiota we know only sparse and mostly questionable records—some indeed were clearly in error (summarized by Cloud, 1976; Schopf, 1976; Schopf and Walter, 1980; and Roedder, 1981). The record of thoroughly believable cyanophyte-like procaryotes, however, can be taken as far back as the 2.8-Gyr-old Fortescue Group (Tumbiana Formation) of the Hamersley Basin, Western Australia (Schopf and Walter, 1980). In a recent paper (Cloud, 1983) I discussed briefly the problem of extending a record of organisms into earlier times. I concluded that the microbial flora from the 3.4- to 3.5-Gyr-old Warrawoona Group of Western Australia, so widely announced in the media, is probably contaminated by the intrusion of younger, perhaps even Cenozoic, iron bacteria, represented by abundant, well-preserved, twisted, ribbonlike stalks and delicate fibrils similar to those of the modern genus *Gallionella* and by *Metallogenium*-like clusters similar to those of modern rock varnishes. Given that this biota is in a pervasively fractured, recrystallized, and siderite-invaded matrix, it cannot be admitted to the select company of persuasively syngenetic fossil microbiotas. Both the Warrawoona and rocks of similar age in Zimbabwe (Orpen and Wilson, 1981), however, do contain wrinkled laminar sequences and vertical buildups that may be of microbial origin. So the evidence, although indirect and at a weak level of confidence, suggests a simple microbial presence that long ago.

Indirect evidence consistent with still earlier life is found in the oldest rocks we know, the 3.76-Gyr-old metasediments of the Isua region of southwest Greenland, where Schidlowski (1981) has found depletion of ^{13}C over ^{12}C, suggesting biologic partitioning. That conclusion is weakened by the evidence from type II diamonds that physical processes can produce a similar depletion (Milledge and others, 1983), but it is defensible and it permits the further conclusion that very simple forms of life, perhaps methanogenic bacteria, were already present when the oldest sedimentary rocks known were deposited. That was about as far back as life could have originated, survived, and become entombed in sediments. Evidence from radiometric ages on the Moon, crater counts on the Moon, Mercury, and Mars, and the pockmarked surfaces of the icy satellites of Jupiter and Saturn implies that a heavy rate of meteoritic and cometesimal infall continued throughout the solar system for hundreds of millions of years after the main aggregation of the planets. Such a bombardment, combined with residual and radiometric heating of the primordial Earth, would have rendered it inhospitable to life, or even to chemical evolution leading toward life, until the planetary surface had solidified and cooled enough to accumulate and retain a permanent atmosphere and hydrosphere, perhaps, 4 to 3.8 Gyr ago.

CONCLUSION

Hutton would be delighted to know that the fragmentary and incomplete record of a Cryptozoic Eon that crops out over only 17% of the rifting and rafted continental surface, but that represents 85% of Earth history, has at last revealed vestiges of a beginning. Combined with evidence from meteoritic geochemistry, this tells of a beginning not only for life on Earth, but also for Earth itself, its sister planets, and even the Sun. Our descendants, if any, 4 or 5 Gyr from now, will see the end as the Sun expands into its next phase of evolution within the galaxy.

REFERENCES CITED

Barghoorn, E. S., and Tyler, S. A., 1965, Microorganisms from the Gunflint chert: Science, v. 147, p. 563–577.

Bonney, T. G., 1895, On the mode of occurrence of Eozoon Canadense at Cte St. Pierre: Geological Magazine (Dec. 4), v. 2, p. 292–299.

Cloud, Preston, 1965, Significance of the Gunflint (Preambrian) microflora: Science, v. 148, p. 27–45.

—— 1968, Pre-Metazoan evolution and the origins of the Metazoa, *in* Drake, E. T., ed., Evolution and environment: New Haven, Yale University Press, p. 1–72.

—— 1973, Pseudofossils: A plea for caution: Geology, v. 1, p. 123–127.

—— 1976, Beginnings of biospheric evolution and their biogeochemical consequences: Paleobiology, v. 2(4), p. 351–387.

—— 1983, The biosphere: Scientific American, v. 249(3), p. 176–189.

Cloud, Preston, and Glaessner, M. F., 1982, The Ediacarian Period and System: Metazoa inherit the earth: Science, v. 217, p. 783–792.

Cloud, Preston, and Hagen, Hannelore, 1965, Electron microscopy of the Gunflint microflora: Preliminary results, Proceedings of the National Academy of Sciences, v. 54, p. 1–8.

Cloud, Preston, and Morrison, Karen, 1979, On microbial contaminants, micropseudofossils, and the oldest records of life: Precambrian Research, v. 9, p. 81–91.

—— 1980, New microbial fossils from 2 Gyr old rocks in northern Michigan: Geomicrobiology Journal, v. 2(2), p. 161–178.

Cloud, Preston, Licari, G. R., Wright, L. A., and Troxel, B. W., 1969, Proterozoic eucaryotes from eastern California: Proceedings of the National Academy of Sciences, v. 62, p. 623–631.

Cloud, Preston, Gustafson, L. B., and Watson, J.A.L., 1980, The works of living social insects as pseudofossils and the age of the oldest known Metazoa: Science, v. 210, p. 1013–1015.

Dawson, J. W., 1865, On the structure of certain organic remains in the Laurentian limestones of Canada: Quarterly Journal of the Geological Society of London, v. 21, p. 51–59.

Frazer, P., 1888, Report of the sub-committee on the Archean: American Geology, v. 2, p. 143–192.

Geikie, Sir Archibald, 1897, The founders of geology: London and N.Y., Macmillan and Co. Ltd., 297 p.

Hofmann, H. J., 1967, Precambrian fossils(?) near Elliot Lake, Ontario: Science, v. 156, p. 500–504.

—— 1971, Precambrian fossils, pseudofossils, and problematica in Canada: Geological Survey of Canada Bulletin 189, 146 p.

Horodyski, R. J., 1981, Pseudomicrofossils and altered microfossils from a middle Proterozoic shale, Belt Supergroup, Montana: Precambrian Research, v. 16, p. 143–154.

Kalkowsky, E., 1908, Oölith and stromatolith im norddeutschen Buntsandstein: Zeitschrift der Deutschen Geologischen Gesellschaft, Bd. 60, p. 68–125.

Knoll, A. H., and Barghoorn, E. S., 1974, Ambient pyrite in Precambrian chert: New evidence and a theory: Proceedings of the National Academy of Science, v. 71(6), p. 2329–2331.

Licari, G. R., 1978, Biogeology of the late pre-Phanerozoic Beck Spring Dolomite of eastern California: Journal of Paleontology, v. 52(4), p. 767–792.

Matthew, G. F., 1890, On the existence of organisms in pre-Cambrian rocks: Bulletin of the Natural History Society, New Brunswick, v. 2(9), p. 28–33.

Merrill, G. P., 1906, Contributions to the history of American geology: Report of the U.S. National Museum for 1904, p. 189–734.

Milledge, H. J., Mendelssohn, M. J., Seal, M., Rouse, J. E., Swart, P. K., and Pillinger, C. T., 1983, Carbon isotope variation in spectral type II diamonds: Nature, v. 303, p. 791–792.

Moore, E. S., 1918, The iron formation on Belcher Islands, Hudson Bay, with special reference to its origin and its associated algal limestones: Journal of Geology, v. 26, p. 412–438.

Orpen, J. L., and Wilson, J. F., 1981, Stromatolites at ~3,500 Myr and a greenstone-granite unconformity in the Zimbabwean Archaean: Nature, v. 291, p. 218–220.

Pickett, John, and Scheibnerová, Viera, 1974, The inorganic origin of "anellotubulates": Micropaleontology, v. 20(1), p. 97–102.

Playfair, John, 1802, Illustrations of the Huttonian Theory of the Earth: Dover Edition of the University of Illinois 1956 facsimile reprint; Dover Publications Ind. (New York), 528 p. (1964).

Roedder, Edwin, 1981, Are the 3,800-Myr-old Isua objects microfossils, limonite stained fluid inclusions, or neither?: Nature, v. 293, p. 459–462.

Schidlowski, M., 1981, Antiquity of photoautotrophy: A frontier revisited, *in* Kageyama, M., and others, eds., Science and scientists: Tokyo, Japan Scientific Societies Press, and Dordrecht, D. Riedel Publishing Co., p. 57–65.

Schopf, J. W., 1968, Microflora of the Bitter Springs Formation, late Precambrian, central Australia: Journal of Paleontology, v. 42, p. 651–688.

—— 1976, Are the oldest 'fossils' fossils?: Origins of Life, v. 7, p. 19–36.

—— 1977, Biostratigraphic usefulness of stromatolitic Precambrian microbiotas: A preliminary analysis: Precambrian Research, v. 5, p. 143–173.

Schopf, J. W., and Oehler, D. Z., 1976, How old are the eukaryotes?: Science, v. 193, p. 47–49.

Schopf, J. W., and Walter, M. R., 1980, Archaean microfossils and "microfossil-like" objects—A critical appraisal: Geological Society of Australia, 2d International Archaean Symposium, Extended Abstracts (Perth, 1980), p. 23–24.

Schopf, J. W., Barghoorn, E. S., Maser, M. D., and Gordon, R. O., 1965, Electron microscopy of fossil bacteria two billion years old: Science, v. 149, p. 1365–1367.

Timofeev, B. V., 1969, Sferomorfidii Proterozoya. [Proterozoic Sphaeromorphyta]: Institute for Geology and Geochronology Precambrian, Akademiya Nauk, Scientific Press (Leningrad), 146 p. (in Russian).

—— 1974, Mikrofitofossilii Proterozoya i rannego Paleozoya, S.S.S.R. [Plant microfossils of the Proterozoic and lower Paleozoic of the U.S.S.R.]: Institute for Geology and Geochronology Dokembriya, Akademiya Nauk S.S.S.R. (Moscow, Leningrad: Izdatel'stvo "Nauka"), 80 p. (in Russian).

Tyler, S. A., and Barghoorn, E. S., 1954, Occurrence of structurally preserved plants in pre-Cambrian rocks of the Canadian shield: Science, v. 119, p. 606–608.

—— 1963, Ambient pyrite grains in Precambrian cherts: American Journal of Science, v. 261, p. 424–432.

Walcott, C. D., 1883, Pre-Carboniferous strata of the Grand Canyon of Colorado, American Journal of Science, v. 26, p. 427–442.

—— 1899, Pre-Cambrian fossiliferous formations: Geological Society of America Bulletin, v. 10, p. 199–244.

—— 1914, Cambrian geology and paleontology, III, No. 2, Pre-Cambrian Algonkian algal flora: Smithsonian Miscellaneous Collections, v. 64, p. 77–156.

—— 1915, Discovery of Algonkian bacteria: Proceedings of the National Academy of Sciences, v. 1, p. 256–257.

Walter, M. R., Oehler, J. H., and Oehler, D. Z., 1976, Megascopic algae 1300 million years old from the Belt Supergroup, Montana: A reinterpretation of Walcott's Helminthoidichnites: Journal of Paleontology, v. 50(5), p. 872–881.

Wang, Fuxing, and Luo, Qiling, 1982, Precambrian acritarcha: A cautionary note: Precambrian Research, v. 16, p. 291–302.

White, G. W., 1956, Introduction and biographical notes to University of Illinois facsimile reprint of John Playfair's Illustrations of the Huttonian Theory of the Earth, originally published 1802: University of Illinois Press (Urbana), p. vii.

Winder, C. Gordon, 1978, Two giants of geology: Joint annual meetings of Geological Association of Canada and Geological Society of America, Toronto, 1978, p. 10–12.

Printed in U.S.A.

Geological Society of America
Centennial Special Volume 1
1985

James Hall's discovery of the craton

Robert H. Dott, Jr.
Department of Geology and Geophysics
University of Wisconsin
Madison, Wisconsin 53706

ABSTRACT

Preeminent nineteenth century American geologist James Hall was not only one of the most competent and ambitious paleontologists and stratigraphers of his day, he also had a firsthand early grasp of midwestern as well as Appalachian geology. In 1841 he joined David Dale Owen for a tour of the Ohio Valley to see how far west the New York State stratigraphic divisions could be extended. He was warmly rewarded. With similar motivation, he also joined the Foster-Whitney-Whittlesey Lake Superior Survey of 1850–51. Then, when the New York legislature terminated his salary during one of its feuds with the irascible paleontologist, Hall accepted an invitation to organize a survey for the young state of Iowa in 1855–59; and in 1856 he assumed an advisory role to the embryonic Wisconsin Survey. He also became a consultant for several other states and had helped to instigate the first Hayden-Meek western Survey to the South Dakota Badlands (1853).

Hall's unique acquaintance with cratonic as well as Appalachian stratigraphy and structure equipped him to recognize the fundamental differences between cratons and orogenic belts. The great thickness and structurally complex nature of Appalachian Paleozoic strata then led to development of his famous theory of mountain building, which was first expounded in the 1857 AAAS presidential address soon after his principal midwestern experiences. Although Hall's inferred simplisitc causal relationship between sediment thickness and mountain building proved wrong, the observation that thick strata are somehow associated with active tectonism was of fundamental importance and is still very much relevant to modern concepts of global tectonics. Among Hall's own priorities, however, the mountain theory was subordinate to his passion to possess and to describe the Paleozoic fossils of North America as well as to extend the New York stratigraphy across the craton.

BACKGROUND

James Hall, first president of the Geological Society of America, was born in Hingham, Massachusetts, on September 12, 1811, to English-immigrant parents of modest means. He attended a public school blessed with a gifted teacher, who inspired Hall in intellectual matters. Young Hall learned of the new Rensselaer School under the direction of Amos Eaton and walked some 200 miles to Troy, New York, in 1830. Apparently he was attracted by the revolutionary "Rensselaer Plan," which emphasized science, rather than the classics, and included both field and laboratory work experiences involving the student in an active role (Clarke, 1923; Fisher, 1978). Besides Eaton, Hall was associated at Rensselaer with such geologic notables as Ebenezer

Emmons, Douglas Houghton, Abram Sager, and others. The Rensselaer School was a remarkable success, for it supplied more nineteenth century leaders of American geology than any other institution (Friedman, 1979).

Upon graduation in 1832, Hall made a tour on foot of the Helderberg Mountains to collect fossils from the great Silurian-Devonian sequence there. A job at Rensselaer as librarian allowed Hall to stay on and obtain the Master of Arts degree in 1833. Later he was appointed an assistant in chemistry (1835–41). Whenever possible, he continued to collect fossils and minerals. In 1836 the New York legislature authorized a geological survey organized in four districts under the initial purview of

William W. Mather, Ebenezer Emmons, Timothy A. Conrad, and Lardner Vanuxem, respectively. James Hall was engaged as Emmons' assistant in the Second District, and his first task was to study the iron deposits of the Adirondack Mountains. A year later the Third and Fourth Districts of central and western New York were revised; Conrad's duties were changed, Vanuxem was given charge of the Third District, and Hall was put in charge of the new Fourth District in westernmost New York (Fisher, 1981). In 1837 Hall's professional career really blossomed as the abundantly fossiliferous Silurian-Devonian strata of this district became his to study intensively. Hall's assistants included George W. Boyd, Eben N. Horsford, and Ezra S. Carr, who were or were to become Rensselaer products like Hall. The Fourth District Survey was completed with publication of the final results in 1843 (Hall, 1843).

Hall was a leading exponent of a then widely held view that an American stratigraphic classification was best for America. European names like Cambrian and Silurian were considered undesirable here, but only partly because of the Sedgwick-Murchison controversy over their validity in Britain. Because of the importance of nomenclature in those early days of investigation in America, there came a proposal from Albany for a meeting of American geologists. Hall was among those who met in Albany in 1838 to discuss the need for a geological forum to deal with problems of mutual interest among American geologists. In 1840 the first formal meeting of the American Association of Geologists occurred in Philadelphia, and out of that organization evolved the American Association for the Advancement of Science in 1847. Still later the Geological Society of America was born (1888) as a sort of stepchild of the AAAS with Hall as its first president (1889) (Fairchild, 1932). Certainly his great international prominence and maturity made Hall a logical choice for president, but also one suspects that his electors considered his great ego and volatile temper in making a prudent as well as logical choice.

Great interest was aroused in Europe by the flow of reports emanating from North America, and most especially from New York. Hall's reputation abroad was growing rapidly; by the latter half of the century, Hall and James Dwight Dana were surely the most famous American geologists. As early as 1837, Hall had been recognized with membership in the Imperial Mineralogical Society of St. Petersburg; and he was one of the first people sought out by Lyell during his visits to North America (1841 and 1845). Lyell, moreover, was not the only one to ask "Which way is Albany?" (Clarke, 1923, p. 321). Later Hall was chosen to be organizing president of the First International Geological Congress at Paris of 1878. He was entertained warmly by many European notables on that occasion and during his other visits to Europe (1872, 1881, 1897). He was elected foreign correspondent of the Academy of Sciences of France in 1884, the only English-speaking member at that time.

Many of Hall's assistants on the New York and other surveys became distinguished men in nineteenth century American geology. Some of the more familiar names of people who came under

his influence—and sooner or later his abuse—include the following:

Charles E. Beecher	Eben N. Horsford	Charles D. Walcott
Ezra S. Carr	Joseph Leidy	Charles A. White
John M. Clarke	W J McGee	Charles Whittlesey
Nelson H. Darton	Fielding B. Meek	Robert P. Whitfield
Grove Karl Gilbert	Charles S. Prosser	Josiah D. Whitney
Ferdinand V. Hayden	Carl Rominger	Amos H. Worthen
	Charles Schuchert	

As Fisher has argued (1978), anyone wanting a career in paleontology needed to come to Albany to apprentice with James Hall because there were no formal programs in colleges or universities; Fisher wrote, "The mere privilege of examining so many fossils in one place was a singular opportunity. . . . The experience gained by learning not only the science but art of (paleontology from Hall) . . . produced an extraordinarily competent scientist" (Fisher, 1978, p. 150).

Other prominent geological and paleontological figures with whom Hall was more than casually associated over the years included:

Louis Agassiz	Joseph Henry	Edward Orton
Joachim Barrande	Edward Hitchcock	David Dale Owen
John J. Bigsby	T. Sterry Hunt	Raphael Pumpelly
Elkanah Billings	J. Peter Lesley	Ferdinand Roemer
Timothy A. Conrad	Leo Lesquereux	Henry D. Rogers
James D. Dana	Sir William Logan	William B. Rogers
Sir William Dawson	Sir Charles Lyell	Benjamin Silliman
Edouard Desor	Jules Marcou	Frank Springer
Amos Eaton	Othniel C. Marsh	Lardner Vanuxem
Ebenezer Emmons	William W. Mather	Eduard de Verneuil
John W. Foster	John S. Newberry	Charles Wachsmuth

The above lists provide terse statistical evidence of James Hall's scientific preeminence, for which there is abundant additional evidence, of course. Not revealed above, however, is Hall's incredible egotism and irascibility. Biographies by J. M. Clarke (1923) and D. W. Fisher (1978) provide abundant details about Hall's character and career from which a few snippets are paraphrased here.

Hall's assistants in particular experienced their mentor's terror tactics, but others were not spared. In addition to awesome verbal diatribes, the hapless objects of Hall's frequent wrath received vigorous pummelings and brandishing threats underscored by either an ever-handy cane or shotgun. Surely the most extreme example of Hall's self-righteousness was an infamous moonlight boat ride down the Hudson River during which Hall allegedly threw overboard the entire printing of an offensive popularized geological chart just published (1849–50) by a North Greenbush school teacher (James T. Foster) which was destined for New York City. Hall, it seems, was planning just such a chart himself (Clarke, 1923, p. 208–209).

Even in his late years, Hall was a formidable opponent. In 1884, when he was 73, there was a claim of a "mongrel fauna" containing two species of spiriferid brachiopods never before seen together. Clarke reported that (p. 513):

The warmth of the contention excited the attention of the reporter for the Philadelphia *Press* which came out the next morning with the story that Hall had said: "If any one will show me these two Spirifers side by side in the same rock, I will sacrifice my life's work. I will give up my reputation, eat my hat and make the person who shows me the rock a present of my coat and boots!" The story was good and was sent broadcast to the newspapers with amplifications, the New York *Tribune* adding that Professor Williams took the next train for Ithaca, sent on a piece of rock containing the offending Spirifers with the message: "You have it now. Please eat your hat and send me your coat and boots by express."

Undaunted, Hall responded in the press that the facts were still as he claimed, that no such specimens had been forwarded to him, and that he was still fully clothed.

Hall's single-minded acquisitiveness for fossils is almost as legendary as his indignation. He stooped to almost every conceivable means short of homicide in attempting to acquire outstanding collections. His most effective technique was to invite adoring collectors like Charles Schuchert to work with him in Albany and to bring their collections. Commonly when the flattered assistant eventually moved on, his collections did not. In characteristic spider-and-fly fashion, Hall once wrote a collector in California that he could "not do better than to furnish me with specimens, sections, etc., of the beds . . . giving me thus the means of bringing the subject more fully and forcibly to our geologists" (Clarke, 1923, p. 299). The fact is that he was correct, too.

Besides having a paranoic streak, Hall was clearly a workaholic. He drove himself and his assistants mercilessly and could rarely say "no" to new schemes. Many times his friends advised him to ease up. For example, no less a personage than Joseph Henry wrote the following to Hall on August 2, 1858 (quoted by Clarke, p. 350):

My own opinion is that you have undertaken too much for your strength and that you will be obliged to share the field of research with others, not as your equals but as working under direction and receiving a proper share of the credit and emolument. * * * *

I would in conclusion impress upon you that life, health and above all, peace of mind are far above scientific reputation and that these ought not to be endangered by an attempt to accomplish more than your own strength or a due regard to the advance of truth will justify.

Truly your friend and servant.

JOSEPH HENRY

Hall seems to have brooded over such advice as evidenced by the following portion of a letter written to Dana on September 10, 1858 (quoted by Clarke, p. 350):

Were I to tell you that I am an unbeliever in science you need not be shocked. It seems to me I can no longer go on and the events of the past year have again and again turned me to resolution that did I know enough of any respectable business or profession to earn an honest living I would at once abandon science forever and could I erase my name from every printed page and annihilate all I have done, it should be done tomorrow.

In the long run, however, advice to take work less seriously was to no avail. One wonders how much longer Hall might have lived beyond the young age of 87 had he heeded Joseph Henry's admonishment.

HALL IN THE MIDWEST

Ohio Valley Excursion

When the Fourth District studies were nearing completion, Hall inevitably became curious to know how far the New York stratigraphic classification could be extended beyond its home state. So in 1841 he made his first of many odysseys westward into the heart of the lowland interior of the continent, now referred to as the tectonically stable North American craton (Fig. 1). He joined David Dale Owen for a boat trip down the Ohio River from Louisville to New Harmony. In a mere two months, he examined strata and collected fossils in Ohio, Kentucky, Indiana, Illinois, Missouri, Iowa, and Wisconsin (Clarke, 1923, p. 94). He returned with ample evidence for extending the New York classification over this entire region, which he documented in the *American Journal of Science* in 1842 as well as on a geologic map of the Middle and Western States contained in the final report of the survey of the Fourth District (Hall, 1843).

Lake Superior Survey

Hall's financial situation seems always to have been tenuous, according to Clarke (1923). Hall himself commonly aggravated his own situation by frequently alienating the New York legislature, by purchasing irresistible fossil collections, by sometimes failing to sell some of his own collections at precarious times, and by investing impetuously in shady ventures. Clarke contrasts the irony of Hall's hypercritical attitude toward his close associates with his gullibility toward complete strangers promoting dubious schemes. Another trait was a frequent failure to repay debts, especially to those whom he did not like very well.

Because of his always marginal and sometimes nonexistent (i.e., suspended) New York income, Hall frequently accepted other temporary employment, which usually led to important expansions of his professional perspective and his burgeoning collections. His second odyssey to the craton was occasioned by such an engagement in 1845 by a private company to examine Lake Superior copper deposits on Isle Royale and the Keweenaw Peninsula of Michigan (Fig. 1). This opportunity was, at least in part, an outgrowth of Hall's acquaintance with fellow Rensselaer alumnus, Douglas Houghton, who had been doing the pioneer geological surveys of Michigan. Hall's party brushed closely with tragedy in crossing the lake from Isle Royale to Copper Harbor in its small sailboat instead of waiting for an overdue schooner that was supposed to fetch them. Nothing was ever published of this trip except Hall's brief account quoted by Clarke (p. 145–148).

Foster-Whitney Survey of Michigan and Wisconsin

In 1847 the federal government authorized a land survey of the Upper Peninsula of Michigan with the talented but erratic Charles T. Jackson in charge. In 1848 he was forced to resign, and his assistants John W. Foster and Josiah Whitney were

CRATONIC ODYSSEYS OF JAMES HALL

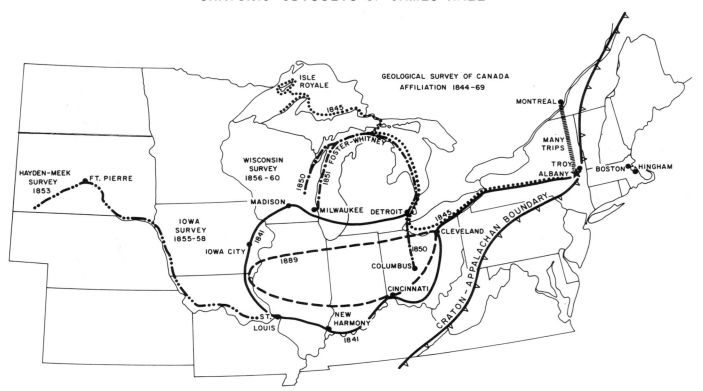

Figure 1. Map of James Hall's various trips into the North American craton.

designated joint directors. Hall, who was already acquainted with both, was engaged to join the survey in 1850 to provide expertise in Paleozoic stratigraphy and paleontology. Colonel Charles Whittlesey was also engaged in study topography and terrestrial magnetism, and a young Swiss protégé of Agassiz, Edouard Desor, was invited to study the drift. Desor, who was already well acquainted with Hall, had accompanied Agassiz and Jules Marcou in 1848 on a birchbark-canoe expedition to Lake Superior. Hall devoted his attention in two brief field expeditions (1850 and 1851) to eastern Wisconsin and adjacent upper Michigan (Fig. 1). It was Foster's and Whitney's intention that Hall would concentrate upon the "Silurian" (including Ordovician of present terminology) and "Devonian" above the Potsdam Sandstone (Upper Cambrian in modern terminology), and the publication generally reflects this division of labor. Nonetheless, we can assume that Hall had a good comprehension of the whole project. For example, in an 1851 letter from Whitney to Hall (quoted by Clarke, p. 224), whimsical reference is made to the Potsdam Sandstone beautifully exposed on the south shore of Lake Superior:

The What d'ye call-ems of the Pitchered Rocks—have you satisfied yourself what they are? I mean the Fucoides duplex (or the Fucoides do perplex) of the Grand Portal. I have not heard from you in so long that for aught I know you may have rolled yourself up in a ball and got petrified.

Whether or not Hall ever visited the Pictured Rocks (Fig. 2), he obviously was well aware of the Cambrian sandstones there and that they contained trace fossils. The Foster-Whitney report of 1851 contains a number of shrewd stratigraphic insights, which appear to be a blend of the thinking of Hall with that of the principal authors. For example, the following quotation reveals advanced thinking about the still-popular notion of universal formations in the context of the red Proterozoic strata of Lake Superior (p. 114):

. . . lithological characters form an uncertain criterion in determining their age, at points widely separated. Yet, we have seen distinguished geologists undertake to identify these rocks, simply from lithological characters, with the *bunter sandstein* of Germany, with the new-red of England, and the trias of the Connecticut valley. Such a doctrine is repugnant to all of our preconceived notions.

The "distinguished geologists" were Owen and Jackson. To underscore belief in lateral variation of lithology (or facies in modern terminology), Foster and Whitney cite the following variations observed by them (p. 114):

The Potsdam sandstone of New York is a Quartzose rock, whose particles are firmly aggregated, while the same rock, on the northern slope of Lake Michigan, is so slightly coherent, that it may be crushed in the hand. The calciferous sandstone of New York, when traced west, passes into a magnesian limestone. Even in that state, according to Hall,

Figure 2. The Grand Portal of the Pictured Rocks along the south shore of Lake Superior in Michigan. The cliffs are eroded from Cambrian "Potsdam" Sandstones (now Munising Formation) (Plate 12, Foster and Whitney, 1851).

groups which, at one extremity, are of great importance, and well characterized by fossils, cannot be identified at the other.

Their account of the identification of the Potsdam in the Great Lakes region also reveals a mature stratigraphic insight (p. 133):

The almost uninterrupted continuity with which this rock can be traced, even from its eastern extension through Canada and along the north shore of Lake Huron to the St. Mary's River, and thence westerly, leaves no doubt as to its true position and identity in age with the Potsdam sandstone of New York. If we were at a loss in thus tracing it continuously, we have still the evidence of the succeeding fossiliferous strata, which show conclusively, the same relations to this sandstone as they do to its equivalent in New York. With both these evidences combined, we cannot hesitate for a moment in our conclusion regarding its age and place in the series.

By 1852 the Foster-Whitney Survey was disintegrating. Besides other clashes within the survey group, Hall had developed one of his characteristic "sooner-or-later" dislikes for Foster. Hall was very much distracted at the time by a threatened libel suit over the jettisoned geological charts, the launching of a pet project for the University of Albany, and completion of Volume II of the *Palaeontology of New York*. He spend only a few weeks in the field but nonetheless provided important counsel on stratigraphy and paleontology. He described the "Silurian" strata and fossils of northern Michigan and eastern Wisconsin, recognized

ancient coral reefs therein, presented a comparison of midwestern stratigraphy with that of New York (Table 1), and made a critical comparison of North American with European Paleozoic terminology. Chapter 17, apparently written by Foster and Whitney, presents a 10-page review of contemporary ideas on the "Elevation of Mountains." It is based chiefly upon the ideas of de Beaumont and contains no premonition of Hall's theory of 1857.

Meek-Hayden Survey of the White River Badlands

By 1852 when *Palaeontology II* was published, Hall's far-flung activities dictated the need of assistance. He already had made use of collectors, such as colorful Colonel Ezekiel Jewett. In 1852 he engaged Fielding B. Meek, who was recommended both for his geologic and artistic skills, and Ferdinand V. Hayden, who was an eager young friend of John S. Newberry of Ohio. In 1853, Hall reluctantly agreed to sponsor Hayden and Meek for an expedition into the White River Badlands of Nebraska Territory (now southwestern South Dakota; Fig. 1) to collect newly discovered Cenozoic nonmarine mammalian and invertebrate fossils. Meek was indispensable to Hall, but like many another assistant, he was glad to be separated from the unpredictable wrath of his mentor if only for a few months. Eventually and with great difficulty he was able to extricate himself permanently from Hall's employment, to which he nonetheless was greatly in debt.

TABLE 1. HALL'S COMPARISON OF THE PALEOZOIC STRATIGRAPHY OF THE MIDWEST WITH THAT OF NEW YORK*
(Lines emphasize major differences between the two regions as recognized by Hall)

WEST Upper Mississippi Valley		EAST New York
Burlington limestone	32.	Great carboniferous limestone
Chemung group	31.	Gray and yellow sandstone
Hamilton group	30.	Sandstone & shale of Catskill Mountains
Upper Helderberg limestones	29.	Chemung group
(Wanting at west and southwest)	28.	Portage group
(Wanting at states north of Ohio River)	27-24.	Hamilton group
Onondoga salt group	23-20.	Upper Helderberg group
(Wanting west of Lake Michigan)	19.	Oriskany sandstone
Niagara limestone	18-14.	Lower Helderberg group
Clinton group	13.	Onondoga salt group
(Wanting in the west)	12.	Niagara group
Hudson-river group	11.	Clinton group
Galena or Upper Magnesian limestone	10.	Medina sandstone
Trenton limestone	9.	Oneida conglomerate
Black-river and Birdseye limestones	8-6.	Hudson-river group
St. Peters or Upper Sandstone of the Northwest	5.	Trenton limestone
Calciferous sandstone or Lower Magnesian limestone	4.	Birds-eye and Black-river limestone
Potsdam sandstone	3.	Chazy limestone
	2.	Calciferous sandstone
	1.	Potsdam sandstone

RELATIVE THICKNESS

*Chiefly after Foster and Whitney, 1851, Pt. II, pp. 288 and 291 with supplementation from Hall and Whitney, 1858, v. 1, p. 46 and Hall and Whitney, 1862, p. 13.

The Iowa Survey

When Dartmouth-educated James W. Grimes succeeded to the governorship of Iowa in 1854, he immediately proposed a geological survey of the state, and through eastern friends he learned of Hall. The new western states typically looked eastward for talent, and the New York Survey was the natural model for geological investigations; at this time it eclipsed all other state and federal efforts. Direction of said survey was offered to Hall in 1855, and he promptly accepted in spite of deep involvement with the Geological Survey of Canada and the New York paleontology. Seemingly there were two reasons for his acceptance: first, his perpetually precarious financial situation (his salary having been suspended since 1850), and secondly, an unquenchable thirst for more fossils. Grimes at once liked Hall, whom he described as "one of the most modest and unobtrusive men I have ever met" (Clarke, 1923, p. 273). This apparent flaw in Grimes' judgment may testify to a kind of Jekyll-Hyde streak in Hall.

Hall's friend Whitney was also engaged for the survey, and Grimes accepted Hall's suggestion that Amos Dean of Albany be appointed first chancellor of the new University of Iowa. Hall himself became the first professor of geology and natural history, but he apparently never presented a single lecture. So Hall had quickly become intimately embroiled in the political affairs of Iowa. Although he could spend only one or two months per year in the state, he orchestrated the survey brilliantly from Albany. He engaged Amos H. Worthen of Illinois as his paleontological assistant, while Whitney concentrated upon mineral deposits. It had not escaped Hall's attention that Worthen possessed one of the finest crinoid collections in the country. A condition of employment to be oft repeated with others was that Worthen would allow Hall to describe his crinoids, which was done in the Iowa reports (Fig. 3). After a brief field excursion in the fall of 1855, Hall returned to Iowa again only for winter meetings of the legislature to politic on behalf of the survey. The younger men did the actual field work, but they were constantly distracted by

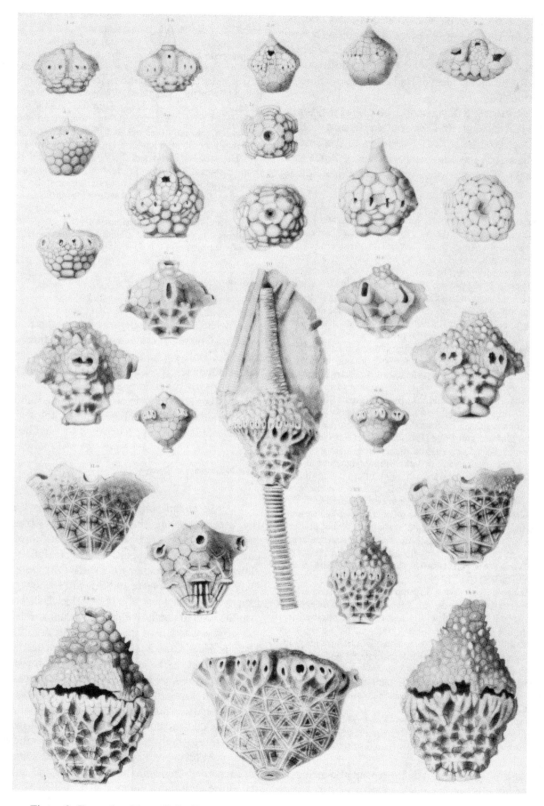

Figure 3. Example of beautifully illustrated crinoids from the Burlington limestone portrayed in the Geology of Iowa (Hall and Whitney, 1858, v. I, Part II, Plate 10).

financial embarassments. Erratic and inadequate tax receipts made it necessary for Hall to borrow funds in Albany. When the Iowa Survey was discontinued in 1859, two volumes had been published (Hall and Whitney, 1858). One that treated the paleontology of eastern Iowa fulfilled most of Hall's selfish ambitions for his Iowa connection (Fig. 3).

Chapter II of the 1858 Iowa report on General Geology carries the subtitle "General Remarks on the Geology of the Northwest, and the Relation of the Formations to those of the East." Several quotations provide vivid evidence of Hall's great stratigraphic insight. Table 1 will complement the following quotations:

In tracing westward the geological formations as known in New York and Pennsylvania, we find them, with one or two exceptions, gradually becoming thinner, until at last several of them are scarcely recognizable, or are so attenuated as to be overlooked in a country deeply covered by modern deposits (p. 38).

This diminution of the lower limestones is in a measure compensated by the accession of the Galena or Lead-bearing limestone, which is an upper member of the Trenton limestone series not developed in the East, but forming, in Wisconsin, Illinois and Iowa, a more conspicuous feature than the Trenton limestone proper.

In the Hudson-river group and succeeding Oneida conglomerate, a most marked change takes place: the former is estimated to have, in Pennsylvania, a thickness of six thousand feet, and the latter a thickness of one thousand eight hundred feet; while the same rocks, along the Green mountain range and in Canada, are in even greater force. This group of strata, consisting of green, blue and red shales, shaly sandstones and some beds of limestone, can be traced westward from Canada and New York, by Lake Ontario and Lake Huron, by Point aux Baies at the northern extremity of Lake Michigan, and thence from below the head of Green Bay and by the eastern side of Lake Winnebago, and at intervals across the State of Wisconsin. Along this line of almost continuous outcrop there is a constant thinning of the beds, until at last the whole group has diminished to a thickness of less than one hundred feet. Thus we see that the group of strata, forming the most conspicuous feature in the Green mountains, with a thickness greater than the actual height of these mountains, and making up a large part of the whole Appalachian chain, has become so insignificant that it had never been recognized along the Upper Mississippi until during the progress of the present survey in the autumn of 1855 (p. 39–40).

The strata which in New York and the east constitute the limestones of the Upper Helderberg series, the Hamilton, Portage and Chemung groups, are all embraced within less than two hundred feet: while in New York and Pennsylvania these formations have a thickness of from four to six thousand feet and form conspicuous topographical features; constituting a great part of the White mountains, and important portions of the Appalachian chain.

The Catskill mountain group, having a thickness of more than three thousand feet, and which forms the conspicuous highlands from which it derives its name, it quite unknown to the west of the Cincinnati axis; and the succeeding conglomerates, sandstones and shales are likewise absent from the series in the west, while their places are in a degree filled by the carboniferous limestones which attain far less thickness (p. 40–41).

In the Hall and Whitney report of 1858, Hall also included a brief discussion of his idea that the thicker Paleozoic strata of the Appalachian region had somehow caused the ultimate deformation of those strata—the theory of mountains that had been the subject of his presidential address to AAAS a year earlier.

This remarkable fact of the thinning out westwardly of all the sedimentary formations, points to a cause in the conditions of the ancient ocean, and the currents which transported the great mass of materials along certain lines which became the lines of greatest accumulation of sediments, and consequently present the greatest thickness of strata at the present time. It is this great thickness of strata, whether disturbed and inclined as in the Green and White mountains and the Appalachians generally, or lying horizontally as in the Catskill mountains, that gives the strong features and the hilly and mountainous country of the east, and which gradually dies out as we go westward, just in proportion as the strata become attenuated.

The subdued features of the West are therefore due, not alone to the absence of great disturbing forces, but to the absence or the great tenuity of the formations, or paucity of materials or strata to be disturbed. The thickness of the entire series of sedimentary rocks, no matter how much disturbed or denuded, is not here great enough to produce mountain features; and the most elevated portions of this region are those where no disturbing force essentially affecting the horizontality of the strata has acted (p. 41–42).

The Illinois Survey—An Aborted Scheme

In 1857 Illinois undertook to reorganize and invigorate its geological survey. Worthen was one of three applicants for its headship. Hall wrote a glowing recommendation for him, but then in a ludicrous breach of tact and judgment, he also supported the other two candidates. His attempt to "ride three horses at once" (Clarke, p. 185) exploded and left Hall the enemy of all. This meddling resulted in Illinois paleontology being inaccessible to Hall—exactly the opposite result to what he had in mind.

The Wisconsin Survey

While already engaged by New York, Canada, and Iowa, Hall accepted yet another affiliation in Wisconsin in 1856. That state undertook a reorganization of its three-year-old survey under a cumbersome partnership of Hall, fellow Rensselaer alumnus and former assistant Ezra S. Carr, now a professor at the University of Wisconsin, and Edward Daniels, a former but ineffective survey director (Clarke, 1923). Hall barely had time to visit the state to confer with the governor and legislature; running the survey was left in the rather ineffective hands of Carr and Daniels. He already had firsthand knowledge of Wisconsin geology, however, as a result of field work for the earlier Foster-Whitney survey, and through correspondence and fossil collections from local naturalists, such as Increase A. Lapham of Milwaukee (see Mikulic, 1983). The employment of Whitney to study the lead and zinc deposits adjacent to Iowa and Illinois and Whittlesey to study mineral deposits in northeastern Wisconsin led to publication of a large volume in 1862, but a hostile legislature regarded the results as insufficient and abruptly terminated the entire endeavor (Bailey, 1980, p. 10). Wisconsin was interested in economically rewarding deposits, and not fossils, so a frustrated Hall published the paleontology of Wisconsin in a New York report in 1867 and separately in 1871.

Other Surveys

To varying degrees, Hall was involved also with other surveys either for advice in choosing personnel, as titular head or as consultant on paleontology. These included the surveys of Missouri (1853 and 1871), California (1853–56), transcontinental railroad surveys (1853–57), New Jersey (1854–57), Ohio (1854–57), Texas (1858), Mississippi (1858), Michigan (1869–70), and Pennsylvania (1870–75) (see Clarke, 1923 for details). He also recommended F. V. Hayden for first director of the U.S. Geological Survey (D. W. Fisher, written communication, 1984). Except for identification and description of fossils for some of these surveys, Hall's contributions were minuscule, but the list nonetheless attests to his prominence.

The Last Trip West

In 1889 at the age of 77 and while president of the new Geological Society of America, Hall mounted his last invasion of the Midwest (Fig. 1). The purpose was to beg, borrow, and buy brachiopods for his latest project to revise this vast group of Paleozoic invertebrates (Clarke, 1923). The trip was rewarding overall, but its culmination was the luring to Albany of young Charles Schuchert of Cincinnati with one of the finest collections ever. Hall's biographer, John M. Clarke, also assisted in the brachiopod studies. The final volume—number VIII of the *Palaeontology of New York* (1894)—resulted, but Hall's last major publication before his death was a long overdue, colored geologic map of the state, which appeared almost anticlimactically in 1894.

During the completion of the *Brachiopoda,* Hall had his last and sweetest wrangle with bureaucracy. The executive secretary of the regents, who oversaw Hall's program, had become overzealous in trying to impose strict accounting and efficiency procedures (there were efficiency experts already in the 1880s). Such a ruckus developed that the legislature stepped in and appointed the aged prima donna as state paleontologist and state geologist for life and with complete independence of management. This must have reminded Hall of an earlier observation by himself on the death of a particularly vicious political enemy that "Providence is usually on my side" (Clarke, p. 285, footnote 5).

Even as an octogenerian, Hall continued to receive honors, including a last trip to Russia in 1897 at the age of 86 to be *ancien president honoraire* of the Seventh International Congress at St. Petersburg. The life of James Hall finally ran out at a summer retreat in New Hampshire in August 1898, just a month short of 87 years of unprecedented scientific production, political intrigue, countless honors, and bitter invective.

HALL AND THE ORIGIN OF MOUNTAINS

In the published version of his famous theory of mountains, first presented orally in his 1857 presidential address to the American Association for the Advancement of Science, he stated that "the greater the accumulation, the higher will be the mountain range" (Hall, 1859, pt. 1, p. 96). This was the crux of his idea that great thicknesses of strata were prerequisites to folded mountain ranges. It is more than ironic that this idea has brought Hall the greatest fame, for it received only a fraction of the attention and time that its author devoted to his fossils and stratigraphy.

Hall was influenced by J.F.W. Herschel's suggestion in 1836 that vertical movements of the crust occur because of changes in pressure and heat at depth, which in turn respond to erosion and deposition at the surface (Clarke, 1923; Dott, 1979). Lateral compression to form folded mountains, as espoused earlier by de Beaumont and the Rogers brothers, was too catastrophic for Hall. Instead, he ascribed to the view suggested by Herschel of quiet gravitational equilibrium (an embryonic isostasy) at a time when the interior of the earth was generally supposed to be extremely pliable—possibly even liquid (Fig. 4). As thick sedimentation occurred, the crust would be bowed downward and eventually the upper layers would become wrinkled to form mountainous structures, said Hall (Fig. 5).

The line of greatest depression would be along the line of greatest accumulation [that is] the course of the original transporting current. By this process of subsidence . . . the diminished width of surface above,

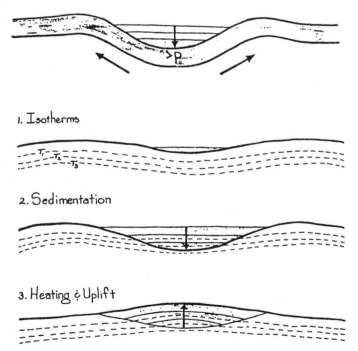

Figure 4. Herschel's concept of subsidence resulting from the pressure caused by sedimentation (upper) and uplift due to thermal expansion (lower) as the downwarping crust encounters warmer isotherms. Herschel anticipated the principle of isostasy in postulating this vertical gravitational buoyancy. (Drawn by Karen Dott Ordemann from Herschel's discussion in two 1836 letters quoted by C. Baggage, 1837.)

Hall's Theory

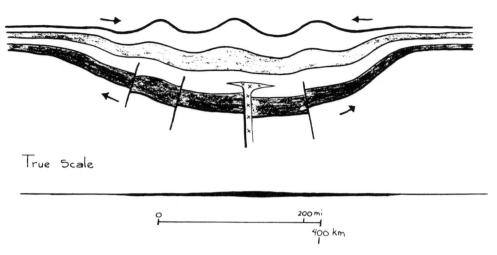

True Scale

0   200 mi

400 km

Figure 5. Hall's theory of downwarping resulting from sedimentation. Note that upper layers are crumpled as their circumference diminishes, whereas lower ones are broken by tension (and dikes are intruded) as their circumference increases. (Drawn by Karen Dott Ordemann from Hall's verbal discussion in the *Palaeontology of New York,* Vol. 6, 1859.)

caused by this curving below, will produce wrinkles and folding of the [upper] strata. That there may be rents or fractures of the strata beneath is very probable, and into these may rush the fluid or semi-fluid matter from below, producing trapdykes, but the folding of strata seems to be a very natural and inevitable consequence of the process of subsidence (Hall, 1859, p. 70 and 73).

Hall was extremely vague about the cause of uplift of mountains, but he distinguished elevation from folding. The uplift, if accounted for at all, was simply ascribed to continental-scale rather than local elevation. He later sought to avoid the issue when challenged on this point by Dana and others by disclaiming that he had ever intended to offer a complete theory of mountain building. Gravitational buoyancy was always implied, however, as the major factor (see quotations above and, for fuller discussion, Dott, 1979; see also Mayo, this volume).

Hall's contribution to mountain building was marginal at best and was soon eclipsed by the much more profound and comprehensive contraction theory of Dana (1873). Nonetheless, Hall's emphasis on some sort of cause-and-effect relationship between mountain belts and very thick strata has had a significant influence upon three generations of geologists. Dana was impressed enough by Hall's observation that Paleozoic strata are 10 times thicker in the Appalachian Mountains than in the lowland of the Midwest (the craton) that he formalized the concept with the term *geosynclinal,* which was later converted to the noun *geosyncline* (Dana, 1873). Emil Haug soon transported the concept to Europe and glorified it as James Hall's Law of mountain chains formed on geosynclines (1911, p. 159–160).

The relationship—if any—of strata to mountain building

has been debated warmly for almost a century. The long-held Hallian view that mountains somehow *result from* thick sedimentation is no longer tenable, for plate tectonic theory shows clearly that the cause and effect is just the reverse. That is, thick sedimentation within orogenic belts is the *result,* not the cause of mountain building. Mountains are cannibalistic, and most of the sediment originates from erosion of the rising mountains (and islands) and is then deposited in depressions within and beside a developing mountain system.

Even if Hall was more wrong than right in his vague causality, he nevertheless was the first person to underscore clearly the great stratigraphic contrasts between mountain belts and what we now refer to as cratons (Dott, 1979). Thus he drew attention at an early stage to large-scale stratigraphic patterns among the larger tectonic elements of the crust and revealed many other shrewd stratigraphic insights that were ahead of the times. By virtue of his breadth of experience in both cratonic and orogenic belt strata, and in no small measure because of his incredible drive, Hall was uniquely equipped to see this fundamental distinction (Dott, 1979).

CONCLUSIONS

As influential as Hall's mountain theory has been, apparently it was not seen by its author as a major part of his endeavors. His priority for it seems evident in the fact that he made no immediate effort to publish his presidential address. Instead, he buried the theory in the introductory pages of *Palaeontology of New York,* Volume III (1859), which appeared two years after the oral presentation. As noted above, a brief but even more obscure statement had appeared in an 1858 Iowa Survey report. The actual text of the address was not published until 1883. Except for an address in New York City in 1868 on the "Evolution of the North American Continent," we find no significant follow-up studies on mountain theory (Clarke, 1923). Indeed, Hall was much more the empiricist than theoretician, and in this respect he stands in sharp contrast to Dana. Although he failed to produce a meaningful mountain building theory, nonetheless Hall had discovered the craton!

It is abundantly clear that Hall's top priority always and unerringly lay with the acquisition and description of fossils and the extension of the New York stratigraphy based thereon. Seemingly, nothing could have seemed of higher importance than to have the best of North America's Paleozoic fossils safely ensconced in his cabinets at Albany, at any cost. Based upon his record, he might well have succeeded had he not died so young, for Providence usually did seem to be on his side.

ACKNOWLEDGMENTS

An oral version of this paper was presented in a History of Geology Division Symposium on the Craton in 1983 at the invitation of Ursula Marvin. Encouragement by Marvin, William Jordan, and Ellen Drake helped me complete this written version. The manuscript benefited substantially from reviews by Donald W. Fisher and Jean Prior.

REFERENCES CITED

Babbage, C., 1837, The Ninth Bridgewater Treatise: A Fragment: London, John Murray, p. 207 and 216–217.

Bailey, S. W., 1980, The history of geology and geophysics at the University of Wisconsin-Madison 1848–1980: Madison, University of Wisconsin, 174 p.

Clarke, J. M., 1923, James Hall of Albany: Albany, 565 p.

Dana, J. D., 1873, On some results of the earth's contraction from cooling, including a discussion of the origin of mountains, and the nature of the earth's interior: American Journal of Science, 3rd. Series, v. 5, p. 423–495; v. 6, p. 6–14, 104–115, 161–172, 381–382.

Dott, R. H., Jr., 1979, The geosyncline—first major geological concept "Made in America," *in* C. J. Schneer, editor, Two Hundred Years of Geology in America: Durham, University Press of New England, p. 239–264.

Fairchild, H. L., 1932, The Geological Society of America 1888–1930: Special Publication of Geological Society of America, 232 p.

Fisher, D. W., 1978, James Hall—patriarch of American paleontology, geological organizations, and state geological surveys: Journal of Geological Education, v. 26, p. 146–152.

——1981, Emmons, Hall, Mather and Vanuxem—the four "horsemen" of the New York Geological Survey (1836–1841): Northeastern Geology, v. 3, p. 29–46.

Foster, J. W., and Whitney, J. D., 1851, Report on the geology of the Lake Superior land district; part II-the iron region: Washington, U.S. Senate, 406 p. (including 4 chapters by Hall).

Friedman, G. M., 1979, Geology at Rensselaer: a historical perspective: The Compass of Sigma Gamma Epsilon, v. 57, p. 1–15.

Hall, J., 1843, Geology of New York, Part IV, comprising a survey of the Fourth District: Natural History of New York, Albany, Carroll and Cook, 683 p.

——1859, Description and figures of the organic remains of the Lower Helderberg Group and the Oriskany Sandstone, New York Geological Survey, Natural History of New York, Part 6, Palaeontology, 3, pt. 1.

——1871, Geological Survey of the state of Wisconsin, 1859–1863. Palaeontology, Part Third, Organic remains of the Niagara group and associated limestones. Albany, published privately, 94 p and 25 plates.

——1883, Contributions to the geological history of the North American continent: American Association for the Advancement of Science, Proceedings, v. 31, p. 24–69.

Hall, J., and Clarke, J. M., 1894, Palaeontology of New York, an introduction to the study of the genera of Paleozoic Brachiopoda: Natural History of New York, v. 8, pt. 2, Albany, C. Van Buthuysen and Sons, Inc.

Hall, J., and Whitney, J. D., 1858, Report of the geological survey of Iowa: Des Moines, State of Iowa, 2 vols.

——1862, Report on the geological survey of the state of Wisconsin: vol. 1, general geology and palaeontology: Albany, Wisconsin Legislature, 455 p.

Haug, E., 1911, Traité de Géologie: Paris, Colin, v. 1, 538 p.

Mikulic, D. G., 1983, Milwaukee's gentlemen paleontologists: Transactions Wisconsin Academy of Science, Arts and Letters, v. 71, p. 5–20.

Geological Society of America
Centennial Special Volume 1
1985

E. O. Ulrich's impact on American stratigraphy

George Merk
Department of Natural Science
Michigan State University
East Lansing, Michigan 48824

ABSTRACT

The controversial concepts of Edward Oscar Ulrich (1857–1944) dominated American stratigraphy during the first decades of this century. His U.S.G.S. mapping experience and extensive knowledge of the paleontology and stratigraphy of North America culminated in his masterwork: *Revision of the Paleozoic Systems* (1911) and several late papers (1916, 1920, 1924). The revision, which served as a guide to a generation of geologists, also contains Ulrich's philosophy of correlation.

Application of his geological concepts to American stratigraphic successions led Ulrich to propose radical changes in existing classifications and correlations. Among his more controversial proposals were the creation of two new lower Paleozoic time stratigraphic units of systemic rank, the Ozarkian and the Canadian; a theory of oscillating troughs and barriers to explain the occurrence of the shelly carbonate and graptolitic shale facies in adjacent strike belts in the Appalachians; and the revision of the Croixan Series of the Upper Mississippi Valley. Each of these proposals was incorporated in various textbooks and correlation charts of the period.

However, as detailed stratigraphic studies in the Ozarks, the Upper Mississippi Valley, and in the Appalachians were completed, it was found that Ulrich's concept of correlation was untenable and it was gradually replaced by one stressing facies relations. But, because of his authority, his disputative nature, and the tenacity with which he held his views, the development of American stratigraphy had been hampered.

INTRODUCTION

The development of American stratigraphy in the early twentieth century was strongly influenced by two conflicting systems of ideas concerning the kind of relations, especially correlations, that exist among stratigraphic units. One group supported the "Wernerian" idea that each stratigraphic unit persists in rock type and fauna laterally until that unit ends in non-deposition. The other group sponsored the concept that stratigraphic and faunal units normally end in a facies change. The result of this conflict was that stratigraphers, depending upon which of the two systems they accepted, were led to widely divergent interpretations of the same stratigraphic data. Although the facies concept was in common use in Europe during the first three decades of this century, its use in America was impeded at this time by the work of E. O. Ulrich (1857–1944), the outstanding proponent of the view stressing lateral persistence while minimizing facies changes. Ulrich's vast experience and encyclopedic knowledge of the paleontology and stratigraphy of North America enabled him to dominate American stratigraphy during this period.

If each new generation is to build upon the structure left it by previous generations, then each new generation must acquaint itself with what has gone before (Mathews, 1974). The purpose of this paper is to examine E. O. Ulrich's impact on American stratigraphy.

EDWARD OSCAR ULRICH (1857–1944)

Edward Oscar Ulrich, internationally known geologist and paleontologist, was the last of the Cincinnati geologists who attained eminence in the science without full university training. His career was typical of those nineteenth century American paleontologists who, beginning as self-taught amateur collectors, reached professional status initially through independent publication.

Bassler (1944b) states that it was by chance that Ulrich's

youthful collecting instinct was directed to fossils when Reverend Henry Herzer, Methodist minister of the family church and geologist in his own right, told the seven-year-old boy the meaning of the curious stones scattered over the local hills. When Ulrich was 15 years old and tall for his age, he secured a job as a rodman in the Cincinnati Water Works. He worked chiefly in the excavations for the Eden reservoir, the type area of the very fossiliferous Eden Shale. During the two years of this work, he collected fossils extensively and studied them. His father finally persuaded him to attend the German Wallace and Baldwin Colleges at Berea, Ohio, in preparation for the ministry; he was also slated later to attend the Ohio Medical College at Cincinnati. His love for fossils, however, conquered the pressure put upon him to continue his scholastic education. Thus, at the age of 20, he was glad to accept his first scientific position, the curatorship of the Cincinnati Society of Natural History. It was in this capacity that he had his first six articles on invertebrate paleontology published in the *Journal of the Cincinnati Society of Natural History.*

Physically, Ulrich had grown into a robust man, 6 feet 2 inches tall, and conforming to the custom of the time, wore a full beard (Fig. 1). The latter was so impressive that his young nephew invariably spoke of him as "der Onkel mit dem grossen schwarzen Bart" (Bassler, 1944b). Ulrich was a born teacher and loved to ask and answer questions. He said he developed his ideas by going out with others and disputing theirs: "If you quit disputing, I will go home," Ulrich said (Ruedemann, 1947). Raymond (1944) wrote, "Although he was a strong personality, and could be blunt and outspoken, he was always good natured. I never knew him to express anger, no matter what he might feel."

Ulrich was a prodigious reader. His remarkable memory, with almost total recall, made recollection of all he read easy. Early in life, he assimilated a great fund of information for later research work. He had a rare photographic mind for the names and characteristics of actually thousands of species of fossils so that he was never at a loss to reproduce, from memory, the geologic sequence and faunal contents of the many rock outcrops studied in North America and Europe (Bassler, 1944b).

Ulrich was an indefatigable worker and zealous about detail. Soon word spread of his ability as a paleontologist, illustrator, and lithographer (Raymond, 1944). In 1885, he combined efforts with Charles Schuchert, another Cincinnatian, to win contracts from the Illinois and Minnesota Geological Surveys for studies of Paleozoic bryozoans, sponges, and mollusks. It was at this time that Ulrich became involved in his first controversy.

Biologists have divided investigators into splitters and lumpers. Ulrich, owing to his keen discerning mind, was, above all, a splitter, both in his stratigraphic work and in his descriptive work on fossils (but he could also become a lumper when he was looking for general principles and conclusions) (Ruedemann, 1947). Ulrich was to find out that being a splitter can have its problems. After the publication of his work for the Illinois and Minnesota State Surveys, he was accused of making far too many species, the innuendo being that he was paid for his labor at so much per new species. Later workers have shown that his species

Figure 1. E. O. Ulrich, in 1896, at age 39, sitting on the Mohawkian-Cincinnati contact where it outcrops next to the Ohio River. From Bassler (1944b).

were good and that he could have made many more if he had had access to more and better material (Raymond, 1944).

Between 1885 and 1897, Ulrich also described Paleozoic faunas and conducted stratigraphic studies for the Ohio and Kentucky Geologic Surveys. Raymond (1944) states that Ulrich's paleontologic work is the foundation for later studies on Paleozoic Bryozoa, Ostracoda, Conodonts, and Ordovician Gastropoda and Pelecypoda. He also contributed much to knowledge of Cambrian and Ordovician trilobites and Mississippian pentremitids. He was an outstanding authority on the strange tooth-like conodonts and Paleozoic ostracoda, two groups as valuable in oil geology as the foraminifera. It was Ulrich's special pleasure to undertake the study of lesser known groups of fossils, such as the sponges and gastropods, usually neglected by paleontologists because of the fossils' commonly poor preservation and consequent difficulty of investigation (Bassler, 1944a).

As one of the pioneers in the study of the stony bryozoa by means of thin sections, his classification and early monographs on the subject were the basis for all subsequent studies. R. L. Anstey (1984, written communication) points out that:

Ulrich championed the use of new technology—namely thin sections—for the study of bryozoans. His contemporaries continued to use only external morphology, and no microscopy. Ulrich salvaged scrap window glass in Cincinnati to cut up to make into microscope slides. In using thin sections, he followed the example set by the Scottish paleontologist, H. A. Nicholson. Nicholson, however, insisted that the trepostomatous bryozoans were corals. Ulrich considered them bryozoans and was eventually the winner of that controversy.

Ulrich's paleontology was innovative, insightful, and exemplary and his detailed descriptions have stood the test of time. His preparation of bryozoan collections for the British Museum (Natural History) and the musuem of the University of Munich led to his contributions on bryozoans and ostracod crustaceans in the first English edition of Karl von Zittel's *Textbook of Palaeontology* of 1896. Ulrich also served for nine years as associate editor of the *American Geologist.*

It was fitting that the productive years in systematic paleontology should be followed by an interval in which their results could be tested. In 1897, Ulrich's reputation brought him a temporary appointment to the U.S. Geological Survey. This post, which became permanent in 1901, marked a significant shift in research emphasis from paleontology to stratigraphy. At the time, the U.S. Geological Survey was carrying on the extensive reconnaissance mapping program that produced 207 geologic folios in 24 years. With the many folios under preparation by the survey, it had become evident that the old-time, large stratigraphic divisions, such as the Shenandoah Limestone and the Trenton Limestone, would have to be subdivided before the finer details of structure could be depicted on a map (Bassler, 1944b).

Ulrich therefore plunged into the problems in stratigraphy and correlation and, true to his nature, "he had to see for himself." He was persistently busy in those years preparing stratigraphic and paleontologic reports for the folios and monographs of the Geological Survey, and he was, so to say, the final authority on all problems that the mapping geologists of the survey encountered in their work (Ruedemann, 1947). His contributions to stratigraphy and correlation were voluminous, not only in what he himself published, but in the aid which he gave to other geologists during the many years of his association with the U.S. Geological Survey (Raymond, 1944). Thus, it is reported by many authorities that Ulrich undoubtedly saw and studied more Paleozoic geology from the Rockies eastward than any other person at the time (Bassler, 1944b).

Thus, Ulrich developed a reputation during the first two decades of the twentieth century as an eminent authority on American Lower Paleozoic stratigraphy. His unfailing memory and his knowledge of fossils and Paleozoic formations in America were absolutely uncanny and sufficient to awe any adversary. To his vast experiences and encyclopedic knowledge, add his strong health and his ever-fiery enthusiasm: the result is an outstanding scientist who dominated his discipline (Dunbar and Rodgers, 1957).

REVISION OF THE PALEOZOIC SYSTEMS

The major conclusions of the many stratigraphic and paleontologic reports prepared for the folios and the monographs of the Geological Survey during its first 30 years are embodied in Ulrich's *Revision of the Paleozoic Systems* (1911). His masterwork explains his principles of stratigraphy and introduces radical changes in classification in an effort toward systematizing stratigraphic correlation and taxonomy in the United States. Ulrich was of the opinion that the American marine Paleozoic succession is the most complete and best determined and, therefore, the

best fitted for recognition as the world standard in classifying Paleozoic events. Ruedemann (1947) states that Ulrich's opinion was based not only upon reading knowledge, but upon six field trips to Europe in the interval from 1922 to 1931, during which he had opportunity to check his results in America with the classic outcrops abroad.

Although some considered Ulrich's *Revision of the Paleozoic Systems* (1911) controversial and a "raw vision," Ruedemann (1947) characterizes it as his "bible on geologic problems" and maintains that it would be "a safe guide for all coming geologists."

E. O. Ulrich's Stratigraphic Principles

In his *Revision of the Paleozoic Systems,* Ulrich defines and discusses his philosophy and methods of stratigraphic correlation. Basic to this philosophy is his concept of the relation of Paleozoic lands and epicontinental seas to Paleozoic diastrophism. Ulrich believed that from the beginning of the Cambrian to the present times, the ocean basins, on the whole, have always been relatively depressed areas; that the continental platforms have maintained essentially the same general form and location; and that the physical characteristics of the Paleozoic lands were in general much as they are today. The borders on the east, west, and south were more or less elevated and subject to orogenic activities, while the great interior parts, including the northern margin, were practically flat and occasionally affected by gentle and periodic oscillations and tilting of the surface.

According to Ulrich's view, the Paleozoic epicontinental seas occupied mostly small, shallow, and often disconnected basins, communicating at one end or side with the nearest ocean. These basins were filled and emptied many times, occasionally receiving their waters from the Atlantic, at other times from the Arctic, and often from the Gulf of Mexico. None of these epicontinental seas ever extended across a continent. In the Appalachian region, the seas were often contained in narrow troughs which connected at some point with the Atlantic.

Using North America as a type, Ulrich observed that the marine deposits of each of the many geologic ages varied in distribution from those of the preceding and succeeding ages. Because the variations in distribution were differential in extent and direction, and not regularly progressive or regressive, generally, Ulrich inferred that the different patterns were caused chiefly by relatively local oscillation or tilting movements of a rhythmic and periodic nature. He said that the effect of the deformative movement was contemporaneously and universally recorded by a corresponding change in the level of the sea, resulting in a shifting of the strandline. In the median areas of the continents, these deformations were relatively gentle and the vertical components of the movements were small; but on account of the low and featureless relief of the interior area, they caused great displacements of the strandline. These changes are usually indicated in the stratigraphic record by evidence of either an advance or a withdrawal of waters from the epicontinental basins and hence by a stratigraphic break.

Because the oscillations of the land were local in origin, each advance of the epicontinental sea spread over only limited areas of the continent, then retreated, generally returning over different areas with each advance. According to Ulrich's stratigraphic view, such movements of the strandline signify basically two things: (1) Each individual advance of a shallow sea over the continental interior is recorded by the deposition of a rock unit of relatively constant lithology throughout and with its own distinctive fauna. Furthermore, each lithographic unit and its fauna is terminated laterally by non-deposition rather than by a facies change (Ulrich believed that facies changes are relatively unimportant in the Paleozoic rocks of North America). Hiatuses and unconformities would be commonplace, yet inconspicuous, because of the low relief of the continental interior, the shallowness of the epicontinental seas, and the frequency of oscillations. (2) Each submergence and the following emergence provide a natural basis for determining stratigraphic correlation and classifications. Furthermore, according to Ulrich's reasoning, since diastrophic movements are rhythmic in operation, periodic in occurrence, and cause the strandline to move in response to these movements and register their stages, then strandline movement offers an easily applied basis for determining the natural divisions of geologic history.

Ulrich did not claim that these ideas were new, and he gives credit to Chamberlain and Salisbury (1907) for advocating their importance insofar as the major divisions of the stratigraphic column are concerned. But it was Ulrich who demonstrated the applicability of these ideas to the minor divisions of the stratigraphic column and to a general, systematic process.

The very backbone of Ulrich's new conception was his view that the continental seas were small and evanescent; that the land oscillated and tilted periodically, frequently, and rhythmically; that the deformations caused withdrawal and shifting of the seas from one basin to another, thus interrupting the process of sedimentation locally; and that hiatuses and unconformities in the record would be commonplace. This was the physical basis for Ulrich's systematic scheme of correlation.

The Role of Fossils in Correlation

Ulrich (1911) maintained that while the great value of fossils in stratigraphic correlation is generally and justly acknowledged, faunas may, nevertheless, lead us astray in cases of very close and detailed age determinations. The practical utility of paleontologic correlation lies chiefly in seeking to take fully into account the many pitfalls that lie in the path of the paleontologic stratigrapher. Greatest among these dangers are those which arise from: (1) assuming that the composition of a marine fauna largely depends upon environmental conditions; (2) relying only on the mere matching of the general aspects of faunas in making paleontologic correlations of strata; and (3) assuming that the first appearance of a species in a stratigraphic sequence marks its inception, and the latest appearances in the stratigraphic column mark its extinction.

In Ulrich's opinion, the rock record is rarely a true representation of the faunal distribution prevailing at any given locality, at any given time, because of the usual slow rates of deposition which prevail, the normal to-and-fro shifting of environments, and the rapidity with which species and fauna migrate from place to place. In addition, the hard parts of animals are obviously carried by the waves and currents widely beyond the areas in which they lived; therefore, the more or less fragmentary remains of the fauna, from a sandy bottom, a calcareous mud bottom, and a clayey mud may thus be mechanically associated in the same layer. Although marine faunas are greatly influenced by the character of the bottom on which they live, such variations in environments are of small consequence to the paleontologist.

The second pitfall in paleontologic correlation is the tendency to rely on the mere matching of the general faunal aspect by comparing the percentage of similarity between two faunas in making age determinations. According to Ulrich (1911, 1933), such usage leads to erroneous correlations. For instance, two locally successive faunas may be totally dissimilar in the majority of their species and their genera, yet the time break between the faunas may be comparatively a short one, if the fauna represent penecontemporaneous transgressions from different ocean realms. Furthermore, radical differences in the faunas of beds that are unlike in lithology, but occupy seemingly equivalent positions in the stratigraphic scale, may have resulted from the alternate capture of a given continental basin by waters invading from distinct ocean basins. In another example, two or more locally successive faunal aggregates may contain in common a large percentage of identical or closely allied species and yet differ widely in age, because the simulating (recurrent) faunas signify recurrent invasions from the same ocean basin. Because there are, in all such faunas, a percentage of holdovers of static forms that are not determinative of exact geologic time, mere matching of species would most likely lead to erroneous correlations.

In his third listed pitfall, Ulrich (1911) cautions that the first appearance of a species in a stratigraphic sequence does not mark its inception, nor does its latest appearance in the stratigraphic column mark its extinction. As a rule, species continued to exist somewhere, and under the stresses incident to sea retreat, modified descendents arose and when opportunity again offered, invaded the re-established seas and left remains that became the guide fossils of the ascending age.

The best safeguard against erroneous faunal correlations is to discriminate between fossils rigorously on the basis of their minute structural differences. It was axiomatic with Ulrich (1911, 1916, 1920, 1924, 1933) that the inherent correlation value of a fossil species is in proportion to the complexity of its unessential structural parts because the greater the detail and the smaller the distinctions, the more exact the correlations. It was by the recognition of such minute structural differentiation that Ulrich actually identified geologic horizons. The rigorous use of this type of discrimination between fauna subjected Ulrich to reproach as a "species maker." But, in Ulrich's opinion, the "preponderance" of generic or broadly specific affinities, however great, should never outweigh the dissenting evidence of two or three exactly identified varieties.

Even when all these complications have been accounted for and the age of a fossiliferous bed has been properly determined, the purely paleontological method of classifying deposits commonly fails in a most important particular. Namely, it does not exactly locate the contact between directly superposed sediments of different ages. To satisfy this demand, Ulrich (1911) reverts to those physical criteria of diastrophism which indicate displacement of the strandline. Shifting of the strandline, the most important criterion in ascertaining diastrophic action, is recorded organically by (1) abrupt changes in the aspect of the faunas, and (2) the sudden appearance or reappearance of species or genera in deposits of continental seas. Strandline shifts are recorded physically by (1) breaks and hiatuses in the stratigraphic column indicating sea withdrawals; (2) changes in the character of the deposits, especially when this involves abrupt transition from biochemical and organic sedimentation to clastics, or a change from land to marine deposition; and (3) overlaps of marine sediments.

Ulrich (1933) reminds the readers that, at best, the fossils mark zones. Therefore, the manifestations of diastrophism—including, besides those of physical nature, also the effects of earth movements on the character of related faunas—should be accepted as the dominant factor in deciding where the eras should be divided into periods or systems and these into groups and formations. Nature had drawn the physical boundaries definitely and sharply, whereas the ranges of species and genera of fossils fall between or extend across the physically defined boundaries.

THE RATIONALE FOR THE OZARKIAN AND CANADIAN SYSTEMS

In the *Revision of the Paleozoic Systems* (1911), Ulrich also proposes radical changes in the classification, the most important of which is the proposition and description of two new systems, the Ozarkian and the Canadian. He made those systems by taking part of the Upper Cambrian and Lower Ordovician, claiming important breaks above and below those systems. Ulrich believed that these new systems were based on criteria similar to those used in separating the well-known Cambrian and Silurian Systems. Unfortunately, the detailed descriptions end with the close of the Canadian, but the stratigraphic revisions embrace the entire Paleozoic era. Ruedemann (1947) states, "I know from his attitudes that he considered the creation of these two systems as the crowning achievement of his career. With these two systems, he advanced into line with Murchison, Sedgwick, and Lapworth."

At the turn of the twentieth century, no one had succeeded in drawing a satisfactory boundary between the upper limit of the range of "Cambrian" trilobites and the lower limit of the "Ordovician" gastropods and cephalopods (Fig. 2). The lower part of this transitional zone contained remnants of the Cambrian fauna,

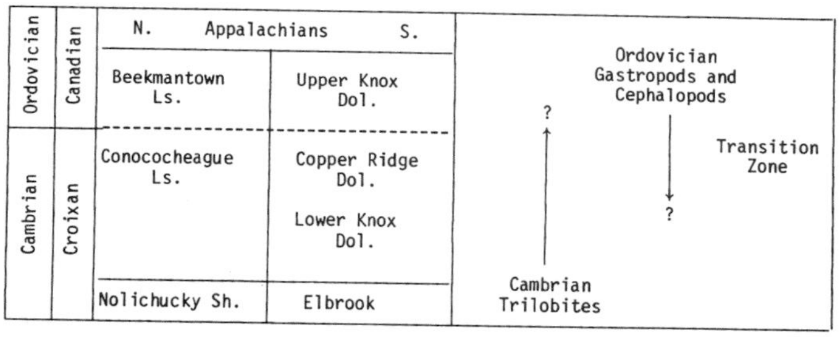

Ordovician / Canadian	N. Appalachians S.		Ordovician Gastropods and Cephalopods	
	Beekmantown Ls.	Upper Knox Dol.	?	Transition Zone
Cambrian / Croixan	Conococheague Ls.	Copper Ridge Dol.	↑ ?	
		Lower Knox Dol.		
	Nolichucky Sh.	Elbrook	Cambrian Trilobites	

Figure 2. Transition zone between upper limit of the range of Cambrian trilobites and lower limit of the range of Ordovician gastropods and cephalopods.

while the uppermost strata held fossils too much like Ordovician species to be interpreted as anything other than post-Cambrian in age. In areas containing such a transitional zone, the boundary between the Cambrian and the Ordovician was either left undecided or it was drawn arbitrarily in the midst of what was thought to be a great, sparsely fossiliferous, transitional series of dolomites and limestones.

It was largely to rectify this conflicting faunal testimony, i.e., trilobites with Cambrian affinities associated with gastropods with Ordovician characteristics, that caused Ulrich (1911) to propose that two new systems, the Ozarkian and the Canadian, be placed in the standard section between the Cambrian and Ordovician Systems. The detached Upper Cambrian formed the Ozarkian System and the separated Lower Ordovician became the Canadian System.

Ulrich cautions the reader, however, that the two new systems, for which the name Ozarkian and Canadian have been selected, are not instituted by mere subdivision of the stratigraphic units previously comprised in standardized Cambrian and Ordovician sections, but they are composed mainly of formations whose stratigraphic positions had been misinterpreted and whose aggregate thickness had been greatly underestimated. The position of the Ozarkian and Canadian Systems, together with the formations referred to them, are shown in Figure 3.

THE OZARKIAN SYSTEM

Ulrich (1911) proposed that the Ozarkian Sequence should have systemic rank because of (1) its great thickness and areal distribution, (2) the distinctive fauna which it contained, (3) the great time span involved in its accumulation, and (4) the conspicuous stratigraphic hiatuses which bounded the rock sequence above and below. His opinion was that these attributes of the Ozarkian Sequence were not inferior to the Cambrian or any other previously established American System.

In the Appalachian Valley, under the name Ozarkian System, Ulrich (1911) included all of the formations that can be shown to be younger than the top of the Upper Cambrian Nolichucky Shale in northeastern Tennessee and the top of the Conasauga Shale in southeastern Tennessee, Georgia, and Alabama, and older than the base of the Stonehenge Limestone of the

		Ozarks	Up. Miss. Valley	Champlain Val.	North Appal.	Tennessee	Alabama
Canadian		Smithville Ls. Powell Fm. Cotter Dol. Jef. City Dol. Roubidoux Fm.	Shakopee Dol.	Beekmantown Ls.	Beekmantown Ls. Stonehenge		Beekmantown Ls.
Ozarkian		Gasconade Dol. Van Buren Dol. Eminence Dol. Potosi Dol.	Oneota Dol. Madison Ss. Mendota Dol. Devils Lake Ss.	Little Falls Dol. Hoyt Ls.	Conococheague Ls.	Knox Dol. Copper Ridge Knox Dol.	Chelpultepec Knox Dol. Potosi Dol. Ketona Dol. Briarfield Dol.
Cambrian		Elvins Gp.	Jordan Ss. Trempealeau Fm. Mazomanie Fm. Franconia Fm.		Elbrook Ls.	Nolichucky Sh.	Conosauga Sh.

Figure 3. Ulrich's classification of the Ozarkian System in the United States.

Beekmantown Group in southern and central Pennsylvania. In the Ozark region of Missouri, Ulrich included as Ozarkian all formations that are younger than the Elvins Group, but older than the Roubidoux Formation. In the Upper Mississippi Valley, Walcott (1914) and Ulrich enclosed the Mendota Dolomite, the Madison Sandstone, and the Oneota Dolomite in the Ozarkian.

According to Ulrich, Ozarkian strata are best developed in the southern Appalachians, where they comprise more than 1,300 meters of dolomite in Tennessee and 850 meters of still older limestone in Alabama, with the total thickness possibly reaching 3,000 meters. This whole sequence is commonly referred to as a single formation, the Knox Dolomite.

Ulrich acknowledged that the described Ozarkian faunas were too few in species to afford a good argument, but the known undescribed species, especially the straight and curved cephalopods, the numerous coiled gastropods, and the true cystoids, all absent in true Cambrian rocks and so different from the many graptolites and the first coiled cephalopods of the Canadian, were certainly sufficiently diagnostic to make the Ozarkian fauna a distinct unit.

In addition, Ulrich notes that an important unconformity, separating the Cambrian from the Ozarkian strata, is shown by the greatly varying age, in different localities, of the beds that form the base of the Ozarkian. For example, in Tennessee, the Copper Ridge Chert (Fig. 3) rests often on the Nolichucky, but more commonly, the Knox begins with an older division 100 to 250 meters thick. In Alabama, three even older formations: the Potosi Dolomite, the Ketona Dolomite, and the Briarfield Dolomite, aggregating 850 meters of dolomite, intervene between the base of the typical Knox and the top unit of the Upper Cambrian, the Conasauga. The hiatus between the two systems, therefore represents over 1,000 meters of known calcareous deposits laid down elsewhere in Tennessee and in central Alabama. And this does not fully measure the time break, for a gap is still indicated between the Briarfield Dolomite at the base of the new systems and at the top of the Upper Cambrian.

The type exposure of the Ozarkian System occurs in the Ozark region of Missouri. Although much thinner than the Appalachian Sequence, the Ozarkian section in Missouri makes a more satisfactory type because the rocks are more fossiliferous and the stratigraphic sequence more complete. The term, Ozarkian, is a modification of the name, Ozark Series, suggested by Broadhead (1891), the first to use the geographic name as a stratigraphic term.

The general stratigraphic succession for the entire Ozark region was developed by Ulrich, who made extensive reconnaissance surveys across the area in the early 1900s. Ulrich's classification as originally proposed has been revised as a result of his own researches and others, but it has never been seriously modified (Bridge, 1930).

As originally defined by Ulrich (1911), the type section of the Ozarkian System comprised all the beds younger than the Elvins and older than the base of the Canadian System, represented in northern Arkansas by the Yellville Dolomite. Ulrich modified this classification in 1915 (Bassler, 1915) and ultimately in 1924 (Willmarth, 1924) (Fig. 4). His ultimate classification of the formations in the Ozarkian System type section assigned the Potosi and Eminence Dolomites to the Lower Ozarkian, the Van Buren and Proctor Dolomites to the Middle Ozarkian, and the Gasconade Dolomite to the Upper Ozarkian, while the Roubidoux and Jefferson City Dolomites were moved up to the Middle Canadian.

Prior to 1908, the entire sequence was classified as Cambro-Ordovician; from 1908 to 1912, it was all referred to as Cambrian; from 1912 to 1930, the practice of the Missouri Geological Survey was to divide the sequence between the Cambrian (Ozarkian of Ulrich) and Ordovician (Canadian of Ulrich) Systems using the Gasconade-Roubidoux contact as the boundary. After 1930, the Van Buren and Gasconade were included in the Ordovician, and the Ozarkian and Canadian designations were not used; the reasons for this change will be discussed later in this section.

Ulrich believed that his Ozarkian System in Missouri was a distinct unit, stratigraphically, lithologically, and faunally. His reasoning is as follows: Stratigraphically, the Ozarkian System in Missouri is 212 meters thick and is bounded, below and above, by significant unconformities—the Elvins-Potosi contact below and the Gasconade-Roubidoux contact above.

Bridge (1930) Dake (1930)		Ulrich (1924)	Formations (Thickness)
Ordovician	Canadian	Canadian	Smithville Ls.
			Powell Fm.
			Cotter Dol.
			Jefferson City (30m.)
			Roubidoux Fm. (67-73m.)
Cambrian	Croixan	Ozarkian	Gasconade Dol. (67-73m.)
			Van Buren Dol. (20m.)
			Eminence Dol. (83-103m.)
			Potosi Dol. (16m.)
		Cambrian	Elvins Gp.
			Bonneterre Dol.

Figure 4. Various classifications of the Ozarkian and Canadian Systems in Missouri.

Lithologically, the Ozarkian System is dominated by cherty dolomites with sandstones and shales appearing only locally. The underlying Cambrian is dominated by sandstones and shales, but carbonates and chert are lacking, whereas the overlying Canadian System is dominated by cherty carbonates, but sandstones appear only locally.

The Ozarkian fauna, according to Ulrich (1911), is distinct and intermediate in character to that of his Canadian and restricted Cambrian Systems. However, his evidence for this, the descriptive Ozarkian faunal lists, were not published until 1930 (Ulrich, Foerste, Bridge), 1938 (Ulrich and Cooper), 1942 (Ulrich, Foerste, Miller, Furnish), 1943 (Ulrich, Foerste, Miller), and 1944 (Ulrich, Foerste, Miller, Unklesbay)—the year Ulrich died.

Bassler (1944b) comments that unfortunately the paleontologic foundation for this work never appeared in print, except in the few cases, listed above, where his association with younger men made publication possible. The results of his paleontologic work are represented by his large private collection of Paleozoic invertebrates with its thousands of type specimens. They include the specimens he collected before joining the federal survey as well as the great quantities of study material he amassed from later official field work. The collections are now housed in the United States National Museum where they are available for comparison and research.

Ulrich's concept of the Ozarkian System was accepted by a number of researchers. Charles Walcott, an authority on the Cambrian System, who had studied Cambrian fossils more extensively than anyone at the time, adopted the Ozarkian System in his later writings, indicating that he believed the term to be justified (Ver Wiebe, 1935). The Ozarkian System concept was also adopted by the U.S. National Museum (Bassler, 1915), the Missouri Geological Survey (Dake, 1921), the New York Geological Survey (Ruedemann, 1929), and the Guidebook to the International Geological Congress (Newland, 1933).

By the late 1920s, however, doubts about the validity of the Ozarkian as a system began to appear. While doing detailed stratigraphy in Missouri, Bridge (1930) and Dake (1930) found evidence of a faunal break in the middle of Ulrich's Ozarkian, between the Eminence and Van Buren Dolomites. Subsequently, a significant unconformity was found at the same horizon. Furthermore, Bridge (1930) cited evidence suggesting that the unconformity between the Gasconade and Roubidoux Formations, separating Ulrich's Ozarkian and Canadian Systems, probably did not exist. The validity of the Ozarkian System concept hinges on the evidence concerning the existence of these features. The elements of the debate follow.

Potosi Dolomite

The Potosi Dolomite marks the base of Ulrich's Ozarkian System and is considered either the lowermost unit of the Ozarkian System or as Upper Cambrian in the conventional classification. Ulrich's claim that the Potosi rests unconformably on older beds ranging from the Derby through the Bonneterre is substantiated by Weller (1928), Dake (1930), and Bridge (1930). The Potosi fauna is not considered of diagnostic value in determining whether the Potosi is of Cambrian or Ozarkian age, primarily because Potosi fossils are so rare (Ulrich reported finding only two Potosi fossils: an undeterminable gastropod and a fragment of a cryptozoon).

Eminence Formation

The next younger unit is the Eminence Formation, the cherty dolomite overlying the Potosi. The Eminence is considered either the highest unit of the Lower Ozarkian or of Upper Cambrian age in the conventional classification. The Eminence fauna consists primarily of trilobites and gastropods, with cephalopods entirely absent. Although the trilobites seem to be more closely related to the Cambrian than to Ordovician varieties, Ulrich believed that the rarity of gastropods in beds older than the Eminence and their abundance higher in the section links the Eminence more closely with the younger Ozarkian units than with the underlying Cambrian units.

On the other hand, Dake (1930) and Bridge (1930) point out that the complete absence of the gastropod *Ophileta,* a genus characteristic of the younger Van Buren and Gasconade Formations tends to link the Eminence more closely with the underlying Cambrian rather than with the younger units above.

Van Buren Dolomite

The third youngest unit in the Ozarkian is the Van Buren Dolomite, considered either Middle Ozarkian or lowermost Ordovician. The cherty Van Buren dolomites may be separated easily from the underlying cherty Eminence and from the overlying cherty Gasconade Dolomites on the difference in the character of cherty constituents. The Van Buren fauna consists almost exclusively of cephalopods and gastropods, with brachiopods and trilobites being almost absent. Dake (1930) and Bridge (1930) conclude that a faunal break exists between the Eminence Dolomite and the Van Buren Dolomite. Their argument is as follows: (1) the presence of the Van Buren cephalopods represents an entirely new faunal element in the Ozark region. Therefore, since the cephalopods are more abundant in the immediately higher units in the sections, but are lacking in the underlying units, the Van Buren must be more closely linked faunally with the younger units than with the underlying Eminence; and (2) the Van Buren gastropod fauna is distinctly different from that of the Eminence fauna. For example, Dake (1930) and Bridge (1930) observe that a gastropod closely allied to *Ophileta*, a typical Ordovician gastropod, is common in the Van Buren but is unknown in the Eminence, once again reinforcing their contention that the Van Buren is linked more closely to the younger units (Lower Ordovician) rather than to the Eminence (Upper Cambrian). The deduced existence of this faunal break was strengthened when Bridge (1930) discovered an unconformity at the Van Buren-Eminence contact, at which the Van Buren laps across truncated older beds extending from the Eminence down to the Bonneterre. The existence of this unconformity, coming in the middle of Ulrich's supposedly coherent Ozarkian System and linked with a faunal break at the same horizon, proved damaging to Ulrich's concept of the Ozarkian as a valid system. Today, this unconformity forms the boundary between the Cambrian and Ordovician Systems in the Ozark region.

Gasconade Formation

The unit above the Van Buren is a cherty dolomite, the Gasconade Formation, which is considered either uppermost Ozarkian or Lower Ordovician in age.

Ulrich (1911) and Weller and St. Clair (1928) state that the Gasconade fauna is large; entirely confined to the formation; and consists chiefly of gastropods, cephalopods, trilobites, and brachiopods. The most abundant and easily recognized species within the Gasconade are as follows:

Eoorthis n. sp
Sinuopea obesa (Whitfield).
Sinuopea regalis Ulrich.
Rhachopea grandis Ulrich.
Gasconadia putilla (Sardeson).
Chepultapecia leiosomella (Sardeson).
Helicotoma uniangulata (Hall).
Ozarkotoma acuta Ulrich.
Ozarkispira typica.
Ozarkispira valida.
Ozarkispira complanata.
Ophileta supraplana Ulrich.
Ophileta, at least six other undescribed species.
Camaroceras huzzahensis Ulrich & Foerste.
Walcottoceras shannonense Ulrich & Foerste.
Buehleroceras compressum Ulrich & Foerste.
Clarkoceras crassum Ulrich & Foerste.
Clarkoceras obliquum Ulrich & Foerste.

A number of tuberculated tribolites of or related to the genus *Hystricurus.*

According to Dake (1930) and Bridge (1930), however, these forms are much more closely related to Ordovician than to Cambrian varieties. As evidence, they cite the abundance in the Gasconade of the gastropod, *Ophileta,* a typically Ordovician gastropod. Although a variety allied to *Ophileta* is found in many Van Buren fauna, it is totally lacking in the Eminence Dolomite. Also as evidence, Dake (1930) and Bridge (1930) state that cephalopods are abundant in the Van Buren and the Gasconade, but are absent in the underlying Eminence Dolomite, again a condition suggesting the existence of a faunal break between the Van Buren and Eminence.

The Gasconade rests unconformably on the Van Buren and is, in turn, overlain unconformably, according to Ulrich (1911) and Weller and St. Clair (1928), by the Roubidoux Formation. The proposed unconformity between the Gasconade and the Roubidoux Formations in Missouri is significant because it is used to divide Ulrich's Ozarkian from his Canadian Systems. The evidence for this unconformity, as cited by Ulrich (1911) and Weller and St. Clair (1928), is the considerable variation in thickness across Missouri of the Roubidoux's basal sandstone, the pronounced changes between the Gasconade and Roubidoux faunas, and the sudden and complete change in the sedimentary character of the two formations. In addition, Ulrich (1911) cites evidence for the widespread erosion of the Gasconade Formation from southeast Missouri before the Roubidoux was laid down.

Bridge (1930) interprets the field conditions differently and states that there is very little evidence of erosion of the Gasconade before deposition of the Roubidoux. In support of his contention, Bridge notes that most of the area over which the upper beds of the Gasconade appear to be missing is where the Gasconade occupies hilltops on which only a thin cap of the Roubidoux is still preserved. Bridge offers the following interpretation. Since the overlying Roubidoux is a very permeable formation, an excellent water gatherer, and rests on the Gasconade (famous in the Ozarks for its development of caves, sinks, and large springs), then leaching certainly must have been active along the Gasconade-Roubidoux contact.

That much of the removal of the Gasconade was brought

about by leaching rather than pre-Roubidoux erosion is corroborated, according to Bridge (1930), by the fact that the Upper Gasconade appears to thin out and disappear in place along the main drainage lines toward the present valleys in the area. Thus, leaching, rather than erosion, could explain the widespread removal of the higher beds of the Gasconade, with the subsequent setting, very unevenly, of the overlying insoluble Roubidoux Sandstone. To show that the Gasconade-Roubidoux unconformity, which served as the upper boundary of Ulrich's Ozarkian System, could be the result of post-Roubidoux solution activity proved very damaging to the concept of the Ozarkian as a valid system.

Ulrich continued his investigations of Ozarkian System stratigraphy and fauna until his death in 1944. This research, resulting in much more detailed information regarding the paleontology of the Ozarkian System, was published in collaboration with others as Geological Society of America Special Papers 13, 37, 49, and 58, entitled, respectively: the Ozarkian and Canadian Brachiopods (1938); Ozarkian and Canadian Cephalopods, Part I (1942), Part II (1943), and Part III (1944).

The first of these papers (Ulrich and Cooper, 1938) was critiqued by Charles Schuchert, an expert on North American Paleozoic stratigraphy and on brachiopods. Schuchert (1939) commented that, although the most important results of these faunal lists are due to the skillful and time-consuming preparation of the material for publication, the use of the fauna in support of the Ozarkian System must be criticized. Schuchert (1939) concluded that of the 23 brachiopod genera that Ulrich and Cooper (1938) listed as being diagnostic of the Ozarkian, 10 have no stratigraphic value, and one is not known to pass into the Ozarkian; and of the remaining 12 genera (with 40 species) which are most useful in chronology, four have come up from the Upper Cambrian, five pass up into the Canadian, and three widely distributed genera are restricted to the Ozarkian. Thus, it did not seem to Schuchert that the brachiopods give strong support to Ulrich's opinion that the Ozarkian Period and System is a valid concept.

Conclusion: Ozarkian System

The Ozarkian System was based on fossil evidence and certain unconformities, and both criteria were found, on detailed analysis, to be faulty. Dunbar and Rodgers (1957) explain that the Ozarkian System has entirely collapsed, for every unit that Ulrich himself classified as Ozarkian now can be shown to be correlative with some other unit that Ulrich himself accepted as either Cambrian or Canadian and of which the "Ozarkian" unit is typically a poorly fossiliferous dolostone facies.

J. M. Weller (1960) states that the introduction of this system accomplished nothing but confusion. The Ozarkian is now impossible to define because of complex correlations, some of which were not well founded, and because the system had different stratigraphic limits, as recognized by Ulrich and his followers, in different areas.

	New York	Normanskill Sh. / Deepkill Sh.	Central Appalachians	Ozarks
Ordovician	Chazyan Gp.	Normanskill Sh.	Carlim Ls.	St. Peter Ss.
Canadian	Beekmantown Ls. Tribes Hill Ls.	Deepkill Sh.	Bellefonte Dol. Axemann Ls. Nittany Dol. Stonehenge Ls.	Smithville Ls. Powell Fm. Cotter Dol. Jef. City Dol. Roubidoux Fm.
Ozarkian	Little Fall Dol.		Conococheague Ls.	Gasconade Dol. Van Buren Dol. Eminence Dol. Potosi Dol.

Figure 5. Classification of the Canadian System in the United State.

THE CANADIAN SYSTEM

Ulrich (1911) restricted and redefined the Canadian System of Dana (1874), making it approximately equivalent to the Beekmantown and Tribes Hill Limestones of Early Ordovician age. Figure 5 shows the formations which make up Ulrich's Canadian System in the United States.

His reasons for proposing that this portion of the Ordovician System deserved systemic rank are as follows: (1) The Beekmantown Limestone represents one of the largest stratigraphic units of the Ordovician in North America. Where fully developed, its thickness of 850 meters represents half of the entire Ordovician Series—and presumably half of Ordovician time; (2) the Canadian System is a lithologic unit, separated from the underlying Ozarkian and overlying Ordovician Systems by significant unconformities, and distinguished by the nature of its rocks, mainly cherty dolostones, carbonates, and dark shales; (3) the Canadian fauna is distinct and different from that of the sub- and superjacent faunas. As evidence, Ulrich (1911) cites the fact that ostracods, orthidae, asaphidae, and graptolites occur for the first time in the Canadian System, whereas the tabulate and rugose corals, cyclostomatous and crystostomatous bryozoans, pelecypods, and crinoids first appear in the Middle Ordovician; and (4) the Canadian is comparable in thickness of deposits, time value, diastrophic history, and faunal differences to either the accepted Silurian or Devonian Systems; thus it should have systemic rank.

Ulrich points out that, at any one locality, many gaps in the record are likely to be found, indicating that the epicontinental sea retreated and permitted erosion of recently deposited materials. Almost always there is a zone of chert and insoluble residues at this level, indicating prolonged solution of the limestone before a new sea covered it. The greatest break of this kind is found, according to Ulrich (1911), at the top of the Canadian System. The exposure on which the Canadian System was originally founded, and which make up the type section, are in the Champlain Valley, the St. Lawrence Valley, and in northwestern Newfoundland. Each area possesses different lithologic and faunal characteristics.

The Champlain Valley exposures, represented by the Beekmantown Limestone, possess a carbonate facies. (Although Ulrich was opposed to the facies concept in correlation, he used the term *facies* in this instance simply as a designation of rock type.) The carbonate-facies-type exposures extend southward from Quebec City and Montreal throughout the Appalachian Trough to Alabama. Throughout its extent, the strata consist of massive gray dolomite containing edgewise conglomerate horizons, mudcracked bedding planes, and oolites. The carbonate facies fauna consist of abundant trilobites, brachiopods, corals, and other littoral, marine, invertebrates with calcareous exoskeletons. Ulrich notes that a similar lithology and fauna occur in the Canadian rocks of the Ozark and Upper Mississippi Valley regions.

The Newfoundland exposures consist chiefly of limestone which contain, besides a much larger number of localized species, just enough Beekmantown fossils to make it reasonably certain that these widely separated limestones are approximately contemporaneous. According to Ulrich (1911), the general composition of this Newfoundland fauna is more closely simulated by the Canadian fauna in Oklahoma and western Texas than by any other Beekmantown Limestone. Ulrich (1911) states that the fossils most useful in identifying these Canadian carbonate formations are the trilobite, *Bathyuriscus amplimarginatus*; the brachiopod, *Dalmanella wemplei*; and the gastropods, *Heliconetona,* and *Ceratopea keithi.*

The Saint Lawrence Valley exposures are represented by dark shales which extend from Gaspé to Quebec City and then southward to eastern New York and New Jersey. The dark shale fauna is composed chiefly of graptolites, which, with only a very few exceptions, are entirely lacking in the carbonate facies fauna. In addition, the dark shale also lacks the benthonic forms so common in the carbonate facies.

The dark shales are found in several zones, each of which is marked by a diagnostic species. As commonly designated, these zones, in descending order, are as follows: (1) *Dictyonema flabelliforme* zone; (2) *Tetragraptus* zone, with two subzones—*Clonograptus* or lower *Tetragraptus* bed and *Dichograptus* bed; and (3) *Didymorgraptus bifudus* or *Phyllograptus* type zones. In New York, these deposits are known as the Deepkill Shales; their graptolite faunas were described and their stratigraphic relations discussed by Ruedemann (1904). In Canada, these deposits have long been called the Levis Shales for extensive exposures opposite the city of Quebec.

Ulrich believed that the dark shale facies were deposited in narrow, synclinal troughs or channels on the inner border of Paleozoic land masses, and that the dark shale was derived from these bordering land masses. From the distribution of the pelagic graptolite fauna, he judged that these channels entered the continent from the North Atlantic between Newfoundland and Labrador, extended southward up to the St. Lawrence to Vermont, then into New York, presumably joining the Atlantic again in New Jersey. Ulrich (1911) noted that, throughout most of this region, the deposits in the channel are buried beneath other sediments, perhaps chiefly of Cambrian age, which have been thrust northwesterly over them; that the shale belt is now part of a great thrust sheet which is limited on the northwest by a post-Ordovician fault (Logan's Line); and that the resultant deformation of the shale belt has added to the difficulties of interpretation. He notes that another intramarginal channel occurs in the Ouachita Trough of central Arkansas and eastern Oklahoma and that similar channels are indicated by graptolite shales in western Nevada.

How did Ulrich explain the occurrences of these contemporaneous faunas in adjacent, but lithologically dissimilar strata? Because the carbonate and black shale belts persist for hundreds of kilometers in separate, but parallel outcrop belts, Ulrich (1911) said that each belt must have been deposited in individual troughs separated by an intervening barrier. The barrier would prohibit the intermingling of the invading epicontinental seas and the mixing of the faunas coming from the various adjacent oceans. The adjacent lithologically dissimilar beds can be explained by assuming that these adjacent troughs oscillated rhythmically and periodically: When one trough was emergent and drained, the other was submerged and receiving sediment. Ulrich envisioned that, because of the oscillational nature of the troughs, the sea would shift from one basin to another, causing the marine deposits to alternate regularly from one trough to another. Thus, when the graptolite-bearing shales were being deposited, the calcareous trough was emergent, and vice versa. During the Taconic orogeny, the two sets of formations were brought into contact by extensive overthrusts, which have completely overridden the barrier that once separated the basins.

Ulrich (1911) denied that the differences in the carbonate and dark shale faunas could be attributed to normal environmental changes within a continuous sea (i.e., a facies change). The chief reason for denying a facies change as an explanation is that the geographic location of the faunas in outcrop do not fit the accepted paleogeographic model of Canadian-Lower Ordovician time: The deep water pelagic graptolite dark shale fauna are found in the easternmost part of the outcrop belt, the area expected to be closest to the Appalachian highlands of Canadian time; whereas, the littoral calcareous fauna are found in the western portion of the outcrop belt, the area located farthest from the Appalachian highlands of Canadian time. Thus, according to Ulrich, to claim that these faunal differences were the result of a facies change, would mean that the carbonate and dark shale outcrop positions should be just the reverse, laterally, of what they are.

Because Ulrich was an authority figure in American stratigraphy, his theory that the dark shale and carbonate facies were deposited in separate, but parallel troughs strongly influenced much of the work in the Appalachian region until after 1920 (Dunbar, 1954). Largely due to research done by R. Ruedemann and G. M. Kay in the Appalachian Province, Ulrich's theory is now discredited. Ruedemann (1901) was among the first to demonstrate the occurrence, in the Middle Ordovician along the bluffs of the Mohawk, of a complete gradation of the carbonate facies (Trenton Limestone) into the dark shale facies (Canajoharie Shale). Kay (1937, 1943) has shown that this kind of lateral change of facies is common in the Adirondack region. In addition, Kay (1937, 1943) has shown that the carbonate facies, in the Middle Ordovician, grade upward into fine clastics that clearly indicate derivation from an eastern source. Dunbar (1954) concluded that there is no known reasonable source for the large volume of the Canajoharie clastics, if they are conceived as having been deposited to the west of a barrier that separates them from a source farther east of an intervening sea. Since the shales and limestones interfinger in a broad transition zone, it is evident that they are the same age and were formed in an open seaway with no barrier in between, the mud settling near shore and the limey sediment farther away.

Kobayashi (1933) points out that there is no positive evidence to suggest the existence of the land barrier within the Appalachian trough. All the facts are much more readily explained by the difference of oceanic currents and rock facies. He further states that after the accumulation of the Paleozoic sediments, the Appalachian trough suffered severe crustal deformation through which the complicated overfoldings of the region were accomplished. If there had been any land barrier in the trough, then the sequence should naturally be different according to the "Decke," and the rock succession of a certain Decke ought to represent the entire absence of the sediment during the time of the land barrier, but there is no such Decke in the Appalachian Valley (Kobayashi, 1933). Jones (1933) asserts that Ulrich has paid too little attention to changes of facies, which are especially marked in the Appalachian Valley and its eastern border or the western border of the Piedmont tract. Prouty (1946, 1948) shows such facies differences to exist in the Upper Cambrian and Lower Ordovician rocks in Virginia and Tennessee on either side of the Tazewell Arch and the Rome Barrier of Ulrich—extensions of the Adirondack Arch. C. E. Prouty (1984, written communication) points out "these lithofacies and biofacies are explained presently as the result of current, geochemical, and faunal barriers. If Ulrich, in his interpretation of the carbonate and black shale facies had accepted facies barriers and left out oscillation theory as a causal mechanism, his explanations would have been on the mark." Jones (1933) noted that equally marked facies changes can be observed in Britain between the faunas on the opposite margins of a single geosynclinal trough. Today it is commonly believed that the carbonate and black shale facies in the Appalachians represent miogeosynclinal and eugeosynclinal depositional sequences, respectively, and that these sequences, beginning in Late Cambrian time, persisted at least into Middle Ordovician time.

Ulrich never accepted these new conclusions, and continued his investigations of Canadian System stratigraphy and fauna until his death in 1944. More detailed information regarding the paleontology of the Canadian System was published in collaboration with others as Geological Society of America Special Papers 13, 37, 49, and 58, entitled respectively: the Ozarkian and Canadian Brachiopods (1938); Ozarkian and Canadian Cephalopods, Part I (1942), Part II (1943), and Part III (1944).

In the first of these papers, Ulrich and Cooper (1938) had listed 31 genera of brachiopods as being diagnostic of the Canadian System. Schuchert (1939) found, on analyzing Ulrich's list, that of the 31 genera, eight had come up from older formations, but 23 genera were indeed restricted to the Canadian System. Schuchert (1939) concluded, on the basis of the brachiopods, that a good case could be made for a Canadian System.

Conclusion: Canadian System

Dunbar and Rodgers (1957) considered that the Canadian of Ulrich, with minor modifications, is a perfectly satisfactory unit; the chief question with regard to it, however, is whether it should rank as a system as Ulrich thought, or as the basal series of the Ordovician. For a time, such a system achieved limited recognition among Ulrich's followers, but this name is now used for the lowest series of the Ordovician. R. C. Moore (1958) comments that if the foundations of the time-rock classification had been based on early geologic studies in North America rather than Europe, it is very probable that what we now class as the Lower Ordovician (Canadian) Series would have been defined as an independent geologic system. It is fairly well distinguished by the nature of the rocks, by distinctness of widespread interruption of sedimentation above it, and by faunal differences from underlying and overlying stratigraphic divisions.

Bassler (1944b) states that today there seems to be a tendency to recognize the Canadian as valid and to abandon the Ozarkian by referring its strata to the Canadian and Cambrian. According to Bassler (1944b), time and more investigations, including particularly the publication of more complete paleontological evidence, will bring the truth. In the meantime, Robert Browning's passage from *Rabbi ben Ezra* must suffice:

Now who shall arbitrate?
Ten men love what I hate,
Shun what I follow, slight what I receive;
Ten, who in ears and eyes
Match me: we all surmise,
They this thing, and I that; whom shall my soul believe?

CLASSIFICATION OF THE CROIXAN SERIES OF THE UPPER MISSISSIPPI VALLEY

Introduction

The philosophy of correlation that lay behind Ulrich's stratigraphic work is readily understood in the light of his experience. New faunas meant new mapping units to be discriminated within the older broad units, and as each new unit was found it was assigned its place in the steadily growing stratigraphic column. Where faunas were absent, the corresponding rock units were presumed to be absent, and absent because of unconformities resulting from non-deposition; indeed, Ulrich held that virtually all rock units are separated by such unconformities. At no place, therefore, could a complete section be found, the standard column being a composite of units from many different areas, dovetailed into a single sequence (Dunbar and Rodgers, 1957).

An excellent illustration of Ulrich's method of correlation can be shown by his involvement in the classification of the Croixan Series of the Upper Mississippi Valley. The modern classification scheme of the Croixan Sequence in the Upper Mississippi Valley was largely developed during the period 1911 to 1939. This classification represents the results of regional studies by Ulrich (1911, 1913, 1916, 1920, 1924) and by teams of geologists from the three states most concerned: Trowbridge and Atwater (1934) for Iowa; Twenhofel, Raasch, and Thwaites (1935) for Wisconsin; and Stauffer, Schwartz, and Thiel (1939) for Minnesota. Twenhofel and his coworkers (1935) interpret the Croixan Sequence of the Upper Mississippi Valley to represent essentially continuous deposition without major unconformity; any stratigraphic breaks are thought to be on the order of diastems. By drawing formational boundaries at places of decided faunal and sedimentary changes, Twenhofel and others have divided the Croixan Series into three formations. On the other hand, Ulrich (1924), by denying the existence of facies changes and by stressing the lateral persistence of strata and the existence of unconformities, divided the same sequence into 10 formations, resulting in a geologic section 40 percent thicker than that produced by his rival researchers.

The Mendota-Madison Controversy

The classification of the St. Croixan Series of Minnesota and Wisconsin, established by Owens (1852), Irving (1877), N. H. Winchell (1886), and with only slight modification by Berkey (1897), had been in use for more than 40 years (Fig. 6). In this classification, the Mendota Dolomite, a local unit occurring in southeastern Wisconsin, was generally accepted as being correlative with the more widespread St. Lawrence Dolomite of southwestern Wisconsin and adjacent Minnesota and Iowa. In addition, the formation overlying the Mendota, the Madison Sandstone, another local unit of southeastern Wisconsin, was correlated with the more widespread Jordan Sandstone of western Wisconsin and adjacent Minnesota and Iowa.

Ulrich (1911) started a controversy that was to endure for some 20 years when he questioned this correlation with these words:

The Mendota Dolomite is commonly identified with the St. Lawrence Limestones of Minnesota, but I am not convinced that this is fact. Unfortunately, I have had no sufficient opportunity to study this problem in the field, and the fossils from beds between the Dresbach below and the Oneota above, now in my hands, are far from satisfactory in kind, number, and exact stratigraphic assignment.

Ulrich thought that a new viewpoint was required with regard to the stratigraphy of the Upper Mississippi Valley. He

West		East
L. Ord.	Shakopee Dol.	Shakopee Dol.
	Oneota Dol.	Oneota Dol.
U. Cambrian	Jordan Ss.	Madison Ss.
	St. Lawrence Dol.	Mendota Dol.
	Dresbach Ss.	Dresbach Ss.

Figure 6. The commonly accepted Upper Cambrian-Lower Ordovician stratigraphic sequence for the Upper Mississippi Valley, circa 1900.

assumed that the true relations of these largely drift-covered exposures would not be fully determined until a more detailed investigation of the bedding planes, the vertical and geographic ranges of the particular faunal species, and the faunal associations of the various beds had been made. And, equipped with his diastrophic principles, this is just what Ulrich intended to do.

But unraveling the stratigraphic relationships in the Upper Mississippi Valley Croixan Sequence would not be easy. The Croixan Series in the Upper Mississippi Valley includes dolomites, dolomitic shales, glauconites, glauconitic sandstones, shaly sandstones and fine-to-coarse sandstones and conglomerates locally. Some of these strata grade laterally into beds of quite different character and in disconnected sections, the same bed may be difficult to trace or correlate. This is especially the case with much of the St. Lawrence Formation and the Franconia Sandstone, although such changes are not confined to them (Stauffer, Schwartz and Thiel, 1939).

The fossils, for the most part, are scarce and poorly preserved, and only a small number had been named, described, and figured (Trowbridge and Atwater, 1934). It was also noted by Stauffer, Schwartz, and Thiel (1939) that the Croixan fauna is particularly sensitive to the changes in sedimentation, and faunal variation is the only reliable means of recognizing the several divisions. However, Croixan faunules may be provokingly elusive, especially in the portions so thoroughly reworked by marine worms or other burrowing forms during sedimentation.

Ulrich began field work in the Upper Mississippi Valley stratigraphy in Wisconsin in 1913. His research was carried on during short periods of a few days or weeks during at least 10 summers, and only when the pressure of other duties with the U.S.G.S. permitted. The result of Ulrich's first season of field work appeared the next year when Walcott (1914) published Ulrich's revised tentative classification of the pre-Ordovician formations in the Upper Mississippi Valley. In this classification, the Upper Cambrian Series in the Upper Mississippi Valley is subdivided into six lithologically and faunally distinct formations, named in ascending order: The Mt. Simon Sandstone, followed by the Eau Claire Shale, the Dresbach Sandstone, the Franconia

Ulrich (1924)		
System	Series	Formation
Canadian	Upper Canadian	Shakopee Dol.
Ozarkian	Upper Ozarkian	Oneota Dol.
Ozarkian	Lower Ozarkian	Madison Ss.
Ozarkian	Lower Ozarkian	Mendota Dol.
Ozarkian	Lower Ozarkian	Devils Lake Ss.
Cambrian	Croixan	Jordan Ss.
Cambrian	Croixan	Trempealeau — Norwalk Mbr.
Cambrian	Croixan	Trempealeau — Lodi Mbr.
Cambrian	Croixan	Trempealeau — St. Lawrence
Cambrian	Croixan	Mazomanie Fm.
Cambrian	Croixan	Franconia Fm.
Cambrian	Croixan	Dresbach Ss.
Cambrian	Croixan	Eau Claire Sh.
Cambrian	Croixan	Mt. Simon Ss.

Twenhofel, Raasch, Thwaites (1935) Bridge (1937)		
Formation		System-Series
Shakopee Dol.		Ordovician (Canadian)
Oneota Dol.		Ordovician (Canadian)
Trempealeau F.	Madison Mbr.	Cambrian (Croixan)
Trempealeau F.	Jordan Mbr.	Cambrian (Croixan)
Trempealeau F.	Norwalk Mbr.	Cambrian (Croixan)
Trempealeau F.	Lodi Mbr.	Cambrian (Croixan)
Trempealeau F.	St. Lawrence	Cambrian (Croixan)
Franconia Fm.		Cambrian (Croixan)
Dresbach Fm.	Galesville Mbr.	Cambrian (Croixan)
Dresbach Fm.	Eau Claire Mbr.	Cambrian (Croixan)
Dresbach Fm.	Mt. Simon Mbr.	Cambrian (Croixan)

Figure 7. Various classification of the Croxian Series of the Upper Mississippian Valley.

Sandstone, the St. Lawrence Formation, and the Jordan Sandstone. Overlying these are the Lower Ozarkian Mendota Limestones and the Madison Sandstone; the latter overlain by the Oneota Dolomite of the Canadian System. This classification was important because it assigned thicknesses to the formations and imposed limits not introduced in earlier discussions (Stauffer, Schwartz, and Thiel, 1939).

Another point of emphasis was that Ulrich now believed that the Mendota Limestone and Madison Sandstone were both younger than the Jordan Sandstone and separated from it by an unconformity. Ulrich's reason for this hypothesis was his belief that the fauna of the "true Mendota" was different from and much younger than that of the St. Lawrence and Jordan faunas and both the Mendota and Madison fauna could be correlated with the Ozarkian Series of his Missouri section.

Ulrich (1924) further changed the classification scheme (Fig. 7) by introducing three new formation names (the Devils Lake Formation, the Trempealeau Formation, and the Mazomanie sandstone), while assigning the St. Lawrence Formation to member rank within his newly created Trempealeau Formation. Ulrich's youngest new formation, the Devils Lake Formation, was formed from the lower portion of the Mendota (Ulrich, 1914) and the upper portion of the Jordan Sandstone. The underlying Trempealeau Formation was made from the lower part of the Jordan Sandstone and the upper part of the St. Lawrence (Winchell, 1874). In addition, Ulrich subdivided the Trempealeau into three members, in descending order: the Norwalk Sand-

stone, the Lodi Shale, and the St. Lawrence (and Black Earth) Dolomite. Ulrich's third creation, the Mazomanie Sandstone, was formed from the lower portion of the St. Lawrence (Winchell, 1874) and the upper part of the Franconia Formation. The creation of the Mazomanie would involve Ulrich in his second major controversy within the Croixan Series in the Upper Mississippi Valley.

Although Ulrich's final reclassification of the Upper Mississippi Valley Croixan did not appear in print until 1924, he used units from it (e.g., the Devils Lake) piecemeal in his 1916 and 1920 publications.

Pertaining to the physical evidence for his classification of the Mendota and St. Lawrence dolomites, Ulrich (1916) made three conclusions: (1) The Mendota Dolomite is of limited distribution, outcropping in eight locations that are arranged in a northwest/southeast-trending belt some 50 miles long and 4 to 5 miles wide; (2) the Mendota intervenes locally along an unconformity which separates the Jordan and Madison sandstones. As evidence of this unconformity, Ulrich cites the presence, between the Jordan and Madison, "of a slightly uneven plain of contact, the distinct nature of which is marked by unmistakable criteria of a reworked sand deposit." He states, "The most convincing of these criteria is the relative coarseness of the quartz grains that make up the first few inches or more of the overlying sandstone—a condition resulting from the washing and sifting by wave action to which the previously exposed and weathered surface of the Jordan was subjected at the time of the early

Ozarkian marine transgression"; and (3) the beds underlying the Mendota outcroppings vary decidedly in age from place to place. As evidence, Ulrich cites the occurrence, near Black Earth, Wisconsin, of a dolomite that he calls the Black Earth Dolomite. It underlies the Lodi Shale and is correlative with the St. Lawrence Dolomite. Ulrich notes that the Black Earth Dolomite thins to disappearance eastward. This situation is depicted on a larger scale in Ulrich's (1920) diagram showing the stratigraphic sequence of the St. Croixan and Ozarkian formations across Wisconsin. Here, the Mendota lies either on the Devils Lake with the Jordan missing, or on the St. Lawrence with the Devils Lake and Jordan missing. This sub-Mendota unconformity suggested to Ulrich (1916) that differential movement, emergence, and locally varying amounts of erosion had occurred before deposition of the Mendota began.

To explain his views, Ulrich (1916) claimed that in south-central Wisconsin, at the end of Cambrian time, a shallow and narrow trough, some 50 miles long and 4 or 5 miles wide, extending from the Baraboo area southeasterly past the Madison area, was eroded through the Jordan Sandstone and into the St. Lawrence Dolomite. Subsequently, during Ozarkian time, this trough was filled with the Mendota Dolomite and the Madison Sandstone, which were of very similar composition to the underlying St. Lawrence Dolomite and Jordan Sandstone, respectively. When the trough was filled, the Oneota Dolomite was deposited over the whole region.

Ulrich (1916) explained the lithologic and faunal similarity of the Mendota and Black Earth Dolomites on the basis of recurrent physical conditions and recurrent faunas. His chief basis for considering the Mendota younger than the St. Lawrence Dolomite is paleontological. The fossils of the "true Mendota" are, he believes, much younger than those of the St. Lawrence.

Dunbar and Rodgers (1957) state that another geologist had collected a fauna of 13 species from a dolomite bed (Black Earth Dolomite) that was admitted by all to be part of the St. Lawrence and that 10 of these species (mostly primitive gastropods, according to Ulrich, 1924) were also known from the Mendota Dolomite. Ulrich (1916) nevertheless maintained that he could discriminate the Mendota from the Black Earth (St. Lawrence) specimens by minute subspecific differences and that the great similarity of the two faunas was due to recurrence after the lapse of a considerable part of a period. He states that the three "Mendota-like" fossil species restricted to the Black Earth Dolomite must be "for the present the real guide fossils for the Black Earth-St. Lawrence Dolomite zone." Twenhofel, Raash, and Thwaites (1935) note that "what these three species are, Ulrich does not reveal."

Ulrich's (1924) classification scheme concerning the pre-Ordovician Paleozoic formations of the Upper Mississippi Valley appeared for some time in influential guidebooks (Thwaites, 1931), in correlation charts (Reeside, 1932; Shimer, 1934; and Bridge, 1937), and in textbooks (Miller, 1952, and partially in Moore, 1958), thus giving his classification scheme credibility.

However, Ulrich's views on the stratigraphic succession in the Upper Mississippi Valley were not universally accepted by all stratigraphers working in the area. Thus, it became important that differences of opinion be recorded, lest the assumptions be made that there was general acceptance of Ulrich's published statements, particularly as this is the area generally regarded as that of the type section of the Upper Cambrian of North America. During the next decade, the pre-Ordovician Paleozoic sequence in the Upper Mississippi Valley was subjected to regional studies more rigorous than any performed previously. These regional studies were performed principally by Trowbridge and Atwater in Iowa; Twenhofel, Raasch, and Thwaites in Wisconsin; and Stauffer, Schwartz, and Thiel in Minnesota.

The first research critical to Ulrich's classification was performed by W. V. Searight (1924) and described in an unpublished (M.S.) thesis at the University of Iowa. Searight made careful stratigraphic, petrologic, and paleontologic studies of the St. Lawrence and Mendota dolomites but found no evidence by which the St. Lawrence of Minnesota, Iowa, and western Wisconsin could be separated from the Mendota in eastern Wisconsin.

Over the next decade, after accumulating data from many measured sections, Trowbridge and Atwater (1934): Twenhofel, Raasch, and Thwaites (1935); and Stauffer, Thiel, and Schwartz (1939) concluded that the St. Lawrence, Black Earth, and Mendota dolomites represent the same stratigraphic unit and are of the same age. The investigators give these reasons why the units should not be separated:

(1) The Mendota has never been observed to occupy a stratigraphic position other than that of the St. Lawrence; the superjacent beds, wherever exposed, are the Lodi siltstones; the subjacent beds beneath the conglomerates at the base of the dolomites are the upper Greensand beds of the Franconia; the Mendota of the type sections lies exactly as far below the base of the Oneota as does the St. Lawrence, a few miles further west (pointed out by Thwaites).

(2) There is little evidence of pre-Mendota erosion, such as Ulrich postulates produced the cut into the underlying deposits (i.e., through the Jordan, Lodi, and St. Lawrence), and it is impossible to trace this postulated erosional depression in either the field or in the subsurface.

(3) The Mendota has never been observed in any exposed section showing an unconformable relationship with any part of the Trempealeau Formation. Even though, locally, there is an irregular bedding plane, and the immediately overlying sand is coarser than that beneath (which may indicate that the lower sand was reworked to form the upper), nothing more than a diastem is needed to explain the situation.

(4) The similarity of the 10 species that occur in the Mendota and in the Black Earth dolomites might be explained on the basis of recurrent faunas, but in light of a considerable mass of opposing evidence, it is difficult to accept this explanation in this particular case. Twenhofel, Raasch, and Thwaites (1935) comment that it appears strange that practically an entire fauna should recur with only slight mutation across a systemic

boundary—especially in the Cambrian of this region where few species have a maximum range of more than 25 feet, and it is readily possible to found zones within a formation on the basis of fossil genera. Further, the meager faunal collections and the none too satisfactory preservation of many of the fossils scarcely seem to warrant recognition of such minor differences. Trowbridge and others (1934), Twenhofel and others (1935), and Stauffer and others (1939) concluded, therefore, that the faunal differences postulated by Ulrich should not be regarded as anything more than the natural amount of disparity to be expected between the fossils from slightly different horizons within a single stratigraphic unit.

The Franconia-Mazomanie Controversy

The second controversy within the St. Croixan Series involving Ulrich concerned his creation of the Mazomanie Formation. These controversial beds lie between the Franconia and St. Lawrence (Trempealeau) formations, possess a thickness of approximately 70 meters, and consist primarily of sandy shales, dolomitic shales and glauconitic sandstones. Ulrich's involvement in the situation began when he placed, in his *Revision of the Paleozoic Systems* (1911), all of the Franconia beds within his Dresbach Formation. Later, however, in Walcott (1914), he recognized the Franconia as a separate formation between the Dresbach and the St. Lawrence.

The boundaries of the Franconia have, henceforth, a history of uncertainty, as various stratigraphers, especially those working in Minnesota, tended to expand the limits of the Franconia upward at the expense of the lower St. Lawrence (Trempealeau) boundary. The Franconia was subdivided by Twenhofel and Thwaites (1919) into five members, in ascending order, as follows: (1) the basal sandstone and overlying calcareous layers, (2) the micaceous shale, (3) the lower greensand, (4) the yellow sandstone, and (5) the upper greensand.

In 1920, a major difference in the interpretation of the beds lying between the St. Lawrence (Trempealeau) and Lower Franconia Formation occurred when Ulrich (1920) concluded that: (1) the glauconitic beds in eastern Wisconsin between these two formations were younger in their entirety than the similar beds in western Wisconsin occupying the same stratigraphic position; and (2) a considerable unconformity, extending over a wide area, existed at the base of the St. Lawrence. This assertion was based upon Ulrich's belief that the Mazomanie was absent in western Wisconsin, and in Minnesota and Iowa; and (3) therefore, the beds in eastern Wisconsin should be separated from the Franconia and designated as a new foundation, the Mazomanie (which Ulrich named after the village of that name in Dane County, Wisconsin). To explain this situation, Ulrich (1920) postulated that the Franconia strata, of his definition, were deposited in a sea invading from the west. This sea was assumed to have retreated after having reached the central part of Wisconsin. The land then tilted to the east, permitting the Mazomanie water to enter and deposit sediments overlapping the Franconia.

Ulrich (1920) stated that he had found clear evidence in the Wisconsin Valley of this overlap in the thinning to disappearance of the Mazomanie on the east side of the Wisconsin Arch. He further stated that on top of the arch the St. Lawrence (Trempealeau) rests directly on the Dresbach. Ulrich concluded that the Mazomanie must be younger than the Franconia, because, not only does the Mazomanie overlie it, but the Mazomanie is also more calcareous than the type of Franconia, further evidence of deposition by a different epicontinental sea.

Ulrich's interpretation of the Mazomanie Formation was granted a measure of acceptance by a number of authors who adopted it in their publications and correlation charts, such authors as Martin (1932), Reeside (1932), Resser (1933), Shimer (1934), and as an alternate interpretation in Bridge (1937). But there was not total acceptance. Ulrich's views received close scrutiny during the next decade as the pre-Ordovician Paleozoic Sequence in the Upper Mississippi Valley was subjected to an increasing number of stratigraphic, mineralogic, and paleontologic studies more rigorous than any performed up to that time.

As a result of these studies, interpretations contrary to Ulrich's views began to appear. The first paper to appear challenging Ulrich's interpretation was a mineralogic study by Pentland (1931). He compared the heavy mineral suites within the Mazomanie and Franconia Formations and concluded that the Mazomanie, at its type locality and elsewhere in eastern Wisconsin, is equivalent to the Franconia of western Wisconsin. Later, Trowbridge and Atwater (1934) reported that they agreed with Ulrich that the Franconia beds should have formational rank, but they disagreed that the Mazomanie is younger than the Franconia. They gave these reasons for their disagreement: (1) the Franconia and Mazomanie are similar lithologically; (2) their thicknesses are very nearly identical; (3) they occupy identical stratigraphic positions between the Dresbach and St. Lawrence Formations; (4) there is no evidence of an unconformity between the Franconia and the St. Lawrence (Trempealeau) from which position the Mazomanie is missing, nor is there evidence of an unconformity between the Dresbach and the Mazomanie from which position the Franconia is missing; (5) the Mazomanie fauna, first reported from the Upper Mazomanie in eastern Wisconsin, has been found also in the upper portion of the Franconia in western Wisconsin; and (6) the Franconia fauna of western Wisconsin has been found below the Mazomanie fauna in the Mazomanie area.

Also, Twenhofel, Raasch, and Thwaites (1935) concluded, after measuring and describing many stratigraphic sections, that Ulrich was in error about the Mazomanie. They gave these reasons: (1) Ulrich's error was due to a mistaken correlation of the calcareous zone of the basal Franconia with that of the St. Lawrence member of the Trempealeau; (2) nothing of the kind described by Ulrich exists; his Mazomanie continues westward to the Mississippi River and beyond, while his Franconia continues eastward over the arch with undiminished thickness; and (3) the Mazomanie does not merit the rank of a formation; all stratigraphic evidence indicates that it is little more than a minor subdivision of the Franconia formation.

Conclusion: Croixan Series

The teams of geologists from the three states most concerned—Trowbridge and Atwater (1934) for Iowa; Twenhofel, Raasch, and Thwaites (1935) for Wisconsin; and Stauffer, Schwartz, and Thiel (1939) for Minnesota—interpreted the Croixan Sequence of the Upper Mississippi Valley to represent essentially continuous deposition from beginning to end, without major unconformity; any such stratigraphic breaks as were present are thought to be of the order of diastems. By drawing formational boundaries at places of decided faunal and sedimentary changes, the research teams divided the Croixan Series into three formations having a combined thickness of 460 meters; these are, in ascending order—the Dresbach, the Franconia, and the Trempealeau (Fig. 8). They agreed that the Mendota Dolomite of Ulrich is a part of the St. Lawrence Dolomite; that the Norwalk, Devils Lake, and Madison sandstones are only local facies of the Jordan Sandstone, and that the Mazomanie Formation is continuous with the Franconia Formation and not lithologically distinct enough to deserve recognition. The consensus is that the Croixan Series of the Upper Mississippi Valley was deposited in shallow water; any differences that exist in the character of the strata, the fossil content, and other features are thought to be the result of differences in depth, changes in source material, and changes in rate of deposition, but not diastrophism.

On the other hand, Ulrich (1924), by denying the existence of facies changes and by stressing the lateral persistence of fauna and strata and the existence of unconformities, divided the same sequence into 10 formations having a combined thickness of 656 meters, thus producing a section 40 percent thicker than that produced by his rival researchers.

DISCUSSION OF ULRICH'S VIEWS

Ulrich (1933) noted that the inferences, deductions, and changes he had proposed in his *Revision of the Paleozoic Systems* (1911) received considerable criticism over the years. Today this criticism is somewhat understandable when viewed from the perspective offered by a knowledge of the historical development of the facies concept. The facies concept was proposed by Gressley in 1838. By the 1880s, it was in common use in Europe, especially by Johannes Walther and his followers in Germany. In North America, use of the facies concept developed more slowly. But, by 1900, its supporters included such influential American geologists as H. S. Williams, C. S. Prosser, J. M. Clarke, J. E. Hyde, and A. W. Grabau. They used the facies concept to clarify stratigraphic relations principally in the Appalachian region. Thus when Ulrich's *Revision of the Paleozoic System* (1911), containing his principles of correlation, was published, it met with opposition promptly. Ulrich (1913) acknowledged this opposition by noting the disparaging term "raw vision" his adversaries used when referring to the *Revision*. Ulrich (1913) tried to rationalize the perceived shortcomings in the *Revision* by noting that:

some of the imperfections which are now so obvious in these pages are due to the necessity of adapting my conclusions to methods and opinions which happened to prevail at such times on the Federal Survey. The greater part of them, however, is to be ascribed to personal errors in observation and judgment, and to change in methods and views incident to the transition from the old to a newer conception of geological conditions. The bonds of convention and early training were too strong to be easily thrown aside. In ordinary cases, simple caution might well pass as a sufficient excuse for delay in presenting evidence. But this has not been an ordinary case. Great and really fundamental changes were contemplated. To begin with, I realized that many sections must be studied before the facts in any one of them can be properly interpreted. It became apparent also very soon that many sections must be studied again and again in the light of later discoveries. It is impossible to note every feature of a stratigraphic section on the first visit. But a delay in publishing would prevent others from the full use of the data so that it could be tested in the field as soon as it came to hand. If the tests proved satisfactory, then the ideas were accepted as good; if they did not apply truly, then we tried to learn why and wherein they failed so that they might be modified to fit the facts.

Dunbar and Rodgers (1957) state that Ulrich made errors of interpretation and, for years, his authority stood in the way of further progress. The reader is thus confronted with a critical question. With all of Ulrich's enthusiasm, talent, tenacity, and

Ulrich (1924)	
Formation	Thickness (meters)
Madison Ss.	13
Mendota Dol.	7
Devils Lake Ss.	33
Jordan Ss.	25
Trempealeau Formation — Norwalk Mbr.	0-17
Trempealeau Formation — Lodi Sh Mbr.	0-8
Trempealeau Formation — St. Lawrence	0-8
Mazomanie Fm.	0-55
Franconia Fm.	0-57
Dresbach Ss.	14-83
Eau Claire Sh.	67-117
Mt. Simon Ss.	33-233
Total = 656	

Twenhofel, Raasch, and Thwaites (1935)		
Formation		Thickness (meters)
Trempealeau Formation	Madison Ss Mbr.	13
	Jordan Ss Mbr.	33
	Norwalk Mbr.	0-17
	Lodi Sh Mbr.	14
	St. Lawrence	0-8
Franconia Formation		17-50
Dresbach Formation	Galesville Ss. Mbr.	25
	Eau Claire Sh. Mbr.	40-67
	Mt. Simon Ss. Mbr.	67-233
Total = 460		

Figure 8. Various Croixan Series composite sections in the Upper Mississippi Valley.

experience in tracing key beds, and collecting and carefully identifying fossils, how could he lack the required vision to break away from his Wernerian-type views? Like all of us, Ulrich's concepts were shaped and limited by his experiences, which in Ulrich's case consisted of his early self-training in the Cincinnati area, his service with various midwestern State Geological Surveys and with the U.S. Geological Survey. In all of these, it appears that he dealt primarily with cratonic-type exposures and had little experience with geosynclinal sequences or with present-day sedimentary or faunal environments.

Bassler (1944b) sheds light on the subject with his comment that

The Cincinnati student, although ideally located for the study of paleontology, was always at a disadvantage in other phases of geology. The uniform distribution of the strata and their many included fossils over long distances, in practically horizontal beds with little changes in lithology, no faulting, no silification of the rocks or fossils, and no marked unconformities, all gave the impression of an unbroken series of deposits. No real bioherms or typical coral reefs with their changes of fauna and rocks at the same level, no development of facies lithology, no cryptovolcanic structures to bring the underlying formations to the surface, and no angular unconformities on a visible scale were present to aid in solving the geologic history. It is no wonder then that the Cincinnatians tended toward paleontology and that they might err when working in another geological province like the folded Appalachians.

Jones (1933) notes that Ulrich has worked largely on lower Paleozoic rocks deposited on the craton or its warped-down margins where the criteria relating to the evidence of diastrophism are generally observable. Although Jones agrees in general with the criteria recommended by Ulrich for the delimitation of formations, he cannot agree with Ulrich's conclusions concerning geosynclinal deposits as compared with those of the craton, because the types of deposits are radically different and the criteria relating to the evidence of diastrophism available in the one are absent or only slightly developed in the other. Also, Ulrich paid too little attention to facies changes which are especially marked in the Appalachian Valley; equally marked changes can be observed in Britain between the faunas on the opposite margins of a single geosynclinal trough (Jones, 1933).

J. M. Weller (1960), in critiquing Ulrich's *Revision of the Paleozoic Systems,* agrees with Jones (1933), Dunbar and Rodgers (1957), and others by noting that Ulrich ignores environmental differences that might affect faunal relations and correlations.

CONCLUDING REMARKS

To summarize, a brilliantly able paleontologist and stratigrapher allowed his ideas, based on reconnaissance studies, to crystallize into a system of concepts that was widely accepted in North America. Dunbar and Rodgers (1957) state that from these concepts Ulrich deduced certain principles by which he reinterpreted many of the correlations that had previously been made. Thus he repeatedly reinterpreted what had been thought to

be a broad facies change as a series of separate deposits in separate basins, each deposit being of limited extent and of different age from the others. For the Appalachians, this led to the concept of troughs and barriers, according to which deposits of, for example, sandstone, shale, shaly limestone, and pure limestone, thought by earlier workers to be an orderly succession of facies, were held to have been deposited separately and successively in one or another of four or five parallel troughs, in such manner that no two deposits were contemporaneous (Dunbar and Rodgers, 1957).

In case after case, where Ulrich had denied facies changes and found disconformities, restudy reversed his conclusion, first in the Ozarks with the work of Bridge (1930) and Dake (1930), then in the Upper Mississippi Valley with the work of Trowbridge and Atwater (1934), Twenhofel, Raasch, and Thwaites (1935), and Stauffer, Schwartz and Thiel (1939), and finally in the Appalachians, with the work of Ruedemann (1901) and Kay (1937, 1943). Ulrich never accepted these new conclusions, but he fought a losing battle against the continually accumulating evidence.

Nevertheless, Ulrich was a geologist of great distinction in his day, upon whom many honors were bestowed. He was an original Fellow of the Geological Society of America and also the Paleontological Society, for which he served as president. He was a member of the National Academy of Sciences and other scientific societies, including Corresponding Member of the Geological Society of London and the Geological Society of Stockholm. Various honors in recognition of his researches came to him. The National Academy of Science, in 1930, awarded him the Thompson Medal for his attainments in paleontology; and in 1932, the Geological Society of America honored him with the Penrose Medal for his contributions to geology. His college, later called Baldwin-Wallace, bestowed on him the honorary degrees of M.A. in 1886 and D.Sc. in 1892.

Raymond (1944) said that Ulrich made great contributions in his field for he provoked, "and I really mean provoked," many geologists to make much more careful studies than would otherwise have been done. Also, if it had not been for him, we should have had no Schuchert or Bassler in geology, and there are many others who owe their start to him.

Bassler (1944b) comments that no outline of Ulrich's accomplishments is complete without mention of his faithful, lifelong assistant, Rector Mesler, who was appointed to the U.S. Geological Survey in 1911 as assistant to Ulrich. A fine collector and skilled preparator, with a devotion to detail, he had just the talents to aid in forwarding Ulrich's researches and to care properly for the great collections the two were soon accumulating. He was active in this work to the last, for during a few moments' rest immediately preceding Dr. Ulrich's funeral service on February 25, 1944, he passed on to join his chief.

ACKNOWLEDGMENTS

I am grateful to Chilton E. Prouty, Robert L. Anstey, Ellen

T. Drake, and the GSA reviewers for their useful comments on this paper.

REFERENCES CITED

Bassler, R. S., 1915, Bibliographic index of American Ordovician and Silurian fossils: U.S. National Museum Bulletin 92, v. 2, p. 1441–1447.

—— 1944a, Edward Oscar Ulrich (1857–1944): Bulletin of the American Association of Petroleum Geologists, v. 28, p. 687–689.

—— 1944b, Memorial to Edward Oscar Ulrich: Proceedings Geological Society of America (1945), p. 331–351.

Berkey, C. P., 1897, Geology of the St. Croix Dalles: American Geologist, v. 20, p. 33–35.

Bridge, J., 1930, Geology of the Eminence and Cardareva Quadrangles: Missouri Bureau of Geology and Mines, v. 24, 2nd ser., p. 1–185.

—— 1937, The correlation of the Upper Cambrian Sections of Missouri and Texas with the section in the Upper Mississippi Valley: U.S. Geological Survey Professional Paper 186-L, p. 233–237.

Chamberlain, T. C., 1909, Diastrophism as the ultimate basis of correlation: Journal of Geology, v. 17, p. 685–693.

Chamberlain, T. C., and Salisbury, R. D., 1907, Geology, v. 2, Earth History: New York, Henry Holt, 692 p.

Dake, C. L., 1921, The problem of the St. Peter Sandstone: Missouri Bureau of Geology and Mines, v. 6, no. 1, 225 p.

—— 1930, The geology of the Potosi and Edgehill Quadrangles, Missouri Bureau of Geology and Mines, v. 23, 2nd ser., p. 1–233.

Dana, J. D., 1874, Geology and natural history, Canadian Period: American Journal of Science, 3rd ser., v. 8, p. 214.

Dunbar, C. O., and Rodgers, J., 1957, Principles of stratigraphy: New York, John Wiley & Sons, 355 p.

Irving, R. D., 1877, Notes on some new points in the elementary stratification of the primordial and Canadian rocks of south central Wisconsin: American Journal of Science, 3rd ser., v. 9, p. 441–442.

Jones, O. T., 1933, Discussion: *in*, E. O. Ulrich, 1933, Principles for the correlation and classification of strata and their application to the Lower Paleozoic: 16th International Geological Congress, Rept. 1, Washington, p. 517–518.

Kay, G. M., 1937, Stratigraphy of the Trenton Group: Geological Society of America Bulletin, v. 48, p. 233–302.

—— 1943, Development of the northern Allegheny Synclinorium and adjoining regions: Geological Society of America Bulletin, v. 53, p. 1601–1658.

Kobayashi, T., 1933, Discussion; *in* E. O. Ulrich (1933), Principles for the correlation and classification of strata, and their application to the Lower Paleozoic: 16th International Geological Congress, Rept. 1, Washington, p. 517–518.

Martin, L., 1932, Physical geography of Wisconsin: Wisconsin Geological and natural history survey, Bulletin 36, p. 4.

Matthews, R. K., 1974, Dynamic stratigraphy: Englewood Cliffs, New Jersey, Prentice-Hall, 370 p.

Miller, W. J., 1952, An introduction to historical geology: New York, D. Van Nostrand, 555 p.

Moore, R. C., 1958, Introduction to historical geology: New York, McGraw-Hill, 656 p.

Newland, D. H., 1933, The Paleozoic stratigraphy of New York: International Geological Congress, 16th, Washington Guidebook 4, p. 1–136.

Owen, D. D., 1852, Lower sandstone of the Upper Mississippi: Geological Survey Wisconsin, Iowa and Minnesota Report, Lippincott & Grambo, p. 48–58.

Pentland, A., 1931, Heavy Minerals of the Franconia and Mazomanie sandstones, Wisconsin: Journal Sedimentary Petrology, v. 1, p. 23–36.

Prouty, C. E., 1946, Lower Middle Ordovician of southwest Virginia and northeast Tennessee: Bulletin of the American Association of Petroleum Geologists, v. 30, p. 1140–1191.

—— 1948, Trenton and Sub-Trenton stratigraphy of northwest belts of Virginia and Tennessee: Bulletin of the American Association of Petroleum Geolo-gists, v. 32, p. 1596–1626.

Raymond, P. E., 1944, Edward Oscar Ulrich: Science, v. 99, no. 2570, p. 256.

Reeside, J. B., 1932, Stratigraphic nomenclature in the United States: International Geological Congress, 16th, Washington Guidebook 29, pl. 1.

Resser, C. E., 1933, Preliminary generalized Cambrian time scale: Geological Society of America Bulletin, v. 44, p. 738–739.

Ruedemann, R., 1901, Hudson River beds near Albany and their taxonomic equivalents: New York State Museum Bulletin 42, p. 489–587.

—— 1904, Graptolites of New York, Part 1. Graptolites of the lower beds: New York State Museum, Memoir 7.

—— 1929, Alternating oscillating movement in the Chazy and Levis Troughs of the Appalachian Geosyncline: Geological Society of America Bulletin, v. 40, p. 409–416,

—— 1947, Biographical Memoir of Edward Oscar Ulrich 1857–1944: Biographical Memoirs, National Academy of Sciences, 24, no. 7, p. 259–280.

Schuchert, C., 1939, Ozarkian and Canadian brachiopods: American Journal of Science, v. 237, p. 135–138.

Searight, W. V., 1924, Correlation of the uppermost Cambrian strata of the Upper Mississippi Valley [M.S. thesis]: State University of Iowa, p. 1–108.

Shimer, H. W., 1934, Correlation chart of geologic formations of North America: Geological Society of America Bulletin, v. 45, pl. 118.

Stauffer, C. R., Schwartz, G. M., and Thiel, G. A., 1939, St. Croixan classification of Minnesota: Geological Society of America Bulletin, v. 50, p. 1227–1244.

Thwaites, F. T., 1931, Description of geologic cross section of central United States—Michigan, Wisconsin, Illinois: Kansas Geological Society, 5th Annual Field Conference Guidebook, p. 66–71.

Trowbridge, A. C., and Atwater, G. I., 1934, Stratigraphic problems in the Upper Mississippi Valley: Geological Society of America Bulletin, v. 45, p. 21–80.

Twenhofel, W. H., 1954, Correlation of the Ordovician Formations of North America: Geological Society of America Bulletin, v. 65, p. 247–298.

Twenhofel, W. H., Raasch, G. O., and Thwaites, F. T., 1935, Cambrian strata of Wisconsin: Geological Society of America Bulletin, v. 46, p. 1687–1744.

Ulrich, E. O., 1911, Revision of the Paleozoic Systems: Geological Society of America Bulletin, v. 22, p. 281–680.

—— 1913, The Ordovician-Silurian boundary: International Geological Congress, 12th, Toronto, Canada, Comptes Rendus, p. 592–667.

—— 1916, Correlation by displacements of the strandline and the function and proper use of fossils in correlation: Geological Society of America Bulletin, v. 27, p. 451–490.

—— 1920, Major causes of land and sea oscillations: Journal of the Washington Academy of Sciences, v. 10, no. 3, p. 57–78.

—— 1924, Notes on new names in table of formations and on physical evidence of breaks between Paleozoic Systems in Wisconsin: Transactions of the Wisconsin Academy of Sciences, Arts, and Letters, v. 21, p. 71–107.

—— 1933, Principles for the correlation and classification of strata and their application to the Lower Paleozoic: 16th International Geological Congress, Rept. 1, Washington, p. 517–518.

Ulrich, E. O., Foerste, A. F., and Bridge, J., 1930, Systematic paleontology of Eminence and Cardareva Quadrangles: Missouri Bureau of Geology and Mines, v. 24, 2nd ser., p. 187–228.

Ulrich, E. O., and Foerste, A. F., 1935, New genera of Ozarkian and Canadian cephalopods: Bulletin of the Scientific Laboratories, Denison University Bulletin, v. 35, no. 17, p. 259–290.

Ulrich, E. O., and Cooper, G. A., 1938, Ozarkian and Canadian Brachiopoda: Geological Society of America Special Paper, no. 13, 323 p.

Ulrich, E. O., Foerste, A. F., Miller, A. K., and Furnish, W. M., 1942, Ozarkian and Canadian Cephalopods, Part 1 (Nautilicones): Geological Society of America Special Paper, no. 37, 157 p.

Ulrich, E. O., Foerste, A. F., and Miller, A. K., 1943, Ozarkian and Canadian Cephalopods, Part 2 (Brevicones): Geological Society of America Special Paper, no. 49, 240 p.

Ulrich, E. O., Foerste, A. F., Miller, A. K., and Unklesbay, A. G., 1944, Ozarkian and Canadian Cephalopods, Part 3, (Longicones and Summary): Geological

Society of America Special Paper, no. 58, 226 p.

Ulrich, E. O., and Schuchert, C., 1902, Paleozoic seas and barriers in eastern North America: New York State Museum Bulletin 52, p. 633–663.

Ver Wiebe, W. A., 1935, Historical geology: New York, John S. Swift, 316 p.

Walcott, C. D., 1914, Cambrian geology and paleontology: Smithsonian Miscellaneous Collections, v. 57, p. 353–362.

Weller, J. M., 1960, Stratigraphic principles and practice: New York, Harper & Brothers, 725 p.

Weller, S., and St. Clair, S., 1928, Geology of Ste. Genevieve County, Missouri:

Missouri Bureau of Geology and Mines, v. 22, 2nd ser., 352 p.

Wilmarth, M. G., 1925, The geologic time classification of the U.S. Geological Survey compared with other classifications: U.S. Geological Survey Bulletin, no. 769, 138 p.

Winchell, N. H., 1874, Geology of the Minnesota Valley: Minnesota Geological and Natural History Survey, 2nd Annual Report, p. 147–155.

——1886, Revision of the stratigraphy of the Cambrian in Minnesota: Minnesota Geological and Natural History Survey, 14th Annual Report, p. 325–337.

Geological Society of America
Centennial Special Volume 1
1985

The Iowa landscapes of Orestes St. John

Jean C. Prior
Iowa Geological Survey
Iowa City, Iowa 52242

Carolyn F. Milligan
Box 171
Window Rock, Arizona 86515

ABSTRACT

Five original field sketches of the Iowa landscape, drawn in 1868 by Orestes H. St. John, were found in 1975 at the California Academy of Sciences in San Francisco. The pencil sketches were meticulously drawn, dated, and signed and included detailed landmark annotations. St. John, whose career paralleled significant developments within the field of geology, was Assistant Geologist of the Iowa Geological Survey from 1866 to 1869. Return of the sketches for the Survey archives revealed that these drawings correspond to five of the thirteen lithographs which illustrate the two-volume 1870 *Report on the Geological Survey of the State of Iowa.* The precise annotations enabled the authors to relocate and document the sites as they appear today. Revisiting these sites provided an opportunity to observe 110 years of historical change in land use from native prairie to intensive agriculture, urban growth, and woodland succession, as well as the effects of surface mining of mineral resources and the protection afforded to land in the public domain. Each site also had a significant geological context: the Missouri River valley and adjoining Loess Hills near Sioux City, the glacial moraines enclosing the Spirit Lake and West Okoboji Lake areas, the gypsum deposits near Fort Dodge, and the limestone resources exposed along the Des Moines River valley near Keosauqua. In relying on his own artistic skills to document these geological features and resources during this reconnaissance era of geological studies in Iowa, St. John joins the 19th Century tradition of documentary artists in America.

DEDICATION

The rugged peaks of the Teton Range in northwestern Wyoming bear the names of several men who first explored these landscapes and studied their geological meanings—Mt. Owen, Mt. Meek, Mt. Moran, and Mt. St. John. Fritiof Fryxell, Emeritus Professor of Geology at Augustana College (Illinois), is an eminent historian of explorer F. V. Hayden, artist Thomas Moran, and artist/photographer W. H. Jackson. Dr. Fryxell was instrumental in the founding of Grand Teton National Park, was its first ranger-naturalist (Love and Reed, 1968), and made first ascents of most of the Teton peaks (Leigh N. Ortenburger, Pers. commun., Dec. 6, 1981), including one (11,412 feet) which he named for Orestes St. John (Fryxell, Pers. commun., Jan. 5, 1977). While researching the early life of F. V. Hayden with his colleague J. V. Howell 30 years ago, Fryxell sought in vain for St. John's papers. This chapter is dedicated to Dr. Fryxell in recognition of his interest in St. John and the special focus of his past research, as well as his teaching and publications which reflect the responsibility he felt to communicate to others the geological heritage of their surroundings.

INTRODUCTION

Geological reconnaissance and artistic tradition travelled together in the Trans-Mississippi West during the 19th Century. The scientific survey rosters included documentary artists and photographers who could accurately record topography, stratigraphic sequences, and geologic structure. For example, Thomas Moran, who travelled with Ferdinand V. Hayden and John Wesley Powell, painted dramatic landscapes that reflected a deep reverence for wilderness, a characteristic of the Hudson River School painters with whom he is linked (Fern, 1976). William Henry Holmes, geologic artist with the Hayden Surveys, as well as with C. E. Dutton's investigations of the Grand Canyon District and G. K. Gilbert's studies of Lake Bonneville, drew panoramic line-drawings that demonstrated an uncanny ability to portray the details and depth of a geologic landscape (Nelson, 1980). William H. Jackson, also an accomplished artist who travelled with the Hayden Surveys, is best recognized as a creative pioneer in photographic art and for the incomparable landscapes he recorded from the 1860s into the 1880s (Fryxell, 1939). The work of these men had widespread popular appeal, for few Americans had seen the far West; their awe-inspiring scenes contributed significantly to Congressional action which set aside the western lands of the Yellowstone, the Grand Canyon, and the Tetons as National Parks.

In the 1970s and '80s, the centennial celebrations of the U.S. Geological Survey and the Geological Society of America have sparked considerable interest in tracing the roots of geological science in the United States and in looking more closely at some of the scientists who carried its concepts forward. The discovery in California of Orestes St. John's sketches of the Iowa landscape, and the reoccupation of his sites to document 110 years of change, provide an opportunity to look more closely at Iowa geology and the geologists who operated within the prevailing 19th Century framework of geological research, traditions of art history, and trends in national experience.

ORESTES ST. JOHN

Orestes Hawley St. John was born in Ashtabula County, Ohio, in 1841 and died in San Diego, California, in 1921 (Fig. 1). The intervening years which span his career mirror the development of geological science in 19th Century United States. In early childhood his family moved to Waterloo, Iowa, and before his twenties, he had made remarkable collections of the fossil fauna from Devonian outcrops along the Cedar River valley. The richly fossiliferous strata exposed here in eastern Iowa, including the bordering Mississippi Valley, was an environment which nurtured many prominent careers in geology (e.g. Charles A. White, W. J. McGee, Frank Springer, Francis M. Van Tuyl, Harry S. Ladd, Merrill A. Stainbrook, and William M. Furnish).

St. John's excellent collections, particularly of fish remains, drew the attention of the renowned Swiss naturalist Louis Agassiz. A professor at the Harvard University Museum of Comparative Zoology, Agassiz was the beacon to which many scholars interested in the natural sciences were drawn, St. John among them. At that time, field geology was largely a discipline of recorded observations, collections of specimens, and broad deductions. Geologists were, in reality, naturalists—curious, observant, and broadly trained in fields which later became distinctly separ-

Figure 1. Orestes H. St. John (1841–1921). From Keyes, 1942.

ate disciplines. St. John's colleagues at Harvard included ornithologists, botanists, zoologists, and physicians, as well as geologists. In 1865 St. John was chosen by Agassiz to accompany him on the Thayer Expedition to Brazil. In addition to significant geological observations, St. John is credited with the collection of numerous specimens of mammals, birds, reptiles, fishes, insects, and crustaceans, as well as detailed notes on their habits, habitats, and anatomy (Keyes, 1942, p. 7). Agassiz and St. John remained close friends until Agassiz's death in 1874. The influence of this European naturalist left an indelible mark on St. John and on the geological science of the period.

The Iowa Legislature, meanwhile, had authorized funding for a state geological survey in 1853. This three-year effort, under the direction of James Hall of Albany, New York, was followed in 1866 by a second funding period, this time under the guidance of Charles A. White, M.D. On the strength of Agassiz's recommendation, White chose St. John to be Assistant Geologist of the Iowa Geological Survey. Between 1866 and 1869, White, St. John, and Rush Emory, a chemist, were the state's geological corps. St. John's field efforts focused on the coal deposits of south-central Iowa and on the geology and mineral resources of the western half of the state. The two-volume summary of this four-year study, *Report on the Geological Survey of the State of Iowa* was published in 1870. In the transmittal letter to Iowa's Governor Merrill, White makes special reference to his admiration of St. John's work and to the fact that the diagrams, maps, and sketches which illustrate the report were done by St. John (White, 1870, v. 1, p. 3). The superimposed letters "S" and "J", which appear on each of the thirteen lithograph illustrations, further mark them as his.

In addition to this work, St. John continued his paleontological pursuits, describing new fossil species and concentrating in particular on Paleozoic fishes. He became well acquainted with Charles Wachsmuth and Frank Springer, trading them crinoid collections for fish collections. In a letter to F. B. Meek in 1867, St. John writes, "I have recently received, as a present, a splendid collection of fishes from the Burlington Limestone—a gift from a much esteemed young friend, Mr. Springer, whom you may recollect having seen at the time of your visit here last spring. Had they been so many nuggets of gold, the gift would not have been half so valued" (Smithsonian Institution Archives). After completing his work for the Iowa Survey, he joined Amos Worthen at the Illinois Geological Survey and completed two elaborate (and illustrated) monographs on the fossil fishes of that state (St. John, Worthen, and Meek, 1875; Worthen, St. John, and Miller, 1883).

These fledgling days of the Iowa and Illinois Geological Surveys were characteristic of changing trends in geological research and American history. The need for rapid reconnaissance, motivated by the search for natural resources, was at the heart of efforts by the states to establish geological surveys within their borders. For the territories farther west, this was an era of discovery and exploration with the granting of federal commissions for expeditions—Hayden, King, Powell, and Wheeler—to assess the geology of the American West. These lands offered bright promise in the post-Civil War period, and the country's attention was gladly diverted to the grandeur of the Western mountains as first seen through the paintings of the documentary artists who accompanied these expeditions. In the middle 1870s, St. John was engaged in geological field work with the U.S. Geological and Geographical Survey of the Territories under Ferdinand V. Hayden. He worked in New Mexico (Hayden, 1876), Idaho, Wyoming, and Colorado, principally in the Teton Ranges, the Gros Ventre, the Snake River, the Caraboo, Blackfoot, and Wind River Mountains (Hayden, 1879, 1883).* He recognized for the first time fossil sequences on the west side of the Continental Divide that were identical to those along the Mississippi Valley at Burlington (Keyes, 1921, p. 38).

At this time, St. John was living in Topeka, Kansas, and at the close of the field seasons with Hayden, he turned to studies of the coal deposits in Kansas, New Mexico, and adjoining states as

*See Authors' Note following References Cited.

a geological consultant. Financiers with the Atchison, Topeka, and Santa Fe Railroad, the Maxwell Land Grant Company, and the St. Louis, Rocky Mountain, and Pacific Railroad recognized his meticulous field work and detailed reports on coal reserves as a sound basis on which to invest large sums of money. His work was responsible for opening the largest coal fields known in Kansas and New Mexico (Keyes, 1921, p. 39). What had started as a "temporary" excursion into the consulting business became a permanent position for the remainder of St. John's professional career, and the mainstay of many a geological career since then.

Orestes St. John is best remembered as a paleontologist, with especially valuable contributions made on the subject of fossil fishes, and with a talent for detailed descriptions of species. He was recognized as a keen observer, a deliberate and dedicated worker, and an outstanding field geologist who made fundamental contributions to stratigraphy and general geology.

RETURN OF THE SKETCHES

In April of 1975, the Iowa Geological Survey received an unexpected package from Ian Campbell, former State Geologist of California, and then with the California Academy of Sciences in San Francisco. Inside were six remarkably executed pencil sketches done in 1868 by Orestes St. John—five landscape scenes of Iowa and one of a scene in Kansas. Each drawing has carefully penciled notations as to location, special landmarks, direction of view, and date. St. John's signature or the superimposed initials "S" and "J" also appear on each drawing. In 1868, St. John was Assistant Geologist with the Iowa Geological Survey. Campbell, in his letter to Iowa's State Geologist Samuel J. Tuthill, explained that the sketches were found among the professional effects of Leo Hertlein, a geologist with the Academy for most of his professional life. Hertlein had died three years earlier, and in sorting through some of his stored materials, Peter Rodda, Chairman of the Academy's Geology Department, and Campbell had come across these sketches, recognized their Iowa connections, and suggested they more appropriately belonged in the archives of the Iowa Geological Survey.

The significance of this unusual find took on added meaning when the sketches were compared with the two-volume *Report on the Geological Survey of the State of Iowa,* published in 1870. These volumes, valuable as summaries of early geological reconnaissance in Iowa, also had been long admired for the thirteen lithographs which illustrate them and which were known to be the work of St. John. Four of the drawings found in California are the original field sketches from which the lithographs were made; the fifth sketch was drawn from one of the lithograph sites, but from a nearly opposite viewpoint.

The surprising discovery of these sketches after so many years, from such an unexpected source, and the possible existence of the other sketches or papers of St. John all remain, even at this date, an intriguing and unsolved mystery. Later communication with Rodda revealed the sketches were found together, in a folder among old U.S. Geological Survey Bulletins, publications, and class notes belonging to Hertlein, and were boxed in the Academy basement for at least the last 25 years. There were no notes or other materials of any kind related to the sketches or to St. John. Hertlein maintained a rather complete separation of his personal and professional life, and any association with St. John was more likely in the course of his career than through any family connection (Rodda, Pers. commun., March 26, 1979).

Hertlein was a Kansas farm boy who travelled west in 1918, entered the University of Oregon, and became interested in geology, specifically paleontology. Following his graduation in 1922, he went to Stanford University to study for his doctorate. Part of the St. John estate is known to have gone to Stanford that summer (Stanford University Archives). Myra Keen, Emeritus Professor of Geology at Stanford, suggests that Hertlein may well have helped with the unpacking of the collections. The books probably became incorporated into the department's Branner Library, with duplicate copies sold to interested students. If St. John had placed what sketches he had in his own copies of the 1870 volumes, and Hertlein had a chance to buy them, he may well have done so (Keen, Pers. commun., Sept. 9, 1979).

When Professor Keen arrived 14 years later, St. John's fossil collections were still unidentified and stowed away in drawers. She later identified and catalogued most of this material, and recalls that the specimens were wrapped in newspaper, with each locality and the date of collection labeled on the inside of the sheet in black ink. The only other material she saw from the original bequest was a collection of fine, imported drafting pens which were doled out to the department's most skilled draftsmen (Keen, Pers. commun., Sept. 9, 1979). It is interesting to note that in the letters written by St. John to F. V. Hayden while in the field between 1874 and 1879 are his expressed concerns about hostile Indian parties, severe snowstorms, the lack of maps, and the need for a bottle of liquid India ink (National Archives, Washington, D.C.).

The Stanford University Libraries could find no further information on the St. John material. Roxanne-Louise Nilan, Stanford University Archivist, pointed out that detailed record keeping of such materials did not begin until the late 1930s at the earliest (Nilan, Pers. commun., Aug. 24, 1979). It has been her experience, and that of others, that manuscripts, sketches, correspondence, diaries, and field notebooks donated to science departments in the early years often ended up buried in closets and attics, and subsequently were lost.

There is no guarantee that the sketches were part of St. John's estate or were ever at Stanford; this is only the closest link found to date between St. John and Hertlein. The connection, if any, may be elsewhere in California, perhaps in Kansas, or a major piece of the puzzle may still be missing. Deaths and loss of records have closed most avenues to further systematic searching for information. Other St. John materials were located at the Smithsonian Institution, the National Archives, Harvard University, the Kansas State Historical Society, and the Iowa State Historical Department.

ST. JOHN'S 1868 FIELD SEASON

On June 3, 1868, St. John wrote to Fielding B. Meek concerning a collection of fossil fishes that the two men and Louis Agassiz were each examining in turn. The collection had been detained, and St. John wrote that he had made arrangements, ". . . to have the package forwarded to me at some point in the western part of the State on its arrival here [Iowa City], that I may prepare my notes in our camp. Dr. White has just returned from Des Moines, and has succeeded—thanks to your important communications—in arranging the matters of the Survey in a very satisfactory manner. We start tomorrow for the western part of the State" (Smithsonian Institution Archives).

Thus began, presumably on June 4th, the trip during which all five of the recently found Iowa sketches were made. The 1868 date on seven of the remaining eight lithographs suggests that all illustrations for the 1870 Report were drawn during this field season.

Insights into Iowa field work during this period are rare, but among the papers of F. B. Meek housed in the Smithsonian Archives is an *Iowa, Nebraska 1867* Notebook. These field notes of Meeks record a trip made the previous year in the company of St. John, White, and Gill which left Iowa City in a wagon on Thursday morning, May 20th, and arrived in Nebraska City Friday afternoon, June 8th. Meek's measured and described stratigraphic sections and his detailed fossil identifications are interestingly interspersed with some of the personal experiences of their journey—a hazardous crossing of the flooded Skunk River valley, adoption of a stray dog, diets of pigeon, dove, squirrel, rabbit, and curlew (". . . one measured 3 feet and one inch across between the tips of the extended wings — bill 6 inches long."), as well as encounters with a community of French settlers and another of New Englanders in southwest Iowa. In addition, Meek comments frequently on the nature of the terrain, the distribution of timber and prairie, the condition of farms and fencing, the weather and travel conditions, and exceptionally beautiful sunsets.

At a pace of 20 to 30 miles per day, early geologists had an understandably close association to the land they viewed. They had days to familiarize themselves with details of the terrain. This "long gaze" imposed by extended travel and living outdoors was important, for the landscape and the occasional exposures of bedrock were the major basis they had for making geological interpretations. St. John's artistic talents and dedication to detail also served them well, for he had the means to document accurately his views of geologically significant landscapes. He drew as a scientist and as an observer, for the purpose of conveying geological meaning and understanding to the readers of his reports.

August 13, 1868, is the date St. John penciled on the bottom of his sketch of the western Iowa hills which border the Missouri River valley south of Sioux City (Fig. 2). The unusual appearance and characteristics of this terrain fascinated those who saw it for the first time. White and St. John wrote in the 1870 Report, ". . . as the moundlike peaks and rounded ridges jut above each other,

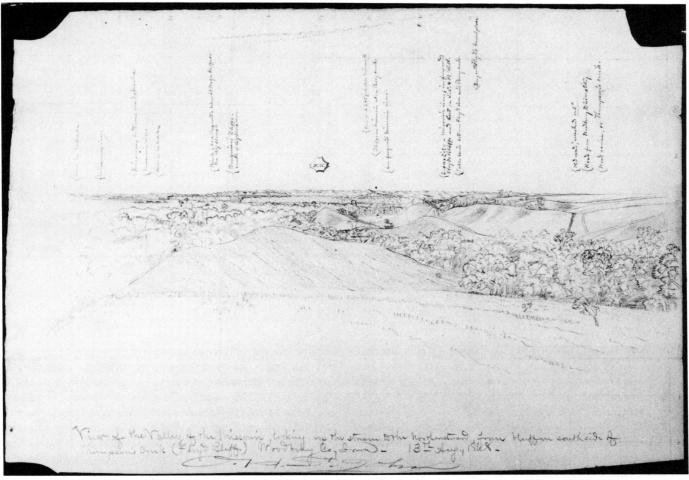

Figure 2. "View of the Valley of the Missouri, looking up the stream to the Northwestward, from bluff on south side of Thompson's Creek (Floyd Bluff) Woodbury Co., Iowa-13th Aug., 1868." Original field sketch.

or diverge in various directions while they recede upward to the upland, the setting sun throws strange and weird shadows across them, producing a scene quite in keeping with that wonderful history of the past of which they form a part" (White, 1870, p. 104). The authors referred to this material as the 'Bluff Deposit,' and noted that it was called "silicious marl" by David Dale Owen, and that it was "similar to that deposit in the valley of the Rhine, known there by the provincial name of 'loess'." They also noted, "A thrifty growth of young forest trees is now spontaneously springing up everywhere in this deposit, and rapidly encroaching upon its hillsides and prairies whenever they are by any means protected from the annual prairie fires" (White, 1870, p. 112).

St. John's sketch from atop the "Bluff Deposit" looks northwest across the great bend of the Missouri River past Sioux City. His marginal notations above the northern horizon direct attention to hills and Missouri River bottomland in Nebraska, bluffs at the mouth of the Big Sioux River, Perry Creek and the Floyd River, buildings at Sioux City, Sergeant Floyd's burial place, and the wooded ravine along Thompson's Creek in the foreground. The precision in the distant landscape is remarkable, and even includes a steam locomotive and adjacent telegraph line threading through cuts in the bluffs, as well as a stern-wheeler on the river just below Sioux City.

By September 25, 1868, St. John had arrived within northern Iowa's lake district, and on that date completed a sketch looking northward along the axis of Lake West Okoboji from a gravelly ridge above its southern shore (Fig. 3). White and St. John referred to the lake as "Minnetonka," and commented that its meaning of "great water" in the Sioux dialect referred to the Indian usage to distinguish this larger and deeper lake from its eastern portion. The 1870 Report, in its discussion of rivers and lakes near the "Great Watershed," stated that Minnetonka is representative of the category "drift lakes" with which north-central Iowa is distinctly marked, and which have their origins in depressions on the land surface left undisturbed since the glacier disappeared. They noted that the shores of these northern Iowa lakes contain some of the most delightful spots to take up resi-

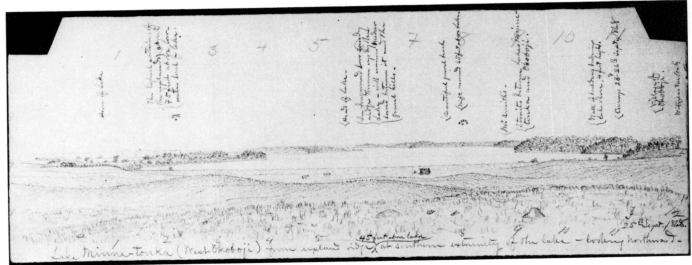

Figure 3. "Lake Minne-tonka (West Okoboji) from upland ridge 45 feet above lake at southern extremity of the lake—looking Northward. 25th Sept., 1868." Original field sketch.

dence in Iowa, with clear water and clean gravelly beaches, abundant fish, and myriads of migratory waterfowl—and ". . . they have already begun to be a favorite resort for those in pursuit of health and pleasure" (White, 1870, p. 73).

St. John's sketch clearly illustrates the rough, irregular character of the land bordering West Okoboji. Numerous embayments and forested projections of land are shown. In marginal notations above the northern horizon, he locates the highest drift mounds which rim the lake, gravel beaches, and a four-foot wall of boulders which borders the shore. "Mr. Smith's," the "Village of Okoboji," an oak witness tree, and a spot on the east shore marked as "Camp 23rd-26th Sept. 1868" are also identified. Signs of settlement include a fenced farmstead above the southwest shore, with grazing horses and a wagon in the foreground meadow.

From this three-day encampment, St. John had good access to the entire lakes area, and to terrain that clearly displays its glacial origins. On September 26, St. John was on another gravelly ridge, overlooking the southern end of Spirit Lake and the upper portion of Lake East Okoboji to the northeast (Fig. 4). Prairie uplands rim the lakes, and as noted in the 1870 text, the shores of Spirit Lake are almost always gravel, and its banks are wooded for a short distance away from the water until the prairie takes over. He noted on the sketch, as well as in the text, that five to six feet of fall occur at the Spirit Lake outlet, and that a small flour mill was located on the spot. He added that the water supply was sufficient for this operation, but because these lakes lie almost directly on the "Great Watershed," they do not receive much water, and thus should be relied on primarily as reservoirs rather than for water power.

By October 3, 1868, St. John had moved on southeastward to the Fort Dodge area and completed a sketch of a gypsum quarry in a ravine three miles south of Fort Dodge (Fig. 5). The text accompanying the corresponding lithograph in the 1870 Report further identifies this site as one of the principal quarries opened on Two Mile Creek. The glacial drift is seen lying directly on the irregular, eroded solution channels which characterize the gypsum surface. Trees are shown on the unexcavated hillside above the quarry and a solitary figure walks along the trail past the quarry face. The gypsum deposits were regarded as important for their use in agriculture, their use in the Fort Dodge area for building stone, and for the manufacture of high quality plaster-of-paris.

In late October, 1868, St. John was in southeastern Iowa, and on October 29 sketched a scene along the Des Moines River one mile above the town of Keosauqua in Van Buren County (Fig. 6). The sketch was drawn from within the river channel during a period of low flow. Sandbars and snags are evident as well as the first few buildings of Keosauqua in the downstream distance. As in the sketch of the gypsum quarry, a lone man stands observing the view from the water's edge. The scene includes the far side of the valley against which the river is eroding. Exposed are strata of limestone and sandstone belonging to the St. Louis Formation. Glacial drift is shown over the bedrock, and the entire valley bluff is covered with timber. The lithograph in the 1870 Report appears within a discussion of the "Carboniferous" System (White, 1870). The authors pointed out that during deposition of this formation, some disturbance and distortion of the strata occurred which is evident in the gentle folds of the exposed rock layers. These carbonate rocks were regarded as an excellent source of lime; some of the thicker more uniform beds were desirable for building purposes, and some layers which had a very compact texture were being tested for use as lithographic stone.

Figure 4. "Spirit Lake, from ridge north of Spirit Lake village at the southern end of lake—looking northward. Upper extremity of East Okoboji in foreground on the right. 26th Sept., 1868." Original field sketch.

Figure 5. "Gypsum Quarry in ravine 3 miles south of Fort Dodge on east side of the Des Moines, Webster Co., Iowa—The largest exposure at this locality—the gypsum 20 to 30 feet in thickness. 3rd Oct., 1868." Original field sketch.

Figure 6. "View on the Des Moines one mile above Keosauqua, Van Buren Co., Iowa. In bluffs on right side of river Saint Louis limestone exposures with intercalated layer of sandstone "a" - 29th Oct., 1868 (Looking eastward from north or left bank of the river.)" Original field sketch.

OUR 1978 FIELD SEASON

The detailed character of these sketches, and their landmark annotations, suggested the possibility of relocating the 1868 sites. The drawings also had attracted the interest of art historians at the University of Iowa, particularly Carolyn Milligan, an artist and then Curator of Visual Materials at the Department of Art and Art History. Together the authors (Prior and Milligan) retraced St. John's travels, revisited the scenes he saw, and reoccupied the vantage points from which he drew, as nearly as we could determine them. Milligan made preliminary field sketches at each site, and from the photographs also taken, she later completed a second set of pencil drawings in the same detailed documentary style of St. John.

St. John's "Bluff Deposit" is now recognized as one of Iowa's major geomorphic provinces, the Western Loess Hills (Prior, 1976). The hills still stand sharp-featured, narrow-crested, intricately divergent, and fascinating in appearance. Much attention is now being given to the inventory and preservation of natural areas and ecological habitat within this region. The encroachment of trees and shrubs, just beginning in the 1860s, has

continued and is one of the obvious natural changes noted in comparing views of this locality.

The site from which St. John sketched is known locally as "First Bride's Bluff," near the northwest end of Ravine Park on the south side of Sioux City (Fig. 7; Woodbury County, NE ¼, Sec. 12, T.88N., R.48W.). St. John's sketch encompasses a view to the northwest, a view now obscured by trees and brush on the foreground ridge, but on gaining a better vantage point, a view with dramatic changes which have accompanied the city's urban development. Entire portions of the bluffs have been removed and are now occupied by U.S. Highway 75, Interstate 29, and a railroad. Residential areas occupy the waves of upland ridges to the north. Electric transmission lines and poles, bridges, cultivated fields, and confining revetments that hold the river in place complete the scene.

This sketch was not made into a lithograph illustration. If an observer turns 180 degrees from this view and look south, however, the scene is exactly that illustrated by the lithograph between pages 104 and 105 of White's Volume 1 (Fig. 8). This view clearly shows the broad crescent-shaped curve in the line of treeless hills stretching south to Sergeant Bluff, dense flood-plain

Figure 7. View northwestward along Missouri River valley toward downtown Sioux City from the site of Figure 2; 1978 photograph.

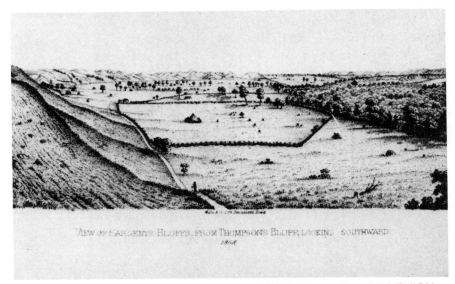

Figure 8. "View of Sargents Bluffs [sic] from Thompson's Bluff, looking southward (1868)." Lithograph from White, 1870, Vol. 1. Opposite view from that sketched in Figure 2.

forest adjoining the river, open fields, a farmstead with fenced pasture, and the now-familiar man with a walking stick making his way along the road at the base of the bluffs. This pastoral setting is now occupied by U.S. Highway 75, Interstate 29, its cloverleaf intersection with Interstate 129 from Nebraska, the Sioux City sewage-treatment lagoons, and the Sioux City Municipal Airport; the floodplain forest is gone, and abundant timber covers the hills instead (Fig. 9). The landscape views here are spectacularly broad, encompassing great distances. The basic elements of hills, valley, and river remain much as St. John saw them. The greatest changes are those brought by settlement and urban growth, and they are more dramatically illustrated here than at any of the other sites.

The Iowa "Great Lakes" have become one of the tourist meccas of the state, a trend observed as early as the 1870 Report. Small villages, summer homes, permanent residences, public beaches, boat docks, amusement parks, and tourist accommodations dot the miles of picturesque shoreline. At the same time, these lakes and nearby marshes, bogs, and fens clustered amid the knobby, gravelly hills are recognized by the state's zoologists, botanists, and geologists as one of Iowa's significant natural regions. The University of Iowa's Lakeside Laboratory on the

Figure 9. View southward toward Sergeant Bluff from the site of Figure 8; 1978 photograph.

Figure 10. View northward toward West Okoboji Lake from the site of Figure 3; 1978 photograph.

shores of West Okoboji has been an important center for teaching and research since 1909. There are a number of state parks, game management areas, and state preserves in Dickinson County. These offer public access to the lakes, as well as protection of the fragile ecological habitats, native prairie remnants, and significant glacial landforms which remain.

St. John's drawing of Lake Minnetonka (West Okoboji) shows the knobby relief characteristic of the glaciated landscapes which enclose the lake. The cobbles and boulders in the foreground are part of the gravelly drift deposited by the slow melting of stagnant, late Wisconsinan glacial ice between 12,000 and 14,000 years ago. These features are now classified as part of the

Altamont Moraine, one of several major crescent-shaped bands of knob-and-kettle terrain that arc across the Des Moines Lobe of north-central Iowa.

St. John had an unobstructed view to the horizon from his upland ridge above the southern extremity of the lake. Our search for the right hummock from which he drew, took us eventually to a spot known locally as "Lilac Hill" (Fig. 10; Dickinson County, SE ¼, Sec. 36, T.99N., R.37W.). We found the view to the lake partially obscured by trees and brush grown up around the homes that occupy all the intervening space to the lower meadow. The hill, however, is a tiny refuge of native prairie plants. The site is poised between the open agricultural

Figure 11. View northward toward Spirit Lake from the site of Figure 4; 1978 photograph.

land to the south and the enclosing rim of trees and homes which focus northward on the lake. Belonging neither to the upland prairies now converted to cultivated fields, nor to the lower meadows now homes and trees, it stands as a solitary remnant of what once was. The detail in St. John's drawing enabled us to identify a number of geographic features recognized today: Brown's Bay, Garlock Slough, Pocahontas Point, Eagle Point, Gull Point, and Smith's Bay.

Locating the site of St. John's other drawing from the lakes area was more difficult. The ridge north of Spirit Lake village and at the southern end of the lake is prominent, easily located, and still strewn with glacial erratics (Fig. 11; Dickinson County, SE ¼, Sec. 32, T.100N., R.36W.). We could find no vantage point, however, which permitted a line-of-sight view with Spirit Lake in the background and East Okoboji in the foreground. Two later discoveries provided a likely explanation.

The shorelines of both Spirit Lake and East Okoboji are indented with a number of irregularly-shaped sloughs. A 1916 publication by the Iowa Highway Commission titled *Iowa Lakes and Lake Beds* shows a long, narrow slough projecting northwestward from the upper end of East Okoboji. This slough is now drained and only some marshy ground remains. St. John's marginal notes also point to "Melonia locality, slough connecting the two lakes" just above his westernmost extension of East Okoboji. So, it is clear the shoreline has been modified in this area, removing, at least in part, the expanse of water St. John saw between his vantage point and Spirit Lake in the background.

Additionally, there is the strong possibility that St. John completed this sketch from two different sites, the second one downslope to the east, nearer the East Okoboji shoreline. There is a considerable horizontal distance covered in this drawing, and to reproduce the perspective he has shown suggests a condensing of two fields-of-view into a single scene on paper.

The shoreline in the vicinity of the isthmus separating these two lakes, rises more gradually than the steeper rim lands at the south end of West Okoboji Lake. These open, expansive views remain today. However, the uncluttered boundary between land and water, and the few scattered signs of civilization, are dominated now by intersecting state highways and beach-front residences along the lake shores, and by cultivated cropland on the higher mounded uplands.

The first three sites revisited still retained their basic natural components of river and valley, lakes and land. Major landmarks could be identified. The Fort Dodge sketch site presented a different set of problems. This close-up view of a gypsum quarry is identified on St. John's drawing as being in a ravine three miles south of Fort Dodge on the east side of the Des Moines Valley. It is further identified in the 1870 text as being on Two Mile Creek, which is also shown on an accompanying map. In the Iowa Geological Survey Annual Report for 1917–1918, Two Mile Creek is identified as being "better known as Gypsum Hollow," and on the map accompanying the report, is marked as "Gypsum Creek," the name that appears today on the USGS Fort Dodge 7.5-minute topographic quadrangle. This Annual Report also describes the exposures of gypsum that occur for nearly a mile upstream and "the numerous old quarries which supplied two mills, long since dismantled, during the nineties" (Wilder, 1918). Convinced that we are in the right valley, we felt that St. John's sketch is probably of one of these early quarries, and in searching, we found remnant natural exposures less than a mile from the mouth of the creek (Fig. 12; Webster County, NW¼, NW¼, Sec. 6, T.88N., R.28W.). One exposure is a portion of the remaining present valley wall, and the other is an outlier, for some reason not taken during earlier quarrying, still standing near the probable position of the original valley wall in the 1860s. These exposures are weathered with the same characteristic patterns as St. John

Figure 12. View of remnant gypsum exposures along Gypsum Creek south of Fort Dodge from the site of Figure 5; 1978 photograph.

described; old tree-stumps are scattered through the overgrowth of younger trees and brush, and an abandoned rail line occupies what may have been the narrow pathway drawn by St. John.

The similarity to the St. John view ends here. The glacial overburden and natural hillslope sketched above the gypsum face are gone, and after climbing to the top of this brushy ravine, the value of the gypsum deposits, barely scratched in the 1860s, becomes apparent. We are in the heart of the U.S. Gypsum Company quarries. Open-pit mining, working faces, spoil piles, haul roads, heavy equipment, graded slopes, and reclaimed areas, all the signs of modern quarrying activity dominate over a square mile of the land surface surrounding Gypsum Creek.

Gypsum Creek was the starting point of this activity; now it is the only relatively unaffected landscape which remains. The exposures along the valley sides were the logical place for early quarrying. Now everything is mined out, except a few isolated remnants adjacent to the natural drainage where recovery was not economical. That anything resembling St. John's sketch remained at all is surprising, but, as in the case of Lilac Hill, the site is at a boundary, at a break in slope, an unused bit of land, by-passed by modern changes because of its marginal position.

The last of St. John's Iowa sketches was drawn along the Des Moines River one mile above Keosauqua. This site is located in the great bend of the Des Moines River in Van Buren County (Fig. 13; SE¼, NW¼, Sec. 2, T.68N., R.10W.). St. John drew from a sandy stretch of river channel exposed by low-flow conditions. His access, and ours, to this shoreline was across the wide point-bar deposit on the inside of the meander loop. The St. Louis Limestone exposures occur along the steep slope of the outer, cut-bank of the meander. On one of our visits to this site, similar sandy flats were exposed above water level, and people were observed walking in shallow water out into the channel.

During other visits, the river filled the channel bank-to-bank and our viewing was restricted to the tree-lined bank at the left of the sketch. The interval of exposed strata appears about the same today as shown in St. John's sketch. This similarity, and the known shallows exposed at low flow, suggest that the configuration of the river channel and the on-going alluvial processes are little-changed along this segment of the valley, even with the additional influence of Red Rock and Saylorville Reservoirs constructed upstream.

This site is the least changed of the five we compared. Hillslope can be matched with hillslope, rock exposure for rock exposure, and ravine for ravine. This fact is a consequence of the establishment of Lacey-Keosauqua State Park in 1921 which preserved much of the scenic, wooded, and dissected terrain on the south side of the river, the very view of St. John's sketch.

COMPARISONS AND CONCLUSIONS

It is important to consider the purpose for which St. John's sketches were done. He, as a scientist and an observer, drew them as a means of accurate geographic documentation, and as illustrations for a scientific report. As he conducted his field work, we assume he was motivated by the interesting natural features he observed, by an understanding of how well a scene conveyed meaning to the geologist, by what various contours and materials revealed about the geologic processes which shaped the land, and by the economic resources associated with these features and deposits. He was not a professional artist nor influenced by artistic traditions. He did not embellish or stylize the drawings as an artist might have done. The pencilled grid on all his drawings even suggests the use of a device—a sketch frame or camera obscura—a standard field technique used to aid in achieving

Figure 13. View of the Des Moines River valley above Keosauqua from the site of Figure 6; 1978 photograph.

proper perspective and proportion. Yet to today's art historians, the drawings have artistic validity and vitality (*Five Iowa Landscapes-Past and Present, 1979*). Despite their intended function, they are St. John's own individual representations of the landscape, influenced by long days spent in the field becoming familiar with its details. They also are set apart by the precision and attention to detail with which they were done. His careful rendering of cultural features, and the presence of the small figure observing the scene, always unobtrusive, seem to go beyond a strictly scientific purpose or the need to include some recognized object for scale; they suggest a genuine enjoyment of the place and a thoughtful, introspective nature. They date from the 19th Century "topographical landscape" tradition, which encompassed not only professional artists but those within the scientific community who documented the land for other than aesthetic reasons. There was a closeness between science and art in this period, a shared reverence of nature, and significance given to its study in any form.

This personal view of the landscape, which these drawings represent, was replaced in succeeding years by the camera and by improvements in travel and field techniques. Similarly, the shift from naturalist to specialist among geologists coincided with a change in orientation from land surface to subsurface geologic investigations. The changing trend in the acquisition of geologic data is underscored by comparing the 19th Century influence of the railroads, in their quest for routes and resources, to the role of the petroleum and water-well drilling industries in the 20th Century. It is interesting to note, however, that as a modern tool for unraveling subsurface conditions, remote-sensing techniques utilizing satellite images focus once again on the land surface (Lowman, this volume). From the pencilled grid which aided St. John's sensing perspective, a parallel may be found in the guiding grid of section-line roads that marks off our perspective in an image of the midwestern landscape taken from 500 miles into space.

St. John's landscape drawings, removed from their original context and placed alongside modern drawings, photographs, and maps of the sites, invite comparison of the effects of 110 years of change. They now provide a context for the consideration of past and present attitudes toward the land as well as consideration of future trends. They serve to bring the land once again to public attention though for a different purpose, and they form the basis of an archival record which future geologists, historians, and artists will find of value.

ACKNOWLEDGMENTS

The search for the remaining sketches and records has been widespread. In addition to those cited in the text, we would like to acknowledge the assistance of Clifford M. Nelson, U.S. Geological Survey; Margaret S. Terry, San Pedro, California, Elizabeth Mead, Orangevale, California, and C. Edward Hertlein, Pratt, Kansas, all relatives of Leo Hertlein; Katherine V. W. Palmer, Ithaca, New York; Earl L. Packard and Leigh N. Ortenburger, both of Palo Alto, California; Olaf P. Jenkins, Pacific Grove, California; W. R. Moran, Los Angeles, California; Alan E. Leviton, California Academy of Sciences; Alan L. Bain, Smithsonian Institution Archives; Michael Goldman, National Archives and Records Service; Ann Blum, Harvard Museum of Comparative Zoology; Larry Jochims, Kansas State Historical Society; Loren Horton, Iowa State Historical Society; Lida L. Greene, Iowa State Historical Museum; the San Diego Historical Society; and the University of Wyoming Library.

This project has benefitted from conversations and exhibition programs presented with Richard Thomas, Professor of History, Cornell College, Mt. Vernon, Iowa; Ellwood C. Parry, III, Professor, and Jean H. Leighton, Ph.D. candidate, both from the

School of Art and Art History, University of Iowa, Iowa City. The support of the Iowa Geological Survey is also appreciated.

REFERENCES CITED

Fern, Thomas S., 1976, The drawings and watercolors of Thomas Moran (1837–1926): Catalogue of an exhibition at the University of Notre Dame Art Gallery, Notre Dame, Indiana, Apr. 4 to May 30, 1976, 152 p.

Fryxell, Fritiof, 1939, William H. Jackson, photographer, artist, explorer: American Annual of Photography, v. 53, p. 208–220.

Five Iowa Landscapes—Past and Present: Notes accompanying an exhibition funded by the Iowa Board for Public Programs in the Humanities, May 15-Aug. 20, 1979.

Hayden, F. V., 1876, Bulletin of the United States Geological and Geographical Survey of the territories, v. 2, no. 4, Notes on the geology of northeastern New Mexico (St. John), p. 279–308.

Hayden, F. V., 1879, Eleventh annual report of the U.S. Geological and Geographical Survey of the territories, embracing Idaho and Wyoming, being a report of progress of the exploration for the year 1877; Report on the geological field work of the Teton division (St. John), p. 321–508.

Hayden, F. V., 1883, Twelfth annual report of the U.S. Geological and Geographical Survey of the territories: A report of progress of the exploration in Wyoming and Idaho for the year 1878, Part 1; Report on the geology of the Wind River district (St. John), p. 173–269.

Keyes, Charles R., 1921, Memorial of Orestes Hawley Saint John: Geological Society of America Bulletin, v. 33, p. 31–44.

——1942, Pioneer geological work of St. John: Pan-American Geologist, v. LXXVII, no. 1, p. 1–18.

Love, J. D. and Reed, John C., Jr., 1968, Creation of the Teton landscape: Grand Teton Natural History Association, 120 p.

National Archives and Records Service, Microfilm Publication M623, *Records of the Geological and Geographical Survey of the Territories ("Hayden Survey"), 1867–79,* Roll 12, (St. John to F. V. Hayden letters).

Nelson, Clifford M., 1980, William Henry Holmes: Beginning a career in art and

science: Records of the Columbia Historical Society, Washington, D.C., v. 50, p. 252–278.

Prior, Jean C., 1976, A regional guide to Iowa landforms: Iowa Geological Survey Educational Series 3, 72 p.

Smithsonian Institution Archives, Record Unit 7154, Orestes Hawley St. John Papers, 1856–1892, Box 2, Folder 11, December 28, 1867 (St. John to F. B. Meek letters).

——June 3, 1868 (St. John to F. B. Meek letters).

——Record Unit 7062, Fielding B. Meek Papers, Box 10, *"Iowa, Nebraska, 1867"* Notebook.

Stanford University Archives, Stanford University President's Report of 1922, and Ray Lyman Wilbur Papers, Geology Department, 1922.

St. John, Orestes, Worthen, A. H., and Meek, F. B., 1875, Paleontology of Illinois: Illinois Geological Survey, v. 6, pt. 2, p. 245–532.

White, Charles A., 1870, Report on the geological survey of the state of Iowa: Mills and Co., Des Moines, v. 1, 390 p.; v. 2, 443 p.

Wilder, Frank A., 1918, Gypsum: its occurrence, origin, technology and uses: Iowa Geological Survey Ann. Rept., v. 28.

Worthen, A. H., St. John, O. H., and Miller, S. A., 1883, Paleontology of Illinois: Illinois Geological Survey, v. 7, pt. 2, p. 53–373.

Authors' Note

The Hayden reports contain numerous sketches which portray landscape and geologic features. The texts do not indicate who made these sketches, but Ortenburger and Fryxell (Pers. commun., Dec. 6, 1981) suggest it certainly could have been St. John. These Iowa sketches demonstrate that he had sufficient artistic skill to have done the sketches in the Hayden volumes. Keyes clearly believed them to be St. John's work (Keyes, 1921, p. 38).

Geological Society of America
Centennial Special Volume 1
1985

Dust in the wind: J. A. Udden's turn-of-the-century research at Augustana

William B. Hansen
Division of Mineral Resources
U.S. Bureau of Land Management
P.O. Box 36800
Billings, Montana 59107

ABSTRACT

As geologist and director of the Texas Bureau of Economic Geology, J. A. Udden gained a reputation as a pioneer geologist from 1911–1932. Less is known about Udden's tenure at Augustana College in Rock Island, Illinois, from 1888–1911. A study of letters to and from his teacher, students, and sons during this period, in light of some of his key turn-of-the-century publications, shows the influence of Udden's research at Augustana.

Many of the concepts which brought him recognition later in life were developed under the guidance of his Augustana teacher, Josua Lindahl. As early as 1891, Udden advocated an actualistic approach to geology much like that of Johannes Walther. Udden's research on wind-blown sediments led to the development of a particle distribution scheme that is used by sedimentologists today—the Udden-Wentworth scale. Working with T. C. Chamberlin, Udden was one of the first geologists to demonstrate that the Pleistocene loess of the Upper Mississsippi Valley was a wind-blown, not water-laid sediment. His interest in wind led to the construction of a working model of a flying machine. He perceived the importance of drill cuttings long before the oil and gas industry realized their value in subsurface geology. An interest in electricity led to Udden's recommendation of seismic reflection as an exploration tool for oil and gas. Despite a full-time teaching load, Udden published 46 papers during his 23 years at Augustana.

INTRODUCTION

In a letter to his son Jon, J. A. Udden (1926) wrote,

"Many years ago I began a practice never to write anything *in the form of criticism* . . . I merely presented the facts as I saw them and understood them."

Presentation of observed facts was a lifelong mission of Johan August Udden. Perhaps this was rooted in his Swedish-American heritage. Born in Lakasa, Sweden, on March 19, 1859, Udden came to America with his parents at the age of two and settled on a homestead in Carver, Minnesota. The family took the name of their Swedish home, Uddabo, and shortened it to Udd for their American last name. On entering school, however, young Johan and brother Svante took matters into their own hands and

changed their last name to Udden after being teased about their short, unusual name (Heiman, 1963).

At the age of 17, Johan enrolled at the fairly new Augustana College, in Rock Island, Illinois, concentrating on Physics, Astronomy, and Botany. At this time, in 1876, the school was quite small (Fryxell, 1924):

The faculty (consisting of the president and three professors) and the students (all boys) together boarded, roomed, and recited in the one college building; when boarding, fuel, light, and one room might be had for $2.50 per week; when as yet no degrees had been conferred by Augustana. . . .The first bachelor's degrees were conferred in 1877.

In 1878, Josua Lindahl joined the faculty at Augustana to

Figure 1. Field geology, circa May 1898, near an exposure of Pleistocene Peorian and Sangamon weathered zones, east of Peoria, Illinois. From left to right: Dr. S. W. Beyer, Iowa State College, Ames; Dr. J. A. Udden, Augustana College, Rock Island; Professor T. C. Chamberlin, University of Chicago; Professor Samuel Calvin, University of Iowa; and Mr. Frank Leverett, U.S. Geological Survey (Leverett, 1899).

teach Natural Sciences. He influenced young Johan's geology career, although he did not offer a formal course in geology until after Udden graduated.

After graduation in 1881, Udden traveled to Lindsborg, Kansas, where he became one of the founders and first instructors of Bethany College. He remained close to Lindahl, however, and it is apparent from the 90-odd letters exchanged between the two men over the next 25 years (Lindahl, 1886a, 1886b, 1888, 1890)[1] that Lindahl had a strong influence on his former student.

Udden married Johanna Kristina Davis in 1882. Only three of their children, Jon, Anton and Svante, survived past childhood. Over the years, J. A. Udden developed strong ties with his sons, and his numerous letters (Figure 2) have provided substantial background information for this paper (Udden, 1915, 1926).

Udden found intriguing features to study in Kansas, wherever rivers exposed the geology; observations made between the Smoky Hill and Little Arkansas Rivers resulted in the publication of his first paper (Udden, 1891). This publication established Udden within the geological community as a meticulous observer and reporter of physical features (Heiman, 1963).

In 1888 Udden returned to Augustana to fill the position left by Lindahl, who was moving on to become curator of the Illinois State Museum. Obviously delighted with Udden's decision to

return to Augustana, Lindahl (1888) wrote,

Friend Udden!
The board has today called you to be my successor in biology and geology whilst you will have Mr. Westlund as assistant in physics, chemistry and astronomy. It was at my suggestion that the subjects were thus divided, it being impossible for any one man to keep himself posted on the development of all these six branches of natural science. I had previously recommended that this change should take place two years hence when Victor Peterson, my best student, would be qualified to take charge of the department now offered to you. Singularly enough, I had never for a moment thought that you would accept a position here with your present bright prospects in Kansas.

Over the next 23 years at Augustana, Udden published 46 papers. As early as 1890, Udden managed to do field work in Texas during the summer as well as carrying a full-time teaching load at Augustana. Heiman (1963) wrote a detailed study about this field work based on a collection of Udden's field notes stored in Texas. Much attention has also been given to Udden's post-Augustana accomplishments as a geologist and later Director of the Texas Bureau of Economic Geology from 1911–1932 (Baker, 1932, Ferguson, 1981, Leighton, 1950), but little has been written on the midwestern geological research Udden performed while at Augustana from 1888–1911. This paper attempts to summarize that research and provide some insight into the man behind it, based on the actual publications and the wealth of letters stored at the Augustana College Archives (Figure 3).

---

[1] A detailed list summarizing over 300 letters to and from Udden, now filed in the Augustana College Archives is available upon request. Many of them were written in Swedish and have not been translated into English.

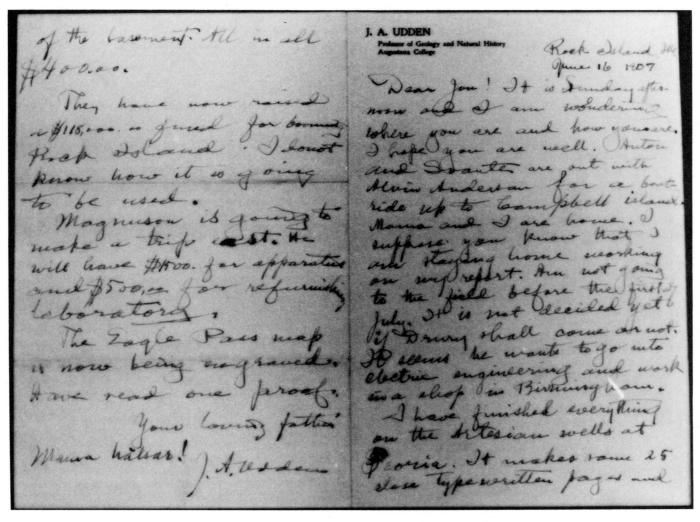

Figure 2. An example of one of the letters J. A. Udden wrote to one of his sons (Udden, 1907). Most of the letters have some reference to geology indicating its importance to Udden.

THE INFLUENCE OF LINDAHL

Josua Lindahl continued to instruct young Udden in the basic sciences long after his graduation from Augustana in 1881. In 1886, when Udden had taken a short leave of absence from Bethany College to do graduate work at the University of Minnesota, Lindahl (1886a) wrote to Udden, on the subject of gravity,

The attraction between the Earth and an apple on the bough determines the weight of the apple expressed in gravity units. But when the apple ripens and its gravity exceeds the cohesion on its stem, the very same attraction will produce a motion which in its turn will be converted into mechanical force in the moment the motion is checked by the ground. This latter force, the violence of the blow, depends on the weight and velocity of the apple, and the velocity again depends on the height of the fall. Thus knowing how great the velocity is, or, which is the same thing, how many seconds the apple has been falling, it can be calculated how great a force (Kinetic) was produced by the gravity of the apple.

The sabbatical from Bethany apparently kindled an interest in electricity for Udden. When he sought advice from his former teacher as to whether he should pursue this interest, Lindahl (1886b) was emphatic in his reply:

Friend August:
My modesty obliges me to acknowledge that there are certain things that fall beyond my knowledge. You have just happened to strike something of that kind in your electrical experiments. You ask whether I regard it worth your while to further inquire into the matter. To this I would answer - no! Not under your present circumstances. There are two classes of men who do such work, and you belong to neither. The one consists of men, possessing an exhaustive knowledge of the present state of the physical science, and who . . . experiment intelligently generally using a skillful artisan to work under their directions, and their aim is a purely scientific investigation into the laws of nature. The other class comprises such men, who have an encyclopedic knowledge of a certain class of phenomena and possess more skill in their fingers than in their brains . . . Both of them must be enthusiastic for their work, and devote *all* their time to it, else they will accomplish nothing. Colonel John Ericson, in comparing himself with Edison characterizes these two classes as respectable *investigators* (or in his Branch constructors), and

Figure 3. The Augustana College Archives as it looks today on the top floor of the Denkman Library in Rock Island, Illinois. More than 300 of J. A. Udden's letters and manuscripts have been preserved in the Archives.

inventors. Of course he was unjust in refusing to acknowledge Edison as an investigator, but otherwise I think his characterization of the two classes is correct. Joseph Henry belonged to the first class. I have read his trial before a committee, appointed by the Smithsonian Institution, at his own request, to investigate into a charge made against him by the inventor of the telephone, Bell . . . In performing these 5000 experiments he has at his disposal a fully equipped physical laboratory, skilled mechanic and all the money needed for pushing the work. You cannot afford to spend a great deal of money and time on random experiments, nor can it be advisable for you to aim at the acquisition of a high scientific standing in the Science of Electricity. The field is already worked up too much, and you cannot expect to accomplish much without spending years of assiduous labor on this topic exclusively. Get a good textbook, and enlarge upon it *to a limited extent* for the purpose of strengthening your own power of observing and interpreting phenomena, but for no other purpose. You have another field in which you should invest your spare time to far better advantage, and that is the study of the mounds. . . .Every little observation may prove a valuable contribution to our fragmentary knowledge of that branch of Archeology.

The "mounds" that Lindahl referred to were a series of Indian mounds that Udden excavated in Kansas and reported on in a later Augustana publication (Udden, 1900). Lindahl was evidently aware of Udden's ability to make meticulous observa-

tions and did not want to see it wasted in a field already "crowded" with investigators.

In the same letter, Lindahl (1886b) stresses to Udden the obligation they both have to their students and schools.

You hesitated, some time ago, to make any particular efforts to accumulate collections for a future museum . . . I have spent considerable thought on that question lately, and I am now of the opinion, that the very best way, in which both you and I can be useful to our respective schools, is just in making good musuems. It is a well-known fact that students, who recite well and make it a matter of overwhelming or sole importance to do so, will rarely make their mark in the world as great men, and institutions, where the classroom is most excellent but no opportunities are given for waking up the boys to observing or thinking beyond the pages of the textbooks, will invariably fail to turn out men of great ability, whereas other institutions, provided with rich museums, libraries, laboratories, etc., but where the classroom is not even as highly developed as in some one-horse-power institutions will hatch out prominent men in all lines of intellectual work. The time allowed for the natural science in our Swedish-American sectarian schools, where two mother tongues and Christianity together with Latin and Greek occupy about 60% of all the class studies, is and must remain entirely inadequate for the demands on a college bred man in America and in this 19th century.

Figure 4. The Fryxell Geology Museum at Augustana College today as it has grown from Lindahl and Udden's initial collections. Many of the original museum pieces are still on display in the museum.

What shall we do to help this sad state of our department and our schools? Make the museum attractive, extensive, carefully arranged, well-labeled, so that the students will find it a pleasure to drop in there as often as possible; and they cannot help to learn a great deal without knowing that they are actually studying.

Udden apparently took Lindahl's advice to heart, for when he arrived some years later to teach at Augustana, he said (Fryxell, 1924, p. 16),

I tried to keep the collections made by Lindahl in good shape, not only because I knew they were valuable, but also because I had a kind of religious loyalty to my old teacher and predecessor in office.

This led at least one of Udden's students (Augustana College Archives, undated) to proclaim,

I recall his geologic museum as I saw it in the early nineties. It was not a collection of rare fossils and costly minerals. It was a model museum once defined as a 'collection of labels illustrated by specimens'. The specimens were good enough, but the labels were so clear and full that in due sequence and with unavoidable gaps filled, they would have made up a very good textbook in elementary geology.

It is worthwhile noting that the Fryxell Geology Museum at Augustana has continued to grow far beyond the expectations of Lindahl and Udden (Figure 4).

One fact that the letters between Lindahl and Udden reveal is that Udden did not make up his mind to specialize in geology until 1890, after he had arrived at Augustana to teach. Lindahl's (1890) power of persuasion apparently convinced him to make up his mind:

Give as much as possible of your time to geology. You have better chances for promotion there than in any other branch of science. But study it systematically - don't just pick it up!

It is also apparent from the letters that Lindahl was preparing young Johan for a teaching position bigger than the one he now held at Augustana. Udden apparently had an opportunity in 1890 to teach at the University of Texas, but passed it by. Writing from his new position at the Illinois State Museum, Lindahl (1890) encouraged him to prepare academically for the next opportunity that came along:

Buy a set of crystal models and study crystallography - you will be required to teach it in a university class. Get a $100 microscope with a polariscope outfit included - (When you are ready to buy it, I wish you would let me help you get it complete and with a good discount - I would ask Rolfe in Champaign to order it; he has a most excellent outfit he got for less than $100); buy a set of microscopic slides of minerals, grind down minerals and study them microscopically. Then take dynamical geology, and last paleontology. . . .the least important for a professor of geology at a University. I omitted chemistry, but don't you omit it. Take a course in blowpipe analysis, that will do to begin with.

Once again Udden took Lindahl's advice seriously and spent $100 dollars for microscopes in 1896. However, with a trait that came to be characteristic of Udden, the money was spent for the benefit of his students rather than himself (Fryxell, 1922, p. 12):

Dr. Udden early realized that in order to accomplish anything of practical benefit to his pupils, he must have the necessary apparatus, and as there was no appropriation or other available funds for this purpose, he took one hundred dollars out of his own account and bought a dozen microscopes and about twice that number of outfits of scalpels, scissors, tweezers, etc. I have the honor to have been a member of the first class to use these precious instruments, so delightfully interesting in their shiny newness, and well do I remember with the affectionate care the doctor guarded them - the look of anguish on his face as perchance he would detect someone of us, untrained in the use and value of such things, polishing away at the delicate lens of a microscope, and better still do I recall the righteous indignation with which he would rescue a keen-edged scalpel from some base misuse!

Lindahl's 1890 letter to Udden continues on with an interesting commentary on the state-of-the-art geology textbooks of that time:

Get the following books:
Brush's Blowpipe Analysis
Brush's Determinative Mineralogy
Rosenbush - Microscopical Physiography of Rock Making Minerals (!!)
Geikie's Textbook of Geology (!!!)
Geikie's Classbook of Geology (smaller)
I heard the leading professor of geology at the Madison meeting declare that there is no American textbook in geology fit for use except as an appendix to Geikie's . . .
Geikie's is used in all the best Universities in the country -Winchell is of no account. Dana's new book (in preparation) will probably be best to use as an appendix to Geikie.

It is clear that Lindahl recognized early that his former student had the potential to make a significant contribution to the science of geology. If, as it has been claimed, Udden's later success in research can be attributed to his encyclopedic knowledge of the basic natural sciences (Baker, 1932), geology has Lindahl to thank for steering him in that direction.

UDDEN'S WORK WITH WIND

Early Observations

Udden published his first professional paper in 1891, three years after he was appointed to Augustana. This paper, like many of his early publications, was based on field observations he made while teaching at Bethany College in Kansas. As with his later research, he took advantage of opportunities he found in an area some would consider geologically boring. He later wrote (Udden, 1898a),

A Western naturalist once said the geology of Kanses was monotonous. In one sense this remark is certainly justified. . . .Occasionally however, some feature of special interest crops out from the serene uniformity, and the very nature of its surroundings then makes it appear all the more striking.

In the case of his first paper that "feature" was a sediment-clogged valley in central Kansas, blanketed with a thick layer of volcanic ash, which had been observed by Udden years earlier. "Megalonyx beds in Kansas" (Udden, 1891) quickly established Udden as a meticulous observer and reporter of facts within the geological community. One geologist from Texas, Robert T. Hill,

sent his compliments to Udden immediately on his "most interesting paper in the last *Geologist.* It is well written and most instructive throughout" (Heiman, 1963, p. 16).

Udden did not just report his observations, he also interpreted the environment of deposition (Udden, 1891):

At another place where it (the ash) occupies a position about 40 feet higher, the particles are not assorted, the bed rests on a thin black seam containing bog manganese and bog-iron, and underneath the jointed clay is dark and carbonaceous. This suggests the bottom of a swamp.

Interpretation of the stratigraphic record using modern environments of deposition is a concept that Johannes Walther was encouraging at this time in Europe. Unlike many other American geologists, Udden was familiar with Walther's writings in the late 1800s. Udden actually referenced Walther's "Die Denudation in der Wüste" as early as 1898 (Udden, 1898b). Both men tended to emphasize "laws of nature" in their later papers (Udden, 1894, 1914; Walther, 1894). One of Udden' students would later remark that he considered Udden's work on wind deposits to be ranked on the same scale as those of Walther (Augustana College Archives, undated). Perhaps Udden's familiarity with the work of Walther was related to his ability to read and speak several languages. He was known to have lectured in German and maintained an extensive collection of untranslated foreign papers, including some from Cambodia and India (Heiman, 1963, p. 10).

An additional early Udden publication at Augustana affords a further glimpse of Udden's observations of wind-blown sediment in Kansas (Udden, 1898a):

This ash occurs in several outcrops in McPherson County in the central part of Kansas, where the writer had an opportunity to study it somewhat in detail a few years ago. Some of the features of the dust at this place reveal the conditions under which it was formed with considerable distinctness. . . .It may be said to consist of angular flakes of pumice, averaging one sixteenth of a millimetre in diameter, and having a thickness of about one three-hundredth of a millimetre. . . .
If these are removed and examined under the microscope, they are seen to be hollow spheres. . . .These are some of the original bubbles that never burst. . . .It is evident that not every droplet of the molten magma would form a single sphere, but that many would swell up into a compound frothlike mass of pumice. . . .The nature of the force which caused the eruption may thus be understood from the study of one little grain of the dust.

These early observations indicate that Udden was becoming intrigued with the study of the wind and its effect upon small particles. As C. L. Baker (1932) would later put it, "He came early to the conclusion that the best contribution he could make, with the limited opportunities of his environment, would be in fields overlooked by others. Thus he came to emphasize the minute. His geologic observations are characterized by their minuteness, as well as by their comprehensiveness."

Udden also recognized sedimentary structures in the ash beds of Kansas. He used them to determine the direction of the prevailing wind at the time of deposition. In a unique demonstration of actualism for that time, he discovered he could trace the path of the storm from the changing ripple marks as he measured up section (Udden, 1898a):

These successive changes are best explained as attendant upon the passage . . . of what our daily weather maps call a 'low area'. . . .One such rotation of the wind generally lasts a day or two. The shower must then have kept on for the same length of time, if not longer.

At the end of the article, Udden prodded the uniformitarian attitude among geologists (Udden, 1898a):

Modern science has taught us that the geological forces are slow and largely uniform in their work, and that most of the earth's features must be explained without taking recourse to theories involving any violent revolutions or general terrestrial cataclysms. While the making of this dust is not any real exception to the law of uniformity, we are reminded that Nature is quite independent in her ways, and that even in her sameness there is room for considerable diversity.

This article appeared in *Popular Science Monthly*. Udden apparently was shy of the negative feedback these unconventional ideas might generate if published in one of the respected professional journals of that day.

Wind As a Geologic Agent

Udden continued his studies of wind-blown particles throughout most of his career at Augustana. He was always aware of those times when dust was in the air. For instance, an average person riding in a railroad coach in the late 19th Century had to deal with the disturbing inconvenience of dust blowing in the open windows. Udden looked at it in a different way. He observed (Udden, 1898b, p. 33), "Quartz particles considerably larger than fine sand are here moved nearest the ground. But the material which is lifted high enough (five or six feet) to come in through the windows and doors of passenger coaches is much finer."

To gather additional samples of dust in the wind, Udden constructed measuring devices out of what materials he had available, since neither he nor the college could afford elaborate labware. He wrote (Udden 1898b, p. 40–41):

One of the devices used in collecting dust directly from the atmosphere consisted of some whisks of broom-corn, smeared with glycerine, and suspended from a pole ninety feet above the ground. The observations were made on a bluff overlooking the Mississippi River at Rock Island in Illinois. The whisks were taken down once a day and washed in water which was allowed to stand until the dust had settled. One series of such samples was secured during the month of March in 1895.

Realizing the shortcomings of using just one type of measuring device, Udden came up with an additional type of measurement (Udden, 1898b, p. 43):

Some dust was collected at the same place and at the same height by suspending two pieces of muslin held horizontally on a frame. The muslin was smeared with glycerine, to which the dust was adhered. This was secured by washing and allowed to settle as before.

No matter which device Udden used, the results matched. Both instruments collected more particles in the 1/32 - 1/64 mm range than any other size. Still not satisfied, Udden constructed a third device for dust collection (Udden, 1898b, p. 44):

Another device for collecting dust from the atmosphere consisted of a hollow cylinder, with apertures on the side for receiving the wind, and with strips of muslin suspended inside. These strips as well as the inner surface of the cylinder were washed once a week, and adhering particles then secured. . . .The cylinder was suspended from the same height and from the same flag pole as the broom corn and the muslin previously mentioned.

The variety of sample data in Table 1 shows Udden's willingness to collect samples wherever they were available. Once he had the samples in hand, Udden organized the particles by grain size with each group "bearing a constant ratio to the diameters in their next group." For instance, the diameter of particles in the largest grade had twice the length of the particles' diameters in the next finer grade, and so on.

The distribution scale which he settled on (Table 2) is still used today and is known as the Udden-Wentworth scale (Blatt, *et al.*, 1972). To separate the particles into the various sizes of this scale, Udden built a series of sieves to handle the coarse gravel through fine sand (Figure 5). Below the fine sand range, Udden depended on particle counts under the microscope (Udden, 1898b).

Udden discovered that at least 70% of each dust sample fell within two of the size ranges. This was a phenomenon that Udden (1914, p. 736) described as the Law of Secondary Maximum:

When a transporting medium is supplied with sufficiently heterogeneous material it will tend to carry and to deposit more of two certain sizes of material than of any other sizes.

Udden also discovered that the masses of the particles in each size range had a fixed ratio to each other. This he called the Index of Sorting. For eolian sediments, he found this index to be near 4.5:1 (Udden, 1916).

Udden, Chamberlin, and the Origin of Loess

With classes recessed at Augustana, Udden went to work for Dr. T. C. Chamberlin of the University of Chicago in the summer of 1893. They set out to examine drift deposits in the upper Midwest.

Monica Heiman (1963, p. 60) surmises that Chamberlin must have had some influence that summer on young geologist Udden, but a letter from Udden (1926) to his son Jon suggests that the influence may have been reciprocal:

TABLE 1. SOME OF THE WIDE VARIETY OF SAMPLES
UDDEN USED IN HIS WIND STUDIES*

Label Number	Sample
265	Dust in a school house four miles from a blown field, North Dakota.
267	Dust collected in a running railroad coach in southern Minnesota.
306	Dust shaken from the trunk of an oak, Rock Island, Illinois.
308	Dust in rain water from the roof of a house, Rock Island, Illinois.
312	Dust washed from some popular leaves, Rock Island, Illinois.
318	Dust washed from foliage of trees, La Salle, Illinois.
319	Dust washed from dry leaves of oak trees, Rock Island, Illinois, February 1895.
327	Shower dust on the ice of the Mississippi, near bank, Rock Island, Illinois, January 1895.
328	Shower dust from a crack in the ice of the Mississippi, near channel, Rock Island, Illinois, January 1895.

*Udden, 1914. Numbers are the original sample labels that Udden used in his 1914 publication on clastic sediments.

Figure 5. The series of sieves that Udden fashioned by hand out of copper and wood. The sieves are now stored in the Augustana College Archives. Each sieve corresponds to a different grade scale on the Udden-Wentworth particle size scale.

TABLE 2. CLASSIFICATION SCALE UDDEN USED IN DESCRIBING WIND-BLOWN SEDIMENTS*

Class	Diameter in mm
Coarse gravel	from 8 to 4
Gravel	from 4 to 2
Fine gravel	from 2 to 1
Coarse sand	from 1 to 1/2
Medium sand	from 1/2 to 1/4
Fine sand	from 1/4 to 1/8
Very fine sand	from 1/8 to 1/16
Coarse dust	from 1/16 to 1/32
Medium dust	from 1/32 to 1/64
Fine dust	from 1/64 to 1/128
Very fine dust	from 1/128 to 1/256

*Udden, 1898b, p. 6. This scale was later modified slightly by Wentworth (1922) and is a primary grade scale in use by sedimentologists today.

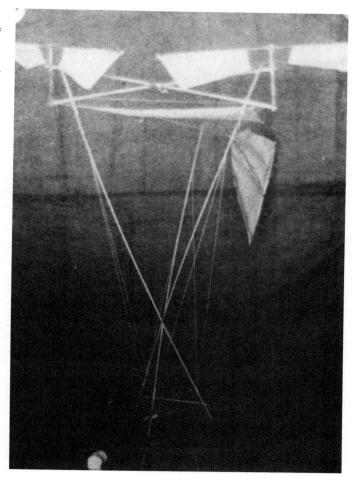

Figure 6. J. A. Udden's working model of a flying machine, called rotopter, as it appeared on its maiden flight in Old Main in 1907.

I think I have corrected even Chamberlin once. That was in my 'Loess As A Land Deposit' (1897) . . . Chamberlin never afterward wrote about the loess as a water deposit. This may have been merely a coincidence. And thereby hangs another tale.

In the past 40 years, T. C. Chamberlin's paper, "Supplementary Hypothesis Respecting the Origin of Loess of the Mississippi Valley" (1897) has been credited with initiating the concept of loess as an eolian deposit (Leighton and Willman, 1950, p. 606). Chamberlin presented that monumental paper on August 12, 1897, before the American Association for the Advancement of Science meeting in Detroit, Michigan (Chamberlin, 1897). Interestingly enough, the paper which Udden refers to in the above quote was read in the same room two days earlier on August 10, 1897, at the Geological Society of America meeting. However, Chamberlin does not appear to have actually attended the session where Udden's paper was read (Geological Society of America, 1897, p. 1, 12).

The two men continued to do field work together the spring following their 1897 talks as shown by a photograph (Figure 1) taken of a field party organized by Frank Leverett in 1898 (Leverett, 1899). It seems that Udden's thoughts on the wind blown origin of loess had some influence on Chamberlin's thinking.

Construction of a Flying Machine

In retrospect, it seems logical that Udden would follow his studies of dust in the wind with the construction of a flying machine. He first sketched in 1893 his "rotopter," which resembles a modern-day helicopter. He went on to build a working model of the rotopter in 1908 (Udden, 1908). Udden's interest in the machine apparently dwindled after exchanging letters with Orville Wright, who expressed his doubts (Wright, 1917). At any rate, Udden makes no mention of the machine in his letters after 1917. Perhaps he remembered the words of his teacher, Lindahl, warning him to stay out of fields which would detract from his work in geology.

ADDITIONAL RESEARCH

One of the major contributions Udden made in his later days in Texas was convincing the petroleum industry to preserve drill cuttings. This was an idea Udden first developed at Augustana; he kept track of as many local well cuttings as he could during his time there.

His work with well cuttings led geologists in the early 1900s to declare that there were few other places in the world where subsurface conditions were as well documented as the Rock Island area (Baker, 1932). Udden's interest in the subsurface also led to the publication of a geologic cross-section across Northern Illinois which was displayed at the 1892 Chicago World's Fair (Udden, 1893).

Udden's expertise in subsurface geology led to his telling the Board of Regents at the University of Texas at Austin that there

Figure 7. Old Main, the building that Udden taught his geology classes in, standing today as it did when first completed in 1894. Many of the cuttings Udden described from the Rock Island area are stored in the attic of the Old Main dome.

were prospects of oil and gas on university-owned lands. This contributed to the discovery of the West Texas oil field.

Several other of Udden's accomplishments in Texas can be traced back to research begun at Augustana. Udden's paper, "The Mechanical Composition of Clastic Sediments" (Udden, 1914) was basically an amplification of research he had accomplished while at Augustana (Udden, 1898b). It was only after strong encouragement from his friends that he decided to include his earlier publication in the 1914 publication. Apparently, his friends felt the 1898 publication would benefit from the increased exposure of being published in a Geological Society of America Bulletin.

In 1920, Udden wrote a paper suggesting that seismic reflection might be a useful tool in oil and gas exploration (Udden, 1920). This is the first time such a geophysical technique was mentioned in the literature, as far as this author knows. Udden's early studies into physics and electricity with Lindahl apparently triggered this idea.

LATE TIES WITH AUGUSTANA

Udden had ties with the Midwest and Augustana College

long after he had arrived in Texas in 1911. Between 1922 and 1931, he wrote at least 20 letters to Dr. Fritioff Fryxell, who was now in charge of the new Geology Department at the school. In 1923, Udden wrote,

I have a number of photos and many more drawn sketches of such structures in the loess of Rock Island. It has been my intention to publish these without trying to give the history of the Rock Island loess, in which the history of three glacial and two interglacial periods are involved, whose general outlines only have been made out.

Udden trusted Fryxell with these observations on the Rock Island loess, but apparently treated other Midwest geologists differently. For in the same letter (Udden, 1923), he asked Fryxell, "Until further I shall ask you not to show these photos to the Illinois geologists. My epistle must be made brief this time."

In 1924, Udden thought so highly of Fryxell as to offer him a job in Texas (Udden, 1924). Fryxell declined, however, deciding to stay on with the small, Midwestern college.

CONCLUSIONS

J. A. Udden's contributions to the science of geology while

he was at Augustana have been relatively unknown or forgotten. Yet his research into wind-blown sediments has been fundamental to our knowledge of eolian deposits and our understanding of loess. His studies of well cuttings opened up a new interest in subsurface geology. Sedimentologists should be grateful that J. A. Udden did not ignore the dust in the wind.

ACKNOWLEDGMENTS

I would like to thank Dr. Fritioff M. Fryxell, Professor Emeritus of Geology, Augustana College, for suggesting this study ten years ago. I would also like to thank Dr. Richard C. Anderson, Chairman, Department of Geology at Augustana College for his relentless support and encouragement throughout this project. Special thanks goes out to the people in the Augustana College Archives for their cooperation. Professor Robert H. Dott, Jr., University of Wisconsin-Madison, is also thanked for his interest and suggestions. Most of all, I wish to thank my wife and children, Chris, Peter, and Amber for their love and patience during the long days that this paper was being written.

REFERENCES CITED

Augustana College Archives, undated, Anonymous personal sketches of Udden's museum by one of his students: Manuscript 17.

Baker, Charles Laurence, 1932, Memorial of Johan August Udden: Geological Society of America Bulletin, v. 44, p. 402–413.

Blatt, H., Middleton, G., and Murray, R., 1972, Origin of sedimentary rocks: Prentice-Hall, Inc., p. 45.

Chamberlin, T. C., 1897, Supplementary hypothesis respecting the origin of the loess of the Mississippi Valley: Journal of Geology, v. 5, p. 795–802.

Ferguson, W. Keene, 1981, History of the Bureau of Economic Geology, 1909–1960: Bureau of Economic Geology, University of Texas at Austin, 329 p.

Fryxell, F. M., 1922, Science at Augustana College: Augustana Book Concern, Rock Island, Illinois.

——1924, The Augustana College museum of natural history: Augustana Bulletin, Series 19, no. 15, p. 16.

Geological Society of America, 1897, Proceedings of ninth summer meeting held at Detroit, Michigan, August 10, 1897: Bulletin, v. 9, p. 12.

Heiman, Monica, 1963. A pioneer geologist: biography of Johan August Udden: Braswell Printing, Kerrville, Texas, Limited edition published by S. M. Udden, 215 p.

Leighton, M. M., 1950, Dedication of Lindahl-Udden memorial: Illinois State Academy of Science Transactions, v. 43, p. 18–23.

Leverett, Frank, 1899, The Illinois glacial lobe: U.S. Geological Survey Monograph 38, p. 187.

Lindahl, Josua, 1886a, Letter to J. A. Udden dated November 13, dealing with concepts in physics: Augustana College Archives, manuscript 17, carton 1, folder 4.

——1886b, Letter to J. A. Udden dated December 21, concerning Udden's research interests: Augustana College Archives, manuscript 17, carton 1, folder 4.

——1888, Letter to J. A. Udden, dated July 11, dealing with Udden's post at Augustana: Augustana College Archives, manuscript 17, carton 1, folder 4.

——1890, Letter to J. A. Udden dated February 28, containing instructions on geology: Augustana College Archives, manuscript 17, carton 1, folder 4.

Udden, J. A., 1891, Megalonyx beds in Kansas: The American Geologist, v. 7, no. 6, p. 340–345.

——1893, A geological section across the northern part of Illinois: Report of the

Figure 8. Udden as he appeared later in life, donning the Order of the North Star presented to him by the King of Sweden in 1911 (Leighton, 1950).

Illinois Board of World's Fair Commissioners, p. 115–158.

——1894, Erosion transportation, and sedimentation performed by the atmosphere: Journal of Geology, v. 2, p. 318–331.

——1897, Loess as a land deposit: Geological Society of America Bulletin, v. 9, p. 6–9.

——1898a, A geological romance: Popular Science Monthly, v. 54, p. 222–229.

——1898b, The mechanical composition of wind deposits: Augustana Library Publications, no. 1, 69 p.

——1900, An old Indian village: Augustana College Library Publication, no. 2, 80 p.

——1907, Letter to Jon dated April 16: Augustana College Archives, manuscript 17, carton 1, folder 3.

——1908, Letter to Dr. Earnest La Rue Jones dated October 23: Augustana College Archives, Manuscript 17, carton 1, folder 17.

——1914, Mechanical composition of clastic sediments: Geological Society of America Bulletin, v. 25, p. 655–743.

——1915, Letter to Anton, dated February 3, concerning Pleistocene deposits in Rock Island: Augustana College Archives, manuscript 17, carton 1, folder 3.

——1920, Suggestions of a new method of making underground observations: AAPG Bull., v. 4, no. 1, p. 83–85.

——1923, Letter to F. M. Fryxell dated January 7, regarding Rock Island loess: Augustana College Archives, manuscript 17, carton 1, folder 6.

——1924, Letter to F. M. Fryxell offering him a job in Texas: Augustana College Archives, manuscript 17, carton 1, folder 6.

——1926, Letter to Jon Udden dated April 24, concerning review of publication: Augustana College Archives, manuscript 17, carton 1, folder 3.

Walther, Johannes, 1894, Einleitung in die geologie als historische Wissenschaft Jena: Fischer Verlag, v. 3, chapter 27.

Wentworth, C. K., 1922, A scale of grade and class terms for clastic sediments: Journal of Geology, v. 30, pp. 377–392.

Wright, Orville, 1917, Letter to J. A. Udden dated February 17, 1917: University of Texas at Austin library.

Printed in U.S.A.

Geological Society of America
Centennial Special Volume 1
1985

Ralph Stockman Tarr: Scientist, writer, teacher

William R. Brice
Geology and Planetary Science
University of Pittsburgh at Johnstown
Johnstown, Pennsylvania 15904

ABSTRACT

R. S. Tarr (1864–1912), a native of Massachusetts and graduate of Harvard University, devoted his very short life to physical geography and glacial geology. Although his first professional work was in the arid regions of the Southwest, he soon returned to the East and began field studies in the Cape Ann area. In 1892 Tarr received an apppointment to Cornell University, where he stayed until his untimely death in March 1912. Once at Cornell, he began investigating the Finger Lake region of central New York. He was especially interested in the origin of the large valleys of Cayuga and Seneca Lakes. Twenty years later Tarr said with conviction that the cause was glacial erosion. He was aided on this quest for the causes of the large Finger Lakes by several journeys to active glacial regions. The first was to Greenland in 1896, and starting in 1905, Tarr made several trips to the Yakutat Bay region of Alaska. His ideas and interpretations were not without controversy, and often he would find himself on opposite sides of an argument from T. C. Chamberlin and W. M. Davis.

In addition to his work as a scientist, Tarr was a master educator and writer. He touched the lives of countless students in his own classes, and, especially, through the numerous textbooks he produced. Tarr was able to write for many audiences: his peers in geology with his many scientific publications, students at all levels of education with his popular textbooks, and an audience beyond the academic world through the more popular science magazines of his day.

INTRODUCTION

Ralph Stockman Tarr, one of the pioneers of glacial geology and physical geography, also had strong interests in teaching and education—education at all levels, elementary and secondary as well as college and university. In addition to his professional writing, Tarr was a frequent contributor to various popular science publications of his day, and he produced textbooks on a wide range of subjects. This article will review these many facets of Tarr's life and his contributions to the science of geology.

Coming as he did from Gloucester, Massachusetts, an area of the country where evidence of glaciation abounds, it is no surprise that Ralph Stockman Tarr (Figure 1) devoted his professional life to the study of ice masses and the effects that they have had on the landscape. He was born in this New England seaport January 15, 1864, and perhaps it was watching the many schooners and other tall ships pass through the harbor of his hometown that gave Tarr the sense of adventure and longing to travel that remained with him until his untimely death at home in

Ithaca, New York. It is both fitting and a little ironic that after facing hazardous ocean voyages and living on glaciers in the wilds of Greenland and Alaska, Tarr died quietly at home on March 12, 1912, after a short illness. He was 48.

In the autumn of 1884, Tarr entered Harvard University where he came under the influence of Professors Nathaniel S. Shaler and William Morris Davis. His classmates included such men as R.A.F. Penrose, Collier Cobb, J. B. Woodworth, George E. Ladd, and Charles L. Whittle, who was his roommate. Later Woodworth described Tarr as:

. . . . perhaps the most brilliant and versatile member of a group of students drawn to the department of geology by the unique personality of Shaler (Woodworth, 1912, p. 30).

Tarr was firmly committed to geology now, and except for a disappointing sojourn at a western ranch in 1887, he stayed at Harvard until he graduated with the class of 1891.

Tarr's life-long work in glacial geology received its formal

Figure 1. Ralph Stockman Tarr at Disenchantment Bay, Alaska. This is one of the few photographs of Tarr without a beard (courtesy of Dr. Arthur L. Bloom, Cornell University).

beginning under his teacher and mentor N. S. Shaler, for his first task as Shaler's assistant was to return to the site of his boyhood to map and describe the geology of Cape Ann Island. Although this project was published in the Ninth Annual Report of the U.S.G.S. under Shaler's name, it was mostly Tarr's work, as indicated in the letter of transmittal where Shaler states, ". . . by far the larger part of the field observations embodied in this memoir have been made by my assistant, Mr. Ralph S. Tarr" (Shaler, 1889, p. 537). Many years passed before Tarr came back to his Cape Ann work under his own name (see Tarr, 1904a).

The year after his graduation from Harvard, Tarr published his first paper in glacial geology describing a large morainal band in central Massachusetts which extends almost continually from Cape Ann to the Connecticut River (Tarr, 1892f). This initial work by Tarr was warmly received by his contemporaries. After reading it, I. C. Russell was so impressed that he sent Tarr copies of his own papers about Mt. St. Elias, a set of papers which Tarr felt were ". . . the most remarkable glacial publications of the decade" (Woodworth, 1912, p. 32; and Tarr, 1891c). Thus, when Tarr received an appointment in the spring of 1892 to the faculty

Figure 2. Cornell Greenland field party, 1896. (L-R) J. O. Martin, T. L. Watson, Professor A. C. Gill, E. M. Kindle, Professor R. S. Tarr, and J. O. Bonsteel (Photo from Department of Geological Sciences, Cornell University.)

of Cornell University, he was already off to a good start in his studies of glacial landscapes.

TARR AS A SCIENTIST

R. S. Tarr's contribution as a scientist can be divided into five sections, three of which are geographically defined: (1) his work in western Greenland, (2) the many trips to and descriptions of Alaskan glaciers, (3) the work he did in the Finger Lakes area of New York, (4) general geomorphology, and (5) his experimental investigations on the behavior of ice, a study which indirectly caused his demise. At times his views brought him in conflict with such people as William Morris Davis and Thomas C. Chamberlin. No one can accuse Tarr of being afraid to put forth views which he knew would be controversial.

Greenland

In 1896 Tarr organized an expedition to the west coast of Greenland where he had his first chance to see glaciers in action (Tarr, 1896g). His was one (Figure 2) of three scientific parties

that were part of Lt. Peary's second attempt to recover a large meteorite and to continue the quest to reach the North Pole. In addition to Tarr and his group, George Barton and a small party landed on Disco Island (Figure 3), and Professor Alfred Burton of MIT went to Umanak Fjord (Anonymous, 1896, p. 335). Information gained on this trip and on Tarr's subsequent expeditions to Alaska played an important role in shaping his thinking about what he saw around him in the Finger Lake region of New York.

The expedition travelled on the steamer *Hope,* and on August 7, 1896, Tarr's small band of hardy geologists was dropped by Lt. Peary on the Upper Nugsuak Peninsula (74° N Lat.), some 100 miles north of the last Eskimo village, where they stayed for 31 days. One of the first duties they undertook, like all good explorers, was to name the heretofore unnamed landmarks. Tarr and his party are believed to be the first non-Eskimos to explore this region in detail. On the southeastern base of the peninsula, they found an extension of the massive continental ice sheet, which they named after their alma mater (Figure 4), and on the northern side was another glacier which became the Wyckoff Glacier in honor E. G. Wyckoff of Ithaca, New York, who

Figure 3. S. S. *Hope* in Disco Island harbor, 1896. Photograph made from the original 8 × 10" glass negative. (Photo from Department of Geological Sciences, Cornell University.)

Figure 4. North Cornell Glacier and Mt. Hope. Photograph made from the original 8 × 10" glass negative. (Photo from Department of Geological Sciences, Cornell University.)

helped finance the project. One large bedrock mass protruding through the ice, a nunatak, was named Mt. Schurman for Cornell President J. G. Schurman.

Photography was a vital part of Tarr's scientific work for it served him, as it does all of us today, as a permanent record of what he observed. In all of his writings, Tarr included as many photographic illustrations as he could, and the expedition photog-

rapher, J. O. Martin, supplied him with an ample number. A 1907–08 department inventory lists over 400 photographs from this Greenland trip, many of which still exist. In addition to the many original photographs, a rock and mineral collection made on this 1896 expedition is still in the department at Cornell. As mentioned earlier, the steamer stopped at Disco Island, where the group collected several samples of native iron which formed part

Figure 5. Pack ice in bay. R. S. Tarr is believed to be the person wearing the hat and standing in the front of the boat. (Photo from Department of Geological Sciences, Cornell University.)

Figure 6. S. S. *Hope* icebound in the mouth of Cumberland Sound in September 1896. (Photo from Department of Geological Sciences, Cornell University.)

of a mineralogical study made at Cornell in 1977, over 80 years later (Bird and Weathers, 1977).

Tarr was very prolific with reports about the Greenland expedition, and soon he became involved in a controversy with Thomas C. Chamberlin and Rollin Salisbury over the degree of glaciation in Greenland during the last ice age. Even before Tarr's first paper was published, shortly after Tarr presented his findings at the GSA meeting December 30, 1896, Chamberlin (1897) responded in an editorial in *The Journal of Geology*. He claimed that there was no real disagreement between them about the former extension of the ice, simply that his own ideas had not yet been published. However, Tarr (1897e), true to his nature of not speculating much beyond his actual data or observations (with one large exception which will be discussed later), quickly pointed out that he was not making any sweeping generalizations about the entire Greenland coast, but that his conclusions were only for his specific study area, the Nugsuak Peninsula:

For the larger question, how far it [the ice-sheet] extended, and how much coast it covered, I believe it is well to wait until further evidence is at hand (Tarr, 1897a, p. 344).

Thus, Tarr was convinced that if he presented evidence that rugged and serrated peaks on the Upper Nugsuak Peninsula had been overridden by ice, it was not wise to use angular topography, as Chamberlin and Salisbury had done, to state that ice did not reach half the Greenland coast. At least in his mind, Tarr felt that his evidence cast some doubt on their conclusions.

This stand was not taken by Tarr just to support his own work over Chamberlin's, because earlier, Tarr had shown that even when he had a distinct preconception and built-in prejudice against an idea, if confronted with facts and field evidence supporting the new idea, he would reverse his position (see Tarr, 1894c, p. 339). Before working in the Finger Lakes region of New York, Tarr had seen glacial erosion only in New England,

and based on his studies there, he had been convinced that there was no connection between rock basins and glacial action. When confronted with the overwhelming field evidence around Cayuga and Seneca Lakes, however, Tarr reversed his position. His apparent obstinacy with regard to the work of Chamberlin and Salisbury, therefore, flowed from his belief in fact and detailed observation rather than in limited observation and speculation.

While exchanging letters and barbs with Chamberlin, Tarr was busy communicating other details from the observations made on the expedition through numerous other papers: valley glaciers of the area around Disco Island and Nugsuak (1897g); the ability of the arctic sea ice to do geologic work (1897c) (Figure 5); evidence of glacial action in Labrador and Baffin Land (1897b); comments on the differences of climates on either side of Davis' and Baffins' Bay (1897h). This last paper tells of the ship *Hope* being icebound for 60 hours at the mouth of Cumberland Sound on 12 September 1896 (Figure 6). At the conclusion of this paper, Tarr asked a series of interesting rhetorical questions:

. . . is there any relation between the down-sinking of Greenland and the uprising of Labrador and Baffin Land? . . . is the load of ice really the cause for the sinking which allows for its [the ice's] withdrawal? . . . but as the answer is not definitely at hand they may well be left as mere queries (Tarr, 1897h, p. 320).

Tarr did not find the large moraines which he had seen in the United States. The moraines he found at the Cornell Glacier looked to him more like eskers. He felt that this was the result of a rapidly retreating ice-front and there was simply no time for the large piles of debris to accumulate (Tarr, 1897-1, p. 149).

Many years passed before Tarr was again able to study the effects of active glaciers. For the next nine years, he confined his investigations to areas that had been ice covered, but were not any more. The short time he spent in Greenland certainly gave

Figure 7. Professor Ralph S. Tarr on the Valdez Glacier, Alaska, 1909. (Photo from Department of Geological Sciences, Cornell University.)

Figure 8. Field party enroute to Alaska, 1906. (L-R) B. S. Butler, J. L. Rich, Professor R. S. Tarr, R. R. Powers, and O. D. von Engeln. (Photo from Department of Geological Sciences, Cornell University.)

him a feeling for the effect of large ice masses on a landscape, experience that proved invaluable as he undertook further study of the Finger Lakes region of New York. His next venture into the far north came in 1905 when he began a study of the Alaskan glaciers.

Alaska

Tarr's work in Alaska began in the summer of 1905 when he led a party for the U.S.G.S. to the region of Yakutat Bay (Figure 7). On this expedition, also partly funded by the American Geographical Society, Tarr had two assistants, Bert S. Butler, who later worked with the U.S.G.S. and was the 1947 Penrose Medalist of the Geological Society of America, and Tarr's brother-in-law, Lawrence Martin, later with the University of Wisconsin. This one field season convinced Tarr that if he wanted to study glaciers in action, Alaska was the best place to do it and Yakutat Bay offered access to several active glaciers. Also, this region had been previously described by several observers going back to 1890 when I. C. Russell visited the area (Russell, 1892).

Immediately upon his return from the first Alaska trip, Tarr published the first of a series of scientific reports on the glaciers in the Yakutat Bay region. Tarr found that some glaciers had retreated from the positions Russell observed in 1891; yet the Turner and Hubbard glaciers had advanced. Just as in Greenland, Tarr discovered evidence of a former ice advance, extending out far enough to cut off Russell Fjord and pond up its waters (Tarr and Martin, 1906b, p. 161). But the discoveries which intrigued Tarr the most were the numerous examples of sudden changes in the relative positions of the sea and the land. Almost everywhere beaches were being either drowned or raised well above present sea level. One shoreline showed evidence of at least 47 feet of uplift (Tarr and Martin, 1906a, p. 64). The most probable cause for these changes in level was the repeated series of earthquake shocks which rocked the region from September 3 to 20, 1899

(Tarr and Martin, 1906a, p. 30-31; 1906c, p. 39). Tarr was able to relate these changes in level with the 1899 earthquake because G. K. Gilbert had visited Yakutat Bay just three months before it occurred; and Gilbert, Tarr felt, was too keen an observer to have missed such obvious features as uplifted shorelines and drowned forests. Yet, no such description occurs in Gilbert's reports (Tarr and Martin, 1906a, p. 44–45).

The next year, Tarr and a small band of students (Figure 8), one of whom paid all of his own expenses, went back to Alaska. Only a few days after the 1906 camp had been established (Figure 9), Tarr noticed that several glaciers had undergone dramatic advances since his visit just one year earlier; in some places the glaciers had advanced as much as two miles in that time. The ice advance was so great at several locations that the party was unable to repeat traverses, ". . . a sea of crevasses, across which travel was utterly impossible" (Tarr, 1907, p. 258). The wave of the advance moved through the glaciers so rapidly that instead of the ice just moving downslope it was badly broken and crevassed. Yet, other glaciers in the area showed no noticeable change in position. These phenomena puzzled Tarr, but he suspected they were somehow related to the same earthquake that had caused the changes in elevation he had seen the previous year. In keeping with his systematic approach to science, Tarr explored several other possibilities, but after careful analysis he returned to the earthquake of 1899 as the best probable cause. The violent shaking of the source area had dramatically added material to the upper levels of the glaciers. The surge generated by this sudden addition of material was transmitted downslope to the foot of the glacier where it caused the cracking and rapid advance (Tarr, 1907). In spite of his analysis, Tarr was not entirely certain of his conclusion, especially with the ". . . briefness of time for the transmission of the great wave" (Tarr, 1907, p. 281–2). The results of these first two trips to Alaska were presented in the U.S.G.S. Professional Paper #64 (Tarr, 1909).

If Tarr's conclusion about the earthquake shaking was valid,

Figure 9. Field camp near Black Glacier, Alaska, 1906. Professor R. S. Tarr is standing just to the right of the flag; next to him is J. L. Rich. Seated fifth from the left is O. D. von Engeln, and seated at the far right is R. R. Powers. The others are unidentified. (Photo from the Department of Geological Sciences, Cornell University.)

then some of the glaciers that had not moved by 1906 could be expected to exhibit the same surge in the very near future. It wasn't until the summer of 1909 that Tarr had the opportunity to test his hypothesis when he returned to Yakutat Bay for the third time, again in the company of his brother-in-law Lawrence Martin. This was the first of several expeditions to the Alaskan glaciers sponsored by the National Geographic Society and conducted by Tarr and Martin. As Tarr had predicted, several of the formerly stagnant glaciers had surged dramatically. The 1909 observations indicated that the surges themselves were short-lived. Sections that were impassable in 1906 were passable in 1909 owing to the healing of the large crevasses (Tarr and Martin, 1910). The fact that the surges did not seem to continue for very long strengthened Tarr's argument that uplift and tilting of the land were not responsible; for if tilting were responsible, then the surges should be part of an overall increase in glacier flow, but that did not appear to be happening.

Some situations, however, did not seem to follow the predicted behavior. The 1909 trip included a short exploration in the Copper River Valley which had experienced a series of earthquakes in October of 1900, just over a year after the Yakutat Bay earthquakes, but the glaciers of the Copper River Valley did not show signs of surging. In fact, a railway had been built between two glaciers and on a third, and the tracks had not moved. Even more puzzling, although these earthquakes were less severe than the ones at Yakutat Bay, avalanches of snow and rocks occurred in the glacial source areas without any surging. Tarr's printed comment was, "These facts may complicate the situation somewhat" (Tarr and Martin, 1910, p. 38); but beyond that he offered no explanation other than to imply that the differences could be explained by the severity of the respective earthquakes.

Tarr felt confident enough about his earthshaking idea to publish a lengthy discourse after the 1909 trip (Tarr, 1910). In this paper, he outlined the observations that had been made in Alaska and put forth various hypotheses that might explain the observations. In keeping with his normal thoroughness, Tarr went to great lengths to show the weakness of each argument. He then listed the support for what he termed "The Glacial Flood Hypothesis":

Figure 10. Camp on Strawberry Island at the Kwik River, Alaska, 1906. Seated (L-R) Benno Alexander (one of the packers hired in Alaska), B. S. Butler (?), R. S. Tarr, R. R. Powers, J. L. Rich (?), J. H. Thompson (?) (another packer hired in Alaska). Photograph is believed to have been taken by O. D. von Engeln, who was the official photographer for the party. (Photo from the Department of Geological Sciences, Cornell University.)

Figure 11. Transportation while at Yakutat Bay, Alaska, 1906. Tarr (with hat) is standing in front of the boat. The others are not identified. (Photo from the Department of Geological Sciences, Cornell University.)

(1) There were a series of unusually vigorous earthquakes in September 1899. (2) Excessive avalanching is known to have occurred in this region during the 1899 earthquakes. (3) Both the earthquakes and the phenomena of spasmodically advancing glaciers are centered in the Yakutat Bay region. (4) Of the seven advancing glaciers, six are known to have begun their advance since 1899, and in the other case, that of Galiano Glacier, the evidence is all but conclusive that the advance was after 1899. (5) There has been a rough progression in the glacier advance from the smaller to the larger glaciers, beginning almost at once with the smallest, then affecting the larger ones, then still larger, while as yet the largest have not responded. (6) The advance was spasmodic, indicating a spasmodic cause. (7) The duration of the advance was brief, indicating a cause of short duration. (8) The cessation of the advance was abrupt, indicating that the cause was terminated abruptly. (9) The extent of the advance was great, involving the rapid transfer of large masses of ice in thickening, spreading, and pushing the glacier fronts forward, proving that the cause involved an intensive addition to the glacier supply. (10) The transformation is without recorded parallel, both in character and extent, which indicates the necessity of seeking an unusual cause (Tarr, 1910, p. 19).

This theory was one of Tarr's major contributions to glaciology, and is one that has drawn much attention over the years. Unfortunately for Tarr's theory, the glaciers around Anchorage did not show any surging phenomena after the 1964 earthquake (see Shreve, 1966; and Walker et al., 1982). Perhaps conditions were not right for the surge to occur, but the fact that nothing happened after such a major earthquake does not speak well for Tarr's idea.

Martin returned to Alaska in 1910 while Tarr attended the International Geological Congress at Stockholm. The Stockholm visit gave him an opportunity to travel with a field party to Spitzbergen and compare glacial action there with what he had seen in Greenland and Alaska. Both men were back in Alaska in

1911; but this was to be Tarr's final trip, for his death occurred early in 1912. Several more reports were produced from their work at Yakutat Bay, some of them published posthumously (see Tarr and Martin, 1913). Two of the most extensive were the U.S.G.S. Professional Paper 69 (Tarr and Martin, 1912) and the National Geographic book on the glaciers of Alaska (Tarr and Martin, 1914). The professional paper had a preface by G. K. Gilbert, who had this to say about Tarr:

His biography when written will be a record of distinguished achievement in physical geography. The present volume [professional paper 69] testifies to his high rank as an investigator (Tarr and Martin, 1912, p. 10).

Finger Lakes

Tarr began his investigations in the Finger Lakes region of New York upon his appointment to Cornell University in 1892 (Tarr, 1896a), and they continued on and off between the various trips to Greenland, Alaska, and other parts of the Northern Hemisphere. His first real work in geology was mapping glacial deposits around Cape Ann, Massachusetts, while he was an assistant to N. S. Shaler, so Tarr was no stranger to glaciated topography, but his curiosity was greatly aroused by the huge north-south valleys now occupied by Cayuga and Seneca Lakes.

At first, Tarr did not believe the valleys were modified by ice at all; he felt they were just large river valleys. As he was able to spend more time mapping the glacial deposits of the region and looking at the tributary valleys, his interpretation underwent a complete change. In 1892, the same year that Tarr went to Cornell, D. F. Lincoln published an article suggesting a glacial origin for the large lake valleys, but from what Tarr said later, his thinking was influenced more by his own field work than by Lincoln's paper (see the footnote in Tarr, 1894c, p. 348; 1894b;

1894f). In an early article about the Finger Lakes, "Lake Cayuga A Rock Basin," Tarr outlined the current theories on their origin and carefully pointed out the fallacies in each, based, of course, on his own field investigations. He stated:

It is not surprising that a region so peculiar as the Finger Lake district should have attracted widespread attention, but it seems extremely strange that until within a year or two no one should have seen the very plain evidence of the origin of these valleys (Tarr, 1894c, p. 342–343).

He then went on to describe the field evidence which led him to agree with Lincoln and others who felt that the valleys were, indeed, greatly modified by glacial erosion; giving such evidence as: (1) buried valleys on the cliff sides, and (2) Salmon Creek and Six Mile Creek both "hanging" some 200 or 300 feet above the present lake bottom. At the time no data existed to support Tarr's estimate of at least 200 or 300 feet of deltaic fill at the south end of the valley under the city of Ithaca. Tarr's powers of intuition were vindicated when a few years later a well drilled on the delta struck bedrock at 430 feet (Tarr, 1904b, p. 273). All of this caused Tarr to suggest that Lake Cayuga occupied a large basin carved into the bedrock by glacial erosion of a much shallower and higher (in elevation) pre-existing river valley. Even though Tarr's paper added strength to the glacial origin argument, not everyone was convinced; Culber (1895, p. 366), for example, stated, ". . . not a single case of a lake basin which can be proven to have been made by ice action has been discovered."

Surprisingly, over the next 10 years, Tarr's position gradually changed with regard to the glacial origin of these valleys. Tarr seemed to retreat from his former positive stand to a much more cautious one. What caused this dramatic switch is not easy to determine, but it seems to have been related to certain facts brought to light by his continued field work around the two large lakes, Cayuga and Seneca. In keeping with Tarr's scientific character, he was not going to make these new facts public until he was sure of a solution; he wrote, "It had been my intention not to publish these facts until it was possible to definitely settle the questions raised [about the glacial origin of the valleys]" (Tarr, 1904b, p. 271). A criticism by Fairchild (1904) of a paper by a former student of Tarr's, Frank Carney, prompted Tarr once again to take up the issue of the origin of the Finger Lake valleys.

In his paper, "Hanging Valleys in the Finger Lake Region of Central New York," Tarr produced a series of cross sections illustrating the relationships between the smaller east-west streams and the main north-south valleys based as much as possible on actual well drilling data. His profiles more than support the designation "hanging valleys." Also, Tarr noted an abrupt change in valley-wall slope at about the 800 or 900-foot contour. Below this point, bedrock is found in outcrop; but above, little bedrock is encountered. All of this forced Tarr to conclude that in the Cayuga valley the glacier must have eroded a minimum of 850 feet and in the Seneca valley from 1400 to 1500 feet, amounts which were much higher than previous estimates. The presence of buried gorges indicated more than one ice advance, and therefore

the erosion need not have been caused only by the last advance (Tarr, 1904b).

The above facts notwithstanding, Tarr looked critically at observations which did not support the glacial erosion theory:

(1) The fact that, while most of the tributaries are in hanging valleys, there is discordance, especially well marked in two cases, both of which, however, may possibly be explained in harmony with the glacial erosion theory; (2) the supposed ice-eroded valleys have the cross section of gorges [formed by river cutting]; (3) the presence of many angular, precipitous cliffs in the zone of supposed ice erosion below the hanging valley levels; (4) the presence, in two places, of evidences of preglacial decay, in situations where such vigorous ice erosion as is postulated should have removed them; (5) the presence of a cavern in a similar position; (6) the existence of an island in Lake Cayuga in the line of glacial erosion; (7) the fact that the ice erosion could have been accomplished only during those periods of time when the ice was thin enough to be deflected from its main course by the valleys which lie transverse to the main ice movement (Tarr, 1904b, p. 288–289).

Given this evidence, Tarr was forced to fall back on the rejuvenation hypothesis. Even though this hypothesis "long seemed improbable," it did allow the deep valleys to be produced by running water, with only very moderate changes by the glaciers.

The statement of my conviction that the glacial erosion theory cannot be accepted as proved, seems especially important in view of the fact that in my earlier paper, overlooking some of the evidence, I published the conclusion that the Lake Cayuga valley owes its great depth to ice erosion. Until the facts opposing glacial erosion are explained, or until the possibility of the rejuvenation theory is eliminated, the current theory for the origin of these valleys by glacial erosion, recently revived, cannot be considered established (Tarr, 1904b, p. 291).

Considering how strongly he had worded the 1894 paper, it must have been hard for Tarr to admit that he might have been premature in his very positive endorsement of the idea of a glacial origin for the valleys, especially as he was normally so cautious with his conclusions.

Tarr's reversal on the glacial erosion theory had an influence on other geologists, for in that same year Dryer (1904, p. 459–60) wrote:

. . . [the] peculiar features by which these [the Finger Lake] valleys differ from normal stream valleys are due to the work of glacial ice, and that during their occupancy by ice-fingers they suffered deepening which is conservatively estimated at 400 feet. . . .In the absence of opposing evidence, this conclusion seems almost too obvious to be debated. But since this paper was read Tarr has brought forward evidence discovered by him in the Cayuga and Seneca Valleys which opposes very grave objections to the theory of glacial erosion as generally applicable to account for the Finger Lake basins. In view of all the facts, the question must remain unsettled.

The following year, Tarr continued to cast doubt on the glacial erosion theory with his paper illustrating some examples of moderate glacial erosion in the Finger Lakes area, and he suggested that the erosion of the Wisconsin Ice Sheet above the present lake level was very slight. This was not the first time Tarr had seen evidence of ice doing very little erosion. While in Green-land he had seen grooves cut into well-decayed rock, yet the glacier had not removed the soft, weathered material (Tarr, 1897d). Even earlier, while working in the Cape Ann region, Tarr had seen examples of decayed granites that were not re-moved by the glacier overriding them. Describing what he found in the Cayuga valley, he again expressed his doubts about glacial erosion in these large valleys:

. . . .but if this evidence [moderate glacial erosion] eliminates erosion in the visible part of the valley by the only known ice-advance in this region, as it seems to do, it throws doubt upon the whole hypothesis of ice erosion for this valley (Tarr, 1905a, p. 163).

The part of this quote about the only known ice advance may seem slightly puzzling in light of what Tarr said earlier where he implied more than one: "during the first advance" and "during as many glaciations as this region experienced" (Tarr, 1904b, p. 284). In other papers as well, he spoke of more than one ice advance (see Tarr, 1905c, 1906a–b). But this is the cautious Tarr speaking, for his field investigations did not reveal direct evidence of more than one ice advance in central New York (Tarr, 1905d, p. 217). Therefore, he was careful to use the term "known advance."

While it took Tarr about 10 years gradually to reverse his position from strong support to a more cautious and skeptical attitude as to how much influence the ice actually had on the Finger Lake valley formation, it took him only a few short months to move back to a position of strong support. In the January-February 1906 issue of the *Journal of Geology,* Tarr began to change his position back to a glacial explanation:

In a paper presenting the newly discovered facts [the hanging valley paper (Tarr, 1904b)] the general question of the origin of the Finger Lake valleys was reconsidered and three hypotheses were proposed as working hypotheses. No attempt was made to establish either of the hypotheses, but the object of the paper, avowedly an unfinished study, was to show that the problem was less simple than formerly believed, and that there were objections to the glacial-erosion theory of sufficient force to warrant a question whether some other theory than glacial erosion might not account for these valleys. The fact that I have been quoted as an opponent of the glacial-erosion theory for these valleys, which has not been the case (Tarr, 1906a, p. 18).

He went on to say that the discordance seen in Six Mile and Salmon Creeks did not necessarily oppose the glacial theory but rather confirmed it because new profile data indicated that both these valleys were occupied by active ice and were therefore scoured, which explains the discordance in elevations. In light of these new data, Tarr changed his mind. He nicely side-stepped any other problem by saying that even if new evidence opposes erosion by the Wisconsin advance, it does not rule out erosion by an earlier ice sheet, even though he could find no deposits left by a previous glaciation. He closed this paper by completely demol-ishing the rejuvenation hypothesis which he had previously championed, and by supporting the idea of more than one major ice advance:

Formation of the lake valleys entirely by Wisconsin ice-erosion is also out of the question. But the striking resemblance of the topography to that of regions of known ice-erosion seems to demand origin by glacial erosion; and in spite of the opposing evidence from a remnant of residual decay, and of other objections of less importance, it is believed that such an origin must be assigned to these valleys, and that the apparent objec-tion to such origin is dependent upon some local condition which permit-ted residually decayed rock to remain, while elsewhere in the valleys there was profound ice-erosion (Tarr, 1906a, p. 21).

Furthermore, in the May 1906 issue of *The Popular Science Monthly,* Tarr came down squarely on the side of glacial erosion, for he wrote (1906b, p. 391):

. . . glacial erosion will explain the conditions in the Finger Lake valleys, and no other theory so far proposed will do so. Moreover, these valleys were a highway for glacial motion, as is proved by the presence of pronounced moraines along their sides and at their heads.

In this latter paper, Tarr still advocated more than just the Wis-consin advance, ". . . for the present we can point with certainty to no greater complexity than that of two periods" (p. 393).

The culmination of his work in the Finger Lakes, which began when he came to Cornell in 1892, was the publication of the U.S.G.S. Watkins Glen-Catatonk Folio Atlas in 1909 for which Tarr prepared the glacial geology (Williams et al., 1909). Even at this late date, Tarr still seemed somewhat unconvinced when he wrote of the glacial origin for the large valleys:

. . . that the ice moved through some of the larger valleys with great power and effect is probable; but even in these valleys the products of residual decay are found, and the topography of the southern half of the quadrangles does not suggest marked ice erosion (Williams et al., 1909, p. 115).

In the summary of the glacial section of the folio, Tarr very nicely skirted the issue by falling back on glacial erosion as the only possible cause. But he advocated multiple ice advances and sug-gested that most of the erosion had been caused by the earlier ones.

That Tarr was finally convinced of the powers of glaciers to modify greatly a landscape can be seen in a draft of an address he planned to give before the American Association of Geographers written only a few months before his death. Tarr (1911, p. 1148–50), wrote:

And glacial erosion takes rank among the most efficient agents of denudation operating in recent times. . . .If glaciers can erode, as seems established, the forms which they would produce are those which have been described in this address. Finding these forms practically coexten-sive with regions of former glaciation a natural explanation is that they are the result of glacial erosion. No other rational explanation has been proposed, and therefore glacial erosion is accepted.

The actual address, entitled "Glaciers and Glaciation of Alaska" (Tarr, 1912), was quite different from what he wrote in his first draft quoted above. Why he changed topics is unknown, but I suspect he felt the glacial erosion concept was well established by 1911 and needed no further words on the subject from him. He therefore switched attention to his work in Alaska.

Thus, Tarr's major contribution to the geology of the Finger Lakes was his mapping of the surficial deposits, descriptions of the large valleys, and eventually, his support for the glacial origin of the valleys, although his own position vacillated. When confronted by what he felt was conclusive field evidence and with the data from his first experimental work on ice flow, Tarr was not afraid to use a platform such as his presidential address to state this conclusion. Unfortunately he died only a few months after he gave the address, so it was left to others finally to convince the last of the doubters. Had he lived but a few more years, their work would have been easier.

Geomorphology

In some respects, all of the preceding sections belong under the general heading of geomorphology, but as they dealt with specific geographic locations, they were handled separately. At the very beginning of Tarr's career he was involved with desert rather than glacial landscapes. From his early work in the Southwest came a series of papers dealing with arid landscapes and drainage patterns in parts of Texas and New Mexico (Tarr, 1890a–d; 1891a–b). But soon he became involved with the glacial deposits of the Northeast and thereafter wrote very little about deserts except in his textbooks.

While he was investigating specific geographic areas, such as the Finger Lakes, Greenland, or Alaska, Tarr ventured into the more general topics of glacial geology. In 1894, he published a small paper on the origin of drumlins (1894e) in which he was in opposition to such colleagues as Geikie in Great Britain, and Davis, Chamberlin, and Salisbury in the United States. Perhaps this early opposition to Chamberlin and Salisbury was a harbinger of the verbal battles that they were to wage later over Tarr's statements after the Greenland trip in 1896. The prevailing theory for the origin of drumlins considered these elongated hills to be of a constructional rather than destructional origin, and Tarr proceeded to offer evidence which more strongly suggested the latter, with a partial working over of morainic or other drift deposits. He wrote:

It is not the author's [Tarr's] intention to insist that this theory is a true explanation, much less that it is the only explanation; but, rather, to urge that it be not too easily put to one side. . . .if this theory does furnish the real explanation, the proof of it will be more easily and more quickly discovered if investigators bear in mind that it is by no means a disproved theory, as some students of glacial geology seem to assume (Tarr, 1894e, p. 405).

This is another example of a major trait of Tarr's where he put more faith on his own observations than on the interpretation of others regardless of who they were.

Tarr was not averse to expressing an opinion on almost any

topic, ranging from glaciers to the origin of coral atolls (Tarr, 1896f) and changing sea level around Bermuda (Tarr, 1897i–j). The work about Bermuda was written while he was on a "vacation to recruit and to see something of midoceanic coral life" (Woodworth, 1912, p. 35). It was his paper on the peneplain (1898b) that, perhaps, drew the most fire because he was in opposition to his old Harvard professor, William Morris Davis. Privately, even before he went to Greenland, Tarr had begun to question the evidence of peneplains. He even sent Davis a manuscript outlining his concerns and objections to the whole concept, but it did not convince Davis (Tarr, 1898b). Not to be cowed by the "voice of authority," Tarr simply looked more deeply into the problem, and the more he investigated it, the more convinced he became that the peneplain concept was in error:

Therefore, notwithstanding the fact that nearly all American geologists have adopted the peneplain explanation, and that no one has publically questioned it, I have decided to at last state my objections in print (Tarr, 1898b, p. 351).

As indicated in a footnote on page 351, the preprint reviewers for this manuscript did not include W. M. Davis.

As indicated by what he wrote to his friend and confidant, J. B. Woodworth, on February 12, 1898, the work on the peneplain was the result of a long and careful study:

Lack of confidence is not one of my failings, but there is a matter, these five years on my mind, upon which I sometimes question my own powers. Do you know that I am a disbeliever in peneplains? Several years ago I wrote an article opposing the theory, but took alarm when I saw every one else believing in them. I pigeonholed the paper. I rewrote it. Again I put it away. Now I have it out again, rewritten. I believe it is right, yet I can not help doubting it, for I seem to have no companions (Woodworth, 1912, p. 35).

His doubts about the peneplain concept seem to stem from his failure to find the evidence of the peneplains he had expected to see in New England and parts of New Jersey.

His old professor not withstanding, Tarr finally had the courage of his convictions and the paper was published in June 1898. He had become alarmed by the frequent announcement of new peneplains with very:

. . . meager evidence, and oftentimes with no statement of evidence whatever. Frequently a new peneplain is mentioned as one might state the discovery of a delta or fossil vertebrate; and not only are single peneplains found in a given district, but oftentimes several of different ages (Tarr, 1898b, p. 351–352).

This statement reflects a situation that sometimes exists when a new concept bursts on the scene. As the bandwagon begins to roll, the new idea is applied in a sweeping manner often without careful study. This paper was Tarr's attempt to start discussion on the peneplain concept, for the situation had reached the point where even the originator of the concept, William Morris Davis, found it necessary to counsel for more careful study of some proposed peneplains.

Tarr did not question the importance of subaerial denudation in shaping the landscapes, but rather, he wondered if too much importance was placed upon it. His own explanation of the topographic features was what he called a "beveling down to mature form" in which the region is just eroded normally with the peaks of approximately the same original height and of similar rock type being reduced at about the same rate, thus, maintaining a more or less common elevation. Even peaks with rocks of less resistance will reach a point where tree cover and soil cover will reduce the denudation rate and allow a more resistant neighbor to get to the same level. In this manner there would be a "beveling of the hill tops," and the highest area of beveling will be in the tree zone where the rocks are protected by soil and tree cover. Above this zone, denudation proceeds relatively rapidly, while below it the denudation is slower. Thus, hills of similar rock type, but different original elevations, will eventually reach approximately the same height. The same attenuation of the denudation rate will allow hills of different rock type to reach a common elevation. It is not necessary to start with a level or almost level plain in the past to develop the concordant hill tops observed in the modern landscape. No long periods of stability are required, and no special cases are evoked to explain the surface irregularities (Tarr, 1898b).

It is interesting to note at this point that although Tarr was not a "true believer" in the peneplain idea, he and O. D. von Engeln, his assistant in Alaska and later colleague at Cornell, have included the following set of questions in one of their physical geography laboratory manuals, published about 10 years later. Students were asked to select 10 different hills on the Centerpoint, West Virginia, quadrangle and then to answer the following:

How do they [the selected hills] compare in height?
What was the original topography of the region as suggested by the uniform height of the hills? (Tarr and von Engeln, 1910, p. 139–140).

Perhaps peneplains could exist in the laboratory exercise, but not in the field, or perhaps Tarr underwent a change of heart during the intervening years as he did on the issue of the glacial erosion of Cayuga Lake valley.

When Tarr's peneplain paper was published, William Morris Davis was travelling in Europe, so there was some delay in his response, at least in the publication of it. In responding to a professional article, most scholars, even Tarr, would reply with a letter or at most a few pages; but not Davis, his response to Tarr was a paper of 37 printed pages!

Davis began by welcoming the discussion which Tarr's paper should generate, for "as he [Tarr] has well said," the peneplain idea is too important to gain acceptance without close scrutiny. Davis took pains to point out that the idea did not originate with him, only the name. In fact, he stated that the idea came from Powell's "Exploration of the Colorado River." Davis then discussed each of Tarr's points in turn, referring to Tarr's subheadings and page numbers lest the reader get lost. For each

objection Tarr raised, Davis countered with four or five positive examples in rebuttal. Then at the end, just to indicate that he had in no way exhausted his supply of counterexamples, Davis concluded:

The completion of the essay in southeastern France has made it impossible to cite, as fully as I should like, certain pertinent examples, to which, however, there may be opportunity of returning if fuller consideration of the peneplain idea is called for by a continuation of this discussion (Davis, 1899, p. 239).

After reading the polite, but nonetheless pointed response from Davis, Tarr wisely decided not to continue the discussion. In this issue he was plainly out of his area of expertise. In his professional publications, however, Tarr stuck by his opposition to the peneplain concept. Even as late as 1910, in his discussion of the topography of the Watkins Glen-Catatonk Quadrangle, he dismissed it with the note ". . . that the application . . . of the phrase 'almost a plain' gives an entirely erroneous idea" (Williams et al., 1909, p. 215). Perhaps in attacking the peneplain concept, Tarr was just ahead of his time.

This rather strong difference of opinion between Tarr and Davis in no way seemed to affect the professional relationship between the student and his former professor. They still attended many of the same meetings and held each other in mutual respect.

In the field of general geomorphology, Tarr often found himself in opposition to many of his colleagues, especially some of the more influential ones. But he was not intimidated. As a result, his contribution was not so much the alternative ideas he presented, but the fact that he forced discussion on several important topics; i.e., Tarr played the role of devil's advocate.

Ice Experimentation

This final aspect of Tarr's scientific contributions is the least complete and the smallest because just as it was beginning to produce results, he died. He first began to experiment with ice movement when his student, E. C. Case, carried out a series of experiments using paraffin dissolved in refined petroleum to simulate ice (Case, 1895). Tarr found these results to be of little value to him when he tried to explain the sudden glacier surges he had observed in Alaska. He realized that there was a lack of knowledge about the physical properties of ice, especially ice under pressure. To remedy this, Tarr had a large slab of ice from a glacier in British Columbia moved to Ithaca by the Canadian Pacific Railroad. Once in Ithaca, the ice was stored in a cold storage vault until it was needed for experimentation. Before experiments could be conducted, however, the necessary apparatus had to be designed and constructed. At first Tarr and his students were forced to improvise and simply put rocks in a bucket to provide the necessary weight needed and then add more rocks to vary the pressure (Figures 12 and 13).

Tarr was not ignorant of some pioneering work that had already been done in this field (e.g., Mügge, 1895), but in keeping with his attitude of doing things for himself, he obtained his own

Figure 12. Ice bending experiments from 1910–11 by Tarr and Rich (from Tarr and Rich, 1912). (Photo from the Department of Geological Sciences, Cornell University.)

Figure 14. Wooden laboratory built behind McGraw Hall on Cornell campus, during the winter of 1911–12. Tarr and von Engeln conducted high-pressure experiments in this laboratory until Tarr became fatally ill in March 1912 after working in this unheated building (from Tarr and von Engeln, 1915). (Photo from the Department of Geological Sciences, Cornell University.)

Figure 13. Stress being applied to ice by adding rocks to the bucket (from Tarr and Rich, 1912). (Photo from the Department of Geological Sciences, Cornell University.)

Figure 15. Inside the wooden laboratory building showing a piece of the high-pressure apparatus that was used to apply stress to the ice blocks (from Tarr and von Engeln, 1915). (Photo from Department of Geological Sciences, Cornell University.)

piece of a glacier and set to work. The first tests were conducted in the winter of 1910–1911 (Tarr and Rich, 1912). Some of the experiments did not work very well because of the warm temperatures inside his laboratory, but there were some very exciting results. By bending various ice bars, Tarr and Rich found that ice bends more easily in some crystal directions than others, and also:

... that ice may be deformed by tension as well as by compression; and they [the experiments] remove absolutely from this part of the phenomenon of ice motion all possibility of appealing to regelation (Tarr and Rich, 1912, p. 241).

In addition, they discovered that ice deformation is in the nature

of plasticity with the plastic yield point very near the breaking point of the ice and that the optical properties are affected by deformation as well.

The following winter Tarr obtained funds to build an outdoor facility to avoid the problem of high temperatures inside the laboratory. A wooden structure was constructed behind McGraw Hall on the Cornell campus specifically for the ice experiments, which were carried out during the winter of 1911–1912. The main thrust of the work this time was a series of high-pressure experiments on snow and pond ice using a device designed to give them both control and measurement of the pressure being applied. Before the work had progressed very far, however, Tarr became fatally ill. Although the work was unfinished, von Engeln published the results up to that time using Tarr's notes (Tarr and von Engeln, 1915). Not only was Tarr's experimental work cut short by his death, but the final results were published in a German journal one year after World War I had started. As a result, it was many years before the paper had very wide circulation.

The direction this line of research would have taken Tarr is impossible to say, but one can be certain that wherever it took him the results would have been interesting, well thought out, and relevant to the movement of glaciers.

TARR AS A WRITER

Tarr's writing activities fell into three categories; his scientific reports and professional papers, the numerous textbooks that he and his co-authors prepared, and what could be called works for a more general, nontechnical audience. He excelled in all three, and as has been demonstrated, he was prolific with his scientific output. He was almost as prolific in the other two categories as well.

Textbooks

Necessity appears to have been the prime motivation which prompted Tarr to begin writing textbooks. As he explained to his friend J. B. Woodworth in a letter October 1, 1893:

I [Tarr] have put in eight solid hours a day, and as a result I have finished a textbook on economic geology, and yesterday sent the manuscript and illustrations complete to the publisher. It was my plan to do a little field work near Gloucester, do a good deal of much needed reading, and prepare a syllabus on economic geology for use in my class in the spring term. I soon found that the syllabus would require as much time as a text-book, very nearly, and the idea dawned on me that I might as well do the latter; so off came my coat, and between July and October I wrote 1,000 pages of manuscript, blue-book size, and then rewrote it, besides preparing about thirty illustrations. I had the subject fairly well in hand and had complete notes and a fairly good library, otherwise I could not have done it (Woodworth, 1912, p. 33).

Eventually, Tarr produced textbooks because of his desire to fund independent field investigations. He hoped that after a few years his writing would put him in a position to finance his own research (Woodworth, 1912). Only the first product was simply a matter of necessity.

He labored over the proofs and the index during the autumn and early winter of 1893, and the book "Economic Geology of the United States" (Tarr, 1894a) was published early the following year. "Naturally I [Tarr] feel a trifle anxious just now as to the way my book will be received. . . . I have chosen my bed and must lie in it" (Woodworth, 1912). Thus, Tarr waited to run the gauntlet of reviews which would follow the release of his book.

To say the book met with mixed reviews is to understate the situation. Some (Anonymous, 1894) were polite and liked the addition of structural geology and physical geography chapters, but the review by Tarr's old classmate at Harvard, R.A.F. Penrose, was a blistering attack, especially on the topics that were included in the book and Tarr's use of the word "ore." Penrose concluded with:

The errors . . . pointed out are only a few among the many that might be mentioned, but they serve to show the want of familiarity with the subject and the inaccuracies prevalent throughout the volume (Penrose, 1894, p. 230).

A less confident person would have backed away from such an attack, but not Tarr, although he did reply with some reluctance:

It is not a very dignified proceeding for an author to reply to a review of one of his own works; but there are times when dignity must be sacrificed in order to place the truth before the public. It seems to me that one of these occasions has been created by Prof. Penrose's review of my Economic Geology of the United States. . . .This review, which occupies six pages. . . , finds the book wholly bad, and so inaccurate that not a word of praise can be found except for the "Publisher's work," the "good language," and "the general scheme in the arrangement of the subject matter," which he says is "logical" (Tarr, 1894d, p. 361).

Tarr went on to point out that most of the errors were those of the reviewer and not his, and he took great pains to support his use of the term "ore":

Had Dr. Penrose studied petrography he would probably not need to be told that the term ore, as I used it, is to be found in Rosenbusch's publications, and in English, in Teall's British Petrography, p. 52. Thus one page of criticism, and the one upon which most energy is devoted, depends rather upon the reviewer's misconceptions than upon the author's inaccuracies (Tarr, 1894d, p. 361).

Thus, Tarr embarked upon a new aspect of his career, namely, the writing of textbooks. To Woodworth he confided in February 1895:

I do not know what I am coming to. Everything points to the fate of writer of books, great and small. I had promised myself a summer of field work, but it now looks as if it would not materialize. A chance to write three books has come to me in the past two months (Woodworth, 1912, p. 35).

One of the three must have been, "Elementary Physical Geography," for it came out the same year (Tarr, 1895b). (For the record, this book sold for $1.40 in 1895.) This second effort met with more favorable reviews, but the third one brought forth rather harsh criticism, this time from W. M. Davis. Tarr was prompted to write the third book, "First Book of Physical Geography" (1897k), because an earlier version, "Elementary Physical Geography," proved too advanced and difficult for many secondary teachers. Davis found fault with most everything, even parts that he had liked, e.g.:

In most respects it presents a good view of the subject, especially where the treatment turns toward the geological side; but in a number of instances it fails to start at the beginning and make everything thoroughly clear. . . .

Under the lands many good lessons are taught, but process receives relatively more attention than form, . . . this seems a mistake in a book that should be essentially geographical" (Davis, 1897, p. 835).

One has to wonder, though, if Davis's review was clouded by the fact that he was in the process of preparing his own geography text which was published the following year (Davis, 1898).

Much space has been devoted to critical reviews of Tarr's early books to show that most of them were not very favorable. The professional criticism, however, did not appear to have much effect on the books' popularity with the public in general and the public schools in particular. Even the first one, "Economic Geology of the United States," went through three editions; others, for example, "Elementary Physical Geography," in spite of being not very "elementary" had nine editions; and the one Davis did not like went through eight editions. Amazingly, Tarr did all of the textbook writing while maintaining a steady stream of scientific papers. He had a facility of putting his thoughts and ideas on paper at high speed without losing his very readable style. When his books were used in secondary and elementary schools, it was said that the parents gained as much from them as did the students; such was the clarity of his writing (Woodworth, 1912).

There were many other texts on geography and on geology, but Tarr's real contribution came from a collaboration with Frank Morton McMurry, professor of education at Columbia Teachers' College. The Tarr and McMurry Geographies became the standard for geography classes around the world, and their popularity lasted long after both men had passed from the scene. (My wife used a Tarr and McMurry Geography in her elementary school classes in Tasmania in the 1940s.) This series was intended to cover the world, starting with North America, "First book; Home Geography and the Earth as a Whole," "Second Book; North America," "Third Book; Other Continents and a Review of the Whole" (Tarr and McMurry, 1900a–c).

This short review of Tarr's contribution as a writer of textbooks has attempted to give a feeling for the prodigious output that he maintained throughout his life. The only major, valid criticism that can be leveled at Tarr was best expressed by two of his critics, Davis and N. H. Winchell:

. . .the book bears too evident marks of hasty preparation (Davis, 1897, p. 835).

There are evidences, though but few, of too hasty composition, or at least of too hasty publication after composition (Winchell, 1897, p. 278).

Given the number of different and varied projects that Tarr conducted simultaneously, it is not surprising that the impression of haste crept into some of his writing. The real wonder is not so much that there is evidence of haste, but that there isn't more. Even at the time of his death, Tarr had several manuscripts almost ready for publication, and some were published posthum-ously, including a new text, "College Physiography," for which Lawrence Martin did the final editing (Tarr, 1914).

General

In addition to reaching the public through his textbooks, Tarr wrote numerous articles for various popular science publications such as *Scientific American, The Popular Science Monthly,* and *Goldthwaithes Geographical Magazine.* In preparing these articles, Tarr used the same care and meticulous attention to detail that he did in his scientific publications, but in these articles he could expand his prose and use much more poetic descriptions.

Before and just after Tarr's appointment to Cornell, most of his writing was published in popular science journals of the day. Between 1891 and 1892, Tarr published eight different articles, including a six-part series on river valleys (Tarr, 1891d; 1892 a–e). Once established at Cornell and heavily involved with research projects and students, one would think Tarr would cease to write for the more "popular press," but this was not the case. After he returned from Greenland in 1896, not only did Tarr prepare various scientific papers for publication, but also he published a lengthy description of the Greenland glaciers in *Scientific American,* using several of the same photographs that later appeared in his professional work (Tarr, 1897f). This was not the first time his work appeared in *Scientific American,* for in 1893, Tarr had one article on extinct volcanoes in the United States (Tarr, 1893b) and another one on the glacial period (Tarr, 1893a).

Tarr seemed to need these nonprofessional journals and magazines as an outlet for the more descriptive material he wrote. Whenever he took a trip, Tarr would prepare several articles, using the more scientific ones for professional meetings and journals, while the more descriptive and "tourist" type material went to popular journals and magazines. Eight years after his experiences with the U.S.G.S. in the West, Tarr was still producing articles about that landscape (Tarr, 1898c–e; 1899). In one of these, he describes a trip to the Grand Canyon which begins:

It is remarkable how few people visit this, the most wonderful bit of scenery in the world. Many traveling Americans do not even know of its existence (Tarr, 1899, p. 824).

Perhaps he would change his mind if he could see the millions of people who visit it today. Possibly, at the time Tarr visited the Grand Canyon, the 12-hour stagecoach ride from Flagstaff may have discouraged visitors. To illustrate how rhapsodic Tarr could be in these publications, here are some excerpts from the piece about the Grand Canyon:

One cannot hear the roar of the torrent that courses madly down the bottom of the cañon . . . the shadows change, the colors assume new tints . . . the geologist sees more than the beauty and grandeur. There before him are thousands of feet of rock layers formed in the sea by the slow process of deposition, requiring for their accumulation a time so great that whole races of animals came and went, leaving only their fossils as proof of their former existence (Tarr, 1899, p. 825, 826).

Tarr was quite willing to use these articles to influence public opinion, such as one in *The Popular Science Monthly* about Watkins Glen and some of the other central New York gorges:

> As one of the most beautiful and most widely known bits of natural scenery in the state, it seems to many of the lovers of this spot that the legislature should take the necessary action for securing the glen as a public park; and there now appears to be reason for hoping that this will be done during the present session of the legislature.
>
> It seems well, therefore, that it should be taken by the state, made better known, and opened freely to the public (Tarr, 1906b, p. 387, 397).

In this appeal Tarr apparently was successful, for today we can walk through Watkins Glen State Park.

Thus, through these "popular press" journals and magazines, Tarr was able to reach a much wider audience than some of his colleagues who published only in the professional journals. Perhaps it was this wider audience cultivated through *The Popular Science Monthly* or *The Independent* that helped make his secondary school texts so popular. It was a means of communication of which he took full advantage.

TARR AS A TEACHER

Tarr's early professional life was shaped by his teacher and mentor at Harvard, Nathaniel S. Shaler, and it was a debt that he freely acknowledged. Not only did Tarr's own teaching style and philosophy reflect traits he learned from Shaler, but he named a son Shaler in honor of his former teacher. Tragically, however, the young boy died in infancy. The true worth of a teacher is measured, not necessarily by popularity, although Tarr was popular, but by the leadership and inspiration that he provides to the students who are under his care. In this department, also, Tarr receives high marks. As an illustration of the effect he had on students, Tarr so impressed one student, not even a geology major, that the young man paid all of his own expenses to take part in Tarr's second trip to the glacier fields of Alaska.

> His life shows what a great teacher does as a fountain of instruction, inspiration, co-operation and friendship. . . .He had broad interest in man and nature and he brought the youth and nature together, confident of the result, heeding little the nice precisions of any particular method of approaching a problem. He was able to see beyond the young investigator's crudities and errors, to preserve the young man's respect for his intellectual self and allow him to evolve his own ways of working without hindrance (Brigham, 1914, p. 96).

His inspiration and love of subject were passed on to a multitude of students for whom he was a "scientific father." Many of them became distinguished geologists in their own right; I will mention only a few of them. At the top of the list should be Lawrence Martin, later of the University of Wisconsin, who was often a co-author with Tarr and a co-leader of several field parties; O. D. von Engeln, who as a student accompanied Tarr on two trips to Alaska, eventually became a member of Tarr's department and continued at Cornell long after Tarr's death. J. L. Rich went on to a long, distinguished, career at the University of Cincinnati. J. O. Martin and J. O. Bonsteel, who accompanied Tarr to Greenland, and F. N. Meeker continued their work with the United States Department of Agriculture, Bureau of Soils. A brother Raymond and Tarr's son Russell were both geologists. T. L. Watson, another member of the Greenland expedition, taught at the University of Virginia. Ross Marvin, who lost his life on Peary's memorable polar expedition, was working on a thesis in glacial geology with Tarr when he went on that fateful journey. And so the list goes on. All were influenced by the vitality of his teaching, and by the man himself. He was always available to his students, whether in the field or laboratory, and he and his wife maintained a most hospitable home which was always open to students (Brigham, 1914).

Immediately after Tarr's death, a fund was started by his students to erect a memorial to him on the Cornell campus. Finally, in 1915 a nine-ton glacial erratic was brought by horse team and railroad from a farm near Slaterville, New York, to the Cornell campus. As the team and wagon approached the campus, much discussion developed as to how to cross the bridge over Cascadilla Gorge. One school of thought was to make a slow crossing so as not to put any undue stress on the bridge, whereas another group advocated a high-speed crossing to decrease the time spent on the bridge. The speed merchants won the day, and the boulder, the wagon, and six horses went across at a gallop while everyone else stood around and held their breath. The bridge held, and the boulder was delivered to the campus without mishap. Eventually, the memorial was completed and dedicated on February 25, 1919 (Figure 16) (Ainsworth, 1972).

The other Harvard teacher who had influenced Tarr was W. M. Davis, and this influence was reflected in Tarr's love of geography and physiography. Very early in his career Tarr started making geography acceptable as a university subject (1896d). He felt that it could and should be included within curricula. H. S. Williams, friend and colleague at Cornell, said that Tarr was one of the first American teachers to bring physical geography to university rank (Williams, 1912). To do this he had to make a break with the past, both in what was taught and the methods of teaching. He summarized some of his thoughts on the subject in a short paper (1895a) dealing with the need to have laboratory instruction in both geology and physical geography. Tarr was highly critical of most instruction he had seen in colleges, for almost without exception it consisted of lecture-recitations, without any laboratory activity for the students, and he felt the situation was not much better at the secondary level. Earlier, an education study group known as the "Committee of Ten" had emphasized the importance of geology and geography as subjects for the secondary schools. In his paper, Tarr not only supported the group's recommendations, but also made a plea for more laboratory activities to be included in the classes (Tarr, 1895a).

At Cornell he set about to put his beliefs into practice. Over the years he built a well-equipped laboratory for teaching physio-

Figure 16. Reproduction of the bronze plaque which is attached to the Tarr memorial at McGraw Hall on the Cornell campus (original photograph by the author). (Photo from the Department of Geological Sciences, Cornell University.)

graphy and physical geography, and developed a "wet lab" approach for illustrating land form development. The equipment, which was a little like a stream table, was more square or rectangular than long and narrow. One of the "boxes" he used was 8 feet square and 16 inches deep, with each corner mounted on threaded cranks for raising and lowering "sea level" (Figure 17) (Tarr and von Engeln, 1908). In order to illustrate the three-dimensional aspect of landscapes, Tarr had several plaster models of various land forms constructed. With these large models, some of which were up to 3 or 4 feet square, the students were able to see what was on the topographic map. And in keeping with his belief that both geology and geography are visual subjects, Tarr made lantern slides an important part of his classroom teaching (Tarr, 1896 b–c; 1905f). Over the years, Tarr amassed several hundred lantern slide illustrations of almost every imaginable land form and geologic feature. A department inventory made by Tarr in 1907–08 lists over 1,000 photographs in the teaching collections, and this did not include his own personal collection. While most of these were black and white illustrations, a few of the more spectacular scenes were hand colored. Many of these

Figure 17. Large box used in Tarr's physiography laboratory to demonstrate land form development (from Tarr and von Engeln, 1908).

old lantern slides are still at Cornell; and almost 100 years after some of them were taken, they are still being used in the way Tarr intended, as an aid for teaching.

As a professional educator, Tarr's interest went far beyond his own classes at Cornell, for he was vitally interested in education at the secondary and elementary levels, especially education in the earth sciences (Tarr, 1905b). He was concerned with university entrance requirements in science and how changing them would affect what was taught at the secondary level (Tarr, 1896e; 1898). It was Tarr's contention, and one that is still valid today, that students need proper instruction in the secondary schools in order to prepare for college so college teachers will not have to teach subjects that should be taught in high school (Tarr, 1896e). Current university educators echo the same concerns. As for the criticism that college preparation courses were suitable for only a few students, Tarr had this to say:

Indeed, if our theories and principles of education are correct, that which is best calculated to prepare students for further study in college is also best calculated to prepare them for any walk in life in which the man is to live by the intelligent use of his mental powers (Tarr, 1896e, p. 60).

Tarr had a much larger vision of the role of secondary education in our society, one that stretched far beyond college entrance:

. . . for its [secondary school's] main object [is] not the preparation of students for the college, but the training of minds which shall be able to be used in the ordinary battle of life (Tarr, 1896e, p. 59–60).

One very good way to upgrade high school teaching, in Tarr's mind, was to demand more for college entrance, and this would put pressure on the high schools to change their curricula to meet the new and stiffer requirements. This was especially true for science:

. . . that science teachers have a right to demand that science, when well

taught, may properly be placed among the list of college entrance subjects (Tarr, 1898a, p. 50).

Tarr did not expect the secondary schools to operate in a vacuum, however. If the colleges were going to demand more, then not only should they offer assistance and guidance whenever and wherever possible, but also they should serve as a model for the secondary schools to emulate. In keeping with this philosophy, Tarr took every opportunity to share his knowledge and teaching techniques with secondary teachers through state and national organizations. Frequently, Tarr and his students would attend meetings simultaneously to present their findings (Tarr and von Engeln, 1908). He offered suggestions as to how best to work laboratory instruction into the high school curricula (Tarr, 1895a), ideas on what a teacher should have in the way of classroom equipment, and even ideas on what books and journals would be best suited for the secondary teacher to use in class and for continuing education (Tarr, 1896b–c). It is interesting to note that Tarr listed "H. A. Ward, Rochester, New York" as the best source for rock and mineral specimens. As expected from Tarr's professional work as a geologist, he emphasized the need for field study, especially at the secondary level:

One of the most important features of the field study is that the student is taught to see for himself (Tarr, 1896b, p. 168).

Nor was all of Tarr's concern with geography, for good geology teaching was very much part of his plan to upgrade science in the secondary schools. Again, he was aided in this effort by the "Committee of Ten" report which suggested a more systematic study of geology at the secondary level. To do this, Tarr advocated the study of geology at all three levels: the grammar school, high school, and college (properly taught, of course), using photographs, models, specimens, and field excursions under the direction of a properly trained teacher.

If geology could be taught as it should be, it would, upon its own merits, create for itself a place in the curriculum, and one that would be recognized as filling a gap in education that few subjects are so well adapted to fill (Tarr, 1900, p. 17).

In much of the above discussion, Tarr seems to have had a low opinion of the secondary school teacher, but such was not the case. In fact, Tarr was very sympathetic to the plight of the school teacher:

As a body they are overworked and underpaid. They are trained to one line of teaching, and then, by the caprice of the superintendent, perhaps, are given some new method—often a fad—of which so many pass over the educational world. . . .To secure better trained teachers, boards of education should be prepared to offer better compensation. Three or four hundred dollars is now commonly offered [yearly?] (Tarr, 1902, p. 55–56).

This support of the public school teacher was not buried in an educational journal but presented in an article about geography teaching in the *National Geographic* magazine (another case where Tarr had the strength of his convictions).

CONCLUSION

When his entire career is examined, Ralph Stockman Tarr is overwhelming. His capacity for work, whether in the field, the classroom, or at the writing desk, rivaled that of any two people. Tarr had catholic interests when it came to science and nature, and he went to great lengths to communicate his love for both on as broad a scale as possible, from professional scientific papers to textbooks for all levels of education, to articles in the *National Geographic* and *Goldthwaites Geographical* magazine. He was a relentless traveller. He would travel to remote corners of the globe in his quest to understand and describe nature, and then for relaxation he would tour Europe or Panama. His approach to life was "to make opportunities," to use the words of a colleague at Cornell, O. D. von Engeln, and to "make as complete account as could be of his [own] observations of the nature and state of the glaciers" (von Engeln, 1964, p. 78, 101).

He was honored by his peers in various professional societies, not only in this country but also broad. He was made a corresponding member of the Royal Geographical Society of Vienna, but the news of this election did not arrive until after his death. Tarr served as associate editor for the Geographical Society and the *Journal of Geology*, and was Foreign Correspondent for the Geological Society of London. He was listed as a Fellow of the Geological Society of America, Association of American Geographers, American Geographical Society, and the American Association for the Advancement of Science.

What makes the story of Tarr's life all the more remarkable, especially Tarr's many field excursions, is the probability that Tarr suffered from a form of hemophilia. O. D. von Engeln felt Tarr's sudden death was not caused by the cold he caught while conducting experiments on ice in the laboratory behind McGraw Hall but from complications brought about by his hemophilia:

Then he had a relapse and that brought on his hemophilic (bleeder) defect and he slowly bled to death (von Engeln, 1964, p. 107).

No written confirmation of von Engeln's diagnosis has been found as yet, but it does seem to explain other descriptions of Tarr's activities in the field. If this condition is true, then Tarr ran tremendous risks each time he ventured into the field, and yet, in this respect, he was irrepressible. At the time of his death, he was planning a trip to Newfoundland to investigate the glacial features there (Williams, 1912, p. 516). Perhaps it was from such an illness that Tarr drew the necessary strength and drive to face life's challenges and tragedies; he and his wife lost two of their four children. Although Tarr lived a relatively short life (48 years), he accomplished more than most who lived longer. As the memorial by Woodworth (1912, p. 40) concludes: "He did with dispatch all he could, and he made every moment count."

ACKNOWLEDGMENTS

The author hereby gratefully acknowledges the assistance and consideration of the following people: Professors Arthur L. Bloom and Donald L. Turcotte, and Emeritus Professor John W. Wells, of the Department of Geological Sciences, Cornell University; and Dr. Ellen T. Drake, College of Oceanography, Oregon State University. Also, a debt of gratitude is owed to the Department of Manuscripts and University Archives, Cornell University Libraries; Gould P. Coleman, university archivist; and Julia Crepeau, secretary. Finally, copies of the original photographs, most of which were in very bad condition, were prepared by the Audio Visual Services at the University of Pittsburgh at Johnstown; Linda D. Leech, supervisor, and Rick Povich, photographer.

REFERENCES CITED

Ainsworth, Earl, 1972, About the Tarr boulder: Cornell Plantations, Summer 1972, p. 30–32.

Anonymous, 1894, Review of "Economic Geology of the United States" by Ralph S. Tarr: The American Journal of Science, 3rd Series, v. 47, p. 151–152.

——1896, The Peary Greenland Expedition of 1896: The American Geologist, v. 18, p. 335.

Bird, John M., and Weathers, Maura S., 1977, Native iron occurrences of Disko Island, Greenland: Journal of Geology, v. 85, p. 359–371.

Brigham, Albert P., 1914, Memoir of Ralph Stockman Tarr: Annals of the Association of American Geographers, v. 3, p. 93–98.

Case, E. C., 1895, Experiments in ice motion: Journal of Geology, v. 3, p. 918–934.

Chamberlin, T. C., 1897, Editorial: Journal of Geology, v. 5, p. 81–85.

Culver, G. E., 1895, The erosive action of ice: Transactions of the Wisconsin Academy of Arts and Letters, v. 10, p. 339–366.

Davis, William Morris, 1897, Tarr's First Book of Physical Geography: The American Journal of Science, New Series, v. 6, p. 835.

——1898, Physical Geography (assisted by William H. Snyder): Ginn & Company, Boston, Mass., 448 p.

——1899, The peneplain: The American Geologist, v. 23, no. 4, p. 207–239.

Dryer, Charles R., 1909, Finger Lake Region of Western New York: Geological Society of America Bulletin, v. 15, p. 449–460.

Fairchild, Herman L., 1904, Editorial: The American Geologist, v. 33, p. 43–45.

Lincoln, D. F., 1892, Glaciation in the Finger Lake Region of New York: The American Journal of Science, 3rd Series, v. 44, p. 290–301.

Mügge, D., 1895, Uber die plasticitat der eiskrystalle: Separat-Abdruck aus dem Neuen Jahrbuch für Mineralogie, Band 2, p. 211–228.

Penrose, R.A.F., 1894, Book review of: The Economic Geology of the United States by R. S. Tarr; MacMillan & Co., New York, 1894, 509 p.: Journal of Geology, v. 2, p. 226–231.

Russell, Israel C., 1892, Mt. St. Elias and its glaciers: The American Journal of Science, 3rd Series, v. 43, p. 169–182.

Shaler, Nathaniel S., 1889, The geology of Cape Ann, Massachusetts; *in* Ninth Annual Report of the United States Geological Survey to the Secretary of the Interior 1887-88, by J. W. Powell, Director: Government Printing Office, Washingtion, D.C. 1889.

Shreve, R. L., 1966, Sherman landslide, Alaska: Science, v. 154, p. 1639–1643.

Tarr, Ralph S., 1890a, Origin of some topographic features of central Texas: The American Journal of Science, 3rd Series, v. 39, p. 306–311.

——1890b, Drainage systems of New Mexico: The American Geologist, v. 5, p. 261–270.

—— 1890c, Superposition of the drainage in central Texas: The American Journal of Science, 3rd Series, v. 40, p. 359–362.

—— 1890d, Erosive agents in the arid regions: The American Naturalist, v. 24, p. 455–459.

—— 1891a, Lake Bonneville: Goldthwaites Geographical Magazine, v. 1, no. 7, p. 468–471.

—— 1891b, Physical geography of Texas: Goldthwaites Geographical Magazine, v. 2, no. 9, p. 627–630.

—— 1891c, Russell's visit to St. Elias: Goldthwaites Geographical Magazine, v. 2, nos. 10-11, p. 706–713.

—— 1891d, River valleys, I: Goldthwaites Geographical Magazine, v. 2, no. 12, p. 778–782.

—— 1892a, River valleys, II: normal development: Goldthwaites Geographical Magazine, v. 3, no. 1, p. 36–40.

—— 1892b, River valleys, III: causes determining the courses of rivers: Goldthwaites Geographical Magazine, v. 3, no. 3, p. 198–202.

—— 1892c, River valleys, IV: floodplains and deltas: Goldthwaites Geographical Magazine, v. 3, no. 5, p. 350–356.

—— 1892d, River valleys, lakes and waterfalls: Goldthwaites Geographical Magazine, v. 4, no. 2, p. 571–576.

—— 1892e, River valleys, Lakes and waterfalls: climatic accidents: Goldthwaites Geographical Magazine, v. 4, no. 3, p. 695–699.

—— 1892f, The Central Massachusetts moraine: The American Journal of Science, v. 43, p. 141–145.

—— 1893a, The glacial period: Scientific American, v. 68, p. 86, 103.

—— 1893b, Extinct volcanos in the United States: Scientific American Supplement, no. 917, p. 14657–14658.

—— 1894a, The economic geology of the United States: MacMillan & Co., New York, 509 p.

—— 1894b, The origin of lake basins: Nature, v. 49, p. 315–316.

—— 1894c, Lake Cayuga a rock basin: Geological Society of America Bulletin, v. 5, p. 339–356.

—— 1894d, Economic geology of the United States: A reply to Dr. Penrose's review: The American Geologist, v. 13, p. 361–363.

—— 1894e, The origin of drumlins: The American Geologist, v. 13, p. 393–407.

—— 1894f, Lake Cayuga a rock basin: The American Geologist, v. 14, p. 194–195.

—— 1895a, Laboratory methods of instruction in geology and physical geography: Regents Bulletin State of New York, v. 32 (Appendix), p. 992–1011.

—— 1895b, Elementary physical geography: MacMillan & Co., New York, 488 p.

—— 1896a, Geological history of the Chautauqua Grape Belt: Bulletin of Cornell Agriculture Experiment Station, no. 109, p. 91–122.

—— 1896b, The teachers outfit in physical geography I: The School Review: A Journal of Secondary Education, v. 4, no. 3, p. 161–201.

—— 1896c, The teachers outfit in physical geography II: The School Review: A Journal of Secondary Education, v. 4, no. 4, p. 193–201.

—— 1896d, Geography in the university: Book Reviews, v. 4, no. 1, p. 1–7.

—— 1896e, College entrance requirements in science: Educational Review, v. 12, p. 57–64.

—— 1896f, A query concerning the origin of atolls: Nature, v. 54, p. 101.

—— 1896g, The Cornell expedition to Greenland: Science, New Series, v. 4, no. 93, p. 520–523.

—— 1897a, Former extension of Greenland glaciers: Science, New Series, v. 5, no. 113, p. 344.

—— 1897b, Evidence of glaciation in Labrador and Baffin Land: The American Geologist, v. 19, p. 191–197.

—— 1897c, The Arctic sea ice as a geological agent: The American Journal of Science, 4th Series, v. 3, p. 223–229.

—— 1897d, Former extension of Cornell Glacier near the southern end of Melville Bay: Geological Society of America Bulletin, v. 6, p. 251–268.

—— 1897e, Discussion and Correspondence: the former extension of ice in Greenland: Science, New Series, v. 5, no. 117, p. 515–516.

—— 1897f, The glaciers of Greenland: Scientific American, v. 76, p. 216–218.

—— 1897g, Valley glaciers of the Upper Nugsuak Peninsula, Greenland: The American Geologist, v. 19, p. 262–267.

—— 1897h, Difference in the climate of Greenland and American sides of Davis' and Baffin's Bay: The American Journal of Science, 4th Series, v. 3, p. 315–320.

—— 1897i, Changes of level in the Bermuda Islands [Abs.]: Quarterly Journal of the Geological Society of London, v. 53, p. 222.

—— 1897j, Changes of level in the Bermuda Islands: The American Geologist, v. 19, p. 293–303.

—— 1897k, The first book of physical geography: MacMillian & Co., New York, 198 p.

—— 1897l, The margin of the Cornell Glacier: The American Geologist, v. 20, p. 139–156.

—— 1898a, What is the consensus of opinion as to the place of science in the preparatory schools?: Journal of Pedagogy, v. 11, p. 40–51.

—— 1898b, The peneplain: The American Geologist, v. 21, p. 351–370.

—— 1898c, The Badlands of North Dakota: The Independent, September 22; 130 Fulton Street, New York.

—— 1898d, The geology of Yellowstone I: The Independent, November 24; 130 Fulton Street, New York.

—— 1898e, The geology of Yellowstone II: The Independent, December 1; 130 Fulton Street, New York.

—— 1899, The Grand Cañon of the Colorado: The Independent, March 23; 130 Fulton Street, New York.

—— 1900, Geology in the secondary schools: The School Review: A Journal of Secondary Education, v. 8, p. 11–17.

—— 1902a, The teaching of geography: The National Geographic Magazine, v. 13, p. 55–64.

—— 1902b, The Physical Geography of New York State: The Macmillan Co., New York, 397 p.

—— 1904a, Postglacial and interglacial (?) changes of level at Cape Ann, Massachusetts. With a note on the elevated beaches by J. B. Woodworth: Bulletin of the Museum of Comparative Zoology at Harvard College, Geographical Series, v. 4, no. 4, p. 181–196.

—— 1904b, Hanging valleys in the Finger Lake region of Central New York: The American Geologist, v. 33, p. 271–291.

—— 1905a, Some instances of moderate erosion: Journal of Geology, v. 8, p. 160–173.

—— 1905b, Results to be expected from a school course in geography: The Journal of Geography, v. 4, p. 145–148.

—— 1905c, The gorges and waterfalls of Central New York: Bulletin of the American Geographical Society (April), 1–20.

—— 1905d, Moraines of the Seneca and Cayuga Lake valleys: Geological Society of America Bulletin, v. 16, p. 215–228.

—— 1905e, Drainage features of Central New York: Geological Society of America Bulletin, v. 16, p. 229–242.

—— 1905f, The use of lantern slides in the teaching of Physiography: New York State Education Department Bulletin no. 356 (Secondary Education Bulletin no. 28), p. 14–20.

—— 1906a, Glacial erosion in the Finger Lake region of Central New York: Journal of Geology, v. 14, p. 18–21.

—— 1906b, Watkins Glen and other gorges in the Finger Lake region of Central New York: The Popular Science Monthly (May), p. 387–397.

—— 1907, Recent advance of glaciers in Yakutat Bay region, Alaska: Geological Society of America Bulletin, v. 18, p. 257–286.

—— 1909, Yakutat Bay region, Alaska (Physiography and glacial geology by R. S. Tarr and Areal geology by R. S. Tarr and Bert S. Butler): United States Geological Survey Professional Paper no. 64, 183 p.

—— 1910, The theory of advance of glaciers in response to earthquake shaking: Zeitschrift für Gletscherkunde, Band 5, p. 1–35.

—— 1911, Presidential address: American Association of Geographers: Unpublished manuscript of first draft, Department of Manuscripts and University Archives, Libraries of Cornell University, Ithaca, N.Y.

—— 1912, Glaciers and glaciation of Alaska: Annals of the Association of American Geographers, v. 2, p. 3–24.

—— 1914, College physiography (published under the editorial direction of Lawrence Martin): MacMillan & Co., New York, 837 p.

Tarr, Ralph S., and Martin, Lawrence, 1906a, Recent changes in level in the Yakutat Bay region, Alaska: Geological Society of America Bulletin, v. 17, p. 29–64.

—— 1906b, Glaciers and glaciation of Yakutat Bay, Alaska: Bulletin of the American Geographical Society, v. 38, p. 145–167.

—— 1906c, Recent changes of level in Alaska: The Geographical Journal (July), p. 30–43.

—— 1910, The National Geographic Society's Alaskan Expedition of 1909: The National Geographic Magazine, v. 21, no. 1, p. 1–54.

—— 1912, The earthquake at Yakutat Bay, Alaska, in September 1899: United States Geological Survey Professional Paper no. 69, 135 p.

—— 1913, Glacial deposits of the continental type in Alaska: Journal of Geology, v. 21, p. 289–300.

—— 1914, Alaskan glacier studies of the National Geographic Society in the Yakutat Bay, Prince William Sound, and Lower Copper River regions: National Geographic Society, Washington, D.C., 470 p.

Tarr, Ralph S., and McMurry, Frank M., 1900a, First book: home geography and the earth as a whole. Tarr and McMurry Geographies: MacMillan & Co., New York, 279 p.

—— 1900b, Second book: North America. Tarr and McMurry Geographies: MacMillan & Co., New York, 469 p.

—— 1900c, Third book: other continents and a review of the whole subject. Tarr and McMurry Geographies: MacMillan & Co., New York.

Tarr, Ralph S., and Rich, John L., 1912, The properties of ice-experimental studies: Zeitschrift für Gletscherkunde, Band 6, p. 227–249.

Tarr, Ralph S., and von Engeln, O. D., 1908, Representation of landforms in the physiography laboratory: Journal of Geography, v. 7, p. 73–85.

—— 1910, Laboratory manual of physical geography: MacMillan & Co., New York, 362 p.

—— 1915, Experimental studies of ice with reference to glacier structure and motion: Zeitschrift für Gletscherkunde, Band 9, p. 81–139.

von Engeln, O. D., 1964, Reminiscences: Unpublished handwritten manuscript, Cornell University Archives, no. 14/15/856, 225 p.

Walker, Bryce, and the Editors of Time-Life Books, 1982, Earthquake, Planet Earth Series: Time-Life Books, New York, 176 p.

Williams, Henry S., 1912, Obituary-Ralph Stockman Tarr: The American Journal of Science, v. 33, p. 515–516.

Williams, Henry S., Tarr, Ralph S., and Kindle, E. M., 1909, Geologic Atlas of the United States. Watkins Glen-Catatonk Folio, Field Edition: United States Geologic Survey Folio no. 169, Government Printing Office, Washington, D.C., 242 p.

Winchell, N. H., 1897, Elementary Geology by Ralph S. Tarr: The American Geologist, v. 14, p. 277–278.

Woodworth, J. B., 1912, Memoir of Ralph Stockman Tarr: Geological Society of American Bulletin, v. 24, p. 29–43.

Geological Society of America
Centennial Special Volume 1
1985

L. R. Wager and the geology of East Greenland

C. K. Brooks
Department of Geology
University of Papau New Guinea
P.O. Box 320
University P.O.
Papua New Guinea

ABSTRACT

Wager is most well-known for his work on the Skaergaard Intrusion of East Greenland, which was a milestone in the history of igneous petrology. It laid the foundation for a much better understanding of magmatic processes and for a multitude of investigations on similar bodies in other parts of the world. Less well known are Wager's achievements in mapping an extensive area of East Greenland in which spectacular examples of Tertiary intrusive and extrusive rocks occur. This achievement, which was a major contribution to the understanding of the geological history of the North Atlantic region, was only possible because of Wager's outstanding abilities as a mountaineer, polar explorer and leader, as demonstrated on his Greenland expeditions of 1930–31, 1932, 1935–36 and 1953, as well as by his feats on Mount Everest in 1933.

Wager's interests extended to a wide range of traditional earth sciences, including tectonics, geomorphology, sedimentology and metamorphic processes. In addition, he played an innovative role in the fields of geochemistry and radiometric dating. It was his appreciation of the importance of laboratory techniques and innovations that set his work apart from that of most other field geologists of his time, and this, combined with his thoroughness, gave him an important place in the history of geological science.

INTRODUCTION

Laurence Rickard Wager (Fig. 1) was a man of considerable achievement. As an explorer and mountaineer he travelled extensively through an area of spectacularly rugged terrain in East Greenland, climbed to the highest point then attained on Everest, as well as making the first ascent of the highest mountain within the Arctic Circle. As a geologist his interests and published works spanned metasomatic processes, tectonics, geomorphology, differentiation of basic and alkaline magmas, layered structures in igneous rocks, mechanisms of crystal nucleation in magmas, the origin of acid rocks and the regional geology of East Greenland, parts of the Himalaya, western Ireland and Scotland. Although not a geochemist himself, he played a leading role in the application of modern methods of chemical analysis to the interpretation of igneous rocks and to the use of radiometric dating techniques in geology. These innovations characterized the Wager era at Oxford, when the Department of Geology and Mineralogy there rapidly assumed a leading role in these fields.

Wager is best known for his work on layered igneous rocks, particularly his studies of the Skaergaard intrusion in East Greenland which he made into a familiar name for generations of geology students and which stimulated much research in other layered intrusions around the world. The original memoir on this intrusion, co-authored by W. A. Deer, remains today as one of the great works of geology: its observational material is acknowledged by subsequent workers to be of impeccable quality, while the deductions and conclusions are both thought-provoking and of world-wide applicability.

During the 1930s when Wager began his career, the earth sciences were in a state of relative quiescence. The theory of continental drift had been rejected and within the fields of igneous petrology and ore geology most workers were locked into seemingly never-ending debates about the origin of granites or the nature of ore-forming solutions. It was at this time that Goldschmidt was laying the foundations of geochemistry and Holmes was applying radioactivity to the dating of rocks, but these developments left most geologists relatively untouched.

Figure 1. Wager at the Skaergaard base hut during the British East Greenland Geological Expedition 1935–1936.

Wager began his work in Greenland in 1930 and for the rest of the decade amassed, first field data and then mineralogical and chemical data that addressed the major problems of igneous petrology. After a hiatus during the war years, he began to apply more refined methods of chemical analysis and to encourage the development of radiometric dating. This innovative work may have encouraged the enormous developments seen in these fields subsequent to about 1960. His work on the Skaergaard Intrusion focussed on one of the major questions of the time, namely the origin of granitic magmas, and this undoubtedly contributed to the interest which his work aroused. In a subsequent paper (Wager and others, 1957) he addressed another current problem, the origin of metallic ores, but curiously this paper does not seem to have gained widespread attention. Similarly, his other considerable achievements in the regional geology of East Greenland are not well known, although he made major contributions to our knowledge of the North Atlantic craton, the origin of syenitic rocks, the history of the Greenland ice sheet and the tectonics of continental margins.

Wager's work extended to other areas (Scotland, Ireland, the Himalayas, the Caribbean and elsewhere). Here, we concentrate on his work in Greenland. The reader is referred to the biography by his co-worker Deer (1967) for a more complete picture.

THE EARLY YEARS (1904–1929)

Wager was born in Batley, Yorkshire, in 1904, although both his parents were from the south. His father was a schoolmas-

ter and his uncle, who lived in nearby Leeds, was a chief inspector of schools but also, perhaps more significantly, an enthusiastic naturalist and a Fellow of the Royal Society. Wager's childhood background was thus academic, but doubtless the close proximity of Yorkshire Dales (now a National Park) also had a strong formative influence on his appreciation and understanding of scenery and its underlying geology.

Wager subsequently went to Cambridge, which at this time was particularly stimulating for geology, with such residents as J. E. Marr, Alfred Harker, Gertrude Ellis and C. E. Tilley. Here Wager obtained a first in geology in 1926 and went on to three years of research on joint patterns in the Great Scar Limestone and metasomatism in the Whin Sill of northern England. At Cambridge he also became a leading mountaineer, climbing in North Wales, Scotland and the Alps.

READING UNIVERSITY (1929–1935)

Soon after being appointed to a lectureship at Reading, an invitation to join the British Arctic Air Route Expedition (B.A.A.R.E.) of 1930–31 gave him his first chance to visit Greenland, which laid the foundations for much of his subsequent work. On this expedition Wager's work was not only geological, he also took part in a number of major sledge journeys: to the Ice Cap Station, an attempt to reach Kangerdlugssuaq, 500 km north of the expedition base, and an attempt on Mt. Forel, at 3500 m, then the highest known mountain in the Arctic (Fig. 2). The failure of these endeavours by no means reflected on Wager's ability but rather were caused by bad weather and other factors

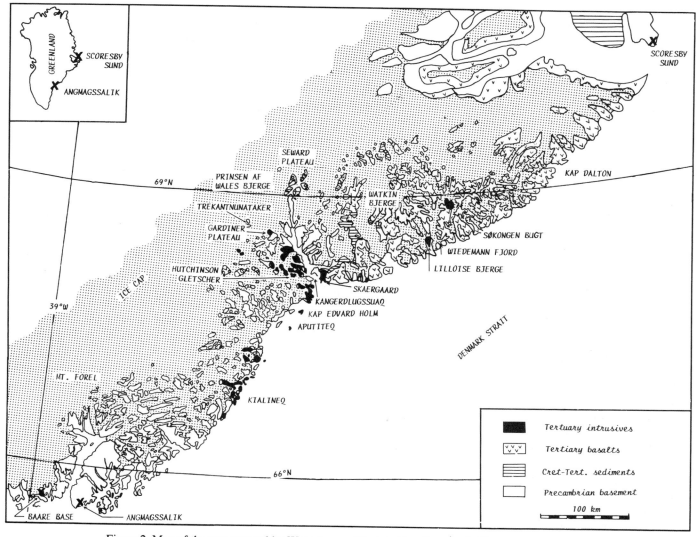

Figure 2. Map of the area covered by Wager's expeditions in East Greenland. These were as follows: 1930–31 in the Angmagssalik district with longer sledge journeys to the Inland Ice and Mount Forel and by ship to Kangerdlugssuaq.; 1932 from Scoresby Sund to Angmagssalik by ship; 1933 from Scoresby Sund by ship; 1935–36 to Watkin Bjerge and in the Kangerdlugssuaq area; 1953, around Skaergaard and Kap Edvard Holm. Angmagssalik and Scoresby Sund are the only settlements in this area.

beyond his control. The journey to Mt. Forel (Fig. 3) led Wager to consider the origin of the Greenland ice sheet, a subject which interested him throughout his life and on which he published an article in 1933.

During sledging journeys in the Angmagssalik area Wager was able to learn a great deal about the Precambrian gneisses and related rocks of the area, whose age was at that time unknown, although some considered them to be Caledonian. It was on the voyage of the expedition's ship "Quest," however, that his most important observations were made. He identified a large number of Tertiary igneous intrusive centres stretching along the coast from Kap Gustav Holm at about 66°N to Kangerdlugssuaq at 69°N (Fig. 2). Wager was fortunate in that ice conditions permitted the exploration of the fjord to its head. He also came to

appreciate the potential of the area and had the tenacity to return subsequently and follow through his discoveries in a way which few others would have done.

It must be remembered that this area is of extreme inaccessibility even today, and in the early 1930s it would have been necessary to travel to the Antarctic to find comparable terrain. Angmagssalik was in those days served only by a single ship each summer, whereas today regular flights are available several times a week all the year round. Kangerdlugssuaq lies 400 kilometres to the North along a coast guarded by great masses of polar pack-ice, constantly drifting southwards from the Arctic Ocean. Special ice-strengthened ships are necessary to sail this coast and even today there is little certainty as to what places might be reached. The land behind this drifting ice is a torn mass of tooth-like peaks

Figure 3. Notebook page dated 25th May 1931 showing view from camp near Mount Forel across the glacier to mountains of Precambrian gneiss with a peneplain preserved in their summits.

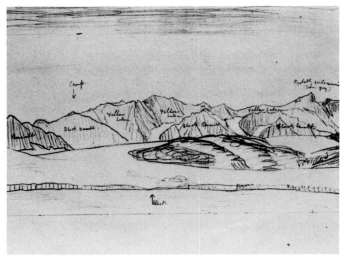

Figure 4. Sheet from 1932 diary showing the Lilloise Range seen from Wiedeman Fjord bearing in a westerly direction about 30 kilometres distant. It was not until 1971 that these mountains were reached and found to be layered basic and ultrabasic rocks and not syenites as believed by Wager.

isolated by great glaciers which tumble down from the interior and the only way to travel long distances overland is by dog sledge along the Inland Ice. This was attempted by the British Arctic Air Route Expedition, who failed due to bad weather. Of course, Wager was uniquely equipped to work in this area, being both a geologist and a mountaineer and there can be little doubt that his love of mountaineering must have been a significant factor in his continued attraction to the area. In fact, the mere problems of living and travelling in such an area take up so much time and energy that it is doubtful that the geological interest alone could have sustained him throughout four expeditions, two of which lasted more than a year. Nevertheless, his recognition of the importance of the Skaergaard Intrusion at this stage was a stroke of genius.

In 1932 Wager took part in the expedition led by Ejnar Mikkelsen and organized by the Scoresby Sund Committee (Mikkelsen, 1933). This expedition carried a number of scientists whose object was to explore and document the coastline between Angmagssalik and the newly established colony of Scoresby Sund, a distance of four degrees of latitude that was very poorly known. Wager was accompanied in the field by his brother, Dr. H. G. Wager, a botanist and Dr. Jens Jensen. They sailed from Scoresby Sund to Angmagssalik in the ship "Søkongen" (which has sailed Greenland waters until quite recently). Wager was a notoriously poor sailor and crossing the North Atlantic in ships the size of "Søkongen" or the Norwegian seal catchers (which were often under 100 tonnes) must have been a nightmare to him. This expedition allowed him to continue his observations already made between Angmagssalik and Kangerdlugssuaq to the North. At Wiedemann Fjord (Fig. 4), a new plutonic body was discovered, but an attempt to reach it by Wager and Captain

Mikkelsen was thwarted by the very heavily crevassed Kronborg Glacier. On this expedition, Wager was also able to study the basalt plateau, which extends from Scoresby Sund all the way to Kangerdlugssuaq covering more than 60,000 km^2. At Kap Dalton, interesting fossiliferous sediments occur which throw light on the age of the basalts and the conditions prevailing at the time they were formed. Wager devoted special attention to these sediments and published a description of them in 1935. As a result of his two expeditions, Wager was, in 1934, able to publish a masterly description of the regional geology of this whole area, which has served as a basis for all further research in the area and remains today little modified by subsequent workers.

On his return from Greenland Wager was selected to take part in an attempt on the summit of Mount Everest in 1933. Together with Wyn Harris, he climbed higher than anyone else until the mountain was conquered twenty years later. Their attempt was made without artificial oxygen supplies and is a supreme testimony to Wager's determination, although it may have contributed to his later heart attacks.

In 1934 he returned to Greenland as a member of Jean Charcot's expedition in "Pourquoi Pas," but because of bad ice conditions little was achieved by this expedition. Wager had realized that an extended stay would be necessary in Kangerdlugssuaq in order to achieve significant results and had accordingly begun to plan a later, more ambitious, expedition. It had been intended to leave equipment and provisions in Kangerdlugssuaq in 1934 for this later expedition, but this was not achieved.

THE 1935–1936 EXPEDITION

A plan was drawn up whereby an expedition was to over-winter in Scoresby Sund, work on the basalts to the South during the autumn and reconnoitre a route to the ice cap leaving a dump there for a journey along the Inland Ice to Kangerdlugssuaq in the following spring. This route had been followed by Martin Lindsey two years earlier, who began on the West Coast, crossed the Inland Ice to the neighbourhood of Scoresby Sund and then turned southwards following the inland margin of the coastal mountains, crossing the huge Kangerdlugssuaq basin and finishing at the 1930–31 base near Angmagssalik. This had been one of the longest unsupported sledge journeys ever made and it was clear that although Wager's plan was feasible, it would leave little time for geology. It had the advantage that the party could travel to Scoresby Sund on the Danish government boat and be picked up by a Norwegian sealer in Kangerdlugssuaq at the end of its shark fishing season. This would save the cost of ship charters, which then as now, were both the most expensive and the most difficult to fund items in an expedition's budget. It was an extremely arduous plan, for which Wager was fully prepared but which few other geologists would have contemplated.

A much more satisfactory solution emerged when Augustine Courtauld suggested a project for climbing Gunbjørn Fjeld (unnamed at this time). This is the highest point of the Watkin Bjerge at about 3700 m, the highest mountain within the Arctic Circle and therefore a powerful attraction for mountaineers. Courtauld proposed to charter a Norwegian sealing vessel to land his party at Wiedemann Fjord and to land the geologists at Kangerdlugssuaq.

In the event, 1935 was an exceptionally bad ice year and the ship which used 17 days on the 400 km voyage from Angmagssalik to Kangerdlugssuaq was several times in danger. Wiedemann Fjord could not be reached and the climbing party began at I. C. Jacobsen Fjord which lies much closer to Kangerdlugssuaq) adding some 150 km to the approach route over unknown country. However 10 days later, thanks largely to Wager's exceptional skill in mountainous country, the top was reached (Courtauld, 1936). Only once has this ascent been repeated: in 1971, again as in 1935, by a combind English-Danish party. Apart from mountaineering, this gave Wager an opportunity to examine at first hand the plateau basalt country which stretches unbroken from Kangerdlugssuaq to Scoresby Sund, and from the coast to the margin of the Inland Ice. Mainly during this journey but also on subsequent ones during this expedition, Wager gained a considerable insight into the nature of the basaltic plateau. He published some of his observations in 1947, but the work originally planned south of Scoresby Sund had to wait until 1965 when this area was visited by an expedition from Oxford University inspired by Wager. During the late 1960s and 1970s the area was subjected to regional mapping by the Geological Survey of Greenland, but this work came no further south than 69°N and at this time is still unpublished.

After the successful attempt on Gunnbjørn Fjeld, the over-wintering party was established at Skaergaard (Figs. 5 and 6). This consisted of Wager as leader; P. B. Chambers, geological assistant and zoological collector; Dr. E. G. Fountaine, medical officer, mountaineer and surveyor; W. A. Deer, geologist; Dr. H. G. Wager, botanist; Mrs. H. G. Wager, assistant botanist; and Mrs. L. R. Wager, responsible for domestic affairs and occasional geological assistance. The Europeans were accompanied by fourteen Greenlanders (Eskimoes) who wished to try out the hunting in the area and were expected to supply a large proportion of the food. The Kangerdlugssuaq area was at this time uninhabited but numerous house ruins testified to earlier inhabitation. The Angmagssalik people had legends relating to the area and it was widely believed to be rich in game. At this time, the population of Angmagssalik had expanded and outstripped the natural resources of the immediate area, leading to the establishment of the settlement at Scoresby Sund. Enterprising hunters still wanted, however, to explore new areas and they readily joined Wager's expedition. In the mid-60s, over-wintering parties of Greenlanders again began to use the area, continuing up to the present day. Their assistance has been a significant factor in the geological exploration of the Kangerdlugssuaq area.

In 1935 it was hoped that the Greenlanders (Fig. 7) would be invaluable sledge drivers, but they were only of limited assistance in this respect, as they were unused to, and frightened of, inland travel over glaciers. Conditioned by their legends to fear the inland regions, which they considered to be inhabited by races of malevolent giants (*tornarssuk*) or dangerous, supernatural outcasts (*qivitoq*), they were not up to Wager's exacting standards on the inland journeys. They were, on the other hand, unparalleled in travelling on the treacherous, ever-changing sea ice.

The base camp on Skaergaard originally consisted of two houses erected by Ejnar Mikkelsen during the 1932 expedition, which were part of a chain of houses along the coast, others being at Kap Dalton, Søkongen Bugt and Mikis Fjord. These were intended as intermediate stations for Greenlanders travelling between Angmagssalik and the newly established settlement at Scoresby Sund. In fact, no parties ever made the journey but the houses stood until recently and, surprisingly, those at Kap Dalton and Søkongens Bugt were still standing in 1980. The British East Greenland Expedition 1935–1936 (as this expedition was officially named) erected a further house for the Europeans which also lasted more than 30 years, until dismantled by hunters, who preferred to live closer to the open water. Wager's house, charmingly situated on Hjemmebugt, was sheltered but had the disadvantage that the winter ice lay long into the summer in the bay. The hunters who began to come in 1966 preferred the site of the wartime American weather station on the neck of the Skaergaard peninsula which was nearer the open water. This weather station was no longer standing in 1966 having been abandoned after World War II in favour of the island of Nordre Aputitêq off Kap Edvard Holm, where the weather patterns were less localised.

Priority for the 1935–36 expedition was given to detailed mapping of the Skaergaard Intrusion, but a regional examination of the whole area was also intended. During the autumn, the lower crevassed regions of the large glaciers were reconnoitred to find a route to the inland areas where the glaciers are snow-

Figure 5. The Skaergaard intrusion and environs from the air. One of the many high-quality oblique air photographs made by the British Arctic Air Route Expedition of 1930–31 and of which the negatives are now destroyed. Precambrian gneisses of Kraemer Island in the foreground, Skaergaard gabbros in the middle distance and basalts behind. Watkin Fjord on left, Denmark Strait in background. View looking eastwards in summer.

covered, relatively flat and provide thoroughfares for travel by dog sledge. Fountaine and Chambers established a supply dump some 60 km up the Frederiksborg Glacier at a height of more than 1000 m which could be used for the long sledge journeys later.

Traditionally, arctic expeditions went into a state of hibernation during the winter, but it is apparent from the diaries and reports that this was not the case on this occasion. Wager kept up field work on the Skaergaard intrusion whenever conditions allowed. During spells of bad weather and the dark period, the time was occupied by thin sectioning of samples, microscope work, mineral separations and collating the results of the field work. Of course, Skaergaard lies only about 200 km north of the Arctic Circle so the dark period is not too exacting. The area is one of unsettled weather with frequent incursions of maritime air and massive snow falls. On one occasion in January rain fell heavily, an event which may have had serious consequences in the Arctic environment.

Early in March, Wager and Fountaine set out on their spring journey to the nunataks on the Seward Plateau and the Prinsen af Wales Bjerge (Fig. 2) where they found the basalts resting on basement gneisses and an upper series of basalts distinct from the basalts at the coast. This major division of the basalts has been confirmed by later workers who have designated them as the "Nunatak Group" and the "Blosseville Group" respectively. It was not until 1982, nearly 50 years after Wager and Fountaine's visit, that these inland regions (Fig. 8) were visited again, this time by helicopter, an indication of the difficulties faced in getting there.

At the beginning of June, Fountaine and Deer travelled by the same route to the Seward Plateau and then across the large funnel-shaped depression which drains the Inland Ice into Kangerdlugssuaq (and is prone to powerful katabatic winds). The crossing of the Kangerdlugssuaq basin from Lindsey Nunatak to Trekantnunatakker was plagued by bad weather and took 23 days, another measure of the magnitude of the operations. These areas have never subsequently been visited. Fountain and Deer's route led them via Gardiner Plateau and through the Kangerdlugssuaq intrusion to the foot of Hutchinson Glacier, where they were taken off by boat. Possibly owing to the bad weather and heavy snow cover, Deer missed the very interesting ultramafic alkaline ring complex at Gardiner Plateau. This was first identi-

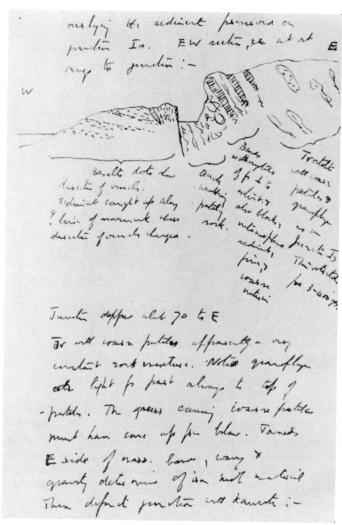

Figure 6. Notebook page showing the contact of the Skaergaard border group (on right) with vesicular basalts (on left).

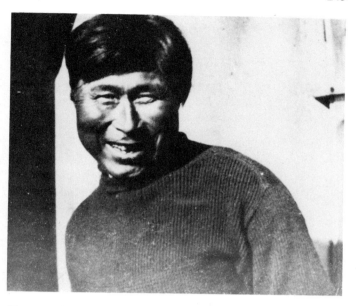

Figure 7. Hans, one of the Greenlanders who accompanied the 1935–36 expedition.

fied by the Northern Mining Co. in 1971 and reported by Frisch & Keusen (1977). Subsequently it has been described in some detail by Nielsen (1980, 1981). That Fountaine and Deer were very close to its discovery is shown by Wager's reference to their finding of "interesting magnetite-tremolite rocks from the basement at Gardiner Plateau" (Wager, 1947). An examination of the relevant material housed in the Geology Department of Oxford shows that it consists of octahedra of magnetite, acicular aegirine and reddish titanoclinohumite: a typical Gardiner Complex assemblage.

When the expedition was taken off in the late summer of 1936 by a Norwegian sealing vessel (Fig. 9), a huge number of samples and field observations had been amassed. They formed the basis for a series of published works, both by Wager and his co-workers and students for more than a generation. The expedition was thus a major landmark, both in the scientific exploration of Greenland and in the history of igneous petrology.

READING UNIVERSITY, 1936–1939

Back at Reading and in close collaboration with W. A. Deer, work progressed on the publication of the expedition's results. First a paper was published in 1937 by the Royal Geographical Society on the Kangerdlugssuaq area, followed in 1938 by an important and much quoted paper on the coastal dyke swarm and flexure, and papers on the Skaergaard pyroxenes (1938) and olivines (1939). Finally, in 1939, the Skaergaard Memoir was published. It formed Part III of the series "Geological Investigations in East Greenland" which ultimately reached Part XI, although after Part IV they were authored by Wager's co-workers. This Memoir is the only number in the series *Meddelelser om Grønland* (which runs to many shelf-metres of volumes) to have been reprinted. This occurred in 1962 with an updated bibliography of Skaergaard works. The memoir was immediately well-received and rapidly became a standard work in the field of igneous petrology, known to generations of geology students worldwide.

In 1940, Wager published some of his observations on epeirogenic earth movements in East Greenland and in 1945 an important paper was published with R. L. Mitchell on the geochemistry of some transition metals in igneous rocks. The last heralded a major thrust in Wager's interests after the war.

THE WAR YEARS AND DURHAM (1939–1950)

At the time of the outbreak of war the Skaergaard work was nearing completion and Wager considered the idea of an expedition to the Ilímaussaq intrusion of South Greenland. This body is an agpaitic nepheline syenite described in a classic work by the Danish petrologist Ussing in 1912. It was no doubt the very

Figure 8. Camp on one of the inland journeys.

spectacular layering in some of the rocks which attracted Wager. The outbreak of war, however, prevented him from carrying out this plan, and just after the war, the newly founded Geological Survey of Greenland began its work in precisely this area. Ilímaussaq was remapped by Ferguson who published his results in 1964.

Wager was commissioned into the Royal Air Force shortly after the outbreak of War in 1939 and served in the United Kingdom, the Middle East, and in Arctic Russia. His geological training was put to good use during this period as he worked in photographic reconnaissance where he could use his knowledge of aerial photographic interpretation. This experience had been gained mainly in East Greenland where the British Arctic Air Route Expedition had taken that country's first aerial photographs using their two De Havilland "Gypsy Moth" aircraft. Wager was apparently involved in the Allied operations on the Lofoten Islands in Norway and many years later he talked of the geology of these islands and the possible igneous layered structures which he had observed on air photographs.

Wager was released from the R.A.F. in 1944 and took up the chair of Geology at Durham, succeeding Arthur Holmes. During his period in Durham, the department was greatly expanded, adding several new members who went on to have distinguished careers.

Also during this period Wager continued his association with the geochemist R. L. Mitchell, who employed emission spectroscopy for the quantitative determination of trace elements. This association led to a major survey of the trace element geochemistry of the Skaergaard intrusion which was published in the

first number of *Geochimica et Cosmochimica Acta.* Wager was one of the initiators of this journal, which rapidly attained a prestigious position in the field, and he was clearly convinced of the coming importance of geochemistry in the earth sciences at a time when most geologists still considered field mapping as the be-all and end-all. Of course, in subsequent years, the situation became completely reversed and field mapping became of secondary importance, especially in Britain and the United States. Wager's strength was in his firm conviction in the value of thorough field work, combined with his perception of the importance of the newly emerging instrumental techniques of chemical and isotopic analysis. The 1957 paper by Wager and Mitchell was thus a pioneering work in a field soon to show explosive growth.

OXFORD UNIVERSITY (1950–1965)

In 1950, Wager was appointed to the chair of Geology in Oxford where he was soon to initiate great changes. Under Wager's leadership, the department in Oxford became a leader in the new phase of development of geology. In particular, the department entered the fields of geochemistry and later radiometric dating and quickly gained worldwide renown, not only in petrology, but also in these two disciplines.

From the early work with Mitchell on trace elements, a collaboration was commenced with the Atomic Energy Research Establishment at Harwell to pioneer the application of neutron activation analysis to the determination of trace elements in igneous rocks. Later this work was carried out at Oxford, and gradu-

Figure 9. "Jopeter" (later "Polarbjørn") of Aalesund (Norway) in Kangerdlugssuaq, 1953. This vessel, typical of the Norwegian seal catchers except that it is larger than most, transported Wager's expedition of that year.

Figure 10. Members of the British East Greenland Geological Expedition of 1953 outside the Skaergaard base hut. Left to right: C. J. Hughes, G. D. Nicholls, P. E. Brown, L. R. Wager, G. M. Brown and W. A. Deer.

ally an impressive series of studies of trace element distributions were published, largely on Skaergaard rocks and minerals, which extended and refined the earlier work with Mitchell. A few examples of this work are Wager, Smit & Irving (1958) on indium, Hamilton (1959) on uranium, Vincent & Bilefield (1960) on cadmium, Vincent & Crocket (1960) on gold, and Brooks (1969) on zirconium and hafnium. The Skaergaard rocks and minerals thus became a natural laboratory where the chemical factors governing the entry of the chemical elements into mineral lattices during crystallization from a magma could be studied. The work formed a bridge between the classic studies of Goldschmidt before the war and the application of these principles to petrological problems, the most spectacular applications of which were to the rocks of the ocean floor (recovered by the D.S.D.P. and I.P.O.D. Projects) and the moon (Apollo Project).

About the same time Wager began to interest himself in radiometric dating of rocks and minerals, and this project was established at Oxford, again with help from Harwell. This activity was probably also inspired by Wager's Greenland work, as he had speculated previously on the age of the basement rocks in the Angmagssalik area, thought by some to be of Caledonian age but believed by Wager to be Precambrian. Wager himself made little use of radiometric methods of dating except when he returned to the problem of the extent of the Caledonian fold belt in East Greenland in 1964 in a short paper with E. I. Hamilton. In the early days Oxford was one of the leading laboratories for isotope studies and has always maintained a high position, thanks to men of the calibre of S. Moorbath and R. K. O'Nions.

Wager's career at Oxford was very distinguished but has only peripheral relevance to the present paper, and the reader is referred to the biography of W. A. Deer (Deer, 1967). He found time for his own research and published often in association with colleagues (although always with the unmistakable Wager stamp) important papers on magma mixing in the Hebridean province, fractionation of Hawaiian, Hebridean and other magmas, igneous layering, cumulate rocks and their nomenclature, and sulphide immiscibility in the Skaergaard magma. In particular, his work on the mechanisms of nucleation and convective processes in magmas foreshadowed a growth of interest in these subjects by fluid dynamicists which is still underway and is leading to a new understanding of igneous processes. His work on layered intrusions culminated posthumously when the accumulated knowledge on the Skaergaard intrusion was brought up to date and a description of other layered bodies worldwide was presented (Wager & Brown, 1967).

Wager's last visit to Greenland took place in 1953 on a combined expedition by the Universities of Oxford and Manchester (Fig. 10). Its object was to augment the collections and observations on some critical areas of the Skaergaard intrusion and complete the field work on the Kap Edvard Holm layered intrusion (which is approximately five times larger in outcrop area than Skaergaard) and the Kangerdlugssuaq alkaline intrusion. Wager regarded the latter as an example of igneous layering in an undersaturated to weakly oversaturated magma and was interested to compare it with his results for basaltic magmas. It has a total outcrop area of over 800 km^2, which gives an idea of the scale on which this short summer expedition was designed to operate. Most workers in Greenland today, used to aeroplanes and helicopters, would not contemplate such an expedition in this difficult country. However, the 1953 expedition was carried out successfully of foot or in a small dingy in spite of adverse ice conditions and meagre provisions. It harvested a considerable collection of material which allowed the laboratory work on Skaergaard, Kap Edvard Holm and Kangerdlugssuaq to proceed, and a number of papers and theses resulted.

In 1963, Wager was in America to present a paper on the mechanisms of adcumulus growth in the Skaergaard intrusion when it was suggested to him by A. R. McBirney that an attempt should be made to take over a drilling barge (as that time the abortive "Mohole Project" was ending) in order to investigate the nature of the unexposed part of the intrusion. This was a problem which had occupied Wager's attention considerably and he was enthusiastic. Nothing happened at once, probably because of the difficulties of locating such a ship in the dangerous ice-filled waters around Skaergaard, but Wager maintained interest in the idea and proposed to mount a major expedition in 1966 to drill a deep borehole into the hidden part of the intrusion and several minor holes with the object of recovering cores specifically designed to study the processes of sedimentation in the Layered Series. In addition, continuing work in several other areas was planned. The areas included the Kap Edvard Holm intrusion, the Lilloise intrusion which lies about 100 km to the E of Kangerdlugssuaq, which had been discovered by Wager in 1932 but never visited, and the Kialineq district, about 100 km to the south, which had only briefly (in 1932) been visited since Wager's discovery of a number of Tertiary plutons there in 1930 during the voyage of *Quest.* Financial support was obtained for this expedition and in September, 1965, Wager and Deer travelled to Scandinavia to discuss plans and arrange a ship charter. But the expedition was carried out without Wager's participation. On a visit to London where he was buying a camera for the coming work, he died suddenly of a heart attack on November 20, 1965. The expedition went ahead, but Wager's drive and enthusiasm was lacking and the drill cores have at this time never fully been examined, nor has any detailed study appeared on the sedimentation processes in Skaergaard such as Wager planned. Work in the Kialineq and Lilloise areas has continued and a new nepheline syenite body was discovered inland from Lilloise. The 1966 expedition erected a small cairn with a brass plaque on the highest point of the Skaergaard Peninsula in memory of this unique man, small in stature, but a giant among geologists and mountaineers (Fig. 11).

FINAL ASSESSMENT OF WAGER'S CONTRIBUTIONS

Geographically, Greenland is a part of North America, although politically part of Europe. Geologically it forms a unique link between these two continents, and in this respect Wager's work in East Greenland is an important contribution to the geological understanding of North America. In particular, Kangerdlugssuaq lies in a critical area. The Precambrian basement, which

Figure 11. Wager in his later years as Professor of Geology at Oxford.

Wager was the first to describe, is part of the ancient North Atlantic craton with close links to West Greenland and Labrador on the one hand and to Northwest Scotland on the other.

The Tertiary plutonic complexes and volcanics are the equivalent on the western side of the North Atlantic of the Faeroe Islands and the classic British Tertiary districts on the eastern side. These are now recognized as having formed from the hot spot now located under Iceland and, as such, give valuable information on the early stages of the break-up of Laurasia. In recent years, attention has focussed on the area as a possible land bridge for the interchange of vertebrates between Europe and North America up to the late Eocene.

Wager's work on the regional geology of the area was impressive. In spite of the exacting nature of the country, large by any standards, the often unreliable weather and demands imposed by the necessity for detailed work, not only on Skaergaard, but also on the Kangerdlugssuaq and Kap Edvard Holm intrusions, the 1935–36 expedition mapped around 10,000 km[2] of country. Called at the time reconnaissance mapping, it has stood the test of time and very few modifications have been made. At the head of Kangerdlugssuaq two new intrusions have been added, the Tertiary Gardiner complex (already referred to) and the Caledonian Batbjerg complex. However, on an earlier map (Wager, 1934), intrusive rocks at these localities are shown, apparently based on visual observations and float found at the head of Kangerdlugssuaq during the 1930 visit with *Quest*. The only other additions to the map in this time are minor. They include a mineralized quartz porphyry stock cutting the Kangerdlugssuaq syenites north of Amdrup Fjord and described by Geyti & Thomassen (unpublished company report) and an ultramafic plug just north of the Skaergaard northern margin described by Kays & McBirney in 1982. The latter may be the origin of the ultramafic blocks observed by Wager and Deer in the Skaergaard Border Group. This plug is largely covered with glacial drift and was only discovered by geophysical means. Further improvements to our knowledge of the area are the discovery of submarine volcanics in Miki Fjord and adjacent areas and the recognition of olivine-rich lavas in the lower part of the Blosseville Group. Wager had observed the pillow breccias, but, surprisingly, was not able to identify them correctly.

Wager's all-round interests extended into geomorphology, and, although never published, he left an abundance of observations and speculations which form the basis of a description of the geomorphology of the area published by Brooks in 1979. Wager's talents in this direction were best displayed by the classic paper on the antecedant drainage origin for the mighty Yo Ri and Arun gorges in the central Himalaya which led him to propose origins for the uplift of the Himalayan chain (Wager, 1937). In a similar way, his enquiring mind led him to consider the processes taking place in the mantle and lower crust which led to the formation of the North Atlantic ocean. These fundamental problems of the North Atlantic areas were inspired by his observations of the coastal dike swarm and flexure in East Greenland. Wager was clearly aware of the relevance which surface features have to processes taking place at depth, and it was seldom, and only recently, that geomorphologists considered their observations in terms of mantle rheology and chemistry, but Wager attempted to do just that.

During the 1935–36 expedition Wager became aware of the considerable extent of sediments in the Kangerdlugssuaq area, and with the help of macrofossils collected at this time and with knowledge gained on previous expeditions he was able to bracket the period of volcanism in East Greenland to rather narrow limits, approximately those of the Eocene. His conclusion, which has not been significantly modified by later work, is interesting as, at that time, geologists were not in agreement as to the age of volcanism in other areas of the North Atlantic Province.

That so much was accomplished in East Greenland, largely by the 1935–36 expedition is without doubt due to Wager's determination, dedication and incredible stamina. These traits are brought out best in two descriptions by his close associates. Of the assault on Everest in 1933, Hugh Ruttledge wrote of Wyn Harris and Wager: "knowing these men as I do they would not turn

aside from a climb within the limits of the possible." Deer (1967) in his biography of Wager published by the Royal Society wrote: "Wager was an exacting leader and sledge companion and gave neither himself nor his colleagues much respite. After the days sledging and evening pemmican, the collections were documented, notes written and problems discussed before the candle was snuffed. The next morning he would take his porridge at scalding temperature, smoke half a cigarette and was impatient to strike camp and begin the day. For lesser mortals there was no time to smoke even half a cigarette unless the porridge was diluted and cooled with a handful of snow." (At that time Wager was a heavy smoker, but those who knew him later in life remember him as fanatically opposed to nicotine. Anyone, no matter how august, who entered his room smoking was immediately sent out and all the windows were opened and the smoke dispersed with much waving of sheafs of paper and histrionic coughing.)

Wager's contribution to North American geology was not only regional. Ironically, after his death, work on Skaergaard has been continued, not by his countrymen, but by North Americans. Perhaps this is because Skaergaard led the way to an understanding, not only of layered intrusions worldwide, but to the excellent North American examples such as the Palisades sill, the Stillwater, Muskox and Duke Island complexes, to name a few. In the 1970s, a series of expeditions instigated and led by A. R. McBirney of the University of Oregon proposed that processes other than crystal settling (double-diffusive convection, liquid immiscibility) had possibly been significant. Also, the gravity and magnetic measurements conducted on these expeditions were not in favour of the subsurface form of the intrusion as proposed by Wager and Deer (e.g. McBirney & Nakamura, 1974; McBirney & Noyes, 1979). T. N. Irvine has carried out extremely meticulous mapping of layered features and backed up his observation by sophisticated laboratory and theoretical studies (e.g., Irvine, 1983 a & b). H. R. Taylor, Jr. and his coworkers (Taylor & Forester, 1979; Norton & Taylor, 1979) further showed that massive isotopic exchanges had taken place with circulating heated ground waters, leading to extensive alteration, so much so that in future the intrusion is in danger of becoming a type locality for metamorphic processes and already gives much insight into fluid flow during mineralization events. All subsequent workers have remarked on the quality of Wager and Deer's observations: the map of the intrusion has been very little modified in the ensuring 40 years and all the basic observations have stood firm.

A curious synchroneity exists between Wager's striving to understand the chemical processes operating in Skaergaard and the rise of modern techniques of trace element analysis. Without doubt Wager was a key figure in the introduction of these techniques to geological science and although he was in no way responsible for the improving technology, he was perceptive enough to grasp the opportunity at a very early stage.

Wager had two outstanding qualities which led to his success as a leader, researcher and explorer. He was both a thinker and a perfectionist, and the two are inextricably mixed and prominent in his doings. A perusal of his manuscripts shows a multitude of alterations, rewrites and polishings. His secretaries were frequently driven to desperation by the long successions of drafts which they prepared for each of his papers, usually working from collaged versions of earlier manuscripts, produced with the liberal aid of scissors and glue and annotated with alterations in an almost illegible hand, usually in pencil. Diagrams were sketched, perused for long periods, redrafted and the whole process repeated many times before he was satisfied. In fact it is doubtful if he ever achieved complete satisfaction, as he continued improving earlier work right up to his death. This is not to say that he was indecisive, it is clear that right at the outset of any piece of work he had a vision of how the finished product should be and this led many of his associates wrongly to construe him as being stubborn, a quality often ascribed to Yorkshiremen.

This tenacity was also a quality which stood him in good stead in his field work and mountaineering. His reluctance to accept a compromise is perhaps borne out by his attitude to the conquest of Everest. In the 60s he was invited to London to give his view on the Chinese claim to have climbed the mountain and filmed their ascent. Pre-war Everest veterans were invited to see the film as only these had knowledge of the Tibetan approach (post-war attempts being through Nepal). On returning to Oxford, Wager was pressed by his colleagues on whether the ascent was genuine. To this he responded brusquely that the mountain was still unconquered, as oxygen had been used both by the Chinese and by Hilary and Tenzing. (His own attempt which had come to within a few hundred feet of the summit had been without any "artificial aids" as he would doubtless have said).

Paradoxically, although an instigator of precise research tools at Oxford (such as neutron activation and mass specrometry), he was inclined to work by intuition. People who asked his criteria for the fitting of a trend line, or the basis for his estimation of the percentage of solidified magma in Skaergaard corresponding to a given stratigraphic location, were often surprised to find it was done by "eyeball" techniques. The lines were frequently based on very few control points and those points which did not fit were rationalized away. His methods, in short, would have been anathema to a person educated in objective scientific assessment. But, surprisingly, his conclusions were generally vindicated by further work. It would undoubtedly be too strong to describe him as an "arm waver," in view of the associations conjured up by this term in the field of global tectonics, but there is little doubt that his striving for the overall picture led him to move faster than the simple work of data acquisition would allow. It is of interest that, although the bulk of his research relied heavily on accurate chemical analyse, he apparently only made one analysis himself. This was in connection with his early work on the Whin Sill of northern England during his time at Cambridge, when such work was extremely time-consuming. Fortunately, his sense of intuition regarding significant directions for research was allied with the ability to collaborate with fastidious co-workers who were able to provide the nuts and bolts of his ideas.

Wager's sense of intuition is also apparent in his method of conducting interviews. He had little time for such documentation as references or C.V.'s. References were preferably taken verbally from respected colleagues. Otherwise, a face to face meeting was preferred, at which Wager could arrive at his own judgment as to the candidate's "soundness," an expression which has apparently now largely gone out of use but which to Wager no doubt described the qualities much admired in earlier times in polar explorers, mountaineers and empire builders.

He was less than perfect as a lecturer, but somehow his students became imbued with a marked degree of motivation. Many of his students went on to become leading earth scientists. They will all remember his strong protective attitudes to his collections. In particular, research students working on Skaergaard materials found it difficult to obtain samples from the inner sanctum where these were kept. In some cases, after several sample-selecting appointments they came away empty-handed and, on occasions, resorted to clandestine methods to obtain material for their projects. In some ways, this attitude was surprising. Wager had otherwise little time for museum collections which lay in drawers gathering dust. He did not generally regard stones as having an intrinsic value, they only became valuable after they had been worked on and documented. He had a number of brushes with the curator of the University Museum, whose main interest was in the prevention of the dispersal of his collections and their orderly storage. Wager, on the other hand, was more interested in extracting this material from its place of storage and crushing it to a fine powder for chemical analysis. Of course, his attitude to the Skaergaard collection was coloured by the fact that a considerable amount of data had already been amassed on these samples, whose numbers he remembered like the names of cherished friends. However, in view of the fact that he intended to return to Greenland and, according to his associates, could always remember, even after a lapse of several years, precisely where each sample had been taken, his extremely protective attitude is perhaps unexpected.

Although his scientific thinking was innovative, his methods were frequently conservative. In the field, he eschewed the use of rubber-shod climbing boots, preferring to stick with the old nailed type which he had grown up with in pre-war days. He regarded the consumption of any foods on expeditions other than porridge and pemmican to be unnecessary luxuries. He would doubtless have looked with scorn on freeze-dried chicken chop-suey or chile-con-carne which are so much favoured by today's outdoor fraternity. In his interpretation of scientific results he never employed any type of statistical evaluation and probable would have had little time for computers, although he might well have embraced remote sensing as a valuable addition, but no substitute, for field work.

Wager, although a small man, was surrounded by an aura of authority and it was generally clear that he would be the one to make the decisions. While not greatly enthusiastic about the administrative side of being a professor (which was at that time in Britain synonymous with being head of the department), he confided at one time that it was "nice to run one's own show." It was this forceful personality which led to the scaling of great mountains, the geological exploration of large areas of previously untrodden country and the perseverance for detailed scientific enquiry. This is the way he will be remembered.

ACKNOWLEDGMENTS

I thank Mrs. P. Wager for indispensable help, both with this paper and in other ways. The manuscript was prepared by Mrs. Druscilla Aruga.

L. R. WAGER'S PUBLISHED WORKS ON EAST GREENLAND

1933. The form and age of the Greenland Ice Cap: Geological Magazine, v. 70, p. 145–156.

1934. Geological investigations in East Greenland: Part I. General geology from Angmagssalik to Kap Dalton: Medd. Grønland, Bd. 105, 46 p.

1935. Geological investigations in East Greenland: Part II. Geology of Kap Dalton: Medd. Grønland, Bd. 105, 32 p.

1937. Kangerdlugssuaq Region of East Greenland: Geographical Jour., v. 90, p. 395–415.

1938. (With W. A. Deer.) A dyke swarm and crustal flexure in East Greenland: Geological Magazine, v. 75, p. 39–46.

1938. (With W. A. Deer.) Two new pyroxenes included in the system clinoenstatite, clinoferrosilite, diopside and hedenbergite. Mineralogical Magazine, v. 25, p. 15–22.

1939. (With W. A. Deer.) Olivines from the Skaergaard intrusion, Kangerdlugssuaq, East Greenland: American Mineralogist, v. 24, p. 18–25.

1939. (With W. A. Deer.) Geological investigations in East Greenland: Part III. The petrology of the Skaergaard intrusion, Kangerdlugssuaq, East Greenland: Medd. Grønland Bd. 105, 552 p.

1940. Epeirogenic earth movements in East Greenland and the depths of the earth: Nature, London, v. 145, p. 938–939.

1945. (With R. L. Mitchell.) Distribution of vanadium, chromium, cobalt and nickel in eruptive rocks: Nature, London, v. 156, p. 207.

1947. Geological investigations in East Greenland, Part IV. The stratigraphy and tectonics of Knud Rasmussens Land and the Kangerdlugssuaq region: Medd. Grønland Bd. 134, p. 64.

1948. (With R. L. Mitchell.) The distribution of Cr, V, Ni, Co and Cu during the fractional crystallization of a basic magma: Geological Congress, 18th Session, Great Britain, Pt. II, p. 140–150.

1951. (With R. L. Mitchell.) The distribution of trace elements during strong fractionation of basic magma - a further study of the Skaergaard intrusion, East Greenland: Geochimica Cosmochimica Acta, v. 1, p. 129–208.

1953. Layered intrusions. Medd. Dansk Geol. For., Bd. 12, p. 335–349.

1957. (With G. M. Brown.) Funnel-shaped layered intrusions: Geological Society of America Bulletin, v. 68, p. 1071–1074.

1957. (With E. A. Vincent & A. A. Smales.) Sulphides in the Skaergaard intrusion, East Greenland: Economic Geology, v. 52, p. 855–903.

1958. (With J. Van R. Smith & H. Irving.) Indium content of rocks and minerals from the Skaergaard intrusion, East Greenland: Geochimica Cosmochimica Acta, v. 13, p. 81–86.

1959. Differing powers of crystal nucleation as a factor producing diversity in layered igneous intrusions: Geological Magazine, v. 46, p. 75–80.

1960. (With G. M. Brown & W. J. Wadsworth.) Types of igneous cumulates. Journal of Petrology, v. 1, p. 73–85.

1960. The major element variation of the layered series of the Skaergaard intrusion and a re-estimation of the average composition of the hidden layered series and of the successive residual magmas: Journal of Petrology, v. 1, p. 364–398.

1961. A note on the origin of ophitic texture in the chilled olivine gabbro of the Skaergaard intrusion: Geological Magazine, v. 98, p. 353–366.

1963. The mechanism of adcumulus growth in the layered series of the Skaergaard intrusion. Mineralogical Society of America Special Paper 1, (Symposium on Layered Intrusions), p. 1–9.

1964. (With E. I. Hamilton.) Some radiometric rock ages and the problem of the southward continuation of the East Greenland Caledonian Orogeny: Nature, London, v. 204, p. 1079–1080.

1965. The form and internal structure of the alkaline Kangerdlugssuaq intrusion, East Greenland: Mineralogical Magazine, v. 34, p. 487–497.

1967. (With G. M. Brown.) Layered igneous rocks: Edinburgh: Oliver & Boyd, p. 588.

REFERENCES CITED

Brooks, C. K., 1969, On the distribution of zirconium and hafnium in the Skaergaard intrusion, East Greenland: Geochimica Cosmochimica Acta, v. 33, p. 357–374.

Brooks, C. K., 1979, Geomorphological observations at Kangerdlugssuaq, East Greenland: Meddeleiser om Grønland. Geosci. v. 1, 121 p.

Deer, W. A., 1967, Lawrence Rickard Wager, 1904–1965: Biographical Memoirs of Fellows of the Royal Society, v. 13, p. 359–385.

Courtauld, A., 1936, A journey in Rasmussen Land. Geographical Journal, v. 88, p. 193–215.

Ferguson, J., 1964, Geology of the Ilímaussaq alkaline intrusion, South Greenland: Meddelelser om Grønland, Bd 172, nr. 4. 82 p.

Frisch, W. & Keusen, H.-R., 1977, gardiner intrusion, an ultramafic complex at Kangerdlugssuaq, East Greenland: Grønlands Geologiske Undersøgelse Bull. no. 122, 62 p.

Hamilton, E. I., 1959, The uranium content of the differentiated Skaergaard intrusion, together with the distribution of the alpha-particle radioactivity in the various rocks and minerals as recorded by nuclear emulsion studies: Meddelelser om Grønland, Bd. 162, nr. 10, 35 p.

Irvine, T. N., 1983a, Observations on the origins of Skaergaard layering: Carnegie Institute Washington Year Bok, no. 82, p. 289–295. 289:

Irvine, T. N., 1983b, Skaergaard trough-layering structures: Carnegie Institute Washington Year Book, no. 82, p. 289–295.

Kays, M. A. & McBirney, A. R., 1982, Origin of picritic blocks in the Marginal Border Group of the Skaergaard Intrusion, East Greenland, Geochimica Cosmochimica Acta, v. 46, p. 23–30.

McBirney, A. R. & Nakamura, Y., 1974, Immiscibility of late stage magmas of the Skaergaard intrusion: Carnegie Institute Washington Year Book, no. 73, p. 348–352.

McBirney, A. R. & Noyes, R. M., 1979, Crystallization and layering of the Skaergaard intrusion: Journal of Petrology, v. 20, p. 487–554.

Mikkelsen, E., 1933, The Scoresby Sound Committee's 2nd East Greenland expedition in 1932 to King Christian IX's Land. Report on the expedition. Meddelelser om Grønland, Bd. 104, nor. 1, 71 p.

Nielsen, T.F.D., 1980, The petrology of a melilitolite, melteigite, ijolite, nepheline syenite and carbonatite ring dike system in the Gardiner Complex, East Greenland: Lithos, v. 13, p. 181–197.

Nielsen, T.F.D., 1981, The ultramafic cumulate series, Gardiner complex, East Greenland. Cumulates in a shallow level magma chamber of a nephelinitic volcano: Contributions in Mineralogy Petrology: v. 76, p. 60–72.

Norton, D. & Taylor, H. P., Jr., 1979, Quantitative simulation of the hydrothermal systems of crystallizing magmas on the basis of transport theory and oxygen isotope data: an analysis of the Skaergaard intrusion: Journal of Petrology, v. 20, p. 421–486.

Ruttledge, H., 1934, Everest: London; Hodder & Stoughton, p. 390.

Taylor, H. P., Jr., & Forester, R. W., 1979, An oxygen and hydrogen isotope study of the Skaergaard intrusion and its country rocks: a description of a 55 m.y.-old fossil hydrothermal system: Journal of Petrology, v. 20, p. 355–419.

Ussing, N. V., 1912, Geology of the country around Julianehaab, Greenland: Meddrelelser om Grønland, Bd. 38, 426 p.

Vincent, E. A., & Bilefield, L. I., 1960, Cadmium in rocks and minerals from the Skaergaard intrusion, East Greenland: Geochimica Cosmochimica Acta, v. 19, p. 63–69.,

Vincent, E. A. & Crocket, J. H., 1960, Studies in the geochemistry of gold - 1. The distribution of gold in rocks and minerals of the Skaergaard intrusion, East Greenland: Geochimica Cosmochimica Acta, v. 18, p. 130–142.

Wager, L. R., 1937, The Arun river drainage pattern and the rise of the Himalaya. Geological Journal, v. 89, p. 239–250.

Watkins, H. G., 1932, The British Arctic Air Route Expedition: Geographical Journal, v. 79, p. 353–367 & 466–501.

Geological Society of America
Centennial Special Volume 1
1985

Dr. Atl: Pioneer Mexican volcanologist

Winston Crausaz
Southwest Missouri State University
Springfield, Missouri 65804

ABSTRACT

Dr. Atl (born Gerardo Murillo in 1875) was the Universal Man of Mexico. In addition to starting the Mexican Mural Movement and launching the careers of such famous artists as Siqueiros, Rivera, and Orozco, he was a revolutionary who, with blazing guns, political intrigue, and innumerable essays, helped to forge the modern Mexican democracy. His first love was volcanoes, and for sixty years he produced an avalanche of sketches, paintings, poems, essays, and monographs devoted to that subject. The mass of paintings and two monographs constitute his main contribution to volcanology.

Between 1911 and 1914 he studied Vesuvius, Etna, and Stromboli under the European volcanologists Perret and Friedlaender. The Mexican volcano Popocatepetl was his favorite mountain. When a crew of sulfur miners awakened it by blasting the crater with dynamite, Dr. Atl prepared a monograph based on decades of personal observations, interviews, Aztec legends, early written reports, sketches, paintings, and historical photographs. He described in detail the first man-made volcanic eruption, which raged in the crater from 1919 to 1938, and formed domes similar to the ones on Mt. St. Helens.

In 1943, when Paricutin erupted from a corn field, Dr. Atl raced to the site and started a seven-year study which culminated in the publication of his second scientific monograph. Although Dr. Atl was 75 at the time, his outlook was surprisingly modern. After discussing the ideas of Alfred Wegener, he declared that the forces of continental drift had formed the volcano.

At the age of 83, he sketched and painted a series of oblique landscapes from airplanes. Six years later, in 1964, while working on three murals in Cuernavaca, Dr. Atl died.

THE LIFE AND DEATH OF DR. ATL

Dr. Atl was born as Jose Gerardo Francisco Murillo (Fig. 1) in the city of Guadalajara in 1875. Murillo especially disliked the romantic religious paintings of the 17th Century Spanish artist Bartolome Murillo. He therefore changed his own last name to Atl, which means *water* in Aztec. Later, the famous South American poet Lugones added the title *Doctor*.

Dr. Atl became a major figure in art, science, literature, revolution, philosophy, poetry, history, politics, and social science.

As an artist, Dr. Atl started both the Mexican Mural Movement and the Revolutionary Art Movement. He launched the careers of Orozco, Rivera, and Siqueirios. His impressionist paintings and sketches captured the essence of Mexico's active volcanoes (Fig. 2).

As a scientist, he published a monograph about the Mexican volcano Popocatepetl, and another about Paricutin. Many of his sketches and paintings realistically record the details of volcanoes and their eruptions (Fig. 3).

As a writer he published several novels and innumerable short stories. One of the latter, "El hombre y la perla," (The man and the pearl) may have been the basis of John Steinbeck's novel *The Pearl* (Adelman, 1976).

As a revolutionary, he persuaded the intellectuals of Mexico City to move to the safe city of Orizaba. There he organized them

Figure 1. Self-portrait of Dr. Atl painted in 1955 and published in color on a Mexican postage stamp.

into an effective propaganda machine that produced newspapers and posters. Dr. Atl is regarded as one of the fathers of the Mexican Revolution, and a father of Mexican democracy.

As a philosopher, he studied philosophy in Italy and Paris and frequently addressed the great problems of man and the universe.

As a poet he wrote, among other works, the magnificent *Las sinfonias del Popocatepetl (The symphonies of Popocatepetl)*. After publication in Mexico, it was translated into Italian and published in Milan. While it contains little of scientific value, it captures the spirit of the wind, the forest, the rocks, the silence, the cold, and the eruptions of a great volcano.

As an historian, he published a six volume work on the churches of Mexico, which included photographs by the painter Frida Kahlo; as a politician, he managed to block a loan of 130 million French francs to the Mexican villain General Huerta; and as a social scientist, he published a two-volume monograph on the popular arts of Mexico.

In later years Dr. Atl terrorized Mexican leaders with his frequent and sharp attacks in the press. By 1920 American photographer Edward Weston wrote that as they walked down Madera Avenue, every second person driving or walking "bowed or called to him" (Weston, 1973).

Early Travels

Before he was 21, Dr. Atl had already spent four months exploring the mountains between Lake Chapala and the Pacific coast, excelled in art, founded a newspaper, and published a novel about Jules Verne. He later went to Mexico City armed with glowing letters of recommendation from his art teacher. These enabled him to obtain a 1,000-peso grant from the dictator Don Porfirio Diaz who was eager to send promising Mexicans to Europe where they could absorb and bring back European culture. Dr. Atl visited New York, Paris, and later Italy, where he studied under the Marxist Labriola. Labriola's lectures were to no avail, because, despite his revolutionary zeal and knack for being in the midst of revolutions (or causing them), Dr. Atl remained opposed to Marxism his entire life. In Lausanne, he frequently argued with a young Marxist upstart . . . Lenin. At various times he also visited Spain, England, Russia, Egypt, Germany, and China.

Return to Mexico

After a seven-year absence, Dr. Atl returned to Mexico and initiated a revolution in Mexican art. At about the same time, 1905, he fell in love with the young niece of the impressionist painter Joaquin Clausell. Although the love affair was a disaster, Clausell took the artist on a mountaineering expedition to Iztaccihuatl (Atl, 1921) and that started his lifelong affair with volcanoes.

Dr. Atl was so impressed with the then unknown artist Diego Rivera that he arranged an exposition of his paintings and encouraged his friends to buy them. With the money from the show, Rivera was able to study art in France and Spain.

Second Visit to Europe

When the Mexican revolution began, Dr. Atl returned to Paris where he arranged an exposition of his Mexican-volcano paintings. The show was a success, and Dr. Atl was greeted as a returning hero. In 1912 a newspaper reported (Luna Arroyo, 1952):

Atl is in Paris. He has arrived—a new Atlas—supporting the weight of a world, a world of luminous mountains and tragedies. All Paris has admired his great canvases, painted with a special method, and his small sketches made with insuperable mastery and with a deep sentiment of nature. The life of this painter, doubled as a thinker, is a wonder, intense and singular, like the life of the great adventurers of history, the conquistadores, or the giant figures in the dreams of great artists. At the present time his life is full of legends, which are the same in his Aztec city as in our city of lights. They form a veil that impedes our ability to see through his brutal realism to the true man, who is more interesting, perhaps, than the man of the legend.

Moving on to Italy, he studied volcanology at the University of Naples and learned from some of the great specialists of the time. These included Frank A. Perret, an authority on Stromboli,

Vesuvius and Pelee, as well as Immannuel Friedlaender, the editor of the *Volcanological Review.* In 1921, Dr. Atl guided Friedlaender on a tour of Popocatepetl.

Revolutionary Mexico

Enroute back to Mexico, Dr. Atl stopped in Washington to confer with President Wilson. Mexico was in the midst of the revolution, and wild rumors preceded his arrival in Mexico City. Orozco heard that he was back "disguised as an Italian, without his beard and speaking Italian exclusively" (Orozco, 1962). The artists and intellectuals could not decide if they should take the side of Carranza, the future President of Mexico, or the side of Pancho Villa. They finally arranged a meeting but could make no progress. In the middle of this chaotic situation, Dr. Atl arrived. In an eloquent speech addressed to the intelligensia he persuaded them to unite behind Carranza, and move to the safe city of Orizaba. According to one account, they just managed to get on the last train out of Mexico City (Cherney and Bennett, 1981). Once in Orizaba, they devoted most of their energy to publishing *The Vanguard,* the newspaper of which Dr. Atl was the editor. In an amusing passage, Orozco (1962) describes the activities of the revolutionary:

Dr. Atl, with rifle and cartridge belt, would be off to Vera Cruz to visit Obregon on the field of battle and collect money for our whole establishment; all the while conducting a ferocious political controversy with the engineer Felix P. Palavicini and resolving a thousand problems and still having time left over in which to write editorials, and books, and even poems, without once neglecting his magnificent collection of butterflies.

Devoted Volcanologist

With the return of peace, Dr. Atl devoted himself to painting, writing, and travel. At times he spent months on the Mexican volcanoes. The timing of the 1919 eruption of Popocatepetl was perfect. Starting in 1905, Dr. Atl had thoroughly explored the volcano, and even visited the depths of the summit crater. In 1921 he wrote in a book of poetry:

For millions of years it slept in the silence of death, for millions of years the wind lashed it, for millions of years the forces of nature have tried to destroy it. They closed its mouth, gnawed at its vertebra, they shook its formidable mass and tore its lips which in other times vibrated from thundering eloquence. But one day in its venerable old age, its entrails stirred, and from the decayed lips of the Colossus, fire shot forth again. Oh marvellous teaching of nature! Nothing is old and nothing has died: in the end of the Destruction is Life.

In February of 1943, Paricutin was born several hundred miles west of Mexico City. When Dr. Atl disappeared shortly after the eruption, his friends were worried. He was, after all, nearly 70. They finally found him living in a shack near the base of the volcano. For months, he circled it, sometimes near, sometimes far, making observations, taking notes, painting and sketch-

Figure 2. Dr. Atl painted this scene of the erupting Paricutin in 1943. The painting was published in color on a Mexican postage stamp.

ing. In 1950, after consulting all the available scientific publications, he published *Como nace y crece un volcan: El Paricutin.* The volume was published under the sponsorship of the president of Mexico, Miguel Aleman.

There was one last way to see the volcanoes. He had already looked at them as a scientist, poet, mountaineer, artist, and historian. At the age of 83, with only one leg, he crawled into airplanes, and repeatedly flew over his mountains. From that vantage point he painted aerial landscapes *(aeropaisajes).* These new paintings also utilized the new curvilinear perspective, which showed the broad curve of the earth's horizon, as it might look from space.

Six years later he was still working, this time on some murals in a Cvernavaca hotel. They were his last project. (Dr. Atl, 1978) [Except as noted, this life of Dr. Atl is based on the biography *El Dr. Atl* (1952) by A. Luna Arroyo.]

BOOKS

Popocatepetl

Popocatepetl may be one of the most famous mountains in the world, but before 1939 it had not been comprehensively

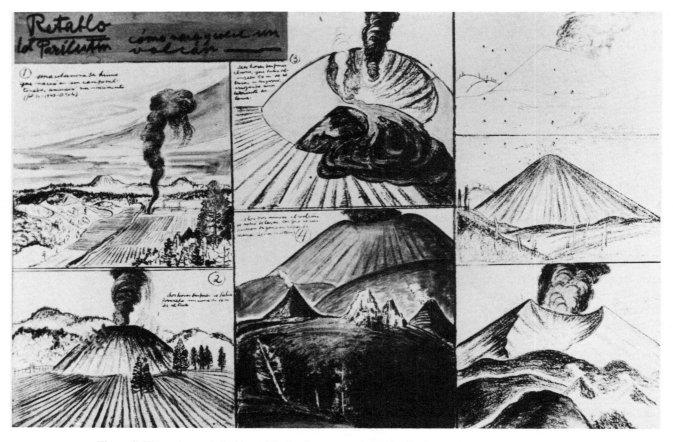

Figure 3. These charcoal sketches of Paricutin were made by Dr. Atl in 1943. (From the collection of the Museo Nacional de Arte Moderno, photographed by Departamento de Artes Plasticas I.N.B.A., Mexico City; courtesy of Museum of Modern Art of Latin America, Organization of American States, Washington, D.C.)

surveyed. That changed with the publication of Dr. Atl's *Volcanes de Mexico: La actividad del Popocatepetl.* Dr. Atl included nearly everything that was known at the time, including statistical data, such as the location, altitude, and composition, and historical studies by the conquistador Cortez and the German explorer Humboldt. There were even a few short articles by such scientists as Aguilera and Ordonez, Friedlaender, Waitz, and Felix and Lenck.

His main contribution to the study of the geology of Popocatepetl was his description of the eruption that started in 1919. It had, Dr. Atl wrote, "acquired an importance of the first order in the history of geology, because it was the result of purely artificial action."

For decades, the natives had mined sulfur from the crater. Near the bottom were three fumaroles which were especially rich. From April to November a broad shallow lake covered the bottom. In order to speed up production, the miners placed 28 sticks of dynamite near several fumaroles in the crater. The resulting explosion threw rocks high in the air, and the bottom and sides of the crater shook violently. The *volcaneros* must have expected that. What they did not expect was the violent snow storm, thick

clouds, and wind so strong they could not walk. The storm lasted six days, and coated the crater with an unseasonally thick layer of snow. The workers also did not expect the man at the winch to flee, leaving them stranded in the bottom of the crater. The cable was impossible to climb, as it was 90 meters long and covered with snow. The walls were vertical and plastered with snow, so they too were impossible to climb. The *volcaneros* were trapped from February 19 to February 25. During that time, two were killed by avalanches, several froze, and two went crazy, ran to the walls, and attempted to claw their way out. The survivor noted that by the last day, the sun had started to melt the piles of snow left by the storm. Bubbles and "thick jets of steam" shot out of the lake or bog in the middle of the crater.

By the end of March, a pile of scoria formed over the ancient chimney, and jets of smoke shot up from glowing cracks near the center. By April, the miners reported that the pile of ashes had slowly converted to a dome. Dr. Atl called the structure an arch, while others called it an upside-down crock. Waitz made the next observations over a year later, in October of 1920. He described the dome as a *tapon* (cork) or *embolo* (piston). The spine of hardened lava, Waitz wrote, was forced up by pressure from below. Dr. Atl disagreed, and noted that there were no recent lava eruptions from the crater of Popocatepetl. The crater was choked with scoria, not lava, from the last eruption. Also, there was no trace of a previous dome in the crater. On November 19, 1920, Dr. Atl climbed to the summit and sketched the new dome. It was 10 meters high and 40 meters across. Two cracks which crossed each other in the form of a cross glowed in the center. The next day the dome was 19 meters high and 50 meters across.

These ratios of height to diameter are nearly identical to those of the domes that formed in the crater of Mt. St. Helens. At one point, the October dome in Mt. St. Helens was 10 meters high and 40 meters wide (Moore, and others, 1981). By March of 1921 the dome in Popocatepetl had increased to a height of 26 meters. Dr. Atl's sketch shows that the shape of the dome was nearly identical to that of the June 1980 dome of Mt. St. Helens (Foxworthy and Hill, 1982). By 1923 it had stopped growing, and explosive eruptions formed a crater in the center. Most activity stopped by 1938, at which time the dome looked like a badly weathered cinder cone.

From 1920 to 1924 a number of central explosions produced huge clouds of thick ash. Dr. Atl published a photograph by the famous German photographer Hugo Brehme, which shows the billowing ash shooting up from the crater. Some of these clouds reached 6,600 meters above the top of the crater. The explosions were preceded by a train-like rumble, followed by clouds of ash that poured out for 30 to 40 seconds.

Dr. Atl was also interested in the Pico del Ventorillo, a 5,000-meter spire on the northwest side of the summit cone. In 1903 he made the first ascent. It is still considered a respectable climb by modern mountaineers. Dr. Atl proposed that some lava came only part way up the main chimney, then squeezed out the side. There is a similar spire on the northwest side of Pico de Orizaba.

Paricutin

The February 1943 birth of the new volcano, Paricutin, spawned innumerable books and articles. The most popular reference is Foshag and Gonzalez (1956) *Birth and development of Paricutin volcano,* which is a set of four volumes. It is cited in such texts as Simkin, and others (1981), Green and Short (1971), Macdonald (1972), and Holmes (1965). Gonzales and Foshag wrote a companion article which is cited in Williams and McBirney (1979), and Macdonald (1972). *El volcan de Paricutin* by Ordonez (1947) is less popular, although Bullard (1976) cites it. The Ordonez book is written in parallel columns of Spanish, English, and French. It contains over 50 illustrations, including several color paintings by Ordonez himself.

None of the above texts mentions *Como nace y crece un volcan: El Paricutin* (1950), by Dr. Atl. This is unfortunate, because in many ways it is the best book on the volcano. It contains twice as many illustrations as any of the others. A number of these are in brilliant color, and preserve the excitement of the events as no photograph could. About one painting of a twisting column of glowing ash and two tornados, Dr. Atl wrote:

From the igneous fountain surged an enormous whirlwind of fire and other whirlwinds of ash accompanied it in a fantastic dance. The glowing column was transformed into a thick cloud, and a suffocating heat invaded the atmosphere.

Other illustrations show, in meticulous detail, the bright red ribbons of lava that flowed out of the central vent. Dr. Atl did not limit himself to strictly scientific illustrations. The book also includes a magnificent portrait of the ragged Dionisio Pulido, the man who actually witnessed the birth of Paricutin.

Much of the text is devoted to a chronological description of the events associated with the first few years of the volcano. However, Dr. Atl was not content merely to describe events. If the first part of the book is representative of the descriptive scientist, the second part is representative of the speculative scientist, the philosopher, and the cosmologist. The main source of the volcanic heat, he wrote, could not have been from the mixing of iron and sulfur, the combustion of petroleum, or the decay of radioactive elements. He followed the suggestion of Professor Gratton of Harvard that the most probable source of volcanic heat was the residual energy left over from the primative globe. To that, only a small amount of heat from radioactive decay or chemical reactions would have to be added to trigger volcanic activity.

Sunspots, he noted, were similar to volcanic eruptions. Since both were manifestations of the same primordial fire, he suggested a cosmological definition of volcanism. It is the direct manifestation of primordial fire in celestial bodies. Had he lived, he would have been delighted by the discovery of volcanism on Io, Mars, and Venus.

In 1950, the geologists associated with Paricutin were unaware of or ignored the possible connection between continental drift (plate tectonics) and volcanism. Moreover, they never wrote about these underlying questions, although they devoted hundreds of pages to minute descriptions of the volcano's history. Dr. Atl was interested in both the details and the general principles. He embraced the theory of continental drift and began chapter one by quoting (in Spanish) from his French edition of *La génèse des continents et des océans* by Wegener. The quote reads:

One thing is certain. The same forces that produce the great folded mountains cause the displacement of the continents. Continental drift, faults and the pushing of mass, earthquakes, volcanism, the alternations of transgressions and polar migrations form, no doubt, a grand system. . . .

If there could be any doubt about the enthusiasm which Dr. Atl showed for the theory of continental drift, it should be put to rest by the following quotation from his chapter "The origin of Paricutin," in which the italics are in the original: *It is possible to conclude, as a consequence, that this abundance of volcanic structures is due to innumerable fractures caused by continental drift. . . .*

CONCLUSION

Dr. Atl signed his name to 1,000 paintings, 11,000 drawings, and 1,200 copies of his book about Paricutin. That accomplishment alone should have been enough to make his presence felt throughout the nation. Today, a large book titled *Dr. Atl* graces the shelves of many Mexican book stores. In 1964, after he died, *Time* magazine noted "Last week Dr. Atl's fire finally went out at the age of 89." It was a fire that burned long and bright, and its afterglow will continue to light the worlds of art and science for years to come.

REFERENCES CITED

Adelman, C. K., 1976, La obra narrativa de Gerardo Murillo, Dr. Atl. [Ph.D. dissert.]: Urbana-Champaign, Illinois, Univ. of Illinois, 288 p.

Atl, Dr., 1921, Las sinfonias del Popocatepetl: Mexico, Ediciones Mexico Moderno, 149 p.

—— 1939, Volcanes de Mexico, La actividad del Popocatepetl: Mexico, Editorial Polis, 73 p.

—— 1947, Un tratado de Geologica dinamica, El Paricutin: Cuadernos Americanos, v. 32, no. 2, p. 104–119.

—— 1950, Como nace y crece un volcan, El Paricutin: Mexico, Editorial Stylo, 152 p.

—— 1974, Dr. Atl, pinturas y dibujos: Mexico, Fondo editorial de la plastica Mexicana, 137 p.

—— 1978, Dr. Atl, Inventor de paisaje: Mexico, Banco Nacional, 55 p.

Bullard, F. M., 1976, Volcanoes of the earth: Austin, Univ. of Texas Press, 579 p.

Cherney, L., and Bennett, M., 1981, Dr. Atl, Father of Mexican Muralism: Americas, v. 33, no. 2, p. 26–33.

Foxworthy, B. L., and Hill, M., 1982, Volcanic eruptions of 1980 at Mount St. Helens: The first 100 days: U.S. Geological Survey Professional Paper 1249, 125 p.

Foshag, W. F., and Gonzales, R., 1956, Birth and Development of Paricutin volcano, Mexico: U.S. Geological Survey Bulletin 965-D, 489 p.

Gonzales, J., and Foshag, W. F., 1947, The birth of Paricutin: Smithsonian Inst. Ann. Rept. for 1946, p. 223–234.

Green, J., and Short, N. M., 1971, Volcanic landforms and surface features: New York, Springer-Verlag, 519 p.

Holmes, A., 1965, Principles of physical geology: New York, Ronald Press, 1288 p.

Luna Arroyo, A., 1952, El Dr. Atl, paisajista puro: Mexico, Editorial Cultura, 247 p.

Macdonald, G. A., 1972, Volcanoes: Englewood Cliffs, Prentice-Hall, 510 p.

Moore, J. G., Lipman, P. W., Swanson, D. W., Alpha, T. A., Growth of lava domes in the crater, June 1980-January 1981: *in* Lipman, P. W. and Mullineaux, D. R., The 1980 Eruption of Mount St. Helens, Washington: U.S. Geological Survey Prof. Paper 1250, 844 p.

Ordonez, E., 1947, El Volcan de Paricutin: Mexico, Editorial "Fantasia," 181 p.

Orozco, J. C., 1962, Orozco, an autobiography [trans. by R. C. Stevens]: Austin, U. of Texas Press, 171 p.

Simkin, T., Siebert, L., McClelland, L., Bridge, D., Newhall, C., Latter, J. H., 1968, Volcanoes of the world: Stroudsburg, Hutchinson Ross Pub. Co., 233 p.

Waitz, P., 1920, La nueva actividad y el estado actual del volcan Popocatepetl: Memoires, Societe sientifique "Antonio Alzate," T. 37, p. 295–313.

Weston, E., 1973, The daybooks of Edward Weston, Mexico [ed. N. Newhall]: Millerton, New York, Aperture Press, 214 p.

Williams, H., and McBirney, A. R., 1979, Volcanology: San Francisco, Freeman, Cooper and Co., 397 p.

Printed in U.S.A.

Geological Society of America
Centennial Special Volume 1
1985

Joseph A. Cushman and the study of Foraminifera

*Ruth Todd**

ABSTRACT

Cushman's interest in forams began with his examination of the *Albatross* collections. In 1906, when he began his systematic study of the Foraminifera for the U.S. National Museum, they were regarded as little more than scientific curiosities of no geologic value because most were believed to range from Cambrian to Recent. Cushman found the existing ten-family classification insufficiently detailed to encompass the myriad distinctions seen in his material. That classification, based primarily upon shape, did not take into consideration the different kinds of wall composition nor the phylogenies of genera.

In 1912, Cushman began an intermittent lifelong association with the U.S. Geological Survey. Two years later he successfully used forams to determine the ages of beds in a water well in South Carolina. In the early 1920s, Cushman took a consulting job with the Marland Oil Company to work briefly in Mexico. His fee gave him the financial means to retire from commercial work at the age of 43 and to build a private laboratory in Sharon, Massachusetts. There he resumed his research studies, accepted students, and started his journal. In the middle 1920s, a controversy developed between Cushman and J. J. Galloway over publication of their respective classifications. Emotions ran high for some years because Galloway believed that Cushman had stolen ideas from his manuscript, which the two had discussed together.

INTRODUCTION

To Dr. Mary J. Rathbun who more than any other person is responsible for starting my serious work on the foraminifera twenty-five years ago. . . .

Sharon, Mass. *June 9, 1928*

—inscription in a presentation copy of the first edition of Cushman's textbook on Foraminifera.

Thus Cushman dated and credited the beginning of a career which was to establish a new field of scientific investigation and a new specialized occupation for thousands of paleontologists around the world.

Cushman's parents were Darius Cushman and Jane Frances (Fuller) Pratt. His father kept a store where he sold and repaired shoes in the small college town of Bridgewater, Massachusetts. When Cushman was 16, his father died and the following few years he and his mother lived under a financial strain. She took college students to board while Cushman earned his way through school by various jobs and obtained a scholarship to Harvard College.

Cushman's brief marriage to Alice Edna Wilson of Fall River ended with his wife's death in January 1912, leaving him with three young children. His second marriage followed in the fall of the next year to Miss Frieda Gerlach Billings of Sharon. The financial straits in which he spent most of his youth were eased by his second marriage as well as by an inheritance for his children from their maternal grandfather. His early employment as curator at the Museum of the Boston Society of Natural History, by which he earned his living for some 20 years, ended in May 1923 when he felt he could earn enough by consulting fees and at the same time have more free time to devote to the study of Foraminifera. His family was all-important to him. His daughter Alice left high school to train in business school and served as his secretary from 1923 until her father's death.

The life of Joseph Augustine Cushman (1881–1949) spanned the time when an obscure hobby was transformed into a science, and his research had a large part in making the study of Foraminifera of practical use. Most of his publications, over 550

---

*Editors' Note: The friends and colleagues of Ruth Todd are deeply saddened by the news of her death on the 19th of August, 1984. We are grateful to Doris Low, Ruth's long-time friend and associate, for completing the minor revisions of this article.

titles (Todd, 1950, p. 40–68), consist of basic data: describing and naming species, organizing them into a natural classification based upon their evolution, and documenting their occurrences in ecologic and stratigraphic frameworks. He published a modern textbook on Foraminifera, started the first journal devoted solely to research on them, and established a laboratory for their study and teaching. His collection of specimens and books on Foraminifera, willed to the Smithsonian Institution, is the largest in the world.

BEGINNING WORK ON THE *ALBATROSS* COLLECTIONS

Cushman's first acquaintance with Foraminifera was in Robert Tracy Jackson's paleontology course at Harvard University; they interested him to such an extent that he changed from his earlier interest in cryptomagic botany to paleontology. In the first winter after receiving his bachelor's degree, he wrote to Smithsonian Institution Secretary S. P. Langley to request a few samples taken by the U.S. Fish Commission Steamer *Albatross,* whose recent biological collecting voyages, begun in 1888 in the Pacific and Atlantic Oceans, had brought back vast collections of bottom sediments, then stored in the U.S. National Museum. This initial request found its way to Dr. Richard Rathbun, assistant secretary in charge of the National Museum. Eleven samples were sent to him, material from the Gulf of Mexico previously studied by James M. Flint, together with a copy of the museum's publication of Dr. Flint's descriptive catalogue of Foraminifera (Flint, 1899).

Sometime during the following two summers (1904 and 1905), which Cushman spent as a "paid investigator" at the Woods Hole Biological Laboratory of the Bureau of Fisheries, he became acquainted with Mary Jane Rathbun, assistant secretary Rathbun's sister, who was a curator of crabs at the National Museum. She encouraged him to take up the study of Foraminifera, and probably through her intervention, he began receiving additional *Albatross* samples. During his early years as curator at the Boston Society of Natural History, Cushman continued his publications on a variety of subjects, mostly botanical. After his second summer (1905) at Woods Hole, during which he worked up some ostracodes of Vineyard Sound and studied the Foraminifera of Woods Hole, he seems to have turned to Foraminifera more seriously.

In the spring of 1906, Rathbun gave Cushman a commission to study and report upon the *Albatross* collections from the entire North Pacific region. It was agreed that the National Museum would publish the complete results of his investigations and, in addition, would publish in advance in the proceedings of the museum any interesting finds and descriptions of any new species found. The museum would also provide photographs and arrange for such drawings as might be necessary. It was further agreed that Cushman would segregate and identify the species and mount four sets of specimens in wooden slides with a black-painted concavity, such as had been used by Dr. Flint whose classic study of the *Albatross* collections from the Caribbean had

Figure 1. Joseph A. Cushman in his laboratory.

appeared only seven years before. The first, most complete set of both old and new species was to be delivered to the National Museum for permanent preservation. In the agreement, Secretary Rathbun pointed out that inasmuch as an act of Congress prohibited the government's acceptance of volunteer assistance, it was necessary for the National Museum to compensate Cushman for his services. The compensation offered was to be one of the duplicate sets of specimens. Rathbun added, "I appreciate that this is a very inadequate return for such services, especially in view of the amount of tedious labor involved, but such is the common reward of science" (R. Rathbun letter, 16 March 1906).

It was a colossal job to offer a young student still three years away from his doctorate. But to Cushman it was an unparalleled opportunity, and he embarked upon a relationship with the National Museum that was to last throughout his life. In having access to the *Albatross* collections, he had available almost unlimited specimens illustrative of the principle of recapitulation. The theory of recapitulation, that the successive stages that an organism exhibits in its life history represent the stages that the genus has passed through in its evolution, had been introduced at Harvard by Louis Agassiz. Cushman's teacher, Robert Tracy Jackson, had applied the principle to the study of echinoids and his teacher's teacher, Alpheus Hyatt, had still earlier applied it to the study of nautiloids. The first two papers Cushman published, while still an undergraduate at Harvard, were on localized stages in common plants in which he gave some examples of recapitulation; namely, that along the stem of a single adult plant may be found in sequence leaves like those of the seedling, then like those of the young plant, and then the adult, and finally regressing to seedling-shaped leaves below the flower. He soon applied this same principle to Foraminifera in his bachelor's thesis on the Lagenidae (Cushman, 1905).

In Cushman's doctoral dissertation, "The phylogeny of the Miliolidae," a study of 20 genera in a single family, Cushman sought to illustrate the principle of recapitulation in Foraminifera. Using or adapting previously illustrated thin sections to supplement his own effective illustrations and diagrams, he gave numerous examples of the existence of structures comprising the youthful stages of one genus that copied the structure of another genus, presumably ancestral to the first genus. Moreover, he found expected discrepancies in that not all species of the genus in question conformed to this pattern, and that not all specimens conformed to the species in question. This discrepancy in one species of a genus was described as acceleration, i.e., skipping or shortening of one or more of the expected stages. In a group of specimens of a species, on the other hand, the discrepancy was described as dimorphism, i.e., two distinctly different shapes (microspheric and megaspheric) attributed to two different types of reproduction (sexual and asexual).

Summarizing the contemporary studies done by Lister (1895) on the biology of Foraminifera, Cushman described the microspheric individual, with a relatively small proloculum, as having been the result of conjugation of two zoospores (gamonts) following withdrawal of protoplasm from the test (or to the outer edge of the test in some genera); and the megaspheric individual, with a relatively large proloculum, as having been asexually produced following withdrawal of protoplasm from the test and its break-up into small masses of protoplasm, each of which secreted a new simple sac-like chamber to become the proloculum of the new specimen.

This dimorphism exhibited itself in several ways: in a larger average size, more numerous chambers, and in the most complete sequence of developmental stages in microspheric forms; and in a smaller average size, fewer chambers, and in shortened or skipped developmental stages in megaspheric forms. This concept of palingenesis in Foraminifera gave form to the evolution of and, consequently, to classification of the 20 genera treated in his dissertation. Although the dissertation was never published, it set for him a pattern for a new classification, as well as a method of going about his work. There was still much to be studied, the vast collections of the steamer *Albatross* lying fallow in the collecting jars stored at the National Museum, now available and awaiting his study.

On 18 June 1909, the day after he passed his doctoral examinations, he wrote to Miss Rathbun: "I have series now of certain things that were so intensely interesting that I nearly neglected my examination on account of them. I never dreamed of having such material but I am sorry to say it will change certain ideas of the classification which some ultra-conservative people may not like (Cushman letter to M. J. Rathbun, 18 June 1909). Yet he pro-

ceeded very slowly with his unorthodox ideas, for he continued to follow the pre-existing classification for nearly 18 years before proposing his own reclassification.

PRE-EXISTING STATE OF KNOWLEDGE ABOUT FORAMINIFERA

At the turn of the century, Foraminifera were regarded merely as obscure objects of curiosity, even by most of the handful of investigators who were occasionally concerned with them. In the British series of volumes, *Cambridge Natural History,* the formal classification of Foraminifera included only the 10 generally accepted families of Brady (1884). Marcus Hartog (1906), the author of the section on Foraminifera, wrote (p. 58): "We must warn the tyro that . . . rigid definitions are impossible: . . . transitional forms making the establishment of absolute boundaries out of the question." And after describing the 10 families and a few genera, he added: "Genera range from lower Cambrian to present day."

Even Cushman's idol, James M. Flint, speculating about the basis of classification, wrote:

The separation . . . where variation is the rule and passage from one type to another is by a sliding scale and not even by a series of steps, is extremely difficult . . . however elastic the definitions of species, or even genera, there will often be a margin of doubt, and the determination of place in the classification must be left to the preference of the individual observer (Flint, 1899, p. 256).

The classic reference for Foraminifera at that time was volume 9 (Zoology) of the *Challenger* expedition reports by Henry Bowman Brady (1884), an immense work of 814 pages and a separate volume of 115 plates of outstandingly beautiful drawings, many in color, by A. T. Hollick, whose high standard of illustration has rarely been equaled. Charles Davies Sherborn's monumental bibliography and index of the Foraminifera (Sherborn, 1888, 1893–96) provided access to virtually all that was known of the described species, including their synonymies. In addition to these works, early works of Soldani, Fichtel and Moll, d'Orbigny, Carpenter, Reuss, Ehrenberg, Terquem, Egger, Schwager, and Stache helped to fill in blank spaces in the geographic and stratigraphic records of species known at the time. Cushman thus began his study of Foraminifera with a rich store of knowledge in the field. The first 15 years of the new century also saw the blossoming of the study of Foraminifera elsewhere, particularly in England, Germany, and the East Indies. Yet the study of Foraminifera was still not something one did for a living, but rather an after-hours hobby or, if independently wealthy, an occupation to fill otherwise vacant hours.

EARLY WORK FOR THE U.S. NATIONAL MUSEUM

The original agreement by Assistant Secretary Rathbun in 1906 specified that the only recompense the museum could offer for work on the *Albatross* collections would be one "good and duplicate series" [of specimens]. Rathbun added, however, that it might be possible to pay for his work at a low rate, perhaps so much a slide, and asked Cushman to make a suggestion in regard to it. Apparently, a price of 15 cents a slide was suggested because the sum of $150 for 1,000 slides was included in a statement of $375 paid to Cushman in 1910, the balance of $225 being for 217 shaded drawings at $1 each and 16 line drawings at 50 cents each. Earlier in a similar account, $750 had been paid for 2,000 slides, 395 shaded drawings, and 110 line drawings done in 1908 and 1909. It was a good start at earning one's living by studying Foraminifera.

With his increased acquaintance with forams, and the constant changing of his ideas as new specimens (that is, specimens unseen before) came under his scrutiny, Cushman realized he needed to go abroad. But his financial reserves were not enough to allow it. In mid-June 1911, he wrote that "circumstances not of my making . . . compel me to think of . . . compensation for my services other than for slides and drawings" (letter to M. J. Rathbun, 12 June 1911). The National Museum tempered its "No" answer with the remark that "you will at least have the satisfaction of being the author of a very valuable contribution to American science" (letter from R. Rathbun, undated in late June 1911).

EARLY EMPLOYMENT BY THE U.S. GEOLOGICAL SURVEY

Cushman's affiliation with the U.S. Geological Survey came indirectly through T. Wayland Vaughan, Geological Survey geologist, whose office was in the National Museum; he had heard of the work being done in Boston on the forams obtained by the *Albatross* expedition to the Philippines. Vaughan's interest was in learning something about the depth and temperature indicated by certain larger forams he was investigating from the Oligocene of the southeastern United States. Believing these fossil shells had modern counterparts in the western Pacific, he sought Cushman's help in identifying some of the larger forams in the *Albatross* collections from the Pacific as well as Cushman's opinion as to the depths and temperatures they lived in. By the following year, Vaughan had decided the survey could benefit from the young man's talent and rapidly increasing familiarity with forams. So on 8 May 1911, he wrote proposing that Cushman undertake for the survey a three-phase program of study of Mesozoic and Tertiary Foraminifera of the Atlantic and Gulf Coastal Plain of the United States, the phases being systematic description, stratigraphic distribution, and ecology. Vaughan felt that since other fossils were available for geologic correlation in the United States, forams were not of critical importance in this phase of the study but might be of help in correlation in areas outside the United States. As far as the survey was concerned, Vaughan's need was mainly information about the conditions under which the foram-bearing beds were laid down.

As his examination work proceeded, Cushman realized he needed more fossil material in order to have a basis for comparison. At Cushman's request, Vaughan devised a quick, inexpensive means of obtaining samples for foram study: fine siftings from collections already in the museum and adhering matrix scraped off megafossils. Vaughan also planned to throw his dragnet for such material farther afield to collections at universities and in state geologists' offices. R. S. Bassler's collections of ostracodes and bryozoans in the National Museum proved especially fruitful as a source of forams.

Being pressed for prompt reports for incorporation into survey manuscripts ready for publication, Cushman responded that he wished to proceed from the more recent material, with which he now felt fairly familiar, down the series to the oldest. Rather than give a complete name (which he would surely have to change when a monographic revision was done), he would use merely the generic name plus a letter to denote the species. Although Vaughan approved Cushman's plan, that is, to begin with the more recent ones and work backward to the older ones, he added: "It may be desirable from time to time to have special reports on collections that do not fit in with the general program."

One of the first duties assigned to Cushman was the study of a series of rotary-drilled samples from a deep water well at Charleston, South Carolina. The samples had already been examined by L. W. Stephenson who had provisionally assigned ages to the various intervals in the well. Cushman's report (1914) was his first attempt to use Foraminifera in stratigraphy, as well as in paleoecologic interpretation. He was successful in finding the Cretaceous needles in the richly fossiliferous Eocene haystack that had resulted from the rotary drilling. In recognizing Cretaceous beds higher in the well than Stephenson had previously done, he gave both himself and his mentor Vaughan a boost and spurred the survey's interest in the usefulness of Foraminifera.

Once Cushman was on the Geological Survey staff, all sorts and ages of material were routed to his office in the Boston Society of Natural History. There was a sudden flowering of interest in forams, and other paleontologists were likewise beginning to realize how convenient and useful they were. Most of the work in those middle years of the decade was kept under cover, however, especially by those few workers who were connected with oil companies. Even the fact that work was going on was not publicized at first.

During Cushman's early affiliation with the Geological Survey, it was T. Wayland Vaughan who arranged for material to be sent to Cushman, gave him work assignments, questioned his identifications, critically reviewed his submitted reports, and supported (although to no avail) his request for more financial aid. Vaughan encouraged his younger colleague, telling him, "The amount of material is so enormous and the need for speedy reports so urgent that we are greatly perplexed," (18 September 1914), and later, "Do all you can to help in the correlation of anything with anything" (4 February 1916). In response to Cushman's complaint that work was not moving along very fast , Vaughan wrote: ". . . .We are building a foundation and this kind of work is always slow. . . .You will have to work perhaps one or two years longer before light will begin to break" (3 April 1916).

ECONOMIC WORK

Geological work in connection with the search for oil in the Tampico area of Mexico was well under way when L. W. Stephenson, one of Vaughan's survey colleagues, went there for a period of six months (January–June 1920). In attempting to work out the stratigraphy of the Tertiary beds, Stephenson found only meager and fragmentary specimens of pelecypods and gastropods (the taxa with which he had been working). But the same beds contained rich assemblages of forams. He wrote to Cushman, "I therefore need all the light I can get on forams" (23 January 1920) and requested copies of as many of Cushman's papers as were available. At the end of 1920, a proposal was made whereby Cushman would do the needed identification work at his office in Boston for some Mexican collections made by Vaughan, his price being $15 per identification. With the large number of identifications to be done, everyone realized the sum would be prohibitively large. There followed some exchanges of proposals—rather high-handed from Cushman and conciliatory from Vaughan—and from Vaughan a classic presentation of the case of pure scientific study versus scientific study done for its practicality. For Cushman, Vaughan weighed the value of the opportunity of making a scientific contribution—not to mention demonstrating the practical value of forams—against the loss of a high salary. For the oil company he weighed the probable, although still unproven, utility of the fossil forams against the cost of the investigation. Vaughan used his own experiences in the field as an example of work that was more of a scientific opportunity than a financial one. He concluded that if the aim of demonstrating the value of paleontological criteria in the solution of problems is not accomplished, "There may be a reaction against the use of paleontologists in petroleum mining and certain kinds of paleontological investigations may suffer" (16 May 1921). Cushman's only response to Vaughan's long exposition of the value of the opportunity being offered him was that they "will get full value for the work that I do" (20 May 1921). An agreement was made for Cushman to do the work.

Exchanges between Cushman and Vaughan were beginning to get a little testy. Vaughan was explicit in telling Cushman what was wrong with the way he was preparing his reports: "They cannot make any application of the lists of species. . . .There are no conclusions . . . which had not been reached by the use of other data" (8 August 1921). Vaughan was justifiably upset because Cushman had not complied with his suggestion that the larger forams be cleared up before going on with the smaller ones. Vaughan was certainly correct that in the field they were more serviceable. Cushman complained that, because the attempt to solve one problem served to unearth other related problems, he was unable to clear up all the larger forams without considering the smaller forams associated with them. Both men were showing increasing signs of irritation. Cushman was turning more and more to the smaller forams and away from the larger, and growing more confident in his use of them.

There were also, at about this same time, three young ladies working in oil company laboratories, who have been credited with being among the first to apply the use of Foraminifera to the search for oil: Alva C. Ellisor, Esther Richards (later Mrs. Paul L. Applin), and Hedwig T. Kniker (see Aldrich, 1982). An excerpt from the presentation of the Distinguished Alumni Award of the Department of Geology of the University of Texas to Alva Ellisor in 1962, quoted by Teas (1965) describes a little of what the beginning of their careers was like:

During the first few years as the work being carried on in the Houston laboratories gradually became known, prejudice and opposition to the validity of this work appeared, not only by petroleum geologists, but by quite a few prominent geologists and paleontologists in the academic world. Some warned the three young ladies they were making a terrible mistake in relying on microfossils. Besides the necessity of *proving* their work to their own companies, it became desirable to convince the rest of the geologic profession. Another difficulty Miss Ellisor faced, and this may be true of the other two, was that she had to work many well samples as 'unknown'—not being given the name and location of the wells. This was not just to test her during the proving stage, but some Humble [Oil Company] geologists wanted to be certain the determinations were based entirely on what was found in the samples, anyd not by any information as to what the age should be . . .

OPINIONS REGARDING VALUE OF FORAMS

At the 11th Annual Meeting of the Paleontological Society held 30–31 December 1919 in Boston, Cushman presented a paper entitled "The value of Foraminifera in stratigraphic correlation." But it was not until two years later, when the Geological Society of America held its convention in Amherst, Massachusetts, that sparks began to be struck. At the Amherst meeting were a number of people who, like Cushman, had already proved to themselves that forams were useful, due to their limited ranges. Also present were some individuals holding the opposite view, that anything having a range from Cambrian to Recent could not possibly be of any value to stratigraphy.

Esther Richards Applin recalled that Amherst meeting and vividly recounted her experiences there. She had been asked to read Cushman's paper on limits of variation in Foraminifera in which he opposed the idea that they were long-ranging or were so plastic that they could not be usefully discriminated. Immediately following Cushman's paper, she also read Dumble's paper which supported the same idea that forams were the most reliable means of correlating well sections. In the open discussion which followed, J. J. Galloway expressed his skepticism about the usefulness of forams in no uncertain terms, concluding: "Gentlemen, you know it can't be done!" Esther Applin recalled that Cushman, standing at the back of the room, was silent, and no one else spoke to oppose Galloway's scoffing remarks. Later Cushman invited Esther Richards and Alva Ellisor, who was accompanying Richards, to visit his office at the Boston Society of Natural History to talk about forams before they returned to their respective oil company jobs on the Gulf Coast.

Also at the fateful Amherst meeting, there was an encounter between Cushman and survey people Vaughan and David White (at that time chief geologist of the survey), during which Cushman took their conversation to mean that since he (Cushman) was an outsider and had done no field work, he could not expect any great amount of financial support from the survey. He felt also that because he had "unintentionally infringed on the rules of the Survey . . . there was nothing further left but to resign" (letter to Vaughan, 12 January 1922). Cushman attempted to see Vaughan again in Amherst and when unable to find him, left him a note, and returned to Boston very much hurt by what he took to be the survey's low estimate of the value of his work. He immediately sent his resignation to take effect the following day, 31 December 1921. He apparently regretted his hasty resignation and wrote to Vaughan about it 12 January 1922, with a tone of farewell: "I have blazed the trail." Vaughan's conciliatory response (15 January 1922)—"If you would accept reinstatement, I am willing to make the move to bring it about" and "I will go over every criticism I have made of you, and will do it in a friendly way"—seemed to have smoothed the ruffled feelings. Cushman's reply, although it stopped short of saying yes, did detail his wish to complete his Eocene and Cretaceous work that had been started for the survey. While his status as employee remained in limbo, Cushman continued to send his reports to Vaughan, mostly the results of his examination of the specimens Vaughan had obtained in Mexico. At this same time, Vaughan stepped down from his administrative post as chief of the section of coastal plain investigations to be succeeded by L. W. Stephenson.

WORK FOR THE MARLAND OIL COMPANY OF MEXICO

Late in October 1922, Cushman was visited by Earl Oliver, general manager, and Earl Trager, chief geologist, of the Marland Oil Company of Mexico. The visit resulted in his agreement to study and report upon the Foraminifera of two complete well sections and not more than 250 areal samples, the work to be finished before the end of the year. The agreed-upon price was $2,500. Before he was very far into the work he recognized "two very definite series . . . and it does not seem . . . that they could possibly be mistaken for one another." He wrote that he had great hopes that soon "as a result of my work it is going to be possible for any member of your staff to definitely place any well sample (letter to Trager, 18 December 1922). As it turned out this was an overly optimistic hope. He was bothered by great uniformity in the foram faunas, which made him suspicious of mixing in the well drillings since the areal samples were distinctive. Earlier geological investigations had suggested the existence of a general east-west zone of fracturing along which petroleum accumulation in the Panuco-Topila region was controlled. If this zone of faulting could be proven to extend across the lease, it was assumed that oil accumulation would also extend across the lease. The field work involved a party of 10 on a four-day trip coasting down through mountains on a flat handcar. Following this excursion, Cushman stayed on in Panuco until the third week of

March, examining samples from the machine borings being done in various parts of the leased area.

Upon his return home to Sharon, Massachusetts, Cushman lost no time in making arrangements for the handling of the expected influx of well samples from Mexico. During the train trip, he had sketched out a design for a small (30 x 40 ft) building, having four rooms, entrance hall, and basement, to be constructed at the lower edge of his orchard. The work of processing and studying the well samples from Mexico was to be carried on there.

Pending completion of the laboratory building, work commenced immediately in the third floor study of the Cushman home. Letters, telegrams, and samples moved at a furious pace back and forth between Sharon and either Panuco, Mexico, or Ponca City, Oklahoma.

After about five weeks of specific information regarding correlations between certain horizons in the shallow machine holes to determine where the stratigraphic highs and lows were, Oliver began to express his doubt: "I do not yet have complete confidence that even Dr. Cushman will be able to make correlations so accurately as he feels that he can do," but added "we must accept Cushman's determinations because there are no better" (Oliver letter to Trager, 10 May 1923).

The following months were rather nightmarish, alternating between pressure from Oliver to recommend precise drilling locations and enthusiasm from Cushman regarding his confidence of eventual results. Cushman felt it was not his business to suggest where the holes should be drilled while Oliver felt just the contrary, that it was indeed his business and if he failed to do so, the company was not getting its money's worth. The work was hard and confusing to Cushman in an area of monotonous shales and complex block faulting and with faunas he had never seen before, and mostly poor faunas at that. The fact that no cores could be taken and everything had to be based upon cuttings added to the insecurity of the interpretations. He began to be suspicious that the company was sending duplicates under different numbers, which is indeed what they were doing, in order to test his work.

In late August, Oliver, Trager, and Cushman met in Sharon for a conference after which Oliver wrote a painfully frank letter in which he threatened an end to reliance upon forams. Nevertheless, Trager was assigned to stay in Sharon at the just-completed laboratory and work with Cushman for awhile. With the two working together, Cushman proposed the making up of sets of the critical specimens that appeared to be of value in correlation in the local field. Neither sets of specimens nor correlation charts were what Oliver had in mind, but rather some specifications as to where the company ought to drill. Oliver went on to mention the $1,000 per day it was costing merely to maintain the organization in Mexico in readiness to drill. Clearly an impasse was near. It was not for lack of effort that Cushman failed to provide what the company needed. In hindsight, it seems that Cushman was getting a lot more out of this Mexican endeavor than was the Marland Company. Working under pressure of an oil company whose interest had to be more financial than scientific was begin-

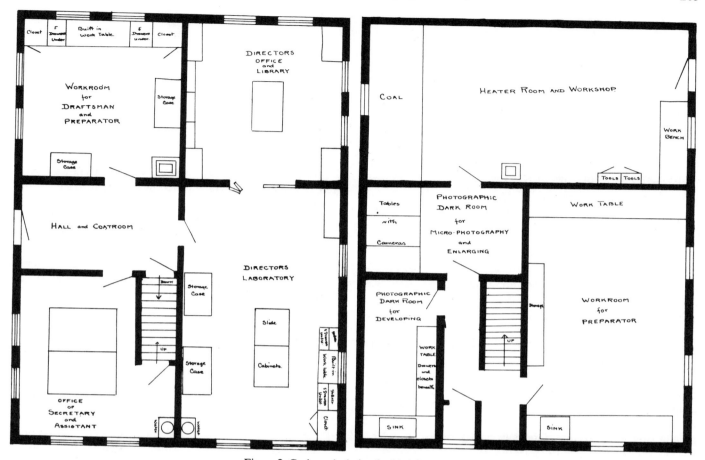

Figure 2. Cushman's design for his laboratory.

ning to tell. Oliver's frustration in his attempts to derive some kind of benefit from Cushman's study of the well drillings was also beginning to tell. In four of the five "deep structure" wells that Marland had drilled, oil had been found in various amounts and these wells were encouraging in a wildcat territory. But the wells had not been satisfactory because of salt water difficulties and low gas pressure. Only one well had yielded 5,000 barrels of oil. On 7 December 1923, Marland's employees at Panuco were notified that they could exchange their shares of Marland of Mexico for Marland of Delaware at eight shares of Mexico for one share of Delaware, Delaware selling for 30–32. The end of Marland's efforts in Mexico was fast approaching.

Cushman had convinced himself that it was futile to hope that the work he had been doing could now be turned over to a successor because too much of it was in his head. He said he still hoped that someone would soon be able to carry on the day-to-day determinations for Marland in Mexico, Texas, and California in order that his contribution could be of greater value to all Marland interests in those fields than if he tried to do it all himself, an obvious impossibility.

With the new year, 1924, Cushman began to tie up the loose ends of the Mexican work while continuing to run a three-ring circus of identification work from Texas and California for the

Marland companies. Donald Hughes was sent by the Marland Oil Company of California to learn the trade (or perhaps it would be better called a craft) by working with Cushman at the laboratory. That spring of 1924, at the joint convention of the American Association of Petroleum Geologists and Society of Economic Paleontologists and Mineralogists held in Houston, it became widely known that many workers were using forams in stratigraphic correlation between wells. Cushman was anxious to present his rebuttal to the "unwarranted criticism of the Foraminifera" given by Vaughan (1923) the previous year in Shreveport, Louisiana. In his rebuttal, Cushman assumed the "blame" for first making a distinction between "larger" and "smaller" Foraminifera, and asked rhetorically, "If these larger species are admittedly of excellent character as regards correlation, what is the matter with the smaller ones?" (Cushman, 1924, p. 489).

Cushman's financial status had improved dramatically with his salary from the Marland Oil Company. His most immediate plan was the beginning of a new journal in April 1925 in which he could have almost immediate publication of the various species and genera he was continually finding. Also, he had decided to open his laboratory, initially restricted to the exclusive use of the Marland people, to students.

Harvard University appointed Cushman to be lecturer in

micropaleontology beginning in the fall of 1926. He had recently added greatly to his library of foram papers, both from European rare-book dealers and by having photographic copies made of library volumes. There being virtually no demand for the old publications on Foraminifera, he acquired most of them for a pittance, such as $10 for the 1803 edition of Fichtel and Moll with original hand-colored plates. He also acquired sets of both d'Orbigny and Reuss models of Foraminifera. His collection of foreign and American foram-bearing samples was growing rapidly. Thus he felt that training young people was one of the best contributions he could make to economic work.

THE GALLOWAY AFFAIR—BACKGROUND

In mid-1924, in response to Waldo Schmitt's request for a brief accounting of the status of his work for the Annual Report of the National Museum, Cushman sent a paper summarizing the biology, the modern and geologic distribution, the methods of handling, and the classification of Foraminifera. The paper was too large for the Annual Report, so it was published separately in July of the next year (Cushman, 1925). The classification followed the ten-family classification of Brady (1884), and the illustrations were all previously published drawings from either his own works or other authors. It was a less than opportune time to republish Brady's existing classification. As it turned out, this paper gave Galloway just the opportunity he needed to assume leadership in the field, and he made the most of it. It spurred him on to work out his own new classification and less than six months later at the meeting of the Paleontological Society at New Haven, 28-30 December 1925, Galloway presented an oral account of the status of his work: "Revision of the classification and nomenclature of Foraminifera" (Galloway, 1926). Cushman was present at Galloway's reading of his revision and he and some others (unnamed) spent half a day in Galloway's hotel room going over the manuscript. Following the New Haven meeting, Cushman reported at length to Mrs. Applin how things had gone there. It had become apparent to Cushman that the necessary comparisons had been done, in many instances, using figures and descriptions rather than actual material. Cushman still did not indicate that he, himself, intended to get into the business of reclassification. It can now only be speculated, but it seems likely, that hearing Galloway's talk about a new revision of classification gave Cushman the idea of preparing a revision of his own without further delay.

The spring after the New Haven meeting Cushman invited Galloway to visit his laboratory in Sharon to talk about forams. Galloway came and spent two days (8-9 May 1926) going over his manuscript and his plans for its completion. Galloway's account of that meeting (letter to Cushman, 29 April 1927) quoted Cushman as replying to his question as to whether he (Cushman) was engaged in any classification work: "No, I am engaged in monographing the species of several of the genera and will leave the classification to you and expect to use your work when it is done." The aftermath of this visit was the origin of the controversy between the two men.

In personalities, they were poles apart. Galloway, as remembered by Alan Horowitz (letter of 16 August 1977), was "a big farm boy that made good, had a very outgoing blunt manner, and said what he thought, . . . almost the exact opposite of Cushman. . . .He was also very generous with his data. . . ." Cushman, despite (or perhaps because of) his humble beginnings, took his role of Harvardman seriously. He was urbane, aloof, and inclined to keep his thoughts and intentions to himself.

Professor Charles Schuchert at Yale, whom Cushman had met while still a student, had kept in frequent correspondence with him. A letter received the same week of Galloway's visit asked, "When are you going to tell us of the phylogenetic and ontogenetic values of forams?" (11 May 1926) and resulted in Cushman's query as to whether he might publish, serially in the *American Journal of Science,* some short papers of a theoretical nature. Accordingly, in the fall, Cushman sent a very brief note (Cushman, 1927a) in which he stated his belief that the material of the test (a function of the physiology of the living animal) must take first rank in classification while form comes second. He gave some examples of identical forms being composed of different material: the arenaceous *Ammodiscus,* the perforate calcareous *Spirillina,* and the imperforate calcareous *Cornuspira.* Cushman stated that a classification could not be worked out without knowing the geologic history of the various genera and, if possible, their phylogenetic relationships. At this time, the perforate *Spirillina* was a stumbling block because its earliest record was, inconsistently, in the Cambrian. In working out his classification, he chose to ignore that record and many years later the original record proved to be Jurassic instead. In Schuchert's acknowledgment of Cushman's manuscript, he indicated some disappointment:

I have long thought that you are afflicted with too much caution—caution is a very good quality but too much of it places the owner into the rear guard and that is where you do not belong. Move to the front, young man, for that is where you should be, and get behind you an army of workers working out the species (17 October 1926).

In short order, Cushman set up phylogenetic trees for five families (Cushman, 1927b, c) taking into consideration the geologic occurrence of the genera and the early stages as represented by young specimens or cut sections of initial stages. He reasoned that because of the alternation of generations—microspheric and megaspheric—in which the microspheric form represents more completely the ontogeny of the species, one must put more weight upon the microspheric forms (which are, unfortunately, much rarer than the megaspheric) in working out the phylogenetic history of a species.

In the more complex forms, the stages representing the early ancestors of the species were pushed back earlier and earlier in the life history until certain of these stages may be lost entirely. One must then look for these primitive species and genera in the present oceans and in the fossil record. Moreover, in the past, certain periods favored evolution of the Foraminifera while others favored long existence of primitive forams.

A simple, single chamber was the earliest and most primitive form of the test. The next stage in development was a second chamber consisting of a long, undivided tube. The division of this long tube into many chambers gave rise to the myriad multilocular forms. The various kinds of coiling of the simple or subdivided tube gave strength and protection to the animal. From this simple outline, one could derive all the myriad forms that forams can assume, while each of these myriad forms could appear in parallel lines having the three different kinds of test wall. Thus an outline for a classification was established, whereby different lines of development would arrive at the same-shaped end form. Such a classification pattern would, of course, produce a multiplication of genera, as well as leave room for still-to-be-discovered genera. Still, many instances of parallelism in Foraminifera, resulting in repeated appearances in the geologic record of morphologically similar genera, awaited recognition. Lines of development needed to be worked out and the stratigraphic positions determined before one could begin to set up a classification.

The Society of Economic Paleontologists and Mineralogists was organized in 1926 during the spring meeting of the American Association of Petroleum Geologists in Dallas. Officers *pro tempore* chosen at the organizational meeting included Galloway to be editor of a proposed new journal. It is clear from letters saved by Cushman that by early 1926 there had built up considerable anti-Galloway feeling, and strings were beginning to be pulled to remove Galloway from any position of influence in the new society. One of the major string-pullers appears to have been Helen Jeanne (Mrs. Frederick B.) Plummer of Fort Worth, Texas.

Mrs. Plummer had first written to Cushman in January 1926 seeking permission to send a specimen for him to examine. Her correspondence style was very formal, courteous, almost mid-Victorian, and Cushman was touched by her charming and ingratiating manner. By her frequent letters, Cushman was kept in touch with behind-the-scenes developments. She left unstated the cause of her opposition to Galloway, but she made frequent reference to the fact that something must be done. She was influential in the political maneuvering that resulted in Cushman's becoming editor while Galloway assumed the presidency the following year, when the *Journal of Paleontology* first appeared. Along with Schuchert she was persistent in pointing out the need for "an orthodox manual as a guide to those who are beginning" (1 April 1926). She was surely right that there was a vacuum to be filled.

With pressure being felt from both Professor Schuchert and Mrs. Plummer, an "Outline of a re-classification of the Foraminifera" was inevitable. It made its appearance as part 1 of volume 3 of the *Contributions* on 28 February 1927 (Cushman, 1927d).

THE GALLOWAY AFFAIR—THE CONTROVERSY

Up until the appearance of Cushman's "Outline," relations between Cushman and Galloway had been cordial. So far as Galloway knew, his visit to the Cushman Laboratory in May 1926, was a pleasant exchange of ideas. Galloway stated he was going back to the figures and descriptions of types and was adhering to the Rules of Nomenclature and, furthermore, that he was teaching his students to do the same.

In January 1925, Cushman received an exceedingly cordial letter from Galloway in which he referred to Cushman as "the dean of foraminiferal study in this country." He acknowledged that Cushman had written him that he (Cushman) was preparing a general paper on the forams. Galloway mentioned "I have in preparation such a work also, but your paper will no doubt fill all the needs in that line" (Galloway letter to Cushman, 15 January 1925). It must have come as quite a shock to Galloway to find, later, that Cushman's general paper (Cushman, 1925), far from proposing any new ideas, merely reiterated Brady's classification. In this same letter (15 January 1925), Galloway said he had come to many unusual conclusions and named some arenaceous families that he believed were not primitive but, on the other hand, were specialized for certain habitats and were mostly degenerate; and some other families and genera that he believed should be put together, ending, "I am not trying to revolutionize the classification, but to help it to evolve into a more reasonable form." He stated that the classification should be based on a comparison of the complete ontogeny and not on adult similarity. With this latter statement, Cushman was in total agreement but gave his reasons for not agreeing with the idea of the early arenaceous groups not being primitive.

During the spring of 1926, Cushman and Galloway continued their cordial exchanges of comments, questions, and views. Galloway stated that it would be another year before he could get his manual out. In the fall, Galloway reported that he had revised his manual once again, "embracing the ideas which I got from you last spring." Cushman carefully answered all of Galloway's specific queries about genera and species and volunteered information and advice where he believed Galloway's ideas were wrong. He was less than forthright regarding his plans to publish his own reclassification.

At last, when only a week remained before his reclassification was due to appear, Cushman wrote (21 February 1927) to Galloway that he was publishing in the next issue of the *Contributions* an outline of his own reclassification. He added the rather lame excuses that "in getting ready for my students . . . it became necessary," and "it will help answer the flood of letters that pour into the laboratory." He added, "I hope you can take over a lot of it into your Manual."

Galloway, in an extensive and carefully thought-out reply (9 March 1927) was frank to say that the outline "came as a distinct surprise" and that it "contains a great deal of truth, for it embraces many of the ideas which I have expressed publicly and to you in our conferences and correspondence." He reminded Cushman that the illustrated phylogenetic trees were the same kind that he was using in his manual and had shown in his slides at the New Haven meeting in 1925. He pointed out many errors in the outline, particularly that Cushman had not carefully followed the Rules of Nomenclature. He believed that with more conferences they might have agreed on one classification, "for there is some one which is the truth." He predicted that Cushman's outline will "only add to the present Tower of Babel," and asked "Can we not get together on this? I do not care for priority or I should have published my preliminary outline long ago, but I do care for the truth and for stability."

In correspondence, the two continued to debate the correct choice of generic names for several of the common genera. In writing to Galloway on 26 April 1927, Cushman made reference to "my larger work . . . on the classification based on developmental stages." Cushman had definite plans for getting over to Europe at long last to examine type specimens and collect from type localities. Cushman's unyielding stand on Montfort's genus *Elphidium* and Lamarck's genus *Discorbis* finally angered Galloway to the extent that on 29 April 1927 he wrote asking Cushman for his assurance that he would not publish his revision until Galloway's had been published. Although his tone was angry, his appeal was perfectly logical. He reviewed in detail the events and dates of their respective presentations of their plans and their publications and concluded:

It seems to me, therefore, and to many others with whom I have discussed the matter, that I have priority as regards a complete study and reclassification of the Foraminifera. We are also convinced that you got the idea from me; the similarity of your treatment to mine is obvious. Now, therefore, we are also convinced that you should abide by the rules of courtesy extended from one scientific man to another when one learns that the other is working on a definite problem, and should refrain from publishing a similar thing until after the work of the first one is published.

What reasons can you have for publishing what is obviously the same kind of thing, differing only in details? That I am incompetent to do it correctly? You have found so far that my work is reliable wherever the facts are obtainable, for instance, the type and definition of *Lagena;* the derivation of *Uvigerina;* and many other facts which I have told you and which you have included in your Outline (but without giving me credit). Do you want the credit yourself? There is enough work connected with the Foraminifera to keep many men busy for many years without duplication, and to furnish credit for all.

In his reply (2 May 1927), Cushman gave no assurance that he would hold back his own classification but said that Galloway's manual would likely appear before his anyway, because it might be a year or more before his textbook would be ready for the press.

In one final effort to delay the publication of Cushman's textbook, Galloway wrote (2 March 1928) to tell Cushman that because John Wiley & Sons, who had intended to publish his manual, had learned about Cushman's proposed book, they were delaying the manual until they learned more about the character of the Cushman book. Moreover, Galloway's promotion at Columbia University depended upon the publication of his manual and therefore Cushman, whether intentionally or unintentionally, was working a grave injustice on him. The appeal was refused in Cushman's lengthy letter of 8 March 1928 in which he defended himself against each of Galloway's contentions, namely, selection of genotypes, use of family trees, and compliance with the International Rules of Nomenclature. Finally he asked:

Why do you not go one step further and take up a general work on Micro-paleontology? You are conversant from your teaching and experience with the various micro-paleontologic groups. Can you not enlarge your book to take in the other groups and make what would be a real text book of micro-paleontology which publishers would compete with one another to get with the growing interest in the subject.

Failing to deter Cushman from his intention of proceeding with publication of his textbook, Galloway appealed to Henry V. Howe, who had been the first president of SEPM. Howe's reply (13 March 1928) summed up the controversy:

The question here hinges around whether Cushman actually informed you that he would give you precedence in the publication of such a revision. If he did then he has treated you very badly. However, if he did, and you cannot prove it by documentary evidence, I wish to extend you my sympathy, but I do not think you will help yourself, and certainly you will do great damage to the society by raising a clamor over it, for it then resolves itself into a question of one man's word against another with the odds all in Cushman's favor, for it hardly seems likely that he would have pledged away his right to publish a textbook upon the group to which he has devoted his life work.

In the same letter, Howe made the point that textbooks are not in the same category with original material when it comes to priority of work. In other words, no one could claim a monopoly upon expressing opinions about other persons' genera, whereas a monopoly could be claimed upon any *new* genus or *new* species that one has found or upon any unmonographed group or fauna upon which one might be working, although in the latter case the time factor could alter that monopoly.

About this time, Cushman also began to seek support from several colleagues for his position that no scientist should be asked to delay a piece of work for the reason that someone else intends to do a similar work some day. At the same time, he was proceeding with the preparation of his textbook "Foraminifera, their classification and economic use." Cushman had the great advantage over Galloway that money for publication was not a limiting consideration. As soon as the textbook was ready, it went immediately to press and appeared in April (Cushman, 1928).

THE GALLOWAY AFFAIR—IN RETROSPECT

It is now of interest to consider whether, indeed, Cushman's textbook was patterned after the Galloway manual, as Galloway believed it was. It is of interest to consider, too, how the development of micropaleontology might have been different had Cushman been generous enough to give Galloway credit, or even an acknowledgment. At that time, it was unusual for one scientist to acknowledge the work of another, or even the help of assistants. We look mostly in vain for such acknowledgment or giving of credit in the Cushman publications, even up to the end of his life. Instead, Cushman's acknowledgment took the form of co-authorship.

Unknown activities behind the scenes and unanswered questions still remain. Why did David White ask to see the Galloway manuscript early in 1926? How did Helen Jeanne Plummer know enough about the Galloway unpublished manual to make a judgment of it in early 1926? How did John Wiley & Sons learn of the prospective Cushman textbook?

In retrospect, neither of the principals emerged wholly untarnished. The current fashion of courteous acknowledgment of the fact that one does not work in a vacuum surely would have lessened the hard feelings and might have made possible even more rapid development of the science of micropaleontology, which owes so much to both men.

COMPARISON OF THE CUSHMAN TEXTBOOK AND THE GALLOWAY MANUSCRIPT

The March 1926 manuscript version of Galloway's manual containing 20 families is the one that Galloway brought to Sharon to talk over with Cushman during that fateful visit in May 1926, and is the one from which Galloway believed Cushman had stolen his ideas. That possibility now seems unlikely because of a number of major differences between the manuscript and Cushman's outline of 45 families that was published less than a year later. Galloway's phylogenetic tree indicated derivation of all families from a simple, imperforate, chitinous-walled, sac-like form, mostly living in warm, shallow, brackish to fresh water, from which three other families, all calcareous, branched out. One of these branches, the globigerinids (a group of generally fragile perforate-walled, floating oceanic genera), was shown to be ancestral, by a major radiation, to five of the remaining families, all calcareous and increasingly complex in numbers of chambers and chamberlets. One of these branches, the rotaliids (a group of generally robust, compactly coiled, bottom-dwelling genera whose habitats were almost universal), was shown to be ancestral, by another major radiation, to five more of the remaining families, one arenaceous and the other four calcareous. The one arenaceous family gave rise to two other arenaceous families while one of the calcareous families, the buliminids (those in which the test was generally elongate, uncoiled, and the chambers were arranged triserially, biserially, or uniserially) was shown to give rise, by still another radiation to the remaining four families, one arenaceous and the other three perforate and calcareous.

Cushman found himself in severe disagreement with most of the indicated derivations. Galloway's placement of arenaceous families as specialized, degenerate forms, rather than ancestral forms, as Cushman regarded them, was a major stumbling block. Galloway's apparent disregard of the occurrence of the families within the geologic framework was a further stumbling block. Cushman could not accept the globigerinids as ancestral to the rotaliids. They both agreed there had been major radiations from both groups, but to Cushman it was the other way around; i.e., globigerinids derived from rotaliids. Moreover, there were numerous other minor disagreements, but most of these were later thrashed out by correspondence.

Seven years later, by the time that Galloway's manual (Galloway, 1933) appeared in print, it was greatly changed from the

manuscript the two men had discussed in Sharon. By that time, the 20 families had been increased to 35, in some instances by levating subfamilies to family status, but it was still ten fewer families than Cushman had in 1927. Now the phylogenetic tree showed globigerinids derived from the rotaliids and at least some of the arenaceous families derived from the primitive chitinous-walled form.

There is now no reason to doubt Cushman's contention that his own classification was, indeed, his own. Even the overall format and the style of enumerating genera were not originated by Galloway but had been in wide use in monographic works from the late 1800s. The passage of time and Galloway's various revisions of his original manual only brought it closer to Cushman's 1927 classification. Galloway's manual included several features that made it a popular textbook. It included a glossary, authorship and dates for each family, and detailed references under each genus, while in Cushman's textbook one needed to search the extensive bibliography, which was arranged under many categories, for the detailed references. Galloway's manual included a historical account of classification, a discussion of principles of classification, and a discussion of organic evolution and its corollaries as illustrated by the Foraminifera. Cushman's textbook, on the other hand, included chapters on the living animal, collection and preparation of material, and methods of study; and described how forams had been used in zonation of well sections that were being investigated in the search for oil.

Cushman's classification, once published, was not much altered through its four editions. Subsequently discovered genera were merely inserted into its framework. In his later years, he left mostly to others the questions of family relationships and devoted himself to the description of faunal assemblages, with new taxa being erected when necessary.

ACTIVITIES AT THE LABORATORY

At the end of the 1920s, the new textbook was in print and meeting more approval than hoped for. The new journal was thriving, both winning praise and gaining subscribers. Cushman had been chosen as new chairman of the National Research Council's Subcommittee on Micropaleontology. He was beginning to teach micropaleontology for Harvard University, having the students come out to the laboratory for instruction—nine students in the spring of 1928 he referred to as a "happy family" of "fine fellows." His long-hoped-for trip to Europe to study types in the museum collections there, to make fresh collections from the many classic localities, and to meet his British and European colleagues had been a great success. He had finally closed the door on the commercial work on which he had spent so much confining effort and time in the preceding years, and he was eagerly anticipating the freedom and time for uninterrupted research. He was actively building up his library of foraminiferal books and papers by purchase at bargain prices of whole libraries from Europe and reprints from rare-book dealers. He was building up his own collection by exchanges with other workers, espe-

Figure 3. The Cushman Laboratory for Foraminiferal Research in Sharon, Massachusetts.

cially from Europe. As the fame of his new laboratory spread, visitors came, students applied for admission, specimens for identification arrived by mail, and requests of many kinds poured in, many in the form of offers of collaboration. He even found time to mount sets of identified specimens to sell to Ward's Natural History Establishment (at their request) and to various individuals.

At the laboratory, he was carrying on concurrently a variety of projects from the U.S. Geological Survey, U.S. National Museum, and the Florida Geological Survey, not to mention working up some rich collections from Fiji. At this time, he also began his work with P. W. Jarvis, an amateur collector at San Fernando, Trinidad, on the exquisite faunas from Trinidad, which within the next 25 years were to bloom into the worldwide planktonic zonation by which the usage of forams was transformed.

With the expansion of students, workers, and projects, an enlargement of space was called for. In the spring of 1929, Cushman designed and had built a steel-sheathed addition to the laboratory for $1295.50, "taking the money out of principle to do it." Into the new "steel room," he moved his own desk and the library and the most valuable of the specimens in the interest of security from fire.

Even as prodigious a worker as he was, the furious pace was beginning to tell. But he had an antidote for his tiredness—"looking over the Trinidad faunas for fresh inspiration." He was

relieved in the spring of 1930 to turn over to Raymond C. Moore the time-consuming, often onerous task of editor of the *Journal of Paleontology*. Students still flocked to his course. It was not formal instruction but rather learning-by-doing with the instructor close by for questions at any time. He was still writing his small papers at breakneck speed: about five weeks from sample to manuscript, then three months to print. He had an incentive: 24 pages and four plates in his *Contributions* needed to be filled every three months.

Forams were being extensively used in California in oil work, and a kind of secondary center for this study had developed there with Hubert G. Schenck at Stanford University assuming informal leadership of the group. Among some of the West Coast workers, Cushman came to be regarded as too much of a dictator. Nevertheless, Schenck and Cushman were in frequent and friendly correspondence, with exchanges of specimens. Several of Schenck's students came to the laboratory as private students for short-term, intensive instruction. A number of the Cushman papers during the early 1930s were collaborative works with Schenck and his California colleagues and students.

Upon the appearance of the Galloway manual in early 1933, Schenck wrote Cushman his appraisal of it: "In some respects it is more useful than yours." It was only a few weeks later that Cushman described his plans, which he requested be kept secret, for a second edition of his textbook with an atlas of plates under a separate cover, and asked Schenck's advice about several points concerning it. In this second edition, he obtained the collaboration of Dunbar and Vaughan to do the sections on Fusulinidae and Orbitoididae, respectively.

Work at the laboratory went on with customary vigor. Survey funds were restored in 1936. Two students from overseas—Shoshiro Hanzawa from Japan and Pedro J. Bermúdez from Cuba—were both at the laboratory that year. It seemed as though Cushman was beginning to feel the pressure of time and demands as he wrote, "I should be three or four people" (letter to Israelsky, 12 August 1936), and he complained, "People seem to think I have only to look at a specimen and immediately give it a name" (letter to Vaughan, 11 February 1936). He expressed suspicion when he found species whose ranges jumped across formation lines and said he needed more continuous core samples. He "religiously avoided working the forams from the point of view of stratigraphy, especially in the first going over. . . .I do not want too much influence from that side" (letter to Stephenson, 10 April 1936). To a prospective student, Steven K. Fox, he wrote "With careful collecting and close study, there is almost no section that contains forams that cannot be zoned by them" (18 February 1936).

The need for depth data for modern species was becoming more pressing. What data were available in the mid-1930s were either imprecise or generalized. He began a quantitative study of the Southern Lines and his assistant, Frances L. Parker, began a similar study of the Northern Lines. Parker's report (1948) on these lines of bottom samples taken by Henry Stetson from shore out across the continental shelf provided some of the first specific information about depth zonation by forams.

In 1937, Cushman began a cooperative study of a line of submarine cores taken by Charles Piggot across the North Atlantic (Cushman and Henbest, 1940). In the quantitative work, he recognized swings between warm and cold species of planktonic forams. The planktonic species, he felt, needed the most work, and he therefore proposed to begin a preliminary study of the Globigerinidae. He put it aside, however, when faced with the utter confusion that had resulted from the lumping of species. Planktonic species were, nevertheless, already being used elsewhere. F. R. Henson wrote him that in Iraq he was using the shapes of cut sections of globigerinids to tell the ages of rocks in stratigraphic sequences between Cretaceous and Miocene.

The difficulty of dealing with the planktonic species in the same way he had always dealt with other species frustrated him. It was clear that those methods would not do. He recognized that the surface of the wall observed under high magnification was critical in the classification of that group of species and at the same time realized his inability to represent the surface adequately, either by photography or by artist's drawings. It was easy to set aside the difficult study of planktonic forams in favor of the numerous requests for work on faunal descriptions, all things that were easy and needed only time and care for their accomplishment. There was no dearth of such requests. Hans Kugler's staff of foram specialists in Trinidad needed help in describing new faunas from the rich fields being investigated there. A drilling into

the Great Barrier Reef of Australia needed study and interpretation. Dr. W. H. Deaderick, a physician in Little Rock, Arkansas, had, as a hobby, amassed a rich and beautiful collection from the Cretaceous of that area and was seeking his help in identification. Three long cores obtained by Charles Piggot of the Carnegie Institution in the Bartlett Deep from deeper water than cores had been obtained before awaited study. Irene McCulloch was attempting to get out a series of papers on the Foraminifera collected by the Allan Hancock Expedition and requested Cushman's collaboration. Thus the intensive work proceeded at the laboratory concurrently on multiple projects.

During World War II, Cushman's major Geological Survey task was to prepare reports on shallow wells currently being drilled in Florida. The reports, needed quickly, were for the purpose of helping the geologists in the field find their way through the structures as new wells were drilled. Always a rapid worker, Cushman returned the reports as identified species plotted on graphs of depths in the wells. He became more and more frustrated as the work went along, feeling that he was working blindly, not knowing relationships between wells, and wondered whether he was doing any good. Eventually, he decided the preparation of an illustrated chart, a "cut and paste job" using his own, already published drawings, would best serve the purpose. Yet he was shocked, as he said, to see, with all that had been published, how much new and unrecorded material there still remained. The expressed need for quick identification of fossils from these shallow wells in Florida, gave him the impetus to begin a project of preparing a series of papers on type localities of formations. This series of papers was still in progress at the time of his death.

The end of the war meant resumption of requests from overseas, and the return of students to the laboratory. Reeside wrote to him, "Your reports on the wells have been like light in darkness" (4 January 1946). Of Cushman's collaboration with the various staff members of Trinidad Leaseholds, Ltd., and other oil companies in South America, H. H. Renz (3 May 1946) expressed the common view that "with your cooperation we are able to . . . [return] to the science a small portion of what we economic paleontologists owe her."

HIS OWN SUMMING-UP

A summing-up of his work may be best stated in his own words written more than 20 years earlier to his British friend Heron-Allen.

As to the old controversy between 'lumpers' and 'splitters' that is a matter largely of personality and point of view. I found in the Pacific work that certain of Brady's figures fitted some of my things but that others of his figures purported to be the same species had very different ranges. I plotted on maps the known records of species in the area and then my own specimens and found they fell into very definite regions and faunal areas. This I found held true when compared with similar work of other authors on crinoids, etc. from the same areas.

Next the comparison of living and fossil forms convinced me that these are much more different than had been supposed when sufficient details were considered. So gradually I was forced into the 'splitter' group. If I had not been I think it might have made a great difference in American work on the group. It was established in my work with the U.S. Geological Survey that the stratigraphic position of well samples could be distinguished by means of the forams and it was a natural step to similar use in connection with geologic studies in the oil industry.

For example, in the Velasco shale of Mexico, a formation of almost uniform gray shale, all looking practically alike for the 750 feet of its thickness and without megascopic fossils of any sort I found the forams could be used. In our work we used about 750 numbered forms to which no names other than generic were applied at that time. When the vertical ranges were carefully plotted and known it became possible to determine the position of any unknown sample very closely in the vertical column, whether an outcrop sample or core from a well.

In the Miocene of California a similar problem was worked out. Collections were made every five feet up the walls of a canyon so that we had about 800 feet of section in foraminiferal shales which looked practically alike and carried practically no larger fossils. After a careful study of the contained forams both my students and myself could place unknown samples in this section without error. This could have been done only by 'splitting.' As a result there are now about 300 workers on the forams on this side of the Atlantic and I imagine another hundred in the universities this year studying forams. The mass of material being accumulated is astounding and as it is gradually made known a great deal of helpful data for evolution and migration of faunas will be available as well as beautiful material for structural and other studies.

While much of the work done will of course be for the development of economic problems nevertheless there will be a certain number of workers who will attack problems from the purely scientific side. Perhaps all this will give you a better idea of my viewpoint.

You may be sure I will welcome any criticism and take them entirely in the spirit in which they are given. None of us can do but a bit in a lifetime of all there is to be done even in one small field of science and most of what we do—perhaps fortunately—will be soon forgotten. Yet I am fascinated by the studies of the forams and am learning a great deal every new day. The unfortunate side is that by the time we really get acquainted with a subject we are cut off just when we might go on and produce some results really worth while. I have tried to get into shape some of my work knowing full well that it is not by any means final nor

above just criticism. It is all very fascinating however and I have found one of the greatest joys of life in working with the forams. Whether it has helped or not I have enjoyed it greatly (Cushman to Heron-Allen, 18 Oct. 1928).

The end of the Cushman era saw a gradual change in the ways forams were studied. The planktonic zonation that had been in wide use for some years by Caribbean economic paleontologists was ushered into print by Bolli, Loeblich and Tappan (Loeblich and collaborators, 1957) and was immediately followed by refinements from many other places around the world. The time was right for the fourth edition of the Cushman textbook to be supplanted. The two-volume work on Foraminifera appeared in the monumental series "Treatise on Invertebrate Paleontology" (Loeblich, Tappan, and others, 1964). The *Glomar Challenger* set forth on its explorations to bring back even greater geological treasures from the strata beneath the floor of the world's oceans than the British sailing ship *Challenger* and the U.S. steamer *Albatross* had done the previous century. There were thousands of people in the allied fields of micropaleontology, biostratigraphy, oceanography, and paleoecology, who were now able to make use of forams in unravelling the complex geological history of the world.

ACKNOWLEDGMENTS

I am grateful to too many colleagues to be mentioned individually. But I must mention three, especially, who gave me freely the benefit of their insights: Richard Cifelli, Frances L. Parker, and Fred B Phleger, Jr. Finally I owe my thanks to the Smithsonian Institution for permitting me access to the Cushman correspondence files now in their care, and to the British Museum (Natural History) for copies of the Cushman/Heron-Allen correspondence.

REFERENCES CITED

Aldrich, M. L., 1982, Women in paleontology in the United States 1840–1960: Journal of the History of the Earth Sciences Society, v. 1, p. 14–22.

Brady, H. B., 1884, Report on the Foraminifera dredged by H.M.S. *Challenger*, during the years 1873–1876: Report on the Scientific Results of the Voyage of H.M.S. *Challenger*, Zoology, v. 9, 814 p., 115 pls.

Cushman, J. A., 1905, Developmental stages in the Lagenidae: American Naturalist, v. 39, no. 464, p. 537–553.

——1914, Protozoa, in A deep well at Charleston, South Carolina, by L. W. Stephenson: U.S. Geological Survey Professional Paper 90-H, p. 79–81.

——1924, The use of Foraminifera in geologic correlation: American Association of Petroleum Geologists Bulletin, v. 8, p. 485–491.

——1925, An introduction to the morphology and classification of the Foraminifera: Smithsonian Miscellaneous Collections, v. 77, no. 4, p. 1–77.

——1927a, Some paleontological evidence bearing on a classification of the Foraminifera: American Journal of Science, v. 13, p. 53–56.

——1927b, Phylogenetic studies of the Foraminifera. Part I: American Journal of Science, v. 13, p. 315–326.

——1927c, Phylogenetic studies of the Foraminifera. Part II: American Journal of Science, v. 14, p. 317–324.

——1927d, An outline of a re-classification of the Foraminifera: Cushman Laboratory for Foraminiferal Research Contributions, v. 3, p. 1–105.

——1928, Foraminifera, their classification and economic use: Cushman Laboratory for Foraminiferal Research Special Publication 1, 401 p., 59 pls.

Cushman, J. A., and Henbest, L. G., 1940, Geology and biology of North Atlantic deep-sea cores between Newfoundland and Ireland. Part 2, Foraminifera: U.S. Geological Survey Professional Paper 196-A, p. 35–56.

Flint, J. M., 1899, Recent Foraminifera. A descriptive catalogue of specimens dredged by the U.S. Fish Commission steamer *Albatross*: U.S. National Museum Report for 1897, p. 249–349.

Galloway, J. J., 1926, Revision of the classification and nomenclature of Foraminifera [abs.]: Geological Society of America Bulletin, v. 37, p. 235–236.

——1933, A manual of Foraminifera: Bloomington, Indiana, The Principia Press, Inc., 483 p., 42 pls.

Hartog, Marcus, 1906, Protozoa, in Harmer, S. F., and Shipley, A. E., eds., The Cambridge Natural History: London, Macmillan & Company, Limited, v. 1, Protozoa to Echinodermata, p. 3–162 (Foraminifera, p. 58–70).

Lister, J. J., 1895, Contributions to the life-history of the Foraminifera: Philosophical Transactions, v. 186, p. 401–453.

Loeblich, A. R., Jr., and collaborators: Helen Tappan, J. P. Beckmann, Hans M. Bolli, Eugenia Montanaro Gallitelli, and J. C. Troelsen, 1957, Studies in Foraminifera: U.S. National Museum Bulletin 215, p. 1–323.

Loeblich, A. R., Jr., Tappan, Helen, and others, 1964, Protista 2, Sarcodina, chiefly "Thecamoebians" and Foraminiferida, Part C, *in* Moore, R. C., ed., Treatise on invertebrate paleontology: New York and Lawrence, Kansas, Geological Society of America (and University of Kansas Press), 900 p., 653 figs.

Parker, F. L., 1948, Foraminifera of the continental shelf from the Gulf of Maine to Maryland: Harvard University Museum of Comparative Zoology Bulletin, v. 100, no. 2, p. 213–253, 7 pls.

Sherborn, C. D., 1888, A bibliography of the Foraminifera Recent and fossil, from 1565–1888; with notes explanatory of some of the rare and little known publications: London, Dulau and Company, 152 p.

—— 1893–1896, An index to the genera and species of the Foraminifera: Smithsonian Miscellaneous Collections, no. 856, p. 1–240; no. 1031, p. 241–485.

Teas, L. P., 1965, Alva Christine Ellisor (1892–1964) [memorial]: American Association of Petroleum Geologists Bulletin, v. 49, p. 467–471.

Todd, Ruth, and others, 1950, Memorial Volume: Cushman Laboratory for Foraminiferal Research, p. 1–68.

Vaughan, T. W., 1923, On the relative value of species of smaller Foraminifera for the recognition of stratigraphic zones: American Association of Petroleum Geologists Bulletin, v. 7, p. 517–531.

Geological Society of America
Centennial Special Volume 1
1985

Wrong for the right reasons: G. G. Simpson and continental drift

Léo F. Laporte
Earth Sciences
University of California
Santa Cruz, California 95064

Wide rejection of a theory now believed correct in essence is a phenomenon of great interest for the history and philosophy of science.
George G. Simpson, 1977

Actually most scientific problems are far better understood by studying their history than their logic.
Ernst Mayr, 1982

ABSTRACT

In the two decades preceding the establishment of plate tectonics theory, G. G. Simpson wrote a series of papers that refuted the paleobiogeographic evidence purportedly requiring direct continental connections. So cogent was his rebuttal that drift proponents thereafter downplayed the past distribution of fossils.

Simpson presented several different arguments. He restricted his discussion to Cenozoic mammals with whose evolutionary history he was most familiar; he challenged the accuracy of the proponents' data; he criticized their methodology, both with respect to its undue appeal to authority and its lack of parsimony; he used his own "coefficient of faunal similarity" to show that present-day mammal distributions on dispersed continents are comparable to those for Tertiary continents; and perhaps most effective was his statistical argument that trans-oceanic dispersal of organisms, while highly improbable for a single event, is practically certain given the many opportunities for such events over geologic time spans. Simpson's objections arose from his own independently developed theory of historical biogeography which, by relying on mobile organisms dispersing across stable continents, was fully adequate to explain the paleobiogeographic data. Simpson thus regarded both drift theory and trans-oceanic land-bridges neither sufficient nor necessary to account for those data.

INTRODUCTION

It is well-known that the theory of continental drift was firmly rejected by the great majority of geologists for a half-century after its initial formulation by Alfred Wegener (1880–1930) in 1912. Following the plate tectonics revolution of the 1960s, various authors suggested the sources and nature of the objections to drift. For example, some emphasized the lack of a suitable mechanism that would permit the less rigid continental sial to move horizontally through more rigid oceanic sima—"you can't push whipped cream through Jello." Others insisted that Wegener and his followers presented evidence for drift that was ambiguous, tenuous, or just plain wrong. Still others argued that

it wasn't a question so much of unconvincing evidence or inadequate mechanism as it was that Wegener was an "outsider," an uninformed astronomer/meterologist whose lack of credentials as a geological specialist undermined his position at the outset. (See, for example, Hallam, 1983, and Marvin, 1973, and references therein).

Pre-eminent among American geologists who opposed drift was George Gaylord Simpson (1902–1984; Figure 1) who, in a series of papers in the 1940s, completely dismantled the paleontological arguments for continental drift as well as for trans-oceanic land-bridges (Frankel, 1981; Marvin, 1973). From 1912 when

Wegener first announced his theory, until the 1940s when Simpson definitively rebutted the paleontological arguments, fossil data were central to the theory of continental drift. After World War II, fossils were either de-emphasized or not mentioned at all in support of drift. On the contrary, when fossils were considered they were used to support a stabilist position (*e.g.,* Axelrod, 1960; Cloud, 1961).

Various reasons have been offered to explain Simpson's strong opposition to continental drift. Both Cracraft (1974, p. 215) and McKenna (1973, p. 295; 1983, p. 475) suggest that Simpson's views resulted from *a priori* stabilist assumptions. Or, in view of the strong influence of William D. Matthew on Simpson's professional education (Simpson, 1978, p. 33–34), we might suppose that Simpson was following the American Museum of Natural History party line, whereby the fixed northern continents were viewed as the centers of origin for most Cenozoic mammalian groups that then migrated into the dispersed and fixed continents of the southern hemisphere (Croizat, 1981, p. 500, 518; Romer, 1973, p. 345). Simpson, of course, knew of Matthew's important work on historical biogeography, *Climate and Evolution* (1939) and he was persuaded by it, but he did not accept Matthew's ideas uncritically (see, for example, Simpson, 1940a, p. 765).

The purpose of this paper is severalfold. First, I will show that Simpson's opposition to trans-oceanic land-bridges and continental drift resulted not from stabilist assumptions but from an earlier-developed theory of historical biogeography. Second, I will demonstrate that Simpson's biogeographic interpretations were independently arrived at, irrespective of his genuine admiration for Matthew. Third, I will dissect the various lines of Simpson's argument to illuminate his style of approach to a scientific problem. Fourth, I will indicate how Simpson's consideration of this issue developed in parallel with other major lines of research he was pursuing at that stage of his professional career. Finally, I will discuss Simpson's belated conversion to plate tectonics and offer my own conclusions why Simpson was misled.

Necessarily, I will go over much of the same ground already well-covered by Marvin (1973) in her excellent book on the evolution of the concept of continental drift (see, especially her discussion of the "verdict of paleontologists," p. 114 ff.), and by Frankel (1981) in his fine essay on the debate over the explanation of disjunctively distributed organisms. Whereas Marvin and Frankel consider the broader issue of stabilist *vs.* mobilist theories and Simpson's role therein, I invert the discussion by keeping the focus on Simpson, the paleontologist, to see how and why he came to oppose drift theory.

WEGENER AND FOSSILS

One year before his death on the inland icecap of Greenland, Alfred Wegener published the last edition that he personally revised of his epoch-making book, *Die Entstehung der Kontinente und Ozeane,* the first edition of which he had published in 1915 (Wegener, 1929). In the initial chapter of that

Figure 1. George Gaylord Simpson, age 42, in the uniform of the U.S. Army, 1944. During World War II, Simpson served in the North African and Italian Campaigns where he attained the rank of Major in military intelligence. It was just before and after the war that Simpson wrote his key papers in historical biogeography.

work Wegener informed his readers how he came to his theory of drifting continents. He recalled that the concept first occurred to him in 1910 when he was struck "by the congruence of the coastlines on either side of the Atlantic." However, he regarded the idea as "improbable" until a year later when he chanced upon a report (Wegener, 1929):

in which I learned for the first time of palaeontological evidence for a former land bridge between Brazil and Africa. . . .The fundamental soundness took root in my mind . . . [and o]n the 6th of January 1912 I put forward the idea [of continental drift] for the first time in an address to the Geological Association in Frankfurt am Main. . . .

Thus, we see in Wegener's own words his assertion that a key to his theory of mobile continents was similarity of fossils on continents that are, today, widely dispersed. In that same 1929 edition, Wegener devoted one whole chapter to "Palaeontological and Biological Arguments," where ". . . we are justified in counting as favourable to drift theory all biological facts which imply that at one time unobstructed land connections lay across today's ocean basins" (p. 98). Among the many facts cited by Wegener was a table of data, attributed to Theodore Arldt, that gave the "percentage of identical reptiles and mammals on each side . . . of the former land connection between Europe and North America" (p. 100). The percentages fluctuated widely and unsystematically. Yet Wegener claimed that land connections between North America and Europe were corroborated when the values were high (29–64% for reptiles; 31–35% for mammals) during the Carboniferous, Triassic, Lower Jurassic, and Upper Cretaceous through the Lower Tertiary. No connections existed during the Permian (12% identical similarity of reptiles) and the Neogene (19–30% mammalian similarity). I will return to these data later when considering Simpson's response to Wegenerian arguments.

Elsewhere in his book, Wegener cited the by now familiar arguments for Gondwana based on the *Glossopteris* flora and the presumed freshwater reptile, *Mesosaurus,* of the late Paleozoic, as well as much evidence from disjunct distributions of living plants and animals, ranging from conifers to earthworms. The point I wish to make is that for Wegener fossil data were by no means a trivial part of his argument for drift. Nor were they trivial for his followers. For example, Alex. Du Toit in his book, *Our Wandering Continents* (1937), also offered paleontological support for his views. Hence, in light of subsequent apparent rebuttal of the fossil evidence for drift by people like Simpson, it is not surprising that the whole concept of continental drift was discredited.

SIMPSON'S RESPONSE TO DRIFT

Simpson's response to those who hypothesized continental drift, like Wegener and Du Toit, or trans-oceanic land-bridges, like John W. Gregory (1929, 1930) and Hermann von Ihering (1927), is contained in a series of key articles written during the 1940s (Simpson, 1940a, 1940b, 1943, 1947, 1952) and summarized in a short book, *Evolution and Geography* (1953). Although Simpson regularly discussed biogeography in most of his other work—whether systematic or theoretical—it is in these writings that we find his theory of historical biogeography best developed.

There are two points to be considered. First, Simpson's arguments for stable continents came out of a prior, well-formulated theory of historical biogeography. Having established his own principles of historical biogeography, he then applied them to the hypothesis of drift and land-bridges. Simpson did not argue, *ad hoc,* against mobilist views; rather, he tested those views in light of much broader, more inclusive concepts of the factors controlling the past distributions and abundances of terrestrial organisms.

Second, because his theory of historical biogeography was based essentially on Cenozoic mammals, his arguments against drift and land-bridges were couched in the data provided by them. Retrospectively, it may seem that Simpson was begging the question by using Cenozoic mammals, because we now know that significant breakup and drift of Laurasia and Gondwana took place in the Mesozoic; by Eocene time the continents were widely dispersed and relatively close to their present-day positions.

In this regard it is important to recall what else Simpson was doing during this stage of his professional career. During the 1930s, Simpson published extensively on South American faunas; in 1930–31 and 1933–34, he spent two years in Patagonia on the Scarritt Expeditions of the American Museum collecting Cenozoic mammals. His South American studies gave him a deep insight into the role of faunal isolation and interchange, on continental scale, in determining the geographic character of a fauna.

During the late 1930s and early 1940s, Simpson was also working on his monograph dealing with the principles of mammal classification, by which he became thoroughly familiar with the world-wide biogeography of Cenozoic mammals (Simpson, 1945). Also at this time, Simpson was writing *Tempo and Mode in Evolution* (1944) in which he made rather original arguments, at least for paleontology, for the crucial role that environment played in evolutionary history. For example, in Chapter VI, "Organism and Environment," Simpson introduced the notion of the "adaptive grid" to explain varying rates and patterns of evolution. Finally, Simpson collaborated with Anne Roe, a clinical psychologist (and later his wife) on *Quantitative Zoology* (Simpson and Roe, 1939) wherein he demonstrated his insistence on biostatistical rigor and robustness—qualities which, as we will see, he found sorely lacking in the Wegenerian arguments. All these projects were either completed or well on their way by the time Simpson attended the Sixth Pacific Congress in San Francisco, in the summer of 1939, where he gave a paper on "Antarctica as a faunal migration route." In this paper he outlined what would become his theory of historical biogeography (Simpson, 1940a).

STRUCTURE OF SIMPSON'S REBUTTAL

Simpson's counterarguments to the drifters and land-bridgers can be grouped into three categories: those arguments that pointed out flaws of critical reasoning; those that emphasized what, today, we would call "evolutionary ecology"; and those arguments that were quantitatively formulated. Simpson used these different types of argument in various combinations in his series of articles dealing with problems of biogeography. Looking back at this body of work, we can see these basic threads throughout; and thus we learn something of Simpson's way of approaching a scientific question and "solving" it.

Flaws in the Drifters' Reasoning and Data

Throughout his rebuttal of the mobilist position, Simpson attacked the citation of evidence that he considered wrong or misleading. He also pointed out the lack of parsimony, especially with respect to moving continents about with free abandon in order to accommodate the presence or absence of some particular group of fossils being discussed, even though some other fossil taxa not under consideration indicated quite the opposite. Simpson also chided drifters for invoking the word of authorities in one field of geology in this area of paleontology in which they were patently not competent. The following example will nicely illustrate the foregoing.

In the early part of this century, two distinguished American paleontologists—James W. Gidley of the U.S. National Museum and Henry Fairfield Osborn of the American Museum—believed that some scrappy equid fossils from the southeastern United States were probably referable to the Old World, late Miocene genus, *Hipparion.* In 1919, a French paleontologist, Léoncé Joleaud, claimed—incorrectly, as it turned out—that these New World *Hipparion* were nothing at all like western North American Miocene horses and, therefore, in order for the eastern North American taxa to have reached the Old World there must have been a trans-oceanic land-bridge from Florida to Spain by way of the Antilles and Africa.

Later, that same year, Joleaud extended the temporal duration of his bridge from late Miocene to include early Miocene to late Pliocene. By 1924, Joleaud not only broadened his land-bridge so that it ran from Maryland to Brazil throughout most of the Tertiary, but he had also embraced continental drift to provide the intercontinental connection rather than a trans-oceanic land-bridge spanning two fixed continents. Moreover, to accommodate variations in faunal similarity between the New and Old Worlds, Joleaud opened and closed the Atlantic Ocean in accordian-like movements, thereby requiring multiple episodes of splitting apart of the continents whereas Wegener only required a single parting. (See Simpson, 1943, p. 30, for Joleaud references).

In 1929, John W. Gregory, a distinguished British physical geologist (after whom the Gregory Rift in East Africa is named), but no paleontologist, cited Joleaud's *Hipparion*-bridge in his Presidential Address to the Geological Society of London on the "Geological History of the Atlantic Ocean" (Gregory, 1929). Gregory himself opposed drift, so he kept only the land-bridge notion of Joleaud without accepting the accordian-like movement that Joleaud assigned it.

Alex. Du Toit, in his book *Our Wandering Continents* (1937), accepted a Brazil/Venezuela to Africa land-bridge, citing Gregory as the authority for this concept, even though as a drifter Du Toit would have preferred direct continental connection. Simpson traced in much more detail than I have given here the complicated path of reasoning from scrappy equid fossil evidence in Florida to the entrenched notion of the *Hipparion* land-bridge throughout much of the Tertiary, with or without lateral movement of the continents. Simpson was particularly chagrined that neither Gregory nor Du Toit formed or sought an independent opinion regarding the paleontological basis for the *Hipparion*-bridge in the first place.

Another briefer example of faulty logic cited by Simpson in the drift argument goes back to Wegener himself. As noted above, Wegener provided a table of "percentage of *identical* reptiles and mammals . . . between Europe and North America" (Wegener, 1929, p. 100; emphasis added). Simpson went back to the original reference and discovered that the percentages referred to families and subfamilies, and thus could hardly be called "identical." Simpson (1943, p. 14) expostulated, "Such looseness of thought or method amounts to egregious misrepresentation and it abounds in the literature of this perplexing topic."

Evolutionary Ecology

However cogent Simpson's criticism of the flaws in the reasoning of the mobilists, the debate certainly would not have gone beyond the "is/isn't" stage if Simpson had not also elaborated a broader theory of those factors that limit the distribution and abundance of terrestrial organisms on continental and intercontinental scales. If Simpson could account for past biogeographies, using established principles of evolution and ecology against a background of stable, dispersed continents, then of course there would be no need to invoke additional hypotheses requiring drifting continents or sinking land-bridges.

As mentioned above, during the time Simpson was rebutting drift he also completed his first major theoretical book, *Tempo and Mode in Evolution,* which has come to be viewed as one of several masterworks of the modern evolutionary synthesis (Simpson, 1944). This book broke with the then-contemporary paleontological opinion about the mechanisms of evolution, coming down on the side of the new genetics with its emphasis on mutational and chromosomal variation, size of populations, nature and intensity of selection, and organism-environment interactions in effecting transformation (morphologic and otherwise) of life forms over the ages—what Darwin called "descent with modification." Not only did Simpson not accept, but he also thoroughly debunked notions of "inherent tendency," "orthogenesis," "aristogenesis," "momentum and inertia," and "racial senescence," among others, that were previously used to explain evolutionary patterns seen in the fossil record. In a sense, Simpson was an iconoclast who broke with existing paleontological tradition and offered original and testable interpretations of the fossil record (Laporte, 1983).

In passing, I would claim that *Tempo and Mode,* alone, refutes any claim one might make that Simpson was a crowd-follower with respect to his position on continental drift. It is unreasonable to assert that he opposed the evolutionary theories of R. S. Lull and H. F. Osborn with *Tempo and Mode,* yet uncritically followed W. D. Matthew with respect to biogeography. Simpson always started from first principles and came to his own, independently derived, conclusions.

In Chapter VI of *Tempo and Mode* Simpson specified the many ways in which organisms can be expected to interact with their environment, biologic as well as physical. Organisms not only evolved within a given "adaptive zone," but the physical parameters defining the zone also changed during geologic time. Thus, ". . . the course of adaptive history (as read from the fossil record) may be pictorialized as a mobile series of ecological zones with time as one dimension" (Simpson, 1944, p. 191–192). This quote indicates Simpson's strongly-held view of the role environment played in the fine details of evolution, as well as in its much broader aspects. In this regard, Simpson staked out an early claim as an important contributor to what is, today, called "evolutionary ecology."

With respect to biogeography, Simpson (1940b, p. 138) addressed what he identified as the "broadest problems of all in

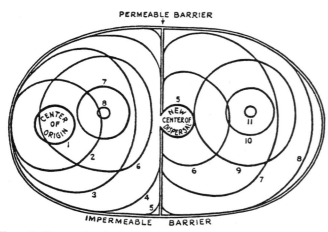

PERMEABLE BARRIER

IMPERMEABLE BARRIER

Figure 2. Simpson's schematic diagram (1940b, p. 145) illustrating his concept of expanding vertebrate populations (1–4) and their breaching of a barrier and further expansion (5–8); during the latter, secondary expansion, the population in the primary area contracts markedly (6–8); later, the secondarily expanded population also contracts (9–11). During the contracting phases, the populations have disjunct distributions.

this field, the general way in which land animals tend to become distributed and in which their distribution tends to change in time." This latter problem included, as well, the more specific one of "the different types of migration routes between major land areas, the way in which one type or another can be inferred from the faunal evidence, and the effect that a given type has on the faunas that use it." Simpson then hypothesized that once a new group of organisms arose in some small geographic area, it spread outward over a much broader area. How far and how fast it spread depended upon the environmental tolerance and the inherent dispersal abilities of the organism in question. Simpson conjectured the expansion and contraction of the organisms over time as environmental opportunities waxed and waned. At some later time, different parts of the populations might well be widely geographically isolated, yielding the classic "disjunct distributions" that were so readily cited by drifters and land-bridgers to support their positions, yet which could be explained in this alternative way (Figure 2).

Simpson acknowledged that terrestrial organisms would, necessarily, encounter barriers of one sort or another eventually, no matter how permissive the environment or inherently dispersable the organisms. Mountain ranges, deserts, seas, and ice would inhibit further migration. Simpson emphasized, however, that barriers to dispersal should not be judged as all-or-nothing propositions. Rather, he recognized that land vertebrates would respond differently to each kind of barrier, according to their environmental adaptability, numbers of individuals, size and bulk of the animals, and the degree of environmental variability across the barrier and its geographic extent. Simpson (1953, p. 20) stated, "Freedom of movement and its restriction are relative.... It is all a matter of degree, of diverse probabilities."

Although Simpson viewed barriers to faunal migration as a spectrum of decreasing opportunity, he thought it convenient to subdivide that spectrum into barriers of three sorts. *Corridors* were weak barriers, so they permitted a rather full and free interchange of animals across them. One might see different proportions of taxa at either end of a long corridor, like between central Asia and western Europe, owing to some attenuation because of the long distances involved (some 8,000 kms in the case of China and France). Evidence of such a corridor in the geologic past would be the widespread distribution of very similar, but not necessarily identical, terrestrial organisms.

At the other end of the spectrum were *sweepstakes routes,* where the barrier was virtually complete, and yet despite low probability, chance events such as rafting across ocean deeps and island-hopping along a broad archipelago resulted in the penetration of the barrier by only a few species. Fossil evidence for such a route would be the presence of a few similar taxa in two regions whose fossils were otherwise completely distinct from each other. Not only did the faunas differ greatly in degree of similarity when comparing corridors *vs.* sweepstake routes, but they also differed in their ecologic balance. That is, faunas crossing a corridor would include organisms spanning a wide range of habits and habitats, whereas sweepstakes-route organisms would produce an ecologically unbalanced fauna and, of course, the migration would be in one direction only. The terrestrial faunas of Madagascar, Australia, and the Hawaiian Islands are examples of such sweepstakes-route migrations.

Filter bridges, according to Simpson, spanned the range between corridors and sweepstakes routes; there were varying degrees of selectivity among barriers that allowed, more or less, different kinds of organisms to pass. Between continents, a filter bridge was typically a relatively narrow isthmus of limited environmental range and therefore a stronger impediment to faunal interchange than a corridor, but much less an impediment than a sweepstakes route. Simpson cited the present-day Isthmus of Panama as one such filter bridge. Evidence for past filters would be intermediate between the kind found for corridors and sweepstakes routes: moderately ecologically-balanced faunas with some migrants from each side of the barrier.

For Simpson, then, present and past distribution of terrestrial organisms was an ecological issue: what was the nature of the barrier to be crossed, and what were the corresponding environmental tolerances of the organisms attempting to breach the barrier. Simpson viewed paleobiogeography, therefore, as a study of the potential mobility of organisms across stable continents (Figure 3). In addition, Simpson approached the study as a dynamic one whose elements were continuously changing, whether they were the nature of the barriers as well as their geographic breadth and temporal persistence, or of the organisms and their range of environmental response. Consequently, Simpson argued for a probabilistic analysis of barriers rather than one that insisted on all-or-nothing statements.

Quantitative Insights

In all of Simpson's scientific work there is a strong quantitative element, no doubt reflecting both his ease with mathematics and his high standards of rigor and reproducibility in scientific argument. In the late 1930s, Simpson collaborated with Anne Roe on *Quantitative Zoology* (1939), a text on biostatistics. It is not surprising, therefore, that in developing his theory of historical biogeography, Simpson would introduce a quantitative point of view.

Because much of the argument for or against direct, physical connection of continents was based on qualitative statements of faunal similarity that were imprecise and poorly reproducible, Simpson invented what he called the "coefficient of faunal similarity," a measure still widely used today. The coefficient was a simple ratio of the number of taxa (species, genera, families, or orders) in common between two faunas divided by the number of taxa in the smaller of the two; multiplication of the ratio by 100 converted it to a percentage, *i.e.,* $100 (C/N_1)$. Although Simpson did not formally propose this measure of faunal similarity until 1947 in a short note to *Evolution,* he had used it earlier (Simpson, 1936), and had discussed its derivation in a footnote (Simpson, 1943).

Simpson's most cogent use of his coefficient of faunal similarity was in "Mammals and the Nature of Continents" (Simpson, 1943) where he provided a table showing his calculations of the coefficient for genera of recent and fossil mammals on several continents. Simpson introduced the data by remarking "how gratuitous are the vague statements and sweeping claims of past faunal resemblances so great as to be inconsistent with the present positions of the continents" (Simpson, 1943, p. 19).

TABLE 1. COEFFICIENTS OF FAUNAL SIMILARITY, $100\ C/N_1$, MAMMAL GENERA

82 Ohio/Nebraska (800 km)	
67 Florida/New Mexico (1600 km)	Recent
64 France/N. China (8000 km)	
24 New Mexico/Venezuela (5300 km)	
45 Eur/NA (8000 km, Greenland) - Early Eocene	
15 Eur/NA (16,000 km, Asia) - "Pontian"	
0 NA/SA (9600 km, Panama) - Early Pliocene	
8 SA/Africa (7600 km) - Triassic reptiles	

With respect to the first value of 82% in the table, Simpson asserts that it recorded such a resemblance (1943, p. 20–21)

as should frequently appear between exactly opposite points on different continents that were, according to drift theory, in former contact. . . .But *no fossil land faunas resembling each other to a degree at all comparable with this have ever been found on continents now separated.* Only evidence of this kind would be more consistent with drift theory than any other, and no such evidence is known [emphasis in original].

Simpson noted that the next two values came from continents with distinct climatic differences and minor geographic barriers (67%) or marked geographic barriers (64%). He claimed that the values were

Figure 3. Sketch by Simpson (1953, p. 53) in which he suggested the "problem of relationship between African and South American porcupines" turns out to be "another example of convergence in separate parallel radiations".

representative of mammalian faunas of distant parts of the same continent. . . .Resemblances of this order of magnitude may be about the least expected between continents united according to the drift theory and about the greatest to be expected between distinct continents. . . .according to non-drift theories. *Resemblances of this degree are altogether exceptional among fossil vertebrate faunas of continents now distinct* [emphasis in original].

The last value for Recent land mammals (24%) came from different faunal realms connected by an isthmian link with marked climatic and geographic barriers. According to Simpson, values "far less than this have repeatedly been given as conclusive evidence for drift or transoceanic continents. The example shows that the evidence leads to no such conclusion."

Turning to the coefficients for fossil land mammals, Simpson calculated a value for Eocene genera (45%):

Of two continents now separated. This is the best fossil mammal evidence . . . of early Eocene (Sparnacian) European mammals also known from beds of the same age in North America . . . for the drift or transoceanic continent theories that has ever been found, but even this really tends more to oppose than to favor those theories. These figures appear to be consistent with full continent union only if the areas in question were very distant, more distant than drift theory postulates.

Simpson was arguing, of course, that one might expect a higher value for these Eocene genera if from the same supercontinent, considering that the generic faunal similarity between China and France, today, is 64%. Although we know now that in fact Laurasia was well split apart by the early Eocene, Wegener (1929, p. 18–19) showed northern Laurasia reasonably intact. Du Toit (1937, p. 219 ff.) also argued for a separated southern Laurasia by Early Tertiary time, but in the Arctic region it was still more or less connected.

The similarity coefficient for the early Pliocene ("Pontian") was clearly lower still (15%), and according to Simpson, argued for a long continental connection via Asia. The final mammalian generic coefficient was zero similarity for early Pliocene faunas from North and South America, just before the isthmian connection was re-established.

Against this background of recent and fossil mammals, Simpson (1943, p. 20, 22) provided a single coefficient for generic similarity (8%) of "Triassic reptiles . . . for comparison and to support the incidental statement that the supposed pre-mammalian resemblances of southern faunas have been grossly exaggerated . . . (and) decidedly inconsistent with any direct union . . . of South America and Africa."

After his return from military service, Simpson (1947) came back to the subject of Eurasian-North American similarities of Cenozoic mammals, *not* to contend with the drifters and land-bridgers, but instead to develop more fully his concepts of historical biogeography. In fact, Simpson remarked that this effort was stimulated by Ernst Mayr, his American Museum colleague, who asked Simpson (1947, p. 616 footnote),

During what times in the Tertiary was a migration route between Eurasia and North America open? How long did it exist in each case? What faunal elements moved from east to west and what others from west to east? What was the climate of the bridge during each of the various times when it was open?

Simpson calculated the coefficients of similarity for scores of families and a like number of genera for 14 intervals of Cenozoic time. His values depended upon the extensive biogeographic records he had accumulated for his monograph on mammal classification (Simpson, 1945). The coefficients fluctuated almost an order of magnitude for genera (5% to 42%) and by more than half for families (52% to 89%). For both families and genera, similarities were highest in early Eocene; lowest immediately thereafter in the mid-Eocene for genera, early Oligocene for families; and generally increased throughout the rest of the Cenozoic. Simpson attributed the strong fluctuations in Holarctic faunal resemblances to a combination of factors, including direct faunal interchange, local extinction of indigenous forms, local evolution and concommitant differentiation of indigenous forms, extinction of migrants in one region, and migration of species into one region from a third (*i.e.*, neither from North America nor Eurasia). Simpson (1947, p. 685–686) concluded that

All the interchanges were selective, and they apparently tended to become more selective as [Cenozoic] time went on. Not the only, but probably the most important, selective factor seems to have been climatic; the migrants generally are those groups tolerant of relatively cold climates. . . .All the faunal evidence is consistent with a single land route, the Bering bridge between Alaska and Siberia, as the sole means of mammalian interchange between Eurasia and America.

Although the 1947 paper did not address continental drift, *per se*, it nevertheless played an important part in the debate, because in it Simpson provided a rigorous set of quantitative data on faunal resemblances between continents, and adequately explained those data in terms of evolutionary and ecological principles, with no recourse at all to drift or trans-oceanic land-bridges.

So convincing and authoritative was Simpson's explanation that his principles of historical biogeography transcended the data upon which they were based, data which, admittedly, postdated significant rifting of Laurasia. What chance, then did the drifters and land-bridgers have with their much scrappier, qualitative data whose meaning was interpreted more or less solely on whether different geographic areas were in direct physical connection or not, with little or no attention paid to evolutionary or ecological factors?

Viewed in this way, I think the fact that Simpson's data were from Cenozoic mammals and not, say, from Carboniferous or Triassic reptiles and amphibians is *not* why the paleontological argument for drift converted so few. Instead, it was that the stabilists now had a *theory* of historical biogeography that could explain *all* important fluctuations in faunal resemblance without having to resort to moving the continents themselves about.

In 1949, at a symposium on continental drift across the South Atlantic Ocean sponsored by the American Museum of Natural History, Simpson (1952, p. 163) gave a paper on still another aspect of quantifying arguments dealing with the issue of drift. He opened his paper noting that

Most of the large and wordy literature of historical biogeography . . . has involved postulates that are commonly unformulated and only rarely made explicit as working hypotheses. These hidden premises frequently are 'all-or-none' propositions, in the form of . . . 'either-or' dichotomies.

Specifically, Simpson was referring to whether the dispersal of a group of organisms required a land connection or not, and the inappropriate conclusion that, if dispersal does require land, then ". . .disjunctive areas occupied by the group have been connected by continuous land." On the contrary, argued Simpson, "There is no group of organisms that cannot be dispersed across water. . . .The predictive or inferential situation is not one of absolute alternatives but one of degree. It is a matter of probability."

Simpson noted that geographic distribution, which is known, must be a function of either land connection or dispersal potential across water, essentially one equation with two unknowns. He argued that too often it had been claimed that because very low probabilities of dispersal across water (or other major geographic barriers) had been assumed to equate to zero, there then must have been some land connection (*i.e.*, transoceanic land bridge or pre-drift contact of two continents). Simpson emphasized that very low probability was not equal to zero probability. In fact, even very low probabilities, if given enough opportunities or trials, could yield high probabilities of occurring at least once.

Simpson then showed from simple probability statistics how this counter-intuitive result comes about, as outlined in Table 2.

If, for example, the probability, *p*, of a single breeding pair of rodents in any one year being dispersed to some oceanic island is one in a million ("a very low probability"), the chances of such dispersal occur-

TABLE 2. DISPERSAL PROBABILITY STATISTICS
(Simpson, 1952)

tp =	1.0	2.0	5.0	10.0
b =	0.63	0.87	0.993	0.99995

p = Dispersal probability for single trial.
t = Number of trials.
b = Total probability of occurrence at least once.

ring after one million years (*t,* trials) is, in fact, better than 60% (*tp* = 1, then *b* = 0.63). After 2 million years, 0.87, and after 5 million years virtually certain (.993)!

Simpson thus demonstrated to his satisfaction that, owing to the previously discounted factor of many multiple trials or opportunities for dispersal because of the long intervals of geologic time involved, very low probabilities of dispersal for a single trial were obviously not equal to zero. On the contrary, they approached one, given an appropriate length of time. That being the case, intercontinental faunal resemblances need not be the result of direct land connection, but rather the almost inevitable ability of organisms to disperse themselves over relatively short intervals of geologic time, irrespective of their inherently very low probability of dispersal in any one trial.

Simpson was aware that at first glance his argument might explain too much, in that if dispersal was eventually inevitable, how come terrestrial organisms were not more or less uniformly spread around the world's far-flung continents? Simpson returned to his previously-developed ideas about the role of environment in limiting the distribution of organisms, as well as the importance of already occupied habitats resisting invasion by newly introduced migrants. Thus, Simpson was again able to emphasize the role of evolutionary ecology in clarifying paleobiogeography.

The simplicity and quantitative rigor of this argument may well have been the last straw in weighing down and collapsing use of faunal resemblances of terrestrial organisms by the drifters and land-bridgers to support their theories.

Simpson consolidated his views on historical biogeography in a small book (Figure 4) entitled *Evolution and Geography: An Essay on Historical Biogeography with Special Reference to Mammals* (Simpson, 1953). Five of the six chapters developed what Simpson saw as the critical factors in determining past and present distributions of organisms. The arguments were essentially identical to the ones worked out in his earlier papers, emphasizing an ecological and quantitative point of view. In the last chapter, he specifically addressed the issue of continental drift and, not surprisingly, Simpson (1953, p. 61–62) asserted that

All the biogeographic features in the known history of mammals are best accounted for on the theory that the continents have had their present identities and positions and that there have been no land bridges additional to those that now exist (North America-South America and Eurasia-Africa) except for a northern Asia-North America bridge."

Simpson's penultimate figure in this volume (Figure 5) con-

Evolution and Geography

An *Essay on* HISTORICAL BIOGEOGRAPHY
With Special Reference to MAMMALS

by George Gaylord Simpson

Figure 4. Cover of Simpson's summary essay (1953) on historical biogeography, *Evolution and Geography,* that illustrates his emphasis on the crucial role of overland dispersal in explaining the past distributions of land mammals.

trasted the Eocene world according to Wegener with one, "according to fossils and related evidence" à la Simpson. Simpson's Eocene world was virtually identical to today's world, except that Simpson showed a sweepstakes route between Southeast Asia and Australia, and a filter bridge/corridor between the New World Nearctic and the Old World Palearctic regions. Simpson's caption for this figure concluded that "Wegener's ideas of Eocene geography were evidently wrong." In his last figure of this volume he showed (p. 63) the geography of the Cenozoic World,

... the world of the Age of Mammals. Major features of the geographic history of mammals are best accounted for by considering the continental blocks and the main sea barriers as constants and three main filter bridges [Bering Strait, Isthmus of Panama, N. Africa-S. Europe] and one main sweepstakes route [island archipelagoes of Southeast Asia] as variables....The conclusion seems to apply not only to the biogeography of mammals but also to that of all contemporaneous forms of life. It remains possible ... that continents drifted in the Triassic or earlier, but there is little good evidence that such was the fact.

As in this last sentence just quoted, and elsewhere, Simpson was certain that what he inferred about the historical biogeography for the Cenozoic Era also applied to the Mesozoic Era, even though the data for his inferences were solidly based only on Cenozoic mammals. But we should not forget that Simpson was a knowledgeable student of Mesozoic mammals—in fact the world's authority (Simpson, 1928, 1929). We must assume that what he concluded for Cenozoic paleobiogeography was conso-

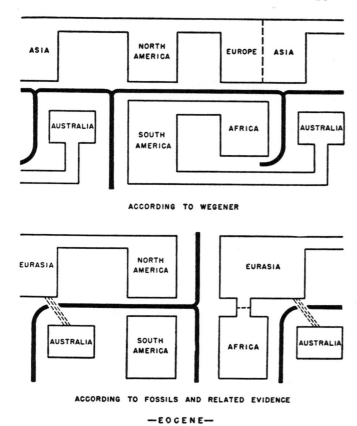

Figure 5. The next-to-last figure in *Evolution and Geography* whose captions reads: "A zoogeographical test of a paleogeographic theory. The upper figure shows the connections and barriers between continental blocks in the Eocene according to Wegener's version of the theory of drifting continents. The lower figure shows the connections and barriers demanded by known faunal resemblances and differences. Wegener's ideas of Eocene geography were evidently wrong."

nant with what he knew of Mesozoic mammalian biogeography. Therefore, we cannot claim that Simpson would have arrived at Wegenerian conclusions if only he had examined a more appropriate set of data.

Simpson's opposition to drift and trans-oceanic land-bridges can be reduced to his preference for mobile organisms and stable continents as against mobile continents and stable organisms. After all, the mechanisms of terrestrial land-animal dispersal were known, whereas the geophysicists were virtually unanimous in denying any suitable mechanism for moving the continents. Although Simpson never cited geophysical authority to support his case—he would not have allowed himself advantages he denied his opponents by appealing to authority—he certainly knew of the opposition of the geophysicists with respect to horizontal continental movement.

THE PLATE TECTONICS REVOLUTION

So effective was Simpson's case against Wegener-Du Toit paleobiogeography, that fossil evidence for drift was thereafter de-emphasized. Du Toit, himself, was ambivalent with respect to their value when responding to Simpson's 1943 paper (Du Toit, 1944). Du Toit acknowledged that "Simpson has exposed several weaknesses in the writer's [*i.e.*, Du Toit] statements and arguments on this subject [of faunal relationships]" (p. 149). Du Toit gave an item-by-item rebuttal to Simpson, but the main freight of his argument was geological, not paleontological. I think the real impact on Du Toit can be read from the concluding remarks in his rejoinder (p. 163): "The acid test of that Hypothesis [of Drift] will, it is felt, depend among other things on its ability simply and logically to account for the harvest of fossil forms yet to be unearthed." Obviously, Du Toit had been pushed by Simpson to the position that fossil forms *in hand* did not provide the acid test for the hypothesis, after all.

Du Toit was not alone in his ambivalence about the value of fossils in demonstrating drift. Throughout the 1950s and early 1960s, fossils played no significant part of the debate. On the contrary, other paleontologists came to Simpson's conclusions using data other than of that Cenozoic mammals. For example, Preston Cloud, reviewing the principles of marine biogeography and Phanerozoic marine invertebrate fossils, concluded (1961, p. 194) that "Certainly no paleobiogeographic evidence known to me requires drift of either poles or crust." Another distinguished paleontologist, Daniel Axelrod (1960, p. 280), came to a similar conclusion after reviewing fossil plant data: "At the present time there does not appear to be any paleontologic evidence that *demands* drifting continents or wandering geographic poles; in fact, the evidence seems to militate against major movement of either sort" [emphasis in original].

Another indication of the lack of enthusiasm of paleontologists for drift might be seen in some of the conferences and symposia debating drift just before the hypothesis of sea floor spreading was proposed. For example, in a collection of articles in the 1962 book entitled *Continental Drift*, edited by the British geophysicist Keith Runcorn, the only article that touched on paleontological data in any way at all is one written by Neil Opdyke, a Runcorn Ph.D. student and paleomagnetician. Opdyke (1962) reviewed the sedimentary rock record for paleoclimatic indicators and included less than one page on recent and fossil reefs. Given that the book was attempting to build the strongest possible case for continental drift, I find it significant that not only were fossil data virtually ignored, but that what little there was was contributed by a geophysicist, not by a paleontologist.

The key papers in the plate tectonics revolution were geological and geophysical in nature. Thus, fossils had no part whatsoever in establishing first-motion at transform faults, developing a magnetic-reversal time scale, determining the mirror-image of magnetic reversals across mid-oceanic ridges, or computer puzzle-fitting of continents. The concept of sea floor spreading, of course, came out of these and related observations. The data from the oceans *compelled* sea floor spreading, which in turn *compelled* continental drift. However much data from fossils may be compatible with, corroborate, or help choose among conflicting ideas about just where and when drift occurred, they are at best only suggestive of drift. Thus, Simpson was right in asserting, in effect, that fossil data do *not* compel drifting continents.

SIMPSON'S CONVERSION

When did Simpson acknowledge the mobilist view of the world? During the plate tectonics revolution, itself, Simpson was aware of the debate that was ensuing between stabilists and mobilists, but he still came down on the side of fixed continents (Simpson and Beck, 1965, p. 32).

During the latter part of geological history at least, the last hundred million years or more [*i.e.*, well back into the Cretaceous], major seas and lands, the oceans and the continents, have had substantially their present identity. . . .Some will disagree with this statement, but it is the consensus.

And again (Simpson, 1966, p. 3):

Even the new paleomagnetic data, which raise serious doubts as regards earlier times, confirm that the southern continents have been at least near their present positions throughout the Cenozoic.

But by 1970 Simpson had recognized that "continental drift" was, indeed, an established fact. He still hesitated, however, in using fossil data alone to determine past continental configurations, as the following quote from a letter to *Science* indicates. Simpson was commenting on an article by Elliott and others (1970) reporting the recent discovery of the mammal-like reptile, *Lystrosaurus,* in Antarctica (Simpson, 1970, p. 678):

. . .the occurrence of *Lystrosaurus* and some other more broadly similar faunal elements in Antarctica, South Africa, and India is evidence that those crustal segments were then continuous, and they are interpreted as parts of pre-drift Gondwanaland. [The authors] also mentioned that *Lystrosaurus* and other generally similar faunal elements likewise occur in Sinkiang [China], that is in what is now central Asia, but they did not discuss the bearing of that fact on paleogeography. Application of their reasoning would indicate that central Asia also was then in continuous (not necessarily direct) connection with Antarctica.

Simpson went on to point out that drifters did not include central Asia within the southern hemisphere supercontinent and, therefore, "we may still have to reconsider the concept of Gondwanaland and some other paleogeographic and biogeographic points." Of course, Simpson was *not* really suggesting that China was once part of Gondwana. Rather, by indirection, he was showing how paleobiogeographic data were not sufficient, in themselves, to compel a unique solution to the problem of ancient continental position. Simpson (1983, p. 115) returned to this point:

There are, however, hints that the situation may not have been as simple as it seems at first sight. *Lystrosaurus* is not known from either South America or Australia, both generally believed to have been parts of a unified Gondwanaland when that reptile lived. Moreover, both the plant *Glossopteris* and the reptile *Lystrosaurus* are known from parts of Asia

well north of India in regions generally believed to have been parts of Laurasia, not of Gondwanaland, when those organisms lived.

Edwin H. Colbert (1971, p. 263), one of the party that discovered Antarctic *Lystrosaurus,* was aware that the fossil did raise biogeographic questions, so he suggested a "long and circuitous route" of travel for the beast from the southern hemisphere to East China. And, if the reptile did range that widely, it would be comparable to the present-day geographic range of species of American and Chinese alligators, which ". . . would seem to represent the end effects of movement between North America and Asia across the trans-Bering filter bridge in middle Tertiary time." Colbert's explanation thus was reminiscent of Simpson's pre-drift arguments.

From the early 1970s, Simpson accepted plate tectonics and made whatever adjustments were necessary in his subsequent writings that involved paleobiogeography, although he certainly did not see plate tectonics as invalidating in any way his principles and concepts of historical biogeography (*e.g.,* Simpson, 1976, p. 8 ff.; 1983, p. 93 ff.). In taped remarks Simpson made in August, 1975, when reviewing his major publications, he alluded several times to his previous, incorrect views about the stability of the continents. In particular (Simpson, 1975, p. 96):

I'm not really very happy about [my book, *The Geography of Evolution* (1965)]. . . .The book is sort of a mish-mash collection of essays. . . .Some. . . .are all right, but some of them are badly outmoded, in fact were already becoming outmoded at the time when the book was published . . . just before it was clear that . . . plate tectonics was going to change the whole background for ideas on ancient geography and ideas on the geography of life.

Simpson's miscalculations in this field continued to be on his mind when writing his book-length autobiography (Simpson, 1978, p. 271). He remarked,

. . .the views that I supported and expounded years ago are still basically valid . . . [but] since reaching my seventies I have myself changed my mind on a basic principle of [a] . . . field of major importance to me: paleogeography.

Simpson (1978, p. 272–273) took pains to clarify the nature of his opposition to Wegener and continental drift.

The theory as Wegener advanced it rested on minimal evidence, much or all of which could also be explained by other theories, including that of stable continents. . . .I soon found (and this is still correct) that most of Wegener's supposed paleontological and biological evidence was equivocal and that some of it was simply wrong. . . .Thus in the 1930s and 1940s after lengthy investigation I concluded that the then-available real evidence of known land mammals not only did not support but opposed any effects of continental drift during the Cenozoic. . . .I did not deny the possibility of earlier effects of drift, but at that time I considered evidence for the drift theory so scanty and equivocal as to make it an unconfirmed hypothesis.

Simpson (1978, p. 273) indicated that the crucial data for

continental drift came from magnetic reversals across ocean ridges. Like other opponents of mobile continents he was converted by the totally different, non-paleontologic data coming from the deep sea.

CONCLUSIONS

Simpson's opposition to the fossil evidence adduced for continental drift or trans-oceanic land-bridges came directly from principles and concepts of historical biogeography that he was developing during the course of his own research on Cenozoic mammalian faunas. That Simpson was so effective in rebutting the drifters and land-bridgers can be directly attributed to the depth and all-inclusiveness of Simpson's notions of what factors were critical in limiting the geographic distribution and abundance of terrestrial vertebrates. Rather than arguing *ad hoc,* point-by-point, against the Wegenerians, Simpson formulated a theory that was both ecologic and evolutionary, and whose application had robust quantitative predictions. The soundness of Simpson's approach is attested to, in that his theory of historical biogeography still stands, today, even in a mobilist world (see, for example, Mayr, 1982, p. 449 ff.; McKenna, 1973; 1983).

Simpson's opposition to drift and land-bridges, therefore, was not the result of inherited tradition, or prior assumptions, or fear of going against the tide of established opinion. Simpson had demonstrated freshness of viewpoint and originality of thought in other areas of his research at this time that preclude the judgment that he was merely "following the crowd." On the contrary, even when the plate tectonics bandwagon was in high gear, Simpson hesitated in climbing aboard until well after the new paradigm was firmly established.

Simpson, after all, was wrong about continental drift. What led him astray? Simpson's specialty was Mesozoic and Cenozoic mammals, so naturally the data upon which he based his theory of historical biogeography were those provided by mammals, especially Cenozoic mammals whose fossil record was so rich as compared to that of Mesozoic mammals. Obviously, these fossil data would be much less likely to argue for mobile continents than, say, Permo-Triassic mammal-like reptiles.

One might wonder, then, if Simpson had worked with those organisms would he have come to different conclusions about drift? I think not. Colbert, a student of Mesozoic reptiles, was aware of the faunal resemblances of Mesozoic terrestrial tetrapods, yet he explained their distribution pretty much according to Simpson's biogeographic principles (see, for example, Colbert, 1952). Alfred S. Romer, another distinguished vertebrate paleontologist, perhaps leaned more toward drift, but his conclusions were always equivocal on this point, as the following statement by Romer (1968, p. 228–229) shows:

With my own interests centered on older periods, however, I have found myself weakening in my earlier beliefs as to continental fixity. . . .Although I do not believe that the evidence is—as yet—strong enough to make a very positive statement . . . For the northern continents, the evidence from the Mesozoic is not at all decisive, and the same is true of

the African-South American relationships. In the south, however, there is evidence strongly suggesting (although not proving) that in the Triassic there was free "trans-Atlantic" faunal interchange between Africa and South America.

The British paleontologist, T. Stanley Westoll (1944, p. 109), concluded from his study of Carboniferous fresh-water fishes that their distribution was "most convincing evidence for the nonexistence of the North Atlantic basin during late Carboniferous times." Yet, he concluded that these fish "can offer no decisive evidence in favor of either" trans-oceanic land-bridges or continental drift, although he thinks the latter more likely than the former. Westoll did not, however, claim much more for his data than that they "provide another small factor in favor of continental drift."

We can never know if Simpson would have been a Wegenerian had he worked with older terrestrial faunas. But if other very capable paleontologists working with these faunas were equivocal or merely lukewarm about drift, on what basis could we conclude that Simpson would have thought differently?

There is one aspect to Simpson's research experience that may have been most decisive in his stabilist views. During the 1930s, much of his research, although by no means all, was on South American mammals, which in the Cenozoic had evolved more or less in isolation from the rest of the world's continental faunas. Simpson's research demonstrated an early Cenozoic connection with North America, presumably by the proto-Panamanian isthmus; long island-continent separation permitting a unique terrestrial mammalian fauna to evolve; and a late Cenozoic inter-American connection ("filter bridge") across which Neotropical and Nearctic faunas crossed and intermingled (see Simpson, 1980, for an up-to-date summary). Both then and now, the history of the South American fauna during the Cenozoic Era is best explained by stable continental land masses intermittently connected by a relatively narrow isthmus, with occasional sweep-stakes dispersal during times of separation (Marshall and others, 1982). Naturally, a theory of historical biogeography based upon these New World terrestrial mammals would inevitably reflect underlying stabilist phenomena. Despite his being strongly influenced by his South American research, it is still moot whether Simpson would have derived a different view of historical biogeography had he worked on, say, intercontinental biotas from pre-drift Gondwana.

There are two, final statements I would make about why Simpson held to his stabilist views. First, and paradoxically, Simpson perhaps put too much confidence in the fossil record in determining past continental configurations. Because fossils did not conclusively demonstrate drift, Simpson was therefore convinced that continental positions must have always been fixed. Ironically, even Simpson was seduced by his own incisive and inexorable logic.

Second, I do not think it was part of Simpson's research style to reconsider a problem once he had seriously thought about it and come up with an answer. This is not to say he was rigid or pigheaded in his views. Rather, Simpson would attack a problem by careful evaluation of the data, by deliberate and logical formulation of his argument, and by construction of whatever theory was necessary to explain his conclusions. Once having done that—and the process might be extended over a number of years, as in the case of historical biogeography—Simpson moved on to new problems and to new areas of research interest. He did not tinker, fuss over, or agonize about issues upon which he had made a well-considered scientific judgment. Therefore, after the

1950s, Simpson did not again address the issue of drift, at least not in the broad terms of his earlier work. Of course, he continued to make biogeographic statements as appropriate when writing about evolution, or describing a fauna, and so on. But he did not reopen the whole question of his views of historical biogeography in the light of the newly established mobilist theory. Quite clearly, Simpson viewed his theory of historical biogeography as essentially unaffected by drift, with considerable justification, I would add.

REFERENCES CITED

Axlerod, D., 1960, The evolution of flowering plants, *in* Tax, S., ed., Evolution after Darwin, v. 1: Chicago and London, University of Chicago Press, p. 227–305.

Cloud, P. E., 1961, Paleobiogeography of the marine realm, *in* Sears, M., ed., Oceanography: American Association for the Advancement of Science, Publication No. 67, p. 151–200.

Colbert, E. H., 1952, The Mesozoic tetrapods of South America: American Museum of Natural History Bulletin, v. 99, p. 237–254.

Colbert, E. H., 1971, Tetrapods and continents: Quarterly Review of Biology, v. 46, p. 250–269.

Cracraft, J., 1974, Continental drift and vertebrate distribution: Annual Review of Ecology and Systematics, v. 5, p. 215–261.

Croizat, L., 1981, Biogeography: Past, present, and future, *in* Nelson, G., and Rosen, D. E., eds., Vicariance biogeography: New York, Columbia University Press, p. 500–523.

Du Toit, Alex. L., 1937, Our wandering continents: Edinburgh and London, Oliver and Boyd, 336 p.

Du Toit, Alex. L., 1944, Tertiary mammals and continental drift: American Journal of Science, v. 242, p. 145–163.

Elliot, D. H., Colbert, E. H., Breed, W. J., Jensen, J. A., and Powell, J. S., 1970, Triassic tetrapods from the Antarctic; evidence for continental drift: Science, v. 169, p. 1197–1201.

Frankel, H., 1981, The paleobiogeographical debate over the problem of disjunctively distributed life forms: Studies in the History and Philosophy of Science, v. 12, p. 211–259.

Gregory, J. W., 1929, The geological history of the Atlantic Ocean: Quarterly Journal of the Geological Society of London, v. 85, p. lxvii–cxxii.

Gregory, J. W., 1930, The geological history of the Pacific Ocean: Quarterly Journal of the Geological Society of London, v. 86, p. lxxii–cxxxvi.

Hallam, A., 1983, Great geological controversies: New York, Oxford University Press, p. 110–156.

Ihering, H. von, 1927, Die geschichte des Atlantischen Ozeans: Jena, Gustav Fischer.

Laporte, L., 1983, Simpson's "Tempo and mode in evolution" revisited: American Philosophical Society Proceedings, v. 127, p. 365–417.

Marshall, L. G., Webb, S. D., Sepkoski, J. J., Jr., and Raup, D. M., 1982, Mammalian evolution and the great American interchange: Science, v. 215, p. 1351–1357.

Marvin, U., 1973, Continental drift: The evolution of a concept: Washington, D.C., Smithsonian Press, 239 p.

Matthew, W. D., 1939, Climate and Evolution: New York Academy of Sciences, Spec. Public. no. 1, 223 p.

Mayr, E., 1982, The growth of biological thought: Cambridge, Massachusetts, Harvard University Press, 974 p.

McKenna, M., 1973, Sweepstakes, filters, corridors, Noah's arks, and Viking funeral ships in palaeogeography, *in* Implications of continental drift to the Earth sciences, v. 1: New York and London, Academic Press, p. 295–307.

McKenna, M., 1983, Holarctic landmass rearrangement, cosmic events, and Cenozoic terrestrial organisms: Annals of the Missouri Botanical Garden, v. 70, p. 459–489.

Opdyke, N. D., 1962, Palaeoclimatology and continental drift, *in* Runcorn, S. K., Continental drift: New York and London, Academic Press, p. 41–65.

Romer, A. S., 1968, Notes and comments on "Vertebrate Paleontology": Chicago and London, University of Chicago Press, 304 p.

Romer, A. S., 1973, Vertebrates and continental connections: An introduction, *in* Tarling, D. H., and Runcorn, S. R., eds., Implications of continental drift to the Earth sciences, v. 1: London and New York, Academic Press, p. 345–349.

Simpson, G. G., 1928, A catalogue of the Mesozoic Mammalia in the geological department of the British Museum: London, British Museum (Natural History), 215 p.

Simpson, G. G., 1929, American Mesozoic Mammalia: New Haven, Yale University, Peabody Museum Memoir 3, Part 1, p. 1–236.

Simpson, G. G., 1936, Data on the relationships of local and continental mammalian land faunas: Journal of Paleontology, v. 10, p. 410–414.

Simpson, G. G., 1940a, Antarctica as a faunal migration route: 6th Pacific Congress Proceedings, p. 755–768.

Simpson, G. G., 1940b, Mammals and land bridges: Journal of the Washington Academy of Science, v. 30, p. 137–163.

Simpson, G. G., 1943, Mammals and the nature of continents: American Journal of Science, v. 241, p. 1–31.

Simpson, G. G., 1944, Tempo and mode in evolution: New York, Columbia University Press, 237 p.

Simpson, G. G., 1945, The principles of classification and a classification of mammals: American Museum of Natural History Bulletin, v. 85, p. 1–350.

Simpson, G. G., 1947, Holarctic mammalian faunas and continental relationships during the Cenozoic: Geological Society of America Bulletin, v. 58, p. 613–688.

Simpson, G. G., 1952, Probabilities of dispersal in geologic time: American Museum of Natural History Bulletin, v. 99, p. 163–176.

Simpson, G. G., 1953, Evolution and geography: Eugene, Oregon, Oregon System of Higher Education, 64 p.

Simpson, G. G., 1965, The geography of evolution: New York, Capricorn Books, 249 p.

Simpson, G. G., 1966, Mammalian evolution of the southern continents: Neue Jahrbuch fur Geologie und Paläontologie Abhandlungen, Band 125, p. 1–18.

Simpson, G. G., 1970, Drift theory: Antarctica and Central Asia: Science, v. 170, p. 678.

Simpson, G. G., 1975, Transcription of tape-recorded personal comments regarding his published works, deposited in the archives of the American Philosophical Society, Philadelphia, Pennsylvania, 135 p.

Simpson, G. G., 1976, The compleat palaeontologist?: Annual Review of Earth and Planetary Sciences, v. 4, p. 1–13.

Simpson, G. G., 1977, A New Heaven and a New Earth and a New Man, *in* Corson, D. W., ed., Man's Place in the Universe; Univ. of Arizona, p. 51–75.

Simpson, G. G., 1978, Concession to the improbable: New Haven and London, Yale University Press, 291 p.

Simpson, G. G., 1980, Splendid isolation: The curious history of South American mammals: New Haven and London, Yale University Press, 266 p.

Simpson, G. G., 1983, Fossils and the history of life: New York, Scientific American Books, W. H. Freeman and Company, 239 p.

Simpson, G. G., and Beck, W. S., 1965, Life: An introduction to biology (2nd ed.): New York, Harcourt, Brace, and World, Inc., 869 p.

Simpson, G. G., and Roe, A., 1939, Quantitative zoology: New York, McGraw-Hill Book Co., 414 p.

Wegener, A., 1929, Die Entstehung der Kontinente und Ozeane: Braunschweig, Germany, Friedrich Vieweg und Sohn, 4th ed. (1966 English translation, The origin of continents and oceans: New York, Dover Publications, Inc., 246 p.)

Westoll, T. S., 1944, The Haplolepidae, a new family of Late Carboniferous bony fishes: American Museum of Natural History Bulletin, v. 83, p. 1–121.

Geological Society of America
Centennial Special Volume 1
1985

Surveying the geology of
a vast, empty, cold country

W. O. Kupsch
Department of Geological Sciences
University of Saskatchewan
Saskatoon, Canada, S7N 0W0

ABSTRACT

Since its inception in 1842, the Geological Survey of Canada (GSC) has had to contend with demographic, geographic, political, and cultural factors peculiar to the country and substantially different from those prevailing in either Britain or the United States of America.

Problems of logistics beset the GSC particularly after Confederation in 1867 and the acquisition of Rupert's Land from the Hudson's Bay Company in 1870, when a vast area, much of it climatically hostile, was added to its exploratory responsibilities. Innovative field exploration methods ranged from track surveys by canoe in the 1880s to helicopter reconnaissance in the 1950s. Although at first geologists were responsible for the topographic mapping of the country, this task was taken over by a Topographical Division, which in time separated from the GSC and became part of the present-day Surveys and Mapping Branch.

Geology departments of universities in Canada were unable to provide sufficient numbers of geologists with a Ph.D. degree, as desired by the GSC. Until after World War II, most professional geologists in either the GSC or the universities obtained their highest degree outside the country, mostly in the United States.

Publication of results on geological studies done by or for the GSC or provincial surveys has been and still is in government documents. Increasingly, periodicals became vehicles of communication. With the establishment of the *Canadian Journal of Earth Sciences* in 1964, Canadians now have a periodical of world standing available at home for the publication of geological papers.

BEGINNING THE TASK

That the task of taking inventory of Canada's minerals, rocks, and fossils would be overwhelming became evident as soon as it was begun.

Alexander Kennedy Isbister was born 1822 in Cumberland House, Saskatchewan. His father was a clerk for the Hudson's Bay Company; his mother was an Indian woman, the daughter of Alexander Kennedy, chief factor of the company. Isbister became a successful barrister in England and a champion of the rights of his half-breed and Métis people (Kupsch, 1977). Among his many interests he included geology. In 1855 he published a review of all that was known about the subject in the "Hudson's Bay Territories, and of Portions of the Arctic and Northwestern

Regions of America." In his paper, which included a geologic map, he acknowledged that the "numberless difficulties inherent in such an undertaking, embracing a range of country so vast and so difficult to explore, or even to obtain access to, must necessarily render any attempt of this nature very imperfect" (Isbister, 1855, p. 497). How right he was in his assessment of the work that lay before as yet unborn generations of geologists. And that, while his scope embraced not even the whole of what we now call Canada.

A VAST LAND

After the U.S.S.R., Canada is the largest country in the world, covering some 9 971 500 km^2 (3 850 000 mi^2). Canada

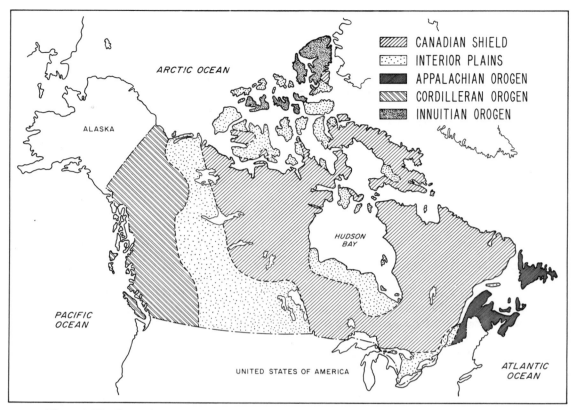

Figure 1. The five major geological and physiographical divisions of Canada (modified after Bostock, 1970, p. 12, and Douglas, 1970, p. 3).

can be divided into five geologic regions that correspond to distinctly different physiographic provinces (Fig. 1). The Appalachian Orogen in the east along the Atlantic Ocean, the Cordilleran Orogen in the west along the Pacific Ocean, and the Innuitian Orogen in the north along the Arctic Ocean are all mountainous. The fourth region comprises relatively undisturbed sedimentary rocks forming the plateaus, plains, and lowlands of the St. Lawrence Valley, the Western Interior, Hudson Bay, and the Arctic. The fifth region is the Precambrian or Canadian Shield.

The Shield comprises about 4 791 500 km² (1 850 000 mi²) of Canada, half of the country. One writer had this to say about it:

Geologically, it is the solid foundation of our country. The same might be said of it historically and culturally as well.
Life on the Shield represents the things that are quintessentially Canadian—things like checked shirts, high-cut boots, toques, parkas, packsacks, snowshoes, and bush aircraft. The human images associated with it are the stuff of Canadian mythology: the Indian trapper, the voyageur, the lumberjack, the prospector, the bush pilot. (Anonymous, 1981)

Rugged impressionistic images of the Shield are very much part and parcel of the Canadian cultural identity. Principally, this has come about through the great popularity, and hence wide distribution, of paintings and their reproductions by the artists known as the Group of Seven. They, more than any others, brought the North, which most Canadians will never experience, into the daily lives of an essentially urban population.

When the Geological Survey of Canada (GSC) started its work in 1842, the country was not as vast as today (Fig. 2). Under the Act of Union, 1840, Lower Canada (Quebec, but smaller) and Upper Canada (Ontario, also smaller than today) had joined into the single province of Canada (Fig. 3). The GSC had been established in 1841 by an act of the legislature primarily to advance the mining economy of this new province of Canada. Ten years later John Richardson (1851, p. 162) would take up the battle cry for a similar inventory to be taken in the vast western and northern region then still outside Canada:

It would be true economy in the Imperial Government, or in the Hudson's Bay Company, who are the virtual sovereigns of the vast territory which spreads northwards from Lake Superior, to ascertain without delay the mineral treasures it contains. I have little doubt of many of the accessible districts abounding in metallic wealth of far greater value than all the returns which the fur trade can ever yield.

The opportunity to expand their work came to the GSC as the country grew.

In the same year that the United States purchased Alaska, Confederation saw New Brunswick, Nova Scotia, and Canada

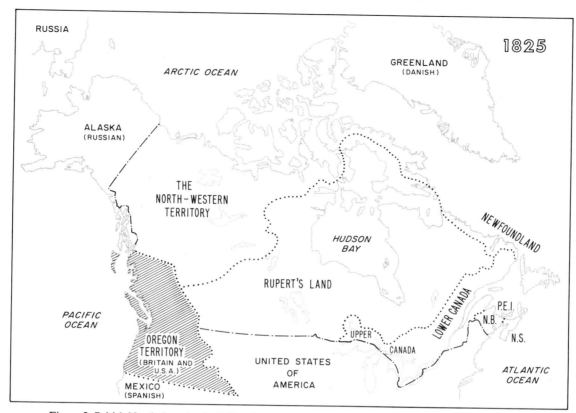

Figure 2. British North America in 1825 (after Howard et al., 1972, v. 5, p. 429; caption from *Atlas of Canada*, 1957, map 109). The international boundary is extended westward along the 49th parallel to the Rocky Mountains (1828). The Oregon Territory is occupied jointly by Britain and the U.S.A. Reannexation: Cape Breton Island to Nova Scotia (1820); Ile d'Anticosti and part of the coast of Labrador to Lower Canada (1825). Agreement between Russia and Britain on the description of Alaska boundary (1825).

united into a federal state, the Dominion of Canada, by the British North American Act (July 1, 1867). The Province of Canada was then divided into Ontario and Quebec (Fig. 4).

A much greater expansion in the GSC mandate for mapping the country occurred in 1870 when the Dominion acquired the North-West Territories (consisting of Rupert's Land, previously the domain of the Hudson's Bay Company which since 2nd May, 1670, had held a charter to the lands "draining into Hudson's Bay" and the North-Western Territory). Only one year later, in 1871, British Columbia joined the Dominion of Canada as the sixth province (Fig. 5). The jurisdiction over the Arctic Islands was transferred from Great Britain to Canada in 1880. Although in subsequent years there were several internal boundary changes, no new land would be added to Canada until 1949 when Newfoundland entered Confederation. The territorial evolution of Canada from 1867 to the present has been aptly summarized by Nicholson (1979).

AN EMPTY HABITAT

More than 95% of the 24 million Canadians live in cities and towns strung out in a strip about 320 km (200 mi) wide along the southern border with the United States.

Canada will always be so infinitely bigger physically than the small nation that lives in it, even if its population is doubled, that this monstrous empty habitat must continue to dominate this nation psychologically, and so culturally. . . .(Lewis, 1946)

The emptiness of the country became less in time, but the effects of filling the void were mainly noticeable in the southern parts of the provinces. The North and the Shield have remained much as they always were: vast and empty. It is estimated that in 1841 the European population of Canada was about 2 million. In 1851, when the first census was taken, this population had grown to 2 436 000. Since then it increased gradually, with a few accelerations on account of waves of immigrants, to the present 24 million (Fig. 6).

Estimates on the numbers of indigenous Indian and Inuit people spread across the top of North America have only low reliability. In 1604 when settlement started on tiny St. Croix Island in the St. Croix River between Maine and New Brunswick on the north side of the Bay of Fundy, there were perhaps 100

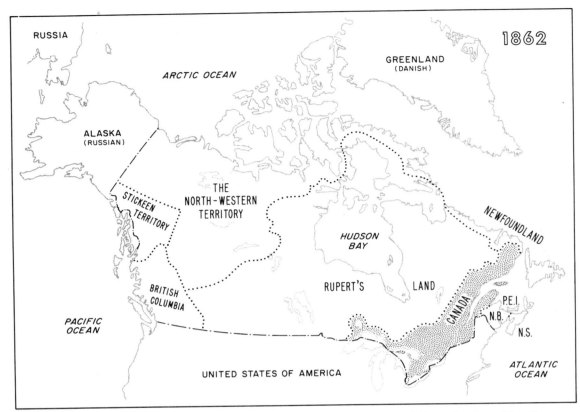

Figure 3. British North America in 1862 (after Howard et al., 1972, v. 5, p. 544; caption from *Atlas of Canada*, 1957, map 109). New Caledonia, with extended boundaries, becomes the British colony of British Columbia (1858). The Stickeen Territory is delimited (1862).

000 indigenous people. Now there are estimated to be some 325 000 Status Indians, 26 000 Inuit and as many as perhaps 1 000 000 Métis and non-Status Indians, all accounting for about 6% of the total Canadian population.

Even compared to Alaska, northern Canada is empty. While Alaskans occupy 4 km^2 (1.5 mi^2) per person, that density is nearly 20 times greater than in the Northwest Territories where one person enjoys 75 km^2 (29 mi^2). Before the coming of large numbers of settlers in the last two decades of the nineteenth century and the first decade of the twentieth century, the same emptiness characterized the Prairie West. Wrote Andrew Lawson (1926, p. 61–62):

It was my good fortune to see something of the northwest of Canada just at the time when the gates to it were being opened, and before the great migration had reached beyond the new Province of Manitoba. In the summer of 1882 I accompanied Robert Bell, of the Geological Survey of Canada, as his assistant, on an excursion across the prairies to the waters of the Athabasca. It was a very simple sort of an expedition. We had two buckboards and two horses. Bell drove one and I the other. The country that we traversed was a wilderness full of charm and wonder to an untraveled youngster like myself, in his third year in college. We journeyed for days and even weeks without seeing a habitation, and the only people met with in a thousand miles were those at the Hudson's Bay Company's posts, or an occasional pioneer making his lonely way from one post to another. A vast solitude possessed the landscape, which, though ever shifting, was ever the same—an endless rolling surface of waving grass and myriads of flowers beneath a bright blue sky, with drifting snow-white clouds. Here and there the poplar bluffs and the gleam of saline waters in small lakes accentuated the glorious monotony of the prairie.

When William Edmond Logan, the director, and his one assistant started the GSC's first field season in 1843, their domain was then the Province of Canada with a population of somewhat less than 2 million. Headquarters were established in Montreal. Travel was by horse and carriage, field work on foot.

Where rivers allowed the use of canoes, these, most Canadian of all vehicles, were the means of transport (Figs. 7, 8). Track surveys of the Dartmouth, York, Malbay, and Grand Rivers were made by Robert Bell in 1863. The technique of using canoes to survey the geology as well as the topography along Canada's many waterways was developed and refined over time. It reached its zenith in the last quarter of the nineteenth century after the acquisition of Rupert's Land, when mapping the whole Precambrian Shield became the responsibility of the GSC.

Water covers 7.6% of Canada's territory. Again, the distribution of this geographical factor is uneven, the greatest concentration of water being on the Shield and along its margins (Fig. 9). As this is also the land that was home to large numbers of fur-bearing animals, particularly beaver, geologic exploration followed the time-honoured water highways used by Indians and

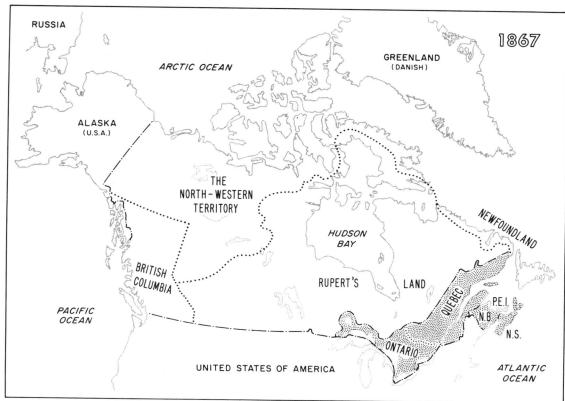

Figure 4. British North America in 1866; Canada in 1867 (after Howard et al., 1972, v. 7, p. 614; caption from *Atlas of Canada,* 1957, map 109). 1866: British Columbia attains its present boundaries by uniting the colonies of Vancouver's Island, British Columbia, and the Stickeen Territory, with a northern boundary along the 60th parallel. 1867: New Brunswick, Nova Scotia, and Canada are united in a federal state, the Dominion of Canada, by the British North America Act (July 1, 1867). The province of Canada is divided into Ontario and Quebec. The United States of America proclaims the purchase of Alaska from Russia (June 20).

traders in their pursuit of game and profit (Kupsch, 1979, p. 124). In the summer of 1879, Robert Bell's party alone mapped a total distance of 2735 km (1700 mi) in track surveys north and northwest of Lake Winnipeg. The Nelson River basin, Lake Winnipeg, the Churchill River, and Hudson Bay were included in this season's work of exploration (Ami, 1927, p. 22).

Even when the rivers were not frozen, travel was arduous. The daily routine of canoe travel was recorded in detail by the biographer of Sir John Richardson (McIlraith, 1868, p. 139–140). The day started at two o'clock, or half past two if the morning was dark. In twenty or thirty minutes at the most, the party was embarked, and the men "after their morning dram" struck up a cheerful song and paddled away at about four miles an hour. Every half hour they rested for about two minutes and lit their pipes; hence, these rests became known as *pipes.* At nine o'clock they put ashore for breakfast, which lasted for three-quarters of an hour. After breakfast the procedure was as before: paddling and alternately singing and smoking. At three o'clock "a *pipe* somewhat longer than usual, affords the men time to take a mouthful or two of pemmican" and John Franklin and John Richardson ate their lunch. At eight o'clock the party put ashore to camp for the night. After supper the men were asleep within

little more than an hour after landing. The officers who sometimes slept in the canoe during the day did not go to bed "till ten or eleven."

The physical labour of boat travel was even greater than the paddling of a canoe. Isbister (1845, p. 337) gave this account:

I would be tedious to describe each day's progress in detail; to-day's was but a repetition of yesterday's struggle against the rapidly increasing current; the men, now straining on the tow-line and dragging the reluctant boat after them—now plunging breast-deep into some river which poured its turbid contents into the main stream in the line of their march—now, when the current, directed by some opposing spit on the other side, bore down with its whole volume upon the bank which, by constant attrition, it had worn into a perpendicular cliff, and sometimes undermined it, embarking and bending to their oars, making for the opposite side with what vigour they might, when the everlasting tow-line was again thrown out, and the same unvarying round of tramping, tugging, and wading had to be repeated.

A COLD COUNTRY

Among Isbister's "numberless difficulties" are a host of meteorological conditions that render geologic observations difficult,

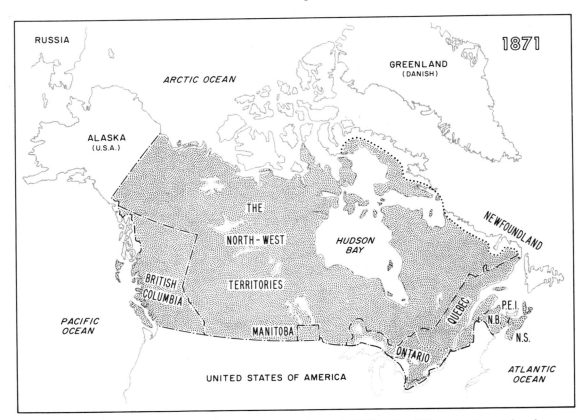

Figure 5. Canada in 1870; Canada in 1871 (after Howard et al., 1972, v. 7, p. 619; caption from *Atlas of Canada*, 1957, map 109). 1870: The North-West Territories (Rupert's Land and the North-Western Territory) are acquired by Canada from the Hudson's Bay Co. From part of them, Manitoba is created as the fifth province. 1871: British Columbia joins the Dominion of Canada as the sixth province.

if not impossible. Principal among these are snow cover and the freezing of navigable waterways.

On his return from Coronation Gulf overland to Fort Enterprise in early September, 1821, John Richardson, surgeon and naturalist to the first expedition led by John Franklin, made few observations on the rocks and minerals because he "travelled over this district when the ground was covered with snow" (Richardson, 1823, p. 534). Richardson then was close to death caused by the starvation his party had to endure on account of a lack of game. In the first quarter of the nineteenth century, British North America was colder and had a longer lasting snow cover than present Canada (Fig. 10). A worldwide trend toward amelioration of the climate, noticeable in Canada as well but less universal and more complex, was, however, to start after Franklin's last expedition (Fig. 11; Ladurie, 1971, p. 85).

Unfortunately, the open-water season in Canada is short and canoes are not of much use when cold sets in and water turns to ice (Fig. 12). In the late nineteenth century the end of the track survey by canoe not uncommonly meant a trip by dogsled or snowshoe back to the nearest railroad station (Fig. 13). In 1893 Joseph Burr Tyrrell (Fig. 14), accompanied by his brother James, travelled from Edmonton, whence they had gone by train from Ottawa, to Lake Athabasca, Black Lake, and Selwyn Lake to the Dubawnt River, which they descended to Baker Lake and

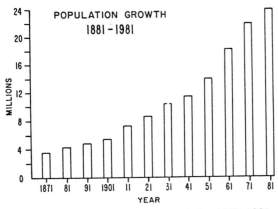

Figure 6. Growth of Canada's population 1871–1981.

Chesterfield Inlet, reaching this place on September 12. From Chesterfield, Tyrrell and his party walked south along the shore of Hudson Bay under appalling travel conditions due to frequent storms, freezing temperatures, and decreasing daylight. Food ran out, illness and frostbite appeared, but they were saved just in time when dogteams sent from Churchill reached them. On November 6 the party left Churchill by dogteam and snowshoe to reach Selkirk, Manitoba, and the railroad by the New Year (Blackadar, 1976, p. 12).

Figure 7. Encampment on the banks of the Red River. Members of the Assiniboine and Saskatchewan Exploring Expedition, June 1, 1858 (credit C-4572 Public Archives of Canada; H. L. Hime, photographer; *see* Huyda, 1975).

Figure 8. Making a portage. Voyageurs and canoemen of the Assiniboine and Saskatchewan Exploring Expedition, June 2, 1858 (credit C-4574 Public Archives of Canada; H. L. Hime, photographer; *see* Huyda, 1975).

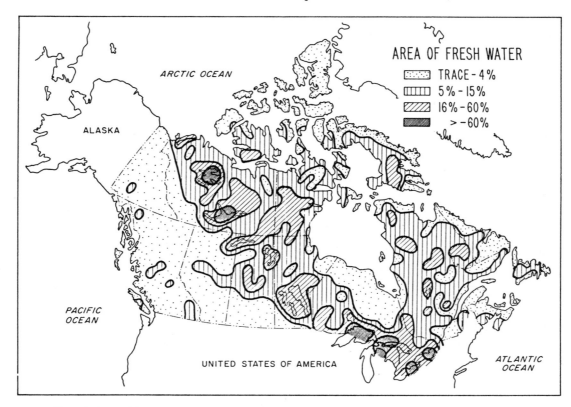

Figure 9. Area of fresh water in Canada (modified after *National Atlas of Canada,* 1973, map 7-8).

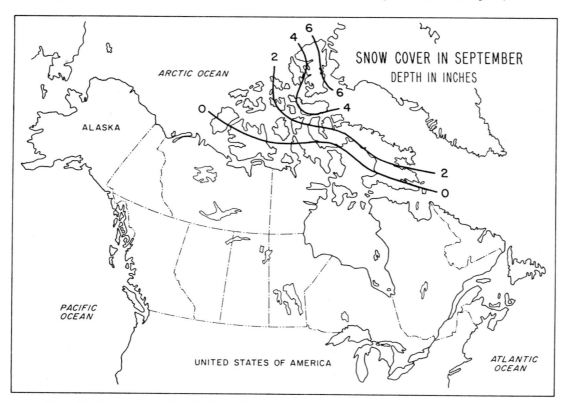

Figure 10. Snow cover in Canada, September 30, under prevailing climatic conditions (modified after *National Atlas of Canada,* 1971, map 13-14).

COPPERMINE MOUNTAINS

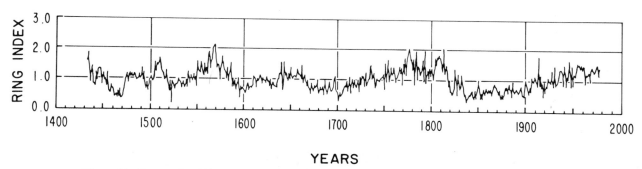

Figure 11. Coppermine Mountains tree-ring index graph, A.D. 1400 to present (G. C. Jacoby, Tree-Ring Laboratory, Lamont-Doherty Geological Observatory, Palisades, New York 10964, personal communication, 1982, reproduced with his permission).

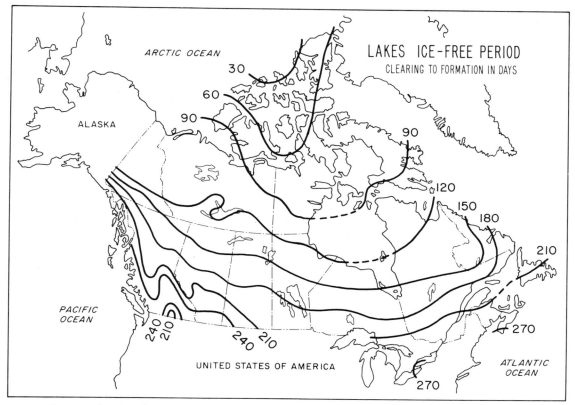

Figure 12. Ice-free period of Canada's lakes in days between break-up and freeze-up (modified after *National Atlas of Canada,* 1971, map 13-14).

A MARITIME NATION

Canada is not only a northern nation but it is also a maritime nation. Bordered by three oceans, it has, at 250 000 km (150 000 mi), the longest salt-water coastline of any nation. It extends over 40° of latitude and covers a wide range of environments and geologic terranes, of which so far only about 3% has been investigated (McCann, 1980, p. iii and ix). On account of the cold climate, most of the coastline is not open to navigation for much of the year. Nevertheless, it was by ship that geographic and geologic exploration of the Canadian Arctic was done, from Frobisher's first visit to southeastern Baffin Island in 1576 to the advent of aircraft in the present century. The story of Arctic geographic exploration has been told many times, but the role geologists played received little attention until publication of Zaslow's (1975) thorough study of the history of the Geological Survey of Canada. More recently, useful summaries were provided by Christie (1982) and Christie and Kerr (1981).

Figure 13. Dog carioles; part of the [Hind] Expedition returning to Crow Wing by the winter road, Tuesday, November 30, 1858 (credit C-20249 Public Archives of Canada; Assiniboine and Saskatchewan Exploring Expedition, 1858; H. L. Hime, photographer; *see* Huyda, 1975).

MOTORS TO HELP MEN

Soon after the hostilities of the First World War had ended, aircraft made their entry into Canada's North (Fig. 15).

On October 17, 1920, the first aerial photograph of northern Canada was taken by Frank H. Ellis, northeast of Hudson Bay, Saskatchewan, on a flight to The Pas, Manitoba, from a height of 915 m (3 000 ft) (Kupsch, 1973, p. 17). The Royal Canadian Air Force in 1932 resupplied a GSC party in canoes on the Yellowknife River (Wray, 1980, p. 46). Soon after, outboard motors would become common. These were some of the early introductions of new tools for the exploration geologist, which would forever change his work by dramatically extending the effective field time.

Another dimension was to be added after the Second World War—the introduction of helicopters. In 1955 the first comprehensive geologic mapping of the Arctic Islands was directed by Y. O. Fortier. In this undertaking, use was made of heavy helicopters as well as single-engine aircraft equipped with oversized, underinflated tires. Light helicopters were used as early as 1953 in mountainous areas of coastal British Columbia.

Figure 14. Joseph Burr Tyrrell (1858–1957) as a young man (credit 201735-A Geological Survey of Canada).

Figure 15. Vickers "Viking" IV flying boat G-CYEZ of the R.C.A.F., 1926. Photo by R. D. Davidson (credit PA 20089, Public Archives of Canada). Note the bow-mounted camera for taking oblique photographs.

EDUCATION

From its beginning the GSC has had a tradition of hiring the best-educated personnel available. An act of 1890 concerning the operation of the GSC recognized this by requiring that "officers who were engaged in scientific work should normally be graduates of a Canadian or foreign university, the Mining School of London, the Ecole des Mines of Paris, other comparable schools, or the Royal Military College" (Blackadar, 1976, p. 15). In a country as empty as Canada, this was a demand that was difficult to meet.

In 1821 McGill College was founded in Montreal by Royal Charter. It became McGill University in 1843. John William Dawson (1820–1899), educator, geologist, naturalist, and a prolific writer on these topics, became the institution's Principal in 1855, a post he was to occupy for the following 38 years (Fig. 16). He had been associated with Sir Charles Lyell in his study of the Maritime Provinces. Especially in the early years of both institutions there was a close connection between the GSC and McGill University. In the nineteenth century the University of Toronto, founded in 1849, also provided the science graduates needed to help build a growing country.

The influence some of the early Canadian geology profes-

sors had on their students, and through them on geological institutions other than the universities where they taught, still needs to be traced and evaluated in detail. Of particular interest is the influence exerted by Sir John William Dawson, a forceful and persuasive speaker and writer, on two of the most intriguing and important scientific controversies in his or any other time—the theories of organic evolution and continental glaciation. Principal Dawson would have none of either. As late as 1935 his second son, William Bell, still under his father's influence, insisted at a convention that the world was created in six literal days (Numbers, 1982, p. 541). Only six years before his death Sir William desisted from advancing any further arguments in favour of his beloved Drift Theory, but he never recanted (Brookes, 1982, p. 143).

Until the twentieth century, geology departments in Canada were not only small but they also offered little in the way of graduate instruction. This void was filled by the GSC which adopted a policy of employing university professors for summer field work. One of the earliest so hired was Henry Youle Hind, professor of chemistry and geology in the University of Trinity College, Toronto, who explored the Prairie West in 1857 and 1858. From early times on, then, the GSC became Canada's chief agency for geologic research. Its professional staff of 2 in Logan's

Figure 16. Sir William Dawson with a group of geologists, 1895 (credit Photographic Archives, McCord Museum, McGill University, Montreal).

first year grew to the present 300, who are supported by 500 others. Summer field work and geologic mapping of areas suitable for a thesis under the supervision of GSC professionals became part of the graduate training of many Canadian geologists whether they were studying at home or abroad, which more often than not was in the United States.

PUBLICATION

As time passed, the number of geological institutions in Canada—provincial geological surveys, bureaus of mines, geology departments of universities—grew, particularly in the past 50 years. As more geologic work was done, more publications appeared. Reports on field work for the GSC were issued as government publications. So were reports written for provincial surveys. Shorter papers appeared in British journals—particularly the *Quarterly Journal of the Geological Society* (London)—during the nineteenth century. Gradually more papers were contributed to American periodicals. At first this was mainly to the *American Journal of Science,* then to the *Geological Society of America Bulletin,* the first volume of which, published in 1890, contains several contributions by Canadians, including Robert Bell, Joseph Burr Tyrrell, and Sir William Dawson. Only since 1964 has Canada had in the *Canadian Journal of Earth Sciences* a multidisciplinary periodical that covers the geological sciences that is of more than local significance, and that covers science rather than technology. One year earlier the companion *Canadian Geotechnical Journal* commenced publication.

PROGRESS MADE

During the past two centuries or so, when matters pertaining to mineral and fuel resources became increasingly important for a fledgling nation, Canada developed the institutions it needed to give these matters the attention they deserved. The birth and growth of these institutions were influenced by both physical and socioeconomic factors peculiar to a northern nation, half of which is lake-dotted Precambrian Shield. The physical factors of vastness and of inclement weather are no longer the detriments to geologic work they were in the past before the introduction of airplanes and helicopters and the transfer of much outdoor work into the laboratory. Socioeconomic factors dependent on a small population, such as the availability of highly educated scientists, on the other hand, will still exert a great influence on Canada's geological institutions. It has recently been estimated that the country will need, in the near future, about 3100 postgraduates per year in the physical sciences and related branches of engineering. The supply, however, will be only 2100 (Cruikshank, 1982). Emptiness will still be with Canada for some time to come.

REFERENCES CITED

Ami, H. M., 1927, Memorial of Robert Bell: Geological Society of America Bulletin, v. 38, p. 18–34.

Anonymous, 1981, The Canadian Shield: The Royal Bank Letter, v. 62, no. 6, November-December, 4 p.

Blackadar, R. G., 1976, The Geological Survey of Canada—Past achievements and future goals: Ottawa, Minister of Supply and Services Canada Catalogue no. M40-38/1976, 44 p.

Bostock, H. S., 1970, Physiographic subdivisions of Canada, *in* Douglas, R.J.W., ed., Geology and economic minerals of Canada: Geological Survey of Canada, p. 10–30.

Brookes, Ian A., 1982, Ice marks in Newfoundland—A history of ideas: Geographie physique et Quaternaire, v. 36, no. 1, p. 139–163.

Christie, R. L., 1982, Geology of the top of the world—Concepts change as exploration advances: Geos, v. 11, no. 1, p. 10–13.

Christie, R. L., and Kerr, J. Wm., 1981, Geological exploration of the Canadian Arctic Islands *in* Zaslow, M., ed., A century of Canada's Arctic Islands: Ottawa, Royal Society of Canada, p. 187–202.

Cruikshank, John, 1982, Ability to fill Canada's research needs in peril: Toronto, The Globe and Mail, October 12, p. 9.

Douglas, R.J.W., ed., 1970, Introduction, Geology and economic minerals of Canada: Geological Survey of Canada, p. 2–8.

Howard, Richard, Lacoursiere, Jacques, and Bouchard, Claude, 1972, A new history of Canada: Montreal, Editions Format, 15 volumes, 1438 p.

Huyda, Richard J., 1975, Camera in the Interior: 1858 (H. L. Hime, photographer, The Assiniboine and Saskatchewan Exploring Expedition): Toronto, The Coach House Press, 55 p.

Isbister, A. K., 1845, Some account of Peel River, N. America: Journal of the Royal Geographical Society, London, v. 15, part 2A, p. 332–345, topographic map.

——1855, On the geology of the Hudson's Bay territories and of portions of the Arctic and North-western regions of America; with a coloured geological map: Quarterly Journal of the Geological Society of London, v. 11, p. 497–521, geological map. Reprinted 1856, in American Journal of Science, ser. 2, v. 21, no. 63, p. 313–338; map omitted.

Kupsch, W. O., 1973, Geological and mineral exploration in Saskatchewan—A *pré*cis of its history to 1970, *in* Simpson, Frank, ed., An excursion guide to the geology of Saskatchewan: Regina, Saskatchewan Geological Society Special Publication no. 1, p. 1–34.

——1977, Métis and proud: Geoscience Canada, v. 4, no. 3, p. 147–148.

——1979, Boundary of the Canadian Shield, *in* Kupsch, W. O., and Sarjeant, W.A.S., History of concepts in Precambrian geology: Geological Association of Canada Special Paper 19, p. 119–131.

Ladurie, Emmanuel, Le Roy (Translated by Barbara Bray), 1971, Times of feast, times of famine—A history of climate since the year 1000: Garden City, New York, Doubleday and Company Inc., 426 p.

Lawson, A. C., 1926, Out of beaten paths: University of California Chronicle, January, p. 61–62.

Lewis, Wyndham, 1946, as quoted *in* Fulford, Robert, 1982, This Canada: Edmonton Journal, September 20.

McCann, S. B., ed., 1980, The coastline of Canada—Littoral processes and shore morphology: Geological Survey of Canada Paper 80-10, 440 + x p.

McIlraith, Rev. John, 1868, Life of Sir John Richardson: London, Longmans and Green, 280 p.

Nicholson, N. L., 1979, The boundaries of the Canadian confederation: Macmillan, 252 p.

Numbers, Ronald L., 1982, Creationism in 20th century America: Science, v. 218, 5 November 1982, p. 538–544.

Richardson, John, 1823, Geognostical observations, *in* Franklin, John, Narrative of a journey to the shores of the Polar Sea in the years 1819, 20, 21, and 22: London, John Murray, App. 1, p. 497–538.

——1851, Arctic searching expedition: A journal of a boat voyage through Rupert's Land and the Arctic Sea in search of the discovery ships under command of Sir John Franklin, with an appendix on the physical geography of North America: London, Longman, Brown, Green, and Longmans, v. 2, app. 1, p. 162.

Wray, O. R., 1980, By canoe up the Yellowknife River in 1932, Part One: Musk-Ox, no. 26, p. 21–47.

Zaslow, M., 1975, Reading the rocks—The story of the Geological Survey of Canada 1842–1972: Ottawa, Macmillan Company of Canada Limited and Canada Department of Energy, Mines and Resources, 600 p.

Geological Society of America
Centennial Special Volume 1
1985

The G. William Holmes Research Station, Lake Peters, northeastern Alaska, and its impact on northern research

J. Thomas Dutro, Jr.
U.S. Geological Survey
Room E-316, Museum of Natural History
Washington, DC 20560

ABSTRACT

The Arctic scientific research station at Lake Peters, northeastern Alaska, epitomized scientific field studies in the north for nearly two decades after its founding in 1958. More than 80 scientists based at the station during the years of greatest activity conducted research in 20 scientific disciplines. Some of the more detailed projects involved: Pleistocene and bedrock geologic mapping, geomorphology, glaciology, meteorology, hydrology, physical and biological limnology, botany, archaeology, ichthyology, mycology and ecology.

Lake Peters, one of the Neruokpuk Lakes, is a large, deep glacial lake, located in the Arctic National Wildlife Refuge in the northeastern Brooks Range. Lake Peters, adjoining Lake Schrader, and the surrounding country are ideally situated for research in various scientific disciplines probing the Arctic environment. Located in one of the more scenic parts of Alaska, the lakes and surrounding mountains also draw a number of visitors each year for recreational camping, hiking and mountain climbing.

The facility was officially named the G. William Holmes Research Station in dedication ceremonies held at the station June 21, 1970. Holmes, a U.S. Geological Survey geologist who died during the winter of 1970, helped establish the station in 1958 and conducted early geological research in the area.

INTRODUCTION

The post-World War II history of Arctic research supported by the United States government is intricately entwined with the rise and demise of the Arctic Research Laboratory (ARL) at Point Barrow, Alaska. The complete story involves complex interactions among several federal and state agencies, the armed forces, at least three universities, and hundreds of individual scientists. Much of this tangled web is explored by Reed and Ronhovde (1971) in their book *Arctic Laboratory,* which documents the life of the facility from its founding in 1947 through 1966.

One of the ancillary offshoots of ARL was the research station at Lake Peters in the northeastern Brooks Range. The story of this field station, in a small way, mirrors the entire U.S. effort to support scientific activities in the North. Initiated during the International Geophysical Year in 1958, use of the station burgeoned during the 1960s, when the U.S. scientific community was in furious turmoil to catch up in response to the Russian *Sputnik* initiative.

A series of events brought about the gradual decline of both ARL and the Lake Peters station (named the G. William Holmes Research Station in 1970). Discovery of a major oil field at Prudhoe Bay in 1968 and the subsequent pipeline project emphasized the need for massive, industry-supported, mainly technical engineering programs in northern Alaska. At about the same time congressional passage of the Alaska Native Claims Act and the National Environmental Policy Act changed forever the pattern of small, serendipitous, mainly academic research projects. By 1970 it became clear that large, multidisciplinary and multiagency-supported programs were the wave of the future.

In 1971, the National Science Foundation was named the lead federal agency to support such ambitious programs as the

Tundra Biome Program and the Research in Arctic Tundra Environments (RATE) which dominated the scene from 1970 through 1977. Two other major efforts in the 1970s were the Arctic Ice Dynamics Joint Experiment (AIDJEX) and the Outer Continental Shelf Environmental Assessment Program (OCSEAP).

Meanwhile, most of northeastern Alaska was placed in a partially restricted category called the Arctic National Wildlife Refuge and managed by the U.S. Fish and Wildlife Service. The facilities at Lake Peters were eventually taken over by the Fish and Wildlife Service, which now uses the station as its summer headquarters for Refuge operations.

Nevertheless, the existence of a stable, well-equipped field station at Lake Peters for more than 15 years provided the essential logistical base to support a number of vital, small research projects in both the physical and biological sciences.

This historical vignette is presented in appreciation of the opportunities provided by the ARL to dozens of scientists. I, and my colleagues on the U.S. Geological Survey, were fortunate to spend four summers at the Holmes Station from 1968 to 1971 while geologically mapping the Mt. Michelson and Demarcation Point quadrangles as part of the early resource assessment of the Arctic National Wildlife Refuge.

EARLY HISTORY

The area of the Neruokpuk Lakes has been geologically intriguing since the time of Leffingwell's explorations early in this century. During the period 1907–1914, Leffingwell made at least five trips into the mountains from the Arctic coast to study the geology and glacial history of the region (Leffingwell, 1919). On one of these traverses, in 1911, he reached Lake Peters and made a reconnaissance of the geology in that area.

The region was examined in a reconnaissance fashion by several parties of geologists of the U.S. Geological Survey during the exploration of northern Alaska for the U.S. Navy Office of Petroleum and Oil Shale Reserves from 1944–1953 (Reed, 1958). In 1952, a tent camp was established near the outlet of Lake Peters and detailed stratigraphic studies were undertaken along the front of the mountains from Coke Creek on the east to Whistler Creek on the west. The general results of this work were published in 1962 (Brosgé and others, 1962).

The actual establishment of a permanent research station on Lake Peters did not take place until the summer of 1958, when a party headed by G. W. Holmes initiated glaciological, meteorological, hydrological, limnological, and Pleistocene geological field work (Fig. 1).

The establishment of the station was actually accidental. Spurred by the activity in the Arctic that accompanied the International Geophysical Year, Holmes began making plans to conduct Pleistocene geology studies on Ellesmere Island. Through his personal contacts with the U.S. Air Force, Holmes approached Colonel Louis DeGoes of the Terrestrial Sciences Laboratory, Air Force Cambridge Research Center, for assistance. Because of

logistical difficulties, the plans for an Ellesmere program were abandoned and in early 1958 Holmes was asked to select a site in Alaska where similar work could be done. Lake Peters was chosen and plans went forward to set up a permanent station there in the summer of 1958 (Fig. 2).

The party arrived on June 21 and field research was conducted during the summers of 1958 and 1959. The U.S. Geological Survey conducted the actual field work, with the financial support of the Air Force Cambridge Research Center, and some logistical support from the Office of Naval Research through the Arctic Research Laboratory (ARL) at Barrow. C. R. Lewis made traverses of the Sadlerochit Valley accompanied by Livingston Chase, geologist, D. G. Anderson, hydrologist, and L. A. Spetzman, botanist, of the U.S. Geological Survey. Fernand de Percin, meteorologist, U.S. Army Quartermaster Research and Engineering Center, Peter Larsson, glaciologist, McGill University, and John E. Hobbie, limnologist, Indiana University, assisted Holmes on several traverses. The base camp at Lake Peters was under the management of Maj. Frank Riddell, Royal Canadian Army (Ret.), from 1958–1961. The results of this work were published by the U.S. Geological Survey (Holmes and Lewis, 1965).

During the initial reconnaissance phase in 1958, Holmes, DeGoes, F. C. Whitmore, Chief of the Military Geology Branch, U.S.G.S., and J. H. Hartshorn, geologist on loan from the U.S.G.S. to the Air Force Cambridge Research Center, met at Lake Peters to formulate plans for continuing research in the area (Fig. 3). It is a tribute to the farsightedness of these men that most of the research discussed at this early stage was actually completed and the results have provided sound bases on which later work in several disciplines was built.

ACTIVITY DURING THE 1960s

Each summer during the 1960s the station served as home base for a variety of Arctic studies. Most prolonged of these was the work of Hobbie on the limnology of lakes Peters and Schrader. Hobbie, who had initiated work on the lakes during his first summers with Holmes, returned to the station in August of 1960 and spent a full year studying the lakes and their immediate physical environment (Hobbie, 1962b).

Bedrock geologic mapping of the Lake Peters area was begun by B. L. Reed of the U.S. Geological Survey during the summer of 1960. Reed, assisted by Peter Workum and James R. Fisher, spent a total of six months in 1960 and 1961 preparing a map and report which was published by the Survey as Bulletin 1236 (Reed, B. L., 1968).

The base camp remained under the supervision of Maj. Riddell during those summers, assisted by Vincent Peabody from ARL, Barrow.

After these first years, by mutual agreement between Britton and DeGoes, management of the station was transferred to the Office of Naval Research, and logistical support for a number of scientific projects was supplied through the Arctic Research Laboratory at Barrow under the direction of Max C. Brewer.

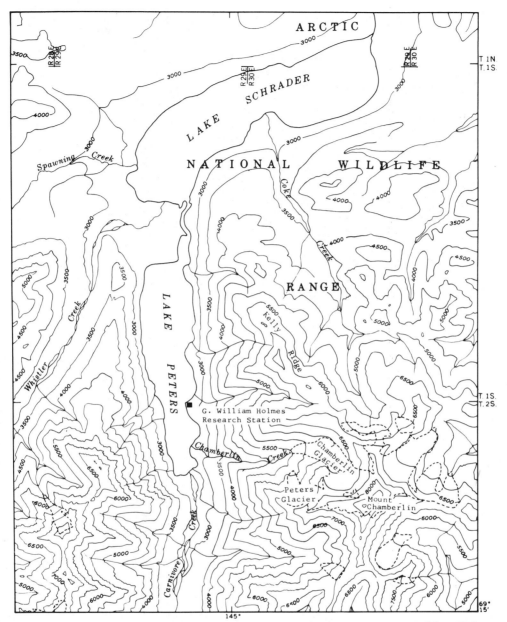

Figure 1. Topographic map of the Lake Peters area showing location of the G. William Holmes Research Station (base from U.S. Geol. Survey Mt. Michelson B-2 quadrangle, 1956, 1:63,360).

Beginning in 1962, most of the research for the next several years was biologically oriented, although some glaciological and geophysical work was undertaken. Studies, mostly sponsored by universities with ARL support, were conducted on early man sites, general botany, lake trout, insects and spiders, singing voles, bryology, mycology and microalgae, mosses and grasses, phytoplankton, Arctic char and lake ecology. During this period, also, the station served as an intermittent base for a number of surveys conducted by the U.S. Fish and Wildlife Service; among the larger animals studied were bear, sheep, musk ox, and caribou. In 1962, a group of scientists undertook a recreational survey of the Arctic Wildlife Refuge for the Fish and Wildlife Service; and in

1968 two scientists examined the area as a part of the National Landmarks Survey.

Beginning in 1968, the station was used as the main base for a major geological program of the U.S. Geological Survey to map most of the Arctic Wildlife Refuge at a scale of 1:250,000. After initial reconnaissance in 1968, three full field seasons involving six geologists, plus support personnel, were accomplished under the leadership of H. N. Reiser. Mapping of the Mt. Michelson and Demarcation Point quadrangles, and small parts of the adjacent Table Mountain, Arctic and Philip Smith Mountains quadrangles, was completed.

In all, more than 80 scientists used the facilities of the

Figure 2. Lake Peters camp, August 1958. View from northeast. Photograph by F. de Percin.

Figure 3. Scientific party and visitors, August 1958. Front (left to right) Maxwell E. Britton, Head, ONR Arctic Program; Frank C. Whitmore, U.S. Geol. Survey; John E. Hobbie, Indiana Univ.; J. H. Hartshorn, U.S. Geol. Survey. Rear (left to right): W. D. Kingery, Mass. Inst. Tech.; Robert Fisher, pilot, ARL; F. de Percin, U.S. Army Natick Quartermaster R. & E. Center; M. C. Brewer, Director, ARL; Maj. Frank Riddell, Station manager; G. William Holmes, Project leader, U.S. Geol. Survey. Photograph by Col. Louis DeGoes, U.S. Air Force.

TABLE 1. LIST OF SCIENTISTS OCCUPYING HOLMES STATION, WITH THEIR
AFFILIATIONS AND PROJECTS, FROM 1958 THROUGH 1971*

Name	Organization	Project
1958		
G. William Holmes	U.S.G.S.	Pleistocene geology
Fernand de Percin	U.S.A.Q.	Meteorology
John E. Hobbie	Indiana Univ.	Limnology
Peter Larsson	McGill Univ.	Glaciology
D.G. Anderson	U.S.G.S.	Hydrology; glaciology
W.D. Kingery	Mass. Inst. Technology	Ice physics
1959		
G. William Holmes, C.R. Lewis, Livingston Chase	U.S.G.S.	Pleistocene geology
Lloyd A. Spetzman	U.S.G.S.	Plant geography
John E. Hobbie	Indiana Univ.	Limnology
Carroll Rock	U.S.A.Q.	Meteorology
David F.Barnes	U.S.G.S.	Geophysics
Ellsworth Clarke	U.S.A.Q.	Meteorology
Frank Leavitt	Dartmouth	Glaciology
1960		
John E. Hobbie	Indiana Univ.	Limnology
Bruce L. Reed, Peter Workum	U.S.G.S.	Bedrock geologic mapping
1961		
John E. Hobbie	Indiana Univ.	Limnology
David Frey	Indiana Univ.	Cladocera adaptations
Bruce L. Reed, James R. Fisher	U.S.G.S.	Bedrock geologic mapping
Gordon Greene	U.S.G.S.	Geophysics
Ralph Solecki, Jerome Jacobsen, Bert Salverson, R.D. Blanchard	Columbia Univ.	Early man sites
Olle Martensson	Univ. Uppsala	Botany, mosses
William C. Steere	N.Y. Bot. Garden	Botany, mosses
Kjeld Holmen	Univ. Copenhagen	Botany, mosses
W.R. Farrand	Columbia Univ.	Geology of archeological sites
R.D. Wood	Nat. Mus. Canada	Botanical photography
Charlotte Holmquist	Riksmuseem, Sweden	Amphipods
Ulf Letteval	Univ. Lund, Sweden	Limnology
R. Bergstrom	Univ. Washington	Lake trout
1962		
J. Muguruma, K. Kikuchi	Hokkaido Univ.	Cloud geophysics, lake ice
R.Ashley	San Diego St.	Insects and spiders
O. Watt, F.B. Day, C.D. Evans, D.R. Klein, L.G. Kenwood	U.S.F.& W.S.	Arctic Wildlife Range recreational survey
1963		
T.J. Cade, E.J. Willoughby	Syracuse Univ.	Zoology, golden eagle and falcons
1964		
R. Tenaza	San Francisco St.	Brown lemming behaviour
H.E. Childs	Cerritos College	Brown lemming behaviour
H. Meguro	Tohoku Univ.	Glaciology
K. Ito	Kyoto Univ.	Glaciology
Y. Tsuru	Tokyo Univ.	Glaciology
1965		
S.W. Greene	Univ. Birmingham, England	Bryological survey
J. Sugiyama, T. Hirata, I. Maruyama	Tokyo Univ.	Mycology, ecology, microalgae
David F. Barnes	U.S.G.S.	Geophysics, gravity survey
William C. Steere	N.Y. Bot. Garden	Botany, bryology
T.A. Smith, H.S. Sears	U.S.F.& W.S.	Pesticide survey
R.D. Biggers	Boy Scouts of Am.	Glaciology
G. Daniels	Boys Life	Photography

TABLE 1. LIST OF SCIENTISTS OCCUPYING HOLMES STATION, WITH THEIR
AFFILIATIONS AND PROJECTS, FROM 1958 THROUGH 1971*
(continued)

Name	Organization	Project
1966		
G.L. Smith	N.Y. Bot. Garden	Mosses, grasses
R.J. Barsdate, F.J. Little, Jr.	Univ. Alaska	Water chemistry, fresh-water sponges
1967		
Andre Journeaux	Univ. Caen, France	Geomorphology
Y. Kobayashi	Nat. Sci. Mus., Japan	Microbiology
1968		
H.N. Reiser	U.S.G.S.	Bedrock geologic mapping
A.K. Armstrong	U.S.G.S.	Carboniferous stratigraphy
I.L. Tailleur, H. Tourtelot	U.S.G.S.	Black shale stratigraphy
G. Phillips	U.S.C.& G.S.	Geophysiology
J.R. Blum	A.D.F.& G.	Land survey
L. Nichols	A.D.F.& G.	Sheep survey
L.P. Glenn	A.D.F.& G.	Caribou project
Peter Lent	U.S.F.& W.S.	Natural landmarks survey
J.E. Hobbie	N. Carolina St.	Phytoplankton
S. Holmgren	Univ. Uppsala	Phytoplankton
J. Kalff	McGill Univ.	Phytoplankton
L.B. Jennings, R.A. Hinman	A.D.F.& G.	Musk-ox range survey
S. Heizer	Univ. British Columbia	Zoology, arctic char
1969		
L.P. Glenn	A.D.F.& G.	Bear survey
J.E. Hemming	A.D.F.& G.	Caribou survey
H.N. Reiser, W.P. Brosgé, J.T. Dutro, Jr., R.L. Detterman	U.S.G.S.	Geologic mapping, Mt. Michelson quadrangle
A.K. Armstrong	U.S.G.S.	Carboniferous stratigraphy
E.G. Sable	U.S.G.S.	Geologic Study, Okpilak granite
H.S. Sears, A. Thayer	U.S.F.& W.S.	Pesticides survey, wildlife range management
Seattle Mountaineers (at least seven)	Seattle, Wash.	Camping and hiking
German Alpine Club (three men)	Munich, Germ.	Mountain climbing
1970		
H.N. Reiser, W.P. Brosgé, J.T. Dutro, Jr., R.L. Detterman	U.S.G.S.	Geologic mapping, Mt. Michelson and Demarcation Pt. quadrangles
A.K. Armstrong	U.S.G.S.	Carboniferous stratigraphy
A. Thayer	U.S.F.& W.S.	Wildlife range, sheep survey
J.M. Lauritzen, R. Wilman	Fairbanks, Alaska	Mountain climbing
1971		
H.N. Reiser, W.P. Brosgé, J.T. Dutro, Jr., R.L. Detterman	U.S.G.S.	Geologic mapping, Demarcation Pt. and adjacent quadrangles
A.K. Armstrong	U.S.G.S.	Carboniferous stratigraphy
Shigeho Rikimaru and associate	Kanazawa Univ., Japan	Mountain climbing

*Note: U.S.G.S. = U.S. Geological Survey; U.S.A.Q. = U.S. Army Quatermaster
Research and Engineering Center, Natick, Massachusetts; U.S.F.& W.S. = U.S.
Fish and Wildlife Service; A.D.F.& G. = Alaska Department of Fish and Game;
U.S.C.& G.S. = U.S. Coast and Geodetic Survey.

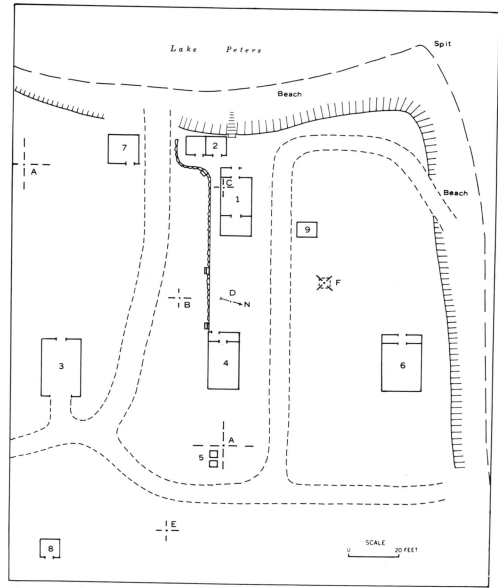

Figure 4. Detailed plan of buildings and facilities at the Holmes Research Station on Lake Peters: 1. Kitchen; 2. Tool-shed and shower; 3. Jamesway, warehouse and garage; 4. Office-laboratory; 5. Surface meteorological station; 6. Sleeping quarters; 7. Generator house; 8. Latrine; 9. Storage tent, on runners; A. Dipole antenna masts; B. Pole for antenna lead-ins; C. Antenna mast; D. Anemometer mast and weathervane (North indicated by arrow); E. Cumulative anemometer pole; F. Clothespole.

Holmes Research Station to a greater or lesser extent. They were involved in about 20 separate research projects in a dozen scientific disciplines. A list of scientists and visitors occupying the station and their projects, as complete as available research records allow, is included as Table 1.

PHYSICAL FACILITIES AT THE STATION

The permanent station consisted of a laboratory building, sleeping and eating quarters and storage buildings. Electricity was supplied by a diesel-powered generator and the buildings were heated so that year-round occupation was feasible, although most of the research activity was confined to the months of May through August. Eight to ten people were comfortably quartered at the station (Fig. 4).

There is ready access to the lake by air, either on wheels or skis during the period when the lake is frozen or on floats when it is ice-free. Large aircraft, including C-47s (R4Ds), C-46s, and Hercules 130s, have regularly landed on the ice in late May and early June.

The critical nature of light conditions, temperature and ice conditions can be appreciated by looking at a work-feasibility

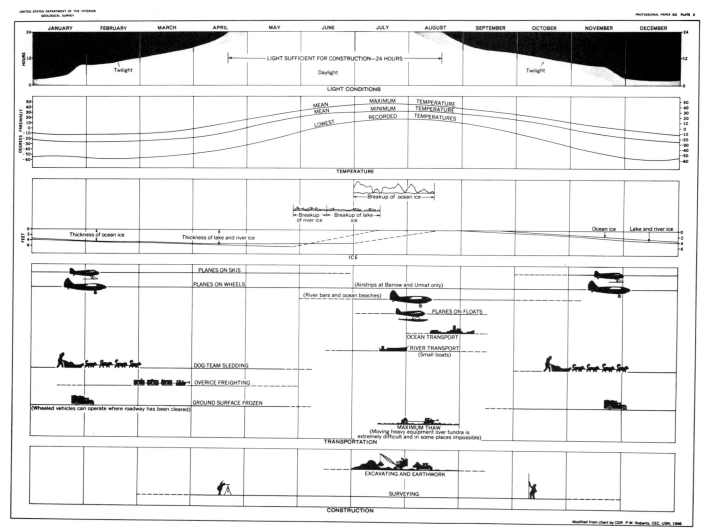

Figure 5. Work-feasibility in northern Alaska as imposed by light conditions, temperatures, and ice conditions (from J. C. Reed, 1958, plate 2).

chart for northern Alaska (Fig. 5). During the year, there is a short time-window when field studies can be carried out efficiently. Logistics must be planned and timed with great care so that supplies and equipment are transported and available to the scientists when field work can begin. Break-up of lake ice on Lake Peters, and similar large glacial lakes in the Brooks Range, begins in early June. Consequently, fuel, heavy equipment and bulky supplies must be moved during May and early June when light and ice conditions are optimal. In July and August, contact by float plane makes possible the movement of people, food, supplies, mail, and sample collections to and from the lake. During the few weeks in late June and early July when the lake ice is melting and breaking up airdrops must be relied upon.

The Holmes Station had all the physical advantages to enable effective staging and execution of scientific field studies in an area where the geological and environmental aspects of an alpine region were close at hand. Therefore the station was an incomparable asset to the scientific community as a whole. All the people and organizations who were responsible for the establishment, care and keeping of the station are to be commended for their persistent faith in the validity of the concept of a permanent research facility at Lake Peters.

DEDICATION CEREMONIES

After Holmes' death in 1970, several of his friends on the U.S.G.S. suggested that the Lake Peters station be named in his honor as a memorial to his work and his part in establishing the station. F. C. Whitmore and Eugene Robertson contacted Max C. Brewer and John Schindler of the Arctic Research Laboratory about the feasibility of such a plan. It was agreed upon and a bronze plaque was secured by Brewer for installation at the station during the summer. We conducted a modest dedication ceremony on June 21, just 12 years from the day that Holmes

Figure 6. Group assembled for the dedication ceremonies at the G. William Holmes Research Station, June 21, 1970. Lower row (left to right): R. L. Detterman, W. P. Brosgé, A. K. Armstrong, J. T. Dutro, Jr., John Kelley, H. N. Reiser, all geologists with the U.S. Geol. Survey, and Tom Mundin, helicopter pilot; above (left to right): Chris Dutro, camphand; Ed Besse, cook; and Ed Ipalook, helicopter mechanic.

and his party first arrived at Lake Peters to establish the station (Fig. 6). The plaque was installed on the west front of the laboratory building at the station. The inscription reads:

"G. William Holmes Research Station
Lake Peters, Alaska
Named for G. William Holmes, Geologist, U.S. Geological Survey, who conceived the idea of this Station, selected the site, and led the party that established it in 1958."

As a matter of historic record, the inscription on the sign that originally was on the north side of the first building (cook shack) read as follows:

"Lake Peters Research Station
Property of the Terrestrial Sciences Laboratory, U.S. Air Force Cambridge Research Center. Operated by the U.S. Geological Survey in cooperation with Arctic Research Laboratory, Office of Naval Research, and USAF Arctic Aeromedical Laboratory, Alaska Air Command."

The sign has since been obliterated, but it provides an insight into the planning and cooperation that went into the establishment of the station.

IMPACT OF THE HOLMES STATION ON ARCTIC RESEARCH

As the only permanent research station in northeastern Arctic Alaska for nearly two decades, the Holmes Station's impact on Arctic scientific studies was considerable. Because of its unique location, a number of studies were completed that otherwise could never have been considered. The relative ease of access and the first-rate physical facilities expedited research immeasurably.

The geological mapping of the Mt. Michelson and Demarcation Point quadrangles, even though supported by helicopter traverses each season, could not have been completed in the time allocated to the program had not the Holmes Station offered a centrally located, permanent base-camp from which to operate. This mapping project proved to be a template for the subsequent major U.S. Geological Survey program to assess the mineral resources of all of Alaska. This Alaska Mineral Resource Appraisal Program (AMRAP) is still active and, in a little more than a decade, has produced resource assessments for more than 35 quarter-million-scale quadrangles totalling more than 400,000 square kilometers.

Similarly, the biological, geophysical and glaciological research done from this station would not have taken place had there not been a station uniquely located where alpine, tundra and glacial-lake environments are so readily available. Resultant scientific publications on the Arctic environment increased our understanding of the natural setting and resources of this northern region, and they are permanent testimony to the value of the station. A list of significant publications, most of them resulting directly from research headquartered at the G. William Holmes Station, follows.

SELECTED REFERENCES ON
NORTHEASTERN ALASKA

(Reports marked with asterisk were direct outgrowths of work at the G. William Holmes Research Station)

*Armstrong, A. K., 1972 (1973), Pennsylvanian carbonates, paleoecology, and rugose colonial corals, north flank, eastern Brooks Range, arctic Alaska: U.S. Geological Survey Professional Paper 747, 21 p.

*Armstrong, A. K., and Mamet, B. L., 1970, Biostratigraphy and dolomite porosity trends of the Lisburne Group, in Geological Seminar on the North Slope of Alaska, Palo Alto, Calif., 1970, Proceedings: American Association of Petroleum Geologists, Pacific Section, Los Angeles, Calif., p. N1–N15.

*Armstrong, A. K., and Mamet, B. L., 1975, Carboniferous biostratigraphy, northeastern Brooks Range, Alaska: U.S. Geological Survey Professional Paper 884, 29 p.

*Armstrong, A. K., Mamet, B. L., and Dutro, J. T., Jr., 1970, Foraminiferal zonation and carbonate facies of Mississippian and Pennsylvanian Lisburne Group, central and eastern Brooks Range, Arctic Alaska: American Association of Petroleum Geologists Bulletin, v. 54 (5), p. 687–698, 4 figs.

*Barnes, D. F., 1970, Gravity and other regional data from northern Alaska, in Geological Seminar on the North Slope of Alaska, Palo Alto, Calif., 1970, Proceedings: American Association of Petroleum Geologists, Pacific Section, Los Angeles, Calif., p. I1–I19.

*Barnes, D. F., and Hobbie, J. E., 1960, Rate of melting at the bottom of floating ice: U.S. Geological Survey Professional Paper 400-B, p. B392–B394, 3 graphs.

*Barnes, D. F., and others, 1976, Gravity data from Mt. Michelson, Flaxman Island, Demarcation Point, and Barter Island quadrangles, Alaska: U.S. Geological Survey Open-file Report 76-258, 4 pls. Scale 1:250,000.

Brosgé, W. P., and others, 1962, Paleozoic sequence in eastern Brooks Range, Alaska: American Association of Petroleum Geologists Bulletin, v. 46 (12), p. 2174–2198.

*Brosgé, W. P., and Reiser, H. N., 1976, Preliminary geologic and mineral resource maps (excluding petroleum), Arctic National Wildlife Range, Alaska: U.S. Geological Survey Open-file Report 76-539, 4 pls. Scale 1:500,000.

*Brosgé, W. P., Reiser, H. N., Dutro, J. T., Jr., and Detterman, R. L., 1976, Reconnaissance geologic map of the Table Mountain quadrangle, Alaska: U.S. Geological Survey Open-file Report 76-546, 2 pls. Scale 1:200,000.

*Cade, T. J., 1960, Ecology of the peregrine and gyrfalcon populations in Alaska: University of California Publications in Zoology, v. 63 (3), p. 151–289, illus.

*De Percin, F. P., 1958, The summer climate of the Lake Peters area, Brooks Range, Alaska, an interim report: U.S. Quartermaster Research and Engineering Center, Environmental Protection Research Division, Research Study Report RER-25, 10 p., illus.

*Detterman, R. L., 1970, Sedimentary history of Sadlerochit and Shublik Formations in northeastern Alaska, in Geological Seminar on the North Slope of Alaska, Palo Alto, Calif., Proceedings: American Association of Petroleum Geologists, Pacific Section, Los Angeles, Calif., p. O1–O13.

*Detterman, R. L., 1974, Fence diagram showing lithologic facies of the Sadlerochit Formation (Permian and lower Triassic), northeastern Alaska: U.S. Geological Survey Map MF-584, Scale 1:500,000.

*Detterman, R. L., Reiser, H. N., Brosgé, W. P., and Dutro, J. T., Jr., 1975, Post-Carboniferous stratigraphy, northeastern Alaska: U.S. Geological Survey Professional Paper 886, 46 p.

*Dutro, J. T., Jr., 1970, Pre-Carboniferous carbonate rocks, northeastern Alaska, in Geological Seminar on the North Slope of Alaska, Palo Alto, Calif., 1970, Proceedings: American Association of Petroleum Geologists, Pacific Section, Los Angeles, Calif., p. M1–M7.

*Dutro, J. T., Jr., Brosgé, W. P., and Reiser, H. N., 1972, Significance of recently discovered Cambrian fossils and reinterpretation of the Neruokpuk Formation, northeastern Alaska: American Association of Petroleum Geologists Bulletin, v. 56 (4), p. 805–815.

*Hobbie, J. E., 1960, Limnological studies on Lake Peters and Schrader, Alaska:

U.S. Air Force Cambridge Research Center, Geophysical Research Directorate, Scientific Report 5, 47 p., illus.

*Hobbie, J. E., 1962a, Aquatic ecology in the Arctic National Wildlife Range: Proceedings Alaskan Science Conference, 1961, p. 31–36.

*Hobbie, J. E., 1962b, Limnological cycles and primary productivity of two lakes in the Alaskan arctic: Indiana University doctoral dissertation, 124 p., illus.

*Hobbie, J. E., 1964, Carbon-14 measurements of primary production in two arctic Alaskan lakes: International Association Theoretical and Applied Limnology, Verhl. 1964, v. 15, p. 360–364, illus.

*Hobbie, J. E., and Cade, T. J., 1962, Observations on the breeding of golden eagles at Lake Peters in northern Alaska: Condor, v. 62 (3), p. 235–237.

*Holmes, G. W., and Lewis, C. R., 1961, Glacial geology of the Mount Chamberlain area, Brooks Range, Alaska, in Geology of the Arctic, v. 2. University of Toronto Press, Toronto, Canada, p. 848–864, illus.

*Holmes, G. W., and Lewis, C. R., 1965, Quaternary geology of the Mount Chamberlain area, Brooks Range, Alaska: U.S. Geological Survey Bulletin 1201-B, p. B1–B32, illus.

*Holmquist, C. M., 1963, Some notes on *Mysis relicta* and its relations in northern Alaska: Arctic, v. 16 (2), p. 109–128, illus.

*Koranda, J. J., 1962, Plant ecology and the Arctic National Wildlife Range: Proceedings Alaskan Science Conference, 1961, p. 37–44.

*Kososki, B. A., Reiser, H. N., Cavit, C. D., and Detterman, R. L., 1978, A gravity study of the northern part of the Arctic National Wildlife Range: U.S. Geological Survey Bulletin 1440, 21 p.

*Larsson, P., 1960, A preliminary investigation of the meteorological conditions on the Chamberlain glacier, 1958: Arctic Institute North America Research Paper 2, 89 p., illus.

Leffingwell, E. de K., 1919, The Canning River region, northern Alaska: U.S. Geological Survey Professional Paper 109, 251 p., illus.

*Lewis, C. R., 1962, Icing mound on Sadlerochit River, Alaska: Arctic, v. 15 (2), p. 145–150, illus.

*Mamet, B. L., and Armstrong, A. K., 1972, Lisburne Group, Franklin and Romanzof Mountains, northeastern Alaska: U.S. Geological Survey Professional Paper 800-C, p. C127–C144.

*Muguruma, J., and Kikuchi, K., 1963, Lake ice investigation at Peters Lake, Alaska: Journal of Glaciology, v. 4 (36), p. 689–708, illus.

Murie, O. J., 1962, Wilderness philosophy, science and the Arctic National Wildlife Range: Proceedings Alaskan Science Conference, 1961, p. 58–69.

*Rainwater, F. H., and Guy, H. P., 1961, Some observations on the hydrochemistry and sedimentation of the Chamberlain Glacier area, Alaska: U.S. Geological Survey Professional Paper 414-C, p. C1–C14, illus.

*Reed, B. L., 1968, Geology of the Lake Peters area, northeastern Brooks Range, Alaska: U.S. Geological Survey Bulletin 1236, 132 p., illus.

*Reed, B. L., and Hemley, J. J., 1966, Occurrence of pyrophyllite in the Kekiktut Conglomerate, Brooks Range, northeastern Alaska: U.S. Geological Survey Professional Paper 550-C, p. C162–C166.

Reed, J. C., 1958, Exploration of Naval Petroleum Reserve No. 4 and adjacent areas of northern Alaska, 1944–53, Part 1, History of the exploration: U.S. Geological Survey Professional Paper 301, 192 p., illus.

Reed, J. C., and Ronhovde, F., 1971, Arctic laboratory: Arctic Institute North America, Washington, D.C., 748 p.

*Reiser, H. N., 1970, Northeastern Brooks Range, a surface expression of the Prudhoe Bay section, in Geological Seminar on the North Slope of Alaska, Palo Alto, Calif., 1970, Proceedings: American Association of Petroleum Geologists, Pacific Section, Los Angeles, Calif., p. K1–K13.

*Reiser, H. N., and Tailleur, I. L., 1969, Preliminary geologic map of the Mt. Michelson quadrangle, Alaska: U.S. Geological Survey Open-file Report. Scale 1:250,000.

*Reiser, H. N., and others, 1970, Progress map, geology of the Sadlerochit and Shublik Mountains, Mt. Michelson C-1, C-2, C-3, and C-4 quadrangles,

Alaska: U.S. Geological Survey Open-file Report, 5 sheets, Scale 1:63,360.

*Reiser, H. N., Brosgé, W. P., Dutro, J. T., Jr., and Detterman, R. L., 1971, Preliminary geologic map of the Mt. Michelson quadrangle, Alaska: U.S. Geological Survey Open-file Report. Scale 1:250,000.

*Reiser, H. N., Brosgé, W. P., Dutro, J. T., Jr., and Detterman, R. L., 1974, Preliminary geologic map of the Demarcation Point quadrangle, Alaska: U.S. Geological Survey Map MF-610, Scale 1:200,000.

*Reiser, H. N., Brosgé, W. P., Dutro, J. T., Jr., and Detterman, R. L., 1980, Geologic map of the Demarcation Point quadrangle, Alaska: U.S. Geological Survey Map I-1133, Scale 1:250,000.

*Rock, C., 1959, Climatic observations during spring and early summer, 1959, Lake Peters, Alaska, *in* Bushnell, V. C., ed., Proceedings of the Second Annual Arctic Planning Session, October 1959: U.S. Air Force Cambridge Research Center, Geophysical Research Directorate, GRD Research Notes 29, p. 95–97.

*Sable, E. G., 1961, Recent recession and thinning of Okpilak Glacier, northeastern Alaska: Arctic, v. 14 (3), p. 176–187, illus.

*Sable, E. G., 1965, Geology of the Romanzof Mountains, Brooks Range, northeastern Alaska: U.S. Geological Survey Open-file Report, 218 p.

*Sable, E. G., 1977, Geology of the western Romanzof Mountains, Brooks Range, northeastern Alaska: U.S. Geological Survey Professional Paper 897, 84 p., illus.

*Spetzman, L. A., 1959, Vegetation of the Arctic Slope of Alaska: U.S. Geological Professional Paper 302-B, p. 19–54, illus.

*Steere, W. C., 1958, *Oligotrichum falcatum,* a new species from Arctic Alaska: Bryologist, v. 61 (2), p. 115–118, illus.

*U.S. Geological Survey Military Geology Branch, 1959, Preliminary report of the Mt. Chamberlain-Barter Island project, Alaska, 1958: U.S. Air Force Cambridge Research Center, Bedford, Mass., 83 p., illus.

*Wood, G. V., and Armstrong, A. K., 1975, Diagenesis and stratigraphy of the Lisburne Group limestones of the Sadlerochit Mountains and adjacent areas, northeastern Alaska: U.S. Geological Survey Professional Paper 857, 47 p., illus.

Geological Society of America
Centennial Special Volume 1
1985

Notes of a packer in the Idaho-Montana line survey, summer of 1899

Ross R. Brattain*

After I had finished my second year at Whitman (Whitman State College), Wink Worthington and I rode our bicycles to Spokane. It seems that Leslie (Leslie Brattain, Ross Brattain's brother) was still reading law with Blake & Post or had gone into some office to get experience. Anyway he had living rooms and Wink and I stayed there while looking for a job. Wink read in the paper that there were some U.S.G.S. men at the Hotel Spokane who wanted to hire some young men to work for the summer on an Idaho-Montana line survey. College men who had some knowledge of such work preferred. He went to see them and took me. He had worked with his brother surveying and I had worked on a job of subdividing a township and also I had experience as a packer. We were both hired. The engineer in charge was a Mr. Rayburn and the head man to start us off was a Mr. Gannett. (Henry Gannett, illustrious geographer and topographic engineer). We were given tickets and told to report at Heron, Montana, which was the closest railroad point to our point of departure. All we had to take with us were clothes to wear on the job, everything else was furnished by the U.S.G.S. We reported as directed and found Mr. Rayburn and Mr. Gannett there and a packer by the name of McQuirck and with him ten pack mules with Mexican aparejos and two saddle horses with men's saddles, one of which was white and the bell mare of the pack train. Why mules prefer a white mare I do not know unless it is the same as gentlemen preferring blondes.

The monument from which the line started was on the summit of the Bitterroot Mountains. The line between Montana and Idaho had been determined from the western boundary of Wyoming to this point. From it the line was to go due north to the Canadian line. There were no trails running straight in any direction and the line had to angle through the Cabinet Range and as it turned out went close to the top of Scotchman's Peak that was over a mile high and on its top was one of the triangle monuments from which readings were taken to other peaks for mapping the country. The starting monument had been set the

summer before and Mr. Gannett was the only one in the outfit that knew where it really was and how to go to it.

Before we started Charley Goodsell, a student at W.S.C. (Whitman State College) whom we knew, joined us. The point was some ten miles southwest of Heron, Montana, and the line crossed the Clarks Fork River about the same distance below the town of the same name. The instrument used to survey the line and measure the distance of each reading was a Theodilite. A big heavy instrument with a telescope twenty-two inches long and stadia rods six inches wide and twelve feet long. They made their measurements by reading the stadia rods. Both the head flagman, or rod man, and the rear rod man carried these rods and readings were double-checked. The instrument was simply covered up and left on line. All brush and trees were cut five feet each side of the line. The crew consisted of Rayburn, his assistant, two rod men, three axe men, two packers and a cook. Worthington was head flagman and Goodsell was the rear flagman. He carried the lunch box of good proportions as it had lunch for the seven men on the line.

The big handicap from the packers' standpoint was that the line went straight and the trails did not. Because I knew something about packing I was made assistant packer. Not only did we have to keep the crew in supplies, but we had to locate camps as close to where the line would come as possible, and we had to have water for the camp and grass for the mules. Mr. Gannett took us up to the starting point. We went up a long ridge, which seemed longer for we had not done much climbing and were soft. McQuirck walked and led the bell mare, which we named Bell. Mc knew the names of most of the mules as he had worked with the string before. We finally reached the monument and I helped the cook, got him wood and we put a kettle of beans to boil so it would be ready to put the pork in and bury in coals for the night so we would have them for breakfast. Rayburn and Gannett were taking readings from various witness rocks and all identification exposures that would identify the location of the monument. The next day they started the line.

We stayed up at the summit camp for three nights but as the line went downhill, which meant that the linemen had to walk farther uphill each day they worked, we had to find a camp below them. We could see what looked like a flat quite a ways below us, so we took a chance and packed the mules, and

*Editors' Note: The manuscript for this chapter came to us through the good graces of Preston Cloud who received it from Walter H. Brattain, Nobel Laureate in Physics and son of the author. It is printed here with the kind permission of Mrs. Walter H. Brattain. The editors believe that inclusion of this piece lends authenticity and color to this section dedicated to the contributions to North American geology as a result of organized effort.

Cabinet Mountains.

McQuirck led Bell and started down a ridge on which the line started. In about three miles we could see the flat and what looked like a small stream. It was on the east side of our ridge. Mc started to work down but there was lots of down timber. Soon Mc, the cook and I all had axes to help our mules. The side of the hill became steeper and we did not want our mules to be jumping downhill in down timber so we decided the best way was to cut a trail as near straight as possible. It was a bigger job than we asked for but we chopped away. Then we heard the boys holler, "Where's camp?" For once my voice was of service and I yelled to bring their axes, which the axe men did and after their arrival we were down and unpacking. Kitchen mule is always first and soon the cook was preparing supper and the men cutting fir boughs to spread their blankets on. We had expected to get down there before lunchtime so we three had no lunch. After that when we moved we carried in our pockets bitter chocolate and cube sugar and that kept us going, as it was not much rest for the mules to stop unless you unpacked them.

We stayed at this camp for three nights and the line was ahead of us. The next move we went off that bench down to the main valley of the Clarks Fork River that the Northern Pacific Railroad follows. The linemen were tying to every culvert and stream, trail or road and when the line crossed the tracks of the Northern Pacific, they tied to it and not far beyond that the line crossed the river. Then I took the two saddle animals and Mr. Gannett to Heron, Montana. We ate supper there and he caught the eastbound train for Washington, D.C. I saw him on the train and stayed all night there and went back to camp the next morning. The line crew had crossed the river so our next move would be to do the same.

We packers would have been willing to stay in the camp all summer and the crew liked it as they had boats to get over the river but each evening their walk was longer. We gathered up the mules and there were two missing, Little Tommy and his mate. They were young mules. Mc said they had gone back to Heron. When the railroad was built there was a big camp and crew there and much equipment was freighted in to be distributed to contractors doing the grading. Many tons of timothy hay had been unloaded and it had seeded and made wonderful feed for our mules. I got on the riding horse and went back and there the two were. On each mule's hackamore was a short rope long enough to go around his neck and be tied to his throat strap. I tried to start towards our camp, but they would turn back to the timothy patch. So I caught them without any trouble, took down the

Mule train. Could be Scotchman's Peak in the distance, but the packers should be going toward it.

ropes on each and tied them together so they would have to walk side by side and used the leather blind as a quirt and soon they were on their way to camp. Finally they came to where a tree had fallen across the trail. Tommy jumped it but his mate did not, so I whacked him across his tail with the blind until he got over. Then we came to another tree and they stopped, twisted their heads until their noses almost touched, then swung their heads in unison and then both jumped at once and went over. I did not hit them any more.

Mc and the cook had the rest of the mules packed and we packed the two and moved to the riverbank. There we took off the packs and the aparejos, leaving only the hackamores. Then we ferried all the supplies and equipment across the river. Then McQuirck tied a thirty-foot rope to Bell's hackamore and had the cook get in the boat and hold the rope. He took the bell off her neck and got in the boat and hung the bell on the left oar, so when he rowed it would ring. He had me push the boat off and the cook let the rope out until the boat was about twenty feet from shore. I was holding Bell by her hackamore close to the water. When Mc was ready he told the cook to tighten the rope and I let loose of Bell's hackamore and slapped her on her hip. She waded into the stream and started to swim and at each stroke

the bell rang. I took hold of Old Tom's hackamore and led him to the edge and said, "Follow" and he did. One by one the others did the same without any urging by me until there was only Little Tommy and Old Modoc on the bank. Old Modoc was named that because some packer said he was old enough to have been in the Modoc Indian War. Modoc was looking at the boat and finally he went in and Little Tommy followed. Some mules, and some horses for that matter, swim very high and others swim so low that only their nostrils and eyes are out of the water. Modoc was a slow swimmer and also deep. He and Little Tommy went down the river about one thousand feet before they reached the other bank. The boat drifted down about two hundred feet. Mc kept ringing the bell and every one came up to where Bell landed; then Mc put the bell on Bell and the string was together again and we packed them and went to our next camp.

The line went within three hundred feet of the top of Scotchman's Peak, which is somewhere near one mile and a half high. Worse than that when approaching its base it went alongside a cliff that scared two of our axe men off the job. Rolly Perkins, a triangulator with the U.S.G.S. came to Clarks Fork and we met him and brought him out to camp. He wished to go up to the top of Scotchman's Peak to do some reading of angles. I took two pack mules and the two saddle horses, his big instrument, bedding, eats and utensils and we went up there. We camped close to the top and also close to a snowbank where we could get water, went with him up to the monument, helped him take the rocks off the copper marker set in a hole in the solid rock and he set up his instrument and read angles for two days. It was high enough so we put the coffee in melted snow water and let it soak all night so our coffee would have strength at mealtime. I tried to be helpful and did what I could in camp, but that was about the limit of my help as he was going to be there only the two days so the why of the way he took the readings are still a blank page to me. Grass was plentiful and our animals did not go far from camp and were there when we were ready to return. When we got back Perkins asked Rayburn to loan him my chum Wink for a few weeks and the two of them left and we did not see Wink again for over a month.

Not far from camp our little stream joined one somewhat larger that had many fish in its pools. McQuirck suggested to me that we try fishing one afternoon and we did. We used some small black gnat flies, about four feet of line on a willow, and fish would bite as soon as the fly hit the water. They would not average much over six inches. The cook helped clean them and we started to count but quit when we got above three hundred. The cook had a Dutch oven about eighteen inches in diameter in which he melted a five pound pail of lard and then laid the fish on a wire screen and then sunk screen and fish in the boiling lard. I helped pass them to the men, and did he fry fish and did they eat. There were some left for us. McQuirck who was going to leave us to do his haying killed a mountain goat and the men ate some, but to me he was somewhat like the bear. Rayburn sent us into Clarks Fork for supplies as each new camp was farther from the source. McQuirck went with us and so did Rayburn's assistant. We put the animals in a pasture that was leased for the summer and gave them a forty-eight hour rest and plenty to eat. As fast as the assistant purchased supplies, I was busy packing them in the aparejos and making up the loads for the mules. The morning of the third day we packed up and returned to camp.

The snow was leaving us and the streams were falling quite fast. I could not help but wonder what happened to the little fish that were caught in a stream that went dry until new snow came. This thought came to me while we were at the camp where we caught so many fish. Rayburn asked me to see what I could find at least a mile ahead. I got on the horse and followed some deer trails for a ways and found a suitable camp with a stream several feet wide and a volume far above our need and blazed a trail back to our camp and reported to Rayburn, who ordered us to move day after tomorrow. So it was announced to the men to make ready. It was up to them to put all little things into a bag and roll them in their bedding. That was just so we would not have a lot of little things to account for at the next camp. Came the day and we moved. When we got there, to our surprise, the stream had no water in it but luckily we did not have to go over a quarter of a mile ahead till we had plenty of water. It happened to be nearer to the line.

They were running along on a ridge that was climbing up towards Scotchman's Peak, and the second night they were almost in sight of our camp. Charley Goodsell, the rear flagman who carried the metal lunch box was the only one who had to carry anything to camp. The ridge was covered with a growth of willows higher than one's head. Besides, if the last sight was long he would be anywhere from a thousand feet to a half mile behind them. He yelled to get his direction and I let out a Nez Perce war whoop and soon we heard him singing. He was the one that was always happy regardless of weather, and everyone was fond of Charley. He got to where the hill pitched off towards our camp, but he was hidden in the willows. A big tree had been uprooted by the wind and it left on the hillside a bank about six or eight feet high all hidden by the willows. Charley had just called and asked the direction and had resumed his song when it ceased and in its place the rattle of the cups and spoons in the metal grub box, and as that noise ceased there came in rhythmic cadence the most expressive assortment of cuss words it was ever my lot to hear. It was so well done that I let it go for two stanzas before I interrupted with the question, "Are you hurt, Charley?" Back came the answer, "H---, no, I lit on my head; you know that wouldn't hurt me, but the strap came off the grub box and the spoons and cups are scattered from H--- to breakfast." Rayburn hollered, "Let 'em go, Charley. We have plenty in camp." Soon Charley was singing again and was in camp.

We were at that camp for quite awhile because we would have to get clear beyond Scotchman's Peak, as the next move would be where the water flowed towards the Kootenai River. To get there we had to take the pack train within a few hundred feet of the top of Scotchman to avoid a box canyon, and then take the animals along under a snowdrift that we did not like the looks of at all. The formation was quartzite and blocky. The layers were from ten to twenty feet thick and the center of the lift was above us to the south. Layers had broken off and slid to the long talus slope below, and above us there were snowdrifts melting. This water running down across where we had to take the mules was what worried us. To get ready we built a fair trail over the talus slope so we could get to where the mules could have good footing. Our difficulty was finding small slabs of rock to fill holes between the big ones. After we got onto soil we had to go north about two miles to get on top of a cliff that ran almost to the top of Scotchman. When we got to where the cliff ended we came to a glacial meadow that was as beautiful as a picture. Here we made our camp, put up the tents, got a pole and hoisted the Stars and Stripes in front of the bosses' tent. This was the nicest camp we had and one we hated to leave. But I am ahead of my story, for when we brought the pack train up the long grassy ridge and finally reached the quartzite slopes with the water running across it, we had an extra man with us. I was walking, leading Bell and let her go to where she was safe, and Old Tom, the lead mule, and Lazarus, with his heavy kitchen load all got over on safe ground. Edith, the mare mule that always wanted to steal Old Tom's place behind Bell, started to go around Little Tommy and the line showed up and quicker than one can tell it, seven mules were

The "mist" camp in a meadow at the foot of Scotchman's Peak.

down. I ran to Edith and grabbed her hackamore and turned her head uphill. The other two helpers did the same and soon we had them safe from rolling down the slope. One thing about a mule that has ever rolled with a pack on his back, if he slips and falls he will lie still until his pack is removed. We unpacked each one and carried the packs off the slope. Then a man at the animal's head and one ahold of his tail would help him up and take him across. It was a slow job but we were glad that we had not lost any animals or had them break a leg and have to destroy them.

The Bureau fed us well. Sometimes I had to go into some town to get food for camp and the lists of what you can buy were much longer than they had to be to satisfy us. One time we brought in a hindquarter of beef and we had everything from tenderloin streaks to boiling beef.

When the line had passed beyond us a mile we moved to go over the ridge. Everyone including the mules would have been willing to stay where we were until snow came but the line had to be finished, so we obeyed orders to move, and went again along the cliff until we could get on top of it, then started to follow along it like we were going to Scotchman and on top of it we hit a patch of broken slabs that we had to cross. It was right along the edge of the cliff, which by this time was over fifty feet high and it was not broken up in small pieces but was in big slabs like those we had to cross getting around Scotchman. The older mules had no trouble because they would not get excited. I was leading Bell,

but Tom did not like the way I went as there was one big slab that rocked when we went over it and he picked out a route that suited him better. He went quite close to the edge of the cliff and of all things got on a slab that rocked. He got off all right but not so Little Barney. He got on the rocking slab and got excited and made a jump; one front foot hit something and down he went and before either Sawdust Jack or I could reach him, over the cliff he went. Mc reached the edge of the cliff and threw his hands in the air and said in explosive language that Barney was hanging in an old snag. Sure enough, the tree had been broken off, a new top had grown, something had killed it and it was sticking straight up. The top of the diamond hitch had caught over this snag and there was Barney hanging in the air about forty feet above the foot of the cliff. Mc hollered to me to get some rope and we could snub around a tree and let him down. But Barney was excited and in his plunging the weight of his load turned his tail down and the broad strip of leather that goes across the animal's hips to keep the load from sliding up on his neck and is laced with whang strings to the back edge of the aparejo, gave way and through the aparejo Barney went, tail first. Within about ten feet of the ground his rump hit a protruding shelf, and then he fell onto the ground and rolled down to almost level ground. We, both thinking we had seen the last of Little Barney, could not believe our eyes when Little Barney got up and shook himself, but his rump looked like it had been skinned. I went down to him by using

Where Little Barney lost his footing and went over the cliff.

projections on the cliff and the old tree. I asked Sawdust Jack to get a can of carbolized vaseline out of Little Tommy's alfoya and to throw it down. He threw it far enough so it dropped on grass-covered ground. The skin across Barney's rump had been cut by that projection and pushed forward about eight inches and twice that distance across his back. I took handfuls of that vaseline and rubbed all over the exposed flesh, then pulled the skin back into place, then greased all around where the skin was broken so gnats would not get at him. Then I did the same with every scratch, but he did not have many. By the time I was back, Sawdust Jack had recovered Barney's aparejo and pack and distributed it to other animals and we went on leaving Barney on the grass.

We made a camp almost on top of the ridge and the linemen ran the line clear down to Lightning Creek which flows into the Kootenai. While they were doing this Sawdust Jack and I tried to find a way to take the pack train down off the north side of the ridge and reach Lightning Creek. We spent three days but every route we started we ended up against a jump down too great for our pack animals. Neither was the route the men went down feasible. It was the quartzite and just one big step after another.

The final decision was that the crew and cook would take food and a blanket apiece and go down, which they could do in a few hours. I was to take the pack train out light to Clarks Fork,

put them in a pasture which the U.S.G.S. had leased. That would give them two nights and a day for rest. Then I'd buy a full load of supplies and come up till my trail hit Lightning Creek and locate them. They told me roughly how to find them, as I would be going down Lightning Creek and they felt they were on a branch which I would have to go up. I was to meet Wink and two axemen in Clarks Fork so I would have plenty of help coming back. The last camp was a poor one for grass; it was mostly spruce which has but little grass.

All set, we started with eleven mules and two horses, as Little Barney was still below the cliff, if still alive. We were not long reaching the snow we crossed coming. It was last winter's snow on a north slope and quite a deep drift, but our trail through it from coming in made it an easy crossing. We were approaching the place where Barney had fallen over, with a question in our minds about how we would find him, when before he could see any of the animals we heard a whinny and then a long bray. My heart jumped and a tear or two came in my eyes. When we came near the edge he whinnied again and was answered by several of the train. They had not forgotten his voice. We had to go on about a mile before we would be down to his level and when we got there he joined in and followed along. I dismounted and looked him over and while not entirely healed I saw nothing wrong in his condition.

Where Barney's pack caught on the tree.

We made good time and crossed the tilted quartzite with snow water running over it as the animals had little but aparejos to carry, but when we got out on the two-mile long ridge with its oceans of wonderful bunch grass on it I needed about three men and a dog to keep them moving. Also I knew they had gone on short rations for several days and I wanted them to fill their bellies. I made the bottom about sundown, but some of them did not know they had eaten enough. I tailed them, but it was not a success as a wind-storm had felled many small trees across the trail and you can't have a string of horses jumping without breaking the string, so I turned them all loose. Then I took Old Tom and put him in the trail headed towards Clarks Fork. I motioned to him and said, "Lead." He looked at me for a moment and then started along the trail. Luckily there were no trees for a few hundred feet and I mounted and slapped every horse I could get behind until he was in the trail. It soon got dark, nothing but the stars and darker. Where the trail was blocked Old Tom would detour and come back and we kept going. Sometimes he made rather high jumps. All the time we were just bringing up the rear. Once Bell stopped and I got off and felt and the obstruction was waist high. I threw the reins over, crawled under it and picked up the reins and told her to jump, and she made it. It was as dark as could be. After about two hours we came out of where the wind

had struck and at eleven o'clock we came to the hotel, restaurant, and bar in Clarks Fork. A voice out of the darkness said, "Ross, is that you?" It was Wink Worthington and two men with him. They had heard the bells on the pack train. We unpacked and took off the aparejos and stacked them along against the wall of the building, then led the bell mare, Bell, to the pasture with the train following. They had been there many times before. We came back and went up to their rooms, one with two beds in it. When we got into the light I saw three men with full beards. I could not tell Wink until he spoke.

There was something I was to pick up at Clarks Fork that had not arrived and caused us to stay an extra day. But Wink and I did not waste our time. The axe men were willing, but little could they do until we got on the trail. We bought everything on the list, but did not buy any fresh meat for fear it would spoil on us if we were delayed in finding the crew. Staple groceries of every kind, canned goods, smoked meats and too many to mention. One thing the cook had ordered was a reflector thirty-six by twenty-four inches which we were to tie on top of old Lazarus' pack and not get it dented. That was what had not arrived, but it did come the second day. It meant hot biscuits for supper. Then we packed everything that we could in the alfoyas; canned goods and small packages went into them. The load for each mule was

Surveyors and cook go without the packers.

Little Barney.

The "road" to Troy.

by his aparejo, so when the departing day came we had an early breakfast, then readied the train, packing the mules and a small pack for the extra riding horse. We left Little Barney in the pasture for a few weeks' additional rest. We had the restaurant make up a lot of sandwiches so we would not have to stop until that night. We all walked most of the way. We had what was called a road for about ten miles, then it gradually became a fair trail and then nearly none. Being near evening we camped, unpacked the horses and mules and there was good grass there.

Next morning we looked for Scotchman's Peak but must have been too close to see its top. Finally we decided that we had gone that way far enough and took a faint trail going up the stream which we felt sure was Lightning Creek. It was not long until we had to get in the creek as there was less brush to cut. We followed it for several hours and came to where it forked. We took the branch coming in from the north and followed it until it had a gravelly bank on one side and a flat covered with buck brush on the other. It was after noon but we did not stop as I was becoming worried that we might have missed them, and we were already nearing two days late. We could not travel the gravelly hillside bank, nor did we like the big boulders in the stream. There was a big yellow pine over two feet in diameter at its small

end. The axe men went to work and flattened the upper side and Bell followed me across and so did Tom and each one and soon we were out in the buck brush and it was not hard going. The valley with low brush and small scattering trees was widening, so we were keeping fairly close to the stream and working up it. We heard Charley Goodsell yell, "Here they are," and he came running through the brush laughing and with tears of joy in his eyes. He told us he had been telling the boys he heard our bells, but they said it was imagination. He had hardly finished saying it when we were swarmed over by the entire crew including Rayburn. They had been hungry two days and some wanted to start out looking for food. They still had little stewed dried apples, and two little fish cooked in with them. Rayburn congratulated us for finding them. The cook asked which packs had canned goods; he knew the kitchen was on Lazarus. We moved in as camp was not far and had a lot of help unpacking and getting anything the cook called for. He almost hugged me when he saw the big reflector, but agreed we would let the biscuits go till evening. We still had six loaves of bread left from what we brought to eat if necessary to save time. Rayburn told the boys to be ready to go back on line in the morning.

The next move we made was a short one as we still had a

heavy load of provisions and so it being nice weather we left all the tents behind. We found a good camping place under some spruce or hemlock trees. We were still close to Lightning Creek. During our supper a thunderstorm hit us with such a heavy rainfall, our plates got so much water in them that we quit eating. What some of the crew said about me for leaving the tents can better be imagined than written. While the gloom was over the camp, Charley had trimmed the limbs off a big tree so we (he and I occupied the same set of blankets) could make our bed there. His voice rang out in song, "It is raining like hell along the Lightning, From the hilltop comes the donkeys' gentle bray &&&" all to the tune of "On the Banks of the Wabash." I don't know how the others fared, but both Charley and I awoke with our heads in a small pool of water on our pillows. It did not give anyone the sniffles. In fact, I do not remember anyone having a cold during the whole season's work, though we slept in wet blankets often.

In a few days we went back and got the tents. With all the clear weather Rayburn asked me to go with him to take a look ahead as he understood the line would cross nearer Troy, Montana, which was on the Kootenai River and we should be able to see indications of its canyon off to the east. Though heavily timbered one could see plainly the apparent depression in tree-tops. So he said to take the pack train and his secretary and one of the axe men, Sawdust Jack, and go along the valley and the stream and see if we could not pick up a trail going east that would take us into Troy. We found the trail and it went by a mining operation and from there in it was a road. I inquired from the boss of the operation and he told me the road would end at the Kootenai River about a mile upstream from Troy and from there to Troy one had to travel on the Great Northern Railroad and that the passenger train from the East came along at one thirty p.m., so not to try it until it had passed. Trains not on schedule were on the lookout for teams, so one was not in much danger. A high bank was on one side and the river on the other. We got there after train time so went down the track and into Troy. We slept, took a room with two beds, and drew lots as to who would get to sleep alone and it fell to me. At first we feared we would have trouble trying to find a rock or bump to curl up around, but we slept through until we were called at the appointed hour. The secretary made the purchases, Sawdust Jack

Railroad creek, lower part.

and I arranged the loads for each mule, and the next morning we packed the mules and went back to camp.

The next move was Troy. Wink had been talking about quitting at Troy and going back to Whitman which opened on September 15th. I decided to go along with him. The old cook did not like the white the taller peaks were displaying and said he would like to follow us. Our paychecks for the previous months would come to Troy and the pay for the fraction of September could be sent to us, so we agreed to stay until Rayburn could get a cook, two for the pack train, and one on the line.

Geological Society of America
Centennial Special Volume 1
1985

Contributions of the state geological surveys: California as a case history

Gordon B. Oakeshott (Retired)
California Division of Mines and Geology
3040 Totterdell Street
Oakland, California 94611

ABSTRACT

At the opening of the nineteenth century, almost none of the sciences were taught in educational institutions. A point of great importance to the early development of geology was the appointment of Benjamin Silliman as professor of chemistry and natural science at Yale College in 1802. Silliman's students initiated a number of state geological surveys in the 1820s and 1830s, starting with North Carolina in 1823, which made major contributions to geological understanding, principles, and applications. There are now 51 state geological surveys—including Puerto Rico—each with staffs ranging from a single "State Geologist" to over 200 people. Of the original state surveys only the California survey, founded in 1880, has survived to the present day with continuous funding. Alabama, Alaska, California, Illinois, Indiana, Kansas, Oklahoma, and Texas maintain their positions as the largest of the state surveys. The U.S. Geological Survey, which does work in all states and in foreign territory, was founded in 1879.

From their beginnings, through the 1950s, state activities were strongly oriented to mineral-resources studies, geologic mapping, and public information. The 1960s saw the burgeoning of concerns for the "environmental" (and engineering) applications of geology.

The decade of the 1970s witnessed greatly-increased emphasis on engineering and environmental geology, especially investigations of "geologic hazards"; development of mineral resources, with due regard to the disposal of wastes, and other environmental concerns; and a renewed consciousness of the importance of public and political participation. Budgets and personnel of the state surveys approximately doubled during the decade. This trend has continued into the '80s, but, lately, growth has been somewhat inhibited by budget restrictions and inflation.

The state geological surveys are economically and politically responsive to *state* authority and therefore have developed a unique capability to serve directly the geological needs of the public.

CURRENT STATUS OF THE STATE SURVEYS

Organization and Funding

In North America today, in addition to the great federal surveys of the United States, Canada, and Mexico, there are 51 state surveys, including Puerto Rico. In terms of personnel and budget, the largest of these are Alabama, Alaska, California, Illinois, Indiana, Kansas, Oklahoma, and Texas (in alphabetical order).

The numbers of personnel in each of the 51 ranges from one state geologist to over 200 people. Budgets range from a few thousand dollars per year to about $6 million for Alaska and Texas, $4 million plus certain University contracts for Illinois, and $7 million for California's Division of Mines and Geology. California also funds geologic work in other divisions and departments. The total budget for all state surveys in 1980 reached about $65 million—still only a fraction of the $550-million annual budget of the U.S. Geological Survey. The state surveys employ about 1,500 professional people, plus an equal number of non-technical staff.

Cuts in state-survey budgets, some severe, began in 1980 and continued into 85, but overall totals in budget and personnel have not been as severely decreased as in 1980.

I have given some figures for the expenditures (budgets) of the state surveys, but strict comparisons of budgets and numbers of personnel are not valid, largely because 1) the state surveys have quite different positions in the hierarchies of state government, and therefore different sources of support, and 2) the surveys perform different, but related, functions in the various states. Some examples follow.

In Texas, the state geological survey is the Bureau of Economic Geology in the University of Texas at Austin. Many of the surveys follow this pattern and most of the state surveys are affiliated with their state universities. In Illinois, the Illinois State Geological Survey is located on the Univeristy of Illinois campus at Champaign, although it is a division of the Illinois Department of Energy and Natural Resources. Close ties with the state university contribute to stability of a geological survey, generally assure a home and funding, help with shared personnel and laboratory facilities, and foster cooperation in research and education.

Nevada is an example of one of the smaller state surveys which has close university connections. A state act in 1866 established a mining school and since those early days, the Nevada Bureau of Mines, and its sister agency, the Nevada Mining and Analytical Laboratory, have been parts of the Mackay School of Mines of the University of Nevada. The Nevada bureau has not been continuously funded from 1866, but the University alliance has enabled it to continue to function over some of the periods of no appropriations. Another of the smaller surveys, the Montana Bureau of Mines and Geology operates within the Montana College of Mineral Science and Technology, at Butte. The New Mexico Bureau of Mines and Mineral Resources is a division of the New Mexico Institute of Mining and Technology at Socorro. The North Dakota Geological Survey finds that its affiliation with the University of North Dakota is advantageous to both the Survey and the University.

Among the state geological surveys which are wholly independent of the state university is the California Division of Mines and Geology (CDMG). Through the years, since its organization in 1880, CDMG has had numerous name and administrative changes from the old State Mining Bureau which reported directly to the Governor, to the present Division which reports to the Director of the Department of Conservation, who in turn is accountable to the Secretary of the Resources Agency, who is one of the Governor's principal agency heads. While CDMG is recognized as the State Geological Survey and its chief is legally the State Geologist, much geological work goes on in other agencies, such as the Department of Water Resources, the Water Control Board, the Division of Oil and Gas, the Department of Energy, the Department of Transportation, and the Department of Forestry. All, or part, of the activities in these agencies are part of the functions of the geological surveys of the other states. No breakdown of the total costs and personnel expended on geological work in California is presently available.

The Washington Division of Geology and Earth Resources operates within the Department of Natural Resources. In Oregon, the Department of Geology and Mineral Industries is an independent state agency. In Pennsylvania, the Topographic and Geological Survey is in the Office of Resources Management within the Department of Environmental Resources. The Utah Geological and Mineral Survey is in the department of Natural Resources and Energy.

Functions

Most of the state geological surveys have operated within their state jurisdiction more or less as a combined U.S. Geological Survey and U.S. Bureau of Mines. Functions vary widely from state to state, depending on local resources and state support. For the most part, state functions are not regulatory, although this is not universally true.

Texas' State Geologist, Director of the Bureau of Economic Geology, University of Texas at Austin, and 1982 President of the Association of American State Geologists, William L. Fisher, says that "The stuff of the state geologic surveys is energy, mineral, and environmental *resources*. State geologists reflect the issues of our times, because we exist at the boundary of science and public policy" (personal communication, 1984).

The statements of other state geologists, although phrased in different words, tend to agree with Bill Fisher that our concern is with natural *resources*. State Geologist Ernest A. Mancini of Alabama says: "The role of the Geological Survey of Alabama has been primarily one that has supported a broad range of cultural, domestic, and industrial development. Since its beginning, the GSA has taken a pragmatic stance. . . ." (personal communication, 1984). Indeed, this emphasis on practical, applied geology, concerned with local problems, has been the forte of the state geological surveys. James F. Davis of California says that "The state geological surveys have established a tradition of mineral resource inventory and development of frame-work geology" (personal communication, 1984).

"The fundamental function of the Iowa Geological Survey is to collect, interpret and report information on geologic features and resources of the state, including surface and ground water," says State Geologist Donald L. Koch (personal communication, 1984). Here is a specific statement of function. Not all the surveys include water (California has a separate State Department of Water Resources, for example), but most do.

State Geologist William W. Hambleton calls attention to the publication of Kansas' Five-Year Plan which details Survey research in specific areas of ground water, energy, minerals, and land use, and describes involvement in the development of new research techniques. For budgetary purposes, not all states acknowledge a research function.

State Geologist Wallace B. Howe leads the Missouri Division of Geology and Land Survey in a major public-service emphasis on the "application of geology to solving problems affecting the citizens and property of the State." New Mexico's State Geologist Frank E. Kottlowski says that the main emphasis of their program is on the "finding and usage of the state's mineral resources" (personal communication, 1984). This state has a severance tax.

Cooperation with the U.S. Geological Survey

Because the charge of the U.S.G.S. is national in scope while that of state surveys, by definition, is restricted, conflict has existed for many years between some state geologists and U.S.G.S. staff. From time to time a state geologist and his staff have felt overwhelmed by the superior funds, equipment, and personnel of the Federal survey.

For example, in California in 1947, Olaf Jenkins, newly-appointed chief of the State Division of Mines and State Mineralogist (Geologist), regarded the first formal work of the U.S.G.S. in California as an intrusion, although he recognized that far more work was needed than the State of California was willing to finance. Later, when the Menlo Park U.S.G.S. center became established, California financed cooperative programs with the U.S.G.S., but Jenkins remained antagonistic to what he considered the U.S.G.S. "superiority." Now, in the 1980s, the U.S.G.S. finances cooperative programs with the Division of Mines & Geology.

In 1950, Joseph Singewald of Maryland told how he had slapped Maryland State decals on U.S.G.S. government field cars to emphasize the Maryland survey's feelings on U.S.G.S. "intrusion."

In Nevada, that state's new 1:250,000-scale geologic map was completed by the U.S.G.S., in a cooperative program with the state. Kentucky has been mapped by the U.S.G.S. on the 1:24,000 scale, the only state so completed.

It is my impression from personal knowledge and attendance at meetings of the Association of American State Geologists (AASG) that most of the State Geologists resented to some degree the U.S.G.S. entering their states with superior finances, more personnel, and greater scientific preparation and knowledge. Many state geologists were and are political appointees, varying greatly in geological preparation and knowledge. Recent actions of the AASG have done much to reconcile differences between the States and the Federal Survey.

The AASG has long followed the practice of holding an annual meeting, prescribed in its constitution and by-laws, which is attended by the state geologists, some of their staff and honorary state geologists. To this meeting are invited officials of the U.S.G.S., U.S. Bureau of Mines, and others. The program of the meeting is designed "to disseminate information concerning activities of state geological surveys to AASG members and to provide representatives of Federal agencies and professional societies and institutes an opportunity to discuss program activities and future plans of their specific organizations. Thus, the AASG annual meeting provides a vehicle for cooperative planning activities between the U.S.G.S. and AASG members" (Haney 1982).

The AASG in 1982 adopted a detailed plan for cooperation between the Association of American State Geologists and the United States Geological Survey. Under this plan the implementation of programs could occur through contracts, grants, cooperative agreements, work-sharing arrangements, and other techniques. All such programs are now operating quite successfully in many states, with a minimum of controversy.

SUMMARY OF SOME CONTRIBUTIONS BY THE STATE SURVEYS[1]

Only 180 years ago, at the beginning of the 19th century, very little science was taught in our institutions of learning. Early public lectures on geology, however, were often popular partly because the speakers followed a natural theology line. Amazingly, in this modern world, we must often contend with creationists who wish to inject fundamental religious dogma into science as taught in our public schools (Skehan, 1983).

When teaching and field studies in geology did begin, the work was done by the educated men of the day—doctors, lawyers, ministers, and teachers self-taught in science—as there were no trained geologists. It was a turning point in science teaching when Benjamin Silliman was appointed professor of chemistry and natural science at *Yale College* in 1802 (see Skinner and Narendra, this volume.) Although a very young man and trained only in law, Silliman shortly made himself a leading geologist. The Sheffield School of Science at Yale trained many geologists of the next half century. Among these, of course, were the early state geologists. By 1812, Silliman had established a formal course in geology at Yale. In 1818, he founded and began editing

---

[1]For a full treatment of developments in geology before 1879, see Rabbitt, 1979.

the *American Journal of Science,* a medium for geological papers ever since. One can follow the development of the science of geology in its pages for the last 165 years.

One of Silliman's students, Amos Eaton, in 1818 gave a course of lectures to the New York State Legislature at Albany, which laid the foundation for establishment of the New York Geological Survey in 1836. Imagine giving a series of lectures to your State Legislature today! In 1821, Eaton made a geological and agricultural survey of Rensselaer County, New York, and in 1824 he published a geological survey of the route of the Erie Canal. Thus, by the early 1820s we had the beginnings of economic geologic mapping and engineering geology. Eaton was responsible, to a large extent, for initiating a number of state geological surveys in the 1820s and 1830s. Among these were North Carolina in 1823, South Carolina in 1824, Massachusetts in 1830 (the first fully-supported state geological survey), New York and Georgia in 1836, and, in all, 20 states east of the Mississippi during the next several years. Their chief functions were locating and describing mineral resources, collecting and exhibiting mineral and rock specimens, and laying out road and canal routes. It is interesting that governments at this early time began to recognize the usefulness of geology and began to give it financial support.

Between 1830 and 1837, fourteen state geological surveys were initiated. In the early days of these state surveys, the idea that geology should be studied continuously to adjust to advances in geological knowledge and changes in societal needs had not occurred to either legislators or scientists. Both groups believed that when a state was investigated geologically and a report written, the job was done. Thus, many state surveys were terminated during the financial recessions of the 1830s and 1840s, not because of lack of money, but because they were thought to be complete. Some state geologists of early days, like Whitney of California and Hall of New York, continued their work for a time without state funding. California now has a "Sunset" law which requires annual review of the need for continuance of all state agencies.

In 1840 the New York geological survey held a meeting of geologists which was the forerunner of the American Association for the Advancement of Science. At the second meeting, in 1841, Professor Edward Hitchcock made it clear that practical geology was not the sole interest of American geologists, but that "careful attention should be given to the scientific geology of the regions examined . . ." (Rabbitt, 1979).

In the last analysis, however, economics was the controlling factor. The gold rush of 1849 in California was followed by a period of economic optimism and expansion in the 1850s; the first state geological surveys west of the Mississippi River were then established. By 1900, the U.S. had 31 state geological surveys. In 1906 H. F. Bain of the geological survey of Illinois organized the Mississippi Valley Association of State Geologists, which was reorganized in 1908 into the Association of State Geologists, and finally became the Association of American State Geologists (Fig. 1). As Oakeshott and others (1971) point out:

To the present day neither geologists nor legislators have been able to determine and delineate precisely what information is most economically useful! Although major contributions to scientific truth and principles have been made by the state geological surveys since their inception, appropriations and stability came, and continue to come, to those surveys and their leaders who have been realistic in their recognition of the economic basis for their existence.

Many years ago G. P. Merrill (1920) published a 549-page quarto volume on the "American state geological and natural history surveys." This book quoted, in very fine print and great detail, the laws providing for the surveys in the individual states, with notes on administration, publications, and financing. Unfortunately, there was little on the scientific accomplishments of the surveys. Most of Merrill's data carry us up to 1885. It is arranged by states in alphabetical order.

Some Basic Contributions

Some examples of the more notable contributions of the state surveys to geologic principles and knowledge are:

1) James Hall's exposition of geosynclinal history.

2) Early midwest survey work on multiple continental glaciation; later work by Leighton and Frye of Illinois.

3) The establishment by Leighton at Illinois of an engineering-geology division in 1927, perhaps the first formal recognition of this specialty in geology.

4) R. C. Moore's concept of cyclothemic sedimentation, and his classic *Treatise on Invertebrate Paleontology.*

5) Minnesota's classic studies of the Pre-Cambrian by geologists Charles P. Berkey, W. H. Scofield, U. S. Grant, C. W. Hall, A. C. Lawson, Leon Lesquereau, Charles Schuchert, A. N. Winchell, and J. E. Spurr. They recognized the area of the Great Lakes tectonic zone as a "major Archean suture."

6) California's initiation in the early 1960s of one of the first cooperative programs on "urban" mapping—engineering and environmental geology. In 1952, the initiation of statewide earthquake investigations, followed by an extensive state strong-motion program, beginning in the early 1970s.

7) The long-continued program on coal research by the Illinois survey, started by H. F. Bain in 1905, and carried on by H. Cady, preeminent authority on coal geology, from 1907 for 50 years.

8) The pioneer work of M. King Hubbert of Illinois on the use of electrical earth-resistivity surveys in ground water investigations in the 1930s.

9) Kansas's pioneer work in computer graphics, mathematical geology, and well-log analysis.

10) Oregon's on-going important summaries of current thinking on ophiolites and generation of magma.

Many state geologists have been strong in their affiliations and activities with national organizations and in university teaching and administration. The state geological surveys have produced, as well as recruited, many distinguished geologists. An outstanding example that comes to mind is Peter Flawn, former

Figure 1. Meeting of the Association of American State Geologists, February 13-14, 1942, Washington, D.C. Standing from left to right: H. E. Culver (Wash.), E. T. Troxell (Conn.), H. B. Kummel (N. J.), M. E. Johnson (N. J.), E. P. Rothrock (S. Dak.), J. L. Stuckey (N. C.), W. M. Laird (N. Dak.), J. A. Carpenter (Nev.), R. E. Esarey (Ind.), Garland Peyton (Ga.), E. F. Bean (Wisc.), W. F. Pond (Tenn.), F. A. Thomson (Mont.), P. H. Price (W. Va.). Seated clockwise around the table: M. M. Leighton (Ill.), A. C. Trowbridge (Iowa), H. A. Buehler (Mo.), E. K. Nixon (Ore.), E. C. Jacobs (Vt.), R. A. Smith (Mich.), Arthur Bevan (Va.), S. J. Lloyd (Ala.), C. A. Hartnagel (N.Y.), R. C. Moore (Kan.), G. E. Condra (Neb.), E. B. Mathews (Md.), R. H. Dott (Okla.), G. H. Ashley (Pa.)

State Geologist of Texas who is now President of the University of Texas at Austin. Historically, R. C. Moore of Kansas was one of the most distinguished; in addition to the cyclothemic concept of sedimentation and his *Treatise,* he was president of the Geological Society of America.

Nationally known, Ian Campbell initiated one of the first programs in "urban" mapping, recognizing the importance of engineering and environmental geology, and had the division name changed to California Division of Mines and Geology. The title of its chief then became "State Geologist." Campbell greatly enhanced the Division's growing reputation in geology. During his decade of California service, he was president of four national organizations: Association of American State Geologists, American Geological Institute, Mineralogical Society of America, and Geological Society of America.

CALIFORNIA: A CASE HISTORY[2]

Early Surveys

Medical Doctor John B. Trask was appointed in 1851 as "Honorary State Geologist," and California's first State Geological Survey was established with $2,000 in 1853 (Shedd, 1933)[3]. Trask had made an excellent impression on the Legislature by submitting an Assembly document titled "On the Geology of the Sierra Nevada, or California Range," plus three other geological

---

[2]Based on my 50 years personal contacts with the California Division of Mines and Geology; summarized in print, 1980.

---

[3]Shedd, 1933. Leviton and Aldrich (1982) say recognition of Trask as first State Geologist was "unofficial."

reports greatly desired because of declining placer gold production and the growth of lode mining. The Trask survey was not continued after 1856, but he had laid the groundwork for an act of the Legislature which followed in 1860.

That act appointed Josiah D. Whitney as State Geologist and appropriated $20,000 for a geological survey. He was a graduate of Yale and quite noted in the East, having worked under the great James Hall of New York. Whitney was not only a leading geologist but was also an excellent organizer and administrator. He soon built up a fine staff, including Gardiner (topographer), Cotter (packer), William Brewer (chief geologist), and Clarence King (volunteer geologist). Clarence King became first Director of the U.S.G.S. in 1879.

Unfortunately, Whitney was tactless and curt with the public and legislators. Disappointed with budget cuts, he told the Legislature in a face-to-face appearance that, "We have escaped perils by flood and field, have evaded the friendly embrace of the grizzly, and now find ourselves in the jaws of the Legislature."

His first report was a treatise on paleontology! Then, dooming the budget for the Whitney Survey in 1874, he bluntly refused to advise Governor Downey on his mining stock speculations! In 1879–80, Whitney used his own funds to publish two informative volumes on the auriferous gravels of the Sierra Nevada, which the Legislature had wanted in the first place.

Fed up with "geologists" the Legislature established the "State Mining Bureau" in 1880 and placed a "State Mineralogist" in charge, in San Francisco, the center of mining in the state. For the next 66 years state mineralogists were mining engineers. In 1929, however, Olaf P. Jenkins (Fig. 2) a young geologist with a Ph.D. from Stanford, was hired as Chief (and only) Geologist. He stayed until 1958, becoming State Mineralogist in 1946. Jenkins is credited with building a modern State Geological Survey. From 1948, he was assisted by Gordon Oakeshott for the next 10 years. Oakeshott then served as State Mineralogist during the year 1958 until the appointment of Ian Campbell in early 1959. Since 1929, all California State Mineralogists (geologists) and their deputies have been registered professional geologists, and all have been under Civil Service regulations since the early 1930s.

The State Mining Bureau, 1880–1928

An Assembly bill in April 1880 established the State Mining Bureau, headed by a State Mineralogist, who was appointed by the Governor. The State Mineralogist's annual salary of $3,000 was set by law and prevailed for many years. His nearest subordinates drew about half that sum.

The Act of 1880, reflecting disillusion resulting from the Whitney surveys, said nothing of geology or a geological survey. Rather, it provided for an analytical laboratory, library, the display of mine models, a museum, and publication and dissemination of information on mining to the citizens of California. After occupying various offices in downtown San Francisco, the Bureau moved into the then-new Ferry Building on the water front at the foot of Market Street in January 1899.

Figure 2. Olaf P. Jenkins, Chief of the California Division of Mines, 1947-1958.

The only mention of the great San Francisco earthquake and fire of April 18, 1906 (of which I am a survivor) by the Bureau was in the 1908 Report of the Board of Trustees. The Ferry Building, on concrete piles in Bay mud, survived, but its tower, in which the Bureau was located, was severely damaged. The Bureau suffered losses estimated at $1,500. This lack of publicity on such an important event probably reflected the efforts of all authorities in San Francisco to minimize the earthquake and pass if off as a fire—a concerted effort that extended for several years.

Activities of the Bureau for many years centered on maintaining its facilities as required by law, publishing statistical reports on mineral production, and preparing reports on mines and mineral production by counties (58 in California). The "Annual Report of the State Mineralogist" was the principal publication of the Bureau. This was published through 1946, but there is a 21-year hiatus between the Thirteenth Annual Report (1895) and the Fourteenth Annual Report (1916). The *Bulletin* series began in 1888 with *Bulletin* 1, "A description of desiccated human remains in the California State Mining Bureau." The bureau's severest critics suggested that these were the remains of the old statistics-loving mining engineers of the Bureau.

Granted that California's old State Mining Bureau was mining-statistics-centered, there were still some valuable geological products of that period. W. A. Goodyear published notable geological papers, beginning in 1887. Harold W. Fairbanks, a Ph.D. student of Andrew C. Lawson at the University of California, did the excellent geological mapping of the San Luis Obispo folio for the U.S. Geological Survey. Curiously, the U.S.G.S. edited out all the faults on his geologic map but left them in the sections. Fairbanks was an Assistant Field Geologist for the Bureau in the 1890s. While at Cal, he described and named the "Golden Gate formation," recognizing its Cretaceous age on the San Francisco Peninsula. Lawson used his student's work, renaming it the Franciscan, a name which has stuck to the present day (see Hsü, this volume). About 1950 the 92(?)-year-old Fairbanks visited us in the Ferry Building to protest omission of his name from Jenkins' *Mother Lode Bulletin* 49. It seems that Fairbanks had published the first definitive geological paper on the Mother Lode with the Bureau in the '90s.

The first of a long line of colored state geological maps was published by State Mineralogist William Irelan, Jr., in 1891. The map covered the entire state on a scale of 1 inch equals 12 miles, and depicted 8 geologic units. It was extremely crude by today's standards; we do not know Irelan's sources—but he initiated one of the state's major functions.

In this period, some classic *Bulletins* were published; among them *Bulletin* 10, a bibliography by A. W. Hodges, 1896; *Bulletin* 18, *The Mother Lode* by W. H. Storms (later he was shot and killed by an irate miner); *Bulletin* 37, *Gems,* by G. F. Kunz, 1905; *Bulletin* 38, *The structural and industrial minerals,* by L. A. Aubury, 1906; and *Bulletin* 19, *The Randsburg quadrangle* by C. D. Hulin, 1925.

The 1916 colored state map, same scale, was better than Irelan's. Compiled by Professor J. P. Smith of Stanford University, it showed 21 units. Olaf P. Jenkins, later to be State Geologist, was one of Smith's student assistants. The whole state was colored in, even though much of the geology was unknown.

In those early days, because of the state's vast size and slow transportation facilities, there were field offices at San Francisco, Los Angeles, Redding, and Auburn; now they are only at Sacramento (Headquarters and the strong-motion program), Los Angeles, and Pleasant Hill, in the East Bay. From time to time short-term field offices are now set up to save travel and money. Santa Rosa (Forestry contract) and various cities for the compilation of county reports are examples.

In 1927 the state Mining Bureau became the Division of Mines and Mining in the newly created Department of Natural Resources. Thus, an administrative layer was introduced between the State Mineralogist and the Governor.

Building a State Geological Survey, 1928–1946, and the State Geologic Map Program

Walter W. Bradley, who had been with the Division as statistician, curator, librarian, and junior mining engineer since 1912, was appointed State Mineralogist in 1928, the last mining engineer in this position.

As the state grew, other name changes took place (Anonymous, 1961). In 1929, the Division became the Division of Mines, and in 1930 it lost the function of petroleum and gas activities which it had held since 1915. The new Division of Oil and Gas became a sister division in the Department of Natural Resources, as did the Division of Forestry.

Bradley was the last permanent State Mineralogist (Geologist) to be appointed by the Governor. Civil Service was introduced in the early 1930s, but he served as State Mineralogist until 1946.

When Bradley retired, the first examination for State Mineralogist was called. Civil Service did much for California; in particular for the Division of Mines. It qualified applicants by written and oral examinations; it stabilized employment; it provided rules and conditions of employment; it tended to equalize salaries and conditions for men and women; and it raised salaries to parity with industry, government, and the universities. Exceptions to the latter have always been the supervisory and administrative salaries and those of the University of California and Federal Government which now far exceed state schedules. In California, compensation for the State Geologist for the last several years has been only 10 percent above that of his Chief Deputy. Retirement compensation for State service is far below that in the U.S.G.S.

In 1928, Theodore J. Hoover (brother of Herbert Hoover), Dean of Stanford University's Mining Department, and geology Professor J. P. Smith persuaded Walter Bradley to start a Geologic Branch and to name Olaf P. Jenkins (Fig. 2) as Chief (and only) Geologist. Jenkins had a brand new Ph.D. from Stanford for his work on the Eocene Kreyenhagen Shale on the west side of the San Joaquin Valley. Thus, by 1929, the state's geological survey had been started. Jenkins' job was confirmed by Civil Service examination in 1933. Two equal branches were set up in the Division: the Geologic Branch (with a budget of $20,000/year) and the Mining Engineering Branch. A modern *geological* survey had been started in California (Jenkins, 1976). Referring to the old statistically-oriented State Mining Bureau, field geologist N. L. (Tucky) Taliaferro of the University of California said that Jenkins had made an "honest woman out of a whore." As we have seen, it wasn't as bad as that!

As its first major project, the new Geologic Branch enthusiastically began compilation of a new State geologic map on one sheet, scale 1:500,000. This was done entirely with volunteer help. One of the most effective of these was Elisabeth L. Egenhoff who was a graduate in geology from Cal, and talented in English and secretarial skills. In 1935 she became a permanent Civil Service editor and for many years did much to assure the excellence of Division publications. The 1930s saw the depths of the depression, and Jenkins had the services of several noted (later) geologists under the Public Works Administration, including E. Wayne Gallagher, Bert Beverly, Harold W. Hoots, R. W. Dibblee, Jr., and Alfred L. Ransome. The resulting colored state geologic map was published on one sheet in 1938. This map left as white areas the unknown geology of the state.

As the usefulness of the '38 map began to play out Jenkins perceived the need for modernization. From 1951 to 1955, Charles J. Kundert compiled the first eight 1:250,000-scale sheets of Geologic Map of California using the Army Map Service (AMS) base. These first AMS base maps compiled from the available topographic sheets were very inaccurate. Beginning in

1956, the AMS and U.S. Coast and Geodetic Survey began compilation of the 250,000-scale base maps using the best available large-scale sheets and photogrammetric methods. Charles W. Jennings, Kundert's successor, then began compilation of a new, far superior 27-sheet series to cover the state, which he named the "Olaf P. Jenkins" edition. The first sheet of this map, Death Valley, was published in 1958. By 1966 this series was "completed."

In 1981, the first of a wholly new 1:250,000 Geologic Map series (Sacramento quadrangle) was published, compiled by Jennings and staff (D. L. Wagner, C. W. Jennings, T. L. Bedrossian, and E. L. Bortugno, geologists, and R. Moor, draftsman). Each quadrangle includes several maps: Geology, using standard formation names; a complete index of sources; and faults and locations of all radiometrically-dated rock samples. This "Regional Geologic Map Series" is unique in the quality and variety of data shown. Because it has been compiled from the best known large-scale basic geologic maps it can be blown up for field use without sacrificing essential accuracy.

In 1943, the Division of Mines published a monumental work (*Bulletin* 118) on "Geologic formations and economic development of the oil and gas fields of California." It was a cooperative work with 126 leading petroleum geologists as authors. It was remarkable for the completeness and specificity of its data, publishing much that had been considered proprietary information by the oil companies. The Division has consistently assisted the petroleum industry, largely through information on the areal geology of the state. The industry has reciprocated by furnishing a great deal of information for use in compilation of the 1:250,000 state geologic maps.

Expansion of Staff and New Programs in Mineral Commodities and Earthquakes, 1947–1958

Olaf P. Jenkins took over as State Mineralogist and Chief, Division of Mines in early 1947, which gave him greater scope to develop his concept of the Division as the State Geological Survey. Mining Engineer, L. A. Norman, Jr., became Supervising Engineer of the Mining Engineering Branch, and Gordon B. Oakeshott was made Supervising Geologist of the Geologic Branch. The function of mining engineering was gradually undergoing diminished activity and, in 1956, Norman left for industry, and Oakeshott was made Deputy Chief of the Division, abolishing the old branches.

Mineral Commodities. One of the great activities of this period was expansion and redirection of the work on mineral commodities. California is fortunate in having the largest variety and value of mineral commodities in the Union. Much has been written about the historical importance of gold in the State's history and need not be repeated here. From 1948 to 1954 California mined 103,000,000 fine ounces of gold valued at over two billion dollars (at $20.67/ounce). It was the most valuable mineral commodity until exceeded by petroleum in 1907. California's maximum petroleum production was in 1953 when it produced over a million barrels a day for the year. A high level of petroleum production has been maintained. Gold remained the most valuable metallic commodity until surpassed by mercury in 1943, by tungsten in 1944 and since 1952, by both tungsten and iron. The non-fuel industrial mineral commodities and also petroleum now far outstrip the value of California's metals. In 1983, California led the nation in the production of non-fuel minerals with an overall value of $1.8 billion. Principal among these were sand and gravel, borates, diatomite, Portland cement, and crushed stone.

Jenkins developed "expertise" on the mineral commodities by assigning a given commodity (or several) to each staff geologist, including the Chief and Deputy. A bulletin on mineral commodities during this period was published as *Bulletin* 156 in 1950, and a much more comprehensive *Bulletin* 176 in 1957, including 736 quarto pages. Geologic occurrences, economic development and utilization were treated for each substance. Staff members were also added for these latter two subjects.

All states now accept their responsibility for work and information on their important mineral commodities. Outstanding examples of mineral commodity emphasis are coal in Illinois, where extensive research has been done; petroleum in Texas, Oklahoma, and other states; salines from Salt Lake in Utah, and copper in Montana and Arizona.

In 1980, CDMG inaugurated a new program on "regional minerals appraisal." The objective is to "provide mineral assessment information that can be readily applied for land-use decisions and planning."

Earthquakes. California is well-known as the state in which most of the damaging earthquakes have occurred and probably will occur in the future. Yet it wasn't until 1952 that the Division of Mines became involved to an appreciable extent. Three *great* earthquakes (magnitude over 8 on Richter's scale) have occurred: at Fort Tejon in 1857, in the Owens Valley in 1872, and at San Francisco in 1906. Twelve were *major* earthquakes (magnitude 7.0 to 7.7) and over 60 were *moderate* shocks (magnitude 6.0 to 6.9). Records have been kept on earthquakes since 1769.

The Fort Tejon earthquake, on the southern part of the San Andreas fault, was felt as far north as San Francisco. It had little effect because of sparse population. The Owens Valley earthquake in 1872, caused by right-lateral oblique slip on the great fault at the eastern base of the Sierra Nevada, was strongly felt in Yosemite Valley, notably by John Muir. The fault zone was visited by State Geologist J. D. Whitney, but Whitney failed to notice the direction of the lateral component. Short summary reviews of these earthquakes were published by the Division of Mines (Oakeshott, 1964; Oakeshott, Greenfelder, and Kahle, 1972; and Bateman, 1961).

San Francisco, April 18, 1906, was California's greatest and most damaging earthquake. This was caused by right-lateral displacement along the San Andreas fault at a substantiated maximum of about 16 feet, north of San Francisco. Epicenter was under the ocean west of the city; strong shaking lasted about 40 seconds. Loss of life was probably at least 1500, but the Army, the City, and all authorities as a matter of policy understated earthquake losses and passed off the event as the great fire which followed.

Governor George C. Pardee, three days after the shock, appointed the State Earthquake Investigation Commission, headed by Professor A. C. Lawson of the University of California and staffed by leading geologists and other scientists of the day. The final report was a monumental pair of volumes published by the Carnegie Institute (Lawson and others, 1908). Noted geologists on the Commission, in addition to Lawson, included J. C. Branner of Stanford; G. K. Gilbert of the U.S.G.S.; and H. F. Reid, professor of geology at Johns Hopkins. A major technical outcome of the Investigation was Reid's elastic rebound theory.

California's third most costly earthquake was at Long Beach in 1933. This caused the Legislature immediately to pass the Field Act requiring the State Office of Architecture and Construction to set up rules and regulations on earthquake safety in the design and construction of school buildings. The Act has been highly successful in reducing damage in later earthquakes (such as El Centro 1940, Arvin-Tehachapi 1952, and San Fernando 1971).

July 21, 1952, is noteworthy for the major magnitude 7.7 Arvin-Tehachapi earthquake. This occurred on the northeast-trending left-lateral, reverse White Wolf fault. Professor Perry Byerly of the University of California and Professor Beno Gutenberg of the California Institute of Technology had a long-standing agreement that any earthquake north of the latitude of the northern boundary of Kern County was "Byerly's" and south of that boundary was "Gutenberg's"—a surprisingly unscientific agreement for two such eminent seismologists to negotiate.

Nevertheless, within a few days, Cal-Tech department chairman J. P. Buwalda; U.S. Coast and Geodetic Survey head W. K. Cloud; and O. P. Jenkins, State Mineralogist; and their staff members, met at Cal-Tech for discussions. The result was field and laboratory investigations by Cal-Tech and the Division of Mines, and a joint publication prepared by Mines (Oakeshott, Ed., 1955). Among the 36 authors were well-known seismologists, geologists, and structural engineers; 16 agencies (federal, state, and industrial) were contributors. This marked the first serious work on earthquakes by the Division of Mines.

From the time of this major earthquake, Division work on earthquakes proceeded apace. Some highlights were: the great Alaskan earthquake of Good Friday, March 27, 1964, visited by two California geologists; two conferences yielding reports on "Earthquake and geologic hazards conference of 1964," and the "Geologic hazards conference of 1965." Great impetus was given to the earthquake program by the Joint Legislative Committee on Seismic Safety, set up in 1969, active through 1974 and headed by Senator Alfred E. Alquist. This Committee generated and passed many legislative bills to abate future earthquake damage. One outcome has been the permanent Seismic Safety Commission presently (1985) chaired by Professor Bruce A. Bolt of the University of California. This was reinforced by the Governor's Earthquake Council in 1972.

Many California state agencies besides the Division of Mines have been involved to some extent in earthquake concerns: the Office of Architecture and Construction, Office of Emergency Services, Division of Highways, Department of Housing and Community Development, Department of Insurance, Council on Intergovernmental Relations (requires that all cities and counties adopt seismic safety elements as part of their general plans), Office of Planning and Research, Department of Real Estate, and Department of Water Resources (Anonymous 1974).

The San Fernando earthquake of February 9, 1971, again greatly stimulated interest and legislative action on abating future earthquake hazards in California. In the early '70s, the Division hired the State's first geophysicist, and the Personnel Board shortly set up Civil Service classes in geophysics and seismology, parallel to the geology classes (Junior, Assistant, Associate [the full working class], and Senior). The principal programs in the earthquake field now are: active fault evaluation including about 10 technical personnel (Alquist-Priolo Act) strong-motion studies with about 20 technical personnel, and the seismological-geophysical program with some dozen technical people.

The active-fault evaluation program was established by the Alquist-Priolo Special Study Zones Act of 1972. The objective of the law is "to provide for the public safety from surface rupture in hazardous fault zones by restricting construction astride active fault breaks" (Davis, 1980). CDMG delineates special studies zones about one-quarter mile wide along potentially hazardous faults.

The office of strong-motion studies was provided for by legislative act in 1971. Under this act, the Division installs, maintains, and monitors a network of strong-motion instruments to "study the nature of strong earthquake motion and to collect data for the design of earthquake-resistant buildings" (Davis, 1980). The program is financed by a tax of .07 of 1% on building permits. By the end of 1980, a total of 264 freefield stations representing 792 seismic data channels, 47 buildings representing 541 data channels, 19 dams representing 217 data channels, and four transportation facilities representing 77 data channels had been installed. The office is headed by a "Chief" (Senior Geologist).

The seismological-geophysical program presently includes monitoring seismic activity, maintaining an earthquake catalog, crustal movement investigations, geophysical surveys and modest earthquake prediction studies (Fig. 3).

Extensive slope-stability (landslide) studies are part of the Division's program in Geologic Hazards, along with the earthquake studies and cooperative studies with county and local governments (Fig. 4). Legislation was introduced in 1982 by Assemblywoman Gwen Moore proposing establishment of a landslide-hazard identification and mapping program with CDMG "to provide technical advice to local government."

CDMG Matures, 1958–1969

Jenkins retired as State Mineralogist in early 1958. He was

Figure 3. Portable microearthquake recording systems capable of recording earthquakes smaller than Richter magnitude 3.0 are used by the California Division of Mines and Geology to collect data for the study of problems pertinent to seismic safety in California. (From *California Geology,* Feb. 1980, p. 27; photo by Charles R. Real).

succeeded by Oakeshott for one year, who was then followed by Ian Campbell (Fig. 5) in February 1959. Campbell shortly "introduced one of the Division's major programs concerning geologic hazards by starting experimental projects on urban geologic mapping in cooperation with Los Angeles and San Diego Counties. He led in introducing legislation which changed the name of the Division of Mines to Division of Mines and Geology, and the designation of its chief from State Mineralogist to State Geologist in 1961, established serpentine as the official state rock, gold as the official state mineral, and the sabertoothed tiger as the official state fossil." He was also a leader in the move to enhance the status of geologists through registration (Oakeshott, 1980). Campbell was involved in the development of Division programs in geochemistry and geophysics before he came to the Division, and afterwards instrumented these programs.

The Department of Natural Resources was renamed the Department of Conservation, and the Resources Agency was set up just above it in 1961. At the end of the 1960s, the Division experienced many changes: 24 people left by transfer or retirement, Division headquarters were moved to Sacramento, under Governor Ronald Reagan; Reagan set mandatory retirement age of State employees at 67, after Campbell retired at age 70. After Campbell, State Geologists changed rapidly: Vacancy Nov. 1-December 16, 1969; Wesley G. Bruer, December 17, 1969-October 30, 1973; Dr. James E. Slosson, November 1973-September 1975; Thomas E. Gay, Jr., October 1975-February 1978; Dr. James F. Davis, March 1978-present.

Nevertheless, during this nine-year period of turnover, major Division program objectives were maintained and advanced. Bruer was a good business-like executive, Slosson was strongly oriented in geologic hazards, Gay understood that his was an interim appointment and maintained the high standards of the Division.

The 1970s to Mid-1980s and the Public Information Program

Public information is the end-product of all Division mapping, research, and investigations. This is true of all state geological surveys. In support of such policy, CDMG maintains information offices in Sacramento, Pleasant Hill in the East Bay, and Los Angeles; an extensive technical library in Pleasant Hill, and formerly maintained a mineral and rock exhibit in San Francisco.

The move of the Division's San Francisco Branch from the Ferry Building to Pleasant Hill in August, 1984, entailed some drastic changes. Space and personnel were reduced; the outstanding mineral, mining, and geologic library was housed in smaller space and some of the historic documents were sent to a Sacramento warehouse. The mineral exhibit was placed in storage (in 1983) in a stone jailhouse in Mariposa, California, a town near the southern end of the Mother Lode. The original intention of the move of this historic collection of industrial minerals, rocks, and artifacts to Mariposa had been to have the county place it on

Figure 4. Huge rhyolite boulder shaken loose from a ridge top during the Mammoth Lakes earthquake sequence of May 1980. The boulder rolled several hundred meters down slope to this resting place, approximately 10 km due east of Mammoth Lakes, California. (From *California Geology*, Sept. 1980, cover; photo by Charles R. Real).

Figure 5. Ian Campbell, State Geologist of California, 1959-1969.

public display in this old gold-mining town. The lease agreement between the State and Mariposa has proven, unfortunately, to be inadequate, and Mariposa County funds have not been forthcoming. The gold exhibit has been protected in storage at another location. The Mineral Exhibit and Library, at this point, remain problems for the Division administration and higher State officials.

CDMG conducts an "Advisory Services Section" with some dozen geologists assigned to these activities. Some of them are: Geologic-seismic review involving the State's interests in Forestry, the Energy Commission, Coastal Commission, Dams (such as the Auburn Dam), the Nuclear Regulatory Commission, hospital sites, Environmental Impact Reports, Alquist-Priolo waivers, and reclamation-plan reviews. Some of these contracts are reimbursed, some not.

Cooperative studies with county and local governments have grown steadily since inaugurated in 1961. Most of these involve geologic hazards (actual and potential) and mineral resources. They are usually supported by cooperative funds.

Many other cooperative studies are conducted with non-State agencies, such as the U.S.G.S., U.S. Bureau of Mines, and

the Earthquake Engineering Research Institute. The latter was founded in 1949 by a small group of noted seismologists to aid the U.S. Coast and Geodetic Survey in its strong-motion earthquake program, based in San Francisco. The Institute has since expanded to about 1,500 members (1985). It shares an office in Berkeley with the Seismological Society of America. According to its Registration brochure, the Institute "is a national professional society devoted to finding better ways to protect life and property from the effects of earthquakes."

Publications

Since information for the public is the end-product of a survey's work, the best way to make a permanent, useful record of that work is through publications. CDMG has prided itself on the number, variety, and high quality of its publications since the old Geologic Branch was started in 1929. State geologist James F. Davis (1984) records,

"Information about the geology, mineral resources, and geologic hazards of California is published by CDMG and distributed to professionals and interested lay persons. The results of studies made by scientific staff members of the Division and others in the earth science professions are published in *California Geology,* the Division's monthly magazine, and in bulletins, special reports, map sheets, special publications, preliminary reports, and geologic maps. All publications are priced at the cost of printing and handling; costs of data-gathering and processing are not included. Revenue from the sale of Division publications during fiscal year 1979-80 was $153,098.

News releases concerning the Division's publications are prepared and distributed to the media. Thousands of inquiries received annually by the Division are answered by supplying copies of *CDMG Notes, California Geology,* and, when warranted, by individual letter from a

Figure 6. The majestic Cascade volcano, Mt. Shasta, dominates the scenery in northern California. (Photo by CDMG).

member of the technical staff. These activities help to inform the public of the importance of geology to the safety and economy of California.

Detailed reference to the subject matter of all publications may be found in the Reports of the State Geologist.

CONCLUSION

The state geological surveys are economically and politically responsive to *state* authority and therefore have developed a unique capability to serve directly the geological needs of the public.

ACKNOWLEDGMENT

My particular thanks to the many state geologists who sent personal letters, photos and other materials. I have incorporated these materials where possible.

SELECTED REFERENCES

Anonymous, 1961, Division changes name and status; the Second Geological Survey; and Past State Geologists; California Division Mines and Geology, Mineral Information Service, v. 14, no. 12.

Anonymous, 1974, Meeting the earthquake challenge: California Legislature, Joint Committee on Seismic Safety, 223 p.

Bateman, Paul C., 1961, Willard D. Johnson and the strike-slip component of fault movement in the Owens Valley, California earthquake of 1872: Bulletin, Seismological Society of America, v. 51, no. 4, p. 483–493.

Davis, James F., 1984, Activities of the California Division of Mines and Geology July 1, 1980 through June 30, 1982: 74th Report of the State Geologist, California Division of Mines and Geology.

Fisher, W. L., 1982 Annual Report, Bureau of Economic Geology: University of Texas at Austin, 43 p.

Haney, D. C., 1982, Implementation plan for cooperation between the Association of American State Geologists and the U.S. Geological Survey: Association of American State Geologists, unpublished report.

Hansen, Michael C., and Collins, Horace R., 1979, A brief history of the Ohio Geological Survey: Ohio Journal of Science, v. 79, no. 1, Jan., p. 3–14.

Hoover, Linn, 1971, State geological surveys: Geotimes, v. 16, no. 7, p. 11.

Jenkins, Olaf P., 1975, Early days memoirs: Ballena Press, Ramona, California.

Jenkins, Olaf P., 1976, Building a state geological survey: Angel Press, Monterey, California and others.

Lawson, A. C., and others, 1908, The California Earthquake of April 18, 1906: Report of the State Earthquake Investigation Commission: Carnegie Institute of Washington, Publication 87, 2 vols., 451 p.

Legget, Robert F., 1963, Geology in the service of man, *in* Albritton, Claude C.,

Jr., ed., The fabric of geology, Addison-Wesley Publ. Co., Reading, Massachusetts, p. 242–61.

Leighton, M. M., 1951, Natural resources & geological surveys: Economic Geology, v. 46, no. 6, p. 563–577.

Leviton, A. E. and Aldrich, M. L., 1982, John Boardman Trask: physician-geologist in California, 1850–1879, *in* Leviton, A. L. and others, eds., Frontiers of geological exploration of western North America, Pacific Division of the American Association for the Advancement of Science, p. 37–69.

Merrill, George P., 1920, Contributions to a history of American state geological and natural history surveys: U.S. National Museum, Bull. 109.

Morey, G. B., and others, 1982, Geologic map of the Lake Superior region, Minnesota, Wisconsin, and Northern Michigan: Minnesota Geological Survey. S-13, 1:1,000,000.

Oakeshott, Gordon B., ed., 1955, Earthquakes in Kern County, California, during 1952: California Div. Mines, Bull. 171, 283 p.

Oakeshott, 1964, Earthquakes in the San Andreas fault zone, Maricopa to Elizabeth Lake: AAPG-SEPM guidebook, p. 41–46 (Fort Tejon).

Oakeshott, Gordon B., and others, 1971, Origin and development of the state geological surveys: Journal of the West, v. 10, p. 133–162.

Oakeshott, Gordon B., 1972, Role of the State geological surveys in education: Journal of Geological Education, v. 20, no. 1, p. 3–8.

Oakeshott, Gordon B., 1980, Inside California's Division of Mines and Geology . . . The first century: California Geology, California Div. Mines and Geology, v. 33, no. 4, p. 75–83.

Oakeshott, Gordon B., 1982, Contributions of the state geological surveys to the development of North American geology: Abstracts with programs, Geological Society America, v. 14, no. 7, p. 579.

Oakeshott, Gordon B., Greensfelder, Roger W., Kahle, James E., 1972, The Owens Valley earthquake of 1872—One hundred years later: California Geology, v. 25, no. 3, p. 55–61.

Rabbitt, Mary C., 1979, Minerals, lands, and geology for the common defence and general welfare, Volume 1, before 1879: U.S. Geological Survey, 331 p.

Shedd, S., 1933, Bibliography of the geology and mineral resources of California to December 31, 1930, p. 5.

Skehan, S. J., James W., 1983, Theological basis for a Judeo-Christian position on creationism: Journal of Geological Education, v. 31, no. 4, Sept., p. 307–314.

State Geologists Journal, various annual reports and unpublished personal letters, through 1984: Association of American State Geologists.

Stuckey, Jasper L., 1965, North Carolina: its geology and mineral resources: North Carolina Division of Land Resources (p. 215–272, Contributions to the geologic history of North Carolina).

Geological Society of America
Centennial Special Volume 1
1985

The role and development of the Smithsonian Institution in the American geological community

Ellis L. Yochelson
U.S. Geological Survey
Washington, D.C. 20560

ABSTRACT

For more than three decades, from its founding in 1846, the Smithsonian Institution served, not only as a focus for what federal geology there was, but also published scholarly papers on earth sciences, and helped inform the general public about the earth and its history. The educational activity of exhibiting collections increased greatly with the opening of a distinct museum building in 1881. Starting from 1879, the growing collections of the U.S. Geological Survey served as a bond between that organization and the Smithsonian. The opening of a new and far larger museum building in 1910 stimulated further growth of this relationship, but there was little increase in scientific staff of the U.S. National Museum for the next four decades; most of the geologists at the Museum were paid by the Department of the Interior. Despite its limited staff, the Museum performed its function well as a national repository; the mineralogical collections in particular grew to worldwide prominence. Surmounting difficulties of little financial or technical support, a few dedicated individuals on the museum staff published significant papers in the 1930s and 1940s. In the post-war decades, the Museum grew dramatically both in physical size for collection storage and in scientific staff. It is now a preeminent research institution; the principal areas of strength in geological research are in meteoritics, volcanology, and the systematic study of many fossil groups, although contributions are being made in many other subjects. The collections of geological objects cannot be duplicated; the institution will continue to provide the raw material for a variety of future investigations.

INTRODUCTION

The Smithsonian Institution was established in 1846 by a bequest from James Smithson (1765–1829), English natural philosopher and chemist, to found in Washington, D.C., "an Establishment for the increase & diffusion of knowledge among men" (True, 1946). The zinc carbonate mineral Smithsonite is named after this man who published a number of mineralogical papers at a time when the chemical composition of minerals was a new field. Along with his monetary bequest, Smithson's cabinet of minerals also came to the new institution. Sadly, it was destroyed in the fire of 1865 that gutted much of the second floor of the "Castle," the first building and present headquarters of the Smithsonian Institution.

Joseph Henry, the first Secretary of the Institution (1846–1878), initially saw the role of the Smithsonian mainly as a center to support intellectual activities by awarding modest sums to assist original research, by lending instruments and collections, by providing advice on scientific work, and by publishing the results of investigation. Although Henry was a physicist, he was aware of the value of geology, having served as field assistant to Amos Eaton, the pioneer geologist from Rensselaer Institute in New York state.

The acquisition and display of collections gradually supplemented the research activities and under the tenure of Spencer F. Baird, the second Secretary (1878–1887), these efforts took precedence over the early roles of the institution. It was the broadly conceived museum function, that largely, but not exclusively, influenced and nurtured the development of geology in North America for about a century. The collections were maintained and enlarged for the benefit of the nation; research by the staff, although important, was secondary because of the small size of the staff and the large size of the collections.

In the last few decades, there has been a major change in

Figure 1. The chemical laboratory of the U.S. Geological Survey in the original National Museum; this is currently the Arts and Industries Building. The lab probably was in the northeast tower of the building. From this beginning arose the U.S. Geological Survey analytical laboratories in Reston, Denver, and Menlo Park. (U.S. Geological Survey, Photo Library "Office Scenes" No. 13).

emphasis. The published research of a greatly increased staff has had significance for a variety of fields in earth science. Increasingly, research is devoted to new fields and new areas outside of the United. States. Meanwhile, the exhibits were modernized half a century after being installed, and these in turn are now being replaced.

EARLY HISTORY OF THE UNITED STATES NATIONAL MUSEUM

The Smithsonian Institution is an "establishment" whose precise status continues to baffle the public. The institution is a quasi-governmental corporation which has as one of its duties the administration of a number of small federal bureaus. The first of these to emerge as a distinct entity within the Smithsonian was the U.S. National Museum, although precisely when it became a separate organization is uncertain.

During the first decade of its existence, the Smithsonian Institution assembled its own collection of art and natural history objects and also had received some scientific material from the

government. For example, specimens obtained by David Dale Owen in his surveys of mineral districts in the Upper Mississippi Valley were transferred in 1854 from the General Land Office to the Smithsonian, as were those collected by Foster, Jackson, Locke, and Whitney from the same region (Merrill, ms. p. 1). It is a tribute to the careful work of many generations of curators that the original fossils illustrated by Owen are still available for study nearly a century and a half after they were collected.

From a practical viewpoint, 1857 is the most likely date to consider for establishment of the National Museum. that year Joseph Henry agreed to house a potpourri of objects and material from various sources that had accumulated in the Patent Office, providing that federal funds were obtained not only to transport them but also to maintain them; with the approval of Congress, the funds were received from the Department of the Interior. Intermittently during its first decade of existence, the Smithsonian had been given small sums of federal money, but since 1857 there has been continuous financial support for the collections by the federal government; however, it was not until 1879 that the term U.S. National Museum was used in a congressional appropria-

Figure 2. The physical laboratory of the U.S. Geological Survey in the National Museum; this is currently the Arts and Industries Building. On the center table is a "decade box" used to increase electrical resistivity, but it is not clear what experiments were being run. This room is a forerunner of the Geophysical Laboratory of the Carnegie Institution of Washington. This lab probably was also in the northeast tower along with the chemical lab, although neither can be located with certainty. Both Figures 1 and 2 were probably taken about the same time, somewhere between 1881 and 1890; a group portrait of the chemists taken in the late 1880s suggests that the photograph may have been taken then (U.S. Geological Survey, Photo Library "Office Scenes" No. 11).

tions bill. The recognition of a national museum as an appropriate expenditure of public funds is significant for all the sciences, not only for geology.

Most of the material collected at governmental expense and housed in the Patent Office came from Lieutenant Charles Wilkes' United States Exploring Expedition (1838–1842). James Dwight Dana's ideas on development of coral reefs were partly based on the collections he made as a member of that expedition. (There is a curious mirror symmetry between the careers of Charles Darwin, who embarked on a long government voyage and eventually became noted for his biological work, and Dana, who embarked on an equally long government voyage, eventually to find his ultimate fame as a mineralogist. Both had earlier training in geology, and after the voyages both published extensively in zoology and geology.)

Among this early material, which is still on exhibit, the

largest object is a 3,000-pound boulder of native copper from northern Michigan to be seen on the second floor of the National Museum of Natural History. The curiosity that the specimen excites as a display item is reflected by its high polish from being touched by thousands of visitors. The cost of moving the rock on numerous occasions far exceeds the market value of the copper. For more than a century, the object has continued to excite curiosity, and as such it represents the public exhibition function of the Smithsonian.

From the time exhibits first opened, they attracted visitors. Despite the difficulty of crossing a canal and swampy ground to reach the red sandstone Castle, the public came in large numbers to see the displays. Whether any congressmen became aware of geology by viewing these objects and thereby voted for the various early explorations and surveys is unknown, but it could have been a factor. Even before there were exhibits, Henry instructed the public with lectures. Henry recorded (1850, p. 21), "A second course of lectures was delivered by Dr. Hitchcock, President of Amherst College, on geology in the lecture-room of the east wing of the Smithsonian building. . . ." These lectures, provided by the Smithsonian before 1865, stimulated considerable interest in science. Oehser (1959, p. 218) noted that "other geological lecturers include the elder Benjamin Silliman and James D. Dana of Yale; Joseph LeConte, then of Georgia; T. Sterry Hunt, of Canada; Fairman Rogers, of Philadelphia: and Louis Agassiz, of Harvard." During the winter of 1853–1854, J. Lawrence Smith lectured on minerals and meteorites; for these talks and for some analyses, he was paid $550, a large sum in those days.

The role of the Smithsonian as a supporter of expeditions and a supplier of scientific advice and assistance has often been noted. As early as 1850, the Smithsonian provided funds for Thaddeus Culbertson to collect in the Badlands; vertebrates he obtained in what is now South Dakota are in the collections. In the mid-1860s, members of Kennitcott's expedition to Alaska, sponsored in part by the Smithsonian, provided testimony on the rich natural resources in Alaska; as a consequence of the purchase of "Seward's Folly," hundreds of American geologists have had this field area to study. It is appropriate to repeat in this context of expedition supplier that Joseph Henry lent the barometers used by John Wesley Powell during his trip down the Colorado River in 1869.

Almost from its inception, the Smithsonian published papers and monographs in geology when few scientific outlets existed. Henry authorized publication of the 1865 "Contributions to Knowledge" paleontologic monograph by Meek and Hayden on the high plains of the upper Missouri River at a time when the War Department was unable to provide funds for more than preliminary reports of exploration. Although the "Contributions," the "Bulletins," and "Proceedings" of the United States National Museum and the "Smithsonian Miscellaneous Collections" are no longer printed, new series have taken their place. The Annual Report of the Smithsonian regularly featured articles on geology for the "intelligent layman," including reprinting of key papers from foreign and domestic journals. If the Smithsonian had done nothing more than act as publisher of geologic literature, its contribution to the science would have been enormous. Although much of this geologic literature is based on studies of North America, papers concerned with all areas of the world, and all subjects of geology, are included.

Display and publication were not the only educational activities undertaken by Joseph Henry. As a regular part of its activities, the Smithsonian gave away duplicate specimens to educdational institutions. For example, from 1846 to 1879, the staff distributed more than 10,000 fossils and more than 21,000 rocks and minerals (Merrill, ms. p. 2). This practice of distributing teaching materials, particularly rocks and minerals, continued for the first quarter of the 20th century.

The Smithsonian has disseminated knowledge, but is also has increased it. Although the Smithsonian supported some early mineralogic investigations (Mason, 1975), the scientific tradition

in geologic research within the institution properly begins with Fielding Bradford Meek, who was trained by David Dale Owen and James Hall (Nelson and Yochelson, 1980). This Ohioan escaped from Hall's domination in Albany and came to Washington, D.C., in 1858. Meek served as a paleontologist for all four of the post-Civil War territorial surveys in the 1860s to 1870s. He also described fossils for the California, Illinois, Missouri, and Ohio surveys. Meek described fossils for the California, Illinois, Missouri, and Ohio surveys. Meek described fossils from all geologic periods and almost all zoologic phyla; he was almost always right on his stratigraphic assignments and on his biologic placement, in spite of the sketchy knowledge of the times and the limited comparative material available. Meek never received a regular salary from the Smithsonian Institution; his work for state and federal surveys was done on contract. He was Joseph Henry's prize lodger; Meek lived in the office Henry gave him in one of the tower rooms of the original Smithsonian building until his death in 1876, the year of the nation's centennial.

THE CENTENNIAL OF THE UNITED STATES AND THE FOUNDING OF THE U.S. GEOLOGICAL SURVEY

Two events, not obviously linked, had a profound effect on the development of geology in America during the last quarter of the 19th century. First, the Centennial Exposition of 1876 in Philadelphia not only celebrated the history of the nation but also spawned a series of American fairs and expositions that persisted for nearly half a century. At most of these events, the Smithsonian sponsored exhibits of natural history, continuing its tradition of diffusion of knowledge. These displays commonly featured geology, both as a basic science and as a practical pursuit that produced valuable ores, minerals, and fuels (Merrill, ms. p. 99–116). To the exposition visitor, geology became a familiar word.

At the Philadelphia exposition, Smithsonian Institution Assistant Secretary Baird assiduously cultivated the various industrial firms and foreign countries which had exhibits, urging them to donate their material to the Smithsonian. Baird was so successful that it was necessary to build another structure in Washington to house the materials. The result was the National Museum which opened in 1881, a colorful, ill-designed brick building just to the east of the Smithsonian Castle (Rathbun, 1905). This structure is still in use under the name of the Arts and Industries Building. Currently, all the exhibits in that building are restored as a replica of the 1876 centennial; and on the east side of the south hall, one may see an old-style exhibit of minerals and ores. This new facility gave the Smithsonian, by American standards of the time, a major display museum. Geology and paleontology played a significant role in the exhibits, occupying nearly a quarter of the public space.

The second event of this decade that profoundly affected development of American geology was the formation of the U.S. Geological Survey in 1879. Directly associated with this, the function of the U.S. National Museum as the official government

Figure 3. The physical geology exhibit in the southwest court of the National Museum. The date is uncertain, but the photograph probably was taken during the 1890s. The Museum is a great believer in recycling, for the specimens shown were used in the original displays in the new National Museum, and some are currently on exhibit in the Physical Geology Hall, East Wing 2nd floor of the present-day Museum of Natural History (Smithsonian Institution, NMNH Photographic Laboratory No. 16841).

repository was formalized. The act of Congress founding the agency included the sentence: "And all collections of rocks, minerals, soils, fossils, and objects of natural history, archaeology and ethnology, made by the Coast and Interior Survey, the Geological Survey, or by any other parties from the Government of the United States, when no longer needed for investigations in progress, shall be deposited in the National Museum" (quoted in Rabbitt, 1979, p. 283–284). John Wesley Powell, second Director of the USGS (1881–1894) played a part in having this sentence written into the act. As a consequence, many federal specimens, particularly fossils, which might have been discarded, continue to be transferred to the Museum and preserved for the benefit of future generations.

One of the earliest activities of Clarence King, the first Direc-

tor of the Geological Survey (1879–1881), was to involve that agency in the tenth census. The 1880 census studied mineral and ore deposits in detail, the only census ever to do so, and valuable collections made in connection with it are still available for examination in the Museum. Ever since its founding, the Smithsonian had taken the view that its charter established it as the national repository, but the legislation of 1879 made this function crystal clear and all tenth census specimens came to the Museum.

The new museum building performed another more basic function when it became the headquarters of the Geological Survey. As originally conceived by King, the Geological Survey operated from several regional centers. When King resigned in 1881, Powell promptly consolidated the staff in Washington. Physical and chemical laboratories were established in the northeast tower of the National Museum and paleontological work was pursued in the south tower.

In 1883 Baird wrote:

The work of the U.S. Geological Survey, also of enormous magnitude—begun under Mr. Clarence King and continued under Maj. J. W. Powell—has resulted in the accumulation of several tons of specimens of fossils, rocks, minerals, ores, and the like. Very few of these can at present be exhibited for want of the necessary space. The survey requires a large number of experts and assistants, and is at present very badly accommodated. Some twenty rooms in the new Museum building have been assigned as quarters for the Director of the Survey and his assistants.

This, however, causes great inconvenience to the other work of the Museum, and as the survey now occupies a large building in Washington for which it pays considerable rent, and for want of quarters in Washington is obligated to scatter its stations over various parts of the United States, it is thought desirable to ask Congress for an appropriation to erect a second museum building corresponding in general character to the first, but on the opposite of the square, along the line of Twelfth Street.

This building it is proposed to devote almost entirely to the mineral department of the National Museum; and when completed to transfer to it everything of a geological and mineralogical nature, and also to prepare a portion of it especially for the accommodation of the Geological Survey which is at present so inconveniently provided for (Rathbun, 1905, p. 265).

In spite of several attempts by Baird, Congress did not appropriate funds for this projected building. As the Geological Survey grew, it expanded to rented quarters in the Hooe Iron Building on "F" Street, Northwest (Yochelson and Nelson, 1979). The chemical laboratories apparently were the last whole unit of the Survey to leave the Museum. Geological Survey paleontologists remained behind, in part because of the difficulty of moving the rapidly increasing collections. Several generations later, some Survey paleontologists continue to occupy offices in the National Museum of Natural History. They have thus formed an integral part of the museum intellectual community dating back more than a century. The close and continuing cooperation between the Smithsonian and the Geological Survey has never been formalized by any written agreement.

DEVELOPMENT OF A MUSEUM STAFF

Although the museum had made earlier appointments of an honorary or temporary nature in the earth sciences, in 1880 George W. Hawes came on the staff as the first paid curator of mineralogy (Mason, 1975). In a sense, he was the first Museum geologist; unfortunately, he died only two years after taking the curatorship. Under Assistant Secretary G. Brown Goode, who succeeded Spencer Baird as the chief officer of the National Museum, this organization had a bewildering assortment of ever-changing departments and divisions. Such a scheme may have appeared to be poor administration, but the appointment of scientists paid by others to the Museum staff permitted these scientists to formally supervise the work of lower-paid assistant curators and aides. There was virtually no money available for

Figure 4. George P. Merrill in his later years, at a moment when he was not writing on geology. Merrill had a reputation of being helpful and friendly. Apparently, he was truly that rare creature, a person beloved by nearly all. (NMHN, Department of Mineral Sciences files).

paid positions, and given this circumstance, Goode took the only appropriate course.

Mason (1975) has clarified who was on the geologic staff at the time and what administrative divisions were used. During the tenure of Goode, essentially every honorary curator had his own separate administrative niche. The year 1883 was a banner year when Frank Wigglesworth Clark, eventually Chief Chemist of the Geological Survey, became an honorary appointee on the staff in the Division of Mineralogy (Mason, 1975, p. 4) and Charles Doolittle Walcott, eventually the third Director of the U.S. Geological Survey, was given a comparable appointment in Paleozoic paleontology.

Lest there be any misunderstanding of the position of honorary curator, this title was used for non-salaried employees, and not for honorific purposes. Many members of the Geological Survey served as honorary curators in the Museum, another tradition which extends to this day under the title of "research associate." Near the turn of the century, the honorary title was abolished, and from the staff listing alone, it is impossible to determine who in the Department of Geology was actually paid by the National Museum. Most of those listed for the next 50

Figure 5. The Smithsonian display showing a *Triceratops* in the foreground, with cases of rocks and minerals behind the skeleton, at the Pan-American Exposition, Buffalo, 1901. This papier-mâché replica is gone, but its spirit lingers on in "Uncle Beazley," a fiberglass *Triceratops* on the Mall outside of the Museum of Natural History. For the Louisiana Purchase Exposition, St. Louis, 1904, a papier-mâché *Stegosaurus* was constructed and this life-size model is still on display (Smithsonian Archives, Record Unit 95, Box 62, NMNH Photographic Laboratory No. 16393).

years as "custodian" were in fact "honorary" members of the staff.

The most notable Museum staff member of this early period was, doubtless, George P. Merrill, who began in 1881 as an aide (Schuchert, 1931). Merrill made significant contributions in several fields. He rapidly became interested in the economic products of geology, taking up work begun by Hawes. Among other miscellaneous items, the Museum had assembled a collection of ornamental building stones from the Philadelphia exposition, and Merrill was the type of individual who could make the most out of what was available to him. He also had available the material collected by the tenth census and obtained additional material

from various stone companies. The subject of building stones was treated by Merrill in several bulletins of the National Museum and in one of the first American books to consider the subject of weathering (Merrill, 1897). The brick museum contained a display of building stones and, in the new building prior to World War I, Merrill filled one enormous hall with blocks and slabs of ornamental stones. As the crowning achievement of this career, Merrill selected the marble for the Lincoln Memorial.

An important early geologic function of the staff of the National Museum at that time, which may be overlooked today, was that of responding to public inquiries about samples. Merrill (1901) indicated that in the 20 years since the museum had been opened, 6,000 packages, or essentially one a day, were received from the general public with requests for identification. Many of the inquiries were of an economic nature, as to whether the taxpayer had found a valuable gold deposit or coal mine. Today the museum staff still responds to a flood of general geologic inquiries.

It is evident from the number of public inquiries alone that the facilities of the National Museum did not allow for a great increase in research activities, over those done prior to 1881. As Merrill (ms., p. 17) noted,

> The years immediately following the death of Dr. Hawes were full of hard work and much trial, the latter due largely to the lack of a single authoritative and directing head to the department, and the former to the enormous mass of material received from the Centennial Exposition and the Tenth Census. Several thousand cubes of building stones remained to be dressed and prepared for exhibition, work was at first carried on in the unfinished rotunda of the Museum and subsequently in the southwest court. Hundreds of boxes of geological specimens, rocks, and ores, were brought from their storage in the Armory and arranged in piles. . . .There was no official catalogue . . . labels, where such existed, had so decayed as to be illegible.

In spite of such a difficult beginning, Merrill continued to build the economic geology collections throughout his long career with the organization. Today, virtually all traces of half a century of staff and exhibit interests in economic geology have vanished from the institution; only one or two exhibits remain in the hall of physical geology. In marked contrast is Merrill's work on meteorites. In its first few years, the Smithsonian had supported investigations of meteorites (Mason, 1975, p. 2) but did not continue this research. After nearly half a century of neglect of the field, Merrill reinstituted work on extra-terrestrial geology, and this effort has continued unbroken to the present. Finally, Merrill also deserves the title of father of the history of American geology by virtue of his National Museum bulletins on contributions of American geology during its first century (Merrill, 1906) and on the history of state geological surveys (Merrill, 1920); these culminated in his well-known book on the history of geology (Merrill, 1924).

In 1897 and 1898, C. D. Walcott, who had his office at the museum building as a Geological Survey paleontologist until he became Chief Geologist of the U.S. Geological Survey in 1892, returned to the Smithsonian Institution on a part-time basis, to serve as acting assistant secretary, in charge of the U.S. National Museum. Almost immediately, he reorganized and systemized the manifold departments and divisions to establish a clear chain of command. In 1897, the Department of Geology was created and Merrill became its first Head Curator. Initially, there were three divisions within the department: physical and chemical geology, mineralogy, and stratigraphic paleontology. Minor readjustments of divisions and a separation of vertebrate paleontology from the invertebrates occurred, but the department endured essentially unchanged for six decades. A summary of the first half-century of geological investigations by the Smithsonian was written (Rice, 1897), and this may be contrasted with the more focused account

Figure 6. Charles Doolittle Walcott quarrying in the Middle Cambrian Burgess Shale, near Field, British Columbia. This photo is posed, but Walcott did much of the day-to-day work, including some dynamiting of loose slabs. Walcott's youngest son Sidney (for whom the first described Burgess Shale fossil *Sidneyia* was named) dates this picture at 1911 or 1912 (Smithsonian Archives, Record Unit RM7004, Box 44, folder 4).

of the Department of Geology written by Merrill (1901). Photographs of some of the displays in the National Museum, given by Merrill, may strike one as crude by today's standards, but they were first class for their time.

Attendance figures of visitors provide perspective on the impact of public exhibits. During Fiscal Year 1888–1889, about the time the Geological Society of America was founded, 374,843 people visited the National Museum. Granted that this was an election year and brought about 100,000 more visitors than normal to the institution, year in and year out the Museum was a magnet for tourists. The collections also grew, and by 1895, the Museum had 25,431 minerals, well over 300,000 fossils, and 63,606 other geological specimens to curate (Goode, 1895). Much of the material was a result of collecting by members of the U.S. Geological Survey.

The career of paleontologist Charles Schuchert further illustrates the intertwining of Geological Survey and National Museum activities. As noted, Charles D. Walcott was the Honorary Curator of Paleozoic invertebrate fossils. When he left the museum building to become chief geologist in 1893, he arranged for Schuchert to be hired by the U.S. Geological Survey to assist in curating the collections, along with other duties. Shortly thereafter, Walcott became Director of the Geological Survey in 1894, and Schuchert transferred to the Museum staff to insure continued care for the collections. The appointment of Schuchert was "of more than passing importance, since for the first time in their history the paleontological collections were placed in charge of an official paid by the Museum" (Gilmore, 1941, p. 314). During the decade Schuchert worked for the Museum, he published numerous papers on invertebrate paleontology, including a significant review of the Brachiopoda (Schuchert, 1897).

After the death of O. C. Marsh in 1899, Schuchert spent some months at Yale sorting and packing the U.S. Geological Survey vertebrate collections. These specimens filled five boxcars when they were sent to Washington. Because of that work, Schuchert was well known at Yale; this acquaintance with the man and his research may have been a prime contributing factor in his appointment to the university faculty in 1904.

The major transfer to the National Museum of vertebrate fossils from the Geological Survey collections which had been at Yale, combined with two earlier shipments from O. C. Marsh, formed the nucleus of the National Museum's extensive fossil vertebrate collections. This gave reason for the appointment of vertebrate paleontologists to the staff. The first two, Charles W. Gilmore and James W. Gidley, were notable workers in the field, and following the pattern of Merrill, they began at the lowest rung of the non-professional ladder. Like other parts of the Department of Geology, the Division of Vertebrate Paleontology, its name changing slightly from time to time, had a history of large collections to curate and small staff to care for them. Gilmore (1941, p. 315) noted that "in the 42 years that paid personnel have been engaged in fossil vertebrate work in the National Museum, 28 persons have been employed, of which only 14 were on a permanent basis." Gilmore's listing includes both Schuchert and Marsh, the only other professional staff being Frederic A. Lucas and C. Lewis Gazin. Despite the small size of the Museum's professional staff, for nearly half a century following Marsh's death, the Geological Survey depended upon it for assistance in the study of vertebrates. Only in the last half of this century has the Geological Survey employed a few vertebrate specialists. The National Museum staff devoted to vertebrate work was by all odds larger than other divisions of the department.

For many years paleobotany was under the Division of Botany of the Museum. In 1898 Albert C. Peale became the first aide paid by the Museum to curate the paleobotanical collections. After his death in 1914, it was almost another half-century before the Museum employed another paleobotanist. Meanwhile, paleobotany in the Museum flourished under U.S. Geological Survey auspices. The efforts of Lester F. Ward, Frank H. Knowlton, C. David White, and Roland W. Brown improved the study of living as well as fossil plants. Cope (1897) gives a summary of the contributions in paleobotany and in other fields of paleontology during the first half-century of the Smithsonian Institution, although his bias naturally enough is toward the vertebrates.

It was not just the Department of Geology that benefited

Figure 7. Ray S. Bassler in his office at the southeast corner of the new museum, surrounded by cases of fossils. This picture in 1920 is typical of Bassler, for in later years his looks changed little, and he never used a much more sophisticated microscope (NMNH, Department of Paleobiology, Bassler photographs in reprint collection).

from the presence of Geological Survey scientists. No history of the Smithsonian Institution would be complete without remarking on the work of William Healy Dall. Dall was the preeminent Tertiary investigator of his day, but he was also the doyen of the study of living mollusks, being the founder of the Division of Mollusks and its Honorary Curator (1868–1927).

In 1901 Ray S. Bassler arrived in Washington, D.C., from Cincinnati, Ohio, to serve as an assistant to Charles Schuchert, and thereafter slowly worked his way upward in the Museum (Caster, 1965). Following Merrill's death in 1929, Bassler became the second Head Curator of the Department of Geology and remained in that position until his retirement in 1948. Bassler's prime interest was in the study of bryozoans. He is best known for his cataloguing work, having produced systematic catalogues of crinoids, ostracodes, corals, and of all North American Ordovician and Silurian invertebrates. E. O. Ulrich, of the U.S. Geological Survey (see Merk, this volume) and Bassler were American pioneers in the study of conodonts. Bassler's survey of the contributions of the Smithsonian to geology, produced for the centennial of the institution, gives an overall assessment of the accomplishments of a century (Bassler, 1946).

Since the shift of Schuchert from Geological Survey to National Museum, there was and continues to be occasional transfer of Geological Survey professional and technical staff to the Museum. During a century, only one or two people have moved from the Museum to the Geological Survey.

THE NEW NATIONAL MUSEUM

From its opening in 1881, the red brick Museum building was so inadequate that in the 1890s the need for another structure became imperative. Before the turn of the century, the Museum collections continued to grow at a rapid rate, even if the staff grew hardly at all. The third Secretary of the Smithsonian, Samuel Pierpoint Langley (1887–1906), is generally given credit for the granite museum built across the Mall from the "Castle." In fact, this may be another example of cooperation with the Geological Survey, for during the time Walcott was Acting Assistant Secretary, in charge of the National Museum, he may have broken down congressional resistance against a new building.

Walcott commonly met "Uncle Joe" Cannon, Speaker of the House, and took him for a social buggy ride. At one of these meetings, "Uncle Joe 'paused, his foot on the step and said: 'Walcott, you may have a building for the Survey or one for the National Museum, but you can't have both' " (Willis, 1947, p. 32). Walcott chose the museum, and it is interesting to speculate what might have been the effect on geology if Walcott had been able to get both; the Geological Survey had to wait 70 years for its building. Ground was broken for the new museum building in 1904. Langley died early in 1906; the following year, Walcott became the fourth Secretary of the Smithsonian.

The new museum quarters proved a boon for the collections, which continued to grow, and for most of its geologists and paleontologists. Because the older brick museum building had become so impossibly crowded, new staff could not be placed there. Edward O. Ulrich (see Merk, this volume) was one of the Geological Survey paleontologists who began his career in the Hooe Building. Certainly his transfer to the new Museum was a noteworthy event. With many fossils available to him and Bassler to dominate, he proselytized vigorously for his Ozarkian and Canadian Systems; in present-day terms these would be approximately Late Cambrian and Early Ordovician as they are currently used in North America. Just possibly, if he had stayed with the field mappers his arguments might have swayed other geologists on the Geological Survey to his interpretation of the geologic column. Ultimately, it was his inability to convince other American paleontologists of the temporal succession of the faunas distinguishing his "systems" that caused his concepts to founder. The Ozarkian and Canadian systems were used for a time in several foreign countries; this usage traces back to paleontologists visiting the Museum who came in contact with Ulrich.

This new museum building also may have been the cause of a different anachronism in American geology. George H. Girty, another paleontologist of the Geological Survey, had bitter words with Schuchert while they were both in the brick museum, and to add to this, he also did not get along with Walcott. When the other paleontologists were consolidated into the new building, Girty did not move with them. This inertia may have been of his own choosing, for in the decade before his death, long after Schuchert and Walcott had left the scene, he did not come to the new building, and he may never have before that. Eventually Girty ended up in the Interior Building, which housed the Geological Survey when that agency left the Hooe Building in 1917. Girty was a strong proponent of the Carboniferous System, and he was close at hand to influence the Geologic Names Committee of the Survey. As a consequence, it was not until the 1950s that the U.S. Geological Survey adopted use of the Mississippian and Pennsylvanian systems, years after most geologists in America used these terms.

Merrill and his staff moved fossils, minerals, and stones to the new building late in 1909 and, for the next several years, supervised work on new display halls. Because of the influx of Geological Survey paleontologists, emphasis in the building shifted to study of fossils and particularly to biostratigraphy. The star performer in earth sciences, however, was not in the new museum building but in the old Castle. In addition to an enormous administrative burden, Secretary C. D. Walcott was a preeminent paleontologist and for two decades, until his death in 1927, practiced his craft at the institution (Yochelson, 1967). During his career with the Geological Survey, Walcott had seen virtually all the outcrops of Cambrian rocks in the United States and eastern Canada. His new position enabled him to undertake much fieldwork in western Canada. His monumental field and laboratory efforts in that region resulted in publication of six entire volumes of the Smithsonian Miscellaneous Collections. Walcott's discovery of the Middle Cambrian Burgess shale fauna in 1909 may have been the most significant fossil find ever made! Modern restudy of this remarkably preserved biota has continued to yield important insights for paleontology.

In his role as Director of the Geological Survey and later as Secretary, Walcott kept his eye on the practical side of geology. In 1904, following the St. Louis Exposition, the National Museum formed a Department of Mineral Technology with Walcott in charge. Merrill (ms., p. 109–110) recorded,

An excessive amount of material was brought from St. Louis. . . .Too liberal a spirit was found to have been manifested in both the gift and acceptance of the material and the trials and earlier experiences with the Centennial collections were repeated. Much of this excess of material it should be stated was solicited by others than members of the Museum force . . . a very considerable proportion bore evidences of a generosity on the part of the donors actuated by an indisposition to pay freight charges for the return of their own material.

Little is known of departmental activities for its first decade, and in all likelihood there was little to report. However, once the exhibits were installed in the new National Museum building, emphasis shifted to the brick Arts and Industries Building and new displays there informed the public about the production and treatment of ores.

In 1910 when the new National Museum opened, new public displays were installed that far surpassed those in the old museum building. In the four display halls on the first floor and three on the second, nearly one-third of all exhibit space was devoted to rocks and fossils. For about a year during the First World War, the display areas were closed and the entire Museum was used by the Bureau of War Risk Insurance as office space. After the war, the exhibits were refurbished and opened to the public. These displays continued to arouse public interest in the earth, its resources, and its former inhabitants. At a time when there was little awareness of science, the importance of the Mu-

Figure 8. The Hall of Rocks and Minerals on the second floor of the east wing, Museum of Natural History; the boulder of native copper is in the foreground. This is now the site of the Mineral Hall, and for those who have seen it, the improvement is obvious. No date is given, but this could have been photographed any time between 1920 and 1955, although it may have been taken in 1921 for Merrill (1923). The absence of visitors may mean that this view was taken before the Museum formally opened that day (NMNH. Photographic Laboratory, No. 24880).

seum cannot be overestimated. In his history of the department, Merrill (1923) estimated that a half-century after the National Museum was clearly established, about 53,000 geological specimens were on exhibit and nearly one and one-half million specimens were in the study series.

World War I involved more than soldiers and guns. The Museum staff supplied mineral specimens for experiments. Merrill was detailed to the Council for National Defense to look for optical quality quartz. Walcott, as president of the National Academy of Sciences and one of the founders of the National Research Council, recognized the importance of natural resources in the struggle, and he saw the implications of involving the

geologic community in the war effort in both expected and novel ways.

Efforts masterminded by Walcott involved some members of the Geological Survey and Museum staff in areas such as terrain study and water supply investigations. After the war, these studies led to investigation by Smithsonian employees of the nation's energy supplies and natural resources. Although this activity was abandoned after a few years, these studies did influence government policies on natural resources and federal development of hydroelectric power. During the 1930s, 1940s, and 1950s, the National Research Council played a major role in development of the earth sciences; it is unlikely that this organization would have developed to the extent that it did without Walcott as one of its co-founders, or would have continued to exist after World War I without his guidance.

THE 1920S, 1930S, AND 1940S

Despite the new building, the lack of funds for the museum and its staff made for great difficulties. The appropriations were small and there was

absolutely nothing in the way of monies for purchases . . . and much discontent arose which was only prevented from becoming serious through the abundant opportunities for research offered and with which the scientific corps busied themselves. The appointment of Walcott as Secretary in 1907 gave hope to the geological corps, which, however, was not fully realized. Indeed conditions became more and more discouraging, taking finally an upward turn for the better only with the appointment of the present incumbent—A. Wetmore—in 1927 (Merrill, ms. p. 117–118).

These words were written shortly before the depression of 1929 when funding became far worse.

The late 1920s were a time of great loss to the staff; both Walcott and Dall died in 1927, and Merrill two years later. There were some replacements. By coincidence, in 1919 both Earl Shannon and William Foshag were hired by the Museum. Through their joint cultivation of Washington Roebling, the Roebling Mineral Collection was donated to the Museum in 1925. This, in turn, led to the Canfield Mineral Collection also coming to the National Museum. Collectively, these specimens gave the Museum a dominant worldwide position in mineralogy, and because the endowment provided by Roebling provided financial support, the Museum has been able to purchase specimens needed for the collections or for display. Foshag developed the Museum as an important center for mineralogy. Although virtually all of his publications are on mineral compositions and collecting localities, Foshag is best known for his work on the Mexican volcano Paricutín.

Because the National Museum is a public institution, it is a servant of the public. Since its inception, the Museum has followed an enlightened policy of allowing qualified investigators to study its geological collections. At a time when some private museums would not permit their minerals to be studied, the National Museum honored requests for both loans and gifts of rocks and minerals. Elements of the Cross-Iddings-Pierson-Washington classification of igneous rocks, details of optical mineralogy investigated by E. S. Larsen, and refinements in the study of opaque ores and minerals by M. N. Short are some of the fruits of this policy.

Before World War II, no type material of fossils was loaned, although other specimens were sent out on request. Occasionally, visitors came to examine the collections, although seldom was there more than one person at a time on the premises. Since the 1940s, the old policy against loan of type specimens has been liberalized, and now hundreds of loans of fossils a year are assembled, packed, and shipped.

Figure 9. William F. Foshag, in a portrait of smoking volcano and smoking geologist. The picture of Paricutín appeared in *Life Magazine* April 17, 1944 (p. 88). Presumably, the montage may have been made at about the same time (NMNH, Department of Mineral Sciences files).

Another less-known aspect of museum policy has contributed significantly to American geology by permitting geologists who have been on the premises and who wish to remain active after formal retirement to retain offices. Several dozen geologists have had their scientific careers extended a decade or more as a result of this policy. Thus, the Museum of the 1920s, 1930s, and 1940s may have included a collection of "old timers," but they were a productive group. For example, Charles Butts in his retirement compiled the geologic map of Virginia, some of the office work being done in the Museum building. This generous policy continues, and more than half a dozen "retired" senior scientists are among those actively engaged in research in the building.

One important event in the interval between the world wars was the employment by the Museum in 1930 of a young paleontologist, G. Arthur Cooper, one of Charles Schuchert's last students at Yale. Cooper remains at the museum after more than half a century. His contributions to geology and paleontology are legendary, and may be divided into several facets. First, he is responsible for the Geological Society of America Devonian Correlation Chart of North America, published in 1942. His

Figure 10. G. Arthur Cooper in the field in west Texas, the Del Norte Mountains in the distance. Much of the 70 tons of Permian limestone was backpacked out of the hills to the roads and jeep trails by Cooper and his colleagues (photograph from a color slide taken by R. E. Grant, co-author of the Cooper-Grant monographs).

earlier studies of Devonian brachiopods led to several major changes in correlation, subsequently confirmed by study of other fossil groups. Second, he has contributed materially to investigations of Ordovician brachiopods. He played a significant part in helping to assemble the North America Ordovician Correlation Chart for the Geological Society of America, a dozen years after his Devonian work. Even more significant, he established the Whiterock Stage as the lowest subdivision of the Middle Ordovician; this stage has been recognized worldwide, and rocks of this age occur on Mount Everest. Third, he instituted a study of Permian brachiopods. A collaboration with R. E. Grant culminated in the largest monograph on brachiopods ever produced. Fourth, since retirement Cooper has written more on living brachiopods than any other person. His significant contributions to geology are recognized by medals from the National Academy of Sciences and the Geological Society of America.

Cooper completed much of his work under difficult conditions. Like many organizations in America, the National Museum suffered during the Great Depression of the 1930s. Cooper recounts being given authority to go into the field one year but being told not to collect large quantities of fossils, for there was no money to pay freight, little space for storage, and no technical support for preparation. In spite of this, Cooper persisted and perhaps 5 percent of the combined Museum and Geological Survey Paleozoic megainvertebrate collections are the result of his efforts in the field.

Walcott left money to support field work and publication on fossils; after the death of his widow, in the 1940s this fund was substantially increased. The Walcott Fund is larger than the Roebling Fund and has been used to good advantage by the Museum, but its overall impact on the field of paleontology has not been as dramatic as that of the Roebling Fund on mineralogy. Extra non-government funds during the depression helped the mineral collection of the Museum become preeminent. Likewise, partly by purchase and partly by fieldwork supported by the Walcott Fund, many outstanding collections of fossils also have been assembled in Washington. But because of the nature of the subject, it is impossible for any organization to have the best collections in all fields of paleontology. Perhaps the administration of these funds demonstrates the difference between the relatively simple inorganic systems of minerals and the far more complex biologic systems represented by fossils.

During World War II, the Museum staff again turned their talents toward the war effort. History repeated itself, and again the Department of Geology helped in the search for piezoelectric quartz crystals and other critical minerals. Foshag had many years of field experience in Mexico, and he was transferred to the U.S. Geological Survey to be liaison to the Geological Survey of Mexico, particularly in mineral resource investigations. Cooper also went to Mexico to conduct basic stratigraphic investigations. He confirmed the discovery of Cambrian rocks by two Mexican geologists, and his other investigations of the Paleozoic sequence in Sonora in 1944 are important chapters in knowledge of western geology. Unfortunately, these investigations never received proper notice at the time, for the Mexican volcano Paricutín chose the same year to erupt, and it received all the attention. Foshag's investigations of Paricutín's growth laid the foundation for a continuing study of volcanology by Museum geologists.

All the Geological Survey paleontologists at the Museum, save one, were transferred to new wartime duties. Some were out successfully searching for mineral deposits, particularly bauxite and fluorite. Others worked for the Military Geology Unit; because of wartime secrecy, the efforts of these paleontologists never received proper recognition. One paleontologist correctly deduced the launching site for Japanese fire balloons which had landed in the northwestern United States by studying the microfossils in the ballast sand of a captured balloon.

Immediately following the surrender in the Pacific, E. P. Henderson and William Foshag were sent to Japan to evaluate gems in the vaults of the Bank of Japan. This turned out to be the largest evaluation of diamonds ever conducted. Unanticipated uses were found for the extensive collections in Washington. At least twice, Geiger counter surveys were run in the Museum to search collections, including those of the tenth census, for uranium-rich rocks that might lead to areas which could be explored for this mineral. The exhibit of rock samples of all geologic periods was used as the basis for the Geological Society of America rock color chart.

Unfortunately, with depleted staff, little space, and obsolete laboratory equipment, the National Museum was in a poor position to conduct much basic research. R. S. Bassler eventually retired, and from 1948 until his death in 1956, Foshag served as the third Head Curator of Geology (Schaller, 1957). He took the first steps of rebuilding the department to its pre-war level.

EVOLUTION OF THE CURRENT STRUCTURE AND STAFF

Between 1948 and 1952, the number of paleontologists on the Geological Survey increased several-fold as a reflection of the post-war growth of that agency. Because there was no additional space on the third floor of the Museum, the "Stone Hall" of Merrill on the second floor was closed and used by the paleontologists for offices and storage for more than a decade. This area is now the exhibit hall of Physical Geology; no trace of its earlier history remains. It was some of the least satisfactory office space in Washington, yet a number of excellent papers were produced by new employees. Cooperation between the staffs of the National Museum and the Geological Survey paleontologists probably reached its acme during the 1950s and 1960s.

Thanks to the efforts of E. P. Henderson, who came on the staff from the Geological Survey in 1929, during the 1930s and 1940s Stuart Perry supported work on meteorites and eventually donated his collection to the National Museum. As in mineralogy, the Museum then came to have the largest collection of meteorites in the world, although meteoritics still remained an esoteric field. The launching of Sputnik in 1957 produced a dramatic interest in the movement of solid bodies through the atmosphere, and information about that subject became a critical factor in national defense and space research. Because the Museum had available this large collection of meteorites, it immediately attracted the attention of investigators from the National Aeronautical and Space Agency and other civilian and military agencies; as might be expected, this activity led to an increase of the staff. After more than half a century at the Museum, Henderson continues his studies of meteorites.

In 1956 Cooper became the fourth Head Curator and began a modest program of increasing the size of the staff. In addition to several new mineralogists, a few paleontologists were hired. After this slight increase in the staff of the Department of Geology, Cooper suggested that the department be divided into two, in broad terms separating the fossils from the minerals. In spite of foot-dragging by some administrators, Cooper pushed his scheme to a successful conclusion. In 1963, shortly after new quarters in the east wing were occupied, Cooper, the last Head Curator of geology, became the first Chairman of the Department of Paleobiology; divisions of paleobotany and sedimentology were added to that department, but internal divisions subsequently were abolished.

More or less coincidentally with the bifurcation of the department, the concept of a rotating chairmanship, commonly for five years, was instituted throughout the National Museum. Cooper was followed by Porter M. Kier, Richard E. Grant, Martin A. Buzas, and Ian G. Macintyre. George Switzer, the first chairman of the Department of Mineral Sciences, was followed by Brian H. Mason, William G. Melson, Daniel E. Appleman, and Robert F. Fudali. All are well known in the geological community for their research in a variety of fields. It is reasonable to say that from the 1960s onward the character of the Museum changed dramatically from that of mainly a repository institution in which research was pursued, to predominantly a research institution that maintains collections of specimens among the finest in the world. One cannot cite the research and significance of one member of the staff to the development of earth sciences without citing that of all. To do so would both expand this summary to inordinate lengths and also bring it into the field of current events rather than history.

Late in 1962 an east wing of the building, devoted entirely to research laboratories and collection storage, was completed. Transfer of the collections and curators to these new quarters in a sense was comparable to the move across the Mall five decades earlier. Both the Department of Paleobiology and the Department of Mineral Sciences flourished with this extra space. The staff of each department is far larger than the size of the old department at its largest.

The Department of Mineral Sciences was lodged on part of the fourth floor and eventually grew to occupy all the offices and much of the storage area. Half the new space was devoted to paleontology, staffed by the Department of Paleobiology and the U.S. Geological Survey. At one time the combined staff of this department and the paleontologists of the Geological Survey was the largest, most diverse group of paleontologists in one institution in the western world. To the outside world, it has never been particularly clear who worked for the National Museum and who worked for the Geological Survey.

As might be anticipated, a few years after the wing was opened almost all the space was filled. The collections of vertebrate and invertebrate fossils filled three entire floors of the new museum wing and occupied additional areas on two more floors and part of the basement. A few years later a west wing was built and the paleobotanical collection was moved there. These cases occupy half of the third floor and may be one of the largest fossil plant collections in the world.

Up until the 1950s, perhaps 90 percent of the collection was a result of transfer from the Geological Survey. Since that time, the Deep Sea Drilling Program, collecting by museum curators, and the foraminifer investigations have added millions of specimens; other programs have added material, although not in such large quantities. There is a constant influx of type specimens given the Museum by outside investigators. Comparison of size is difficult, but overall, the collection of fossils at the Museum is by far the largest in North America and is comparable to the largest in Europe. Probably, there are as many invertebrate fossils in the building as in all other collections in the United States east of the Mississippi River combined. The collection of Foraminifera attracts more visiting investigators than any other comparable collection in the entire Museum.

The diversity of paleontological specialties is a great strength of the Department of Paleobiology. In the exploitation of two techniques, it has been a trailblazer. One of its prime contributions has been the development and exploitation of hydrochloric acid solution of limestone to yield silicified invertebrate megafossils. More than 70 tons of Permian limestone have been dissolved on the premises. Similarly, although thin-sectioning and peel techniques were not invented at the Museum, they have been developed to a high state and applied particularly to the elucidation of both Recent and fossil bryozoans.

The Department of Mineral Sciences has been heavily involved in extra-terrestrial geology. Partly as a consequence of the activity in meteorite studies, the Museum also became a center for those interested in tektites. Although there is somewhat less intense activity in tektite studies since the theory of their origin as moon fragments has been abandoned, interest in meteorites remains unabated. The discovery of meteorites in Antarctica and the agreement with the National Science Foundation that all specimens found on that continent by American scientists are to be part of the national collection insure that meteoritics will continue to be a major focus of the Department of Mineral Sciences. Museum scientists were included among those who studied the original collection of moon rocks. It was a technician from the Department of Mineral Sciences, Grover Moreland, who cut the first thin section of an extra-terrestrial rock.

With more space available in the new wing, the Museum was able to provide better facilities for visiting investigators. Even more important, the Smithsonian Institution began a series of pre- and post-doctorate awards in the mid-1960s. A significant number of American and foreign geologists are "alumni" of the two departments. With so many more people around the building for a limited time and an increasing number of students on the premises, the concept of a staid and stodgy museum has vanished.

In the late 1950s, the Museum began an ambitious exhibits program. The exhibits, virtually unchanged since the early 1920s, were not simply refurbished but were completely redesigned. There is no need to dwell on anything but the finished product, except to note that this has not been a simple task! The dioramas of past invertebrate communities and the murals of Cenozoic mammals along with other displays were striking. The Mineral Hall is marked by a sign indicating that it is the world's largest, most complete mineral collection, and this claim has never been disputed. The Hope diamond which came to the building in 1958—by regular mail—has supplanted Lindberg's airplane, *The Spirit of St. Louis,* as the single display object most inquired about by visitors.

Because of the opening and closing of various halls, it is difficult to gove any exact dates, but one may characterize the 1960s and 1970s as the era when exhibits again finally received their proper share of attention. It may be helpful to place the public exhibits in perspective. Each person entering the building is counted; although multiple visits by the same person cannot be distinguished, these are not significant. The current figures indicate that each year about 2 to 3 percent of the population of the United States view the exhibits. Occasionally, during the summers, it is necessary to close the building until some of the crowd leave.

In the early 1960s, the U.S. National Museum, which had never really been legally founded, was subdivided. In 1969, the name became the National Museum of Natural History, to coordinate with other national museums administered by the institution. Of the five Museum of Natural History directors, in the last two decades, two, the paleontologist P. M. Kier (1973–1979) and the volcanologist R. S. Fiske (1980–1985) have been earth scientists.

In 1984, the centenary of the National Gem Collection was celebrated and in a sense this event summarizes the development of the collections and the exhibits. Shortly after Frank Wigglesworth Clark was appointed as an honorary curator in 1883, he was given $2,500 in appropriated funds. With this he purchased 1,000 gemstones and arranged a display for the 1884 Cotton Exposition in New Orleans; following, it was on display for two years in Cincinnati. The three gem cases in Washington attracted considerable interest and this was considered the best public display of gemstones on the continent. Before the turn of the century, a few private donations and another modest appropriation for an exposition was given. A private fund allowed for purchases, but little more of significance happened to the collection for 50 years. In the mid-1950s, development of a new gem and mineral hall and the donation of the Hope diamond increased interest in the gem collection. Entirely by private donation and private funding, the National Gem Collection has now grown to be the finest public exhibit of gems in the world. Although some institutions have more spectacular individual stones, this is the most balanced collection on display.

OTHER SMITHSONIAN ACTIVITIES

During Samuel Pierpont Langley's day, a tiny astrophysical observatory was established behind the Castle. This remained a limited activity primarily concerned with non-geological phenomena, although the work done on the solar "constant" has interest for geologists. Langley's new astronomy ultimately led to a better perspective on the stars and planets. The observatory was moved to Cambridge, Massachusetts, in 1956, and metamorphosed into the Harvard-Smithsonian Center for Astrophysics; scattered among its 500 employees are half a dozen earth scientists. After the dawn of the "space age" in 1957, astronomical and astrophysical activities of the Smithsonian were expanded and have played a role in the extension of the term *geology* to include investigations throughout the solar system. One example of new technology was the use of the Schmidt Camera network, whose continuous search of the sky was able to track several meteors and to lead to immediate recovery of a fallen meteorite.

Development by the Smithsonian of the Center for Short-lived Phenomena at Cambridge, in 1968, has been a boon to the geologic community. This activity evolved into the Scientific Events Alerts Network in 1975 and is now headquartered in the Museum of Natural History Building. The reports issued by the network may be the most important service performed by the institution for seismologists and volcanologists throughout the

world. A few years ago, no one would have predicted the development of a global network to respond almost immediately to volcanic eruptions, meteorite falls, and earthquakes.

In the mid-1960s, Congress authorized the Smithsonian Institution to participate in the "Public Law 480" program. Money owed to the United States by various foreign countries, but restricted to use in that country in its own currency, became available for joint scientific investigations by American nationals and scientists of a country in the program. As a consequence, the Smithsonian has funded geologic investigations in other parts of the world by members of the American geological community and has fostered exchange visits for foreign workers.

As a follow-up to the NASA Apollo program, a Center for Earth and Planetary Studies was formed in 1972. The opening of the National Air and Space Museum in 1976 provided this center with a permanent home and it is an integral part of that museum. Earlier work of the center was concerned with the moon, but interests have shifted to the earth and its desertification and lineation. Emphasis is being given to spectral characteristics of rocks in remote sensing (see Lowman, this volume). Comparative planetology is also studied, particularly in the fields of cratering and tectonics. Recognition of the importance of collections has not diminished, and the center is one of nine NASA repositories for images of planetary data. Although preservation of the lunar samples is not under the jurisdiction of the Smithsonian, the National Museum of Natural History has several moon rocks on display as does the National Air and Space Museum. History tends to repeat itself, for the public's interest in touching a moon rock in the Air and Space Museum is like their interest in touching the native copper boulder.

THE FUTURE

Because it is the height of folly to speculate on the course of reseach, only a few general comments should be made. Given the realities of space and money limitations in the Museum of Natural History, it is likely that the Museum research staff will remain more or less at its current size, although that of the Geological Survey may diminish. Indeed, the present strength is most exceptional when viewed against the history of the institution. For three-quarters of a century, there were seldom more than seven active workers in geology during any one decade. Currently, the combined scientific staff of the two Museum departments is about 40.

As a result of the construction in 1983 of a Museum Support Center about 10 miles from the museum, there is considerable relief from the space limitations for storage of collections. Whether this new facility will be essentially a warehouse or will provide the opportunity for novel investigations remains to be determined; the moves of collections in the past have been both upsetting and stimulating.

The Museum has always emphasized collections and ought to continue to do so. In Mineral Sciences, a few choice additions will continue to be made to the mineral collection. The synthetic mineral collection will be strengthened and more gems will certainly be donated to the nation. Seventy-five years ago no one could have anticipated the significance of x-ray analysis, and 25 years ago no one could have predicted what would be learned from the electron probe. Whatever new instruments become available, the Museum will have the collections to which these tools can be applied.

Fossils are not a renewable natural resource, and the collection of megafossils assembled at the Museum can never be duplicated. Thus, it will continue to attract investigators interested in systematic studies and related fields. The day of great growth of vertebrate and invertebrate fossil collections has long passed. Sadly enough, some institutions once prominent in paleontology have relinquished their holdings, and the Museum has been involved in salvaging collections. During the early 1970s, the number of paleontologists employed by the Geological Survey outside of Washington exceeded the number on the premises. In 1979, the headquarters of the Paleontology and Stratigraphy Branch left the Museum, and since then the number of Geological Survey paleontologists in Washington has declined; it is unlikely that this trend will be reversed. On a more positive note, many American invertebrate paleontologists now deposit their type specimens at the National Museum, and this trend is likely to continue. While there are many facets to paleontology, the strength of the Museum probably will continue to be in systematics.

The internal history of the Smithsonian and the Museum is filled with what appears to the outside world as anachronisms. Thus for two decades, the Department of Paleobiology has housed a group studying sedimentology. In the future this activity possibly may be strengthened and made a separate department. Judging from what has been accomplished to date, emphasis will be on fieldwork in shallow marine waters, in part in association with reefs. Laboratory studies of deep-water sediments and fossils will continue, but it seems unlikely that the Smithsonian will operate its own oceanographic vessels.

A third generation of public displays in paleontology is now being opened, and some of the geology exhibits are scheduled for change. In this age of mass media and easy travel, the exhibits and the general area of public education in science will never again be allowed to fall into the doldrums. Popularization of science and research do not always fit well in the same person, but there are now sufficient scientists at the institution to do both in an effective manner.

In a fashion, one may summarize the significance of the National Museum-Geological Survey community in Washington by noting the past presidents of the societies who were working in the building as of 1984. Michael C. Fleischer and Brian H. Mason have both served the Mineralogical Society of America; and G. Arthur Cooper, Richard E. Grant, Porter M. Kier, William A. Oliver, Jr., Norman F. Sohl, and Ellis L. Yochelson have been presidents of the Paleontological Society. In earlier times, older generations of workers were similarly honored by their colleagues. Three members of the Museum community—Walcott, David White, and Timothy W. Stanton—left to serve as Chief Geologists of the U.S. Geological Survey and then returned to pursue research.

The Museum staff cannot encompass all research in the earth sciences and never will be able to do so. Its record in the past suggests: that which it does, it does well. Curatorial functions are not glamorous, but they have been invaluable in the past and will be equally invaluable in the future, preserving samples of the earth's history for generations to come. The Museum has done its fair share for the "increase and diffusion of knowledge" in the earth sciences.

REFERENCES CITED

Bassler, R. S., 1946, The Smithsonian: Pioneer in American geology: Science, 104: p. 125–219.

Caster, K. E., 1965, Memorial to Ray S. Bassler (1878–1961): Geological Society of America, Bull. 76: p. 167–174.

Cope, E. D., 1897, 679–696, *in* Goode, G. B. (ed.), The Smithsonian Institution 1846–1896, the history of its first half century: City of Washington, 856 p.

Gilmore, C. W., 1941, A history of the Division of Vertebrate Paleontology in the United States National Museum: Proceedings of the United States National Museum 90: p. 305–377.

Goode, G. B., 1895, An account of the Smithsonian Institution: its origins, history, objects, and achievements: City of Washington for distribution as the Atlanta Exposition, 38 p.

Henry, J., 1850, Fourth annual report of the Board of Regents of the Smithsonian Institution: 31st Congress, 2nd session, Senate Miscellaneous Document 120, 64 p.

Mason, Brian, 1975, Mineral sciences in the Smithsonian Institution: Smithsonian Contributions to the Earth Sciences 14: p. 1–10.

Merrill, G. P., 1897, A treatise on rocks, rock weathering, and soils: MacMillan and Company, New York, 411 p. [2nd edition, 1906].

—— 1901, Editorial Comment: The department of geology in the National Museum: American Geologist 28: p. 107–123.

—— 1906, Contributions to the history of American geology: U.S. National Museum, Annual Report for 1904: p. 189–733.

—— 1920, Contributions to a history of American State geological and natural history surveys: U.S. National Museum Bull. 109, 549 p.

—— 1923, The department of geology of the United States National Museum: Smithsonian Institution Report for 1921: p. 261–301.

—— 1924, The first 100 years of American Geology: Yale University Press, New Haven, Connecticut, 773 p.

—— An historical account of the Department of Geology in the U.S. National Museum, 130 p. (This incomplete manuscript is housed in the library of the Department of Mineral Sciences and includes some events through 1927).

Nelson, C. M., and Yochelson, E. L., 1980, Organizing Federal paleontology in the United States, 1858–1907: Journal of the Society for the Bibliography of Natural History 9 (4): p. 607–618.

Oehser, P. H., 1959, The role of the Smithsonian Institution in early American geology: Journal of the Washington Academy of Sciences 49: p. 215–219.

Rabbitt, M. C., 1979, Minerals, lands, and geology for the common defence and general welfare; volume 1, before 1879: U.S. Geological Survey, 331 p.

Rathbun, Richard, 1905, The United States National Museum: an account of the buildings occupied by the National collections: Report of the United States National Museum for 1903: p. 177–309.

Rice, W. N., 1897, Geology and mineralogy, 631–646, *in* Goode, G. B. (ed.), The Smithsonian Institution 1856–1896, the history of its first half century: City of Washington, 856 p.

Schaller, W. T., 1957, Memorial of William Fredrick Foshag [1894–1956]: American Mineralogist 42: p. 249–255.

Schuchert, Charles, 1897, A synopsis of American fossil Brachiopoda, including bibliography and synonymy: U.S. Geological Survey Bulletin 87: 464 p.

—— 1931, George Perkins Merrill (1854–1929): Geological Society of America Bull. 42: p. 95–122.

True, W. P., 1946, The first hundred years of the Smithsonian Institution 1846–1946: Smithsonian Institution, 64 p.

Willis, Bailey, 1947, A Yanqui in Patagonia: Stanford University Press, 152 p.

Yochelson, E. L., 1967, Charles Doolittle Walcott 1850–1927: National Academy of Sciences Biographical Memoirs 39: p. 471–540.

Yochelson, E. L., and Nelson, C. M., 1979, Images of the U.S. Geological Survey, 1879–1979: U.S. Government Printing Office 1979-281-363/7, 56 p.

Geological Society of America
Centennial Special Volume 1
1985

Rummaging through the attic;
Or,
A brief history of the geological sciences at Yale

Brian J. Skinner
Department of Geology and Geophysics
Yale University
P.O. Box 6666
New Haven, Connecticut 06511

Barbara L. Narendra
Peabody Museum of Natural History
Yale University
P.O. Box 6666
New Haven, Connecticut 06511

ABSTRACT

Commencing with the appointment of Benjamin Silliman as Professor of Chemistry and Natural History in 1802, the history of instruction and research in the geological sciences at Yale can be conveniently divided into seven generation-long stages. Each stage was characterized by a group of faculty members whose interests and personalities imparted a distinct flavor and character to the institution; as those faculty members left, retired, or died over a decade-long period of change, responsibility for geological studies passed to a new generation.

The first stage began with the appointment of Silliman; the second started in 1850 as Silliman's career drew to a close and J. D. Dana, his son-in-law, was appointed to the faculty, and brought the first Ph.D. degrees in the United States. The third stage commenced in 1880, and the fourth beginning in 1900, brought the first faculty appointments specifically for graduate instruction. The fifth and sixth stages saw the formative moves that welded different administrative units together, leading to today's Department of Geology and Geophysics. Stage seven, commencing in 1965, includes the present (1984), but holds the seeds of stage eight.

The increasing diversity of research activities in geology has led to a doubling of the number of geological faculty employed at Yale approximately every 50 years. The number of Ph.D's awarded has increased at a parallel rate. We suggest the size of the faculty will probably double again by the year 2035 and that production of Ph.D's will probably rise to a rate of 12 to 15 a year.

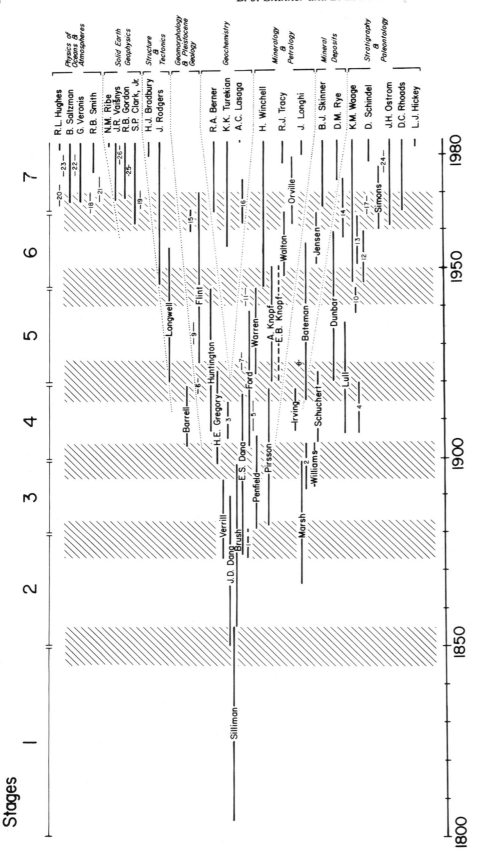

Figure 1. Sequence of faculty members teaching in the geological sciences at Yale. Distinct breaks in faculty appointments occur about 1850, 1880, 1900, 1920, 1945, and 1965. Names associated with heavy lines are mentioned in the text or are current members of the faculty. After 1900, instructorships are included only for those who progressed to higher rank. 1) G. W. Hawes; 2) C. E. Beecher; 3) Isaiah Bowman; 4) G. R. Wieland; 5) Freeman Ward; 6) J. P. Buwalda; 7) W. M. Agar; 8) K. C. Heald; 9) Helmut de Terra; 10) G. E. Lewis; 11) George Switzer; 12) J. T. Gregory; 13) J. E. Sanders; 14) A. L. McAlester; 15) A. L. Washburn; 16) R. L. Armstrong; 17) A. W. Crompton; 18) T. D. Foster; 19) N. L. Carter; 20) J.C.G. Walker; 21) H. T. Rossby; 22) M. E. Fiadeiro; 23) R. E. Hall; 24) D. R. Pilbeam; 25) Harve Waff; 26) Emile Okal. Designations of fields of study are matters of convenience and are open to question; for example, Rye and Skinner could equally well be listed under Geochemistry, Lasaga and Gordon under a new entry "Materials Science," and Saltzman and Turekian under Pleistocene Geology because of their work in paleoclimatology.

INTRODUCTION

Yale was one of the first institutions in the United States where geology was taught; the subject has been offered continuously since 1804, and as a separate discipline since 1812—the longest unbroken run, so far as we know, of any institution in the country. A study of the history of geology at Yale shows that any account must focus on the individuals who have worked at the institution—how they interacted and how their interests and activities changed the science and the institution. The story of the development of geology at Yale shows how individuals shape an institution, and could apply to dozens of other distinguished universities.

Geology was first taught at Yale in the first decade of the 19th century, but records show that even earlier there was research and teaching by Yale people in topics that would today be considered geology. In the 18th century, for example, it was a popular occupation for the mathematically inclined to calculate almanacs estimating the positions of the Moon and the planets and, especially, predicting the dates of eclipses. Today this would be called planetology, much of which is housed under the title of geology. Evidence of a marked interest in another aspect of planetology—meteors, fireballs, and comets—appears in a pamphlet prepared by Yale President Thomas Clap and published posthumously (Clap, 1781), which records his conclusion that fireballs are a class of comets that circle the Earth in highly eccentric orbits. Clap's conclusion was incorrect, but his enthusiasm for science and improvements in the teaching of mathematics were influential on the future of the sciences at Yale.

In 1801 President Timothy Dwight appointed a professor of mathematics and natural philosophy (physics)—the first of several faculty appointments that would place science on a secure and permanent footing at Yale. For his second appointment in 1802, he chose a recently graduated student who was then entering on a legal career; Dwight prevailed on him to abandon the law and become a teacher. The student, Benjamin Silliman (1779–1864), had never had a course in chemistry, mineralogy, or any other subject allied directly to geology, but he accepted Dwight's offer and, at the age of 23, was appointed Professor of "Chymistry" and Natural History. Silliman immediately set to work learning the subjects he was to teach by attending lectures in chemistry given by Professor James Woodhouse in the Medical School of the University of Pennsylvania. While in Philadelphia, he also attended lectures by Caspar Wistar on anatomy and surgery, took a private course in zoology given by Benjamin Barton, and made social contact with Joseph Priestley. Beyond some professional advice and assistance offered by Dr. John Maclean, Professor of Chemistry at Princeton, the Philadelphia experience was the only formal training in science that Silliman had when he presented his first course in chemistry to Yale College students in 1804–1805. The course consisted of 60 lectures, with mineralogy introduced at appropriate points. Silliman knew only too well that he was not really prepared for a career in either chemistry or natural history, so in 1805 he journeyed to England and Scotland where he spent a year inspecting mines, visiting various institutions, and studying in Edinburgh. While in Edinburgh, he became interested in the Vulcanist-Neptunist debate, which was then at its peak. He attended lectures by Dr. John Murray, an avowed Wernerian-Neptunist, and Dr. Thomas Hope, an avowed Huttonian-Vulcanist. Though more impressed by Murray, he struggled with the conflicting philosophies, remaining ". . . to a certain extent, a Huttonian, and abating that part of the rocks which the igneous theory reclaims as the production of fire, . . . as much of a Wernerian as ever" (Fisher, 1866, v. 1, p. 170). His earliest paper on the geology of New Haven (Silliman, 1810), written soon after his return from Scotland, makes interesting reading because it reflects the conflict in his mind and his attempts to resolve it. As he carried out this earliest geological investigation of the New Haven region, he was accompanied on horseback by interested local citizens, including Noah Webster, whom Silliman described as being "in the meridian of life" and "among the most zealous of my companions. . . ." (Fisher, 1866, v. 1, p. 216).

By the fall of 1806, Silliman was ready for a full-time teaching role. He was the founder of both the geological and chemical sciences at Yale, and more important for our story, one of the founding fathers of geology in North America.

Geological activities at Yale can be readily divided into seven stages: the first started about 1800 and covered the long career of Silliman and the early work of his distinguished student and son-in-law, James Dwight Dana (1813–1895). The second stage began about 1850, as Silliman's teaching career drew to a close. Two events of major importance marked the opening of this second stage. The first was the founding of a scientific school under the direction of Benjamin Silliman, Jr. and John Pitkin Norton; the other was the appointment of Dana to the faculty of Yale College. The third stage opened 30 years later, about 1880, by which time G. J. Brush had directed the Sheffield Scientific School to considerable prominence and the Peabody Museum of Natural History had been founded. Subsequent steps came in more-or-less generation-long gaps of 20 to 25 years. Each step began with a group of distinguished faculty members who were appointed over a period of about eight to ten years, and imparted to the institution a special flavor determined by their particular interests. As members of the group retired or died, responsibility passed to a new generation and a new pattern started to emerge. Distinct changes occurred about 1880, 1900, 1920, 1945, and 1965. It is now apparent that the Department of Geology and Geophysics has entered yet another stage of generational refurbishing in the 1980s; future histories will probably mark 1985 as the midpoint of the change. The dates of change are not exact—"about 1900" really means the time span from a few years before 1900 to a few years after 1900. An examination of Fig. 1 (adapted from an earlier diagram by Jensen, 1952) reveals that the seven steps are quite distinct.

STAGE 1: THE YEARS BEFORE 1850

When Benjamin Silliman started teaching chemistry at Yale, geology was not even a recognized discipline in most of the major academic institutions of Europe and North America. As illustrative material became available, Silliman expanded his classes in geology and mineralogy. His first course, initiated in 1807, was a private one based on early mineralogical acquisitions. In 1812, with the famous mineral collection of Colonel George Gibbs available for his use, he separated the geology and mineralogy lectures from his chemistry course and started a new course required of all Yale seniors (Narendra, 1979). By the time Silliman died in 1864, geology had risen to such prominence that 33 states had founded geological surveys (Merrill, 1920), and many geological topics—such as continental glaciation—captured the public imagination. The growth of the subject in North America was in no small measure aided by Silliman's elegant bearing (Fig. 2), powers of persuasion, and gifted teaching. But Silliman had acquired respectable scientific skills as well. His student, Amos Eaton, said, "Silliman . . . gives the true scientific dress to all the naked mineralogical subjects which are furnished to his hand" (Eaton, 1820, p. ix). Silliman was a founding officer, in 1819, of the first national organization for geologists, the American Geological Society, and he attracted to Yale many students and postgraduates who became leaders of the fledgling science. In 1818 Silliman founded the *American Journal of Science,* providing a publication outlet for scientists in all fields. That journal has been published at Yale, without a break, to the present day.

Silliman's effect on Yale as an institution was enormous. His leadership, forceful ideas, and organizational skills provided the momentum in the development of the physical sciences that led directly, albeit some years after his death, to Yale's becoming a university rather than a college.

STAGE 2: THE YEARS BETWEEN ABOUT 1850 AND ABOUT 1880

Two events mark the opening of stage 2. The first was the appointment, in 1850, of James Dwight Dana (Fig. 3) as Silliman's successor. Dana graduated from Yale College in 1833 and grew to great prominence during the next 20 years as a result of his scientific papers and contributions arising from his participation in the United States Exploring ("Wilkes") Expedition (1838–1842). Dana was also famous for his *System of Mineralogy,* first published in 1837, and his *Manual of Mineralogy,* which first appeared in 1848. By 1849, as Rossiter (1979) has observed, he was only 36 but had already accomplished far more than most geologists did in a lifetime. Aside from being one of the most distinguished geologists in North America, Dana was also Silliman's son-in-law. His appointment as the Silliman Professor of Natural History in 1850 (changed to Silliman Professor of Geology and Mineralogy in 1864) was necessary to keep him in New Haven rather than lose him to Harvard, and it is not hard to imagine the role Silliman might have played behind the scenes in

Figure 2. Benjamin Silliman. A portrait painted by Samuel F. B. Morse in 1825. (Courtesy, Yale Art Gallery).

order to bring about the newly endowed chair. Silliman continued to teach geology until Dana had completed his expedition reports and was ready to take over the lecturing role in 1856.

The second event that marked the opening of stage 2 was a result of Silliman's practice of accepting postgraduates for specialized instruction (for which no degree was given), a practice he started before 1820. His son, Benjamin Silliman, Jr., continued the practice by teaching applied chemistry to some of his father's special students, starting in 1842. This in turn led to the opening in 1847 of what was initially called the Yale School of Applied Chemistry, directed by the younger Silliman and John Pitkin Norton. The school received no financial assistance and little encouragement from Yale. Later named the Sheffield Scientific School after a wealthy benefactor, the institution developed into a successful technical college that awarded its own undergraduate degree for professional training in the applied sciences. In 1852 the new degree, the Bachelor of Philosophy (Ph.B.), was awarded to a group that contained members destined for greatness in geology. Perhaps the most distinguished was George Jarvis Brush (1831–1912; Fig. 4), Professor of Metallurgy from 1855 to 1871, then Professor of Mineralogy and Director of the Sheffield Scientific School. Another was William P. Blake (1826–1910), a prominent mining engineer, whose reports about Alaska are supposed to have had a considerable influence on U.S. Secretary of

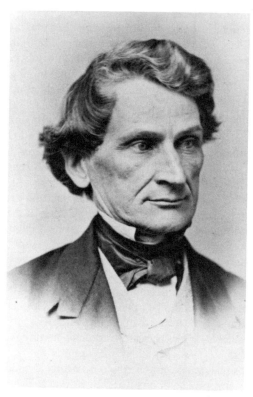

Figure 3. James Dwight Dana. A photograph from an album of a member of the class of 1865, Yale College.

Figure 4. George Jarvis Brush. A photograph from the archives of the Peabody Museum.

State Seward when he considered the purchase of Alaska. Blake later became Professor of Geology and Director of the School of Mines at the University of Arizona. Yet another of the illustrious 1852 degree recipients was William H. Brewer (1828–1910), who worked on the California State Survey, was briefly Professor of Natural Science at the University of California, and then became Professor of Agriculture in the Sheffield Scientific School from 1864 to 1903.

When the School of Applied Chemistry was founded, it was administratively enclosed within a new "Department of Philosophy and the Arts," Yale's first formal graduate department. Master's degrees had been awarded on a somewhat casual basis since the first commencement in 1702, and degrees were also given for professional training in the Schools of Medicine, Law, and Divinity. The new Department, forerunner of today's Graduate School of Arts and Sciences, differed in that it awarded degrees for research. In 1861 the first three Ph.D. degrees in the country were awarded—one went to A. W. Wright (later a Professor of Chemistry and Molecular Physics at Yale), whose topic of research was the same as that of President Clap a century earlier—the velocity and direction of meteors entering the Earth's atmosphere. In 1863 J. Willard Gibbs (later Professor of Mathematical Physics at Yale) received his Ph.D. for a thesis on the form of teeth in spur gears. The thesis has not had any influence on geology or geologists, but Gibbs's later work in chemical

thermodynamics has had a profound impact on most branches of geology. The first Ph.D. in geology was awarded to William North Rice in 1867 for a thesis discussing the Darwinian theory of the origin of species.

Among the many students attracted to Yale by its scientific atmosphere was Othniel Charles Marsh (1831–1899; Fig. 5), a member of the Massachusetts Peabody family, whose sons traditionally went to Harvard. Arriving in 1856, he was graduated from Yale College in 1860, did two years of graduate study in the Sheffield Scientific School, then went to Germany where, with the encouragement of the younger Silliman and J. D. Dana, he pursued his interests in vertebrate paleontology. Marsh managed to convince his wealthy uncle, George Peabody, to provide a gift of $150,000 to found a museum of natural history at Yale. The gift was awarded in 1866, the same year in which Marsh was appointed Professor of Paleontology, the first such professorship in America. The first building to house the Peabody Museum of Natural History was completed in 1876. Marsh had started his paleontological expeditions to the western states (Fig. 6) several years earlier—the first was in 1870—so by the time the museum was opened it housed not only collections of materials previously acquired by Yale, but a wealth of new material ready for study. Marsh's famous studies on extinct reptiles and other animals arose from this material, as did his work on toothed birds and the evolution of the horse, which particularly interested Darwin and

Figure 5. Othniel Charles Marsh. A photograph taken in 1872 by William Notman. From the archives of the Peabody Museum.

Thomas Huxley (Schuchert and LeVene, 1940). Marsh's work marked the beginning of a new line of geological activities at Yale—vertebrate paleontology.

Another among the students who were attracted to Yale sciences during the period from 1850 to 1880 was Clarence King (1842–1901), who graduated from the Sheffield Scientific School in 1862. King's early geological activities were mainly in the West, and from 1867 to 1877 he was Director of the U.S. Geological Survey of the Fortieth Parallel, working in the area now embraced by the states of Nevada, Utah, and Colorado. His greatest scientific interest was probably the origin and geological history of the North American Cordillera, but more important for geology as a profession was his appointment, in 1879, as first Director of the newly founded U.S. Geological Survey. He was the first of many Yale geologists to join and serve that distinguished institution. King stressed the scientific as well as the practical side of geology, and the U.S. Geological Survey follows his tenets to the present day. The importance of the U.S. Geological Survey to the development of geology in both North America and the rest of the world can hardly be overestimated.

As the second stage drew toward its close in 1880, a change of major importance occurred within the complex institution called Yale. Some years earlier, J. D. Dana and others had be-

come champions of the move toward making Yale a university rather than an undergraduate college with appendages in professional schools. However, there was continued resistance in the conservative administration to any plan which would place other units of the academic community on an equal footing with the College and its rigorous discipline of young male minds through the traditional memorization and recitation of classical subjects. Perhaps this regimentation had been necessary in earlier years when some of the students were really children and the faculty had to act *in loco parentis*; Benjamin Silliman, for example, was 13 when he entered Yale in 1792. But times had changed, and Harvard, under President Charles William Eliot, was directing the change. Eliot presented his ideas for a curriculum revision in his inaugural address in 1869; in the preparation of that address he was advised by G. J. Brush and Daniel Coit Gilman (B.A., 1852*), both officers of the Sheffield Scientific School (Morison, 1936). Gilman later became first president both of the University of California and of The Johns Hopkins University. Eliot set forth a plan whereby Harvard would become a university in which graduate degrees were to be offered in many departments and in which training and research in the sciences were to be given special emphasis. When Harvard made such a move, could Yale fail to react?

Eliot's plan temporarily led to a near-total abandonment of required courses for undergraduates at Harvard. Yale's curriculum reform began in the 1870s, when electives were allowed in Yale College for the first time. By 1887 an extensive elective system existed and Yale had officially become a university (Pierson, 1952).

STAGE 3: THE YEARS FROM ABOUT 1880 TO ABOUT 1900

By 1880, Yale's activities in the geological sciences were located in four administratively separate units—Yale College, the Sheffield Scientific School, the Graduate School, and the Peabody Museum. In the College, a general geology course was offered by J. D. Dana using his *Manual of Geology* (first edition in 1862) as a text. It is interesting to look at the exam Dana gave students in 1884 (Fig. 7). The questions have a decidedly modern ring to them and one wonders if students today could handle such an exam in two hours. It is especially interesting to see Question 4 concerning the sources of heat that cause geological changes. This was some years before the discovery of radioactivity, when questions such as the heat generated by gravitational compression were being widely debated. Dana obviously taught a course that was current.

In the Sheffield Scientific School, students followed one of several prescribed programs of study, the choice depending on their proposed profession. Basic geology was taught, oddly enough, by Addison Emery Verrill, Professor of Zoology (using Dana's textbook) and was required for most of the programs.

_____

*Degrees are understood to be Yale degrees unless otherwise indicated.

Figure 6. O. C. Marsh's 1873 student expedition to the West. Marsh can be seen sitting at the top of the small hill in the foreground, directly above the reclining figure looking for fossils on the side of the hill. Photograph by C. R. Savage. From the archives of the Peabody Museum.

More advanced instruction in geological topics was given by G. J. Brush and George W. Hawes. Much of the geological teaching on campus took place in the rooms of the Peabody Museum, where O. C. Marsh held sway as unofficial, but very real director. Marsh himself was what today we would call a research professor, and did almost no teaching. He was not paid a salary for most of his career, except for the last few years of his life when he needed money.

As the third stage opened, the winds of change were still blowing. In Yale College, geology became one of many electives, and in 1887, for the first time in 75 years, it was no longer a required subject for undergraduates. Dana, Marsh, and Brush were still active, but responsibility was passing to a new generation. J. D. Dana's son, Edward Salisbury Dana (1849–1935), was appointed Curator of Mineralogy in the Peabody Museum in 1874; in 1879 he became Assistant Professor of Natural Philosophy in Yale College, where he also taught mineralogy and crystallography, and in 1890 he was appointed Professor of Physics. Shortly thereafter Samuel Lewis Penfield (1856–1906) was

appointed to the Sheffield Scientific School. Penfield had studied mineralogy with Brush and became his successor. With the two Danas, Brush, and Penfield, mineralogy at Yale was, for a few years, very strong and even pre-eminent in the country; it was in this period that E. S. Dana published the famous 6th edition of his father's *System of Mineralogy* (1892). Petrology, a subject considered in those days to be an offshoot of mineralogy, had been taught for some years in the Sheffield Scientific School by George W. Hawes (1848–1882; Ph.B., 1872), who left in 1880 to become Curator of the Geological Department of the U.S. National Museum. Hawes was succeeded by Louis Valentine Pirsson (1860–1919) who had studied chemistry at Yale (Ph.B., 1882), worked as an analytical chemist, and then had served as a field assistant to J. P. Iddings (Ph.B., 1877) and W. H. Weed of the U.S. Geological Survey in their work in the Yellowstone area. Pirsson (1918, p. 255) wrote of Hawes that he was "the earliest of the petrographers in this country." Pirsson himself could also lay claim to being one of the small group of people who developed petrology as a science in its own right. He is widely remembered

YALE COLLEGE.

1884.

DECEMBER EXAMINATION—SENIOR CLASS.

Geology.

TIME, 2 HOURS.

———

1. What are fragmental rocks and the sources of their materials ?

2. The chemical constitution and organic sources of limestones.

3. The relation between the transporting power of rivers and their velocity ; the geological effects of transportation.

4. Sources of the heat concerned in producing geological changes.

5. The distribution of the dry land of North America at the close of Archæan time ; the mountains that then existed.

6. Evidence as to the time of elevation of the Rocky Mountains.

7. When appeared the first fishes ; the first amphibians ; the first mammals ?

8. What are mountains of circumdenudation and how were they made ?

9. Distribution of coal areas in North America ; the age of the coal beds of the Rocky Mountains.

Figure 7. The examination of 1884 given to the class in introductory geology in Yale College by J. D. Dana. The numbers at the bottom of the page were written by the student who took the exam.

as the P of the CIPW system of normative calculations [Cross, Iddings, Pirsson, and H. S. Washington (B.A., 1886)]. In many respects, however, Pirsson reinforced the mineralogical strengths of Yale's geological activities so, as they neared the ends of their careers, J. D. Dana, a generalist, and O. C. Marsh, a paleontologist, were the two who gave breadth to the field. Fortunately, Pirsson became increasingly interested in some of the larger geological problems and in 1893 he was relieved of his teaching in mineralogy so he could take over the course of basic geology from Verrill (Cross, 1920). In 1915, Pirsson and Charles Schuchert published their course notes, which became the first in a long series of physical and historical geology texts authored by Yale geologists. The forefather of them all was Dana's *Manual of Geology,* used both at Yale and around the country for more than a generation. Verrill and Pirsson not only used Dana's text, they

knew the man himself and taught the line of reasoning used in his book. In a sense, the many physical and historical geology texts from Yale faculty are one of the institution's major products.

The staff at Yale may have been small during the final years of the 19th century, but its prestige was high. Both J. D. Dana and Marsh were on the committee of the National Academy of Sciences that recommended the founding of the U.S. Geological Survey to Congress. Marsh was President of the Academy for 12 years (1883–1895), and Dana became the second President of the Geological Society of America, succeeding James Hall in 1889.

J. D. Dana stopped teaching in 1890 and a new cycle of change was underway. Unfortunately, the first steps were faltering. Henry Shaler Williams (1847–1918) was appointed in 1892 and succeeded Dana in 1894 as Silliman Professor of Geology in Yale College. After Williams graduated from the Sheffield Scientific School in 1868, he completed a Ph.D. (1871), taught briefly in Kentucky, and was then called to the chair of geology at Cornell, where he was responsible for the founding of the society of Sigma Xi. He became famous for his studies of the Devonian strata of the eastern United States and for his development of methods of stratigraphic correlation. He would seem to have been an ideal appointee, but unfortunately his few years at Yale were unhappy ones. Reportedly Williams was not as interesting a lecturer as Dana, and he suffered by comparison (Cleland, 1919). Problems caused by the assignment of faculty to different administrative units on campus also seemed to get in Williams's way, as later correspondence from H. E. Gregory to Williams suggests: ". . .it is certainly due to you that the agitation was started and that our faculty began to see clearly the danger of the relationship between the schools as they did" (Gregory, 1906, p. 148).

In 1904, Williams returned to Cornell. Only one other appointment was made before a great faculty expansion started at the turn of the century—that of Charles Emerson Beecher (1856–1904), appointed Assistant in Paleontology in the Peabody Museum in 1888. Beecher, an invertebrate paleontologist, had studied at the University of Michigan and worked in Albany with James Hall; after he came to Yale he completed a thesis on fossil sponges and was awarded a Ph.D. degree in 1889. This led to his appointment as a faculty member (Professor in 1897) in the Sheffield Scientific School, and he succeeded Marsh as unofficial director of the Peabody Museum. Beecher seems to have been the first senior faculty appointee to teach paleontology on a regular basis; unfortunately he died suddenly and unexpectedly in 1904, the same year in which Williams resigned. Marsh had died in 1899, so, as Yale entered its third century of instruction, it was rich with paleontological collections but devoid of faculty members to teach from them.

The number of students trained at Yale during the 20 years from 1880 to 1900 was not especially large, but those Yale graduates went on to play major roles in the development of science in the United States. One of the most significant was Arthur L. Day, who graduated from the Sheffield Scientific School in 1892 and then received his Ph.D. in physics from the

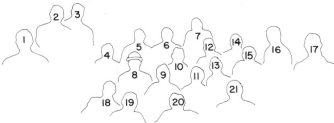

Figure 8. Faculty members and graduate students in front of the Peabody Museum of Natural History, 1911 (the names of faculty members are italicized): 1) Walter R. Barrows; 2) Charles W. Drysdale; 3) Alan M. Bateman; 4) *Richard S. Lull*; 5) *Herbert E. Gregory*; 6) *Freeman Ward*; 7) Donnel F. Hewett; 8) *Louis V. Pirsson*; 9) Henry G. Ferguson; 10) *Charles Schuchert*; 11) Ralph D. Crawford; 12) *William E. Ford*; 13) *Joseph Barrell*; 14) *Isaiah Bowman*; 15) *John D. Irving*; 16) Morley Wilson; 17) Charles C. Evans; 18) unidentified; 19) Kirk Bryan; 20) John J. O'Neill; 21) Bruce Rose.

Graduate School in 1894 for a thesis on "The seconds pendulum: determination for New Haven." Day gained experience in Germany and, on his return to the United States, became an assistant to G. F. Becker, the U.S. Geological Survey's Chief Physicist. The Geophysical Laboratory was born in the laboratories of the Survey, and from here Day was appointed as the first Director of the Geophysical Laboratory of the Carnegie Institution of Washington. Following Becker's intellectual lead, he started a tradition of careful and systematic measurements that have had an enormous impact on the geological sciences. In subsequent years, the Geophysical Laboratory became a working home for many Yale scientists, among them the remarkably productive J. F. Schairer [B.S., 1925; Ph.D. (Chemistry), 1928].

Three other graduate students of this era were destined to play major roles during the next stage of Yale's geological history. Two received their Ph.D.'s in 1899—Herbert Ernest Gregory and

Charles Hyde Warren; the third, Joseph Barrell, received his degree in 1900.

STAGE 4: THE YEARS FROM ABOUT 1900 TO ABOUT 1920

The years from about 1900 to 1920 are among the most distinguished in the history of geology at Yale (Fig. 8). But they were also years of great tragedy and, at times, of turmoil.

In 1904, the geologists of the Sheffield Scientific School acquired a building of their own—Kirtland Hall—newly built of red New Haven Arkose. It was part of an expanding group of laboratories specifically built for the School. With the retirement of Brush in 1898, William Ebenezer Ford (1878–1939) was appointed Assistant in Mineralogy to S. L. Penfield. Ford had graduated from the Sheffield Scientific School in 1899, and, while an Assistant, earned his Ph.D. in 1903 for a group of mineralogical studies. Penfield, a great analytical chemist and determinative mineralogist, died in 1906 at the age of 50. From 1903 onward Ford assumed teaching duties and advanced steadily in academic rank. He made a number of original contributions, but is best known for his revision of E. S. Dana's *Textbook of Mineralogy*, first published in 1877. J. D. Dana authored two mineralogical texts that have continued, through many editions and revisions, to the present day. They are the famous *System of Mineralogy*, and the *Manual of Mineralogy*, the latter now in its 19th edition under the authorship of C. S. Hurlbut, Jr., of Harvard and C. Klein, Jr., of New Mexico. But it was from the pages of the famous fourth edition of E. S. Dana's *Textbook*, published in 1932, that an entire generation of geologists learned mineralogy and crystallography. Popularly known, even today, as "Dana-Ford," its stature and reliability are such that the book is still kept in print by its publishers, John Wiley & Sons, and, as recently as 1980, sold as many as a thousand copies in a single year. While working on the revision, Ford maintained a reference file for all mineral species. When he died, the file passed to a former Yale student, Michael Fleischer [B.S., 1930; Ph.D. (Chemistry), 1933] who kept it current during his distinguished career at the U.S. Geological Survey, and used it for four editions of his very useful volume, *Glossary of Mineral Species*, the most recent dated 1983.

After the untimely death of Beecher in 1904, an unusual but inspired appointment was made in paleontology. Charles Schuchert (1858–1942) had left school at age 14 to enter his father's furniture business in Cincinnati. He was an amateur fossil collector who became so proficient in paleontology and such a recognized authority on brachiopods that, without formal training, he was appointed first by James Hall as his assistant, then by N. H. Winchell to the Minnesota Survey, and eventually as Assistant Curator under C. D. Walcott at the U.S. National Museum. In 1897 Schuchert published the *Synopsis of American Fossil Brachiopoda* and this, along with his other publications, led him to be invited in 1904 to be Professor of Paleontology and Historical Geology and Curator of the paleontological collections in the Peabody Museum. Schuchert was 46 at the time of his appointment, and local legend has it that the first lecture he gave before a Yale class was the first time he had ever attended an undergraduate college lecture. What a difference a new approach can make! While he attempted to improve the instruction of stratigraphy, Schuchert developed a way to plot the thickness and location of strata on base maps. Later he plotted the distributions of marine and nonmarine strata, which led to the development of the sophisticated paleogeographic maps for which he became famous. A firm opponent of Alfred Wegener's ideas of continental drift, Schuchert became the major geological spokesman against the concept at the famous symposium sponsored by the American Association of Petroleum Geologists in New York in 1926 (Schuchert, 1928). Schuchert, like Marsh, devoted his life to Yale and the Peabody Museum. Neither married, and each used his personal funds to enrich Yale's collections. Schuchert also acted

as another of the unofficial directors of the Peabody Museum until his retirement, and on the occasion of his 80th birthday, in appreciation of his long service, he was bestowed with the title "Director Emeritus."

Marsh's successor as vertebrate paleontologist on the faculty was Richard Swann Lull (1867–1957). He brought a new outlook and new interests and quickly became a major figure on the campus. When G. G. Simpson prepared his memorial to Lull he wrote: "The names Marsh, Lull, and Yale are so strongly linked in the history of paleontology that it is almost a shock to recall that Marsh and Lull never met and that Lull was nearly 40 when he began his association with Yale" in 1906 (Simpson, 1958, p. 128). Lull studied zoology at Rutgers College, graduating in 1893, then joined the faculty of the Massachusetts Agricultural College (now the University of Massachusetts) in Amherst. Nearby Amherst College had a major collection of fossil footprints from the local Triassic redbeds; they drew Lull's attention and aroused his interest in vertebrate paleontology. He returned to studies under the direction of H. F. Osborn at the American Museum of Natural History and in 1903 he was awarded a Ph.D. by Columbia University. After three more years in Amherst, Lull came to Yale. He brought with him a love for research and a keen instinct for collecting, both strong Marsh attributes; and he also brought a love of teaching and a flair for innovation in museum exhibit design, neither of which had held much interest for Marsh. Lull taught a course on evolution to Yale undergraduates that was tremendously popular and year after year drew hundreds; "Lull's impressive bearing, his skilled delivery, and his complete command of the subject made each session unforgettable" (Simpson, 1958, p. 128). Outside of Yale, Lull became most famous for his widely read text, *Organic Evolution*, first published in 1917, for his extensive studies of horned dinosaurs, and for his classic volume, *Triassic Life of the Connecticut Valley* (1915) which was a pioneering study of paleoecology. Lull's career at Yale spans all of stage 4 and most of stage 5, because even though he retired in 1936, he remained active in his work in the Peabody Museum until he was nearly 80.

A fourth long-lived member of the faculty was Herbert Ernest Gregory (1869–1952). A member of Yale's class of 1896, Gregory completed his Ph.D. in 1899 and was immediately appointed Instructor in Physical Geography. In 1901 he was promoted to Assistant Professor, and in 1904 he succeeded H. S. Williams as the third Silliman Professor of Geology. Gregory's interests were varied—stratigraphy, structural geology, hydrology, and geomorphology—and in many respects he seems to have been more closely allied to the interests of J. D. Dana, the first Silliman Professor, than to those of H. S. Williams. With William Morris Davis of Harvard, Gregory founded the New England Intercollegiate Geological Conference, an annual field conference held in a different geographical and geological locality of New England each year. The 75th NEIGC took place in 1983. Gregory was also responsible for the founding of the Connecticut Geological and Natural History Survey in 1903. While returning from a trip to Australia and New Zealand, Gregory visited

Figure 9. Joseph Barrell. A photograph published originally in the *Bulletin of the Geological Society of America* (v. 34, 1923, Pl. 2).

Hawaii and found that the Bernice P. Bishop Museum in Honolulu was in need of both a new director and a major revitalization. Welcoming reaffirmation of the historic ties forged by Yale-educated missionaries in the early 1800s, the Yale Corporation and the Bishop Museum Trustees designed an arrangement whereby the Director of the Bishop Museum should also hold the rank of Professor at Yale. Gregory became Director of the Bishop Museum in 1920, and for a few years divided his time between New Haven and Honolulu, until he established his residence permanently in Hawaii. He continued to hold the Silliman Professorship until his retirement in 1936, but by then he had not taught at Yale for at least a decade.

Another paleontologist, who began his Yale association by collecting fossils in the West for Marsh, was George Reber Wieland (Ph.D., 1900). Lecturer in Paleobotany from 1906 to 1920, with nonteaching research appointments thereafter, Wieland is best known for his work on fossil cycadophytes, but he was also active and productive in the study of dinosaurs and fossil turtles (Nelson, 1977).

The careers of E. S. Dana, Ford, Lull, and Schuchert were all long and distinguished; they spanned the time from the end of the 19th century to the middle years of the 20th. It was these four men, together with Pirsson and Gregory, who carried forward the traditions and methods of their Yale forebears of the 19th century. It is fortunate that they were so long-lived and so active,

because the careers of two other faculty members were sadly cut short—those of Joseph Barrell and J. D. Irving.

According to H. E. Gregory (1923), Joseph Barrell (1869–1919; Fig. 9) was the first person appointed to the geological faculty at Yale to carry out the program of organized graduate course work that had been authorized by the University in 1902. Prior to that time, graduate instruction had apparently been largely an extension of undergraduate instruction. From 1902 onward, the Ph.D. degree would not only entail completion of research and a satisfactory thesis, but, increasingly, formal courses as well. Barrell received a B.Sc. from Lehigh University in 1892 and a degree in mining (E.M.) in 1893. He then instructed in mining and metallurgy at Lehigh for four years and completed a geological study of the highlands of New Jersey, for which he received an M.Sc. in 1897. This prepared him to work with Pirsson, Penfield, and Beecher at Yale from 1898 to 1900, when he received a Ph.D. for a thesis on the geology of the Elkhorn District, Montana. He then returned to Lehigh for three more years. He was appointed Assistant Professor of Structural Geology at Yale in 1903 and Professor in 1908, a post he retained until his tragic early death from spinal meningitis in 1919.

Barrell's interests were eclectic, and he published major papers on such topics as isostasy, geologic time (he believed the earth to be at least 1.5 billion years old), the influence of climate on the nature of stratified rocks, the nature and relationship of marine and continental environments of deposition, volume changes during metamorphism, and the planetesimal hypothesis of the origin of planets. He was apparently an extraordinarily acute and demanding teacher of graduate students, but too demanding for most undergraduates. Among his peers and colleagues he seems to have been held in great affection but also in great awe; in Barrell's memorial, Gregory wrote (1923, p. 22) that "he possessed many attractive human traits, but his intellectual power was so obvious and so continuously displayed that 20 years of intimacy has left me an impression of a mind rather than of a man." Geologists often overlook Barrell because he died young and because his interests ranged so widely, but as G. L. Thompson (1964, p. 11) remarked, "many modern ideas are essentially his but since these ideas involved the basic fundamentals of geology, few people realize that Barrell was the originator."

The second important appointee primarily for graduate instruction was John Duer Irving (1874–1918), a petrologist and economic geologist who was appointed Professor of Geology and Mineralogy in the Sheffield Scientific School in 1907. Irving had studied at Columbia, where he received his Ph.D. in 1899. He was teaching at Lehigh University when a small group of people gathered in Washington, D.C. to form a not-for-profit membership corporation in order to publish a new journal, *Economic Geology.* Irving became its first editor, and when he moved to New Haven he brought the editorial responsibility with him. Though the journal has no formal connection with Yale, it has been housed and edited there ever since—after Irving, by A. M. Bateman and today by B. J. Skinner. In July 1917, having obtained a leave of absence from Yale, Irving left for France to serve with the Eleventh Regiment of Engineers. He died a year later, a victim of influenza, while on duty at the Flanders front (Kemp, 1919).

Irving gave an inspired course, and some of his lecture notes remain. His successor, Bateman, continued to use many of the same notes. Eventually, the course grew into Bateman's famous volume, *Economic Mineral Deposits,* first published in 1942 by John Wiley & Sons. Bateman once pointed out to one of us (B.J.S.) how the structure of Irving's course could still be seen in parts of his book.

Mention must be made of the activities of three Yale faculty members who made important contributions to geology but who were not professional geologists. The first two were the geographers, Isaiah Bowman (Ph.D., 1909) and Ellsworth Huntington (Ph.D., 1909). Both were Harvard trained—Bowman as an undergraduate, Huntington as a graduate student. Both were strongly influenced by William Morris Davis, and both went on to make lasting contributions in geomorphology. Bowman was at Yale from 1905 to 1915, then became Director of the American Geographical Society (1915–1935) and eventually President of The Johns Hopkins University (1935–1948). While at Yale he wrote his most important scientific book—*Forest Physiography* (1911); the volume is really the first comprehensive account of landforms of the United States. Huntington's career at Yale was much longer than Bowman's. With the exception of a two-year gap from 1915 to 1917 while he was in military service, he was a member of the Yale faculty from 1907 to 1945, though from 1917 onward his position was that of Research Associate in Geography, and he did little teaching. An intrepid explorer and prolific writer, Huntington made contributions to knowledge about the geomorphology of the Near East, China, India, and Siberia, but he is most famous for his extensive studies of climatic changes and their influence on civilizations.

The third person who made important contributions to geology, though not a geologist, was Bertram Borden Boltwood (1870–1927) who graduated from the Sheffield Scientific School in 1892. Following studies in chemistry in Germany with Krüss and Ostwald, he received a Ph.D. in chemistry from Yale in 1897. In 1906 he joined the Yale faculty as Assistant Professor of Physics. After a year's leave (1909–10) when he worked in Manchester with Ernest Rutherford, he became Professor of Radiochemistry. Boltwood was both a superb chemical analyst and an accomplished physicist. Between 1900 and 1906, when he and the geologist J. H. Pratt (Ph.B., 1893; Ph.D., 1896) were working as consulting mining engineers and chemists, Boltwood became interested in radioactive minerals. When Rutherford and Soddy theorized in 1903 that a radioactive element disintegrates spontaneously, emitting energy and forming a new element that may in turn disintegrate, Boltwood found his life's work—identifying the daughter products. He published extensively and made many contributions—some with Rutherford, with whom he worked closely—but most important for geology was his demonstration (Boltwood, 1905) that lead is the end product of uranium decay. From this came his suggestion (Boltwood, 1907) that simple

lead/uranium ratios in minerals should give an estimate of the time of crystallization of the mineral. Radiometric dating was born; one of the specimens he analyzed was a uraninite crystal from a pegmatite near Glastonbury, Connecticut, for which he estimated an age of 410 million years. Boltwood was not aware that two isotopes of uranium were present and that two different daughter products were included; fortuitously, the errors involved compensated each other and the date he calculated is surprisingly close to the age we would assign today (265 m.y.).

STAGE 5: THE YEARS FROM ABOUT 1920 TO ABOUT 1945

At the beginning of stage 5, a university-wide administrative reorganization occurred and the present-day departments of study were formed. Unification of all the geology faculty of the three schools—Yale College, the Sheffield Scientific School, and the Graduate School—created a geology department that was a distinct budgetary unit with the power to govern its own faculty appointments, responsible to a central university administration. Charles Schuchert was named the first Chairman of the new Department of Geology in 1920.

The Peabody Museum building was demolished in 1917 to provide space for a new Yale dormitory complex, the Harkness Quadrangle. Almost immediately thereafter the United States entered World War I, and construction of the promised new museum building was not begun until 1923. In his annual report as Geology Chairman in 1921, Schuchert spoke for all the anguished curators whose collections were inaccessible: "The paleontologists, with the grand collections in their charge scattered in nine different and strange places, find that their effectiveness in teaching and in extension work is almost nil. The geologists of Yale College are also in temporary quarters, and since 1917 the various members of the Department have been housed in four different buildings. The natural history museums of our country are growing with leaps and bounds, but dear old Peabody Museum is boxed up and the Lord (and the University Corporation) alone know when we shall be allowed to emerge. A more depressing state of affairs no department at Yale was ever subjected to, and all this is being observed by the spirit eyes of Benjamin Silliman, James D. Dana, Othniel C. Marsh, and George J. Brush!" (Schuchert, 1921, p. 180). At long last, under Lull as first official Director, the Museum and its new displays were finished. Eight scientific societies—one of them the Geological Society of America—held their annual meetings in New Haven in December 1925, and 800 of their members attended the dedication of the new building.

In this stage, long-lived Lull, Ford, Schuchert, and E. S. Dana were still present and active; Schuchert and Dana also assumed important roles in university affairs. But increasingly prominent was a faculty who trained some of today's most distinguished geologists. The new faculty members were A. M. Bateman, C. R. Longwell, C. O. Dunbar, A. Knopf, C. H. Warren, and R. F. Flint. With the exceptions of Flint and Knopf, each of these appointees had received all or a major portion of his professional education at Yale. It would seem that Yale found it difficult to look beyond its own graduates when new appointments were to be made.

The first appointment of the new era actually occurred before 1920. Alan Mara Bateman (1889–1971) was a Canadian who graduated from Queen's University and had already gained a good deal of field experience when he arrived at Yale in 1910 to study with Irving. During the summers of 1911 and 1912 he worked for the Geological Survey of Canada on the Fraser River, British Columbia, and from this work came his thesis on the "Geology and ore deposits of the Bridge River district, British Columbia," for which he received a Ph.D. in 1913. Bateman was then invited to join the famous Secondary Enrichment Investigation inspired by L. C. Graton, then at the U.S. Geological Survey, which included members of academia, industry, and the Geophysical Laboratory. The work continued for a number of years and led to many important papers—especially concerning the deposits at Kennecott, Alaska (White, 1974). In 1915 Bateman was appointed Instructor in Geology; when Irving left for military training in 1916, Bateman was appointed Assistant Professor. He continued to work at Yale until his death, rising through the ranks and finally becoming the Silliman Professor of Geology in 1941.

Following Bateman's appointment and the deaths of Pirsson, Irving, and Barrell, there were three appointments in 1920: Longwell, Dunbar, and Knopf. Chester Ray Longwell (1887–1975) entered Yale as a graduate student in 1915 but service in the U.S. Army interrupted his studies, and he did not complete his thesis on the geology of the Muddy Mountains of Nevada until 1920. He was appointed to the Yale faculty in the same year and advanced steadily, becoming Professor in 1929. Longwell's main interest continued to be western geology, especially the western overthrust belt which his thesis did much to define. He was an excellent teacher and such major figures in the geological world as W. W. Rubey and James Gilluly were among the graduate students he influenced. He was also an excellent teacher of undergraduates and taught elementary physical geology at Yale. Following Pirsson's death he inherited the physical geology portion of the Pirsson and Schuchert *Textbook* (1929), which eventually became the famous Longwell, Knopf, and Flint version of *Physical Geology* (1948).

The second appointee of 1920 was Carl Owen Dunbar (1891–1979; Fig. 10), successor to Beecher and Schuchert. Dunbar studied geology at Kansas where he came under the strong influence of W. H. Twenhofel (B.A., 1908; Ph.D., 1912), who was a great admirer of Schuchert and encouraged Dunbar to do graduate work under him. Dunbar received his Ph.D. in 1917, after which he did a year of postgraduate study with Schuchert and then spent two years as an instructor at the University of Minnesota. In 1920 he returned as Assistant Professor of Historical Geology and Assistant Curator of Invertebrate Paleontology, becoming Professor of Paleontology and Stratigraphy in 1930. Dunbar's broad professional interests centered on the fusuline foraminifera and their use in stratigraphic correlation. As a

Figure 10. Carl O. Dunbar and John Rodgers. A photograph taken in 1956. (Courtesy of K. M. Waage).

teacher he was demanding and meticulous and had a great influence on the students he came in contact with—especially the graduate students. He also was an important textbook writer. He collaborated first with Schuchert in revision of the historical side of the *Textbook of Geology* (Schuchert and Dunbar, 1933), which he later rewrote as the highly influential *Historical Geology* (1949). He then revised this book again with K. M. Waage (1969). Perhaps even more influential at the graduate level was the book Dunbar co-authored with John Rodgers in 1957, *Principles of Stratigraphy,* a classic text that grew out of a jointly taught course. As Director of the Peabody Museum he initiated the construction of the museum's first dioramas. Probably the best known exhibit produced during his directorship is the 110 foot-long Pulitzer Prize-winning mural, "The Age of Reptiles," painted by Rudolph Zallinger.

The third appointee of 1920 was Adolph Knopf (1882–1966). Knopf, a petrologist, was successor to Pirsson. All of his training was at the University of California at Berkeley where he was strongly influenced by the teaching of Andrew Lawson. Knopf received a B.S. in Mining Geology in 1904 and throughout his life made important contributions to economic geology. First and foremost, however, he was a petrologist and field geologist. In 1905 he started working with the U.S. Geological Survey and spent six field seasons in Alaska, during which time (1909) he completed his Ph.D. thesis. Between 1910 and

1920 Knopf worked with Waldemar Lindgren, F. L. Ransome, L. C. Graton, W. H. Emmons, J. B. Umpleby, and others as they studied the ore districts of the western United States; he published a number of papers and contributed to many others. One has to wonder how the arrival of such an accomplished economic geologist on the scene at Yale must have seemed to the already appointed A. M. Bateman. By 1920, salaries at the U.S.G.S. were so low compared to those at universities that Graton left for Harvard, Lindgren for M.I.T., Emmons for Minnesota, and Knopf for Yale (Coleman, 1968). Knopf, like Barrell, was a scientist with eclectic interests and a demanding mind. He challenged students mightily and is remembered by them as an inspiring teacher. From Boltwood he developed an interest in radiometric methods of measuring geologic time and while he did not, unfortunately, seize the opportunity to start making such measurements at Yale, he played a major role in bringing geologic insight to the measurements made so enthusiastically by those in the laboratory.

Shortly before Knopf joined the faculty, his wife Agnes died during the influenza epidemic of 1918, leaving him with three children. In 1920, he married Eleanora Bliss (1883–1974), a distinguished Bryn Mawr geologist then working at the U.S. Geological Survey. Unfortunately, neither the geology faculty nor the Yale University administration was ever bold enough or sensitive enough to appoint Mrs. Knopf to the faculty or to allow her to teach, though she worked in her husband's office. A long generation of students remembers with affection and gratitude her intellectual stimulation and advice (Rodgers, 1977). She became the leading structural petrologist in the country and was an active, productive scientist until her death in 1974.

Two other appointees of this stage were C. H. Warren and R. F. Flint. Charles Hyde Warren (1876–1950) graduated from the Sheffield Scientific School in 1896, and received a Ph.D. in mineralogy under the direction of Penfield in 1899. He also spent a summer in the field in Yellowstone with Iddings, as Pirsson had done. After a short term as Instructor in Mineralogy at Yale he moved to M.I.T. and advanced to Professor of Mineralogy in 1912. At the retirement in 1922 of Russell M. Chittenden, Brush's successor as Director of the Sheffield Scientific School, Warren was recalled to replace him. His title was changed from Director to Dean and he was simultaneously appointed Sterling Professor of Geology. Warren was an able and fair administrator—he not only served as Dean but also as Chairman of the Department of Geology from 1923 to 1938. Most of Warren's activities were taken up in administration so he did little teaching and little research. His administrative leadership was vitally important, however, and to his credit can be recorded the final solution of the difficulties caused by the continued maintenance of two different undergraduate schools within the university, and the inevitable duplication of course offerings, among other things. Under his guidance, the Sheffield Scientific School ceased to exist except as a legal title to cover administration of the School's endowment funds.

The last member of the famous group of faculty of stage 5

to be appointed was Richard Foster Flint (1902–1976). Like Knopf, Flint brought new blood and ideas to Yale, for he had received all of his education at the University of Chicago (B.S., 1922; Ph.D., 1925). He joined the Yale faculty the year he received his Ph.D. and became both a masterful teacher and the most distinguished glacial geologist of his age—indeed, he was widely and affectionately known as the "Pope of the Pleistocene." Flint joined with Longwell and Knopf as an author of *Physical Geology,* then published subsequent editions with different co-authors until the time of his death.

To those who were privileged to get to know him, Flint was a warm and responsive friend, but unfortunately he found it difficult to lower a barrier of reserve and to most people he came across as autocratic and forbidding. This did little to endear him to the geological community of North America, so in spite of his scientific status he received few honors in his homeland, but a great many honors from abroad. He was also the losing protagonist in a long-continued debate over the origin of the channeled scablands. Flint ridiculed J Harlen Bretz's suggestion that the unique topography resulted from a catastrophic flood and he was overly dogmatic in his rejection of the idea.

A faculty member appointed to the Department of Zoology during stage 5, G. Evelyn Hutchinson, must also be mentioned. Hutchinson was educated at Cambridge University and worked in South Africa prior to his appointment to the faculty in 1928. From the moment he arrived at Yale he started to influence geology and geologists through his many contributions in biogeochemistry and limnology. Indeed, Hutchinson can be considered the first geochemist appointed at Yale. In 1946, as stage 6 opened, a second geochemist was appointed to the faculty, this time in the Department of Astronomy. Rupert Wildt, who had studied with V. M. Goldschmidt, is more correctly called a cosmochemist, and like Hutchinson he was so eminent in his field that when the new journal *Geochimica et Cosmochimica Acta* was founded in 1951 (American editor, F. E. Ingerson, Ph.D., 1934), the names of Wildt and Hutchinson appeared among the distinguished members of the Editorial Advisory Board.

Institutions are sometimes blessed by a flux of students who for a time catalyze each other's interests and stimulate each other to heights of great achievement. A necessary ingredient for success is that the students have the right temperaments, but even more essential is that they have an outstanding faculty with whom to interact (Rubey, 1974; Gilluly, 1977). All these conditions were met at Yale during the period 1920 to 1945 (Figs. 11 and 12). It was a period when the Graduate School at Yale flowered under the leadership of Dean Wilbur L. Cross and Yale President James Rowland Angell, a period during which, according to the respected historian of Yale, George Pierson (1955), there was a revival of scholarship among Yale undergraduates.

The list of Yale students of this time who went on to gain distinction in geology reads like an honor roll for the profession. Among the geology Ph.D.'s were Henry G. Ferguson (1924), Donnel F. Hewett (1924), Thomas B. Nolan (1924), James Gilluly (1926), George Gaylord Simpson (1926), J. B. Stone (1926),

Figure 11. Faculty members and graduate students in front of Kirtland Hall, ca. 1922 (the names of faculty members are italicized): 1) unidentified; 2) William W. Rubey; 3) Joseph E. Hare (?); 4) James Gilluly; 5) Thomas B. Nolan; 6) unidentified; 7) unidentified; 8) Ludlow Weeks; 9) Carle H. Dane; 10) *Ellsworth Huntington*; 11) *Alan M. Bateman*; 12) *Chester R. Longwell*; 13) *Adolph Knopf*; 14) Waldo S. Glock; 15) *Carl O. Dunbar*, 16) unidentified; 17) J. Doris Dart; 18) *Charles Schuchert*, 19) *Richard S. Lull.*

Wilmot H. Bradley (1927), Carl Tolman (1927), G. Arthur Cooper (1929), Philip B. King (1929), Aaron C. Waters (1930), A. K. Miller (1930), Carle H. Dane (1932), Norman D. Newell (1933), M. N. Bramlette (1936), Paul D. Krynine (1936), Yang Tsun-Yi (1939), Preston E. Cloud (1940), A. L. Washburn (1942), and John Rodgers (1944). Joining this remarkable group were W. W. Rubey, who did not complete his Ph.D. requirements, but who was later awarded an honorary degree (D.Sc., 1960), Harry H. Hess (B.S., 1927) who was also awarded an honorary degree (D.Sc., 1969), and Ralph E. Grim (B.A., 1924). Including those who taught and those who studied, the extraordinary Yale group from this period has been awarded 13 Penrose medals to date; they are (in order of receipt) Schuchert, Simpson, Gilluly, Knopf, Rubey, Hewett, King, Hess, Bradley, Cloud, Rodgers, Waters, and Cooper. Two others, A. L. Day and M. E.

Wilson, have also been awarded the medal, which was first presented in 1927. No fewer than nine of the same group have served as President of the Geological Society of America; they are, in order of appointment, Schuchert, Knopf, Gilluly, Longwell, Rubey, Nolan (when he was also Director of the U.S. Geological Survey), Hess, Bradley, and Rodgers. In addition, J. D. Dana, Arnold Hague, A. L. Day, George S. Hume, Peter T. Flawn, and Paul A. Bailly, recipients of Yale degrees in other periods, have also served as president.

STAGE 6: THE YEARS FROM ABOUT 1945 TO ABOUT 1965

The faculty appointed around 1920 continued essentially unchanged through the 1930s and to the end of World War II in

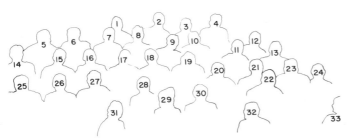

Figure 12. Faculty members and graduate students in front of Kirtland Hall, 1938 (the names of faculty members are italicized): 1) *Alan M. Bateman*; 2) *Richard F. Flint*; 3) *Carl O. Dunbar*; 4) *Charles Schuchert*; 5) J. David Love; 6) *G. Edward Lewis*; 7) *Adolph Knopf*; 8) *Chester R. Longwell*; 9) *Eleanora B. Knopf*; 10) J. Danvers Bateman; 11) Wendell G. Sanford; 12) *William E. Ford*; 13) Joseph P. Jennings; 14) Esa Hyyppä; 15) Chauncey D. Holmes; 16) Hugh H. Beach; 17) Preston E. Cloud; 18) John C. McCarthy; 19) William A. Rice; 20) Allen J. Turner; 21) John S. Shelton; 22) Earl M. Irving; 23) S. Warren Hobbs; 24) Shu Chen; 25) A. Lincoln Washburn; 26) Yang Tsun-Yi; 27) John Rodgers; 28) Gladys M. Galston; 29) Mrs. Lewis; 30) Jean M. Berdan; 31) Mrs. Washburn; 32) Mrs. Hyyppä; 33) Mrs. Shelton.

1945. E. S. Dana died in 1935, breaking the last direct link that went back through his father to his grandfather, Benjamin Silliman. Schuchert died in 1942, at the ripe old age of 84; Ford died in 1939, before retirement; and Huntington died in 1947. Lull retired from teaching duties in 1936, but continued to be active in research. Gregory retired from the Silliman professorship of geology in 1936 and was succeeded in the chair by Knopf, who was, in turn, succeeded by Bateman and then in 1962 by a student of this era, John Rodgers (Fig. 10). There have only been six Silliman Professors from 1850 to the present day. Four additional endowed professorships in geology have been established in this century. C. H. Warren retired in 1945 so by the end of the war there was a small senior faculty—Longwell, Dunbar, Flint,

Bateman, and Knopf. New appointments were needed, not only to maintain faculty strength, but also to handle the anticipated increases in teaching loads as military personnel returned to catch up on delayed educations (Fig. 13).

Following Ford's death, George Switzer joined the faculty as mineralogist, but he was quickly called to the Smithsonian Institution and his place was taken in 1945 by Horace Winchell, son of A. N. Winchell, grandson of N. H. Winchell. He had studied as an undergraduate at the University of Wisconsin and completed a Ph.D. at Harvard. Just prior to Knopf's retirement in 1951, Matt S. Walton was appointed as a petrologist, and started a program of study on the petrology of the igneous and metamorphic rocks of the Adirondacks. In vertebrate paleontology, Lull

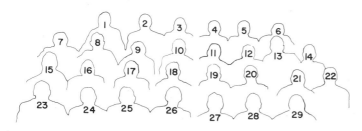

Figure 13. Faculty members and graduate students in front of Kirtland Hall, 1950 (the names of faculty members are italicized): 1) *Adolph Knopf*; 2) *Carl O. Dunbar*; 3) *Karl M. Waage*; 4) *Joseph T. Gregory*; 5) Heikki G. Ignatius; 6) William H. Hays; 7) Dwight R. Crandell; 8) *Chester R. Longwell*; 9) *Eleanora B. Knopf*; 10) *Horace Winchell*; 11) *John Rodgers*; 12) John Imbrie; 13) Walter W. Wheeler; 14) Charles P. Thornton; 15) Spencer S. Shannon, Jr.; 16) Colin W. Stearn; 17) George White; 18) E. R. Ward Neale; 19) John E. Sanders; 20) E. Carl Halstead; 21) J. Thomas Dutro, Jr.; 22) Sanborn Partridge; 23) *Richard F. Flint*; 24) Charles H. Smith; 25) John A. Elson; 26) James W. Clarke; 27) Grant M. Wright; 28) Henry W. Coulter; 29) Richard V. Dietrich.

was succeeded by G. Edward Lewis (Ph.D., 1937), who left for the U.S. Geological Survey in 1945, then in 1946 by Joseph T. Gregory.

Karl M. Waage, a recent Princeton Ph.D. (1946), was appointed in invertebrate paleontology and stratigraphy. John Rodgers (Ph.D., 1944) was appointed in sedimentary geology in 1946, but was then asked to switch to structural geology in anticipation of Longwell's retirement. In 1951, Mead LeRoy

Jensen received his Ph.D. from M.I.T. and was appointed in economic geology. At Yale he quickly organized a laboratory for measuring sulfur isotope ratios, starting what later became a strong trend toward geochemistry. John E. Sanders, who received his Ph.D. in 1953 for a thesis on "Geology of the Pressmen's Home Area, Hawkins County, Tennessee," was appointed Assistant Professor in 1956. Also in 1956 Karl K. Turekian was appointed Assistant Professor in the field of geochemistry. Before the beginning of the next stage in 1965, J. T. Gregory, Walton, Jensen, and Sanders had all departed, and several new appointments had been made. Vertebrate paleontology became very strong with three new men: Elwyn L. Simons, John H. Ostrom, and Alfred W. Crompton. A. Lee McAlester was appointed in invertebrate paleontology, Sydney P. Clark, Jr., in geophysics, and Philip M. Orville, in petrology. In effect, the careers of these people belong with the next stage, but all were appointed prior to 1965.

The biggest change during the stage from 1945 to 1965, however, was the erection of a new building. Since 1904, most of the geologists (except for those who had offices in the Peabody Museum) had been housed in Kirtland Hall. The building was two long blocks away from the new Peabody Museum as well as from the buildings that housed the Departments of Chemistry, Physics, and Zoology. Having outgrown the long-occupied quarters, the Department of Geology needed to move. Through a generous gift from C. Mahlon Kline (Ph.B., 1901), an expanded science complex was created on the north end of the Yale campus, where the Peabody Museum and the existing chemistry, physics, astronomy, and biology facilities already were situated. Three new Kline buildings were erected, one each for biology, chemistry, and geology. In 1963, the Department of Geology moved into its new quarters adjacent to the Peabody Museum. Soon after the move a diversification of activities and an expansion of the faculty took place.

STAGE 7: FROM ABOUT 1965 TO THE PRESENT

Bateman retired in 1957 but remained active as Editor of *Economic Geology,* the position he had held since 1919, until a very few years before his death in 1971. Flint also retired in 1970, and died in 1976; his colleague and close friend, A. Lincoln Washburn (Ph.D., 1942), who had been appointed Professor in 1960, relinquished all teaching duties in 1966 to become a Senior Research Associate. In 1970 Washburn left for the University of Washington. With the departure of Flint and Washburn, the 75 years of studies in the areas of classical geomorphology and glaciology came to an end, although the closely related area of Pleistocene studies continues today through research in climatology, hydrology, and marine geochemistry.

Geochemistry, which had grown strongly under the guidance of Turekian, started to occupy more attention. Richard L. Armstrong (Ph.D., 1964) began a radiometric dating laboratory soon after his appointment as Assistant Professor in 1964. Minze L. Stuiver came from Holland in 1959 to operate the Yale Radiocarbon Laboratory, a project initiated and strongly supported by Flint, but left for the University of Washington in 1969.

The most important change that has occurred during the current stage, however, is an expansion of the Department to become the Department of Geology and Geophysics (1968), and the appointment of a geophysics faculty. Appointments have been made in two areas: (1) the physics of oceans and atmospheres, begun with the appointment of Theodore D. Foster (who

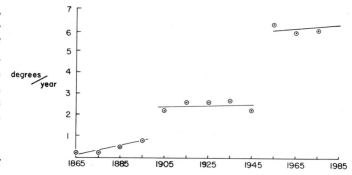

Figure 14. Numbers of Ph.D. degrees awarded in geology at Yale each year averaged over decade intervals. Each decade is plotted at the midpoint of the interval. Prior to the turn of the century, fewer than 10 students received degrees in each decade; the succeeding 45 years, to the end of World War II, saw an average of 25 students per decade; during the third period, from 1950 to the present, about 60 students graduated each decade, giving averages per year of less than 1, 2.5, and 6, respectively.

subsequently left) and then led by George Veronis (1966) and Barry Saltzman (1968), and (2) solid earth geophysics, led by S. P. Clark, Jr. (1962). There have been a number of subsequent appointments in both areas.

GRADUATE DEGREES IN GEOLOGY FROM YALE

We have mentioned some of the students who were at Yale and the roles they played later in life as faculty members in the development of the science. A complete accounting of all Yale students who have made important contributions to geology is beyond the scope of this paper. A total of 351 people had received Ph.D.'s in geology by December 1984. The first was awarded in 1867 to William North Rice, but completion of degrees was sporadic for many years, and until the beginning of the 20th century Yale averaged fewer than one degree per year. Starting about 1900, a noticeable increase in the production rate occurred (Fig. 14); the award rate rose to about 2.5 degrees per year and remained at that level until the end of World War II. The rise actually started in 1903, and coincided with the first appointments made specifically for graduate instruction. Starting about 1950, another distinct jump occurred, to an average award rate of 6 degrees per year— a rate maintained to the present day. This second jump was apparently generated by the increased size of the graduate student body as returned servicemen swelled its ranks after World War II, and by the maintenance of the increased student-body size by an inflow of government funds supporting the research carried out by the students.

DEGREES HELD BY YALE FACULTY

During the first century of geological instruction the instructors were, by and large, educated at Yale. This is not surprising, considering the small number of institutions where geologists

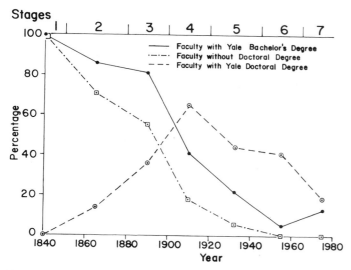

Figure 15. The percentage of Yale geology faculty members educated at Yale was very high during the early stages but has declined steadily as has the percentage of the faculty without doctoral degrees.

were trained; as the number of institutions increased, the percentage of faculty appointees who held Yale bachelor's degrees decreased. The fraction of the faculty educated at Yale has declined steadily throughout the twentieth century and appears to have bottomed out at about 10 percent (Fig. 15; Table 1).

During the first stage of geology at Yale, up to the year 1850, the sole faculty member, Benjamin Silliman, did not hold a doctoral degree. During the second stage, 1850 to 1880, the American Ph.D. degree was introduced and the first holders of doctoral degrees were appointed to the faculty. As shown in Figure 15, the percentage of faculty members not holding earned doctoral degrees dropped steadily and reached zero during stage 6. There was a marked preference for Yale doctoral degrees through stage 4, when 65 percent of the total faculty had Yale doctorates while an additional 18 percent lacked doctorates. From the fourth stage onward, the number of faculty members holding Yale doctorates has declined steadily. With the retirement of John Rodgers in June 1985, Yale geology will be in an unusual situation, at least temporarily—for the first time in more than a century, it will have only one faculty member with a Yale doctorate.

WOMEN GEOLOGISTS AT YALE

Yale's record of awards of Ph.D. degrees in geology to women is unimpressive. Women were admitted for graduate studies at Yale as early as 1892, but it was not until 1903 that the first woman geologist, Mignon Talbot, enrolled to work in paleontology with Beecher. Talbot received a Ph.D. in 1904—near record time—and spent her professional career teaching geology at Mt. Holyoke College. The second woman to receive a Ph.D., in 1908, was Ruth Sawyer Harvey (Jones), who completed a thesis on the "Drainage and glaciation in the central Housatonic basin"

of Connecticut, under the direction of H. E. Gregory. This was the first Yale thesis that specifically focused on the question of glaciation. For a few years Miss Harvey taught physical geography at the Philadelphia High School for Girls, but, apart from preparing her thesis for publication (Harvey, 1920), she seems to have done no further professional work after her marriage in 1917.

The third woman to receive a Ph.D. was Gladys Mary Wrigley, in 1917, who wrote on the "Roads and towns of the Central Andes" under the guidance of Bowman. Wrigley's work was done at a time when geography was administratively considered one of the "geological sciences" and the geological profession can only claim her on the basis of an institutional technicality. Wrigley later worked with Bowman in New York where she became editor of the *Geographical Review*.

Not a single Ph.D. degree in geology from Yale was awarded to a woman for the next 32 years. These were the years when Eleanora Bliss Knopf carried out her studies at a desk in her husband's room, and when photographs taken of department members (Figs. 11 and 12) included Mrs. Knopf and occasionally wives of graduate students or visitors, but only a very few female graduate students, and none who completed the requirements for a doctoral degree. We can offer no explanation for this lamentable hiatus. Following World War II, the situation started to change, albeit rather slowly. Jean Milton Berdan, daughter of a famous Yale professor, received her Ph.D. for a thesis on the "Brachiopoda and Ostracoda of the Manlius Group of New York State" in 1949, and Elizabeth Jean Lowry (Long) completed a thesis on the "Geology of part of the Roan Mountain Quadrangle, Tennessee" in 1951. Then another hiatus occurred, until 1968, when Rosemary Jacobson Vidale was awarded a Ph.D. for her thesis on "Calc-silicate bands and metasomatism in a chemical gradient." Following Rosemary Vidale, 12 more women have received Ph.D.'s in geology (for a total of 15 since 1949) and the number of women students enrolled has risen to a point where an award rate of 1.5 to 2 per year was reached by 1983. It is interesting—and sad—to note, however, that despite 182 years of geology at Yale, no woman has yet been appointed to the regular teaching faculty in the discipline. This situation will surely change during the next step of Yale's geology history, because women students now play an increasingly important role in the Department.

Women were first admitted as undergraduates at Yale in 1969. The first major to graduate from the Department of Geology and Geophysics was Margaret B. Coon, in 1974. The following year there were three women graduates and the number has risen to a point where eight women graduated in 1983 and six in 1984. From 1974 to 1984, the Department graduated 168 majors in all, of whom 58 were women.

CLOSING REMARKS

The history of teaching and research in the geological and geophysical sciences at Yale is one that records a slow, steady

TABLE 1. SOURCES OF UNDERGRADUATE AND DOCTORAL DEGREES
OF THE YALE GEOLOGY FACULTY
(Members are listed by order of their appointments to the faculty.)

Stage	Name	Bachelor's degree	Doctoral degree	Date of Appointment
1	B. Silliman	Yale	---	1802
2	J.D. Dana	Yale	---	1850
	G.J. Brush	Yale	---	1855
	O.C. Marsh	Yale	---	1866
	A.E. Verrill	Harvard	---	?1873
	G.W. Hawes	Yale	Heidelberg	1873
	E.S. Dana	Yale	Yale	1874
3	S.L. Penfield	Yale	---	1881
	L.V. Pirsson	Yale	---	1882
	C.E. Beecher	Michigan	Yale	1891
	H.S. Williams	Yale	Yale	1892
	H.E. Gregory	Yale	Yale	1898
4	J. Barrell	Lehigh	Yale	1903
	W.E. Ford	Yale	Yale	1903
	C. Schuchert	---	---	1904
	I. Bowman	Harvard	Yale	1905
	R.S. Lull	Rutgers	Columbia	1906
	G.R. Wieland	Penn. State	Yale	1906
	F. Ward	Yale	Yale	1907
	E. Huntington	Beloit	Yale	1907
	J.D. Irving	Columbia	Columbia	1907
	A.M. Bateman	Queen's (Ont.)	Yale	1915
	J.P. Buwalda	California (Berkeley)	California (Berkeley)	1917
5	A. Knopf	California (Berkeley)	California (Berkeley)	1920
	E.B. Knopf	Bryn Mawr	Bryn Mawr	---
	C.R. Longwell	Missouri	Yale	1920
	C.O. Dunbar	Kansas	Yale	1920
	C.H. Warren	Yale	Yale	1922
	W.M. Agar	Princeton	Princeton	1923
	K.C. Heald	Colorado College	Pittsburgh	1924
	R.F. Flint	Chicago	Chicago	1925
	H. de Terra		Munich	1930
	G.E. Lewis	Yale	Yale	1938
	G. Switzer	California	Harvard	1940
6	H. Winchell	Wisconsin	Harvard	1945
	J. Rodgers	Cornell	Yale	1946
	K.M. Waage	Princeton	Princeton	1946
	J.T. Gregory	California	California	1946
	M.S. Walton	Chicago	Columbia	1948
	J.E. Sanders	Wesleyan (Ohio)	Yale	1951
	M.L. Jensen	Utah	M.I.T.	1951
	K.K. Turekian	Wheaton College (Illinois)	Columbia	1956
	A.L. McAlester	Southern Methodist	Yale	1958
	A.L. Washburn	Dartmouth	Yale	1960
	E.L. Simons	Rice	Princeton, Oxford	1960
	J.H. Ostrom	Union College	Columbia	1961
	P.M. Orville	Cal. Tech.	Yale	1962
	R.L. Armstrong	Yale	Yale	1962
	S.P. Clark, Jr.	Harvard	Harvard	1962
	A.W. Crompton	Stellenbosch	Stellenbosch, Cambridge	1964
7	D.C. Rhoads	Cornell (Iowa)	Chicago	1965
	R.A. Berner	Michigan	Harvard	1965
	T.D. Foster	Brown	California	1965
	N.L. Carter	Pomona	California (L.A.)	1966
	B.J. Skinner	Adelaide	Harvard	1966
	J.C.G. Walker	Yale	Columbia	1967
	J.R. Vaišnys	Yale	California (Berkeley)	1967
	G. Veronis	Lafayette	Brown	1968
	B. Saltzman	City College (N.Y.)	M.I.T.	1968
	H.T. Rossby	Royal Inst. Technol. (Sweden)	M.I.T.	1968
	R.B. Gordon	Yale	Yale	1968
	D.M. Rye	Occidental	Minnesota	1973
	M.E. Fiadeiro	Lisbon	California (San Diego)	1975
	R.E. Hall	Cal. Tech.	California (San Diego)	1975
	H.S. Waff	William and Mary	Oregon	1975
	D.R. Pilbeam	Cambridge	Yale	1975
	R.B. Smith	Rensselaer	Johns Hopkins	1976
	R.J. Tracy	Amherst	Massachusetts	1978
	D. Schindel	Michigan	Harvard	1978
	E.A. Okal	Paris	Cal. Tech.	1978
	J. Longhi	Notre Dame	Harvard	1980
	H.J. Bradbury	Liverpool	Liverpool	1980
	R.L. Hughes	Melbourne	Cambridge	1982
	L.J. Hickey	Villanova	Princeton	1982
	N.M. Ribe	Yale	Chicago	1983
	A.C. Lasaga	Princeton	Harvard	1984

expansion covering almost two centuries. Starting in 1802 with one man, Benjamin Silliman, the teaching faculty has grown to a total of 22 full-time members in 1984, corresponding to approximately a doubling in the size of the faculty every 50 years. While no one can predict the future with certainty, we believe there are several reasons to expect that the faculty size may increase still further in the future. One reason for such an expectation comes from the broadening involvement of geologists—perhaps we might better say earth scientists—in many aspects of society and the economy. As Paul Bailly (1984) pointed out in his presidential address before the Geological Society of America, in medium-sized and large countries, there is a strong correlation between the Gross National Product (GNP) of a country and the number of geologists employed. For every $50 million of GNP, in 1980 U.S. dollars, one geologist is employed. Bailly also pointed out that in economically developed countries, among which the United States certainly belongs, there is approximately one geologist employed for every 7,000 people. Because both the GNP and the population are expected to grow over the next 50 years, the number of geologists employed will presumably grow too, and it is not unreasonable to conclude that Yale's involvement in that growth will require additions to the faculty. Numbers of students will presumably increase too, so that the production rate of Ph.D's could increase to a level of 12 to 15 a year, while undergraduate majors might increase from today's figure of about 15 to as many as 30 to 35 a year.

A second reason for expecting that growth patterns of the past will continue in the future is the growing diversity of geological activities. During the first stages of the history of geology at Yale, the principal activity of geologists was a descriptive natural history of the Earth. During the century from about 1880 to 1980, geology developed, as Bailly (1984) remarked, "into a body of scientific disciplines with explanatory power and nascent predictive power." In a very real sense, we are now poised on the threshold of a new era in geology when large-scale testing of predictions will occur as a result of major investigative programs. Planetary studies, ocean-floor drilling, global geodynamic studies, and other large programs have already exposed geologists to ways of working and thinking that could barely have been imagined a generation or two ago. New, possibly larger and possibly even more exciting programs have already been spawned. Examples are the U.S.S.R.'s Deep Drilling Program, the Continental Scientific Drilling Program in the United States, Canada's Lithoprobe, and the U.S.'s COCORP programs of deep, crustal seismic sampling. No doubt other such exciting ventures lie ahead, and they will inevitably stimulate the field. Such programs provide a guarantee that the rate of development of the geological sciences will continue to increase and to diversify in the future. Surely the geological and geophysical activities at Yale will also continue to grow and diversify in the future.

ACKNOWLEDGMENT

The authors are grateful to John Rodgers and Karl Turekian for reminiscing to us about former Yale personnel, and to them and Karl Waage, Horace Winchell, Danny Rye, John Ostrom, Catherine Skinner, and John Morgan for reading the paper and providing helpful suggestions, and to Robert Tucker for preparing the illustrations.

BIBLIOGRAPHIC NOTE

The brief nature of this overview of Yale geology has precluded detailed reference to, and discussion of, the large body of relevant manuscript material in the Yale University Archives. This material includes the papers of Benjamin Silliman, the two Danas, Brush, Marsh, Schuchert, Huntington, and other faculty members. Constant sources of reference were the three volumes of the *Historical Register of Yale University* (1939, 1952, 1969), which contain official records of faculty appointments. Sarjeant (1980) greatly simplified the task of finding obituaries where contemporary observations were recorded, particularly in regard to the reasons for the appointment of our faculty members, the effectiveness of their teaching, and the degree of influence they had on their colleagues as well as on their students.

REFERENCES CITED

Bailly, P. A., 1984, Geologists and GNP—future prospects: Geological Society of America Bulletin, v. 95, p. 257–264.

Bateman, A. M., 1942, Economic mineral deposits: New York, John Wiley & Sons, 898 p.

Boltwood, B. B., 1905, On the ultimate disintegration products of the radioactive elements: American Journal of Science, 4th Ser., v. 20, p. 253–267.

—— 1907, The disintegration products of uranium: American Journal of Science, 4th Ser., v. 23, p. 77–88.

Bowman, I., 1911, Forest physiography: New York, John Wiley & Sons, 759 p.

Clap, T., 1781, Conjectures upon the nature and motion of meteors, which are above the atmosphere: Norwich, Conn., John Trumbull, 13 p.

Cleland, H. F., 1919, Memorial of Henry Shaler Williams: Geological Society of America Bulletin, v. 30, p. 47–65.

Coleman, R. G., 1968, Memorial of Adolph Knopf: American Mineralogist, v. 53, p. 567–576.

Cross, W., 1920, Louis Valentine Pirsson: American Journal of Science, 4th Ser., v. 50, p.173–187.

Dana, E. S., 1877, A text-book of mineralogy, with an extended treatise on crystallography and physical mineralogy . . . On the plan and with the co-operation of Professor James D. Dana: New York, John Wiley & Sons, 485 p.

—— 1932, A textbook of mineralogy, with an extended treatise on crystallography and physical mineralogy . . . (fourth edition, revised and enlarged by William E. Ford): New York, John Wiley & Sons, 851 p.

Dana, J. D., 1837, A system of mineralogy, including an extended treatise on crystallography: New Haven, Conn., Durrie & Peck and Herrick & Noyes, 452 + 119 p.

—— 1848, Manual of mineralogy: including observations on mines, rocks, reduction of ores, and the applications of the science to the arts . . . designed for the use of schools and colleges: New Haven, Conn., Durrie & Peck, 430 p.

——1862, Manual of geology: treating of the principles of the science with special reference to American geological history, for the use of colleges, academies, and schools of science: Philadelphia, Theodore Bliss & Co., 798 p.

——1892, The system of mineralogy of James Dwight Dana 1837–1868. Descriptive mineralogy (sixth edition, by Edward Salisbury Dana . . . entirely rewritten and much enlarged): New York, John Wiley & Sons, 1134 p.

Dunbar, C. O., 1949, Historical geology: New York, John Wiley & Sons, 567 p.

Dunbar, C. O., and Rodgers, J., 1957, Principles of stratigraphy: New York, John Wiley & Sons, 356 p.

Dunbar, C. O., and Waage, K. M., 1969, Historical geology (third edition): New York, John Wiley & Sons, 556 p.

Eaton, A., 1820, An index to the geology of the northern states (second edition): Troy, N.Y., Wm. S. Parker, 286 p.

Fisher, G. P., 1866, Life of Benjamin Silliman, M. D., LL. D., 2 vols.: New York, Charles Scribner and Co., 407 p. and 408 p.

Fleischer, M., 1983, Glossary of mineral species, 1983 (fourth edition): Tucson, Ariz., Mineralogical Record, 202 p.

Gilluly, J., 1977, American geology since 1910—a personal appraisal: Annual Review of Earth and Planetary Sciences, v. 5, p. 1–12.

Gregory, H. E., 1906, January 29, Letter to H. T. [sic] Williams *in* "History of the Geological Department at Yale University" (a scrapbook compiled by Gregory ca. 1915): Yale University. Records of the Department of Geology, Yale University Archives.

——1923, Memorial of Joseph Barrell: Geological Society of America Bulletin, v. 34, p. 18–28.

Harvey, R. S., 1920, Drainage modifications and glaciation in the Danbury region, Connecticut: Hartford, State Geological and Natural History Survey of Connecticut Bulletin 30, 59 p.

Historical Register of Yale University, 1701–1937, 1939: New Haven, Conn., Yale University, 589 p.

Historical Register of Yale University, 1937–1951, 1952: New Haven, Conn., Yale University, 327 p.

Historical Register of Yale University, 1951–1968, 1969: New Haven, Conn., Yale University, 878 p.

Jensen, M. L., 1952, One hundred and fifty years of geology at Yale: American Journal of Science, v. 250, p. 625–635.

Kemp, J. F., 1919, Memorial of J. D. Irving: Geological Society of America Bulletin, v. 30, p. 37–42.

Longwell, C. R., Knopf, A., and Flint, R. F., 1948, Physical geology (third edition): New York, John Wiley & Sons, 602 p. [The first (1932) and second (1939) editions were published as *Part I* of *A textbook of geology.*]

Lull, R. S., 1915, Triassic life of the Connecticut Valley: Hartford, State Geological and Natural History Survey of Connecticut Bulletin 24, 285 p.

——1917, Organic evolution. A text-book: New York, The Macmillan Company, 729 p.

Merrill, G. P., 1920, Contributions to a history of American state geological and natural history surveys: U.S. National Museum Bulletin 109, 549 p.

Morison, S. E., 1936, Three centuries of Harvard, 1636–1936: Cambridge, Mass., Harvard University Press, 512 p.

Narendra, B. L., 1979, Benjamin Silliman and the Peabody Museum: Discovery, v. 14, no. 2, p. 12–29.

Nelson, C. M., 1977, Wieland, George Reber, *in* Dictionary of American Biography, Suppl. 5, 1951–1955: New York, Charles Scribner's Sons, p. 742–744.

Pierson, G. W., 1952, Yale College, an educational history, 1871–1921: New Haven, Conn., Yale University Press, p. 48–49, 65–94.

——1955, Yale, the university college, 1921–1937: New Haven, Conn., Yale University Press, Chapters 7 and 8.

Pirsson, L. V., 1918, The rise of petrology as a science, *in* A century of science in America: New Haven, Conn., Yale University Press, p. 248–267.

Pirsson, L. V., and Schuchert, C., 1915, A text-book of geology, for use in universities, colleges, schools of science, etc., and for the general reader. Two vols.: Part I, Physical geology, by Louis V. Pirsson; Part II, Historical geology, by Charles Schuchert: New York, John Wiley & Sons, 1051 p. (also published as one volume, 1915).

——1929, A textbook of geology. Part I, Physical geology, by Louis V. Pirsson (third edition, revised by W. M. Agar [and others]. Revision edited by Chester R. Longwell): New York, John Wiley & Sons, 488 p.

Rodgers, J., 1977, Memorial to Eleanora Bliss Knopf: Geological Society of America Memorials, v. 6, 4 p.

Rossiter, M. W., 1979, A portrait of James Dwight Dana, *in* Wilson, L. G., ed., Benjamin Silliman and his circle: studies on the influence of Benjamin Silliman on science in America: New York, Science History Publications, p. 105–127.

Rubey, W. W., 1974, Fifty years of the earth sciences—a renaissance: Annual Review of Earth and Planetary Sciences, v. 2, p. 1–24.

Sarjeant, W.A.S., 1980, Geologists and the history of geology. An international bibliography from the origins to 1978. 5 vols.: New York, Arno Press, 4517 p.

Schuchert, C., 1897, A synopsis of American fossil *Brachiopoda:* U.S. Geological Survey Bulletin 87, 464 p.

——1921, *in* Reports of the President, Provost, and Secretary of Yale University and of the deans and directors of its several schools and departments for the academic year 1920–1921: New Haven, Conn., Yale University, p. 177–181.

——1928, The hypothesis of continental displacement, *in* Theory of continental drift: a symposium on the origin and movement of land masses both intercontinental and intra-continental, as proposed by Alfred Wegener: Tulsa, Okla., The American Association of Petroleum Geologists, p. 104–144.

Schuchert, C., and Dunbar, C. O., 1933, A textbook of geology. Part II, Historical geology (third edition, largely rewritten): New York, John Wiley & Sons, 551 p.

Schuchert, C., and LeVene, C. M., 1940, O. C. Marsh, pioneer in paleontology: New Haven, Conn., Yale University Press, p. 230–247.

Silliman, B., 1810, Sketch of the mineralogy of the town of New-Haven: Connecticut Academy of Arts and Sciences Memoir, v. 1, part 1, p. 83–96. The paper was read to the Academy on 1 September 1806.

Simpson, G. G., 1958, Memorial to Richard Swann Lull (1867–1957): Geological Society of America Proceedings for 1957, p. 127–134.

Thompson, G. L., 1964, Barrell, Joseph (1869–1919), *in* La Rocque, A., ed., Contributions to the history of geology, 1. Biographies of geologists: Columbus, Ohio, The Ohio State University Department of Geology, p. 9–11.

White, W. S., 1974, Memorial to Alan Mara Bateman, 1889–1971: Geological Society of America Memorials, v. 3, p. 15–23.

Printed in U.S.A.

Geological Society of America
Centennial Special Volume 1
1985

A history of geology at the University of Pennsylvania: Benjamin Franklin and the rest

Carol Faul
Department of Geology
University of Pennsylvania
Philadelphia, Pennsylvania 19104

ABSTRACT

A survey of the study of geology and geologists at the University of Pennsylvania up to the end of the nineteenth century provides a kaleidoscope of the development of American geology. The topics emphasized and the work of Penn faculty elsewhere show the evolution of state and national surveys and organizations.

Geology at the University of Pennsylvania began in 1749 when Benjamin Franklin aided in establishing a formal academy of study. At first, geology was taught peripherally to other fields, mainly by botanists, chemists, and physicians. There was discussion about a professorship of geology and mineralogy in 1816 but the subject was diffused among other professorships. In 1835 geology was formally established at the University of Pennsylvania and Henry Darwin Rogers was appointed Professor of Geology and Mineralogy.

A succession of geologist-administrators dominated Penn geology through much of the nineteenth century; among them, Alexander D. Bache, who left Penn to head the U.S. Coast Survey in 1843; Henry Darwin Rogers, director of the New Jersey Geological Survey and the First Geological Survey of Pennsylvania while a Penn professor; Ferdinand V. Hayden, director of the U.S. Geological and Geographical Survey of the Territories; and J. Peter Lesley, director of the Second Geological Survey of Pennsylvania. Joseph Leidy and Edward D. Cope were renowned vertebrate paleontologists who worked on Hayden's Survey material.

Beginning around the turn of the century, students succeeded their teachers as heads of the Department and geology declined, with intermittent, unsuccessful attempts to revive it, until 1963, when the present reconstruction began. It is a task which Henry Faul, one of its architects, estimated would take 25 years to complete.

INTRODUCTION

In the middle of the eighteenth century, Philadelphia was a large and rich city, with a population of about 12,000 and growing rapidly. There was a strong intellectual element; a large number of books were published, advertised, sold, and read. But the city lacked a school for higher education. A Charity School was begun in 1740; in 1749, mainly through the efforts of Benjamin Franklin, the School became an Academy and thus higher education began in Philadelphia. In 1755, the Academy received a new charter which allowed it to grant college degrees and the College was born. The Academy, the College, and the University are referred to as Penn in this paper.

Benjamin Franklin was a major influence in making Penn the first colonial college free from the control of any one religious sect (Hindle, 1956, p. 66). The trustees were conspicuous for their liberal views. The teaching of science became prominent after 1754 when William Smith was appointed Provost. He required science for all undergraduates and called for nearly 40 percent of the students' classroom time be devoted to scientific subjects (Hornberger, 1945, p. 29). The Medical School, whose faculty taught much of the geology offered during Penn's early days, awarded its first degrees in 1768 (the first medical degrees given in America).

The institution struggled to survive during the Revolutionary War when its halls were used by Americans and British as barracks and hospital, respectively. After the British departed, the school revived and the University of Pennsylvania formally emerged in 1791.

THE EARLY YEARS: 1749–1835

Benjamin Franklin ruminated on diverse geological matters from speculating why fossil shells are found in rocks on mountain tops to the constitution of the Earth's crust and core. He was sure that there had been great geological changes in England after observing coal mines under the sea and oyster shells in the Derbyshire mountains (Smyth, 1905–1907, II:p.312; VIII:p.598-601). Franklin's fascination with geophysics is well known (Pomerantz, 1976); he even suggested using balloons to gather geophysical data. He addressed the energy problem by calculating how much candle wax and tallow could be saved by the citizens of Paris if they changed their waking hours.

Other pre-Revolution scientists in Philadelphia had geologic ideas. John Bartram, the botanist and naturalist, proposed a geological survey of the colonies which, by borings, would determine the mineral resources present (Darlington, 1849, p. 393). He kept informed about the lamented destruction of the vertebrate fossils discovered at Big Bone Lick, Kentucky, in 1739 that were being scattered throughout the Colonies and Europe (Simpson, 1942). The writings of David Rittenhouse, Professor of Astronomy at Penn, 1779–1780, occasionally touched on geology. He followed Franklin as president of the American Philosophical Society. It was there that the scientific life of Philadelphia then centered—Penn was too small and traditional to fully assimilate the plethora of scientific ideas being generated around it.

From the post-Revolutionary War restructuring of the University, Rev. John Ewing emerged as Provost and Professor of Natural and Experimental Philosophy. Because of his interest in astronomy, he tended toward the study of geodesy; he described a method of calculating the size of the Earth, as described in a communication to the American Philosophical Society in 1786 (Phillips, 1884).

Benjamin Smith Barton was a scientist with multiple interests. Professor of Botany and Natural History (appointed in 1789 to the first such professorship in the United States), he included geology and mineralogy in his courses, as described in a student notebook dated 1805 (Smallwood, 1941). He lamented: "Of all the branches of natural history, there is not one which has been cultivated with so little attention, zeal, or success, in the United States, as Geology, or Mineralogy" (Barton, 1807, p. 51–52). He was so inspiring a teacher that a student named his son after him—William Barton Rogers, founder of M.I.T. Thomas Jefferson was well aware of Barton's knowledge of natural history and his ability to communicate it. He wrote Barton (February 27, 1803) asking him to train Jefferson's secretary Meriwether Lewis for the impending expedition with Clark that was to explore the Missouri River and beyond.

Jefferson turned more than once to University of Pennsylvania faculty about scientific matters. When writing Caspar Wistar in 1807 concerning the education of his grandson, he commented: "There are particular branches of science which are not so advantageously taught anywhere else in the United States as in Philadelphia" (Thorpe, 1893, p. 7). Jefferson knew Wistar as a fellow member of the American Philosophical Society (Wistar followed Jefferson as president) and because of a mutual interest in fossil bones, including those Pleistocene remains at Big Bone Lick, Kentucky. Starting in 1789, Wistar taught at Penn, most notably anatomy. His knowledge of contemporary bones was of enormous help in interpreting the vertebrate fossils. He published a few papers, inspired other workers, and might be called America's first professional vertebrate paleontologist.

Charles Willson Peale, besides painting portraits, was impressario of one of the first natural history museums in America, which opened in Philadelphia in 1786. In 1801 Peale organized what was possibly the first American geologic field expedition, to excavate two mastodon skeletons from the swamps of Orange County, New York. Thomas Jefferson deposited some specimens collected by Lewis and Clark in Peale's museum, there being no national museum in which to store them. Peale was associated with the University quite informally, using its facilities. He advertised an "Introduction to a Course of Lectures on Natural History: Delivered to a Course of Pennsylvania, Nov. 16, 1799" (Sellers, 1969, p. 289). In turn, one University professor, B. S. Barton, asked to hold lectures in Peale's museum, then housed in the building now called Independence Hall (Sellers, 1969, p. 305).

Benjamin Rush was the first Professor of Chemistry in America, presenting lectures in the Penn School of Medicine in 1769 (Newell, 1976). Rush was never much interested in mineralogy, but James Woodhouse, succeeding him in 1795, was of a mineralogical orientation. (Woodhouse was offered the job when Joseph Priestly declined it.) By 1795, geology, through mineralogy, through chemistry, through medicine, was working its way into the curriculum.

Woodhouse (see Figure 1) was born in Philadelphia and received a medical degree under Rush. A few years before his professorship, he had founded the Chemical Society of Philadelphia, one of the first such societies known. Membership was drawn largely from the School of Medicine and usually met in Woodhouse's laboratory. The analysis of minerals was a major activity; the members announced that they were willing and able to identify specimens "free of charge" for people throughout the United States, declaring: "By this means many valuable discoveries may be made and we may become acquainted with the operations of nature in this part of the globe" (Smith, 1914, p. 46–47). Some geology appeared, as when Woodhouse attempted to prove, by circuitous analyses, the origin of a "wall" as an eroded basalt dike, ". . . a product of nature and not of art." (Smith, 1918, p. 86–90).

When Benjamin Silliman was appointed Professor of Chemistry and Natural Science at Yale in 1802, the young lawyer wrote

Figure 1. James Woodhouse, Professor of Chemistry, 1795–1809, taught mineralogy in his University of Pennsylvania courses.

interest in spiritualism, a topic that seemed to fascinate a number of Philadelphia geologists and mineralogists in later life.

Adam Seybert, probably the best trained mineralogist and geologist in Philadelphia at the end of the eighteenth century, graduated from the University of Pennsylvania in medicine. He then studied in Edinburgh and London, under Werner in Freiberg, and under Häuy in Paris. Around 1795 he brought home a collection of minerals that William Maclure recalled in a letter to Samuel Morton, dated April 3, 1830, as ". . . a small cabinet of mineralogy which Dr. Seybert brought from Göttingen which cost him 15 dollars was the only one we had to run to in Philadelphia" (Reingold, 1964, p. 36). Silliman had brought Yale's mineral collection in a candle box to Seybert to be identified when he came to Philadelphia for his quick course.

As a result of university politics, Seybert was denied the professorship of chemistry left vacant by Woodhouse's death in 1809, forcing him to turn to politics and business. His son Henry started in his father's geological footsteps but his interest waned. In 1883 he presented a large number of railroad bonds to Penn to endow the Adam Seybert Chair of Moral and Intellectual Philosophy, especially to investigate spiritualism (Smith, 1919, p. 51–52). But for academic politics 74 years earlier, it might have been a chair in geology.

William Maclure was a sometime Philadelphian who examined the geology of the eastern United States, crisscrossing the land and examining the rocks, becoming perhaps America's first field geologist. In 1809 he published a colored geologic map with *Observations on the Geology of the United States.* He had no formal association with the University and directed his scientific interests to the Academy of Natural Sciences which he had helped found in 1812. Starting in 1817, he was president for 22 years.

Philadelphia physician Archibald Bruce in 1810 founded the *American Mineralogical Journal,* the first American publication primarily for geology and mineralogy. An ambitious endeavor, it published the works of many prominent geologists but it survived only until 1814.

At a University trustees meeting in 1816, during efforts to organize a "Faculty of Physical Science and Rural Economy," a motion was proposed for a professorship of geology and mineralogy. After two years of discussion, among the professorships that emerged was one of natural history, including geology, zoology, and comparative anatomy, and one of mineralogy and chemistry as applied to agriculture and the arts (Thorpe, 1893, p. 328–329). Thomas Cooper became Professor of Chemistry and Mineralogy and Charles Caldwell received Natural History and its parts. It was a time of an academic financial crunch and the professors likely received no salary. Caldwell and Cooper resigned in 1820 and 1821, respectively.

Thomas Cooper had introduced a course of geological and mineralogical lectures at Penn in 1816, advertised in the *Analectic Magazine* of October, 1817. He was a prototype of the modern political dissident. He had left England after various skirmishes and settled in Northumberland, Pennsylvania, with

in his diary: "Of course I must resort to Philadelphia, which presented more advantages in science than any other place in our country" (Cheyney, 1940, p. 210). Off he went to Philadelphia to study chemistry and probably mineralogy under Woodhouse and to learn from Penn professors Robert Hare, Benjamin S. Barton, and Caspar Wistar. Later he voiced unkind comments about Woodhouse's teaching, but one senses a clash of personalities— Yankee versus Philadelphian.

Inspired by his teacher James Woodhouse, Robert Hare's great contribution to mineralogy was made in 1801 when he reported to his colleagues at the Chemical Society his invention of the oxyhydrogen blowpipe, essential for mineral identification in fusibility tests and the reduction of metals from their minerals. Hare became Professor of Natural Philosophy and then Professor of Chemistry at the Medical School, from 1810 to 1847. His interest in mineralogy continued and in 1814 he attended the now erudite Silliman's lectures on mineralogy at Yale, becoming ". . . well informed in geology . . ." according to Silliman himself (Fisher, 1866, p. 265–266). As he grew older, Hare developed an

Joseph Priestley in 1793. It was not long before he was serving a six-month sentence in jail for sedition and libel. When Jefferson succeeded Adams as President, however, Cooper became a judge. By the time Cooper had settled down at Penn, his radical ideas were directed toward scientific education. He saw minerals as a tool to understanding the Earth, and geology as a tool to understanding the Earth's mineral wealth; he rejected the rockhound philosophy of minerals as intrinsically valuable objects (Smith, 1919, 69–71). After Cooper left the University, geology was taught more casually, as evidenced by the listing of Robert M. Patterson, Vice-provost and Professor of Mathematics, in a curriculum schedule as also providing geology instruction for sophomores (University Archives, 1824?).

Samuel Morton learned chemistry, mineralogy, and geology at a private school conducted by Cooper. He graduated in medicine from Penn and practiced medicine, physical anthropology, vertebrate and invertebrate paleontology, and geology. Using former Philadelphian Lardner Vanuxem's notes, in 1828, he presented a correlation of the Atlantic Coastal Tertiary strata by stratigraphic paleontology (Morton, 1829).

William Hippolytus Keating and Thomas Say were both members of the Long-Keating expeditions (1823–1825) to survey the Great Lakes area and sources of the Mississippi River. Keating was mineralogist and Say the zoologist. Before, during, and after this time, Keating was Professor of Mineralogy and Chemistry at Penn. He was a founder of the Franklin Institute in 1824. Say was Professor of Natural History, which included geology, at Penn and a charter member of the Academy of Natural Sciences. Both their positions were likely more honorary than remunerative at the University because of stringent finances (Cheyney, 1940, p. 237). Say was perhaps the first American to point out the value of fossils in stratigraphic correlation, stating that the progress of geology ". . . must be in part founded on a knowledge of the different genera and species of reliquiae which the various accessible strata of the earth present" (Say, 1818, p. 382).

With the American Philosophical Society, the Academy of Natural Sciences, and the Franklin Institute all active, Philadelphia geologists were diffused about the city. Most of Isaac Lea's geology centered on the activities of the Academy of Natural Sciences of which he became president. Lea did pioneer work on the Atlantic coastal Tertiary fauna; later he turned to microscopic mineralogy. Richard Harlan succeeded Caspar Wispar as the principal authority on vertebrate fossils at the American Philosophical Society. A Penn M.D., he was the first American to spend a major part of his time on vertebrate paleontology and the first American to apply Linnaean names to American vertebrate fossils (Simpson, 1942).

THE PHILADELPHIA GEOLOGIST-ADMINISTRATORS

Philadelphia produced a succession of effective geologist-administrators; many contributed to national and state geologic surveys. Most were associated with the University of Pennsylvania during or before their other geological endeavors.

Alexander Dallas Bache (Figure 2) was the first of the great geologist-administrators. He was Professor of Natural Philosophy and Chemistry at Penn intermittently from 1828 (at age 22) until he left for Washington to head the U.S. Coast Survey in 1843. Like his illustrious great-grandfather, Benjamin Franklin, he was fascinated by geophysics, especially terrestrial magnetism. He tried to establish an American system to fit into the world network of magnetic observatories. The first observatory was at Girard College in Philadelphia, a school he had founded in 1836 (Odgers, 1947, p. 115–118). Bache had leading roles in the American Philosophical Society and Franklin Institute and in 1863 he became the first president of the National Academy of Sciences.

Peter A. Browne organized geology on the state level. A Philadelphia lawyer with an amateur's zeal for geology, he felt that the field should be more accessible to the people than the limited access provided by the University. In 1826 Browne called a public meeting in Philadelphia, ". . . Intended to Promote a Geological and Mineralogical Survey of Pennsylvania, the Publication of a Series of Geological Maps, and the Formation of State and County Geological and Mineralogical Collections (Browne, 826). He pointed out the value of geology and mineralogy and the progress being made in spreading a knowledge of the field, citing that as late as 1811 a Mr. Cooper had been charged with lunacy for his mineralogical excursions. Browne was distracted by business promotions and the project died. His interest revived, however, and in 1832, largely at his instigation, the Geological Society of Pennsylvania was organized. Its major purpose was to agitate for a state geological survey, and it dissolved in 1836 after its goal was achieved.

Geology arrived formally and permanently at the University of Pennsylvania in 1835 with the founding of the Department of Geology and Mineralogy and the appointment of Henry Darwin Rogers (Figure 3) the first Professor of Geology and Mineralogy. The deciding factor may have been that suggested by historian Edward P. Cheyney: ". . . Henry D. Rogers was in a position to offer his services without salary, and the Faculty, attracted by an opportunity to expand their number without expense, petitioned the Trustees for his election" (1940, p. 225). Rogers was neither a stranger to the city nor to geology. His father had received an M.D. from Penn and Rogers had been born in Philadelphia. During a visit to England he spent much time in London scientific circles and returned home well trained in geology. In 1834, and again the next year, the Franklin Institute let Rogers use a lecture room to teach a course on geology, providing him with exposure to the scientific community. According to *A Guide to A Course of Lectures on Geology Delivered in the University of Pennsylvania* (see Figure 4), Rogers was a thorough teacher: "First, the study of the materials which compose the Earth; Secondly, an investigation of the causes of change now in operation; Thirdly, a study of the several formations in regular series, beginning with the oldest" (Rogers, n.d., p.1).

Figure 2. Alexander Dallas Bache, the first of a series of geologist-administrators at the University of Pennsylvania.

Rogers was active. In 1835 he was appointed to make a geological survey of New Jersey and in 1836, became director of the first Pennsylvania Geological Survey. His major geologic contribution, with his brother William, was unraveling the structure of the Appalachians, the first study of a very large mountain belt.

Charles B. Trego was pleased to see the Pennsylvania Geological Survey succeed. When elected to the Pennsylvania House of Representatives from Philadelphia, he almost immediately introduced and helped guide into law the bill creating the Geological Survey. He may have been offered the post of State Geologist and he did become one of Rogers' assistants (Roberts, 1875). In 1852 Trego became Professor of Geology, Mineralogy, and Paleontology at Penn and was still on the faculty during its geological pinnacle almost 20 years later.

John Fries Frazer was the next geologist-administrator. His mentor was Alexander D. Bache, under whom he had studied and for whom he had been lab assistant at Penn. He was assistant to Henry Darwin Rogers on the Pennsylvania Survey in 1836,

and in 1844, when Bache went to the Coast Survey, Frazer became Professor of Natural Philosophy and Chemistry. He was a joiner, active in the Franklin Institute, the American Philosophical Society, and a founder of the National Academy of Sciences. In 1852 he proposed the School of Arts, Mines, and Manufactures in which the applications of science would be emphasized (Sinclair, 1972). He hired Charles Trego as Professor of Geology, Mineralogy, and Paleontology, and J. W. Alexander as Professor of Engineering and Mining, and asked James Curtis Booth, who had joined the faculty the year before as Professor of Chemistry as applied to the Arts, to join them.

James C. Booth had studied under Robert Hare and William Keating and graduated from Penn in 1828. Possibly the first American to study chemistry in Germany, he returned to Philadelphia to establish a private laboratory in Philadelphia where he instructed students in techniques of analytical chemistry and did assays and other commercial work. As a professor in Frazer's new school, he and J. W. Alexander weren't paid, didn't teach,

Figure 3. Henry Darwin Rogers, founder of the Department of Geology and Mineralogy at the University of Pennsylvania in 1835.

A GUIDE
TO
A COURSE OF LECTURES
ON
GEOLOGY,
DELIVERED IN THE UNIVERSITY OF PENNSYLVANIA.

By H. D. ROGERS,
Professor of Geology and Mineralogy.

GEOLOGY is the science which investigates the mineral structure of the earth, and the causes of the successive changes and revolutions which the organic and inorganic kingdoms of nature have undergone, or are undergoing.

The plan of the course embraces, *First*, the study of the materials which compose the earth ; *Secondly*, an investigation of the causes of change now in operation ; *Thirdly*, a study of the several formations in regular series, beginning with the oldest.

The ultimate materials or chemical elementary substances at present known are 54 in number, 12 being non-metallic, and 42 being metals.

Those which principally enter into the chemical composition of our planet's surface may be classed in the following order, according to the respective importance of each :

Simple non-metallic substances.

1. OXYGEN.	3. NITROGEN.	5. SULPHUR.	7. FLUORINE.
2. HYDROGEN.	4. CARBON.	6. CHLORINE.	8. PHOSPHORUS.

Metallic bases of the alkalies and earths.

1. SILICIUM.	3. POTASSIUM.	5. MAGNESIUM.
2. ALUMINIUM.	4. SODIUM.	6. CALCIUM.

Metals whose oxides are neither alkalies nor earths.

1. IRON.	2. MANGANESE.

Figure 4. Syllabus of a geology course taught by Henry Darwin Rogers, 1842.

and were asked to resign. They did in 1866 and 1865, respectively (Cheyney, 1940, p. 239).

Fairman Rogers studied under John F. Frazer at Penn and then worked with A. D. Bache on the Coast Survey. A Philadelphian of wide interests, varying from magnetism to polo, he replaced J. W. Alexander on the University faculty, receiving the title of Professor of Civil Engineering, Geology, Mining, Surveying, Art of Mining and Mining Machinery. In 1855, under his deanship, he reorganized the technical school started by Frazer (Cheyney, 1940, p. 240). Rogers and Trego each received $62.50 for teaching. Students paid $5.00 per course; tickets for courses were purchased from the janitor (Cheyney, 1940, p. 241). The school was suspended in 1861 during the confusion of the Civil War.

An active and prolific organizer, J. Peter Lesley graduated from Penn in 1838, and at A. Bache's suggestion, worked for Henry D. Rogers and the first Pennsylvania Geological Survey surveying the bituminous and anthracite coal regions. He was then minister of a Congregational Church, an ex-minister, an original member of the National Academy of Sciences, a Pennsylvania Railroad geologist, a consulting geologist, Penn Professor of Mining starting in 1859, Professor of Geology and Mining Engineering in 1872, Dean of the Science Faculty, Dean of the

Towne Scientific School (which had replaced Frazer's technical school). In 1858 Lesley had become incensed that H. D. Rogers had omitted his name from the large geologic map of Pennsylvania that accompanied Rogers' *The Geology of Pennsylvania*, although he had done most of the drafting and coloring. Their friendship ended (Jordan, 1984). Perhaps it was just as well that Rogers had gone to the University of Glasgow in 1857 to occupy the Regius professorship of Natural History.

Lesley is credited with initiating the first extensive studies of the coal, oil, and gas resources of Pennsylvania. As in Rogers' State Survey, a large number of geologists, including C. Ashburner, C. Beecher, Leo Lesquereux, and I. C. White, received their basic training on the Second Survey and went on to enrich American geology.

James Hall, New York State Paleontologist and first president of the Geological Society of America, was a friend and admirer of Lesley. He wrote a strong letter in support of Lesley as Pennsylvania State Geologist (Merrill, 1969). It seems to have been through Lesley's negotiation that Hall sold a large collection of fossils to Penn in 1872. (The latter was making excuses to the University more than 22 years later for not having sent all that was due them (Hall, 1894).).

Ferdinand Vandiveer Hayden (Figure 5) strode upon the University scene in 1865 as Professor of Geology and Mineralogy. Early in Hayden's geological career he came under the tutelage of James Hall, who sent him on a collecting expedition to the Dakota Badlands, starting a lifelong love of western exploration. Hayden had early associations with Penn through Professor Joseph Leidy, who identified many of the vertebrate fossils he collected. (Also, Hayden knew Philadelphia because he had been stationed there for a short time during Civil War service as an army surgeon.) Later, after he began working for the U.S. Government as a geologist in 1867, besides Leidy, Penn faculty members Persifor Frazer and E. D. Cope were associated with his surveys. While a Penn professor, he continued western explorations. The Philadelphia Academy of Natural Sciences published many of his early papers and he went to the Black Hills in 1866 under their auspices. He left the University of Pennsylvania in 1872 when the directorship of the U.S. Geological and Geographical Survey of the Territories became too time-consuming; he returned to Philadelphia when illness halted his work. In 1887 he died and was buried in Woodlands Cemetery, next door to the University.

Figure 5. Ferdinant Vandiveer Hayden, Professor of Geology and Mineralogy, 1865–1872. (Photograph courtesy of National Portrait Gallery, Smithsonian Institution, Washington, D.C.)

THE LATE NINETEENTH CENTURY YEARS

In 1872 the University of Pennsylvania moved from Ninth Street to a new campus in West Philadelphia. A new Department of Science was created. It differed from the School of Arts in providing practical scientific courses during junior and senior years. The Department of Geology and Mineralogy offered a "Course of Geology and Mining Engineering" (University Announcement, 1872, p. 13–15). In 1875 the Department of Science was renamed the Towne Scientific School.

Although the Department of Geology and Mineralogy had existed since 1835, geology was also still taught in other schools and departments of the University. Many of these schools changed their names, merged, divided, or went out of existence. Only the most relevant to the development of geology at Penn are mentioned here. Assistants, instructors, lecturers, and assistant professors appeared throughout the succeeding decades. They are listed in an appendix at the end of this paper.

Persifor Frazer, son of John Fries Frazer, trained at the Frieberg School of Mines and then returned to America to work for the Hayden Survey in 1869 as a mining engineer and geologist. The next year he became Professor of Natural Philosophy and Chemistry at Penn; he assumed his father's job at his death in 1872. (John F. Frazer had died of heart failure while showing visitors College Hall, the first building on the new campus, the day after it had been dedicated.) Persifor Frazer next headed the

Chemistry Department, then, tired of academic life, resigned to become an assistant on Lesley's Second Pennsylvania Survey. He became a consulting mineralogist in Philadelphia and applied his scientific training in diverse fields. He had an avid interest in mind reading and he became a handwriting expert, devising a process of detecting forgeries by composite photography. Vivid newspaper accounts describe his activities as an expert witness at famous murder trials (University Archives, 1909).

Joseph Wharton, Philadelphia geologist, mineralogist, and financier, in 1881 donated $100,000 to the University of Pennsylvania. He is remembered by environmental geologists for an imaginative plan to supply Philadelphia with pure water. Between 1876 and 1880 he gradually acquired a huge contiguous tract of the New Jersey Pine Barrens in which he planned 33 shallow reservoirs. Connected by a network of canals to a huge reservoir in Camden, an aqueduct under the Delaware River would transport the water to a grateful Philadelphia. The plan was thwarted by the New Jersey legislature which prohibited such use of their resource. The money given to Penn went not for geology but to establish the Wharton School, the first school of business associated with a university.

Frederick A. Genth and George A. Koenig both received extensive training in Germany and both immigrated to America, Genth in 1848 and Koenig in 1868. Genth was a commercial chemist before becoming Professor of Chemistry at the new De-

Figure 6. George Augustus Koenig (in black beard and miner's uniform) teaching a mineralogy class in 1887.

partment of Science in 1872 and, two years later, Professor of Chemistry and Mineralogy. The dean was J. Peter Lesley and Genth was employed as a chemist for his Second Pennsylvania Survey. In 1888 he left academia after acrimonious dissension with the Trustees and returned to consulting and research. A foremost pioneer mineralogist, he wrote 54 papers describing 215 mineral species, 24 of them new minerals.

George A. Koenig (Figure 6) was a commercial chemist before joining the Department of Science in 1872, first as Assistant Professor of Chemistry and Mineralogy, then Professor of Mineralogy and Geology until 1892. He discovered diamonds in meteoric iron and described 13 new minerals.

According to a catalog of 1880, the Towne Scientific School had over 10,000 geological specimens. Some of these were likely part of the James Hall collection, acquired by Lesley in 1872, but still only partially received. The University had, of course, acquired geological specimens throughout its history. The earliest mention of a collection is of the Keating suite of rocks from the Great Lakes Long-Keating expedition, given to the University around 1819. Collections assembled by both Genth and Koenig were donated to the Department of Geology and Mineralogy. The Genth collection of German minerals, fossils, and rocks, with a catalog in elegant German script, is dated 1847. Although

catalogs dating from 1832 on are in the Department archives, this is the oldest collection extant. The George Koenig collection of about 3000 minerals and rocks, with a fund to curate them and a bronze plaque to commemorate them, was given to Penn in 1925 by his daughter and son-in-law. The largest and most valuable collection of minerals in the Department is from alumni Joseph A. Clay and his brother J. Randolph Clay, presented a few years after Joseph's death in 1881. It contains many spectacular crystals from localities around the world.

Although the Medical School had ceased to be the primary habitat of mineralogists and geologists around 1816 with the establishment of a Faculty of Natural History, its faculty members often included in their curriculum geological topics related to medical matters. Samuel B. Howell was a practicing physician and Professor of Mineralogy in the Auxiliary Department of Medicine at Penn, 1872–1889, a department organized to supplement the specialized training of medical students. Howell was interested in air quality; he studied the composition of the air in Atlantic City and determined that along the beach it contained up to 50% more oxygen. A newspaper reported that because he decided the air quality was so superior he moved from Philadelphia to a hotel on the beachfront.

Two renowned vertebrate paleontologists reigned in the

Medical School in the latter half of the nineteenth century. Joseph Leidy earned an M.D. at the Medical School, became Professor of Anatomy in 1853, and stayed for 38 years (with time out as a surgeon in the Civil War). Active in the Philadelphia scientific community throughout his career, he became president of the Academy of Natural Sciences of Philadelphia and the Wagner Free Institute of Science, founded in 1852. He was deflected from a career in medicine partly at the urging of Charles Lyell that he pursue paleontology; he was also influenced by his friend, William E. Horner, Professor of Anatomy and fossil bone collector at the Medical School. Leidy's important vertebrate paleontology work began in 1847 with a memoir on the fossil horse of North America. His meticulous descriptions gave vertebrate paleontology scientific respectability. Distressed by the explosive politics of Hayden and caught in the crossfire of the violent feud between Cope and Marsh, Leidy gradually withdrew from the study of vertebrate fossils. He had neither the financial resources nor the taste for battle necessary for survival in the field he described as no longer fit for a gentleman.

Edward Drinker Cope (Figure 7) had learned anatomy from Leidy as a young man; later they both worked on the fossil vertebrate material of the Hayden Survey, Cope acting officially as geologist and paleontologist. (Figure 8 illustrates the kind of envelope Cope used to store specimens from the Hayden Survey.) When his feud with O. C. Marsh of Yale accelerated, Cope was able to hold his own until 1879 when Marsh won the coveted position as head of vertebrate paleontology at the newly created U.S. Geological Survey. This effectively eliminated further participation by Cope in federally-sponsored paleontological projects. About that time, Cope lost his financial resources in a mining hoax. In 1889, Cope became Professor of Geology and Mineralogy at the University of Pennsylvania, giving him financial security and a respected affiliation. In 1890, he was secure enough to inaugurate a mighty public effort to oust arch-rival Marsh from the U.S. government job. This futile exercise brought Cope glaring publicity and came close to costing him his University position. He weathered that storm, partly due to his continuing blizzard of publications, 50 papers appearing in 1890 alone (Osborn, 1931, p. 399). Cope died in 1897; his skeleton is preserved in the University Museum.

DECLINE

For the following 70 years, the destiny of geology at the University of Pennsylvania was in the hands of the head professors who chose their favorite students to succeed them. (The title of department chairman did not appear until about 1950.) The most competent of these was Amos P. Brown; he studied under Lesley at Penn and then worked for him on the Second Pennsylvania Survey. He became Instructor in Mining and Metallurgy in 1889 and received his Ph.D. from Penn in 1892. Brown worked with E. D. Cope in the Dakotas, Kansas, Texas, and Oklahoma in 1893 (Penrose, 1918). In 1892 he was appointed Professor of Geology and Mineralogy in the Auxiliary Department of Medi-

Figure 7. Edward Drinker Cope, Penn Professor, 1889–1897. (Photograph courtesy of Library, Academy of Natural Sciences of Philadelphia.)

cine until that department was abolished in 1898. He was head professor of the Department of Geology and Mineralogy from 1908 until 1917. Brown wrote on geological, mineralogical, and paleontological topics, but his major scientific work was on the crystallography of hemoglobin: with E. T. Reichert of the University he wrote a definitive work, published by the Carnegie Institute of Washington in 1909.

Frederick Ehrenfeld was Amos Brown's favorite student. He was appointed an instructor in geology and mineralogy in 1897 and received a Ph.D. in 1898. Ehrenfeld was an assistant professor, 1907–1923, and put in charge of the Department in 1917. In 1923 he was appointed Professor of Geology and Mineralogy and remained until his death in 1940. His major geologic interest had been the coastal rocks of Maine and he collected many beach sands. His student, Paul J. Storm, had joined the faculty in 1924 as an instructor in geology and mineralogy and received a Ph.D. in 1930. In 1936 he was made Assistant Professor of Geology and Mineralogy and retained the title until his retirement in 1963, even though he had been head of the Department since Ehrenfeld's death in 1940.

University historian Edward P. Cheyney ignores Penn geol-

Figure 8. Edward Drinker Cope used an envelope from the Geological Society of America to store a specimen.

ogy after the Leidy-Lesley-Cope era, except for two sentences about a 1935 cooperative program (p. 428). Why did the University of Pennsylvania administration choose to let geology at Penn inbreed and wane, providing it with minimal moral and financial support? No clear reasons are obvious. The decline seems not to have been so much the result of deliberate policy as of neglect.

The strongest individual attempt to revitalize geology at the University of Pennsylvania was the effort of Philadelphian R. A. F. Penrose, Jr., mining geologist and patron of the Geological Society of America and American Philosophical Society. His father had been a Penn professor and he was long interested in the welfare of the University, becoming a trustee in 1911. In 1920 he declined an offer to become Provost. In an undated manuscript, *Geology and its Scope in University Education,* prepared for a University trustee, he showed a strong interest in and had very definite opinions on the teaching of geology (Fairbanks and Berkey, 1952, p. 628–630). As a trustee he seems never to have earmarked specifically for geology money donated to Penn although, according to Ehrenfeld, he did give geology many words of encouragement. Penrose attempted to persuade John Clarke, New York State Paleontologist, to accept a professorship of geology at the University (Fairbanks and Berkey, 1952, p. 730) but Clarke didn't want to leave Albany. In 1924, Penrose disagreed with the methods the administration planned to increase its endowment. In 1927 he disagreed with plans on the administrative reorganization of the University, resigned from the Board of

Trustees, and seems to have had nothing further to do with the University of Pennsylvania.

Geology existed at a subsistence level until University President Thomas S. Gates entered into an agreement with the Academy of Natural Sciences of Philadelphia called the "Four-Year Course in Earth Sciences" (Dowlin, 1940): the academy would offer University students elementary and advanced courses in paleontology and advanced courses in geology and mineralogy. When the program got underway in 1938, geology was renamed the Department of Earth Sciences. Besides Penn faculty Ehrenfeld, Storm, and Assistant Professor A. Williams Postel, at various times the students learned from Benjamin F. Howell, Edwin H. Colbert, Erling Dorf, and Horace Richards from the Academy, all of them with other academic affiliations. (Geology at the Academy had gradually declined in importance since Cope's time.) The association gradually weakened until geology was taught only by Ehrenfeld, Storm, Richards, plus one or two assistants.

In 1941, the University celebrated with a series of bicentennial conferences, recognizing geology with lectures and a booklet on shifting of sea floors and coast lines (Bowen and others, 1941).

THE RECONSTRUCTION

Early in the 1960s, under a physicist President, Gaylord Harnwell, a biologist Provost, David Goddard, and a Germanist

Figure 9. The Department of Geology is located in the old Dental Hall, constructed in 1897, now named Hayden Hall after Ferdinand V. Hayden.

Dean, Otto Springer, a committee of faculty from science and engineering decided that geology was integral to the composition of a contemporary university. It was reasoned that geology not only had a right to life as a field of study and research at Penn, but it was an essential part of the education of students in other sciences and engineering.

To initiate the reconstruction, Howard A. Meyerhoff, distinguished former senior editor of *Science,* became Chairman in 1963. Paul Storm retired. Five assistant professors were hired to join him and part-time lecturer Horace Richards. Geology was renamed the Department of Geology. After occupying rooms in College Hall since 1872, at first in the second floor east wing, then in 1930 acquiring the former chapel for a museum, geology moved. It relocated in the former Dental Hall (Figure 9), a red brick building completed in 1897 as the first edifice designed and constructed as a school of dentistry, and occupied since 1913 by the School of Fine Arts. Geology occupied newly reconstructed basement laboratories and offices until Fine Arts vacated, belatedly, in 1967; then it took over all but the art studio, once the dental clinical operating room. The building was renamed Hayden Hall in honor of Ferdinand V. Hayden.

In 1966, when Meyerhoff approached retirement age, geochronologist Henry Faul (Figure 10) was chosen as Chairman and Professor of Geophysics to accelerate the recovery and strengthen its program of research. The faculty was upgraded with the acquisition of superior teachers and prolific researchers. Faul guided and goaded geology into a spectrum of research

Figure 10. Henry Faul, Professor of Geophysics, 1966–1981, an architect of the current reconstruction of the Department of Geology.

including geochronology, geochemistry, Pleistocene geology, paleoclimatology, paleobotany, paleobiology, metamorphic petrology, high temperature petrology, and history of geology. Close association with faculty from other schools, including the Department of Landscape Architecture and Regional Planning, the Veterinary School, and the School of Engineering, added breadth to the teaching and research of geology. Research associates, visiting professors, and a strong seminar program contributed to geology's new vitality.

Through support from the administration and externally generated funds, desperately needed research facilities were constructed in 1972 by gutting the south section of Hayden Hall and building three floors of laboratories and offices and space for the map library.

R. Ian Harker became Chairman in 1972 to guide geology in its reconstruction. The seven tenure-track faculty, associates, visiting and adjunct faculty, and graduate students increased the department research productivity.

Robert F. Giegengack became Chairman in 1978. Under his leadership, research has been grouped primarily into three core areas: geochemistry/geochronology, paleobiology/paleoecology, and environmental geochemistry/biogeochemistry as the reconstruction of the proud tradition of geology at the University of Pennsylvania continues.

ACKNOWLEDGMENTS

Joan G. Heubusch thoughtfully and critically read the manuscript. Margaret Fanok helped compile the Appendix.

APPENDIX

Geology Teachers/Researchers at the Academy-College-University of Pennsylvania

Benjamin Franklin: 1747–1790, Founder.

John Ewing: 1779–1802, Provost; Professor of Natural and Experimental Philosophy.

Benjamin Smith Barton: 1789–1791, Professor of Botany and Natural History; 1791–1815, Professor of Natural History.

Caspar Wistar: 1789–1792, Professor of Chemistry; 1792–1810, Adjunct Professor of Anatomy, Surgery, and Midwifery; 1810–1818, Professor of Anatomy.

James Woodhouse: 1795–1809, Professor of Chemistry.

Robert Hare: 1810–1812, Professor of Natural Philosophy; 1818–1847, Professor of Chemistry.

Thomas Cooper: 1816–1821, Professor of Applied Chemistry and Mineralogy.

Robert M. Patterson: 1814–1828, Vice-provost; Professor of Mathematics.

William H. Keating: 1822–1828, Professor of Mineralogy and Chemistry.

Thomas Say: 1822–1828, Professor of Natural History (honorary).

Alexander D. Bache: 1828–1836, 1842–1843, Professor of Natural Philosophy and Chemistry.

Henry D. Rogers: 1835–1846, Founder, Department of Geology and Mineralogy; Professor of Geology and Mineralogy.

John F. Frazer: 1844–1872, Vice-provost, Professor of Natural Philosophy and Chemistry.

James C. Booth: 1851–1856, Professor of Chemistry.

Charles B. Trego: 1852–1855, Professor of Geology, Mineralogy, and Paleontology; 1855–1872, Professor of Geology.

J. W. Alexander: 1852–1855, Professor of Engineering and Mining.

Fairman Rogers: 1853–1861?: Professor of Civil Engineering, Geology, Mining, Surveying, Art of Mining and Mining Machinery.

Joseph Leidy: 1853–1891, Professor of Anatomy.

J. Peter Lesley: 1859–1872, Professor of Mining; 1872–1875, Professor of Geology and Mining Engineering; 1875–1885, Dean of Towne Scientific School.

Persifor Frazer: 1870–1871, Instructor in Natural Philosophy and Chemistry; 1871–1872, Assistant Professor in Natural Philosophy and Chemistry; 1872–1873, Professor of Natural Philosophy and Chemistry; 1873–1874, Professors of Chemistry.

Frederick A. Genth: 1872–1874, Professor of Chemistry; 1874–1888, Professor of Chemistry and Mineralogy.

George A. Koenig: 1872–1879, Assistant Professor of Chemistry and Mineralogy; 1879–1892, Professor of Mineralogy and Geology.

Samuel B. Howell: 1872–1890, Professor of Mineralogy and Geology.

Thomas M. Chatard: 1872–1873, Assistant in Analytical and Applied Chemistry and Mineralogy.

John H. Marden: 1874–1879; Linwood O. Towne: 1879–1881; Hermann A. Keller: 1881–1883; Winchester Dickerson: 1882—Professor's Assistants in Geology and Mining Engineering.

Henry A. Wasmuth: 1883–1886, Professor's Assistant in Geology; 1887–1891, Instructor in Mining.

Marshall F. Pugh: 1888–1889, Professor's Assistant in Mining and Metallurgy; 1889–1891, Instructor in Mining and Metallurgy.

Edward D. Cope: 1889–1896, Professor of Geology and Mineralogy; 1896–1897, Professor of Anatomy.

Amos P. Brown: 1889–1892, Instructor in Mining and Metallurgy; 1892–1917, Professor of Geology and Mineralogy; 1908–1917, Head Professor of Geology and Mineralogy.

Frederick Ehrenfeld: 1897–1907, Instructor in Geology and Mineralogy; 1907–1923, Assistant Professor of Geology and Mineralogy; 1923–1940, Professor of Geology and Mineralogy; 1917–1940, Head Professor of Geology and Mineralogy.

Charles Travis: 1906–1917, Instructor in Geology and Mineralogy.

Paul J. Storm: 1924–1936, Instructor in Geology and Mineralogy; 1936–1963, Assistant Professor of Geology; 1940–1963, Head Professor of Geology and Mineralogy.

Frederick M. Oldach: 1925–1934, Instructor in Geology and Mineralogy.

A. Nelson Sayre: 1926–1929, Instructor in Geology and Mineralogy.

A Williams Postel: 1934–1941, Instructor in Geology and Mineralogy; 1941–1946, Assistant Professor of Earth Sciences.

Edwin H. Colbert: 1939–1941, Lecturer on Geology and Paleontology.

Benjamin F. Howell: 1939–1943, Lecturer on Geology and Paleontology.

Robert G. Chaffee: 1940–1941, Assistant Instructor in Geology.

Erling Dorf: 1942–1943, Lecturer on Paleobotany.

Paul A. Dike: 1947–1952, Instructor in Earth Sciences.

Horace Richards: 1950–1972, Lecturer on Paleontology.

S. Francis Thoumsin: 1941–1942, 1953–1960, 1958–   , Instructor in Geology.

Bruce K. Goodwin: 1960–1963, Instructor in Earth Sciences.

Allen Keller: 1961–1963, Instructor in Earth Sciences.

The following held or hold position in the Department of Geology since the reconstruction began:

Howard A. Meyerhoff: 1963–1967, Professor; 1963–1966, Chairman.

Frederick G. Layman: 1963–1966, Instructor; 1966–1968, Assistant Professor.

Peter Fenner: 1963–1966, Instructor; 1966–1968, Assistant Professor.

George D. Klein: 1964–1966, Assistant Professor; 1966–1969, Associate Professor.

Alan Lutz: 1964–1965, Visiting Lecturer.

Reginald Shagam: 1965–1968, Assistant Professor; 1984–1985, Visiting Professor.

Patrick Butler: 1965–1967, Instructor; 1967–1969, Assistant Professor.

Elysiario Tavora: 1965–1966, Visiting Professor.

Henry Faul: 1966–1981, Professor of Geophysics; 1966–1973, Chairman.

Carol Faul: 1966– , Associate.

Robert F. Giegengack: 1968–1975, Assistant Professor; 1974– , Associate Professor; 1978– , Chairman.

Arthur J. Boucot: 1968–1969, Professor.

J. Granville Johnson: 1968–1969, Assistant Professor.

John B. Lyons: 1968, Visiting Professor.

Ronald Hartman: 1968–1969, Lecturer, part-time.

Allan M. Gaines: 1969–1976, Assistant Professor.

Gunther Wagner: 1969–1971, Instructor.

Subir K. Banerjee: 1969–1971, Lecturer, part-time.

Irving Friedman: 1970– , Adjunct Professor.

R. Ian Harker: 1970–1971, Visiting Professor; 1971– , Professor; 1973–1978, Chairman.

Rene Eastin: 1970–1972, Post Doctoral Fellow.

Charles Schnetzler: 1970–1976, Adjunct Associate Professor.

Jan Burchart: 1971–1972, Visiting Associate Professor.

Charles W. Thayer: 1971–1972, Lecturer; ;1972–1978, Assistant Professor; 1978– , Associate Professor.

Mary Emma Wagner: 1972– , Lecturer.

Hermann W. Pfefferkorn: 1973–1974, Visiting Lecturer; 1974–1979, Assistant Professor; 1979–1983, Associate Professor; 1983– , Professor.

Kenneth Foland: 1973–1980, Assistant Professor.

Elizabeth K. Ralph: 1973–1985, Research Associate.

Lawrence H. Soderbloom: 1973–1976, Associate Assistant Professor.

James G. Bockheim: 1973–1975, Adjunct Assistant Professor.

Michael Welland: 1974, Lecturer.

Clive M. Barton: 1975–1976, Lecturer.

H. N. Kritikos: 1975–1978, Adjunct Associate Professor; 1978–1983, Adjunct Professor.

Grant M. Skerlac: 1975, Lecturer.

Peter Dodson: 1975–1981, Adjunct Assistant Professor; 1981–19 , Adjunct Associate Professor.

Arthur H. Johnson: 1975–1980, Adjunct Assistant Professor; 1980-19 , Adjunct Associate Professor.

Yoshikazu Ohashi: 1976–1984, Assistant Professor.

Robert A. Weeks: 1976– , Adjunct Professor.

Paul W. Levy: 1976– , Adjunct Professor.

David Gaskell: 1977–1982, Adjunct Associate Professor.

Timothy M. Lutz: 1980–1982, Lecturer; 1982– , Assistant Professor.

Barry P. Kohn: 1982–1984, Visiting Professor.

Stephen P. Phipps: 1982–1984, Lecturer; 1984– , Assistant Professor.

Leo J. Hickey: 1982– , Adjunct Professor.

Joseph C. Liddicoat: 1983, Lecturer.

REFERENCES CITED

Barton, B. S., 1807, A discourse on some of the principal desiderata in natural history . . .: Philadelphia, Denham and Towne, 3 p.

Bowen, N. L., Cushman, J. A., and Dickerson, R. E., 1941, Shifting sea floors and coast lines: University of Pennsylvania Bicentennial Conference, Philadelphia, University Press, 30 p.

Browne, P. A., 1826, An address . . . to Promote a Geological and Mineralogical Survey . . .: Philadelphia, printed by P. M. LaFourcade, 8 p.

Cheyney, E. P., 1940, History of the University of Pennsylvania 1740–1940: Philadelphia, University of Pennsylvania Press, 461 p.

Darlington, W., 1849, Letter from John Bartram to John Garden, March 14, 1756, in Memorials of John Bartram and Humphry Marshall: Philadelphia, 585 p.

Dowlin, C. M., ed., 1940, The University of Pennsylvania today: Philadelphia, University Press, p. 20–21.

Fairbanks, H. R., and Berkey, C. P., 1952, Life and letters of R. A. F. Penrose, Jr.: Geological Society of America, 765 p.

Hall, J., 1894, James Hall letter to Joseph Willcox, May 28, 1894, University of Pennsylvania Department of Geology Archives.

Hindle, B., 1956, The pursuit of science in revolutionary America 1735–1789: Chapel Hill, University of North Carolina Press, 410 p.

Hornberger, T., 1945, Scientific thought in the American Colleges 1638–1800: Austin, University of Texas Press, 108 p.

Jordan, W. M., 1984, J. Peter Lesley (1819–1903), pioneer in the geology of Pennsylvania anthracite: Neuvième Congrés International de Stratigraphie et de Géologie du Carbonifère, Compte rendu, v. 1, p. 121–127.

Maclure, William, 1809, Observations on the geology of the United States, explanatory of a geological map: American Philosophical Society Transactions, v. 6, p. 411–428.

Morton, S. G., 1829, Geological observations on the Secondary, Tertiary, and Alluvial formations of the Atlantic coast of the United States: Academy of Natural Sciences of Philadelphia Journal, v. 6, p. 59–71.

Merrill, G. P., 1969, Letter from James Hall to William A. Ingham, June 8, 1874, in The first one hundred years of American geology: New York, Hafner, p. 693–694.

Newell, L. C., 1976, Chemical education in America from the earliest days to 1820: Journal of Chemical Education, v. 53, no. 7, p. 402–407.

Odgers, M. M., 1947, Alexander Dallas Bache: scientist and educator, 1806–1867: Philadelphia, University of Pennsylvania Press, 223 p.

Osborn, H. F., 1931, Cope: master naturalist: Princeton University Press, 740 p.

Penrose, R. A. F., Jr., 1918, Memorial of Amos P. Brown: Geological Society of America Bulletin, v. 29, p. 13–17.

Phillips, H., Jr., comp., 1884, Early proceedings of the American Philosophical Society 1744–1836: American Philosophical Society Proceedings, v. 22.

Pomerantz, M. A., 1976, Benjamin Franklin—the complete geophysicist: EOS, v. 57, p. 492–505.

Reingold, N., 1964, Science in nineteenth century America: New York, Hill and Wang, 339 p.

Roberts, S. W., 1875, An obituary notice of Charles B. Trego: American Philosophical Society Proceedings, v. 14, p. 356–358.

Rogers, E. S., ed., 1896, Life and letters of William Barton Rogers: Boston, Houghton Mifflin Co., 2 vols., 427 p. and 451 p.

Rogers, H. D., n.d., A Guide to a Course of Lectures . . .: in author's collection, p. 1.

Say, T., 1818, Observations on some species of Zoophytes: American Journal of Science, v. 1, p. 381–387.

Sellers, C. C., 1969, Charles Willson Peale: New York, Charles Scribners' Sons, 370 p.

Simpson, G. G., 1942, The beginnings of vertebrate paleontology in North America: American Philosophical Society Proceedings, v. 86, p. 130–188.

Sinclair, B., 1972, The promise of the future in G. H. Daniels, ed., Nineteenth-century American science: Evanston, Northwestern University Press, p. 249–272.

Sinclair, B., 1974, Philadelphia's philosopher mechanics: Baltimore, Johns Hopkins Press, 353 p.

Smallwood, W. M., 1941, Student notebook dated 1805 in the Library of Congress, in Natural history and the American mind: New York, Columbia University Press, p. 291.

Smith, E. F., 1914, Chemical Society of Philadelphia notice in *Medical Repository* 3, 1800, p. 68, in Chemistry in America: New York, D. Appleton and Co., p. 46–47.

Smith, E. F., 1918, James Woodhouse letter to Rev. James Hall in *Medical Repository,* 1798, *in* James Woodhouse: a pioneer in chemistry, 1770–1809: Philadelphia, John C. Winston Co., p. 84–90.

Smith, E. F., 1919, Chemistry in old Philadelphia: Philadelphia, J. B. Lippincott Co., 106 p.

Smyth, A. H., 1905–1907, Benjamin Franklin letter to Jered Eliot, July 16, 1747, *in* The writings of Benjamin Franklin: New York, v. II, p. 312.

Smyth, A. H., 1905–1907, Benjamin Franklin letter to Abbe Soulavie, September 22, 1782, *in* The writings of Benjamin Franklin: New York, v. VIII, p. 598–601.

Thorpe, F. N., 1893, Benjamin Franklin and the University of Pennsylvania: Washington, Bureau of Education, circular 2, 450 p.

University of Pennsylvania Announcement, 1872, Organization and courses of study of the new Department of Sciences, 22 p.

University of Pennsylvania Archives, 1824?, Curriculum list.

University of Pennsylvania Archives, April 9, 1909, Newspaper articles in *The Public Ledger* and *The Evening Bulletin.*

University of Pennsylvania catalogues and directories, 1829–1984.

Geological Society of America
Centennial Special Volume 1
1985

A chronology from Mohole to JOIDES

Elizabeth N. Shor
2655 Ellentown Road
La Jolla, California 92037

ABSTRACT

In the late 1950s, when the competition was strong among oceanographic institutions, an idea arose that was larger than any of them could handle alone. This was the concept of drilling a hole through the crust of the earth to the mantle that came to be called the "Mohole Project." It was a radical approach: to make a gigantic leap forward in exploration of the earth's interior through the sea floor, going far beyond the technology of the day to the limits of what was possible, but not tested and proven. Although the project acquired funding early, the impetus was lost when an unexpected contractor was selected with whom the originators did not work smoothly.

Simultaneously, some geologists were urging the drilling of many deep sedimentary sites in the ocean. The geologic community was divided over the merits of drilling a single hole to the mantle *versus* drilling a number of shallower holes. The Mohole Project was cancelled by Congress as a line item when cost estimates soared. The program to drill at many ocean sites survived, briefly as LOCO and CORE, and finally in 1964 as JOIDES, the scientifically productive Joint Oceanographic Institutions Deep Earth Sampling.

INTRODUCTION

"The ocean's bottom is at least as important as the moon's behind," said Athelstan F. Spilhaus in the 1950s.[1] Oceanographers then were becoming annoyed at the costs of the burgeoning space program. Such feelings may have led them to think big and to converge on the idea of drilling a hole to the mantle of the earth.

Prior to 1950, the few geologists who thought about the sea floor believed that sediments had accumulated and remained undisturbed in the ocean basins. In those basins should be many kilometers of sediments that recorded the geologic history of the earth since the oceans had been formed. A few geophysicists were pondering the nature of the earth's crust as determined from earthquake records; was it like that under the continents, or was it thinner, or was it quite different? Seismic-refraction studies at sea, which began in the late 1940s and early 1950s chiefly by Maurice Ewing, J. B. Hersey, Maurice Hill, and Russell W. Raitt, revealed some unexpected but consistent observations:

(1) Material with velocities appropriate to sediments or clastic sedimentary rocks was, on the average, only about 300 meters thick, an amount of sediment that would have accumulated only since the beginning of the Cenozoic era. To geophysicists this was Layer 1.

(2) Beneath those materials that were clearly sedimentary was a horizon with rather variable seismic velocities that might be appropriate either to consolidated sediments or to extrusive volcanics. This was called Layer 2.

(3) The oceanic crust was thin and had a uniformly high seismic velocity, which was significantly different from that of continental rocks. This was called Layer 3.

(4) The Mohorovičić discontinuity (the demarcation between the crust and the mantle) was much shallower beneath the ocean basins than beneath the continents.

Certainly, some classical geologists had recognized that only by very deep drilling could a true picture of the earth's interior be acquired (summarized in Hess, 1959; Bascom, 1961:35–47). Charles Darwin wrote to Alexander Agassiz in 1881 of his yearning to have cores from ocean atolls. G. K. Gilbert proposed a deep boring into plutonic rock in 1902. T. A. Jaggar, founder of the Volcano Observatory in Hawaii, had proposed to Richard M. Field in 1943 that, after the war ended, a group effort should be made "to drill one thousand core-producing holes in the deep ocean bottoms each one thousand feet deep . . . with worldwide drill-hole distribution." In 1953, Maurice Ewing, speaking in a Distinguished Lecturer series, promoted drilling a hole through the oceanic sediments. Frank. B. Estabrooke presented a plea in a letter to *Science* on October 12, 1956, for "a geophysical penetration of the earth's crust."

So much for "prehistory." This account is for the generation of marine geologists who did not know the world before JOIDES. It is derived from contemporary minutes and correspondence, *not* from reminiscences of the participants.[2]

MOHOLE'S EARLY SUCCESS

The story begins on March 23, 1957, at the end of a meeting of the Earth Science review panel of the National Science Foundation. Hess (1959) said:

Many of the suggested projects were excellent and most were of high caliber. Walter Munk remarked, however, that not a single one of them was of such a nature that a really major advance in Earth Science would result. He suggested the panel invent a project which might strike directly at the roots of a major problem and forthwith suggested a hole to the Moho[rovičić discontinuity]. If the writer [Hess] deserves any credit for past or future association with this project, it is that he took Munk seriously and prevented momentarily adjournment of the panel. In the few minutes remaining he proposed referring the project to the American Miscellaneous Society (AMSOC) for action. This was not a joke. It was done for several very good reasons.

AMSOC has no constitution, no officers, no members, only founders, many of whom are distinguished scientists. Here was a society which could act immediately—no need to wait for the next council meeting to have the proposal referred to a committee which would report to the council a year later.

The American Miscellaneous Society had been christened by Gordon G. Lill and Carl O. Alexis of the Geophysics branch of the Office of Naval Research in the summer of 1952, over a pile of proposals to that office that could be categorized no more closely than one at a time; i.e., they were all miscellaneous. Oceanographic geophysicists comprise most of its claims to be founders.

A few AMSOC participants met at Munk's house in La Jolla, California, on April 20, 1957, for breakfast, and discussed drilling a hole to the Mohorovičić discontinuity. As Willard Bascom (1961, p. 48) recalled,

They were not certain about the minimum depths to the Moho or of the maximum depths that had been reached in the search for oil, so they could not even make a good guess whether or not such a hole was possible. What they could do was talk about past experience and who should be consulted and what such a hole might find.

In short order they agreed on a "deep-drilling committee," of which Gordon G. Lill (Geophysics branch of ONR) was to be chairman, and the members were Joshua I. Tracey and Harry S. Ladd (U.S. Geological Survey), Roger R. Revelle and Walter H. Munk (Scripps Institution of Oceanography), and Harry H. Hess (Princeton). Scripps Director Revelle wasn't present, so "that

afternoon a delegation called on [him] to inform him about the grand new idea that had blossomed on his campus" (Bascom, 1961:49).

A week later the new committee met at the Cosmos Club in Washington, D.C.; they added William W. Rubey (U.S. Geological Survey) to their number, and when Maurice Ewing of Lamont (now Lamont-Doherty) Geological Observatory chanced by, they invited him to join. They agreed to ask the National Science Foundation for funds for a feasibility study, and on July 15 chairman Lill submitted proposal P-3831 to that agency, requesting $30,000.

The concept was moving quickly in various circles, which illustrates the impact of the geologists/geophysicists who formed that original group. The International Association of Physical Oceanography and the International Association of Seismology and Physics of the Earth's Interior passed resolutions in favor of the deep-drilling project in September 1957. Even more significant was the endorsement by the International Union of Geodesy and Geophysics that month at its meeting in Toronto, Canada. Bascom (1961:49-50) said:

There, in several of the discussion groups, the subject of deep drilling arose again, prodded by AMSOC members and by Dr. Tom Gaskell, a British geophysicist. Finally on September 14, 1957, resolution number eleven was adopted.

It read:

The I.U.G.G.
Considering that the composition of the earth's mantle below the Mohorovičić discontinuity is one of the most important unsolved problems of geophysics,
And that, although seismic, gravity and magnetic observations have given significant indications of the nature of this material, actual samples that could be examined petrographically, physically and chemically are essential,
And that modern techniques of drilling deep wells are rapidly developing to the point where drilling a hole ten to fifteen kilometers deep on an oceanic island may well be feasible,
And that the crustal material above the Mohorovičić discontinuity is also of prime interest
Urges the nations of the world and especially those experienced in deep drilling to study the feasibility and cost of an attempt to drill to the Mohorovičić discontinuity at a place where it approaches the surface.

This document opened the door to international competition, for Soviet scientists declared that they were considering such a project—and American scientists were smarting over the first successful man-created satellite, U.S.S.R.'s Sputnik.

On December 6, 1957, AMSOC participants decided "that the problem of drilling to the mantle should be broken down into three parts: (1) a 'practice' hole on the continent to 35,000 feet (an idea that was soon discarded), (2) a 'sedimentary-section' hole in the deep ocean basin, (3) a mantle hole beneath the ocean" (Bascom, 1961:51).

The National Science Foundation (NSF), according to Hess (1959), found itself unable to grant funds to a society that had no officers or a membership list. Hess approached the National Academy of Sciences-National Research Council (NAS-NRC), which "very kindly, and, I might say, courageously, gave the AMSOC committee a respectable home" on December 8, 1957. Proposal P-3831 was withdrawn, and in April 1958, the Division of Earth Sciences of NAS-NRC submitted proposal P-4773 to the National Science Foundation, requesting $30,000 toward a feasibility study. That division formally established the AMSOC Committee, consisting of Gordon G. Lill (ONR, chairman), Carl O. Alexis (ONR), Maurice Ewing (Lamont), Harry H. Hess (Princeton), Harry S. Ladd (U.S.G.S.), Arthur E. Maxwell (ONR), Walter Munk (Scripps), Roger Revelle (Scripps), William W. Rubey (U.S.G.S.), and Joshua I. Tracey (U.S.G.S.).

At the annual technical conference of the Earth Sciences Division of NAS-NRC on April 26, 1958, about 75 scientists gathered at the invitation of Hess and Lill to hear and discuss the idea of a feasibility study and eventually a series of drilling tests. According to Lill and Maxwell (1959), the arguments presented against the project were:

(i) The cost will be so great that funds will be drained from other worth-while work in the earth sciences. (ii) One hole will not suffice. Since the mantle is not homogeneous, it will be necessary to drill many holes to determine its nature. In this case, drilling costs will become prohibitive. (iii) It is known from geochemical experiments that a phase change occurs in the physical nature of rocks under the conditions of pressure and temperature which must exist in the region of the Mohorovičić Discontinuity. (iv) It is known from laboratory experiments that the mantle is composed of peridotite, eclogite, or dunite. (v) The publicity given large projects is bad for science.

Bascom noted that another objection was that the capability of deep-ocean drilling did not then exist. However, A. J. Field (Union Oil Company) showed movies of the ship *Cuss 1* with a full-sized oil rig drilling at sea. Bascom wrote (1961, p. 53–54), "Almost until that moment the capabilities of floating drilling platforms had been kept closely guarded commercial secrets and virtually no one present had seen or even heard of such equipment before."

Those present at the conference unanimously endorsed a deep-drilling project.

The National Science Foundation granted $15,000 to NAS-NRC for the feasibility study on June 26, 1958, and two weeks later the AMSOC Committee asked Willard Bascom to become its executive secretary on a part-time basis; he was then on the staff of the National Academy of Sciences. The term *Mohole* surfaced about this time; it appeared first in print in the "Science and the Citizen" column of *Scientific American* in December 1958, and second as the title of an article by Bascom in *Scientific American* in April 1959.

Several panels of the AMSOC Committee were defined and appointed from the fall of 1958 to the spring of 1959: Drilling (later Drilling Techniques) Panel (William B. Heroy, Sr., chairman); Site Selection Panel (Harry H. Hess, chairman); Scientific Objectives and Measurements Panel (Harry S. Ladd, chairman).

At a June 1959 meeting of the executive board, a tentative budget of $14 million dollars for the entire project was established, and it was decided that the AMSOC Committee should manage the project—not merely advise. The latter concept was accepted by the Governing Board of NAS.

In mid-1959 oceanographers were still congratulating themselves on their participation in the International Geophysical Year of 1957–58, and they were eyeing the nearly unstudied Indian Ocean. They gathered 800 strong in New York City at the First International Oceanographic Congress at the end of August 1959, where the concept of drilling to the mantle was formally announced to reporters. Simultaneously it was presented in NAS-NRC Publication 717: "Drilling Thru the Earth's Crust."

Shortly after that, Hess (1959, p. 254–255) redefined the objectives of what he called "the AMSOC hole to the earth's mantle":

In 1850 Boisse proposed that the Earth's interior might be analagous in composition to meteorites. It was a brilliant proposal for its day and in a general way it is probably correct. In detail, perhaps, it is today being taken too seriously. . . .

The uncertainties in the meteorite analogy cannot be resolved by further speculation. A hole or holes are required if we are to know something more definite about the chemistry of that 84 per cent of the volume of the Earth called the mantle. . . .

A half-ton sample of mantle would probably give more specific information than the 1400 odd meteorites now in collections. . . . It seems foolhardy to put an enormous effort into attempts to sample the moon or even the planets without finding out what is 5 km below the sea floor. The information from the hole is necessary to attack in a well-reasoned manner what we wish to find out about the moon. Compared to space exploration the cost is small.

EXPERIMENTAL MOHOLE

Things were moving fast: in October, Bascom (1959) described a dynamic-positioning system for a drilling ship (which has proven highly successful), and in November the Drilling Techniques panel concluded that the drilling ship *Cuss I* would be a suitable vessel for experimental holes. That ship was owned by a consortium (Global Marine Exploration Company) that included Continental, Union, Shell, and Superior oil companies—hence the ship's name. Although it had what John Steinbeck (1961) called "the sleek race lines of an outhouse standing on a garbage scow," it had successfully drilled in 200 feet of water.

NSF in 1960 increased the funds available to the Mohole Project, and by October Bascom was its full-time director. In February 1961 *Cuss I* was modified, at a cost of one and a half million dollars, to drill in deep water. The modifications included the first use of Bascom's dynamic-positioning system, in the form of "four 200-horsepower outboard motors located around the hull and operated by a central joystick which activated them to compensate for shifts caused by wind and currents" (Greenberg, 1964:118).

The first test hole, in March, was drilled 25 miles off La

Jolla, California, and it broke records: the 3,111 feet of water over which the ship floated was 10 times greater than water depths previously attempted. Five holes were drilled in two days, the deepest to 1,035 feet in sediments of the San Diego Trough (Shor, 1978:304).

Next came the Experimental Mohole, for which *Cuss I* was towed slowly and laboriously by a tug to a site 40 miles off Guadalupe Island, Baja California, Mexico. The Guadalupe site had been selected on the advice of two of the members of the Site Selection Panel, George G. Shor, Jr., and Russell W. Raitt (Scripps Institution), as a geologically interesting spot in an accessible range for the drill ship (which was based in Los Angeles) but well off shipping lanes. It was hoped that the drilling could reach the "second layer," a hard reflecting horizon presumed to be volcanic. William R. Riedel (Scripps Institution) was in charge of the scientific program.

Cuss I held its position in waves to 14 feet high and wind gusts to 20 knots. In 11,672 feet of water, it took half a day for the drill pipe to reach the floor of the ocean before it could even begin drilling. On April 2, 1961, the drill bit reached the "second layer" below 600 feet of sediments and recovered a 44-foot core of basalt (AMSOC Committee, 1961b). President John F. Kennedy telegraphed his congratulations to the National Science Foundation.

The President's Scientific Advisory Committee urged drilling the ultimate hole, but in mid-June 1961 the Mohole Committee and its panels began the long debate: an intermediate ship versus an ultimate ship. For example, on June 12 the Drilling Techniques Panel agreed that the ultimate ship should be developed immediately, and on June 13 the Mohole Committee strongly recommended that "an intermediate drilling program be initiated during fiscal year 1962" (AMSOC Committee, 1961a).[3] At the same time the committee "agreed that the major scientific objective of Project Mohole is to drill to the earth's mantle through a deep ocean basin." It also recommended that "the AMSOC Committee should, in the future, concern itself with matters of scientific policy, engineering review, and budget," that "the operational and engineering future of Project MOHOLE be entrusted to a prime contractor," and that "the AMSOC Committee desires to retain the staff intact on the project."

Bascom advocated in a memo of October 1, 1961, to the AMSOC Committee and its panels that "the proper way to proceed is to build an experimental drilling ship of modest proportions . . . the intermediate ship concept." From its testing, with continuous improvements in design and construction, the ultimate ship would be developed. A Naval Architecture Panel was formed in October 1961, chaired by Harold E. Saunders, and it too joined the debate over one ship or two.

With the major program about to go out to bid, two significant people resigned from the AMSOC Committee for conflict of interest because of their organization's possible intent to bid: Chairman Gordon Lill, who was an advisor to Lockheed Aircraft Corporation, and Roger Revelle, director of Scripps Institution. Harry Ladd served as acting chairman from July 19 to December

9, 1961, when Hollis D. Hedberg, of Gulf Oil Corporation and Princeton University, assumed the chairmanship.

The committee knew of 11 organizations that had expressed interest in bidding. Requests for bids to "provide the necessary services, facilities, materials, and equipment for the Mohole Project" were sent to those companies and to other possible bidders on July 27, 1961. Twelve organizations replied that they wanted to be considered. The AMSOC Committee had no part in the final bid selection. By the end of December, a specially appointed NSF selection panel had narrowed the bidders to five, ranked as follows:

Socony Mobil Oil Company
Global-Aerojet Shell
Zapata Off-Shore Company
General Electric Company
Brown & Root, Inc.

On February 28, 1962, NSF announced the selection of Brown & Root, of Houston, Texas, as the successful bidder. (Lill and Revelle returned to the AMSOC Committee at this time.)

Competition was arising in another aspect: the Soviet Union and Canada separately announced plans to drill five deep holes on land. These rumors shook up the planners of Mohole; shouldn't they be moving faster?

For those who like a choice of endings, you may proceed with Mohole, Concluded; or LOCO and CORE.

MOHOLE, CONCLUDED

What were the qualifications of Brown & Root, a construction and engineering firm, for this new undertaking? It had not been on the original invitation list from NSF for bids, but it did submit a proposal within the deadline. It moved rapidly on the selection list from fifth to first, with—at the end—such criteria considered as "competence and policy factors" and whether an oil company should be selected. "Talk about oil companies grew like Topsy. We early discussed whether an oil company would deflect the project from its scientific objectives," said William E. Benson, head of NSF's earth sciences section (Solow, 1963:198). Solow (1963) and Greenberg (1964:226) pointed out the close association between the chairmanship (Albert Thomas of Texas) of the House appropriations subcommittee that reviewed the NSF budget and the state in which Brown & Root was located.

The selection of Brown & Root was challenged by Senator Thomas Kuchel of California, but the Comptroller General, after investigating, stated that "While the records are not as clear as might be desired . . ., [we] are unable to conclude that the award to Brown & Root was not in the public interest" (Greenberg, 1964:225).

The contract signed to Brown & Root by NSF on June 20, 1962, stated:

This project . . . has as its ultimate aim the drilling of a hole to the Mohorovičić discontinuity. . . .It may prove desirable to expand this broad scope of work . . . to include other geophysical surveys, additional shallower holes in other selected oceanic or continental areas....If such is deemed advisable by the Foundation . . . it would be accomplished through subsequent agreement with the contractor (Greenberg, 1964:225).

Cost estimates by NSF at the time of selection of Brown & Root were $35 to $50 million dollars, to reach the Mohorovičić discontinuity in three to seven years (National Science Foundation, 1962).

Various scientists who had been early participants in Mohole quickly became disenchanted with the selected contractor. Bascom and several other members of the AMSOC Technical Staff formed an independent corporation, Ocean Science and Engineering Corporation (OSE), which they hoped would become a consultant or subcontractor to the contractor or to NSF. The new firm received a 60-day subcontract from Brown & Root, which was not renewed. Soon afterward several of the OSE members received a one-year contract with NSF for engineering tasks and other duties assigned by the managing coordinator of NSF for Mohole, C. Don Woodward.

The outspoken chairman of the AMSOC Committee from December 9, 1961, Hollis D. Hedberg (1962), announced his position:

> It seems to me that the over-all ultimate purpose of the project can be simply stated as *to contribute to the determination of the nature and characteristics of the as yet unknown portions of the Earth's Crust and Mantle.* . . .I should perhaps reiterate that the drilling to the Upper Mantle—the Mohole—should be the central theme of the project because it is the ultimate goal.

He advocated two ships: one for oceanic drilling in many areas, the other for the Mohole or in fact moholes. Not all members of the AMSOC Committee agreed with him, and some feared that the ultimate goal would be lost as shallower drilling used up the impetus and funds of the project. For a vessel for intermediate drilling, the Naval Architecture Panel had recommended a C1-M-AV1, a cargo ship 338 feet long; a number of them had been built for World War II under the Maritime Administration, so a choice among those was available.

Creighton Burk (then completing his Ph.D. at Princeton) became the AMSOC Scientific Officer on August 1, 1962. He strongly favored an intermediate ship as necessary for the Mohole Project.

The Site Selection Committee pursued its original charge to find a suitable location for drilling to the Mohole: The site must be where the Mohorovičić discontinuity was known to be shallow but still of "average" characteristics, not too far from a well supplied port, and in a region of long-term reasonable weather for what was expected to require three years of drilling. In the spring of 1962, the sites under consideration were off Puerto Rico, the Barracuda fault (200 miles east of the Lesser Antilles, east of the south terminus of the Puerto Rico Trench), and somewhere near the Hawaiian Islands. At the request of the AMSOC Committee, a 1,000-foot hole into serpentinite was drilled near Mayaguez,

Puerto Rico, in October 1962, to provide information on rocks similar to the deep crustal rocks that could be expected just above the mantle (Burk, 1964).

Hedberg established the Oceanic Sediments Drilling Panel of the AMSOC Committee in September 1962, with Maurice Ewing as chairman. Hedberg also soon expressed his concern that the prime contractor had not produced tangible accomplishments by early 1963, nor had Brown & Root accepted and commenced plans for the intermediate drilling program. President of the NAS Frederick Seitz commented that he "hoped that in spite of the many problems which are arising, the Committee will be steadfast and unified in working toward a solution" (AMSOC Committee, 1963a). Seitz also took note of the rising cost estimates: Brown & Root had estimated $68 million for drilling the ultimate hole. The head of the Bureau of the Budget also expressed concern over the rising estimates and on March 1, 1963, recommended to NSF "a thorough reappraisal of the second phase of Project Mohole."

The role of the AMSOC Committee was reassessed by NSF and NAS-NRC, and it was agreed in mid-March 1963, that the committee would study all scientific aspects of the project and make specific recommendations to NSF, which would be subject to final review and approval by that foundation. A meeting of the committee on May 11, 1963, was critical. Dissension was appearing in the entire project to the extent that William E. Benson (NSF) felt it necessary to urge "that we express our differences of opinion openly among ourselves but not in the corridors," for the sake of the Mohole, the National Science Foundation, and all large U.S. science projects (AMSOC Committee, 1963b).

At that meeting Mohole-originator Walter Munk noted that "after an auspicious beginning, Project Mohole has passed through two years of administrative difficulties that seem to dwarf the technical difficulties." Brown & Root had, he said, "strongly objected to the desire for, the necessity for, or the usefulness of an intermediate vessel."

Munk, Revelle, and two other committee members (Maurice Ewing and J. Brackett Hersey) left the meeting briefly because of conflict of interest (see LOCO and CORE below); while they absented themselves, the rest of the committee voted unanimously that NSF consider another prime contractor for the intermediate program of the Mohole Project. The drilling programs that would become LOCO, CORE, JOIDES were entering into the Mohole Project. At the May 11 meeting, these items were noted:

> There was concern expressed over the apparent lack of depth in understanding by the Prime Contractor of all the factors associated with the Project. . . .many important aspects of the Project are still in the rough concept stage . . . yet no real engineering plan, and an over-all budget has not been available for us to consider.

In an amicable end that day, however, the committee voted that Brown & Root should continue the design and engineering for an ultimate vessel to drill to the Mohorovičić discontinuity.

In another confrontation, Bascom's company, OSE, reviewed the Engineering Plan Report of Brown & Root in a report of June 1, 1963, and found it "neither a clear plan nor a sound basis for proceeding."

Hedberg was still searching for a consensus among his AMSOC Committee, and in August 1963, reported on his poll of members as to their preference for an intermediate drilling vessel for deep-ocean cores *versus* a vessel to drill solely to the Mohorovičić discontinuity. Twelve members voted for the intermediate vessel first, with taking chances on getting an ultimate vessel later; five favored the ultimate vessel first, with taking chances on getting the intermediate vessel later; two did not return their ballots.

Alan T. Waterman, who had been the only director of NSF from its founding in 1950, stepped down and was replaced by Leland J. Haworth on July 1, 1963. Waterman's decision must have been somewhat affected by the complications of the Mohole Project. Haworth gave his opinion to the AMSOC Committee in September: the Mohole Project had not begun in the right way; one should feel one's way in scientific advance and take steps that one is reasonably sure of. He noted that cost "looms largest to the Federal Executive Branch," also that

he and Dr. [Jerome] Wiesner [the President's Science Advisor] believe that [the AMSOC Committee] probably will have to think of ocean drilling as two programs—those things that lead directly to the very deep drilling and mantle piercing and a more general shallower drilling program with its own scientific objectives (AMSOC Committee, 1963c).

Haworth recommended that there should be "some mechanism in which the operating responsibility for this program can be placed": a single university or a group of universities. He observed that the Mohole Project was the only line item of NSF; he advised that general drilling should be divorced from it, and put in its own program of supporting good research, and so not debated "as a specific item with its advocates and its enemies in the Congress."

AMSOC Committee chairman Hedberg continued to espouse an intermediate program and ship, and was sharply criticized by NAS President Frederick Seitz on October 30, 1963, for his statements before the House of Representatives Subcommittee on Oceanography of the Committee on Merchant Marine and Fisheries (Lomask, 1976: 195). Hedberg resigned on November 8.

Seitz, in January 1964, advocated that a new committee replace the AMSOC Committee for the scientific aspects of oceanic drilling. This was voted favorably by that committee and the new organization was named the National Academy of Sciences Committee on Oceanic Drilling, NASCOD. Gordon G. Lill, the original chairman, accepted the newly created position of project director of Mohole for NSF on February 10, 1964. Creighton Burk resigned as scientific officer.

Brown & Root continued with plans for the ultimate drilling. A news release of January 21, 1964, by NSF announced that a large drilling platform, designed by Brown & Root, would be built, which would first be equipped and used for intermediate drilling "prior to its use for the deep hole through the earth's crust . . . to the mantle." Finally, on January 27, 1965, NSF presented the spot chosen by the Site Selection Committee for the first attempt to reach the mantle: 120 miles north-northeast of Maui, Hawaii.

On September 28, 1965, NSF announced that Brown & Root would award a contract to National Steel and Shipbuilding Company in San Diego for a drilling platform to reach the Mohole; the vessel was expected to be completed early in 1968. Bids on the platform had been higher than expected. But the Executive Committee of the National Science Board had adopted a resolution on August 4, 1965, favoring the continuation of the Mohole Project in spite of rising costs. From the rough estimate of $14 million dollars by the AMSOC Committee in 1959, the estimate in 1966 had risen to $127.1 million.

The first legislative move against Mohole was by the Committee on Appropriations of the House of Representatives, which on May 6, 1966, denied a request from NSF for $19.7 million to continue construction of the drilling platform. (Chairman Albert Thomas, who had championed the project, had died that February.) An appeal to the Senate was "immediately mounted by a handpicked rescue party" (Greenberg, 1966).

Emiliani (1981:1713) finished the story:

A subcommittee of the Senate Committee on Appropriations restored the funds, but on August 18, 1966, the full House of Representatives . . . voted 108 to 59 not to go along with the restoration of funds. . . . Project Mohole was dead. A total of $25,100,000 had been spent.

LOCO AND CORE

The recommendation of June 13, 1961, by the AMSOC Committee that "an intermediate drilling program be initiated during fiscal year 1962" opened a door to geologists with interests in seafloor sediments. So did the success of the Experimental Mohole in April 1961.

Hedberg and Burk (1964) later summarized their own point of view, which represented part of the geological/geophysical community of the time. After noting that "one Mohole would in itself provide little basis for broad generalizations about the earth's mantle," they continued:

The direct information we now have about the composition and structure of the oceanic crust has been obtained from a very limited area at the ocean margins, from a few oceanic islands, and from scattered dredged samples and cores of the upper few scores of feet of sediment. But much additional and very valuable indirect information has been gathered by oceanographic institutions from topographic, seismic, gravity, magnetic, and heat-flow studies in the oceans.

This mass of indirect geophysical data (as well as the cost of getting it) is enormous and growing rapidly—yet we are not significantly better equipped to interpret these data now than we were many years ago. The *one* single thing that the earth sciences need now more than anything else to reduce the subjectivity and speculation in interpretation of such geophysical data, is a sound program of widespread drilling in the ocean crust—to measure the actual properties of the rocks and fluids in the holes and, most important of all, to recover samples of the crustal rocks. . . .

It is only from the rocks that we can determine such critical geological properties as geologic age, fossil content, mineralogy, stratification, thermal conductivity, chemical composition, and remanent magnetization, as well as the equally important lateral (geographic) and vertical (time) variations of these properties within the crust.

Of course, in the long-distant end, it is the answers to basic questions about the earth and life upon it which interest us—answers that only knowledge of the combined chemical, physical, and geological properties of the rocks can lead us to.

The first to take action on intermediate drilling was Cesare Emiliani of the University of Miami, who yearned for sediment cores longer than the 25-meter ones that could be obtained with the piston corer (Emiliani, 1981). He approached NSF in February 1962, about the possibility of funding for obtaining long cores, and was encouraged to organize an interinstitutional working group. John Lyman and William E. Benson of NSF consistently urged that the several oceanographic laboratories put together a joint proposal for operating a sediment-coring ship. They suggested corporate entities similar to Associated Universities, Inc. (which operate Brookhaven), Associated Universities for Research in Astronomy (which operate Kitt Peak Observatory), and Universities Corporation for Atmospheric Research (which operate the National Center for Atmospheric Research).

Emiliani drafted an "Outline of a Research Program on Deep-sea Sediments," under the name LOCO Committee, with himself as chairman. The proposal was to contract with Global Marine Exploration Company for a drilling vessel to occupy four stations in the Venezuelan Basin and four in the Nares Basin (southeast of Bermuda in the north Atlantic). The expected yield would be five to ten kilometers of core material. Emiliani circulated the outline widely for comments, suggestions, and possible participation.

Sedimentary geologists at other institutions responded favorably and immediately pointed out other regions of interest to themselves. For example, Riedel at Scripps suggested a dozen holes in the first three years, half to be in the Atlantic and Caribbean, half in the Pacific.

In Memorandum No. 1 of the LOCO Project,[4] Emiliani (1962a) identified LOCO as an acronym for Long Cores, and allowed that taking cores in the Pacific would "considerably reduce the prospects of success for the LOCO Project."

Because of the Mohole Project, the NAS Committee on Oceanography and the AMSOC Committee sponsored an ad hoc meeting in Miami on June 10, 1962, with representatives from many interested oceanographic institutions and agencies. The group appointed a committee, consisting of James R. Balsley (U.S. Geological Survey), K. O. Emery (Woods Hole Oceanographic Institution), Cesare Emiliani (University of Miami), Maurice Ewing (Lamont Geological Observatory), Hollis D. Hedberg (AMSOC), Bruce C. Heezen (Columbia University), Harry H. Hess (Princeton), Fritz F. Koczy (University of Miami), William R. Riedel (Scripps Institution of Oceanography), and

George G. Shor, Jr. (Scripps). Emiliani was elected chairman of the LOCO Committee.

The group decided that day that the LOCO Committee should be independent of the AMSOC Committee and that the LOCO Project be independent of the Mohole Project, but that there should be overlapping membership. Thus, Hedberg, as chairman of the AMSOC Committee, became a member of the LOCO Committee. Maurice Ewing, who was not at the meeting, had recommended on the previous day that LOCO should be under the AMSOC Committee. Hedberg agreed with Ewing in a letter on July 2, and he then made several efforts to combine LOCO and Mohole. Emiliani (1981) dismissed these as "confusion," but he did become a member of the AMSOC Committee in July 1962.

The University of Miami in the summer of 1962 bought a small oceanographic ship, renamed R/V *Pillsbury,* for geophysical site studies, and submitted to NSF a proposal for one year of support for organizational and administrative costs of LOCO. This proposal was funded.

The LOCO Committee met next on September 13, 1962, at the University of Miami, at which time,

following a suggestion by John Lyman [NSF], it was decided to ask the four leading institutions (Woods Hole, Columbia University [Lamont], University of Miami, and University of California [Scripps]) to establish a nonprofit interuniversity corporation to build, own, and operate a vessel especially designed to drill and core oceanic sediments." [However], "should the Corporation fail to materialize within 3½ months (i.e. by Dec. 31, 1962), the LOCO Committee should proceed at once to make plans for using a commercially available drilling vessel" (Emiliani, 1962b).

At the September 13 meeting, eight locations for coring were agreed upon: seven in the Atlantic and Caribbean, one in the Gulf of Mexico. The LOCO Committee was not in agreement as to whether the project should build a special vessel or contract with an available one. Ewing and Riedel strongly favored building a special vessel.

Emiliani prepared a draft of a charter for a four-institution corporation, Associated Oceanographic Institutions, Inc., which by January 23, 1963, had been signed by the members of the LOCO Committee from Lamont, Scripps, and the University of Miami and by the directors of those institutions (Emiliani, 1963a), but not by the members from Woods Hole Oceanographic Institution. Director Paul M. Fye of WHOI preferred that the corporation "be formalized by an incorporation act among the Trustees, Regents, etc. of the 4 Institutions, rather than among individuals employed by these Institutions." Such a corporation could handle other major oceanographic efforts, not only the LOCO Project. Fye recommended that Scripps Institution serve as an interim grantee and administrator for LOCO.

Creighton Burk, who, as noted above, had become AMSOC scientific officer on August 1, 1962, had been attempting to draw the oceanographic laboratories together in a general drilling project that would have side benefits for Project Mohole. He agreed with Hedberg in strongly favoring the construction of an intermediate vessel.

Although Emiliani considered LOCO unrelated to Mohole, he was almost alone in that regard. The delay on a decision concerning an intermediate vessel for Mohole affected the advice by NSF to various inquiries on drilling programs during 1962 and 1963. A ground-swell movement had begun within the geologic community for "intermediate" drilling, both by paleontologists who were seeking longer sequences for age-dating and by seismologists and other geologists who wanted to know the nature of layers 2 and 3. NSF officials were encouraging the submission of all proposals for oceanic drilling, but they were not encouraging the possibility of two ships for intermediate holes. Emiliani (1962c) noted (with a "tsk, tsk") in LOCO Memorandum 14 that Maurice Ewing (i.e., for Lamont) had submitted a proposal alone to NSF for a drilling program to be carried out by a commercial contractor (it was soon withdrawn).

On January 4, 1963, Roger Revelle, Maurice Ewing, and J. Brackett Hersey (WHOI) informed Emiliani that they, as individuals, intended "to form a non-profit Corporation for the purpose of converting and operating a C1-M-AV1 vessel to be used for drilling and coring through the sediments, the 2nd layer, and possibly below" (Emiliani, 1963a).

The Articles of Incorporation of this Consortium for Oceanic Research and Exploration, Inc. (CORE), dated February 7, 1963, listed as its purposes:

to engage in geological, oceanographic and engineering research; to conduct drilling operations into the crust of the earth; to design, acquire, equip and operate vessels for such research, projects and operations; to contract with other organizations, associations, corporations, or individuals for the conduct of such research, projects and operations; to accept grants, gifts and bequests from any source for the support and financing thereof; and to engage in any activities relating to any of the foregoing purposes.

CORE submitted a proposal to NSF in February 1963, to carry out among other projects "the NAS-NRC experimental-exploratory drilling program recommended by the AMSOC Committee as necessary to the success of the Mohole Project" and especially the "programs recommended by such groups of scientists as the 'LOCO Committee.'"

The LOCO Committee next met on February 18, 1963, at NAS offices in Washington, and agreed to use an intermediate vessel of the Mohole Project for LOCO drilling. They also voted unanimously that a corporation associating the four major oceanographic institutions be established at the earliest possible date (Emiliani, 1963b). The group met again on April 18, 1963, in Washington, and, after several disparate actions, voted to dissolve the committee (Emiliani, 1963c).

Emiliani received NSF funding to carry out drilling from Global Marine's *Submarex* later that year under the LOCO name

(Emiliani, 1981), and then the acronym disappeared. Following the formation of CORE, Emiliani was asked to join the consortium. He demurred, but proposed Fritz Koczy of the University of Miami in his stead.

Throughout the rest of 1963, the directors of the four institutions maneuvered and corresponded. Each of them, except perhaps Woods Hole Oceanographic Institution, wanted to manage a major drilling program.

On March 19, 1964, Gordon Lill, project director of Mohole for NSF, brought together Maurice Ewing (Lamont), Paul Fye (Woods Hole), Fritz Koczy (University of Miami), and Roger Revelle (Scripps Institution), along with NSF personnel Richard Bader, William E. Benson, and John Lyman. Lill's message— seven years after he had first become involved with the Mohole idea—was that the four institutions must together underwrite an agreement for pursuing major oceanic drilling projects and then advise NSF of that accomplishment.

So, on May 10, 1964, a Memorandum of Agreement was signed by the leading officers (not some individuals) of those four oceanographic institutions. Revelle proposed the name: Joint Oceanographic Institutions Deep Earth Sampling: JOIDES. Paul Fye became the first chairman. The first JOIDES drilling project was conducted under Lamont's auspices off the east coast of Florida from the ship *Caldrill* in April-May, 1965, one year before the Mohole appropriation faltered in the House of Representatives. The *Glomar Challenger* was built for JOIDES in 1968 and promptly began work for the Deep Sea Drilling Project, for which Scripps Institution was designated the operating institution.

In the long run, the "intermediate" drilling concept won, perhaps because it appealed to a larger spectrum of geologists than did the Mohole, but more likely because it did not become a line item in Congress.

Explanatory Notes

[1]Spilhaus told the author in June 1983, that he was probably the first to use this statement as such, to a congressional committee during the 1950s. It was often repeated in that decade by other oceanographers when they were speaking for their science. I heard Roger Revelle say it many times.

[2]A recent example of reminiscences is by Munk (1984:9), who said: "Ten years after the MOHOLE demise, the Deep-Sea [sic] Drilling Project initiated an enormously successful ocean-wide sediment drilling program." Actually the drilling program of JOIDES began one year before Congress ended the Mohole Project.

[3]Copies of the minutes of the AMSOC Committee and considerable correspondence on Mohole are in the Archives of the Scripps Institution of Oceanography; similar records are probably in several other oceanographic archives. The author of this article organized the available material in the Scripps Archives several years ago.

[4]Copies of most of the Memoranda of the LOCO Committee (mimeographed or photocopied) are in the Archives of the Scripps Institution of Oceanography, filed with Mohole (for historical convenience).

REFERENCES CITED

AMSOC Committee, 1961a, Minutes of June 13, 1961.

—— 1961b, Experimental drilling in deep water at La Jolla and Guadalupe sites: NAS-NRC Publication 914, 183 p.

—— 1963a, Minutes of February 27, 1963.

—— 1963b, Minutes of May 11, 1963.

—— 1963c, Minutes of September 28, 1963.

Bascom, Willard, 1959, Precise station-keeping in deep water, unpublished report from NAS-NRC. (Copy in Archives of Scripps Institution of Oceanography.)

—— 1961, A hole in the bottom of the sea: Garden City, N.Y., Doubleday & Co., 352 p.

Burk, C. A., editor, 1964, A study of serpentinite: the AMSOC core hole near Mayaguez, Puerto Rico: NAS-NRC Publ. 1188, Washington, D.C., 175 p.

Emiliani, Cesare, 1962a, LOCO Memorandum 1, April 5, 1962.

—— 1962b, LOCO Memorandum 12, September 14, 1962.

—— 1962c, LOCO Memorandum 14, October 12, 1962.

—— 1963a, LOCO Memorandum 15, January 23, 1963.

—— 1963b, LOCO Memorandum 17, February 21, 1963.

—— 1963c, LOCO Memorandum 20, May 3, 1963.

—— 1981, A new global geology, p. 1687–1716, *in* Emiliani, Cesare, editor, The Sea, v. 7: The oceanic lithosphere: New York, Wiley-Interscience.

Greenberg, Daniel S., 1964, Mohole: the project that went awry: Science, v. 143, p. 115–119, 142–144, 334–337.

—— 1966, Mohole: Senate is asked to restore funds: Science, v. 153, p. 38–39.

Hedberg, Hollis D., 1962, Report of chairman to AMSOC Committee, January 17, 1962. Also March 10, 1962.

Hedberg, Hollis D., and Burk, Creighton A., 1964, Drilling the ocean crust: International Science and Technology, October 1964, p. 1–10.

Hess, H. H., 1959, The AMSOC hole to the earth's mantle: Transactions American Geophysical Union, v. 40, p. 340–345. (Reprinted 1960 in American Scientist, v. 48, p. 254–263.)

Lill, Gordon G., and Maxwell, Arthur E., 1959, The earth's mantle: Science, v. 129, p. 1407–1410.

Lomask, Milton, 1976, A minor miracle: an informal history of the National Science Foundation: Washington, D.C., National Science Foundation, p. 167–197 of 285 p.

Munk, Walter H., 1984, Affairs of the sea, *in* A celebration in geophysics and oceanography—1982, Scripps Institution of Oceanography Reference Series 84-5, p.3–23.

National Science Foundation, 1962, News release, February 28.

Shor, Elizabeth Noble, 1978, Scripps Institution of Oceanography: Probing the Oceans, 1936 to 1976: San Diego, Tofua Press, 502 p.

Solow, Herbert, 1963, How NSF got lost in Mohole: Fortune, v. 67, p. 138–141, 198, 203–204, 208–209.

Steinbeck, John, 1961, High drama of bold thrust through ocean floor: Life, v. 50, p. 111.

Geological Society of America
Centennial Special Volume 1
1985

North American paleomagnetism and geology

John Verhoogen
University of California at Berkeley
Berkeley, California 94720

ABSTRACT

This paper sketches out developments in North American paleomagnetism that have markedly affected geological thought, to the exclusion of those related mostly to the past character and properties of the earth's magnetic field or to magnetic properties of rocks. Early work (1948) at the Carnegie Institution's Department of Terrestrial Magnetism was mainly directed, as was also that of P.M.S. Blackett in England, at elucidating the past behavior of the geomagnetic field and its origin; proving or disproving continental drift was not a primary objective, and the concept of sea-floor spreading was, of course, still unheard of. The first surprise came with the discovery of numerous reversely magnetized igneous and sedimentary rocks. It took about 10 years to establish that most of these reversed rocks reflected field reversals rather than self-reversal contingent on mineralogical composition. The development of a magnetic time scale based on the chronology of field reversals was of paramount geological significance as it led to confirmation of the Vine-Matthews hypothesis and to plate tectonics; it is now widely applied to worldwide stratigraphic correlations. Unexpected "abnormal" directions of magnetization in pre-Tertiary rocks were first interpreted in terms of polar wander, or, in the case of deformed rocks, in terms of magnetostriction and stress-induced anisotropy; they became accepted by British paleomagnetists as evidence for continental drift a few years before their American colleagues reached the same conclusion. Paleomagnetism has further affected geological thinking by establishing, through determination of the earth's paleoradius, the improbability of Carey's expansionist views and, most important, in revealing the complex nature of continental margins formed by accretion of exogenous terranes.

BEGINNINGS OF PALEOMAGNETISM IN THE U.S.A.—POLAR WANDER VERSUS CONTINENTAL DRIFT

It is not clear who first used the term "paleomagnetism" which came into current use in the early 1950s. Before that, the common term was "rock magnetism" which, as Irving (1964) has remarked, lacks the historical connotation of the word "paleomagnetism." It is, of course, this historical connotation that relates paleomagnetism to other geological disciplines and explains the impact of the former on the latter. It is interesting, however, that the earliest work was directed towards nothing more than establishing the history of the earth's magnetic field itself; very little thought was given to geological applications such as a test of continental drift, even though Mercanton in Switzerland had noted the possibility of this application as early as 1926. (For early references, see Irving, 1964).

Paleomagnetism in the United States started at the Department of Terrestrial Magnetism (DTM) of the Carnegie Institution of Washington around 1936. In 1938, McNish and Johnson reported on the magnetization of varves and marine sediments. All previous reports on rock magnetism, coming from Germany, France, Japan, and elsewhere, had dealt only with igneous rocks, bricks, pottery, or clays baked by lava flows. McNish and Johnson's demonstration of remanence in sediments thus broadened considerably the field of inquiry.

After an interruption of several years, the work at DTM was resumed shortly after the war and resulted in three important papers published in 1948 and 1949. The 1948 paper by Johnson, Murphy, and Torreson is characteristically entitled "Prehistory of the Earth's Magnetic Field"; the 1949 paper by Torreson, Murphy, and Graham is similarly titled "Magnetic Polarization in Sedimentary Rocks and the Earth's Magnetic History." The purpose these authors have in mind is not geological; they, like P. M.

S. Blackett in England at the same time, wanted facts about the ancient magnetic field in order to test hypotheses on its origin, which was then still quite obscure. The 1949 paper by Torreson, Murphy, and Graham deals with the magnetization of some flat-lying sedimentary rocks in the western United States, most of which range in age from Eocene to Pliocene. The authors conclude that for the past 50 million years the geomagnetic field has maintained its present polarity while the earth's magnetic axis has had an average orientation coinciding with the geographic axis. The authors, however, offer no proof of the stability of the remanence in the sediments they studied.

This matter of stability is the focus of the third DTM paper by J. W. Graham (1949). This study, Graham wrote, was undertaken to investigate whether sedimentary rocks can retain their initial direction of magnetization over long periods. If so, "it should be possible to determine how the direction of the earth's magnetic field has varied during geologic time. . . . the property could be useful in the solution of structural problems" (Graham, 1949, p. 132). One test of stability proposed by Graham is the celebrated fold-test: If samples taken in various parts of a folded bed show a scatter in magnetic directions that is significantly reduced when the magnetic directions are corrected for the tilt of the bed, it may be surmised that the magnetization was acquired prior to folding.

One bed that Graham found to satisfy the fold-test very nicely is the Rose Hill ferruginous sandstone of Silurian age, which was deformed at the end of the Paleozoic. From his observations Graham concluded that its magnetization had been retained for at least 200 million years. There was, however, an odd fact about this formation: At several sites spread over a distance of about 50 miles, the inclination of the magnetization was quite abnormal, with the south-seeking end of the polarization pointing downward (normally, it is the north-seeking end that points downward in the northern hemisphere); the magnetic declination was also rather abnormal. Graham makes no comment on these facts except to state that there is no known mechanism to produce a magnetic field of this kind over such a distance. He apparently did not think that the magnetization of the Rose Hill formation reflected the direction of the earth's field at the time of deposition. Indeed, he and O. W. Torreson read a paper at the 1951 meeting of the American Geophysical Union (AGU) in which they contrast the magnetizations of flat-lying and folded Paleozoic sediments (Graham and Torreson, 1951). These differences are attributed to deformation; they stated:

Inverse magnetizations are present in the deformed geosynclinal sections but have not been found in contemporaneous flat-lying beds. It is believed that these anomalous magnetizations resulted from the physical and chemical conditions peculiar to the geosynclinal sections, and that they do not have any direct bearing on the past directions of the Earth's magnetic field.

Graham (1954) wrote, "We think there is, in our data, a fair case for believing that since Cambrian time the earth's field has remained essentially in its present sense and roughly in its present orientation." But a note of caution is sounded, for he added: "The word 'think' is used advisedly, for we do not have a proof that is certain, and the dikes, sills, and sediments having inverse magnetizations still are not completely understood" (Graham, 1954, p. 215).

But a year later the situation had changed. Prompted by the work of British paleomagnetists on Permian and Triassic rocks in Great Britain, Graham sampled extensively formations of that age in the southwestern United States and found their magnetization to be essentially consistent with the British results. In the abstract to his 1955 paper, he wrote:

It is demonstrated that an impressive number of the observations are in essential agreement simply on the basis of assuming that the rocks were magnetized by a geomagnetic field about like the one today, the essential difference being that this field was in a significantly different orientation. This result is discussed in terms of slip of an outer shell of the earth relative to the axis of revolution (Graham, 1955, p. 329).

A similar conclusion was reached at the same time by R. R. Doell. In his Ph.D. thesis, completed at Berkeley in September 1955, he reported on his sampling in the Grand Canyon of the Colorado River, and located a Permian (Supai) paleomagnetic pole slightly to the west and south of the British Triassic pole of Clegg et al. (1954). Doell also located a Precambrian paleomagnetic pole in the eastern Pacific. Doell, like Graham, interpreted these findings in terms of polar wander. He drew a generalized polar wander path with eight points on or about it, extending from the early Precambrian to the Eocene. Graham's Silurian pole lies on it. In view of the scarcity of data, no attempt was made to draw separate polar wander paths for the several continents. This was done a year later, by Irving (1956) and by Runcorn (1956), mostly on the basis of the difference between Permian paleomagnetic poles from Europe and North America; and by 1957 most British authors had adhered to the view, early propounded by Blackett and his co-workers, that paleomagnetic data were best interpreted in terms of polar wander *and* continental drift (Creer, Irving, and Runcorn, 1957). But American workers remained skeptical about drift. In their influential review, Cox and Doell (1960) appear unconvinced that discrepancies between poles from different continents could not be explained by inaccuracies in stratigraphic correlation or dating, by changes in the configuration of the magnetic field which may not have remained essentially dipolar through all geologic times, or by episodes of unusually rapid polar wander. In a later review (Doell and Cox, 1961) these same authors, though impressed by the evidence for drift of Australia (Irving and Green, 1958), still seem unconvinced with regard to India, and to the separation of Europe and North America in Permian times. They do not doubt, however, the reality of polar wander.

Since it is now generally believed that polar wander has contributed less to the distribution of Phanerozoic poles than plate (and microplate) displacements, it is interesting to consider what prompted early investigators to prefer polar wander to continental drift. This preference was in fact quite logical. Remember that initially it was not known whether the geomagnetic field had always had its present, predominantly dipolar character; to establish that required comparison of contemporaneous directions of the field at widely spaced sites, a comparison that would have been invalid if the sites had suffered relative displacements since acquiring their magnetization. Rough agreement between Permian paleomagnetic poles from Europe and North America could be interpreted as evidence for a dipolar field and small relative displacements. It was thus rather logical to suppose that if these two continents had not moved much relative to each other, neither had they moved much relative to the present geographic poles; but since Permian poles were significantly displaced from the present pole, it must have been the magnetic poles that had moved. The close association of magnetic poles with the rotational pole had been demonstrated from the start by work on Recent, Pleistocene, and Tertiary formations, while the emerging dynamo theory of the origin of the magnetic field required a close association of the earth's rotational and magnetic axes. Clearly, the simplest conclusion to be drawn from the data was that the rotational pole had, in the past, moved significantly with respect to fixed geographic features.

This was by no means a new idea. Polar wander curves began to appear in the geophysical literature as early as 1902; Koeppen and Wegener in 1924 had drawn such a curve on the basis of paleoclimatic data (for early references, see Gutenberg, 1939). In the mid-1950s, influential papers by Gold (1955) and Munk (1956) had revived the old controversy as to whether displacement of the earth relative to its spin axis is possible or not; Gold concluded that it is, but much depends on the mechanical properties of the mantle which were then—and still are—imperfectly known.

The conversion from polar wander to continental drift as the main factor in explaining the pattern of dispersion of paleomagnetic poles came about mostly from a comparison of polar wander curves for separate continents. Grommé, Merrill, and Verhoogen (1967) compared polar wander paths for Africa, Australia, and North America, in which they found a strong suggestion that apparent polar wander during the Mesozoic was entirely due to continental drift rather than to true polar wander. Jurdy and Van der Voo (1974) used reconstructed plate positions and a mean paleomagnetic pole to show that true polar wandering since the early Tertiary amounts to only 2°, less than the uncertainty (4°) of the mean paleopole, and is therefore not significant. Wells and Verhoogen (1967) showed that the scatter of Permian poles for continents bordering on the Atlantic could be significantly reduced by applying to them the rotations that best closed that ocean (Bullard and others, 1965); there was, however, no noticeable reduction in scatter for Carboniferous poles. Hospers and Van Andel (1968) noted that the same analysis carried out on fewer but more reliable* paleomagnetic data produced much better results. If North America is rotated through an angle of 38° about an Eulerian pole at 88.5°N, 27.7°E so as to close the North

Atlantic, agreement between North American and European paleomagnetic poles from the Silurian to the Triassic is virtually exact (McElhinny, 1973, p. 253). These results notably increased general confidence in the validity of paleomagnetic data and in the reality of continental drift, at least with respect to the opening of the North Atlantic. The most compelling evidence for drift had come, of course, from Irving's and Creer's paleomagnetic studies on Mesozoic rocks from the southern continents, particularly Australia, but these studies fall outside the scope of the present review.

Continental drift, which considered only displacement of continents, has now been superseded by the broader concept of plate tectonics which includes, of course, displacement of the oceanic floor. Paleomagnetism now offers the possibility of determining apparent polar wander paths for totally oceanic plates. This is done by combining and weighing data such as magnetic inclinations in deep-sea cores, magnetic anomalies over seamounts, skewness of linear anomalies, and paleoequator determinations from sediment-core analysis. Results have been recently summarized by McWilliams (1983).

No undisputed paleomagnetic evidence exists at present for subduction, collision, spreading, or any other plate tectonic phenomenon before middle Proterozoic time, with the exception of cratonal displacement with respect to the geomagnetic axis (McWilliams, 1983; Van der Voo, 1982). Whether these cratonal displacements represent plate motion or true polar wander or both is not known. The difficulty in establishing true polar wander is the absence of a fixed frame to which polar shifts could be referred. In plate tectonics everything moves, ridges and trenches as well as the plates themselves. The only remaining reference axis is the spin axis itself, with which the average paleomagnetic axis appears to coincide. Determination of polar motion relative to the lithosphere as a whole requires instantaneous knowledge of the relative motion of all plates, including purely oceanic ones which sooner or later vanish without leaving a trace. The analysis cannot therefore be carried back much further than late Mesozoic. For the Cenozoic there seems to be no evidence of true polar wander relative to the lithosphere (McElhinny, 1973; Jurdy and Van der Voo, 1974 and 1975; Jurdy, 1981; for a recent review see McWilliams, 1983). There does, however, seem to be significant polar displacement relative to a frame defined by hotspots (Jurdy 1981) and to the level in the mantle at which hotspots originate (Cox and Gordon 1983); and Gordon (1983) suggests that a rapid shift of about 13 degrees of the spin axis relative to the hotspots of the Pacific Ocean basin occurred between 90 and 81 Ma. ago.

REVERSALS

The greatest contribution of paleomagnetism to plate tectonics, and to geology in general, was not so much the demonstration of continental drift as the establishment of a magnetic stratigraphic scale based on the chronology of reversals of the geomagnetic field. For it was this magnetic time scale which, when compared to the magnetic lineations of the ocean floor, led

---

*i.e., better methods for magnetic "cleaning" of samples.

to the Vine-Matthews-Morley interpretation of sea-floor spreading and to plate tectonics. It has since been used extensively for stratigraphic correlation.

Once again, an unexpected discovery (demonstration of sea-floor spreading) was made in the course of research intended for a quite different purpose, in this case to establish whether the occurrence of reversely magnetized rocks was due to reversals of the magnetic field or to a subtle solid-state phenomenon related to the mineralogy of reversed rocks. No greater contrast could be expected than that between the two interpretations, one of which (field reversal) involves magnetohydrodynamic forces operating on a planetary scale (10^{+8} cm) while the other (self-reversal) involves quantum-mechanical forces operating on the atomic scale (10^{-8} cm). The story has recently been told in great detail by Glen (1982) and can only be summarized briefly here. Although many instances of reversed magnetization had been reported prior to 1949, it was Graham's (1949) discovery of reversely magnetized sediments in the Rose Hill (Silurian) formation of Maryland that really started the controversy, for Graham wrote to Louis Néel, the discoverer of antiferromagnetism and ferrimagnetism, to inquire whether a rock could acquire a remanent magnetization opposed in sense to the field that induced it (self-reversal). Yes indeed, it could, said Néel, who promptly described four different mechanisms by which it could do so (Néel, 1951). Shortly thereafter, Uyeda (1958) found on a laboratory shelf a specimen of dacite from Mount Haruna in Japan which is demonstrably self-reversing. Balsley and Buddington (1954) noted a correlation of reverse magnetization with occurrence of certain minerals (hematite-ilmenite solid solutions) in Precambrian gneisses in the Adirondack Mountains of New York. On the other hand, work by Hospers and others, in Iceland and elsewhere, had shown a frequent parallelism of remanence (either normal and reverse) in igneous rocks and their baked contact rocks, and some stratigraphic ordering (e.g., absence of reversals in Holocene and upper Pleistocene rocks). (Excellent accounts of this work are to be found in Irving, 1964, and in Glen, 1982). The case for self-reversal was made stronger when Uyeda identified the carrier of reversed remanence in the Haruna dacite as a hematite-ilmenite solid solution very similar to the one responsible for reversed magnetization in the Adirondack gneisses. The mechanism for self-reversal in this mineral depends intricately on relative rates of cooling, exsolution, and ordering of Fe^{3+}, Fe^{2+}, and Ti^{4+} ions, and is not one of those originally postulated by Néel. Still other possible mechanisms for self-reversal were proposed, some of them based on progressive oxidation of iron-titanium oxides, and it appeared at one time that there might be a correlation between the polarity of igneous rocks and their state of oxidation. The evidence for and against this correlation is summarized by McElhinny (1973, p. 112–114). Lurking in the background was a conceptual difficulty against field reversals. Rough calculations on the relative magnitude of possible forces involved in the terrestrial dynamo invariably assigned a dominant role to the Coriolis force arising from the earth's rotation, and paleomagnetic evidence linking the earth's magnetic and rotation axes was rapidly accumulating; the earth's magnetic field was clearly strongly coupled to its rotation. How then could the coupling at once be strong enough to account for the observations and weak enough to allow reversing the sense of the field without reversing the sense of rotation?

As Matuyama (1929) and Mercanton (1926a; 1931-32) had noted long before, field reversals, if they occur, should be worldwide. In the absence of self-reversals, all contemporaneous rocks should be magnetized in the same sense. The difficulty in applying this test lay in establishing contemporaneity on a time scale appropriate to magnetic events. The work of Hospers in Iceland had shown that field reversals must be very sudden events requiring no more than a few thousand years and that intervals of either polarity could last perhaps as little as 100,000 years or less. Usual correlation methods based on stratigraphy and paleontology were hopelessly inadequate. Fortunately, improvements at Berkeley in the precision of K-Ar dating methods offered hope of increased resolution. In 1963, Cox, Doell, and Dalrymple (1963) and almost simultaneously McDougall and Tarling (1963) presented polarity time scales based on igneous rocks, for the latest 3 million years. Those time scales underwent numerous corrections and additions, culminating in the 1966 time scale of Doell and Dalrymple (1966) which included the recently discovered Jamarillo event (a short-lived episode of normal polarity which occurred towards the end of the reversed Matuyama epoch). In the meantime, much work was being done, mainly at the Lamont-Doherty Geological Observatory of Columbia University, on the magnetization of deep-sea cores; this work followed from Harrison and Funnel's (1964) initial observation of reversal in sediments from the equatorial Pacific. The discovery of the Jaramillo event inNorth Pacific and Antarctic cores (Ninkovich and others, 1966; Opdyke and others, 1966) finally clinched the argument in favor of field reversals and, by allowing a correct interpretation of the magnetic lineation of the ocean's floor, to a spectacular confirmation of the Vine-Matthews-Morley hypothesis of sea-floor spreading. It was only a short step from there to plate tectonics.

Since then an enormous amount of work has been done on magnetic polarity stratigraphy. Part of this work has been carried out on land and has led to considerable refinement in the exact placement of stratigraphic boundaries, e.g., the Mesozoic-Cenozoic boundary. Much work has also been done on marine magnetic anomalies, which, when coupled to deep drilling through the ocean floor and rock-magnetic studies of dredged specimens, has greatly improved our understanding of the nature of the oceanic crust, of the evolution of ocean basins, and of the mechanism of sea-floor spreading. Comprehensive bibliographies may be found in quadrennial reports of the American Geophysical Union (see, for instance, Larson and Helsley, 1975; Blakely and Cande, 1979; Butler and Opdyke, 1979; Chan and Alvarez, 1983).

TERRANES

Perhaps one of the least expected contributions of paleomagnetism to tectonics is the discovery that continental borders commonly consist of a jumble of small units, known as microplates or "terranes," which may have drifted over large distances and suffered what now appears as intracontinental rotation of large amplitude. One of the earliest cases of this type is that of the Siletz River lavas of western Oregon (Cox 1957). Cox had been led to examine these rocks of Eocene age by the work of Irving (1956) and of Clegg, Deutsch, and Griffiths (1956) on the Deccan traps of India (Cretaceous to Eocene), which showed paleomagnetic poles in the north Atlantic not far from Florida; these poles were interpreted to demonstrate a large northward drift and rotation of India since Eocene time. Cox, who was then still skeptical of continental drift, noticed a lack of reliable paleomagnetic data for North America to which the Indian data could be compared and proceeded to Oregon to sample the Siletz River volcanic series of lower Eocene age. These rocks, which satisfied all tests of reliability, yielded a paleomagnetic pole in the Atlantic not far from the Indian poles; Cox accordingly suggested that perhaps India had not drifted at all but that it was the pole that had made a transient excursion over some 60 degrees of latitude. Irving (1959), on the other hand, interpreted the Siletz pole as being due to a 60° clockwise tectonic rotation of Oregon. Cox (1959), carefully evaluating the data, pointed to several possibilities: drift of India and rotation of Oregon; a rapid excursion of the pole during the late Cretaceous, Paleocene, and Eocene down to latitude 30°N; or a non-dipolar configuration of the field; or a combination of these. "My preference," he wrote, "is for the second hypothesis of rapid change of a dipole field because it fits the data from all continents reasonably well (within the limits imposed by stratigraphic uncertainties) and because it is the simplest" (Cox, 1954, p. 116).

Although a few more cases of suspected tectonic rotation were found at about that time (Newfoundland with respect to North America, Spain with respect to Europe, bending of Japan), it was only after about 1975 that the subject began to receive much attention; but then it literally exploded. Microplates have been discovered or suspected in almost all orogenic belts. (For a review through 1978, see Van der Voo, 1979; for the period 1979–1982, see McWilliams, 1983.) A new vocabulary appears, e.g., "accretion" history of component "terranes" which form a "tectonic collage." Investigations extending through Alaska and the Canadian Cordillera, western United States, Mexico, Central America and the Caribbean, Appalachians, and in many areas outside North America, particularly the Mediterranean area, have revealed geologic histories far more complex than, and of a different nature from, what classical tectonic theory had led older generations of geologists to believe. Displacements of 20 degrees or more in paleomagnetic latitude relative to a neighboring craton, and rotation over large angles, now appear commonplace in orogenic belts and continental margins. The flavor of the subject and its present vigor may be caught by perusing the abstracts of papers presented at the December 1983 fall meeting of the American Geophysical Union. In the Geomagnetism and Paleomagnetism section, no fewer than 22 papers were read on "Ap-

plications of Paleomagnetism to the Tectonics of the Western U.S." A few more paleomagnetic papers were read in the Tectonophysics section at a session devoted to the Franciscan geology of the San Francisco Bay area. It is, incidentally, in this area that one of the first instances of paleomagnetic evidence for rotation was found (Grommé and Gluskoter, 1965).

PALEORADIUS OF THE EARTH

A number of geologists and geophysicists have proposed, at different times and for different reasons, that the earth's radius increases with time. Egyed (1957), for instance, suggests a radius increase at the rate of about 0.05 cm/year, while Carey (1958) calls for a much faster rate of about 0.5 cm/year since the late Paleozoic. Egyed (1960) was also the first to point out that the earth's radius can, in principle, be determined from paleomagnetic data: given two contemporaneous sampling sites at which the paleomagnetic inclination (and hence the paleolatitude) is known, it suffices to determine the distance between two parallels of paleolatitude through these sites. Success of the method depends on three conditions: 1) the purely dipolar nature of the field recorded at the sites, 2) the perfect contemporaneity of the sites on the time scale of apparent polar wander, and 3) the distance between sites having remained unchanged since the magnetization was acquired. Cox and Doell (1961) first applied the method to Permian sites in Europe and Siberia, which may not satisfy exactly condition 3; they concluded that "the average Permian magnetic field, as seen from two sampling areas almost 5,000 km apart on the Eurasian landmass, shows no significant departure from the dipolar configuration and is perfectly consistent with a Permian earth radius equal to the present one," although "the accuracy is probably not great enough to confirm or reject expansion at a rate as small as the 0.4 to 0.8 mm/year figure suggested by Egyed" (Doell and Cox, 1961, p. 301). The same conclusion was subsequently reached by other investigators using different methods. The latest study on this subject, to this reviewer's best knowledge, is that of Van Andel and Hospers (1968) who conclude that the hypothesis of fast expansion put forth by Carey can be rejected at the 95 percent confidence level, but the method cannot resolve slow expansion at a rate of the order of 1 mm/year or less (McElhinny 1973, p. 279).

CONCLUDING REMARKS

The foregoing account completely neglects North American contributions to paleomagnetism that have only indirectly influenced geologic thought but have nevertheless been of prime importance in the development of the subject itself. Among these contributions, one should mention progress in instrumentation, which now allows accurate measurement of the extremely weak magnetization of a vastly increased variety of rocks; progress in magnetic mineralogy; and progress in the understanding of the origin of magnetic remanence, which now allows, for instance, separation of the natural remanence into distinct components acquired at different times, such as the overprinting that commonly marks subsequent orogenic episodes. Much remains to be done, however, particularly on the interpretation of the magnetization of metamorphic rocks without which Precambrian geologic history would be hard to unravel.

REFERENCES CITED

Balsley, J. R., and Buddington, A. F., 1954, Remanent magnetism of the Russell belt of gneisses, Northwest Adirondack Mountains, New York: Journal of Geomagnetism and Geoelectricity, v. 6, p. 178–181.

Blakely, R. J., and Cande, S. C., 1979, Marine Magnetic Anomalies: Reviews of Geophysics and Space Physics, v. 17, p. 204–214.

Bullard, E. C., Everett, J. E., and Smith, A. G., 1965, The fit of the continents around the Atlantic, in A symposium on Continental Drift: Philosophical Transactions Royal Society of London, v. A258, p. 41–51.

Butler, R. F., and Opdyke, N. D., 1979, Magnetic Polarity Stratigraphy: Reviews of Geophysics and Space Physics, v. 17, p. 235–244.

Carey, S. W., 1958, The tectonic approach to continental drift, in Carey, S. W., editor, Continental Drift, a symposium: University of Tasmania, Hobart, p. 177–358.

Chan, L. S., and Alvarez, W., 1983, Magnetic Polarity stratigraphy: Reviews of Geophysics and Space Physics, v. 21, p. 620–626.

Clegg, J. A., Almond, M., and Stubbs, P.H.S., 1954, The remanent magnetism of some sedimentary rocks in Britain: Philosophical Magazine, v. 45, p. 583–598.

Clegg, J. A., Deutsch, E. R., and Griffiths, D. H., 1956, Rock Magnetism in India: Philosophical Magazine, v. 1, p. 419–431.

Cox, A., 1957, Remanent magnetization of lower to middle Eocene basalt flows from Oregon: Nature, v. 179, p. 685–686.

—— 1959, The remanent magnetization of some Cenozoic volcanic rocks [Ph.D. thesis]: University of California, Berkeley, 193 p.

Cox, A., and Doell, R. R., 1960, Review of Paleomagnetism: Geological Society of America Bulletin, v. 21, p. 645–768.

—— 1961, Paleomagnetic evidence relevant to a change in the earth's radius: Nature, v. 190, p. 36–37.

Cox, A., Doell, R. R., and Dalrymple, G. B., 1963, Geomagnetic polarity epochs and Pleistocene geochronometry: Nature, v. 198, p. 1049–1051.

Cox, A., and Gordon, R. G., 1983, Absolute plate motion and true polar wander [abstract]: EOS, v. 64, p. 844.

Creer, K. M., Irving, E., and Runcorn, S. K., 1957, Geophysical interpretation of paleomagnetic directions from Great Britain: Philosophical Transactions of the Royal Society of London, v. A250, p. 144–156.

Doell, R. R., 1955, Remanent magnetism in sediments [Ph.D. thesis]: University of California, Berkeley, 156 p.

Doell, R. R., and Cox, Allan, 1961, Paleomagnetism: Advances in Geophysics, v. 8, p. 221–313.

Doell, R. R., and Dalrymple, G. B., 1966, Geomagnetic polarity epochs: A new polarity event and the age of the Brunhes-Matuyama boundary: Science, v. 152, p 1060–1061.

Egyed, L., 1957, A new dynamic conception of the internal constitution of the earth: Geologische Rundschau, v. 46, p. 101–121.

—— 1960, Some remarks on continental drift: Geofisica Pura e Applicata, v. 45, p. 115–116.

Glen, W., 1982, The Road to Jamarillo, Stanford University Press, 459 p.

Gold, T., 1955, Instability of the earth's axis of rotation: Nature, v. 175, p. 526–529.

Gordon, R. G., 1983, Late Cretaceous apparent polar wander of the Pacific hotspots with respect to the spin axis: Geophysical Research Letters, v. 10, p. 709–712.

Graham, J. W., 1949, The stability and significance of magnetism in sedimentary rocks: Journal of Geophysical Research, v. 54, p. 131–167.

—— 1954, Rock magnetism and the Earth's magnetic field during Paleozoic time:

Journal of Geophysical Research, v. 59, p. 215–222.

—— 1955, Evidence of polar shift since Triassic time: Journal of Geophysical Research, v. 60, p. 329–347.

Graham, J. W., and Torreson, O. W., 1951, Contrasting magnetizations of flat-lying and folded Paleozoic sediments [abstract]: Transactions American Geophysical Union, v. 32, p. 336.

Grommé, C. S., and Gluskoter, H. J., 1965, Remanent magnetization of spilite and diabase in the Franciscan formation, Western Marin County, California: Journal of Geology, v. 73, p. 74–94.

Grommé, C. S., Merrill, R. T., and Verhoogen, J., 1967, Paleomagnetism of Jurassic and Cretaceous plutonic rocks in the Sierra Nevada, California, and its significance for polar wandering and drift: Journal of Geophysical Research, v. 72, p. 5661–5684.

Gutenberg, B., 1939, Hypotheses on the development of the earth' crust, in B. Gutenberg, editor, Internal constitution of the earth: McGraw Hill, New York, p. 177–217.

Harrison, C.G.A., and Funnell, B. M., 1964, Relationship of palaeomagnetic reversals and micropalaeontology in two late Cainozoic cores from the Pacific Ocean: Nature, v. 204, p. 566.

Hospers, J., and Van Andel, S. I., 1968, Paleomagnetic data from Europe and North America and their bearing on the origin of the North Atlantic Ocean: Tectonophysics, v. 6, p. 475–490.

Irving, E., 1956, Paleomagnetic and paleoclimatological aspects of polar wandering: Geofisica Pura e Applicata, v. 33, p. 157–168.

—— 1959, Paleomagnetic pole positions: A survey and analysis: Royal Astronomical Society Geophysical Journal, v. 2, p. 51–59.

—— 1964, Paleomagnetism and its applications to geological and geophysical problems, Wiley, New York, 399 p.

Irving, E., and Green, R., 1958, Polar wandering relative to Australia: Geophysical Journal, v. 1, p. 64–72.

Johnson, E. A., Murphy, T., and Torreson, O. W., 1948, Pre-history of the Earth's magnetic field: Terrestrial Magnetism and Atmospheric Electricity, v. 53, p. 349–372.

Jurdy, D. M., 1981, True polar wander: Tectonophysics, v. 74, p. 1–16.

Jurdy, D. M., and Van der Voo, R., 1974, A method for the separation of true polar wander and continental drift, including results for the last 55 m.y.: Journal of Geophysical Research, v. 79, p. 2945–2952.

—— 1975, True polar wander since the Early Cretaceous: Science, v. 187, p. 1193–1196.

Larson, R. L., and Helsley, C. E., 1975, Mesozoic Reversal sequence: Reviews of geophysics and space physics, v. 13, p. 174–176, 209.

Matuyama, M., 1929, On the direction of magnetization of basalt in Japan, Tyôsen, and Manchuria: Proceedings Imperial Academy Japan, v. 5, p. 203–205.

McDougall, I., and Tarling, D. H., 1963, Dating of reversals of the earth's magnetic field: Nature, v. 198, p. 1012–1013.

McElhinny, M. W., 1973, Palaeomagnetism and Plate Tectonics: Cambridge University Press, 358 p.

McNish, A. G., and Johnson, E. A., 1938, Magnetization of unmetamorphosed varves and marine sediments: Terrestrial Magnetism and Atmospheric Electricity, v. 53, p. 349–360.

McWilliams, A. G., 1983, Paleomagnetism and the motion of large and small plates: Reviews of geophysics and space physics, v. 21, p. 644–651.

Mercanton, P. L., 1926a, Aimantation de basaltes groenlandais: Comptes Rendus Académie Sciences, v. 182, p. 859–860.

—— 1926b, Inversion de l'inclination magnetique terrestre aux âges géologiques: Terrestrial Magnetism Atmospheric Electricity, v. 31, p. 187–190.

—— 1931, 1932, Inversion de l'inclinaison magnetique aux âges géologiques: Comptes Rendus Académie Sciences, v. 192, p. 978–980, v. 194, p. 1371.

Munk, W. H., 1956, Polar wandering—a marathon of errors: Nature, v. 177, p. 551–554.

Neél, L., 1951, L'inversion de l'aimantation permanente des roches: Annales de Géophysique, v.7, p. 90–102.

Ninkovich, D., Opdyke, N. D., Heezen, B. C., and Foster, F. H., 1966, Paleomagnetic stratigraphy, rates of deposition and tephrachronology in North Pacific deep-sea sediments: Earth and Planetary Science Letters, v. 1, p. 476–492.

Opdyke, N. D., Glass, B. P., Hays, J. D., and Foster, J. D., 1966, Paleomagnetic study of Antarctic deep-sea cores: Science, v. 154, p. 349–357.

Runcorn, S. K., 1956, Paleomagnetic comparisons between Europe and North America, Proceedings of the Geological Association of Canada, v. 8, p. 77–85.

Torreson, O. W., Murphy, T., and Graham, J. W., 1949, Magnetic polarization of sedimentary rocks and the Earth's magnetic history: Journal of Geophysical Research, v. 54, p. 111–129.

Uyeda, S., 1958, Thermo-remanent magnetism as a medium of paleomagnetism, with special reference to reverse thermo-remanent magnetism: Japanese Journal of Geophysics, v. 2, p. 1–123.

Van Andel, S. I., and Hospers, J., 1968, A statistical analysis of ancient earth radii calculated from paleomagnetic data: Tectonophysics, v. 6, p. 227–235.

Van der Voo, R., 1982, Pre-Mesozoic Paleomagnetism and Plate Tectonics: Annual Reviews of Earth and Planetary Sciences, v. 10, p. 191–220.

Wells, J. M., and Verhoogen, J., 1967, Late Paleozoic paleomagnetic poles and the opening of the Atlantic Ocean: Journal of Geophysical Research, v. 72, p. 1777–1782

Geological Society of America
Centennial Special Volume 1
1985

The early development of archaeological geology in North America

John A. Gifford
Anthropology Department
University of Miami
Coral Gables, Florida 33124

George Rapp, Jr.
Archaeometry Laboratory
University of Minnesota at Duluth
Duluth, Minnesota 55812

ABSTRACT

As in Europe, in North America the origins of archaeological geology stemmed from discoveries of artifacts and fossil mammal bones together, and the bearing of such associations on the antiquity of humans in the New World. After 1870 the expansion of American and Canadian federal government surveys into the western portions of North America fostered a tradition of geologists undertaking archaeological field studies as a contingent part of their primary work. This informal collaboration ended around the turn of the century (with a few notable exceptions), as both sciences crystallized into more strictly defined fields. Debate over the antiquity of humans in the New World continued into the 1920s, only ending when definitive geological and paleontological evidence was presented to date artifacts to the end of the last glaciation.

While the initial link between archaeology and geology in North America concerned the question of human entry, other fields of research during the late nineteenth and early twentieth century included petrography of stone artifacts as a means of determining their origin, and geomorphic studies of archaeological sites as an indication of landform modification and environmental change. From such "informal" interdisciplinary research, often accomplished by individual initiative, the connection between geology and archaeology grew, slowly in the pre-World-War-II period and almost explosively in postwar decades due to a general realization by archaeologists of the natural environment's importance in understanding cultural development.

INTRODUCTION

Prior to 1977, the geological profession had not recognized formally a subdiscipline devoted to geological aspects of prehistoric culture. In the following account chronicling the history of this "implicit" research area, we are not striving to create a more tangible entity than actually existed. The fact remains, however, that for more than a century some geologists have devoted greater or lesser amounts of their professional work to problems of an archaeological nature, and it is this substantial phenomenon that we intend to document.

This history will be conceptual as well as institutional and biographical. The underlying idea is that through a series of research problems evolving over the past century, geologists and archaeologists have discovered a genuine and to some extent inescapable link between the course of a culture's development and the constraints imposed or opportunities offered by the surrounding physical environment. This is not the classic notion of determinism, but is more like Vita-Finzi's concept of geological *opportunism* in human affairs (1978, p. 11). We will generally avoid more consideration of that particular philosophical debate.

To us, archaeological geology implies a cross-disciplinary or interdisciplinary approach to research problems that either lie in an area of overlap between the two disciplines, or, if basically unidisciplinary, are amenable to methodological or technological assistance from the other discipline. We will document some of the scholarly questions that over the past century have appealed to practitioners from both disciplines. Our anthology will show that the vitality of archaeological geology has depended not only on controversial issues such as the "early man" controversy, or well-focused methodological developments like petrographic techniques for provenance studies, but also on an almost intuitive perception of the fundamental interrelationship between archaeology and geology.

Two general research themes are suggested naturally by classification of the past century's publications dealing with both disciplines. By far the more prevalent theme concerns the so-called "early man" controversy, which dominated archaeological geology from the 1860s through the 1930s. The second involves research directed toward reconstructing physical environments contemporary with prehistoric human activity at a given time and location. Both of these, along with less common research activities such as provenance studies, remain to the present day vigorous research areas in archaeological geology.

THE FIRST HUMANS IN NORTH AMERICA

"No subject has lately excited more curiosity and general interest among geologists and the public than the question of the Antiquity of the Human Race . . ." Thus did Charles Lyell open his book on this topic (1863), specifically the geological evidences for human antiquity.

There were antiquarians in America long before 1863 (the American Antiquarian Society was founded in Worcester in 1812) but studies of artifacts and skeletal remains in North America generally were dependent on developments in Europe or were conducted in America by Europeans. The German naturalist Albert Koch, for instance, in 1838 excavated on behalf of his private museum a spring deposit on the lower Pomme de Terre River in southwest Missouri. He recovered blackened mastodon bones that he contended had been burned by Indians. This discovery and later ones by Koch (fully covered by McMillan, 1976) reinforced a growing interest in North American human prehistory. Well-known geologists such as J. D. Dana later evaluated Koch's geological evidence and dismissed it (Dana, 1875), but the contemporaneity of humans and mastodons in North America continued to stir scientific and popular interest. Great antiquity for human remains could be deduced if they were found in association with the bones of extinct mammals that were known to have lived during the European glacial age.

On both of Lyell's visits to North America (1841–42 and 1845) he was shown sites where human bones had been found in contexts suggesting great age. In *The Antiquity of Man* he reviews the discovery of bones of extinct mammals together with a human pelvis in a gulley near Natchez, Mississippi. At the time of his second visit Lyell was not disposed to put much credence in

the association, and even in 1863, four years after revelations in France and England concerning artifacts in the glacial-age Drift, Lyell (1863, p. 203) could not admit more regarding the Natchez discovery than " . . . so long as we have only one isolated case, and are without the testimony of a geologist who was present to behold the bone when still engaged in the matrix, and to extract it with his own hands, it is allowable to suspend our judgment as to the high antiquity of the fossil." Evidence for human antiquity (i.e., glacial age) cannot be accepted without *correct* interpretation and documentation of *primary association* between human traces and paleontological or geological data.

This need to document primary associations was also appreciated by contemporary researchers. The Smithsonian Institution, which was to dominate anthropological research in the United States throughout the nineteenth century, published an appeal soon after the beginning of the Civil War (Gibbs, 1862) for collection of archaeological data. The appeal first deals with obtaining skulls of American Indians ("Where without offense to the living, acquisitions of this kind can be made, they will be gladly received as an important contribution to our knowledge of the race."). The second category of acquisitions, "Specimens of Art, etc.," includes "those articles found under conditions which connect archaeology with geology, and which may be classified as follows . . .".

The Smithsonian's list of archaeological *desiderata* includes (Gibbs, 1862, p. 395) 1) "The contents of shell beds of ancient date," (which are noted to have produced evidence of a stone-age culture in Denmark); 2) "Human remains, or implements of human manufacture . . . found in caves beneath deposits of earth and more especially of stalagmites . . ."; 3) Spears and arrowheads . . . discovered in connexion with bones of extinct animals . . ."; and 4) "Implements of the same description found in deposits of sand and gravel . . . such as have recently attracted the attention of European geologists."

Most interesting is the following passage (Gibbs, 1862, p. 395):

In all these cases the utmost care should be taken to ascertain with absolute certainty the true relations of these objects. In the case of the shell banks, the largest trees, where any exist, should, if practicable, be cut down and the annual rings counted. Next the depth of the superincumbent deposit of earth should be measured and its character noted, whether of gravel, sand, or decomposed vegetable matter, as also whether it has been stratified by the action of water. Next the thickness of the shell bed should be ascertained and the height of its base above present high water mark, as also whether it exhibits any marks of stratification. Finally, the face of the bed having been uncovered, a thorough examination should be made, commencing at the top and carefully preserving everything which exhibits signs of human art, and noting the depth in the deposit at which they were discovered. Specimens of each species of shell should be collected, and all bones or fragments of them saved. Evidences of fire having been used should be watched for and recorded.

In the search of caverns the same system should be used. . . .

In theory, at least, there was no question about how archaeological data should be collected—with maximum recording of context and associations. Most investigators, however, did not confine themselves to objective recording of potential early human discoveries; they rather naturally desired to place these in some temporal perspective. What American archaeologists lacked for several decades was a geological framework of post-Tertiary strata in North America. Until the 1880s, their discoveries had to be related somehow to a European context. This problem is apparent in Whittlesey's 1869 review of evidence for the antiquity of man in the United States. He notes that the extinct mammals found in the glacial drift of the U.S. are generally similar to those of the European drift, and assumes, in fact, a "general synchronism" between the Quaternary system of both regions. A logical implication follows (Whittlesey, 1869, p. 273): "... since human relics have certainly been found in this formation in the old world, it is reasonable to expect them here."

Other traces of prehistoric habitation in North America were the shell-heaps along the Atlantic coast, which Whittlesey equated with the Danish kitchen-middens, but considered to be of a younger age. A third class of prehistoric sites included caves, rock shelters, and springs. Whittlesey describes investigations at three such sites, Elyria Shelter in north Ohio (examined in 1851), a cave near Louisville, Kentucky (1853), and High Rock Spring, Saratoga, New York (1866). Traces of human remains or activities were found at all three, but he does not consider any one to be of exceptional antiquity. He briefly mentions Koch's discovery of mammoth bones on the Pomme de Terre River in Missouri, and the claim that a stone projectile point was found beneath them (Whittlesey, 1869, p. 278): "If these statements are reliable, they tend to extend the antiquity of the occupation of red men much beyond that of the American shellheaps, in which are no remains of extinct animals. The statement of Mr. Koch is, however, contradicted by one of the men who assisted him in exhuming the skeleton."

Whittlesey (1869, p. 281ff.) presents a very approximate chronology for the prehistoric habitation of North America, assigning to the American Indians 2,000 years and before them another 2,000 years to the mound builders. "It is highly probable, and is in accordance with the analogies of Europe, that since the glacial period, there was a people here, more ancient than the mound builders, but I know of no remains of such a people except the charcoal beds at Portsmouth, Ohio, in the ancient valley alluvium." Whittlesey also mentions the human pelvis from Natchez, and a stone knife from Grinnell Leads, Kansas found beneath 14 feet of sediment, "... but how ancient the overlying deposits are, cannot be considered as well settled." Here is implicit recognition of the lack of a geochronological context for the artifacts.

As American scientists were looking to Europe for some theoretical context for their discoveries, European scientists seem to have manifested a great interest in the North American fossil record because they perceived it as having extended to more recent times than their own. This is illustrated by a statement of S. F. Haven (1867, p. 35), archaeologist and secretary of the venerable American Antiquarian Society:

... while the progressive changes in nature, especially in animated nature, have been so retarded here [in North America] that the phenomena which have in the eastern hemisphere been buried from sight by the physical revolutions of untold centuries, are often in this country more evident and more accessible ... In these inquiries, which far transcend the ordinary limits of antiquarian research, Archaeology, Geology, and Physiology, joining hands, have adopted a new designation [Anthropology] for their labors.

The notion of the recency of geological events in North America was coupled with a belief in a superior antiquity for humanity here. Both these ideas were to figure in geological and archaeological research during the 1870s and 1880s.

It is not suprising, given this intellectual environment, to find by 1870 reports of human bones and artifacts exhumed from Tertiary strata of the western United States. Perhaps the most notorious example of many reports was the Calaveras Skull, which supposedly had been found in 1866 at the bottom of a mine shaft beneath Tuolumne Table Mountain in the Sierra Nevada Mountains of California. The gold-bearing gravel matrix at Table Mountain had previously yielded Late Pliocene-Pleistocene fossils, and was buried by more than one hundred feet of pre-Late Glacial lava. State geologist J. D. Whitney (1867, p. 277) was convinced of the veracity of this and other cultural finds attributed to the Tertiary gravels, including some ground stone mortars and pestles: "... assuming the correctness of Mr. Matson's statements [on the provenance of the Calaveras Skull], this relict of human antiquity is easily seen to be an object of the greatest interest to the ethnologist as well as the geologist ..."

Unfortunately for him, Whitney had ignored Lyell's first dictum—"to behold the bone when still engaged in the matrx." Whitney evidently was sincere in his belief that it represented Tertiary Man in California (Merrill, 1924, p. 493) but the Calaveras Skull was rather like a nineteenth-century Piltdown Man: for 25 years it lay in the Peabody Museum at Harvard, until Powell and Holmes showed in the 1890s that it was a hoax perpetrated on Whitney by local miners, and Hrdlička (1907), p. 21–28) dismissed any claims of its geological antiquity. Exposure of the Calaveras Skull as a fraud demolished much of the credibility for either Tertiary or Quaternary humans in North America.

To understand what happened to the "early man" controversy during the last two decades of the nineteenth century we must turn to the geologists of the United States Geological Survey and the archaeologists of the Bureau of American Ethnology (BAE). Both agencies were created by the efforts of J. W. Powell in 1879. The BAE, in Powell's conception, was to be a permanent anthropological survey for the United States (Hinsley, 1981, p. 147).

The nineteenth century history of the USGS has been told by Manning (1967), who limits discussion of the archaeological aspects of its work to a single paragraph (p. 79). It may be that this brevity accurately reflects the relative importance of archaeological studies in the Survey's overall research activities. However, a somewhat more extensive review of the political and, to a lesser extent, the geological questions surrounding these disputes is presented in volume 2 of Rabbitt's (1980) history of the U.S. Geological Survey.

The key to the early man controversy lay in defining a relative sequence of glacial-age deposits for North America—one of the goals of the Survey's geologists in the 1880s. By then European glaciologists had recognized the existence of glacial and interglacial stratigraphy, and Americans also began to discern regional sequences as large areas of the Midwest and Rocky Mountains were mapped by state surveys and by the USGS (Flint, 1971, p. 17). Inevitably in the course of this fieldwork some artifacts were discovered by geologists able to appreciate the implications of their context. For example, I. C. Russell (1885, p. 246–247) describes and illustrates the obsidian projectile point found by W J McGee of the USGS in lacustrine clays of glacial Lake Lahontan, "associated in such a manner with the bones of an elephant, or mastodon, as to leave no doubt as to their having been buried at approximately the same time." Fieldwork by geologists of the Geological Survey of Canada in the second half of the nineteenth century also produced some of the first archaeological material from the western provinces of that country (Nobel, 1973). In Canada, as in the U.S., a governmental department of archaeology ultimately was to grow out of the federal geological survey (Trigger, 1981, p. 78).

But McGee, initially a proponent of paleolithic man in North America (McGee, 1888), in 1889 disavowed the Lahontan find as a primary association, and therefore dismissed its relevance to the early man question. What happened to his opinion and that of other USGS geologists during the 1880s seems to have involved a complex interplay of discoveries, personalities, national and state politics, and sociological trends in American science, which together polarized geological and archaeological opinions (although not along disciplinary lines). As knowledge about the Quaternary deposits of North America grew through publications of the federal and state surveys, and as the American landscape increasingly was built upon (and into), the quantity of reported "early man" discoveries increased dramatically, while the quality of their documentation in general did not. Some portion of these finds must have been authentic examples of what today would be termed Archaic or Paleoindian remains, but the majority were misinterpretations: they were not really human bones or not artifacts, or they were quite recent human bones and artifacts in contexts that were misunderstood by the discoverers, or, occasionally, the finds were bones and artifacts knowingly misrepresented.

An example of the first category is illustrated as figure 3 in Foster's *Pre-historic Races of the United States*. Titled "Stone hatchet (?) of finely-crystallized sienite" from the Modified Drift of Jersey County, Illinois, it almost certainly represents a frost-shattered, differentially weathered cobble. The putative hatchet was first described by its discovered, W. McAdams, before the Jerseyville Scientific Association in 1872. He forwarded it and a description of its discovery to Foster, a geologist and past president of the AAAS (his presidential address in 1869 was entitled

"On recent discoveries in ethnology as connected with geology"). Foster (1874, p. 68–69) wrote of the find, "On the whole, I will not positively assert that this specimen is of human workmanship, but I affirm that if it had been recovered from a ploughed field, I should have unhesitatingly said that it was an Indian hatchet." This does not speak well of his familiarity with stone artifacts.

Far more notoriety was attached to other discoveries in the second category—misinterpreted contexts. D. J. Meltzer (1983) places the claims for Paleolithic man that were made by F. Putnam, C. C. Abbott, and G. F. Wright in a wider anthropological context than is possible here. Putnam, as director of the Peabody Museum of Harvard University, was the *de facto* head of the paleolithic man faction in American archeology; their opponents were the geologists of the USGS and the archaeologists of the BAE in Washington. Within the USGS, G. F. Wright, whose two volumes (1891, 1904) summarized the evidence connecting early man finds with the glacial period, fell afoul of the increasing polarization of geological opinion and became engaged unwillingly in an exchange of acrimonious publications with several erstwhile colleagues, among them R. D. Salisbury (1893), W J McGee, and Survey director Powell (1893).

The critical role of geologists in the founding and early development of the Bureau of American Ethnology should not be underestimated. Powell founded it and served as its Director until 1902, when the geologist W. H. Holmes succeeded him in that position until 1910. Holmes led the BAE into the important new fields of physical anthropology (he hired Ales Hrdlička for the Smithsonian) and cultural resource preservation (Noelke, 1974). In Noelke's view Holmes had the geological and archaeological talents to be a leader in the development of modern theory and method, but unfortunately became bogged down in distasteful administrative duties.

During the 1870s and 1880s Powell's interest in American Indian linguistics and ethnography largely kept both the USGS and the BAE removed from growing popular interest in the notion of an American Paleolithic period. It was Holmes who established a link between contemporary American Indians and the crude stone tools that were coming to light at several sites on the East Coast and in the Midwest (Hart, 1976). Regrettably, it is beyond the scope of our chapter to attempt an analysis of all the characters involved in the loci of the Paleolithic man controversy: Dr. C. C. Abbott, the physician on whose farm were found the first evidences of paleolithic artifacts in the Trenton Gravels; the Rev. G. F. Wright, professor at Oberlin Theological Seminary, field geologist with the USGS in Ohio, and author of several controversial books on man and the glacial period; W. H. Holmes, artist, USGS geologist and later BAE archaeologist; W J McGee, also of the USGS and later the BAE. Meltzer (1983) covers in particular the contentious exchanges in the scientific journals concerning these finds.

From a geological viewpoint it is more interesting to examine the part W. H. Holmes played in the controversy. He concerned himself with integrating good field geology with an appreciation of its ethnological implications. Hinsley (1981,

p. 105–107) details Holmes' involvement during the 1880s and 1890s: "Over the years Holmes marshalled additional criteria for challenging every claim for Paleolithic man—uprooted trees, mudslides, the integrity of the investigator . . . His published reviews of Paleolithic finds were masterpieces of cautious apparently open-minded treatment that nonetheless systematically destroyed every claim [e.g., Holmes, 1893a]."

Holmes' approach was a combination of astute field observation and experimental archaeology—recreating the techniques of manufacturing stone tools from blanks, and the production of some percentage of rejects therefrom. His paper giving a geological interpretation for the distribution of stone tool types in the highlands of the Chesapeake Bay region vs. the lowlands (Tidewater country) exemplifies his powers of field observation and persuasive argument (Holmes, 1893b, p. 2):

> The geology of the tide-water country is wholly unlike that of the highland, and the rocks available to the aborigines in the two regions were not only different in distribution but peculiar in the shape they took and in other features that affect the character of the utensils made and employed.

He proceeds to show how different rivers draining different geologies have distributed nonuniformly certain sizes of rock types, thereby determining the location of specialized quarry and activity sites where stone tool blanks were prepared. These unused blanks, Holmes (1893b, p. 2) argues, were often mistaken as paleoliths by those who may not have perceived the true nature of the deposits.

> Each province, district and site, here and elsewhere, is supplied with art remains brought together by the various agencies of environment—topographic, geologic, biologic, and ethnic—and the action of these agencies is to a large extent susceptible of analysis, and this analysis, properly conducted, constitutes a very large part of the science of prehistoric archaeology.

In his crusade against unsophisticated acceptance of the geological evidence offered by an archaeological site, Holmes fell into the same trap himself. Half a century later, a geologist with even more archaeological acumen reinvestigated Holmes' work at several quarry sites and published a far more reasonable interpretation of the evidence they contain about prehistoric quarrying activities (Bryan, 1950). Finally, all controversy regarding the Trenton gravels was laid to rest (Richards, 1939).

The early man controversy, focusing as it did the attention of geologists on details of sedimentation, colluvial processes, animal disturbances, frost action, and so forth, at least had the salutary effect of increased understanding of these phenomena. And geologists also became aware of how natural processes could generate pseudomorphic stone artifacts. N. H. Shaler played this small but interesting role in the Paleolithic debate. During his geological survey of Nantucket in 1891, Shaler (1893, p. 184) collected from a deposit of outwash gravel numerous flint pebbles that appeared to be artificially flaked:

Figure 1. Photograph of N. H. Winchell from the University of Minnesota Archives.

> It became evident to me that if one searched these deposits of washed drift with the eye prepared to find implements, an unconscious choice was made of those having forms which would place them in this category; if, on the other hand, every chipped stone was taken the variety thus gathered was so great that it soon became at once both embarrassing and instructive. . . .It is clear that there are perfectly natural processes by which pebbles may be chipped in such a manner that now and then one of them may have a very artificial aspect. . . .It is clear that just here we have a pitfall most dangerous for the unwary.

His was an early recognition of the possible random natural production of stone "artifacts," a topic that continues to arise in studies of certain early human sites in North America.

Many other leading figures of North American geology in the late nineteenth century found the geological aspects of the early man controversy compelling and made significant contributions to the resolution of questions about the glacial context of human and artifactual remains. A good example is the first state geologist of Minnesota, N. H. Winchell (Fig. 1).

While working out the glacial geology of that region, Winchell became interested in the ancient remains of the indigenous Indian tribes, much as Powell had for the American Southwest.

Winchell's interest culminated in a large volume, *The Aborigines of Minnesota,* (1911).

Winchell had earlier been an outspoken adversary of Powell and of the USGS, believing both did little to serve the best interests of American geology. This opposition to Powell was one of the major reasons for Winchell's founding the *American Geologist* in 1888. No geologist affiliated with the USGS was appointed to the editorial board of this journal, and most were proponents of the American Paleolithic theory. Manning (1967, p. 232) notes that it was not of the same caliber as the *Geological Society of America Bulletin,* but that "excellent material sometimes appeared." The *American Geologist* was guided through the end of the century by N. H. and Alexander Winchell and published many of the important papers in archaeological geology. (Perhaps in partial response to its pro-paleolithic stance, T. C. Chamberlin appointed Holmes the "archaeologic geology" editor of his newly-formed *Journal of Geology* in 1894, and both men published therein papers refuting evidence from several early man localities (e.g., Chamberlin, 1902).

N. H. Winchell summarized many of the questions about the geological context of early humans in North America in his GSA presidential address of 20 December 1902 (Winchell, 1903). In this address he discusses the pre-Quaternary landscape, suggests a possible date for the first invasion of Pleistocene ice (200,000 years before the present, B.P.), opts incorrectly for an aqueous rather than eolian origin for loess, considers the glacial context of the Lansing (Kansas) skeleton, and concludes that ". . . man existed in North America at the time of the Iowan epoch of the ice sheet . . ." This was rapidly becoming a minority view.

In the broader controversy, it is apparent that in response to the claims of Wright, Abbott, Putnam and others for Glacial Man, after the mid-1880s the scientists of the USGS and the BAE became more cautious and conservative in handling geological evidence relating to human antiquity. There was also an element of professionalization, if not outright elitism, involved. Hinsley's book (1981) is much concerned with the professionalization of American anthropology in the second half of the nineteenth century as demonstrated by scientists in the Federal Government. He notes (p. 154) that Geological Survey men like Powell and McGee believed themselves to be best suited to the task of synthesis since their experience crossed several specialties of the sciences—they were generalists, and proud of it. It is apparent in the exchanges between Powell and G. F. Wright, (e.g., Powell, 1893) that the Survey director did not really believe Wright capable of correctly interpreting both the geological and archaeological evidence regarding paleolithic traces in the Quaternary deposits of the U.S.

Ultimately, notes Hinsley (p. 151–152), a schism developed between the BAE, with its nineteenth-century scientific approach of natural history survey by individuals (typified by Powell himself), and the "Boas-trained students that formed the profession of anthropology after the turn of the century." Franz Boas, German-born American anthropologist, was Columbia University's first

professor of anthropology, having taught there from 1899 until his death in 1942. Natural scientists like Powell, McGee, Winchell, and Holmes could undertake anthropological research only as long as there was not a professional cadre of scholars formally trained in that discipline, and that began to happen after the turn of the century. N. H. Winchell (1901, p. 248) eulogized E. Claypole, who was another of this type: "The scope of his studies, and of his writings, were broad. As a scientist he belonged to the old school, to which belonged most of the geologists of fifty years ago. He did not content himself with specialization in any line, but participated in physics, paleontology, glacial geology, prehistoric anthropology, and in all general geology, biology, and botany."

This does not mean that the new generation of anthropologists were all arrayed on the side of an American Paleolithic. On the contrary—Ales Hrdlička shared with his mentor W. H. Holmes an avowed disbelief in this idea (or, more accurately, humans as contemporary with the last glaciation in North America—see Hrdlička, 1907, p. 10). What apparently did happen was that anthropology students under Boas, and later, Alfred Louis Kroeber, simply came to regard the Paleolithic man question as a nonquestion, irrelevant to the new anthropological paradigm of historical particularism. It was left by default to Hrdlička, as a professional anthropologist, to pass judgment not only on the physical anthropological and archaeological evidence for human antiquity in the New World, but the geological evidence as well.

From our present-day perspective, it appears that most of the American paleolithic controversy at the turn of the century could have been avoided had archaeologists (and most geologists) appreciated the distinction between the relative chronologies of prehistoric sequences in Euruope and North America. At that time one of the few who did was the "old-school" natural scientist E. Claypole, who in 1898 wrote on the distinction between paleolithic and neolithic time on the two continents. By the end of the nineteenth century, geologists in Europe and in England had come to realize that their paleolithic period extended at least back to the most recent interglacial, while neolithic habitation was exclusively post-glacial. Claypole (1898, p. 338) applies this notion to the American archaeological record, and states clearly that, "No case has yet been brought forward in which the tools or weapons of man have been found in such circumstances as to allow the belief that they were of interglacial age . . . Even the argillyte implements from New Jersey [the Trenton Gravels], probably the oldest yet described, cannot claim an antiquity greater than early post-glacial." He notes that American archaeologists, if they wish to persist in using the term "paleolithic" in reference to artifacts, *must divorce it from any chronological connotation* and strictly limit it to morphology (p. 339): "Strong confirmation of the doctrine which assigns all the hitherto published 'paleolithic(?)' finds of North America to a much later date than that which is justly claimed for those of England is found in the comparative recency of their geological setting." Claypole ends his paper by observing (p. 340): "It is, of course, not to be expected that archaeologists should also be geologists, but it is absolutely necessary in order to avoid serious error that each should be familiar with the discoveries of the other when they touch upon his own particular province. The study of early man is the most important meeting point of the two sciences and here there should be harmony and mutual understanding."

Claypole's astute interdisciplinary attitude unfortunately was ignored, and D. J. Meltzer (1983, p. 23) is correct when he observes that ". . . the proponents of each [side of the early man issue] spent much of their time talking past one another. . ."

And thus events continued into the early twentieth century, due primarily to the character of Aleš Hrdlička of the Smithsonian Institution. He took on the mantle of W. H. Holmes as judge of any discovery involving human remains claimed to be of Glacial man. Krieger's assertion notwithstanding (1964, p. 24–25), Hrdlička did assume (1907, p. 10) the Glacial (= Pleistocene) period to have begun a gradual decline about 20,000 years B.P. in the northern U.S. and to have ended some 10,000 year B.P. (the Champlain substage) at the latitude of the St. Lawrence River valley. Hrdlička therefore was not surprised that all the early human remains he investigated were of "modern," or (to him) postglacial appearance.

It remains a fact that Hrdlička overtly intimidated other professionals into abandoning research on early humans in the western hemisphere. His three BAE publications (1907, 1912, 1918) are models of excellent physical anthropological documentation combined with devastating imputations and innuendo regarding the circumstances of the finds and the dependability of the finder, all in his acid prose. The most infamous of these cases involved E. H. Sellards (Fig. 2), who was state geologist first of Florida (from 1907–1918) and then of Texas (1932–1945). Although he personally found the controversy deeply disturbing (Krieger, 1961), Sellards carried on a two-decade dispute with Hrdlička over the correct interpretation of human remains and artifacts in apparent primary association with extinct Pleistocene fauna at Vero and Melbourne, on the east central coast of Florida.

In Hrdlička's discussion of the Vero find he questions both the primary association of human bones with extinct vertebrate remains and the physical anthropological evidence for the pre-modern identification of the human bones, observing that (Hrdlička, 1918, p. 35):

It is scarcely safe for the geologist or the paleontologist to assume that the problem of human antiquity is his problem. Although it is only just to acknowledge that geology and especially paleontology can be, on occasion, of the greatest aid to anthropology in determining the age of human remains, yet these branches are not adequate in themselves to deal with the subject. In all cases where the remains of man are concerned, be they cultural or skeletal, there enters a most important factor into the case which does not exist for the geologist or the paleontologist, namely, the *human element*, the element of man's conscious activities.

In his conclusion to this publication Hrdlička gives his general opinion regarding interdisciplinary activity: " . . . Our colleagues in collateral branches of science will be sincerely

Figure 2. Photograph of E. H. Sellards from the Texas Bureau of Economic Geology Archives.

thanked for every genuine help they can give anthropology, but they should not clog our hands."

Sellards continued his interest in documenting direct associations between human remains and Pleistocene fauna (Krieger, 1961). He published two extensive bibliographies and indexes to early human localities in North America (1940, 1947), and summarized his views in *Early Man in America* (1952). But the battle had ended a quarter-century earlier with the discovery of projectile points in primary, undisturbed association with bones of an extinct species of bison at a site near Folsom, New Mexico (Figgins, 1927).

Figgins begins his 1927 article thus (p. 229):

When we analyze the technical opposition to the belief that man has inhabited America over an enormous period of time, we find it is not only restricted to an individual minority, but it also appears to be traceable to the results of a too circumscribed viewpoint—a failure to appreciate properly *all* the evidence, and a seeming unwillingness to accept the conclusions of authorities engaged in related branches of investigation.

Figgins goes on to explain how geology provides the bridging argument between human skeletal remains and arguments of Pleistocene antiquity. The Folsom point found in direct physical contact with a rib bone at the type site on the Cimarron River in New Mexico represents an association made famous by a photograph published in Figgins' article (p. 234). The geological details of the Folsom site's associations and those of two others— Frederick, Oklahoma and Colorado, Texas—were discussed in the same issue of *Natural History* by H. J. Cook (1927). He concentrates on descriptions of the position of artifacts and extinct mammals in the local geological sequences, which in all three cases represent Pleistocene stream channel or valley fills. The New Mexico and Texas cases are typical bison kill sites, but the Frederick site, more controversial because of the presence of a spear point in association with earlier Pleistocene material, was subsequently disavowed (Evans, 1930).

It was appreciated by most contemporary archaeologists and geologists who were concerned with the question of human antiquity in North America that the Folsom finds ended serious opposition to at least a late-glacial human presence here. In a remarkably short time enough geological and archaeological evidence was accumulated from sites in the Southwest U.S. to allow a comprehensive multidisciplinary study and review (Howard, 1935) of the early human question.

Howard's archaeological excavations at Burnett Cave and at Carlsbad, New Mexico were done as part of his dissertation under F. Ehrenfeld, Professor of Geology at the University of Pennsylvania; the Carnegie Institution supported his field studies. At several commercial gravel pits between Clovis and Portales, Howard described and sampled the Late Pleistocene ephemeral lake sediments—blue-gray silty sands—which contained mammoth teeth, the bones of extinct horse and bison, and traces of human activity. Grain-size analysis was performed on samples of the lake sediments, and their included diatoms were studied by K. E. Lohman of the USGS. Charcoal fragments from the lake sediments were identified by R. W. Chaney of UCLA, and the vertebrate fauna was described by C. Stock of Cal Tech. An unsuccessful attempt was made by P. B. Sears to extract pollen grains from the sediments.

Howard's synthesis places all the accumulated paleoenvironmental data into a reconstruction of the late Pleistocene environment of the Llano Estacado. He also presents a review of the typology and distribution of known paleo-Indian projectile points in North America and a review of the various contemporary theories regarding the chronology and climatic history of the Late Glacial. He ends with an exhaustive consideration of all known associations of human skeletons and extinct mammals, concluding (p. 151) "... ten thousand years is the most satisfactory date for the length of time man has been in America." This was the datum for early human studies in the mid-1930s. The elegant geological work of K. Bryan and his colleagues and students at sites such as Lindenmeier (Bryan and Ray, 1941), Sandia Cave (Bryan, 1941), and Ventana Cave (Haury, 1950) reinforced 10,000 yr b.p. as the latest possible date and suggested occupation might extend back to the late Pleistocene. With the advent of radiocarbon dating, occupation has been extended further back into the Late Pleistocene by several millennia, thought with increasingly contentious debate associated with earlier sites. In a recent review of early human sites in North America, D. Stanford (1983, p. 65) states:

The major problem concerning most of these sites is that they do not meet certain criteria that would provide clear evidence of pre-Clovis culture. These criteria are: (1) a clearly defined stratigraphy, (2) reliable and consistent radiometric dates, (3) consonance of data from relevant interdisciplinary studies, and (4) the presence of unquestionable artifacts in an indisputable primary context.

His last and most crucial criterion was proposed by Lyell some 140 years ago, and, like the other three, will always necessitate the collaboration of geologists and archaeologists to satisfy it.

PALEOENVIRONMENTS AND ARCHAEOLOGY

An appreciation of the physical environment as a framework within which human societies operate and evolve was slow in developing among archaeologists, and could only be expected to do so after some body of knowledge of prehistory had accumulated. Such appreciation also could come only with the recognition of past human interaction with the natural environment, a subject first covered by G. P. Marsh in his 1864 (first edition) volume *Man and Nature, or Physical Geography as Modified by Human Action* (Marsh, 1874). Strangely, Marsh's classic had little substantive effect on either the concepts or the course of American prehistoric studies, despite the fact that he was a correspondent of Lyell's and that his eclectic examples of human modification of geomorphological phenomena such as soil erosion and river alluviation could have been applied to the archaeological record. This lack of impact from Marsh's book may have been due to the prevailing atmosphere of natural determinism in late nineteenth century science, which ran counter to his thesis of human effects on nature. Also, Marsh's references to the prehistoric (as opposed to the historic) record are limited, although his recognition that shell middens often mark ancient shorelines (1874, p. 640f) shows an awareness of their potential for coastal paleogeographic studies.

More appreciation of the changing environments of prehistory arose among European archaeologists and geologists, possibly because the contrasts between modern and ancient late Quaternary environments were greater there than in North America. J. Steenstrup's excavations of shell middens and peat bogs revealed that pine trees had grown in Denmark during prehistoric times, and A. Morlot's work on the lake dwellings of Switzerland indicated that they were built during a period of much lower water levels. Between approximately 1860 and 1890, European geologists studying the Quaternary glacial record came to realize that the different vegetation and climate of Upper Paleolithic times would have affected, in unspecified ways, the activities of contemporary hunters and gatherers (Geikie, 1895, p. 643).

To a lesser degree the concept of a substantially different prehistoric environment affecting contemporary societies was also appreciated by American geologists and archaeologists of the late nineteenth century, but it was developed fully by a remarkable American archaeological geologist named Raphael Pumpelly. Because Pumpelly's fieldwork was not in North America, we will not treat it in detail, but the novelty and modern aspect of his conception of archaeological research warrants some description. His two-volume autobiography (Pumpelly, 1918) is a fascinating record that reads almost like a fictional adventure story covering half a century.

As a mining geologist, Pumpelly visited and travelled through the interior of Chinese and Russian Central Asia during the American Civil War, and then developed a fascination with the idea of an ephemeral inland sea in that region produced by Late Quaternary climatic fluctuations. According to this theory, the progressive desiccation of this body of water after the third century B.C. forced the outward migrations of Aryan nomadic tribes westward into Europe and eastward into China—an example of environmental determinism on a grand scale. In 1904, at the age of 67 and after a successful geological career, Pumpelly was able to devote all his considerable talents to archaeology and undertake an expedition to Russian Turkestan to investigate the evidence for the theory, thanks to the support of the newly formed Carnegie Institution (Pumpelly, 1908).

The scope of the project, and Pumpelly's imagination in its execution and in the interpretation of results, are astounding for such an early date. Pumpelly's collaborators included ceramicists, vertebrate paleontologists, ethnobotanists, experts on early metallurgy, physical anthropologists, geomorphologists, and climatologists. The team at the Anau site cumulatively reconstructed the physical and human environment of the 4th millenium B.C. Their research was far ahead of its time; it compares favorably with multidisciplinary archaeological research that was done thirty or more years later. Pumpelly clearly understood that in the study of prehistory, which involves spatio-temporal change, physiography and biology are essential factors.

Pumpelly's reconstruction of the chronology at Anau, to take only one of many possible examples, illustrates how his geological reasoning led him to a "working hypothesis," based on several explicit assumptions not listed here (Pumpelly, 1908, p. 50):

This is applying to archeology simple rules of geological reasoning. We know the thickness of the strata of each of the cultures of the three neighboring sites [at Anau], and we know the aggregate existing thickness of the cultures of all the sites. If we take the duration of each culture to be proportional to the thickness of its accumulated strata, the duration of the entire series will be represented by the aggregate existing thickness of all of the strata plus any culture-gaps between different cultures, and minus any overlaps of the cultures of neighboring sites. Having the standing thickness of the different cultures, two additional factors are needed to convert the stratigraphic record into a chronological one:

(a) The assignment of values to the observed four intervals that existed between the different cultures.

(b) The rate of accumulation of the culture-strata.

His final "Stratigraphic Chronology of the Cultures" is reproduced here (Fig. 3); the depositional lacuna inferred to exist between the Copper and Iron Age strata is the first ever portrayed on an archaeological stratigraphic section.

Before returning to Europe and more fieldwork in December 1905, Pumpelly, as retiring president of the Geological

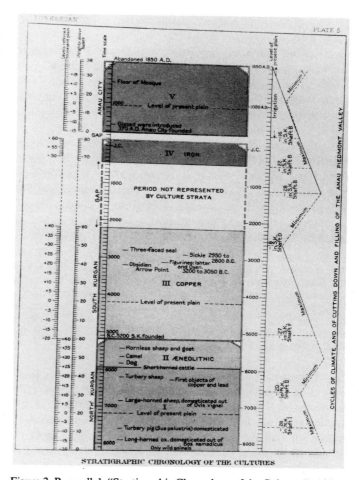

Figure 3. Pumpelly's "Stratigraphic Chronology of the Cultures." (1908)

Society of America, delivered a public lecture on the results of his Anau excavations. He mentions modestly in his autobiography (1918, p. 748) that "The attendance was very large, not only of members but also of citizens, and I was made to feel that the address was liked."

During the first two decades of the twentieth century, little was done in North America to apply geological techniques to archaeological problems. One exception is the work of E. Huntington (who had been a field assistant on Pumpelly's expedition) on climatic change in the American Southwest during Pre-Columbian times (Huntington, 1914). Although a geographer, Huntington's physiographic studies of alluvial terraces as a record of climatic "pulsations" must have been read by and influenced the next generation of geomorphologists who were to investigate archaeological sites in their natural settings. The second exceptional development was a refinement in stratigraphic excavation techniques. Willey and Sabloff (1980, p. 84–93) note that the "stratigraphic revolution" in American archaeology was instigated by two young anthropological archaeologists (M. Gamio and N. C. Nelson) who had excavated at European Paleolithic sites and had been fundamentally influenced by the more geologically-oriented notion of prehistory that already was developing there.

For the quarter-century between 1925 and 1950, research and teaching at the interface between geology and archaeology in North America was largely dominated by one person, Kirk Bryan, who was first a geologist with the USGS but was then for most of his professional career Professor of Physiography in the Department of Geology at Harvard University.

In the early 1920s Neil Judd, director of the excavations sponsored by the National Geographic Society at Pueblo Bonito in northwestern New Mexico, asked Bryan to investigate the local geologic record for evidence of any changes in environment that may have affected the population of Chaco Canyon. Bryan's manuscript report was posthumously published in 1954. As noted by Johnson (1951),

This famous series of studies [at Pueblo Bonito] demonstrates his ability to collaborate with colleagues in different fields, and to select with acumen and with a sense for practical value, facts and hypotheses contributing to the solution of a problem. As the studies progressed they became of increasing significance to archaeological work because the cycles of sedimentation and erosion were synchronized with cultural events, the whole forming a setting for human prehistory.

During the 1930s Bryan was active in assessing the geological setting of sites of the newly recognized Folsom culture, summarizing his work at the original Folsom site and at the Lindenmeier site in a symposium on early humans and their paleoenvironments; the symposium papers were edited by G. G. MacCurdy, Director of the American School of Prehistoric Research (Bryan, 1937). This volume is of great interest, not only for its presentation of paleoenvironmental studies by P. B. Sears (pollen analysis), H. B. Richards (Pleistocene molluscs), E. Antevs (paleoclimate), and M. M. Leighton (weathering profiles), but also because it contains the last statement in print by the indomitable Hrdlička disputing any evidence for glacial-age humans in North America. It therefore marks the end of one era in American archaeology and the beginning of the modern one.

Bryan, born and schooled in New Mexico, devoted many months of field work to the Southwest, studying both the Quaternary geology and archaeology of the region. The Quaternary stratigraphic studies done with C. C. Albritton, Jr., in the Davis Mountains of Trans-Pecos Texas (Albritton and Bryan, 1939), were supported by a grant from the Penrose bequest of the GSA, and made extensive use of associated Paleoindian and Archaic artifacts to establish field relationships among the several mapped units. Bryan's subsequent geological studies at Sandia Cave (Bryan, 1941) and at Ventana Cave (Haury, 1950) are also models of careful application of geological inference and interpretation to the reconstruction of past natural environments.

At Harvard, Bryan assigned to several of his graduate students thesis research that might today be characterized as archaeological geology. L. L. Ray worked out a detailed late glacial chronology for the Cache la Poudre drainage basin in northern Colorado as a means of geologically dating the Lindenmeier site

(Bryan and Ray, 1941). Another student, J. T. Hack, worked with Harvard's Awatovi Expedition in the Jeddito Valley of Arizona, documenting (Hack, 1942) the prehistoric strip mining of coal by the Hopi residents during the Pueblo III-V periods (ca. 13th-17th centuries A.D.). Hack's work was encouraged by the archaeological director, who wrote in the preface: "It was the basic policy of the expedition to encourage the cooperation of students in other branches of learning who believed that their techniques could appreciably augment the strictly archaeological approach to the problems of prehistory. The policy has been justified beyond all expectation." Hack, along with another of Bryan's Harvard students, L. B. Leopold, worked for the USGS after graduating from Harvard.

Yet another of his students was S. Judson, who carried out under Bryan's supervision a very detailed geological study of the Protoarchaic San Jon site in northeastern New Mexico, on the southern High Plains (Judson, 1953). Judson describes the Pleistocene stratigraphy and sedimentary history of the site relative to the bison remains and Plainview points found there. He presents an alluvial chronology modeled after Bryan's for Chaco Canyon, and documents the historic-period development of arroyos in the area due to a change in stream regime (steepened gradients), which he, along with Hack (1942), attributes to increased run-off from destruction by grazing of the vegetation cover, combined with a period of aridity beginning in the 1880s.

Clearly Kirk Bryan influenced much of the American work in archaeological geology shortly after WWII. His work was not limited to the New World; one project that remained unfinished at his death in 1950 involved a geological study of the rock shelter at La Colombière in southeast France. It was completed by two of his junior colleagues, H. L. Movius and S. Judson (1956), and dedicated to his memory.

We believe that by the end of Bryan's career he had effected a convergence of Pleistocene studies and archaeological geology through an explicit paleogeographic approach to archaeology that he stated thus (Bryan, 1951, p. 3): "As an idea and as an objective we need to acquire a complete picture of the landscape at each interval of the past. In this framework man and his economy should fit, and on this background his social structure was built."

Of course others were engaged in geological studies related to archaeology during the pre- and immediate post-WWII period. One such prolific researcher was E. Antevs, who extended the varve chronology method of dating glacial features from Fenoscandia to the northeastern U.S. The study of glacial varves and climatic change ultimately led him to an interest in the relationship of human artifacts to late Quaternary geological events. In 1934, as a research associate of the Carnegie Institution of Washington, he began a study of the type site for the Clovis culture. From this came Antevs' classic "Age of the Clovis Lake Beds" (1935), which integrated well the geological and archaeological settings of the site.

Antevs employed a climatic interpretation of sediment rates to develop a late Quaternary chronology indicating stages when warmth, dryness, coolness, wetness or combinations of these predominated in North America. Dates for these stages were extrapolated with caution from the New England varve chronologies. His interest in the Desert Archaic cultures of the Southwest U.S. is evidenced by a study of the Double Adobe School site on Whitewater Draw in southeast Arizona (Antevs, 1937). He also assisted in the excavation of other prehistoric sites in Arizona and New Mexico (Smiley, 1977; Antevs, 1948, 1949a, 1949b). Antev's work provided an important new dimension in geology's contribution to prehistoric archaeology—the ability, under optimum conditions, to date certain geological deposits and, indirectly, cultural remains in calendar years. Towards the end of his active scientific career the technique was to a large extent superseded by radiocarbon analysis, which has demonstrated the validity of the late Quaternary chronology for North America constructed by Antevs, Bryan, and others.

By about 1950 both geologists and archaeologists had arrived at a mature conception of the potential benefits to be derived from interactive collaboration. This is exemplified by Movius' article (1949), which offers an environmental-deterministic viewpoint of Old World paleolithic archaeology, stressing the nearly inseparable bond between it and the natural sciences. While acknowledging that prehistoric archaeology is a social science, Movius emphasizes that its goal of explicating humanity's progressive ability to cope with the natural environment can only be realized through natural science studies— primarily geological—that are concerned with the sequence and correlation of Pleistocene events.

From this level of conceptualization stems the very sophisticated multidisciplinary projects of R. J. Braidwood and his natural science colleagues in Iraqi Kurdistan during 1947–1955 (Braidwood and Howe, 1960). We assume it is in contrast to his own experiences that Braidwood (1957) comments on the general lack of cross-disciplinary understanding and cooperation between archaeologists and natural scientists. After pointing out that cultural interpretations within an environmental framework demand maximum knowledge of paleoenvironments he observes (p. 16):

> It seems to me that in reaching for the goal, a new field, perhaps 'Pleistocene ecology,' might come into focus, allied to archaeology (and human paleontology) but setting the archaeologist himself somewhat more free to deal with matters of culture. This field or axis of interrelated disciplines (perhaps 'Pleistocene ecology' or 'paleo-environment' or 'Quaternary geography'—I shall not attempt to name it!) would definitely include Man as an element in, and a factor acting upon, the environmental scene.

Eight years later, K. Butzer's *Environment and Archaeology: An Introduction to Pleistocene Geography* was published; it marked the first book-length scholarly treatment of the interdisciplinary field proposed by Braidwood. Much of contemporary field and laboratory research in archaeological geology is directed towards elucidating the paleoenvironments contemporary with a site or region under multidisciplinary investigation; many of the methods applied are explained in Rapp and Gifford (1985).

SUMMARY AND CONCLUSIONS

In choosing to focus our history on the dominant themes of early human and paleoenvironmental studies we unavoidably will be deficient in treating the related research areas that figure in archaeological geology, particularly the subject of provenance studies of the inorganic raw materials used in prehistory. North American geologists perhaps played a lesser role here than did their European co-workers, with the usual notable exceptions (e.g., Washington, 1898, 1923). Artifact provenance studies in North America indeed have focused during the past two decades on obsidian, flint, and other lithic materials, and more recently on galena and native copper; however, much of this analytical work has been done by archaeologists working with chemists or physicists rather than with geologists (e.g., Farquhar and Fletcher, 1980).

The archaeologist W. K. Moorehead (1910, p. 357), in the hope of determining the relative age of stone artifacts based on patination of their surfaces, submitted a collection of samples to "... Prof. John D. Irving of Lehigh University, who is Secretary of the Geological Society of America, and an expert in such matters. Some of these specimens are found to be old, a few very old, and others more or less recent." Irving identified substantial weathering of the worked surface of artifacts such as a syenitic gneiss with oxidized hornblende and concomitant staining of the surface by limonite. He could only inform Moorehead that this degree of chemical weathering probably required more than a few hundred years, but this was as much an indication of the lack of laboratory analytical data on rock weathering rates as a shortcoming of the archaeological application. When we reach the recent period of more sophisticated collaboration, Judson (1957) used the example of how a geochemical study of flint, chert, and obsidian might reveal some facts of mutual interest to geologists and archaeologists.

We now return to our opening statement that prior to 1977 the geological profession had not officially recognized a distinct subdiscipline devoted to geological aspects of human activities through time. Archaeological geology became a bureaucratic reality at the Geological Society of America's annual meeting of that year, in Seattle, when the society's Archaeological Geology Division was formally convened. However, as we have tried to exemplify in this chapter, the record of activity in this subdiscipline extends back to the middle of the nineteenth century. At the turn of the century archaeological geology had the attention and interest of leading geological scientists.

In the early twentieth century factors such as the shift of archaeology from a natural to a more social science (Daniel, 1975, p. 239–242), and the erection of more specific disciplinary boundaries resulted in a decline in overall effort, but archaeological geology activities were still carried on by a few individuals. More work began in the 1930s, with the realization that natural environments of the newly recognized Paleoindian cultures could best be determined through natural science investigation of the associated physical record. Following WW II the level of multi-disciplinary activity accelerated concomitantly with the appearance of new theories of cultural materialism (e.g., Steward, 1949).

The body of published materials on archaeological geology from the last four decades is extensive, but much of it may be classed as "gray literature," i.e., reports and publications of site studies that did not appear as papers in refereed journals or in books and monographs widely available in academic libraries. Any thorough review of this material would be extremely difficult, and we do not pretend to have covered it adequately here. Even less have we exhausted the more recent vast and widely available literature of the past two decades. To do so would require a book-length effort. What we have tried to demonstrate instead is the genuine link that has existed, relatively undocumented, between geology and archaeology, which will form the basis for continued productive cooperation.

REFERENCES CITED

Albritton, C. C., Jr., and Bryan, K., 1939, Quaternary stratigraphy in the Davis Mountains, Trans-Pecos Texas: Bulletin of the Geological Society of America, v. 50, p. 1423–1474.

Antevs, E. V., 1935, Age of the Clovis Lake beds: Proceedings of the Academy of Natural Science, Philadelphia, v. 87, p. 304–312.

——— 1937, Climate and early man in North America, in MacCurdy, G. G., ed., Early Man: Philadelphia, J. B. Lippincott, p. 125–132.

——— 1948, Climatic changes and pre-white man, in The Great Basin, with emphasis on glacial and postglacial times: Univ. of Utah Bulletin, v. 38, p. 168–191.

——— 1949a, Geology of the Clovis sites, in Wormington, H. M., ed., Ancient man in North America: Denver Museum of Natural History Popular Series No. 4, p. 185–192.

——— 1949b, Age of the Cochise artifacts on the West Leggett: Fieldiana-Anthropology, Field Museum of Natural History, v. 38, p. 34–57.

Braidwood, R. J., 1957, Means toward an understanding of human behavior before the present, in Taylor, W. W., ed., The identification of Non-Artifactual Archaeological Materials: NAS-NRC Pub. No. 565, p. 14–16.

Braidwood, R. J., and Howe, B., 1960, Prehistoric investigations in Iraqi Kurdistan: Oriental Institute, Studies in Ancient Oriental Civilization, no. 31, 184 p.

Bryan, K., 1937, Geology of the Folsom deposits in New Mexico and Colorado, in MacCurdy, G. G., ed., Early Man: Philadephia, J. B. Lippincott, p. 139–152.

——— 1941, Correlation of the deposits of Sandia Cave, New Mexico, with the glacial chronology: Smithsonian Miscellaneous Collections, v. 99, no. 23, p. 45–64

——— 1950, Flint quarries—the sources of tools, and, at the same time, the factories of the American Indian: Harvard University, Peabody Museum Papers, v. 17, no. 3, 40 p.

——— 1951, Foreword, in Moss, J. H., ed., Early man in the Eden Valley: Philadelphia, University of Pennsylvania Museum, p. 1–4.

——— 1954, The geology of Chaco Canyon, New Mexico, in relation to the life and remains of the prehistoric peoples of Pueblo Bonito: Smithsonian Miscellaneous Collections, v. 122, no. 7, 65 p.

Bryan, K., and Ray, L. L., 1941, Geologic antiquity of the Lindenmeier site in Colorado: Smithsonian Miscellaneous Collections, v. 99, no. 2, p. 1–76.

Butzer, K., 1964, Environment and Archaeology: an Introduction to Pleistocene Geography: Chicago, Aldine, 524 p.

Chamberlin, T. C., 1902, The geologic relations of the human relics at Lansing, Kansas: Journal of Geology, v. 10, p. 745–777.

Claypole, E. W., 1898, Paleolith and neolith: The American Geologist, v. 21, p. 333–344.

Cook, H. J., 1927, New geological and paleontological evidence bearing on the antiquity of mankind in America: Natural History, v. 27, p. 240–247.

Dana, J. D., 1875, On Dr. Koch's evidence with regard to the Contemporaneity of Man and the Mastodon in Missouri: American Journal of Science, v. 9, p. 335–356.

Daniel, G., 1975, A hundred and fifty years of archaeology: London, Duckworth, 410 p.

Evans, O. F., 1930, The antiquity of man as shown at Frederick, Oklahoma: a criticism: Journal of the Washington Academy of Sciences, v. 20, p. 475–478.

Farquhar, R. M., and Fletcher, I. R., 1980, Lead isotope identification of sources of galena from some prehistoric Indian sites in Ontario, Canada: Science, v. 207, p. 640–643.

Figgins, J. D., 1927, The antiquity of man in America: Natural History, v. 27, p. 229–239.

Flint, R. F., 1971, Glacial and Quaternary Geology: New York, John Wiley, 892 p.

Foster, J. W., 1874, Pre-historic races of the United States of America: Chicago, S. C. Griggs, 415 p.

Geikie, J., 1895, The Great Ice Age and its relation to the antiquity of man (3rd ed.): New York, Appleton & Co., 850 p.

Gibbs, G., 1862, Instructions for archaeological investigations in the United States: Smithsonian Institution, Annual Report for 1861, p. 392–396.

Hack, J. T. , 1942, Prehistoric coal mining in the Jeddito Valley, Arizona: Harvard University, Peabody Museum Papers, v. 35, no. 2, 24 p.

Hart, K. R., 1976, Government geologists and the early man controversy: The problem of "official" science in America, 1879–1907: Ph.D. Dissertation, Kansas State University, 351 p.

Haury, E. W., 1950, The stratigraphy and archaeology of Ventana Cave, Arizona: Albuquerque, Univ. of New Mexico Press, 599 p.

Haven, S. F., 1867, Report of the Council to the Society, Proceedings of the American Antiquarian Society, Annual Meeting, p. 9–63.

Hinsley, C. M., Jr., 1981, Savages and scientists: Washington, Smithsonian Institution Press, 319 p.

Holmes, W. H., 1893a, Are there traces of man in the Trenton Gravels?: Journal of Geology, v. 1, p. 15–37.

——— 1893b, Distribution of stone implements in the Tidewater country: American Anthropologist, v. 6, p. 1–14.

Howard, E. B., 1935, Evidence of early man in North America: University of Pennsylvania, Museum Journal, v. 24, no. 2–3, 176 p.

Hrdlička, A., 1907, Skeletal remains suggesting or attributed to early man in North America: Smithsonian Institution, Bureau of American Ethnology, Bulletin No. 33, 113 p.

——— 1912, Early man in South America: Smithsonian Institution, Bureau of American Ethnology, Bulletin No. 52, 405 p.

——— 1918, Recent discoveries attributed to early man in America: Smithsonian Institution, Bureau of American Ethnology, Bulletin No. 66, 67 p.

Huntington, E., 1914, The climatic factor as illustrated in arid America: Washington, D.C., Carnegie Institution Publication No. 192, 341 p.

Johnson, F., 1951, Kirk Bryan—1888–1950: American Antiquity, v. 16, p. 253.

Judson, S., 1953, Geology of the San Jon site, eastern New Mexico: Smithsonian Institution, Miscellaneous Collections, v. 121, n. 1, 69 p.

——— 1957, Geology, in Taylor, W. W., ed., The identification of non-artifactual archaeological materials: NAS-NRC Pub. No. 565, Washington, D.C., p. 48–49.

Krieger, A. D., 1961, Elias Howard Sellards—1875–1961, American Antiquity,

v. 27, p. 225–228.

—— 1964, Early man in the New World, *in* Jennings, J. D., and Norbeck, E., eds., Prehistoric Man in the New World: Chicago, University of Chicago Press, p. 23–81.

Lyell, C., 1863, The geological evidences of the antiquity of man (2nd ed.): Philadelphia, G. W. Childs, 526 p.

Manning, T. G., 1967, Government in science: Lexington, Univ. of Kentucky, 184 p.

Marsh, G. P., 1874, The Earth as modified by human action, a new edition of Man and Nature: New York, Scribner, Armstrong & Co., 656 p.

McGee, W J, 1888, Paleolithic man in America: his antiquity and his environment: Popular Science Monthly, v. 34, p. 20–36.

McMillan, R. B., 1976, Man and mastodon, a review of Koch's 1840 Pomme de Terre expeditions, *in* Wood, W. R., and McMillan, R. B., eds., Prehistoric Man and his environments: a case study from the Ozark Highlands: New York, Academic Press, 271 p.

Meltzer, D. J., 1983, The antiquity of man and the development of American archaeology, *in* Schiffer, M. B., ed., Advances in archaeological method and theory, Vol. 6: New York, Academic Press, 359 p.

Merrill, G. P., 1924, The first one hundred years of American geology: New Haven, Yale University Press, 773 p.

Moorehead, W. K., 1910, The stone age in North America (2 vol.): Boston, Houghton Mifflin, 768 p.

Movius, H. L., 1949, Old-World Paleolithic archaeology: Bulletin of the Geological Society of America, v. 60, p. 1443–1456.

Movius, H. L., and Judson, S., 1956, The rock-shelter of La Columbière: American School of Prehistoric Research, Peabody Museum of Harvard, Bulletin No. 19, 147 p.

Nobel, W. C., 1973, Canada, *in* Fitting, J. E., ed., The development of North American archaeology: Garden City, New York, Anchor Books, p. 49–83.

Noelke, V.H.M., 1974, The origin and early history of the Bureau of American Ethnology, 1879–1910: Ph.D. Dissertation, The University of Texas at Austin, 338 p.

Powell, J. W., 1893, Are there traces of man in the glacial gravels?: The Popular Science Monthly, v. 43, p. 316–326.

Pumpelly, R., 1908, Explorations in Turkestan (2 vol.): Washington, D.C., Carnegie Institution Publication No. 73, 494 p.

—— 1918, My reminiscences (2 vol.): New York, Henry Holt, 844 p.

Rabbitt, M. C., 1980, Minerals, lands, and geology for the common defense and general welfare. Volume 2, 1879–1904. U.S.G.S. and U.S. Government Printing Office, 407 p.

Rapp, G. R., Jr., and Gifford, J. A., 1985, Archaeological geology: New Haven, Yale University Press, 435 p.

Richards, H. G., 1939, Reconsideration of the dating of the Abbott Farm site at Trenton, New Jersey: American Journal of Science, v. 239, p. 345–354.

Russell, I. C., 1885, Geological history of Lake Lahontan, a Quaternary lake of northwestern Nevada: U.S. Geological Survey, Monograph 11, 288 p.

Salisbury, R. D., 1893, "Review of Man and the Glacial Period," by G. F. Wright: The American Geologist, January, p. 13–20.

Sellards, E. H., 1940, Early man in America, index to localities and selected bibliography: Bulletin of the Geological Society of America, v. 51, p. 373–432.

—— 1947, Early man in America, index to localities and selected bibliography, 1940–1945: Bulletin of the Geological Society of America, v. 58, p. 955–978.

—— 1952, Early man in America: Austin, Texas, University of Texas Press, 211 p.

Shaler, N. S., 1893, Antiquity of man in eastern North America: The American Geologist, March, p. 180–184.

Smiley, T. L., 1977, Memorial to Ernst Valdemar Antevs, 1888–1974: Geological Society of America Memorials, v. 6, p. 1–7.

Stanford, D., 1983, Pre-Clovis occupation south of the ice sheets, *in* Shutler, R., ed., Early man in the New World: Beverly Hills, Calif., Sage Publications, 223 p.

Steward, J. H., 1949, Cultural causality and law; a trial formulation of the development of early civilizations: American Anthropologist, v. 51, p. 1–27.

Trigger, B. G., 1981, Giants and pygmies: the professionalization of Canadian archaeology, *in* Daniel, G., ed., Towards a history of archaeology: London: Thames & Hudson, p. 69–84.

Vita-Finzi, C., 1978, Archaeological sites in their setting: London, Thames & Hudson, 176 p.

Washington, H. S., 1898, The identification of the marbles used in Greek sculptures: American Journal of Archaeology, v. 2, p. 1–18.

—— 1923, Stone adzes of Egypt and Hawaii: Journal of the Washington Academy of Sciences, v. 13, p. 377–383.

Whitney, J. D., 1867, Notice of a human skull, recently taken from a shaft near Angel's, Calaveras County: California Academy of Natural Sciences, Proceedings, v. 3 (1863–67), p. 277–278.

Whittlesey, C., 1869, On the evidence of the antiquity of man in the United States: American Association for the Advancement of Science, 17th Annual Meeting, Proceedings, Section B (Natural History), p. 268–288.

Willey, G. R., and Sabloff, J. A., 1980, A history of American archaeology (2nd ed.): San Francisco, W. H. Freeman, 313 p.

Winchell, N. H., 1901, Edward Waller Claypole, The American Geologist, v. 27, p. 247–247.

—— 1903, Was man in America in The Glacial Period?: Bulletin Geological Society of America, v. 14, p. 133–152.

—— 1911, The aborigines of Minnesota: St. Paul, Minnesota Historical Society, 761 p.

Wright, G. F., 1891, The Ice Age in North America and its bearing upon the antiquity of man (3rd ed.): New York, D. Appleton, 472 p.

—— 1904, Man and the Glacial Period (2nd ed.): New York, D. Appleton, 385 p.

Printed in U.S.A.

Geological Society of America
Centennial Special Volume 1
1985

The anticlinal theory of oil and gas accumulation:
Its role in the inception of the natural gas and modern oil industries in North America

John T. Galey
Galecrest R4
Somerset, Pennsylvania 15501-8679

ABSTRACT

The anticlinal theory of oil and gas accumulation was conceived in the Appalachians by recognition of the relationship of the structure of the strata to the occurrence of gas and oil. A New England M.D. who had migrated to southern Ohio made the first recorded observation of this relationship before 1836, but pronounced no theory. A quarter century later, in 1861, another New Englander pronounced the first theory from observations on the Ontario Peninsula, Canada. Four months later, an Ohio college professor made observations in West Virginia and independently pronounced essentially the same theory. It was employed successfully there until 1865, after which the concept languished until 1878 by reason of its misapplication. The first successful application of the theory in Ontario had followed its pronouncement by only a year.

Prior to the inception of the natural gas industry in 1884, the anticlinal theory was successfully utilized in the search for oil by knowledgeable prospectors. But they often discovered gas instead of oil, and gas was not then a saleable commodity. Nevertheless, this gas was soon to provide the impetus for the inception of the natural gas industry. The anticlinal theory was reintroduced in 1885 at a propitious time to provide the most viable means, the interpretation of the structure of the rocks, of obtaining a continuing supply for a rapidly growing gas industry. By 1901, its understanding aided the discovery at Spindletop in East Texas of the greatest oil gusher the world had ever known and "began a new era of civilization."

INTRODUCTION

Before wells were drilled specifically to find oil, brine from which salt could be produced was the most commercially valuable product that could be obtained by drilling. Such wells had occasionally encountered oil and gas but the only value of oil then was as a domestic remedy for almost every known disease, and it could be obtained only as a by-product from wells drilled for brine or from oil springs. Gas was considered a nuisance and without value. When the value of oil for illumination and lubrication was recognized in the mid-19th century, incentive to secure it in larger quantity was created. Wells were therefore drilled for this purpose, just as they had been for brine.

No sooner had some of these wells been found productive than prospectors sought a means of locating oil. Since successful wells had been drilled near oil seeps, locations for new wells were located along creeks in valleys where crevices containing oil were believed to exist in the rocks below. Wells located in such manner were said to be located on the basis of "creekology." They were also located by such other means as a peach fork or a small phial of oil suspended from a switch in the hands of an oil wizard, at other times by a fortune teller, a will-o-the-wisp or along belts projected with an orientation connecting or paralleling established production, but the majority of prospectors simply trusted to luck.

More scientific means of locating oil and gas were conceived by S. P. Hildreth, M. D., based on his observations of the relationship of their occurrence to geologic structure. He described his observations in a major article, personally solicited by Benjamin Silliman, Sr., for the January 1836 issue of the *American Journal of Science.* Henry Darwin Rogers, however, was the first scientist to attempt to explain the conditions for the accumulation of oil in North America. These observations, however, did not help the prospectors in the region soon to become the major oil producer of the nation. This area of northwestern Pennsylvania is devoid of obvious structural disturbance, and accumulation there is governed by stratigraphy.

The first theory of anticlinal occurrence of oil was advanced before March 1861 by T. Sterry Hunt, former pupil of Benjamin Silliman, Jr., at Yale. Hunt was then a principal assistant to Sir William Logan, Director of the Geological Survey of Canada. He made the observations for this theory on the extension of the Cincinnati Arch into Ontario. Four months later, a similar theory was independently formulated by Professor E. B. Andrews of Marietta (Ohio) College, working on the Horseneck/White Oak/Burning Springs Anticline, the West Virginia "Oil-Break." Their anticlinal theory, briefly stated, is that gas, oil and water from an existing source, separated by gravity, concentrate in impermeably covered porous strata of a trap created by structural disturbance.

Both Hunt's and Andrews' theories are based on the occurrence of oil in anticlinal traps (Fig. 1). Even though other types of productive structural traps have been found to exist, the original terminology is continued here.

Two new oil pools in Ontario were discovered within a year following Hunt's pronouncement of his theory. Numerous failures, however, resulted from hastily inferred conclusions in utilizing the theory in West Virginia. Structural contouring, which had been introduced several years earlier to depict the form of geologic and topographic features, had not yet been employed. It would have doubtless prevented these failures.

These failures together with antagonism to the theory by one of the most prominent geologists of the day, J. Peter Lesley, Director of the Second Pennsylvania Geological Survey, caused the concept to be in disfavor from 1865 until 1885. This situation existed despite the discovery by F. W. Minshall, in 1878, of the reason for the West Virginia failures. Nevertheless, several knowledgeable operators utilized the theory before 1884 with the expectation of finding oil. They found, instead, unusually large flows of gas, which then had no market. Such large volumes of gas, which were being wasted by venting to the atmosphere, impelled George Westinghouse, Jr., the electrical genius, to recognize its commercial value and to initiate the natural gas industry in 1884.

The anticlinal theory was reintroduced in 1885 by Professor Israel C. White, nine months after he had been advised of the manner by which John H. Galey had located an enormous gas well he drilled near Tarentum, Pennsylvania. Reintroduction of the theory came at a most propitious time, for it was to provide a viable means of securing a continuing supply for a new and rapidly expanding natural gas industry. White found ready support for his theory in Edward Orton, the highly regarded State Geologist of Ohio. Even though Ohio contains a paucity of struc-

tural disturbance, Orton's endorsement of the theory vastly accelerated its acceptance.

The most spectacular success of the theory, however, was its aid in discovering in 1901 at Spindletop, Texas, the greatest oil gusher the world had ever known. With the finding of such an abundance of oil, the modern oil industry began, just as the enormous gas discoveries at Murrysville, Pennsylvania, enabled George Westinghouse, Jr., to initiate the natural gas industry in 1884 and give credence to the anticlinal theory. These discoveries resulted from the application of a geological theory, the anticlinal theory of the occurrence of oil and gas. There should be no doubt that it has played the major role in founding the natural gas industry and beginning the modern oil industry in North America.

S. P. HILDRETH, M.D. AND PIONEER GEOLOGIST—THE FIRST NOTED ANTICLINAL OCCURRENCES OF OIL AND GAS

Shortly after S. P. Hildreth migrated to Marietta, Ohio, from New England in 1806, he began geological observations without official position or support while practicing his medical profession (Owen, 1975, p. 30). He examined numerous stratigraphic sections, searching for coal and minerals; in the process he located many oil and gas seeps and studied the salines from which salt was produced, together with their by-products, gas and oil. Much of his stratigraphic and structural information was obtained from drillers and operators of brine wells (Hildreth, 1836, p. 26, 72, 74).

Hildreth's observations were published as occasional articles in the *American Journal of Science and Art.* They attracted the attention of Benjamin Silliman, Sr., who conducted the *Journal* and personally solicited a major article from Hildreth for the January 1836 issue (Owen, 1975, p. 38), entitled "Observations on the Bituminous Coal Deposits of the Ohio and the Accompanying Rock Strata." This is a summary of Hildreth's wide-ranging field investigations of the geology, natural history and mineral resources of the area that he describes as "the southwest termination of that immense valley which lies between the Rocky Mountains on the west and the Allegheny Range on the east" (Hildreth, 1836, p. 1). This article, years ahead of its time, antedates the work of geological surveys of three states, and is one of the most important contributions to knowledge of the occurrence of oil and gas.

One of the early notations of the anticlinal occurrences of oil and gas that Hildreth made resulted from his investigation of the salt works on the Great Kanawha River in western Virginia. He describes the salt works there as follows (Hildreth, 1836, p. 101):

"... from the mouth of the Elk (River), (Charleston, Kanawha County) up the Kenawha (sic) (River) to the salt wells, a distance of nine miles by the river, the rock strata rise at the rate of nearly 50 feet to the mile. Above this point to the extremity of the salines, the strata dip to the S.E. at an angle somewhat less, or about 33 feet to the mile for a distance of six or eight miles—the anticlinal line (axis) of the strata—at the great

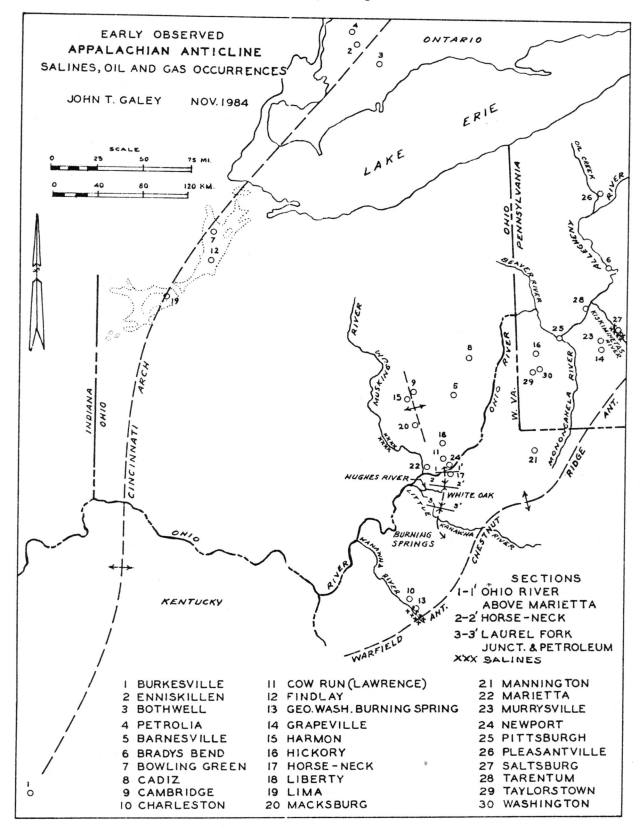

Figure 1. Early observed Appalachian anticline salines, oil and gas occurrences.

salt deposit being near the center of the works—they dip from the line N.W. and S.E. at an angle of three or four degrees."

The George Washington Burning Spring is located near the anticlinal line.

Hildreth illustrates this anticline by a sketch showing the dip of the strata to the northwest and southeast from the anticlinal line elevated 450 feet above a base of 14 miles and accompanies it with a stratigraphic section of the exposed rocks (p. 104). This feature, now known as the Warfield, is the extension to the southwest of the Chestnut Ridge of Pennsylvania. He states that mineral oil rises in nearly all the wells and that immense quantities of gas are continually rising with violent discharges when the whole volume of water in a well is sometimes thrown a hundred feet high with tremendous force and noise (p. 122).

In his description of the Kiskiminetas River salines five miles below "Saltsburgh" and eighteen miles west from the foot of Chestnut Ridge (an anticline) in Pennsylvania, Hildreth describes the structure as follows: "The rock strata rise or dip in conformity with the ridges of mountains and hills, in some instances at an angle of almost twenty degrees." He illustrates this by a stratigraphic section showing the dip of the strata and draws analogy to it with the Kanawha River structure (Hildreth, 1836, p. 72) where,

. . . there is similar elevation of the rock strata but at a much smaller angle—the base line (on the Kiskiminetas) being about ten miles and the elevation three hundred feet—The main muriatiferous deposits—lie at a depth of from 500 feet to 700 feet according to the rise or dip of the rock, corresponding in this respect to the strata on or near the surface.

The salt wells here he notes (p. 74):

. . . are singular in one respect, they afford no petroleum but an abundant supply of carburetted hydrogen—the absence of petroleum seems to indicate some peculiar state of the coal beds in consequence of which they afford the gaseous and not the fluid products—."

The presence of gas without oil in the more intensely folded rocks provides a clue for the carbon ratio theory.

Hildreth (1836, p. 72) states that,

At the salines on the river Muskingum (Ohio) the same arrangement is found to take place. Rock deposits at the depth of several hundred feet at one point on the river, are found on the surface at another place a number of miles above. But all the strata being of secondary origin—the change produced on the surface is not apparent to common observers, and is detected only by close inspection, and the observation of those engaged in penetrating the earth in quest of salt water, who, operating at different points on the river, take notice in their downward passage of some rock, remarkable either for its composition or extraordinary hardness, and noting its distance from the main salt rock, can trace its progress to the surface with great accuracy and certainty.

The saliferous strata on the Muskingum, he states (p. 26), show "the same upward tendency—that has been observed on the Kiskiminitas and Kenawha, but whether this elevation was

caused by the immense evolution of gases—remains as yet unknown." No mention is made of the presence of oil. The foregoing description indicates that the folding is gentler and more obscure here than that on the Kanawha and the Kiskiminetas. It also demonstrates the utility of marker beds in subsurface geology.

Although Hildreth observed and provided the essence of the anticlinal accumulation of oil and gas, he pronounced no theory. If prospectors had followed his precepts, however, they could have utilized them successfully.

HENRY DARWIN ROGERS AND WILLIAM BARTON ROGERS

The accomplishments of the Pennsylvania Geological Survey under the direction of Henry Darwin Rogers (see Faill, this volume) from 1836 to 1842 were prodigious. Preliminary geologic reconnaissance of the entire state was made without benefit of topographic maps or adequate geodetic control. The final report was not published until 1858, and then it was necessary for Rogers himself to fund part of the cost because of the State's financial straits. No mention was made of petroleum or of Hildreth's early observations, but the report indicates that Rogers had a keen sense of geologic structure (Owen, 1975).

William Barton Rogers (see Faill, this volume), brother of Henry, was State Geologist of Virginia from 1835 to 1841. He made a preliminary report of his reconnaissance in Virginia to the House of Delegates in January, 1836, the same month Hildreth's major article was published in the *American Journal of Science*. William Rogers mentioned the petroleum springs near the Hughes River tributary of the Little Kanawha River in West Virginia, as did Hildreth, but each failed to note the occurrence of the White Oak anticline. In his 1839 report to the delegates, William Rogers (1839) described the larger anticline at the famous salt works and Burning Spring on the Great Kanawha River but attached no special significance to it with respect to petroleum.

Before 1840 both of the Rogers brothers knew of the relationship of petroleum to anticlinal structure. Henry states this in an address given at Glasgow in March 1863 which was published by *Harper's Monthly Magazine* in America, July, 1863. In it he states (p. 259–264) that,

Where the general flatness of the coal rocks is only at wide intervals interrupted by narrow but long and sometimes rather sharp anticlinal waves, the more copious emissions of the rock-oil and the native gases is found to be chiefly restricted to the tracts occupied by the crests and sides of these local billows in the strata. It was long ago, before 1840, noted by my brother, Prof. William B. Rogers in his Geological Survey of Virginia and was observed and made known by myself in my own similar exploration of Pennsylvania, that nearly all the localities of abundant and comparatively permanent Artesian salt wells or artificial brine springs, with their almost invariable concomitants, the liquid and gaseous hydrocarbons were situated upon or nearly coincident with the artificial arching of the strata. This was seen on the Kanawha by him and on the tributary river of the Allegheny and Ohio by myself.

Earlier, when Henry Rogers was in Glasgow, Scotland, oil was discovered at Titusville in 1859 by Colonel Drake. The geology of Pennsylvania was still fresh in Henry's mind. He recognized the significance of anticlines without declaring specifically that the new wells were located on one, which they were not. He became the first scientist to attempt to explain the habitat of oil in North America when he addressed the Philosophical Society of Glasgow on May 2, 1860. In this address he stated (Rogers, 1860, p. 355–359),

"Superficially the gas and oil emitting springs within the coal fields are found chiefly upon the anticlinal structures of the strata, in proximity to the outcrops of the black bituminous slate subjacent to the formation in localities where the gas evolving shale lies at the depth of a few hundred feet below the oil . . . imperviousness of the argillaceous strata except where these are fissured, as they are apt to be on the anticlinal axes, serves to hold down the elastic and volatile products."

T. STERRY HUNT

T. Sterry Hunt was a principal assistant of Sir William E. Logan, first Director of the Geological Survey of Canada. Hunt was a native of Connecticut and educated at Yale, where he had been trained under Benjamin Silliman, Jr., as a chemist and mineralogist, but his interest turned to oil.

Oil had been utilized from the Oil Springs that were known since 1830 at Enniskillen on the Ontario Peninsula, and wells were dug about a hundred feet into the gravel above the limestone in 1858 to augment the supply (Harkness, 1924, p. 76–92). The Geological Survey of Canada, notably Hunt, gave particular attention to this occurrence. Since Hunt knew of Logan's observation of oil on an anticline on the Gaspé, at the mouth of the St. Lawrence River, in 1844, he investigated the structure. This was accomplished by having pits dug and well records collected from the time drilling began. Stratigraphic information was sought because the outcrops were masked by surficial deposits. This activity yielded information that enabled him to observe that the productive wells were located along a low, broad east-west-trending anticlinal axis, and he deduced that the oil occurred because of the anticlinal and that fissures in the rock caused by their disturbance provided the reservoir.

Hunt made, as a result, the first definite statement of an anticlinal theory of oil accumulation at a public lecture which was reported by the Montreal *Gazette*, March 1, 1861, as follows (Hunt, 1862, p. 319–389):

. . . he supposes the oil to come from the Silurian and Devonian limestones at considerable depth below, which are in many cases, very full of bitumen in one place . . . containing 8 or 10 percent and he thinks we are warranted in expecting those oils in different places over a great portion of the western peninsula, . . . which had been shown was underlain by these limestones. It required, however, a peculiar arrangement of the strata to allow the oil to accumulate and flow out and this will only be met with . . . along lines of folding and disturbances, which the geologists have shown to exist in various parts of that region.

This statement with accompanying description of structural conditions in the province afforded a useful guide for exploration. Ontario was probably the first area in America where petroleum was influenced by geological guidance and Hunt was never hesitant in providing it. Canada's first commercial oil field was developed here at Enniskillen in 1861 by flowing wells at depths of 350 to 400 feet. Oil was discovered at Bothwell in 1862 and at Petrolia within the year (Logan, 1863, p. 379).

Hunt had become not only the Canadian Geological Survey's oil expert but the world's first authority on petroleum geology. His subsurface studies, largely through his stratigraphic knowledge, had become so sophisticated by 1866 that he was able to locate the axis of the Cincinnati Arch "which divides the coal field of Pennsylvania from that of Michigan" (Hunt, 1866, p. 257). He had also by this time simplified his theory by stating (p. 260),

The existence, in any oil-bearing region, of available sources of petroleum, depends upon a combination of many circumstances: (1) the proper attitude of the strata, (2) the existence of suitable fissures, which may act as reservoirs, and (3) such an impermeability of the surrounding and overlying strata as will prevent the outflowing and wasting of the accumulated oil.

He knew of the "Great American Well" near Burkesville, Kentucky, that had been drilled for brine on Renox Creek in Cumberland County in 1829 and that flowed oil 40 miles down the Cumberland River before catching fire and burning back to its source. It flowed an estimated 50,000 barrels in the first three weeks and was bailed and pumped until 1860 (McFarlan, 1943, p. 283). This oil came from rocks of Trenton age on the Cincinnati Arch at a structural location similar to that of the oil on Manatoulin Island, Ontario (Hunt, 1866, p. 254). This great anticline Hunt foresaw becoming a major oil province, and the Trenton a drilling objective, in view of his favor of indigeneous limestone source beds. Unfortunately for Ohio and Indiana, as a Canadian, he could not apply his talents there; this was eighteen years before the 1884 discovery of the Lima-Indiana Field, based on gas seeps, at Findlay, Ohio.

EBENEZER ANDREWS—GEOLOGICAL RELATIONS OF ROCK OIL

While he was teaching Geology at Marietta (Ohio) College in 1851, Ebenezer B. Andrews became interested in the oil and gas seeps and oil producing brine well at Burning Springs on the Oil-Break of West Virginia, located across the Ohio River south of Marietta (Owen, 1975, p. 62). He recognized this anticline, overlooked by both Hildreth and Rogers, as a habitat of petroleum. This feature extends for about 40 miles from 4 miles north of the Ohio River and east of Marietta to about the same distance south of the Little Kanawha River at Burning Springs (Minshall, 1881, quoted by Peckham, 1884, p. 52). Its axis is nearly north-south about 40° off trend to the northeast-southwest orientation of Appalachian folding. It has a maximum structural relief of about 1500 feet.

When Andrews completed his investigation of the Oil-Break, he published the results in an article entitled "Rock Oil, its Geological Relations and Distribution" in the *American Journal of Science and Art,* July 1861. It is he states (1861, p. 91–92),

Along this line (the axis) the productive wells are found—only in a narrow range extending north and south and little more than 120 rods wide . . . The oil fissures are struck at different depths . . . consequently, there is no such thing as an oil rock . . . The oil is found in any kind of stratum, each oil fissure doubtless extends vertically, or nearly so, through many different strata . . . in the broken rocks as found along the central line of a great uplift we meet with the largest quantity of oil. It would appear to be a law that the quantity of oil is in direct ratio to the amount of fissures.

In another article he wrote (Andrews, 1866, p. 33)

Fissures serve two purposes, one to give space for the formation and expansion of the hydrocarbon vapors and the other to furnish receptacles for the oil when condensed. These fissures must connect with the deeply seated sources of oil.

Andrews' exposition followed Hunt's by four months, but they were entirely independent of each other, even though each invisioned fissures to be the oil reservoir. Andrews, however, believed fissures to be the primary cause of accumulation rather than the anticline itself.

After service in the Civil War, Andrews returned to his teaching at Marietta in time to participate in the oil boom of 1864–65. His published exposition of the anticlinal fissure theory of 1861 had earned him wide repute and he had become the acknowledged oil authority in southeast Ohio and adjacent West Virginia. He had also become one of the first academic geologists to engage in consulting.

By 1865 Marietta had become a regional petroleum center. Its newspaper, the *Register,* had an oil column as a regular feature, and Andrews received his share of publicity. When he was quoted in the column that an area was promising, business people agreed with him and they purchased it. An unfavorable report was almost as effective as a dry hole in condemning a property (Owen, 1975).

Leasing and drilling activity prior to 1864 was confined largely to stream beds on the Burning Springs, the southernmost portion of the Oil-Break, where many wells failed to produce. (Burning Springs is located in Wirt County about 50 miles north of the George Washington Burning Spring.) Discovery of gas seepages attracted the attention of General Adoniram J. Warner, a practical geologist affiliated with Andrews, to the summit of the White Oak section immediately to the north, and the Horseneck north of the White Oak. Andrews and Warner abandoned their idea of locating wells along stream beds when they mapped the White Oak area and found that (Andrews, 1866, p. 35),

. . . all the productive oil wells . . . group themselves along the anticlinal line . . . being the one of greatest fissuring of the rock. . . .Toward the

northern and southern extremities, this line presents a simple anticlinal with the rocks dipping on either side of the axis at angles varying from 5° to 25°. . . .But in the middle part (White Oak), there is a double fracture, the lines of dislocation enclosing a somewhat elliptically shaped area about ten miles long by one wide . . . the dislocated strata inclining in opposite directions varying from 30° to 60°.

They hastily inferred that the axis would be valuable territory to both the north and south. Numerous wells were drilled based on this inference but a high percentage were failures (Minshall, 1881, quoted by Peckham, 1884, p. 52). This high incidence of dry holes damaged Andrews' reputation as an oil finder and he joined the Second Geological Survey of Ohio under John S. Newberry in 1869, prior to F. W. Minshall's discovery in 1878 of the reason for the failures on the axis to the north and south of White Oak (Minshall, 1881).

OTHER EARLY CONTRIBUTORS TO THE ANTICLINAL THEORY

In addition to those mentioned above, three geologists and a petroleum engineer made significant contributions to the anticlinal theory in the mid-1860s. They were John S. Newberry, Alexander Winchell, Charles H. Hitchcock and E. W. Evans.

John S. Newberry, who had gained early experience with each of two groups of scientists exploring the west, became interested in petroleum in 1859, the year before he published an article on the oil wells at Mecca, Ohio (Newberry, 1860a, p. 325–326). He was later to become a geological consultant of the widest experience and prestige and State Geologist of Ohio. He recognized the importance of structure in oil occurrence by 1866 when he wrote (Newberry, 1866, p. 284):

The "oil belt" seems to cover an area on the line of the great anticlinal axis which runs parallel with the Appalachians and separates the eastern and western coal fields of Kentucky. . . .Over this area at numerous localities oil had been found. . . .The region about Burkesville, Cumberland County is one of the proved oil centers.

Alexander Winchell, the articulate and colorful State Geologist and Professor of Geology at the University of Michigan, became widely recognized as an expert on petroleum during the exploration boom of 1864–1865. He had been among the first to expound on source beds, reservoirs, accumulation and overlying protective formations (Owen, 1975, p. 71). He conducted a number of surveys between 1864 and 1867 to determine oil prospects in widely scattered areas of Michigan, Ohio, Ontario, and West Virginia, for private companies. In one of his reports he explains reservoirs as follows (as quoted in Howell, 1934, p. 8):

Where the bituminous shales are covered by an impervious layer, as of shale or plastic clay, the oil and gas elaborated are retained in the rocks, filling cavities by driving the water out by elastic pressure, and saturating porous strata embraced in the formation or intervening between it and the impervious cover above.

Lesley did not believe this explanation twenty years later. Winchell specifically states that both gas and oil are formed in the same source beds and migrate upward into the crest of the anticline under the same factors he describes.

At about this time, a Mr. Biggs, during a meeting of the American Philosophical Society in the spring of 1865, made a statement (as quoted in Howell, 1934, p. 8) suggesting that oil wells flowed because,

Petroleum is an intimate mixture of the gases into which petroleum decomposes, with the petroleum fluid, like that which exists between the carbonic acid and the water in a soda fountain.

Charles H. Hitchcock was State Geologist of Maine in 1861–1862. He had previously served on a State Survey of Vermont as assistant to his father, Edward Hitchcock, a Professor of Geology at Amherst College. There was substantial interest about 1865 in the beauty and value of Albertite that occurred in New Brunswick. Hitchcock investigated this occurrence and produced a paper, "The Albert Coal or Albertite of New Brunswick," which gave most persuasive support to the anticlinal-fissure theory. He discovered that the Albertite occupies an irregular fissure, he called a vein, along the anticlinal line (Hitchcock, 1865, p. 270) and believed it was originally analogous to petroleum, injected into the fissure in a liquid state and was subsequently hardened. His basis for this belief was that liquid petroleum was contained in the coal and shale associated with the vein deposit of Albertite. He states that borings for petroleum in Ohio and West Virginia are most successful along lines of fracture, particularly on anticlinal axes, and agrees with Andrews that oil occurs along anticlinal faults. But he states that the immense yields of wells in such locations suggest the existence of more than cavities filled with liquid. He summarized the conditions of occurrence of petroleum in 1867. These were in accord with those understood by geologists, but he added (Hitchcock, 1866, as quoted in Howell, 1934, p. 8): "Besides petroleum, brine and gas are commonly, if not universally, discharged from the orifice of a well, and we may suppose that before the tapping of the cavity they were arranged according to their specific gravities, the gas uppermost and the brine beneath the others."

E. W. Evans was a Professor of Mathematics who assisted Marietta College in becoming an important petroleum engineering institution. He made one of the earliest statements on extensive lateral oil migration in 1866, after working on the West Virginia Oil-Break. He stated (Evans, 1866, p. 334–343) that,

This large accumulation of oil in the crevices of the anticlinal would seem to be owing not solely to a direct connection by a continued line of fractures with the original sources of the oil in strata underneath, but in great part also to the collection, from a wider area, of oil that has come up elsewhere, as through the crevices of adjacent synclines. For being lighter than water, it would gradually work up between the strata of the slopes.

This is the means by which the reservoir becomes charged.

F. W. MINSHALL—SELF TAUGHT PRACTICAL GEOLOGIST

F. W. Minshall was a self taught practical geologist of remarkable ability, who had worked with Andrews on the West Virginia Oil-Break. He had done considerable work north of it in Ohio, had been field superintendent for an oil company in the Cow Run Field, near Marietta, Ohio, in 1865 and, afterwards, in the refining business at Parkersburg, West Virginia (Owen, 1975).

Minshall's investigation of the Oil-Break showed that it "is a fold or wrinkle in the bottom or a great trough . . . that there are undulations in the axial line which divide the line into three sections . . ." This was the reason for the well failures to the north and south of White Oak. The lows of the undulations contained water, the highs oil, and (Minshall, 1881, p. 52):

. . . had there been no erosion of the surface [the anticline] would have presented three different peaks of different altitudes; that of Horseneck would have been about 500 feet higher than that of Burning Springs and that of White Oak about 300 feet higher than that of Horseneck and the summit of White Oak peak would have been about 2,000 feet above the level of the Ohio River. Under each of these peaks, the rocks lie in the form of a table, say 4 miles long and from 3/4 to 1 mile wide. From the ends and sides of these tables, the rocks dip at certain angles. Taken as a whole, the rocks form inverted basins with flat bottoms and sloping sides. In these inverted basins, nature, for thousands of years, has been collecting gases as the chemist collects them in inverted bottles over the pneumatic cistern.

Minshall determined this by laborious detailed leveling of the surface rocks from which he produced geologic sections across (Fig. 2) and along the entire length of the Oil-Break. He agrees that (p. 51): "The statement which Professor Stevenson makes—is accurate and true for the whole length of the line; there is no evidence of faulting on either side."

Minshall had, unfortunately, misinterpreted Stevenson who wrote (Stevenson, 1875, p. 393),

The rocks along Dunkard Creek (Greene County, Pennsylvania)—show no signs of disturbance worthy of comparison with that found in the interesting Oil-Break of West Virginia. There, as I have shown elsewhere, the strata are so crushed and dovetailed, that no two borings show even approximately the same section; but here (along Dunkard Creek) there is no break deserving the name.

Stevenson further states (p. 394),

The great Oil-Break of West Virginia is merely an anticlinal . . . which exhibits violent faulting and crushing within the fold. . . .East and west from this violently disturbed region, which is very narrow, the rocks are almost horizontal for a considerable distance, until on each side they are cut off by a fault on the east side—being one of several hundred feet. Outside of the narrow break the rocks are regular and in no wise distorted. Within the break the borings become productive wells, whereas outside no productive wells have been obtained, though many borings have been carried down to the oil-bearing strata. There are no crevices, but the rock contains petroleum, just as within the break.

SECTION ON THE OHIO RIVER ABOVE MARIETTA.

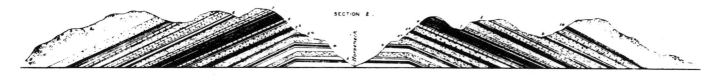

SECTION AT HORSE-NECK. WEST VIRGINIA.

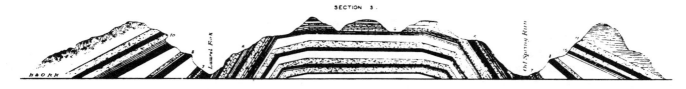

SECTION BETWEEN LAUREL FORK JUNCTION AND PETROLEUM.

WEST VIRGINIA.

ON THE BALTIMORE & OHIO R R

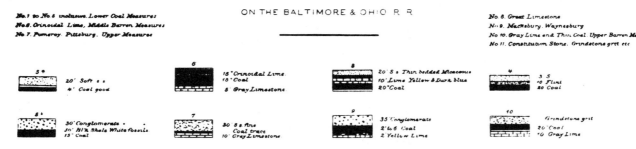

Figure 2. West Virginia Oil-Break cross-sections drawn by F. W. Minshall in 1878. (From U.S. 10th Census Report 1884, v. 10).

Oil operations north of the Ohio River had been going on since 1860 at locations determined by surface indications and usually along stream beds. No drilling beyond the creek beds was done at Cow Run until 1869. In the words of Minshall (1888, p. 465), "Not even the ghost of a geological theorist was seen there by a practical operator prior to 1870." By this time it was discovered by drilling, presumably by Minshall, that the productive area of the Cow Run Field was confined to the crest of a small dome (Figs. 3 and 4). Subsequent exploration for new fields was thereafter based on the contour of the surface rocks as well as surface shows.

Surface shows of oil and gas were known to exist in a relatively narrow belt extending from the mouth of Newells Run on the Ohio River east of Marietta, northwest about 21 miles to Macksburg, a variance from the northeast "belt" orientation to which Pennsylvania operators were accustomed. In 1870, two lines of levels were run to show the structure of both the surface and subsurface rocks in drilled wells. These revealed three anti-clines, named the Liberty, Lawrence and Newport (Minshall, 1888).

The first well was drilled in 1870 by John B. Kigans on the crest of the dome of the Cow Run Field, along the Lawrence Anticline. This well encountered such a strong flow of gas at 1200 feet that it was abandoned because it could not be drilled further. The Liberty Arch was found productive in 1885 when the Bradish Oil Company #1 - C. F. Epler Well, located on its crest, discovered a very strong flow of gas with only a small amount of oil in the Berea sand (Mississippian).

The Newport Anticline, that was found to be the north extremity of the West Virginia Oil-Break, was first drilled in 1890. Both gas and oil were discovered in four different sands, the deepest of which was the Berea. Subsequent drilling revealed the presence of two separate pools, one on the west flank of the structure, called the Newells Run, and the other on the east, the Bosworth. The axis of the feature is between these pools and is nonproductive (Bownocker, 1903).

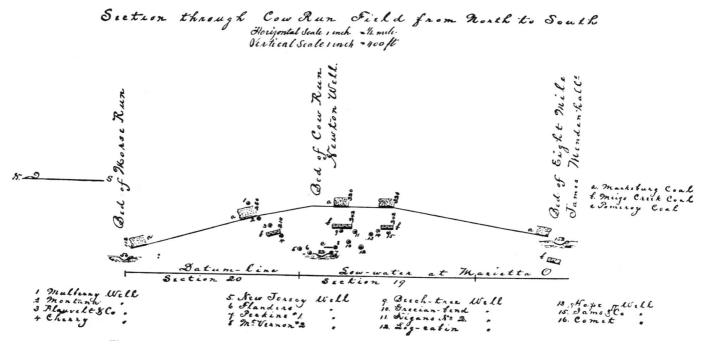

Figure 3. Section through Cow Run Field from north to south drawn by F. W. Minshall in 1870. (From Minshall, 1888).

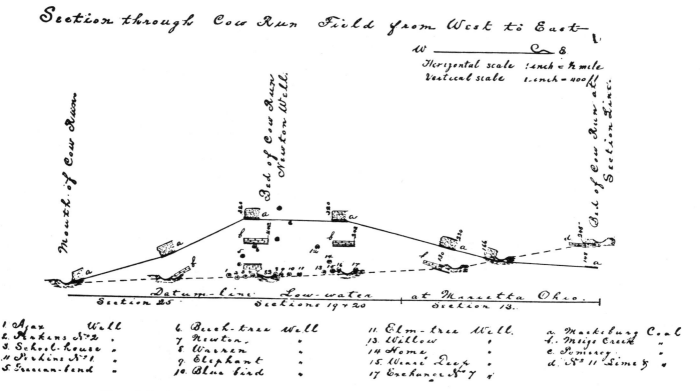

Figure 4. Section through Cow Run Field from west to east drawn by F. W. Minshall in 1870. (From Minshall, 1888).

In 1878, Minshall proved the existence of the Berea sand on the north edge of the Macksburg Oil Field by taking over a shallow abandoned well, the structural position of which he understood, deepened it and found a strong flow of gas.

The well was drilled wet and was standing full of salt water when the first bit was run in the sand after it began to show gas, and when the second bit developed enough gas to throw the brine over the top of the derrick—it was necessary to place an oil saver on the casing head to prevent the driller being pickled (p. 155).

Orton credits Minshall for recognition of the structure of the Macksburg Field which he surveyed by levels on the surface rocks, checked by well levels. Minshall had determined this field "to be a terrace on which, at its higher edge, dry gas was found in good volume, on the terrace itself a valuable oil field has already been worked out, while on the lower border was an ocean of salt water" (Orton, 1890, p. 51). Orton calls the terrace "a suppressed anticline."

Minshall continued to demonstrate his talents as an explorationist when he was called upon for geological advice by a local company at Barnesville, Belmont County, in southeastern Ohio in 1899. The three wells that had been drilled to supply gas to the town had failed. When Minshall made a survey of the surface structure, he discovered a weak axis northwest of the town. A well was promptly drilled and gas found in the Berea sand. The second well was even more successful. Development drilling was carried on sporadically over a period of at least twelve years during which time the gas-bearing area was enlarged and a small oil pool developed (Bownocker, 1903, p. 214–217).

J. PETER LESLEY—ANTAGONISM TO THE ANTICLINAL THEORY

The anticlinal theory had the most antagonistic opponent it could have had in J. Peter Lesley from 1865 until his health broke in 1894, during which time he often attacked it with vigor. Prior to this time, in 1862, he had stated (as quoted in De Golyer, 1961, p. 9),

In some parts of the western coal-field the dip is as high as five degrees and the basins from five to ten miles (sic). Sharp features make local dips of thirty degrees or more, and a central subanticlinal is sure to subdivide the basins. In the secondary basins thus formed the wells are more perfectly artesian as to salt water; but it is upon the subdividing anticlinals that the gas and oil collect. In such regions it is asserted that all the blowing and many of the spouting wells are ranged along the summits of such anticlinals.

He knew that in Pennsylvania the most productive oil fields in 1865 were stratigraphic, in areas of little or no structural disturbance, and that no oil had been found on the prominent anticlines further east. In a discussion at a meeting of the American Philosophical Society in December, 1865, the year he was reputed to know more about the geology of Pennsylvania than any other man, he repudiated his 1862 statement (Lesley, 1865):

Stress has been laid by some geologists of note upon a supposed genetic connection between the accumulation of petroleum and the anticlinal axes but there are no anticlinal axes in the Pennsylvania oil regions of the French and Oil Creek wells, nor in the Pennsylvania and Ohio oil regions of the Beaver River. . . .The only well defined anticlinal among oil fields is a mere upsqueeze crossing the Ohio River near Marietta, bringing the oil rocks near enough to the surface to be tapped, and thus, only affecting the finding of oil. . . .The true relationship of petroleum with surface springs seems to be one of simple hydrostatics. Every natural oil spring is an artesian spring, without regard to the existence of anticlinals or profound earth-crust faults.

This same year, Lesley did consulting in connection with the depth at which the Brady's Bend Iron Works #1 Well, a wild-cat, would reach the Venango Sands. He states (Lesley, 1880, p. XII):

I calculated the vertical distance from the oil sand on Oil Creek up to Coal A; then I calculated the dip of the measures between Oil Creek and Brady's Bend; and then I identified Coal A at Brady's Bend. I reported that the Venango oil sand, if it extended underground as far as Brady's Bend, ought to lie at 1100 feet beneath water (River) level.

When the well was drilled, it found gas and oil at 940 and 1089 feet, although he states, in 1885, that he had fixed their depths in 1865 at 700, 900 and 1100 feet (Lesley, 1886, p. 58). The means he employed to predict the depths are essentially the same as those used by White in 1889 to locate his discovery at Mannington, West Virginia.

Lesley worked on the First Geological Survey of Pennsylvania under Henry D. Rogers and later complained bitterly about the meager credit Rogers gave him and his co-workers on the Survey. Afterwards, he gained wide prestige professionally for his work on the geology of coal and iron deposits (Owen, 1975, p. 53). He did topographic contour mapping across the most rugged terrain of Western Pennsylvania to provide the route for the Pennsylvania Railroad between Centerville (New Florence) in the northern Ligonier Valley, west across Chestnut Ridge to Latrobe, Westmoreland County. Four sections of this map, showing the outcropping rocks and the Chestnut Ridge anticline, were published in the final Pennsylvania Survey Report (Lesley, 1895).

He made an outstanding contribution to the understanding of topographic expression of structural and stratigraphic features in his "Manual of Coal and its Topography," published in 1856, in which he put special emphasis on the superiority of contouring over other means of representing surface form. This resulted from his construction of contour maps of the Broad Top, Pennsylvania coal field in 1855–1856. There is no indication that contouring a layer of rock instead of a land surface had occurred to Lesley before this time (Owen, 1975, p. 80). This work impressed Lesley with the need of a means to determine elevations for rapid reconnaissance and, in collaboration with an expert instrument maker around 1860, he invented a very sensitive barometer (Ames, 1909).

Benjamin Smith Lyman, Lesley's former student and his wife's nephew, proved the feasibility of using subsurface contours to depict the shape of coal, iron and lead deposits in southwest

Virginia in 1866 and 1867. When Lyman was employed by the British government to investigate the oil prospects of the Punjab, India, he produced the first printed structure contour maps in 1870. They were of the Barra Katta Oil Lands (Howell, 1934).

No less an authority than Sir Archibald Geikie credits Lesley with being the innovator of "Topographical Geology" in a memorial address a year after his death in 1903 (Owen, 1975).

When the Second Geological Survey of Pennsylvania was established in 1874, Lesley was appointed its Director. He had been by this time, in addition to being a consultant, a Professor of Geology at University of Pennsylvania and a Congregational minister. This was a critical time in the growth of the oil industry and for the development of petroleum geology but most of the effort of the staff of district geologists, which by 1875 included Ashburner, Chance, F. and W. G. Platt, Stevenson, and White among others, with appropriate personnel, was devoted to coal, iron, and limestone, on which mapping was emphasized. This staff provided such information on oil as it could gather to John F. Carll, who was assigned to the oil region. Henry E. Wrigley was temporarily employed to prepare an historic and economic report on petroleum. Such a staff constituted, beyond question, the most powerful state geological organization in the country.

Policy required prompt presentation of each year's investigation under the name of the geologist who made it, but did not provide for expression of the author's unqualified views, in order to assure no bias of theory would distract from the facts. The policy "imposes on the State Geologist unceasing labor as an editor every day of the entire year. Every sentence of every report must be revised. . . ." (Lesley, 1876, App. p. xx). It seems reasonable to assume that Lesley's keeping the reports "factual" was influenced by his prejudice against the anticlinal theory.

Wrigley's Special Report on Petroleum in 1875 offered little help to prospectors for new fields. He calls attention to an American geologist (Lyman) being employed by the British government to investigate the oil prospects of the Punjab Province of India and to the fact that petroleum in Ohio and West Virginia occurs on plainly marked belts of geological disturbance (anticlinal) which are illustrated on maps accompanying his report. There is no mention made of the presence or absence of anticlines in the Pennsylvania oil region. He does point out the limitations of the belt theory.

The Survey's first report on the Venango Oil District, done by Carll, was published in 1875. It emphasized the fallacy of the notion of the relationship of topographic features to the presence of productive sands. It is most important, however, as an example of subsurface structural mapping which demonstrated monoclinal dip to the south-southwest. Carll's demonstration of monoclinal dips across the Venango oil country, in contrast to the steep dips of the prominent anticlines further east which are barren of oil, strengthened Lesley's disbelief in the anticlinal theory. Lesley's appendix to this report points out particularly the *"undisturbed condition* of all this part of Pennsylvania, yet the most productive oil fields of Venango, Clarion, Armstrong and Butler Counties are *equally undisturbed.* The popular notion that petroleum wells are

dependent on anticlines, faults, or other disturbances is a pure fancy of the imagination. And it is very desirable to get rid of it . . . The district of greatest oil production in Pennsylvania is precisely the district where there has never been any structural disturbances (sic) whatever. . . ." (Lesley as quoted in Randall, 1878, p. 54–55; Lesley italics). This was a correct statement for parts of Venango, Clarion, Armstrong and Butler Counties where 88 percent of the world's oil was then being produced.

Lesley requested Carll to write an article for the leading newspaper of the oil region, the Oil City Derrick, and it was published July 24 and 25, 1876, and again in the 1885 Annual Report of the Survey with the following preface by Lesley (Carll, 1886, p. 59–60):

This article is and was intended to be, as accurately defined and complete a summary of the knowledge obtained by the Geological Survey up to that date, and was published in advance of the regular Reports of Progress, to supply an urgent demand from all classes of people interested in petroleum, for the best practical information respecting its geological relationships. . . .After ten years of universal exploration, it still stands as the truest statement that can be made. No essentially new truth in Oil Geology has been discovered to add to it . . . But what is more to the present purpose, it contains predictions that have come true. . . .

Carll's (1886, p. 73–74) statement on anticlines contained in this article was,

. . . the anticlinals and synclinals now seen on the surface should not be taken as guides in searching for the oil sands. These undulations were produced by movements in the earth-crust long after the oil sands were deposited—as is shown by the fact that the coal measures are equally affected by the same waves—consequently they could have had no agency whatever in controlling and directing the currents which had already laid down the oil-sands thousands of years before. Insofar as these anticlinals and synclinals affect productiveness of the oil sands by affording an opportunity for the gas to collect at the crowns of the arches and salt water to settle in the depressions between them, just so far ought they to have an influence in the selection of the location of an oil well and no farther.

This statement contains the essence of the anticlinal theory and Carll also makes it clear that the anticlines and synclines had no effect on the deposition of the oil sands in which he knows production is controlled by stratigraphy.

Oil production had been extended far beyond the original Venango district by 1880 when Carll's report on "The Geology of the Oil Regions of Warren, Venango, Clarion and Butler Counties" was published. Lesley's unusually long "Letter of Transmission" (Carll, 1880, p. XVI) included the statement that,

The supposed connection of petroleum with anticlinal and synclinal axes, faults, crevices, cleavage planes, etc. is now a deservedly forgotten superstition. Geologists well acquainted with the oil region never had the slightest faith in it, and it maintained its standing in the popular fancy only by being fostered by self-assuming experts who were not experienced geologists.

This appears to discredit Carll's 1876 statement on anticlines contained in the Oil City Derrick article Lesley requested him to write. It also appears that Lesley does not distinguish between the habitat of oil and the ability of a reservoir to produce. Carll had, at the same time, 1880, discredited the idea of fissures or crevices being responsible for large wells by proving that the porosity of a sand was sufficient to account for the most productive wells yet discovered (Carll, 1880, p. 245–255).

Carll's 1880 report contained abundant well data with elevations and topographic maps of the Clarion-Butler trend which were included in the accompanying Atlas, but only local or unrevealing cross sections are shown. It would have been an exhaustive report had it not been for lack of structure contour maps which were an essential part of his 1875 Venango report and in this new area structure of the oil bearing rocks may have had some significance. A fine print footnote appears to make this evasive apology (Carll, 1880, p. 106):

On the theory generally accepted that the anticlinal and synclinal structure of the country was accomplished as a whole, after the close of the carboniferous era, any reference to the Brady's Bend anticlinal and synclinal in this discussion is unnecessary, since they did not exist when the Oil Group and the conglomerate rocks were deposited; but there is a feeling with some geologists that the beginning of the crust movements may have taken place in the coal era, and been only consummated at its close.

Carll completed his Warren County report in 1882. He had collected 685 well records for use in a structural study. Such a study might have remedied the lack of an exploration method oil prospectors needed and the Survey had thus far failed to provide. However, he had found insufficient records useable because of secrecy or misrepresentation.

Carll terminated his connection with the Survey in 1882 in a spirit of frustration after eight years of dedicated service in which he could see no tangible effect of his accomplishments. As a parting tribute to Carll, Lesley wrote in the letter of transmittal for the 1883 Survey Report (Carll, 1883, p. XII),

I desire to publish, in the most unequivocable manner, the fact that Mr. Carll was the first geologist to comprehend the structure of the Oil Regions and to furnish a reliable exposition of its features. . . .It is not too much to say that, however much was previously guessed at, or suspected to be true, the Geology of Petroleum has been virtually created by him [Carll] and that his services to science in sweeping away popular fallacies have equalled those which he had conferred by his discoveries and demonstrations.

Carll returned to the Survey in 1885 for a second tour and, in 1886, gave the most erudite treatment that existed to the principles of petroleum geology as applied to Pennsylvania fields.

John J. Stevenson was the most prestigious of Lesley's staff when he joined it in 1875. He was then Professor of Geology at University of West Virginia, had served under Newberry on the staff of the Geological Survey in Ohio where his investigations revealed several structural features (Orton, 1888; 1890) later found to be productive, and had worked and published on the

West Virginia Oil-Break (Stevenson, 1875). His first work for the Pennsylvania Survey was in the southwestern corner of the State for Report K on the Greene and Washington District. Israel C. White, who had just completed his Master's Degree under him, was his assistant.

During the course of his work in Washington County, Stevenson located the Washington anticline (1876) where White claims he made the first confirmation of his anticlinal theory. Stevenson also located the Claysville anticline where, what was said in 1882 to be the world's largest gas well, was drilled on the Hickory Dome by a Bradford oil operator (Carll, 1887). He also located the Belle Vernon anticline in eastern Washington County where White claims he had another confirmation of his theory (Stevenson, 1876).

In Report K, he states in connection with the oil in the Dunkard Region, Greene County (Stevenson, 1876, p. 394–395):

Experience having shown in West Virginia and southwest Pennsylvania that oil could be procured in economical quantities only where there seems to have been some decided disturbance of the strata, those engaged in procuring this oil came to regard this disturbance as necessary to the original production of the oil itself. But a few considerations, I think, will suffice to show that the oil in no wise owes its origin to disturbance of the strata and the only effect of the disturbance has been to provide reservoirs for the oil in the rock already oil-bearing.

His reason for this belief was that:

In the Dunkard region, the Morgantown and Mahoning sandstone are saturated with oil, for no well was bored without finding greasy, strongly odorous rock at each horizon though, as previously stated, no supply of oil was obtained unless a cavity was struck.

He considers this to be analogous to that on and adjacent to the West Virginia Oil-Break (See Minshall above). He further states (p. 394),

. . . it seems unnecessary, as well as in accordance with observed facts, to regard them (cavities and crevices) as due to the action of some disturbing force exerted along the line (of disturbance) but without sufficient energy to produce permanent effects visible on the surface. . . .

How far southward the productive area extends has never been ascertained, as the creek bed coincides with the anticlinal only in the vicinity of Bobtown. Southward from this, no borings have been made along the anticlinal. The prejudice was in favor of "bottom" lands, not only because they are at a lower level than the upland, but also because it was believed that oil had a tendency to flow into hollows. It was held, also, that the extreme abruptness of the hills along the creek proved that there the disturbance was most violent. For these reasons, those borings (wells) followed the creek and occupied barren ground and no wells were sunk along the line of disturbance. On Dunlap's Creek, the borings made near the crest of the little axis there obtained oil. Borings made away from it, though piercing the same rock, found none.

This statement expressed his view of the anticlinal occurrence of oil. Since Stevenson, an ardent advocate of the anticlinal fissure theory, had expressed disagreement with Lesley's view of oil occurrence on anticlines, he qualifies his statements as applying to no area other than his own in the south by stating (Stevenson, 1876, p. 397):

It must be understood that the above statement is confined in its application to the Southern Oil District. It may not at all apply to the loose-grained and gravel oil belts in the Allegheny River region, which can hold large quantities of petroleum without the aid of cavities or fissures produced by structural disturbances of the strata. In the southern area the rocks are fine-grained and compact, and seem to require lines of disturbance to enable them to hold stores of oil.

Stevenson located the Murrysville anticline two years before it was drilled and the Grapeville anticline ten years before Guffey had it drilled. Description of these and other features is included in his 1877 Survey Report KK on Fayette and Westmoreland Counties.

Oil activity had shifted to Bradford from the Butler-Clarion "Oil-Belt" when H. Martyn Chance, formerly Carll's assistant, was given responsibility for a Survey Report on northern Butler County and Clarion County in 1878 and 1879. Little was known of structural features west of the major anticlines when his work commenced. He made an unusual effort to determine structure by use of well elevations and subsurface data he had on the "Butler-Clarion Oil-Belt" which extended across his district. He located a number of minor flexures but found such gentle dips that the features were difficult to trace. He noted that such effort "must prove futile and be productive of much confusion with but little resulting benefit" (Chance, 1879, p. 9). No attempt was made to relate the minor flexures he found to accumulations of petroleum. Had more adequate maps and mapping facilities been available, he might have had better success.

Several years prior to his commencing work in the district, however, it had been observed that "the oil wells drilled southeast of Millerstown (Butler County) reached the oil sand at a less depth than those nearer town. A careful leveling of the well mouths and a comparison of the relative depths . . . showed that the oil sand certainly dipped to the north. . . ." (Chance, 1879, p. 10). This is the setting of the Millerstown anticline, the next west from Brady's Bend. It is described in Report V on northern Butler County. Since it was not detected in the surface rocks of Clarion County, it was, therefore, omitted from the map and text (Chance, 1880).

The "Clarion Oil-Belt" is shown about three miles west of, and roughly parallel to, the Millerstown axis but no indication of structure is drawn within the "belt." Chance (1879) had observed that prospectors had become converted to the unreasoning dogmatic belief in the "belt-line" theory and have run lines of all imaginable bearings in vain attempts to trace out locally productive pools. This resulted from the Pennsylvania Survey's failure to provide prospectors with a method of exploration.

Ashburner (1880) discovered the northernmost coal outcrop in the Appalachians north of Marilla summit in northwestern Bradford Township during his work for the McKean County Report R. The fact that this coal is younger than rocks to its southeast is indicative of the anticlinal structure of the Bradford Oil Field. But he does not call attention to it. Later during his investigation of the anthracite coal in northeastern Pennsylvania, Ashburner produced beautiful structure contour maps which could serve as models for maps of an oil or gas reservoir.

Lesley was an enigma. For example, while he credited Carll with "virtually creating the geology of petroleum" (Lesley, 1883, p. XII), he discredits this statement in the article he requested Carll to write for the Oil City Derrick in 1876 by his 1880 statement on the connection of petroleum with anticlinal and synclinal axes (Lesley, 1880, p. XVI). Lesley's staff, working in the oil region, knew the stratigraphy and had the capability of doing both surface and subsurface structural contouring with the aid of the sensitive barometer of which he was co-inventor. Such mapping might have furnished a method of exploration for new deposits of which oil prospectors were in dire need and the Survey had yet failed to provide. When Lesley assigned Ashburner to do the McKean County Report in 1880, the Bradford Oil Field there was the greatest in Pennsylvania for it was producing more than 83 percent of the state's total (Ashburner, 1880) and its geology required examination. However, Ashburner was instructed (Ashburner, 1880, p. 282) "to examine and report on the general geology and the value and extent of the coal measures, leaving the examination of the oil district for a special survey which should be made in the future." Did he want an historical or a geological study? Ashburner could not have failed to recognize that his discovery of the northernmost coal in the Appalachians had some meaning. Did Lesley edit out this fact? When White pronounced his theory, Lesley promptly issued a broadside against it (Lesley, 1887, p. 654–655),

. . . the location of anticlinal lines in the Pittsburg region has become a sort of popular mania, produced by a theory. . . .It is impossible . . . that any arrangement of water, oil and gas can occur in the deep oil rocks such as occurs in a bottle. . . .It, therefore, seems to me irrational to assign any importance whatever to the extremely gentle anticlines of the gas-oil region.

IMPETUS FOR THE NATURAL GAS INDUSTRY

In 1878, seven years before Lesley made this statement, however, the first two large natural gas wells were drilled at almost the same time, 9 miles apart, about 18 miles east of Pittsburgh on the Murrysville anticline, in Westmoreland County, Pennsylvania, where flank dips rarely exceed one degree (Stevenson, 1877). The operators were looking for oil but each well discovered gas in volume considered to be the largest yet encountered. One, the Haymaker #1, Remaley, had such a large flow it could not be controlled, was flared and lit the countryside for miles at night. When it was controlled eighteen months later, its open flow was 35 million cu. ft./day. The other well, the Mehaffey et al., #1 Beaver Valley, was considered to be of the same capacity and was also vented to the atmosphere (Carll, 1886). A well of the size of either, according to Carll, would relieve a larger reservoir capacity than an oil well flowing

100,000 barrels/day (Carll, 1890). In March 1882, four years after these two wells, another, said to be the largest yet drilled in the world, was found on a gentler anticline, the Claysville of Stevenson's 1876 report K. This well was the Niagara Oil Company #1 Alexander McGuigan (Carll, 1887), located 2.5 miles southwest of Hickory, Mt. Pleasant Township, Washington County, Pennsylvania on a closure, called the Hickory dome, some 33 miles west and south of Murrysville and about 18 miles from Pittsburgh.

These three wells, located such a distance apart, but in close proximity to a large manufacturing and population center, focused attention on the use that could be made of energy being wasted and the prospects for additional supply that could exist over a wide area. They should have also demonstrated the effectiveness of the anticlinal theory in locating gas which had, only a few years before, been described as "a strange vapor of the earth" and without commercial value.

GEORGE WESTINGHOUSE, JR., FATHER OF THE NATURAL GAS INDUSTRY

George Westinghouse, Jr., the fabulous inventor, recognized that natural gas had commercial value and that it had been utilized locally in a small way. He had become imbued with the idea that low cost natural gas could make Pittsburgh the greatest industrial city in the nation and he conceived a grand plan to furnish gas to the city and its mills (Garbedian, 1943). He organized the Philadelphia Company in July, 1884, and had gas flowing from Murrysville 18 miles to Pittsburgh, by November 1st of the same year. Others who tried and were called "impractical visionaries" gave up the attempt. The scope of his enterprise was, for the first time, conceived and executed with such a high degree of professional skill that Westinghouse must be considered the father of the natural gas industry. The Philadelphia Company, alone, by 1885, among the dozen others, was supplying about 350 million cu. ft./day, much of which came from its 52 wells of the 74 total (Carll, 1886) in the Murrysville Field, through its 600 miles of pipelines, exclusive of connections to 25,000 homes and 700 mills. It employed three to four hundred men working on its lines, not including small armies of from two to five thousand men laying new gathering and transmission lines (Thurston, 1888). The transformation this produced was in Thurston's words "almost like a page out of the Arabian Nights." Pittsburgh was "metamorphosed by natural gas—the fires of hell as it were—into a city of delightful homes, an industrial paradise" (McLaurin, 1896, p. 376).

JOHN H. GALEY

A native of the Pennsylvania oil regions and early oil prospector, John H. Galey (Fig. 5), who had taught school at age 16, achieved prominence in 1860 by drilling the famous Maple Shade gusher at Pleasantville, Venango County, Pennsylvania, when he was 20 years old. He played a large part in opening the

Figure 5. Portrait of John H. Galey (1840–1918), pioneer oil prospector. Photo, Pittsburgh, Pa., 1892.

Clarion and Armstrong County fields in the 1860s and the Butler and McKean County fields in the 1870s (Boyle, 1898). He was a voracious reader of geological reports, had the reputation of being an astute observer, had been practicing the precepts of John F. Carll and was "a close student of sands and their showing" (National Cyclopedia of American Biography, 1922, p. 18). He had talked with von Hofer, the noted geologist and delegate of the Austrian Commission to the Centennial Exposition in Philadelphia when he visited the Pennsylvania oil region in 1876. Hans von Hofer convinced Galey of the relationship of oil occurrence to structural disturbance and the anticlinal nature of oil "belts" in Ontario and West Virginia. He pointed out that the prominent Chestnut Ridge and Laurel Hill anticlines shown on the Pennsylvania Survey's maps paralleled the Pennsylvania oil "belts" which he believed were gentle anticlines that would require careful study to discover. There were no anticlines apparent to Galey, however, in the area he was exploring. His success had resulted from his innate ability of understanding where to look for oil. Because he had experienced the thrill of discovery, had the urge to explore, and had already decided to go west, he became interested in gold and silver and temporarily gave up oil exploration. By this time, 1880, he had made his fortune in oil and formed a partnership with Col. James M. Guffey in which Guffey was the financial promoter and he the explorer. He soon opened the

Texas Mine and Smelter at Galeyville, Arizona, about 60 miles from Tombstone, on the east slope of the Chiricahuas in 1881. The mine prospered at first but by 1883 Galey had lost the fortune he had made in oil and returned to Pennsylvania (Heald, 1967).

Following this reverse, Galey had to get a fresh start. He recalled that von Hofer had, some years earlier, described to him the value of anticlines in locating oil, that brine wells on the Allegheny River near Tarentum had produced oil, and that some wells in the vicinity had found gas. In addition, I. C. White had done a report for the Pennsylvania Geological Survey covering that area (White, 1878). He knew also that two unusually large gas wells had been drilled not far from Tarentum, near Murrysville, and that the geology of that area had been done by Stevenson in his 1877 Pennsylvania Geological Survey Report KK. Upon reviewing it, Galey found that these wells were located on an anticline. The report on the Tarentum area revealed that north of the Allegheny River an anticlinal axis crossed Bull Creek and the Upper Freeport coal outcropped along it (White, 1878). This was the opportunity to utilize von Hofer's theory of locating a well on the crest of an anticline. Galey rode his horse along the outcrop of the coal and compared its elevation with that of the creek. He made the first well location where he believed the coal reached a high elevation along Bull Creek for Guffey and Gailey (sic) #1 Wehrle. It was considered to be a good gas well when completed in the Murrysville sand (Mississippian) at a depth of 1179 feet, in July, 1884. This well was promptly sold to the Pennsylvania Salt Manufacturing Company for use in its plant at Natrona, about 2 miles distant. They soon made a location for a second well on a Bandy farm on the east side of the creek, where he believed the coal was at an even higher elevation, about 1.5 miles upstream from the mouth of Little Bull Creek. The well was completed in September, 1884 (Carll, 1887).

One of his younger brothers, Jim Galey, was the drilling contractor for the well. John visited it when it had started making considerable gas and he told his brother not to drill further because the Murrysville Sand, which when penetrated too deeply, would encounter water and spoil the well. When John left the drilling rig, Jim told his tool dresser he thought this was going to be a big well and he was going to make another run. If they did get water, he said that John wouldn't know they had not already encountered it. After the bit made about a "dozen licks," the drilling cable and tools blew out of the hole and leveled the rig at the walking beam. The roar was such it could be heard for miles. It is said to be one of the largest (gas) wells ever found (Weeks, 1885) and the "driest gas of any well in the field" (Carll, 1887, p. 684–685).

WESTINGHOUSE'S ACCEPTANCE OF THE ANTICLINAL THEORY

Westinghouse was, at this time, in the process of completing his pipeline from Murrysville to Pittsburgh. He had also been drilling for gas on his estate, "Solitude," in the Homewood sec-

tion of Pittsburgh (Garbedian, 1943). The result had at first given great encouragement but ended in disappointment (Carll, 1886). When Galey got word of his own well, he immediately induced Westinghouse to see and hear it, telling him that it would make more gas in a few minutes than his in several years. When Westinghouse saw and heard the well blowing, he was so impressed he bought it on the spot for $100,000, a large price in those days, even for such a large gas well with no pipeline.

Galey told Westinghouse how this well had been located at the highest coal elevation he could find on the anticline, the axis of which White had indicated crossing Bull Creek (White, 1878). Westinghouse was concerned about a continuing supply for his new company and agreed to back Guffey and Galey in finding gas for him. Since White's geology was a clue for their discovery, they engaged him, divulged the manner in which the new well was located and requested him to find more situations of the same kind for them to lease. White (1904, p. 54) states "Guided by this principle (structure), the writer pointed out and located all of the great oil pools of West Virginia for a Pittsburg syndicate in 1884 and 1885. . . .The "Pittsburg syndicate" was Guffey, Galey, *et al.* The lease play which began in 1884 lasted into 1886 and covered a half million acres in 11 counties (Thoenen, 1964).

White's anticlinal theory was published in *Science,* June 26, 1885, in an article "The Geology of Natural Gas" (White, 1904). He claimed that he formulated his theory independently as a result of a suggestion by a Wm. A. Earesman, in 1883, that many great Pennsylvania gas wells were located along anticlinal axes as shown on the Geological Survey's maps. This was subsequently verified over a period of two years by the successful location of several great gas pools including the Grapeville and Washington.

GRAPEVILLE GAS POOL

The Grapeville Gas pool was the first to be located to fulfill Guffey and Galey's request to White and was one of the practical tests White used to verify the theory. White points out in his 1885 *Science* statement that "the geological horizon that furnishes the best gas reservoirs in Western Pennsylvania . . . is the gas rock at Murrysville, Tarentum. . . ." (White, 1885, p. 52). No mention is made of an equally competent reservoir at Grapeville. It was not opened up until the following year, 1886, (Johnson, 1925), the year following publication of the theory in *Science,* June 26, 1885. The Grapeville gas pool is located in an area where seepages existed and a shallow brine well drilled nearly fifty years earlier had encountered gas in such volume it burned down the drilling rig (Thurston, 1888). Guffey opened up and controlled the Grapeville gas pool until it was taken by a corporation (Burgoyne, 1892).

WASHINGTON—TAYLORSTOWN FIELD

The first practical test White states he used for verification of his theory was the Washington, Pennsylvania gas pool. It was discovered by the Peoples Light & Heat Company #1 Hess Well, completed April 30, 1884 (Clapp, 1907). This was four months prior to the date Galey had advised White of the discovery and the manner in which the location for the big well had been made at Tarentum. The Hess discovery was at first considered an excellent well. Its capacity, however, was soon found to be inadequate for the town's supply even with the addition of four wells (Carll, 1886). The location of this well, as shown by the USGS structure map of the Claysville Quadrangle (Munn, 1912) is west of the axis and on the southwest plunge of the Washington Anticline to which Stevenson had called attention in 1876 (Rept. K, p. 30). Had this well been drilled in an identical structural position on the Murrysville Anticline, to the Murrysville Sand, it would have found water rather than gas. Since the well location was on a creek, the Chartiers, the major drainage for that part of southwest Pennsylvania, it's been said it was based on "creekology." Perhaps it was a matter of splitting bets between geology and "creekology." Subsequent drilling revealed that this pool is a part of a large stratigraphic accumulation that became known as the Washington-Taylorstown Field, extending from the Nineveh Syncline on the east about 7 miles west to Washington and about the same distance from it west through the "celebrated" Taylortown (sic) oil field. These are two of the many discoveries based on structural locations that development proved to be stratigraphic. (Amity and Claysville USGS Folios, Clapp 1907 and Munn 1912). The Taylorstown oil field is the one White located and mapped out for Guffey (White, 1904).

There was nothing new in White's theory, as has been previously shown, which had not already been accepted by geologists familiar with the subject. However, it succeeded at the propitious tme by bringing to the attention of those who wanted gas the most viable means of finding the volumes required to fill the need created by Westinghouse. It seems most incredible that White should have claimed original creation of the theory, in view of the other earlier observations that had been made and particularly because of the widespread sensational news of the large flows and enormous waste of gas from wells with anticlinal locations at Murrysville and Hickory. J. J. Stevenson, his mentor and close associate, had noted these anticlines in his Second Pennsylvania Survey Reports and had also written an article on the West Virginia "Oil-Break" while White was studying under him for his Master's Degree. In addition, debate was raging over the theory in the Pennsylvania Survey while White was working there. The irony of White's formulation of the theory on the suggestion of Earesman is that neither Earesman nor J. J. Vandergrift, President of Forest Oil Company and United Pipelines, who funded the investigation that resulted in the theory, were ever known to have put it to practical use. The successful test of the theory at Tarentum, of which White had been informed ten months earlier, and which could not have failed to be a significant factor in its verification, was ignored.

The most legitimate objection to White's theory was based on his statement that "all great gas wells are found on anticlinal axes," which they are not. The condition of the sand of the reservoir has a most important effect on a gas well being "great."

Also, as Minshall found on the West Virginia Oil-Break, axes usually contain undulations. These may not be discernible without structural contouring, which was not then employed, and drilling outside closure would likely be a failure.

There are great gas wells found in stratigraphic traps even without any suggestion of anticlines as, for example, the New Sheffield Field in Beaver County, Pennsylvania, discovered in 1884. The Washington-Taylorstown Field in Washington County, Pennsylvania, mentioned above, is a good example of a stratigraphic accumulation that was discovered because of structural indications.

Chance, who had probably more experience in the Pennsylvania oil region than any other geologist excepting Carll, held the opinion that "belt-lines" were more valuable than anticlines in determining the location of great gas wells (Chance, 1887). The "belt-line" hypothesis, however, is not applicable in locating a new pool because subsurface information is required for its use.

When Carll (1886, p. 45–51) wrote: "The diversified conditions under which all minerals exist make it absolutely certain that no infallible rule ever will be discovered," he accurately summarized the use of theories for prospecting.

MANNINGTON WEST VIRGINIA DISCOVERY

White reconfirmed and publicized the anticlinal theory October 11, 1889, when he and his associates of the T. M. Jackson Company discovered oil in their #1 Hamilton well near Mannington, Marion County, West Virginia. The discovery was made on one of the half million acres of leases in eleven West Virginia counties Guffey and Galey had acquired (Thoenen, 1964) after requesting White to find situations for them to lease similar to where they had drilled the enormous gas well at Tarentum. The leases carried drilling commitments for more than forty wells and since the capital was not available to meet them, many of the leases, including some in the vicinity of Mannington, were surrendered and subsequently acquired by the T. M. Jackson Company.

The oil discovered by the #1 Hamilton was in the Big Injun sand (Mississippian) and while the well was never a significant producer, the field it opened was substantial (Thoenen, 1964). Two hundred wells, of which only five percent were failures, were drilled along the twenty-five mile Northeast trend between Mannington and Mt. Morris, Greene County, Pennsylvania. White (1904, p. 54) claimed to have located all the great oil pools of West Virginia for Guffey, Galey *et al.* in 1884 and 1885, "long before the drill finally demonstrated the correctness of his conclusions." This was one of those pools.

White's description of the manner by which the location was made for this discovery involves the practical application of the principles of the anticlinal theory and was essentially the same as that employed by Lesley in calculating the depth of the Venango sands in the Brady's Bend Iron Company's well as early as 1865.

EDWARD ORTON ACCEPTS WHITE'S ANTICLINAL THEORY

The highly regarded Edward Orton of Ohio was probably the first State Geologist to accept and give prominence to White's anticlinal theory. Orton (1888, p. 93) wrote:

. . . his [White's] applications of it are bold, original and, best of all, successful and they mark a new period in our study of the geology of oil and gas . . . but, as has been already shown, anticlinals are of infrequent occurrence in Ohio. . . . Even if anticlinals are held to account for the fact of oil and gas accumulation, the theory would have but limited application to our geology.

However, Orton related the theory, after the fact, to the November 1884 Findlay, Hancock County gas discovery in the Trenton (Ordovician) limestone at a depth of 1648 feet. The location of the discovery was near a known gas seep and only later determined to be close to the axis of the anticline. This gave confirmation to Hunt's prediction made twenty years earlier.

While the yield of the Findlay discovery was very small compared to the large Pennsylvania wells, the gas was flared and at night could be seen for miles. This spectacular sight created an excitement that precipitated a boom which produced numbers of wells in northwestern Ohio and east central Indiana. Some produced oil of gusher proportions and others that found gas provided the supply for establishment of a local industry only a few months after Pittsburgh's. This development became known as the Lima-Indiana Field.

Just as it was on the Ontario Peninsula, surface structure in the vicinity of the field was difficult to determine. Drilling, however, revealed sharp dips as, for example, at Findlay where the Trenton drops 150 feet in less than 1760 feet and Orton produced structure contour maps in 1889 (Orton, 1890). His study enabled him to predict, in 1890, that oil would be found at Bowling Green, 30 miles northwest of Findlay, prior to the time drilling was done there (Bownocker, 1903).

A local company had made several unsuccessful attempts to find gas for the town of Cadiz, Harrison County, Ohio, before 1887, and Orton was requested to investigate what could be done to find a supply. Stevenson had described a small anticline near Cadiz during his work for the Ohio Second Geological Survey in 1874 (Orton, 1888; 1890). Orton discovered two additional anticlinal features only a short distance from town. All three of these were subsequently successfully exploited by small oil and gas wells in the Berea sand (Mississippian) at relatively shallow depth (Bownocker, 1903).

CAMBRIDGE ARCH—DISAPPOINTMENT

When word of the anticlinal theory reached prospectors in east central Ohio, their activities were directed to the Cambridge Arch which Stevenson had described in his Second Survey Report as a conspicuous uplift, the axis of which lies immediately west of Cambridge (Orton, 1888; 1890). The fact that the axis

was not level, and that the structure was off trend to Appalachian folding did not occur to them. Most wells drilled prior to 1887 were, as a result, failures and those that found gas or oil had sufficient water to seriously impair their productivity. No better result was obtained when lines of levels were run across the feature and operators therefore concluded that the structure was too flat to permit gas, oil and water to separate. The difficulty was that the wells were drilled outside effective closure, just as were wells on the West Virginia Oil-Break.

By 1892, however, the Pebble Rock Oil & Gas Company had levels run on the prominent coal of the area and discovered a dome about 4 miles south from Cambridge. The first well drilled was successful and a valuable gas pool, known as Harmony, had wells with flows as large as 2 million cu. ft./day. The productive horizon is the Berea Sand (Mississippian) at a depth of approximately 1100 feet (Bownocker, 1903).

Orton predicted that oil or gas would be found on an anticline that crosses the Ohio River at Moores Junction, Washington County, 3 miles southwest of Marietta. Four noncommercial wells were drilled to the Cow Run Sand at a depth of less than 750 feet between 1894 and 1898. Coincidental with the realization of failure of the last of the wells, oil was found south of the river. Drilling to the north was thereby encouraged and rewarded with a 150 bbl/day well. Subsequent wells flowed as much as 500 bbls/day.

THE DISCOVERY THAT CHANGED THE WORLD

At about the same time, Captain Anthony F. Lucas believed there was big oil at Spindletop, a solitary mound four miles south of Beaumont, Texas, that contained gas seeps and that he suspected was a salt dome, protruding above the flat Gulf Coastal Plain. He had become infected by Patillo Higgins' enthusiasm over the prospect for oil there, even though Higgins had exhausted his capital in an attempt to find it, and outstanding experts had unhesitatingly condemned the possibility. Yet Lucas tried and found sufficient oil to give him encouragement but he, too, had depleted his resources before he achieved any better result. When he attempted to obtain financing from Standard Oil in New York, they sent their top oil expert to visit his Spindletop prospect and Lucas was told that oil could not be produced there in quantities that would pay for boring the wells.

Lucas was discouraged until Professor William B. Phillips, a University of Texas geologist, came to Beaumont to see him and told him that he believed there was something to the dome idea, that the prospect deserved to be tested and that Guffey and Galey was the firm he should approach to do it. Phillips offered to write a letter to Galey, the explorer of the partnership, who might then be in Corsicana, Texas, 200 miles to the northwest, where Galey had developed the first oil field in 1895.

Galey's oil and gas operations were well known prior to Spindletop; Boyle (1898, p. 670) wrote,

Many of the greatest fields developed since the inception of the oil business owe their existence and life to the ambitious and determined character of this man, [Galey]. Some of the largest producers of oil the country has ever known have been the product of his keen foresight in pushing operations in what might look to another eye like unproductive territory.

Galey was a modest, uncommunicative man who was considered to be a dreamer by his contemporaries. Mellon (1948, p. 159) wrote of him:

Whether he was looking for oil, or gold and silver in his prospecting, Galey did not seek money nearly so much as his own peculiar kind of excitement. The finding of treasure seemed more important to him than anything the treasure could buy.

He had the established reputation for finding more oil than anyone else and "possessed nerve, energy and endurance" (McLaurin, 1896, p. 222). Galey had drilled wells in California from Coalinga to the middle of the Mojave Desert, across the country to Washington, D.C., and had established production in all of the Appalachian States as well as Kansas and Texas, but was unsuccessful in securing drillable leases in the Osage and Indian Territory until after Spindletop. Some of his major oil discoveries included the Turner gusher which flowed 8,000 barrels per day in Lewis County, West Virginia, and the famous Matthews gusher in the McDonald Field, Washington County, Pennsylvania, which flowed 860 barrels of oil per hour and has probably never been equalled east of the Mississippi River.

It was on this man that Lucas called to seek financial support for his prospect. Galey was in Corsicana at the time and was awaiting his arrival. He told Lucas he had heard of his dome theory, what others thought about it, and that such experts were not necessarily right. When Lucas explained his salt dome theory, Galey comprehended that it would provide the same kind of trap as the fold expressed at the surface by the coal on Bull Creek at Tarentum where he had drilled the monstrous gas well sixteen years earlier. The sight of the mound with its seeps heightened his interest. He conceived its form as an expression of the configuration of the strata below, which formed a trap that was the source of the seeping gas and smelled like rotten eggs, similar to that which he had found in Kentucky.

An agreement was made on the spot to drill three wells to a depth of 1200 feet, contingent on Lucas acquiring additional acreage and Guffey and Galey's ability to borrow the money necessary to finance the operation. Their extensive production should have permitted them to do this themselves, but they could not, what with the demands Guffey made on the partnership funds for the purchase of West Virginia coal land and his political activity, in addition to a restricted cash flow caused by flush new production creating a price decline.

Galey invited Lucas to come to Pittsburgh to explain his dome theory to Guffey and their bankers, the Mellons. They approved the loan on the basis of Galey's recommendation. Meanwhile, Lucas had secured additional leases. Guffey approved the land holdings although some vacancies existed which others could drill if production was discovered. The agreement was finalized and Lucas returned to Texas. When Galey reached Beaumont to make the location for the initial well, Lucas had gone to Corsicana to secure the Hammils' services for drilling (Hammil, 1957). Mrs. Lucas was at home and Galey took her to see where he would drive the stake. He drove it into a hog wallow and said, "tell that Captain of yours to start that first well right here."

A few months later, January 10, 1901, after overcoming seemingly insurmountable drilling obstacles before reaching its depth of 1140 feet, the Spindletop gusher blew its top with a deafening roar through the hurtling pipe-splintered derrick, spewing 100,000 barrels of black gold 200 feet to the sky (O'Connor, 1933).

This was the most spectacular event in oil history, the first giant gusher in America, and it resulted in the first salt dome oil field. Little did the eminent geologists of the day realize that Spindletop was drilled with benefit of geology. When he saw the well, Galey's only recorded comment to the press was, "This great gusher marks the beginning of the liquid fuel age." (Clark, J. A., 01/04/57, Houston Post Sec. 6, p. 5). Galey was interested and had invested in the development of the internal combustion engine.

Luck always plays a part in any discovery, but the Spindletop discovery was the result of an understanding of the anticlinal theory and its application to a salt dome. The discovery "constituted a fitting climax [for Galey] to the most illustrious career of wildcatting in the annals of the oil industry" (Clark and Halbouty, 1952, p. 133).

Spindletop had the world's first great oil well and it was also at Spindletop in Texas "where oil became an industry" (Spindletop 50th Anniversary Commission, 1951, p. 4). On the Murrysville anticline in Pennsylvania, twenty-three years earlier, two of the world's first great gas wells were drilled by knowledgeable operators seeking oil, who found gas instead. A single well having the flow of either would have relieved a larger reservoir capacity than the Spindletop discovery. While flowing gas does not provide as spectacular a sight as gushing oil, these wells inspired the initiation of the natural gas industry. The anticlinal theory provided the basis for the location of these wells, as it did for the Spindletop discovery. This makes it abundantly clear that the anticlinal theory, with its long history prior to its formal publication, has played the major role in the establishment of both the natural gas and modern oil industry. The kind of men whose ingenuity made the modern oil industry possible also created the "beginning of a new era in civilization" (Spindletop 50th Anniversary Commission, 1951) by the founding of the natural gas industry seventeen years earlier.

REFERENCES CITED

Ames, Mary Lesley, 1909, Life and Letters of Peter and Susan Lesley: New York, Knickerbocker Press, 2 v.

Andrews, E. B., 1861, Rock Oil, its Geological Relation and Distribution: American Journal of Science & Arts, v. 32, p. 85–93.

—— 1866, Article IV, American Journal of Science, Second Series, v. 42, no. 124, July, pp. 33–43.

Ashburner, C. A., 1880, The geology of McKean Co., and its connection with that of Cameron, Elk and Forest: Second Geological Survey of Pennsylvania, Report R, McKean Co., 371 p.

Bownocker, J. A., 1903, The occurrence and exploitation of Petroleum and natural gas in Ohio: Geological Survey of Ohio, Fourth Series, Bulletin 1, 325 p.

Boyle, P. J., 1898, Derrick Handbook of Petroleum, 851 p.

Burgoyne, A. G., 1892, All Sorts of Pittsburghers, 303 p.

Carll, John F., 1875, Report of Progress in the Venango County District: Second Geological Survey of Pennsylvania, v. I, 1878 ed., p. 1–49.

—— 1880, The Geology of the Oil Regions of Warren, Venango, Clarion and Butler Counties, Second Geological Survey of Pennsylvania, v. 3, no. 24, 482 p., Atlas.

—— 1883, Geological Report on Warren County—: Second Geological Survey of Pennsylvania, v. I-1, no. 21, 439 p.

—— 1886, Preliminary Report on Oil and Gas in Report of State Geologist for 1885: Second Geological Survey of Pennsylvania, Annual Report for 1885, p. 1–81.

—— 1887, Report on the Oil and Gas Regions in Lesley, J. P., Second Geological Survey of Pennsylvania, Annual Report for 1886, pt. 2, p. 575–586; p. 684–685.

—— 1890, Seventh report on the oil and gas fields of western Pennsylvania: Second Geological Survey of Pennsylvania, Oil and Gas Fields, Vol. I-5, no. 8, 356 p.

Chance, H. Martyn, 1879, The Northern Townships of Butler County, Second Geological Survey of Pennsylvania, v. V, no. 17, 248 p.

Chance, H. Martyn, 1880, The Geology of Clarion County, Second Geological Survey of Pennsylvania, Report of Progress, v. V, no. 15, 232 p.

—— 1887, The Anticlinal Theory of Natural Gas: American Institute of Mining Engineers Transactions, v. 15, p. 3–13.

Clapp, F. G., 1907, Amity Quad., U.S. Geological Survey Bulletin 300, 145 p.

Clark, J. A., 1957, No Wildcatter Ever Greater Than Galey: Houston Post, Texas, Section 5, Jan. 6, p. 8

Clark, J. A., and Halbouty, M. T., 1952, Spindletop, Random House, New York, N.Y., 306 p.

De Golyer, E., 1961, Concepts on occurrence of oil and gas in history of petroleum engineering: Dallas, American Petroleum Institute, p. 15–33.

Evans, E. W., 1866, On the Oil-Producing Uplift of West Virginia: American Journal of Science, 2nd Series, v. 42, p. 334–348.

Garbedian, H. G., 1943, George Westinghouse, Fabulous Inventor: Dodd, Meade & Co., New York, N.Y., 345 p.

Hamill, C. G., 1957, We Drilled Spindletop: Spindletop Foundation for Medical Research, 40 p.

Harkness, R. B., 1924, Petroleum in 1922: Ontario Department of Mines, Annual Report, v. 32, pt. 5, p. 76–92.

Heald, W. F., 1967, The Chiricahua Mountains: University of Arizona Press, Tucson, Arizona, 164 p.

Hildreth, S. P., 1836, Observations on the Bituminous Coal Deposits of the Valley of the Ohio and the Accompanying Rock Strata: American Journal of Science, v. 28, no. 1, p. 154.

Hitchcock, C. H., 1865, The Albert Coal, or Albertite, of New Brunswick: American Journal of Science, ser. 2, vol. 39, pp. 267–75.

—— 1866, The Geological Distribution of Petroleum in North America, Report of the British Association as quoted in Howell, J. V., 1934, p. 8.

Howell, J. V., 1934, Historical Development of the Structural Theory of Accumulation of Oil and Gas: in Wrather, W. E. and Lahee, F. H., eds., Problems of

Petroleum Geology: Sydney Powers Memorial Volume, American Association of Petroleum Geologists, p. 1–23.

Hunt, T. Sterry, 1862, Notes on the History of Petroleum or Rock Oil: Smithsonian Institution Annual Report for 1861, p. 319–329.

—— 1866, Petroleum: Geological Survey of Canada, Report of Progress, 1863 to 1866, p. 233–262.

Johnson, M. E., 1925, Mineral Resources of the Greensburg Quadrangle, Westmoreland County, Pennsylvania: Pennsylvania Geological Survey, 4th series, Bulletin 37, 162 p.

Lesley, J. P., 1856, Manual of Coal and its Topography: Philadelphia, J. B. Lippincott & Co., 224 p.

—— 1865, On Petroleum in the Eastern Coal Field of Kentucky and Records of Borings in Pennsylvania: American Philosophical Society Proceedings, v. 10, p. 33–68, 187–191.

—— 1876, Historical Sketch of Geological Explorations in Pennsylvania and Other States: Second Geological Survey of Pennsylvania, v. A, 1878 ed., 200 p.

—— 1880, in Carll, J. F., The Geology of the Oil Regions of Warren, Venango, Clarion and Butler Counties: Second Geological Survey of Pennsylvania, v. 3, no. 24, 482 p.

—— 1883, in Carll, J. F., Geological Report on Warren Co., Second Geological Survey of Pennsylvania, v. I-4, no. 21, 439 p.

—— 1886, in Carll, J. F., Preliminary Report on Oil and Gas in Report of State Geologist for 1885: Second Geological Survey of Pennsylvania, Annual Report for 1885, p. 1–81.

—— 1886, The geology of the Pittsburgh coal region, American Institute of Mining Engineers, Transactions, v. 14, p. 618–656.

—— 1887, Report on the Oil and Gas Regions: Second Geological Survey of Pennsylvania Annual Report for 1886, pt. 2, p. 575–586; 684–685.

—— 1895, Geology of Pennsylvania, v. 3, pt. 1, Carboniferous Formation, p. 1629–2153.

Logan, William E., 1863, Report on the Geology of Canada: Geological Survey Canada Report of Progress from its commencement to 1863, 983 p.

McFarlan, A. C., 1943, Geology of Kentucky: University of Kentucky, Lexington, 531 p.

McLaurin, J. J., 1896, Sketches in Crude Oil: Harrisburg, publ. by the author, 406 p.

Mellon, W. L., 1948, Judge Mellon's Sons: Pittsburgh, privately printed.

Minshall, F. W., 1881, as quoted from State Journal, Parkersburg, West Virginia in Report on the Production, Technology and Uses of Petroleum and its Products, U.S. 10th Census Rept., 1884, v. 10, 319 p.

—— 1888, History of the Development of the Macksburg Oil Field: in Geological Survey of Ohio, Vol. 6, p. 443–75.

Munn, M. J., 1912, Description of the Claysville quadrangle (Pa.): U.S. Geological Survey Geological Atlas Folio 180, 14 p.

National Cyclopedia of American Biography, 1922.

Newberry, J. S., 1860, The Oil Wells of Mecca: Canadian Naturalist, v. 5, p. 325–326.

—— 1866, Prospectus of the Indian Creek and Jacob's Knob Coal, Salt, Oil, etc. Company with a geological report on the lands by J. S. Newberry (abs): American Journal of Science, 2nd Ser. v. 41, p. 284.

O'Connor, H., 1933, Mellon's Millions: New York, John Day Co., 443 p.

Orton, E., 1888, The Geology of Ohio considered in its relations to petroleum and natural gas: Geological Survey of Ohio, Vol. 6, p. 1–59.

—— 1890, Geological Survey of Ohio, First Ann. Rept., 323 p.

Owen, E. W., 1975, Trek of the Oil Finders: American Association of Petroleum Geologists, Tulsa, Oklahoma, Memoir 6, 1647 p.

Peckham, S. F., 1884, Report on the production, technology, and uses of petroleum and its products: U.S. 10th Census, v. 10, 319 p.

Randall, F. A., 1878, Observations of the geology around Warren, Pa.: 2nd Geological Survey of Pennsylvania, v. I, p. 50–55.

Rogers, Henry D., 1860, on the Distribution and Probable Origin of the Petroleum or Rock Oil in Western Pennsylvania, New York and Ohio: Philosophical Society of Glasgow Proceedings, v. 4.

——1863, Coal and Petroleum, Harper's Monthly, v. 27, p. 259–264.

Spindletop 50th Anniversary Committee Official Proceedings, 1951, Spindletop Where Oil Became an Industry (1901-1950): Beaumont, Texas, 201 p.

Stevenson, J. J., 1875, Notes on the geology of West Virginia, no II, Proceedings of the American Philosophical Society, v. 14, p. 370–401.

——1876, Report of progress in the Greene and Washington district of the bituminous coal fields of western Pennsylvania: Second Geological Survey of Pennsylvania, Report K, 419 p.

——1877, Report of progress in the Fayette and Westmoreland district of the bituminous coal fields of western Pennsylvania: Second Geological Survey of Pennsylvania, Rept. KK, 437 p.

Thoenen, E. D., 1964, History of the Oil & Gas Industry in West Virginia, Educational Foundation Inc., Charleston, W. Va., 429 p.

Thurston, G. H., 1888, Allegheny County's Hundred Years, A. A. Anderson and Son, Pittsburgh, Pa., 313 p.

Weeks, J. D., 1885, Natural Gas: U.S. Geological Survey, Mineral Resources, 1883–4, 154 p.

White, I. C., 1878, Report of progress in the Beaver River district of the bituminous coal fields of western Pennsylvania: Second Geological Survey of Pennsylvania, Report Q, 328 p.

——1885, The Geology of Natural Gas: Science, v. 6, p. 43–44.

——1892, The Mannington Oil Field and the History of its Development: Geological Society of America Bulletin, v. 3, p. 187–216.

——1904, Petroleum and natural gas; precise levels: West Virginia Geological Survey, vol. 1A, 625 p.

Wrigley, Henry E., 1875, Special Report on the Petroleum of Pennsylvania: Second Geological Survey of Pennsylvania, Vol. J, p.1–78.

Geological Society of America
Centennial Special Volume 1
1985

Application as stimulus in geology: Some examples from the early years of the Geological Society of America

William M. Jordan
Millersville University
Millersville, Pennsylvania 17551

ABSTRACT

The Geological Society of America was conceived in 1881 and born in 1888 as part of an era of unprecedented mineral-based prosperity that has continued, with wartime and business cycle fluctuations, to the present. From its earliest days, geology and its practical applications, particularly industrial, have been intimately connected. This paper examines the mutual relationship of practical need and geology during the first half-century of the GSA, focusing particularly on how practical needs stimulated theoretical developments. Three specific examples of this relationship are given: G. K. Gilbert's studies of hydraulic mining-induced aggradation of streams in California; the influence of iron and other mining in the Lake Superior region on the early development of Precambrian stratigraphy; and geology's rediscovery of the concept of facies as a consequence of exploration for coal in Utah and the search for petroleum in the Permian of West Texas.

INTRODUCTION

Geology's practical origins can be found in medieval and even older mining, especially in Saxony (Adams, 1938). The intellectual development of the discipline came later—beginning with the natural philosophy of 17th century Europe and crystallized in the late 18th century as an independent scientific discipline under the influence of the industrial revolution. Beginning in Britain during the mid-18th century practical needs played an ever increasing role in the historical development of geology. Although as Porter (1977, p. 132) cautions, "any immediate relationship between industrialization as such and the making of geology is more problematic" than, say, that between industry and chemistry, it is clear that practical concerns provided the original impetus to organization of the local and national surveys that put geology on its modern footing (Fuller, 1969; Jordan, 1979).

In the early 19th century geology blossomed with the explosive growth of transportation-communication networks—first canals and later the railroads—serving industry and commerce. This happened at the same time as the development of a standard geologic time-scale in Europe and establishment of the Geological Survey of Great Britain. In North America as in Europe, practical interest in exploiting natural resources, especially coal

and iron but also agriculture, led to establishment of numerous state geological surveys (see Oakeshott, this volume) and the Geological Survey of Canada. In Pennsylvania, for example, the First Geologic Survey (1836–42) under Henry D. Rogers conducted its work with a mandate to determine the uses to which geologic resources "can be applied in the arts, and their subserviency to the comforts and conveniences of man" (Millbrooke, 1976). The state reestablished its Second Geologic Survey (1874–1889) upon the call of mining and manufacturing interests, and to satisfy demands for public knowledge of Pennsylvania petroleum, first exploited on a large scale in 1859 (Pierce and Jordan, 1982).

On the national scene in the United States, economic geology was fostered by western mining interests, by the great territorial geologic surveys of the immediate post-Civil War years, and through organization of the U.S. Geological Survey in 1879. By the time the Geological Society of America was established, petroleum production was shifting westward and by the turn of the century that industry emerged from its "Age of Illumination" into the "Age of Energy." Anthracite coal production peaked in 1917 as new industrial (and military) forces were beginning to fashion the natural resource based symbiosis of geology and industry that exists today, and by the start of World War I the Geological Society of America (GSA) had just completed its first quarter century. Thereafter practical and particularly industrial influences on geology accelerated greatly. Conversely, as Mary Rabbitt (1980) pointed out, the institutionalization of geology in governmental agencies and in professional organizations such as the GSA was directly beneficial to industry and, through it, society in general. It can be said that the modern world would not have become what it is today without geology.

In this context and within the time frame of the first half-century of the GSA (1888–1938), three specific examples of practical need as stimulus to geology are presented: G. K. Gilbert's California stream studies of hydraulic mining-induced aggradation, the development of Precambrian stratigraphy in the Lake Superior region as a consequence of iron and other mining activity, and geology's rediscovery of the concept of facies stemming from exploitation of coal resources in Utah and petroleum exploration within the Permian of West Texas. These three examples show the influence of concurrent developments, especially within the mineral industries, upon the conceptual progress of geology in its "early mature" phase. To place these examples in context and to show that this period of time does not involve "ancient history," however, we note that Nikola Telsa built his electric motor in 1888, contemporaneous with incorporation of the GSA, and in the same year George Eastman perfected his Kodak box camera. These developments may *seem* remote because at the time Oklahoma was still "Indian Territory," not opened to general settlement until 1889.

ECONOMIC BACKGROUND

This paper does not deal with economic geology, but some background is relevant to the examples cited. In the decade prior to the first meeting of the Geological Society of America at Ithaca, N.Y., in December, 1888, the eastern states totally dominated the U.S. economic and industrial scene. The 1880 census shows Pennsylvania to be the leading mining state, with its production of both coal and iron totalling more than that of the next three states, Colorado, California, and Nevada combined (Rabbitt, 1980). Pennsylvania, Michigan, New York, and New Jersey, in that order, produced 75% of U.S. iron ore. Most copper came from Michiga, while virtually all zinc was mined in New Jersey, Pennsylvania, and the Mississippi Valley. Petroleum production in 1880 was almost entirely from northwestern Pennsylvania. Only gold, silver, and lead production were significant in the West.

The west's climb to eventual prominence in mineral production started with John Marshall's discovery of gold in California in 1848, the Comstock Lode, Nevada, gold and silver strike of 1859, and the concurrent discovery of gold in Colorado. Gold was found in the Black Hills in 1874 to the Indian's (and eventually Custer's) regret, and finally rich silver-bearing lead carbonates were first exploited at Leadville, Colorado, in 1878, in the same year that copper was discovered at Butte, Montana. The

Cripple Creek district, exploited in the post-1888 period, was the last of the great Colorado "gold camps" and served as nurturer of GSA's benefactor R.A.F. Penrose, Jr. (Guilbert, 1983). The most romantic gold-rush of all, however, to the Klondike in the Yukon territory, began late in the century in 1897. Discussion of western mines and miners to about 1910 can be found in Young (1970, 1976) and Wyman (1979), while Barger and Schurr (1944) cover the 20th century U.S. mining picture to the beginning of World War II.

Actually, as shown by the 1890 census and enunciated by Turner in 1893, the western frontier had closed by the last decade of the 19th century. In 1890 Montana copper (amounting to 43% of national production) became economically important in the West, a first for a non-precious metal in that region. Gold production declined, especially in California because of an 1884 ban on hydraulic mining, whereas silver production increased, putting stress on contemporary monetary policy. By 1900, however, gold production was up again, owing both to new discoveries and the adoption of a national "Gold Standard." Petroleum production doubled by 1890, but was still confined mostly to the east, in Pennsylvania, West Virginia, and Ohio (Galey, this volume). However, by 1900 not only had petroleum production continued its increase but more than half was from west of the Mississippi. Similarly, copper and aluminum production were also sharply up by 1900 (Rabbitt, 1980). Thus the Geological Society of America was established during a time of intensive exploitation of mineral resources despite the 1893 depression, the most severe slump of the 19th century.

CALIFORNIA STREAM STUDIES OF G. K. GILBERT

The first of several examples illustrating the complex interplay between utilitarian activity and conceptual progress in geology (even in areas not directly "economic") involves the downstream effect of hydraulic gold mining (Figure 1) on the rivers of the western Sierra Nevada in California. In the fifty years following Marshall's discovery of gold at Sutter's Mill, one and a half billion cubic yards of sediment (a volume eight times greater than that moved during construction of the Panama Canal) had been overturned by mining in the Sierra (Ritter, 1968). As early as 1884, agricultural interests in the Sacramento Valley obtained a federal court injunction to prevent further addition of tailings to the river transport systems, thus effectively killing the hydraulic mining industry. In 1904 the California Mining Association appealed to President Theodore Roosevelt to "re-assess the whole situation, and judge from a scientific point of view whether mining could be revived" (Kelley, 1959). The President gave the task to the U.S. Geological Survey which assigned G. K. Gilbert, who was nearing the end of his illustrious career, to investigate. Gilbert's extensive previous scientific accomplishments are well-summarized in Yochelson (1980).

The first step in Gilbert's investigation was to begin flume experiments (Figure 2), experimentally measuring the transport-

Figure 1. "Malakoff Diggings" hydraulic mine, Nevada County, California. Photograph by C. E. Watkins, c. 1877. (Courtesy of the Bancroft Library, Berkeley, California.)

ing power of running water, supplemented by extensive field observation. Stream processes were for Gilbert a theme that had occupied at least part of his attention since he first proposed the concept of a "graded river," as early as 1877, during his days on the Colorado Plateau as a member of the Powell Survey. His California studies resulted in twin reports, the 1914 *The Transportation of Debris by Running Water,* USGS Professional Paper 86, and the 1917 *Hydraulic-Mining Debris in the Sierra Nevada,* USGS Professional Paper 105. These are well known classics in process geomorphology and pioneering works in sedimentology. Ironically for the mining industry, however, Gilbert's eventual recommendation was against resumption of hydraulic mining and it came *not* because of the aggradation and farmland flooding, the original complaint, the effects of which could be ameliorated, but because the consequences of resumption would be prejudicial to navigation. Siltation had already, at the time of Gilbert's investigations, resulted in an observable migration of the San Francisco baymouth bar, threatening eventually the commerce and prosper-

ity of the entire region. In California, the consequences of investigating a practical problem thus led directly not only to Gilbert's pioneering studies in sedimentology but to some of the earliest observations made in estuarine oceanography. Hagwood (1981) presents a comprehensive history of the hydraulic mining industry in the western Sierra Nevada as part of his study of the work of the California Debris Commission, while details of Gilbert's investigative techniques and his interpretations can be found in Leopold (1980).

LAKE SUPERIOR IRON MINING AND PRECAMBRIAN STRATIGRAPHY

A second example of the utilitarian-geological relationship involves the exploration for metallic ores and extensive turn-of-the-century iron mining in the Lake Superior region. While high-grade iron ore had been produced by underground mining in Wisconsin and Northern Michigan since 1852, Minnesota depos-

Figure 2. Flume built by G. K. Gilbert for hydraulic experimentation at the University of California, Berkeley. (Courtesy of U.S. Geological Survey Photographic Library, Gilbert No. 3408.)

Figure 3. Portrait of Charles R. Van Hise. (Courtesy of U.S. Geological Survey Photographic Library, Portrait No. 58).

its were not opened until the 1890s. Because of their richness, the efficiency of open-pit mining, and ease of transport to smelting locations, by 1895 the United States was the world's largest producer of pig iron and steel (Gregory, 1980).

The development of a Precambrian chronology in North America began early in the investigations of the Geological Survey of Canada under William Logan, well to the east of the Lake Superior region. In 1882 the Canadian Survey set Andrew C. Lawson to work (initially as an assistant to Robert Bell) in the area between Lake of the Woods and Lake Superior "owing to the discovery of precious metals at Lake of the Woods" (Zaslow, 1975, p. 184). Lawson's revolutionary report defining the Keewatin, based on this mapping, appeared in 1885 and his Rainy Lake Region paper, which redefined "Laurentian" as used in the Kenoran district, appeared shortly thereafter in the Geological Survey of Canada's Annual Report for 1887.

Lawson was the first to apply the basic principles of stratigraphy to Archean rocks, but the opening of the Mesabi Range to production in 1890 brought the attention of the U.S. Geological Survey (USGS) to the overlying Proterozoic rocks of the region. Two men affiliated with the University of Wisconsin, Charles R. Van Hise (Figure 3) and Charles K. Leith became the most prominent of a multitude of survey workers. Initially Van Hise

came to the study of the Lake Superior Precambrian as assistant to Roland D. Irving of the Lake Superior Division of the U.S. Geological Survey. Irving, with Thomas C. Chamberlin, pioneered in Precambrian geology while with the Wisconsin Geological Survey (Vance, 1960).

Upon Irving's premature death in 1888, Van Hise became Chief of the Lake Superior Division and subsequently seven USGS Monographs were published. Principal among Van Hise's many associates in this effort was Charles Leith who, while still an undergraduate at the University of Wisconsin and serving as a general assistant to Van Hise, first encountered the Precambrian in 1892 (Bailey, 1981). Leith joined the USGS in 1897 and was assigned to study the Mesabi as Assistant Geologist in 1900; his Monograph 43, *The Mesabi Iron-Bearing District of Minnesota*, appeared in 1903. Between 1903 and 1907, a joint U.S.-Canadian International Committee on Geologic Nomenclature (Zaslow, 1975) conferred on stratigraphic matters. The resulting decisions were incorporated into the comprehensive Lake Superior summary volume, USGS Monograph 52, *The Geology of the Lake Superior Region,* issued in 1911, which was written primarily by Leith but was based mostly on Van Hise's data. Previously Van Hise and Leith had also collaborated on USGS Bulletin 360, *Pre-Cambrian Geology of North America,* published in 1909.

Beside the obvious observation that mining brought USGS attention to the region, which improved understanding of the Precambrian, active development of the area stimulated the personal and "extracurricular" involvement of geologists such as Van Hise and Leith. With the discovery of the rich ore deposits of the Mesabi Range in 1890, geological field work went forward during an intense mining boom (Figures 4 to 6). Close contact was maintained between USGS field workers and the mining companies for exchange of information. In 1892, with the drastic cut in Survey appropriations (Rabbitt, 1980) and Powell's resignation, Van Hise relinquished his regular position with the Survey but continued on a per diem basis, which allowed him to undertake other work as a consultant. In 1901, while Leith was working on the Mesabi, the consolidation of the new United States Steel Company was in progress and several of the companies involved attempted to hire Leith as a consultant. Leith declined because as he stated in a letter to Van Hise, to do so "would be a grave breach of faith, and would not be to my credit in the long run" (McGrath, 1971, p. 232). Van Hise supported Leith's stand, particularly since in 1900 he himself had resumed full-time Survey duties and was by then Leith's supervisor as Chief of the Precambrian and Metamorphic Geology Section.

USGS regulations did not prohibit work for commercial interests in Canada, however, and, starting in 1902, Leith and Van Hise did so together. In 1903, as an advisory member of the geology committee of the newly organized Carnegie Institution of Washington (Carnegie had recently sold his iron and steel interests to J. P. Morgan's U.S. Steel), Van Hise visited Carnegie in Scotland to discuss establishment of a geophysical laboratory, and in the process also obtained $100,000 of Carnegie's money for the purchase of options on mineral lands in Canada, fifty percent of any profits to go to Van Hise! In 1905, Leith himself went on a per-diem basis with the Survey and into partnership as field operative for Van Hise, now president of the University of Wisconsin, sharing half of Van Hise's profits. Eventual publication of the 1911 Lake Superior summary monograph virtually ended the work of both men for the USGS. Van Hise died suddenly in 1918, the result of complications following minor surgery, leaving a comfortable estate, and Leith, by the time of his death in 1956, had amassed a fortune of over $1.7 million (McGrath, 1971).

Memorials to both men can be found in the memoir series of the National Academy of Sciences (Chamberlain, 1924; Hewett, 1957) and Leith's eulogy of Van Hise appears in the 1920 Bulletin of the Geological Society of America. Leith's own memorial, by R. J. Lund, is in the 1956 GSA Proceedings Volume.

GEOLOGY AND THE EARLY PETROLEUM INDUSTRY

In the 1880s, while the iron and steel industry was booming (Lewis, 1976), especially owing to construction of the more than 88,000 miles of railroad then in operation (a network expanded to over 250,000 miles by the end of the century), the petroleum industry was still in its "Age of Illumination" (Williamson and Daum, 1959). Even when John D. Rockefeller organized the Standard Oil Trust in 1882, the industry's main product was kerosene. In 1900 petroleum accounted for only 2.4% of aggregate U.S. energy consumption, and by 1920 when oil accounted for 12.3% of all energy used, almost half of that was in the form of fuel oil, and the remainder was largely still kerosene. Only 2.2% (gasoline) was used as motor fuel (Williamson and others, 1965; Giebelhaus, 1983). Nevertheless, Standard Oil "grew exceedingly large" and, partly because of Ida Tarbell's, *The History of the Standard Oil Company* (1904), the Supreme Court in 1911 ordered the company, as then constituted, to be dissolved.

In terms of petroleum exploration, the discovery at Spindletop a decade earlier in 1901 moved activity definitively southwest of the original Pennsylvania-West Virginia-Ohio heartland (Galey, this volume). At the time, the science of petroleum geology was slowly advancing beyond the "creekology" of the early days in the east. Eventually explorationists came to realize that anticlinal structures were potential traps, an insight that resulted in the fielding of geologic parties on a large scale in the early 20th century as structural closure became a principal object of exploration. The history of the early petroleum industry and its growing use of geology is recounted by DeGolyer (1947), Knowles (1959), and especially Owen in his monumental *Trek of the Oil Finders* (1975). The year 1917 marks the beginning of professionalization of petroleum geology with the establishment of the American Association of Petroleum Geologists (AAPG) during the boom days of World War I, and at the start of the new automobile age (Morley, 1966).

Geophysical prospecting techniques appeared in the 1920s, as recounted by Bates and others (1982). The torsion balance was imported from Hungary, the seismometer from Germany, and electric-logging from France. The first U.S. production to be discovered by seismology was at Sugarland, Texas in 1927. Micropaleontology as an exploration tool also came in the 1920s when the first industrial laboratory for paleontology, that of Humble Oil under the direction of Alva Ellisor, was established in 1921. Joseph Cushman (Todd, this volume) was applying micropaleontology in the field for the Marland Oil Company of Mexico in 1922. Ernest W. Marland, founder of what was to become Conoco, was one of the first managers to employ geologic specialists on a large scale. By 1926, when the Society of Economic Mineralogists and Paleontologists was founded, paleontologic technique was well enough established that the organizing committee chose to broaden the membership base to include all microscopist specialists then active in the petroleum industry (Russell and Tener, 1981).

The career of Wallace E. Pratt, who joined Humble Oil in 1918 as its first professional geologist and is now best known perhaps for his statement that "oil is found in the minds of men," is exemplary of the early interrelationship of science and practical need. Pratt, in 1921, discovered, for example, that fault planes can serve as hydrocarbon traps, and it was also under his direction that the 1927 Sugarland seismic discovery was made. Pratt is

Figure 4. Panorama of open pit Shenango Iron Mine, Mesabi District, Minnesota. Photo by E. F. Burchard, c. 1911. (Courtesy of U.S. Geological Survey Photographic Library. From USGS Monograph 52, Plate 11B.)

Figure 5. View of Mountain Iron Mine, showing steam shovel "bucking" bank of ore. (From USGS Monograph 43, Plate 26 B.)

Figure 6. First train load of Mesabi ore delivered to the Duluth, Missabe, and Northern Dock, 1892. (From USGS Monograph 43, Plate 3.)

Figure 7. Aerial view of McKittrick Canyon, Guadalupe Mountains National Park, Texas. (Courtesy of *The Lamp*, Exxon Corporation.)

Figure 8. P. B. King on first day of field work mapping the southern Guadalupe Mountains, January, 1934. (Courtesy of W. L. Hiss and Permian Basin Section, SEPM.)

to be remembered also for his personal purchase, and eventual donation to the U.S. government, of the 5,632 McKittrick Canyon acres (Figure 7) that now form the core of Guadalupe Mountains National Park (Copithorne, 1982; Salvador, 1983).

However, it is the rediscovery of the facies concept by the petroleum industry in the 1920s, based on reef structures marginal to the Permian basins of West Texas which the Guadalupe Park displays so well, that provides the best example illustrating the role of societal need in the nurturing of concepts in geology. While the original recognition of sedimentary facies in ancient rocks is credited to observations in the French Jura by Gressly and Prevost in 1838 (Moore, 1949; Nelson, 1985), the concept lay fallow for many years (Merk, this volume), until rediscovered in the course of exploration for minerals early in the 20th century. For example, although Frank R. Clark, working for the U.S. Geological Survey in the western Book Cliffs coal region of Utah, discovered and documented well-developed facies intertonguing in 1911 (Clark, 1928), this fact was not widely known. This Cretaceous sequence was investigated subsequently on an inten-

sive basis by Spieker and many others beginning as early as 1921, but wide recognition of the importance of its theoretical implications did not come until the 1949 publication of GSA Memoir 39, *Sedimentary Facies in Geologic History* (Longwell, 1949).

It was the effort to understand the West Texas Permian that had the greatest impact (in North America at least) on the revival of geologic thinking with regard to facies. The story of recognition of the West Texas Guadalupian reefs is well-told by P. B. King (Figure 8) in *The Evolution of North America* (1959). As a young college graduate in 1924, King was sent by the Marland Oil Company to West Texas. There, where the oil play was being developed in the Permian lying unconformably below Cretaceous cover, drillers' logs were of vital importance but were difficult to interpret because, as King states, the "real difficulties were with the rocks themselves, which obeyed no laws with which we were familiar at the time" (King, 1959, p. 35). After leaving Marland in 1929 for graduate work at Yale, which involved field studies in the Permian of the Glass Mountains, King discovered the same stratigraphic puzzles he had observed in West Texas. Charles

Schuchert, supervising his work, mentioned the possibility of reefs and urged King to look into it. As King puts it: "I was unimpressed, and to bolster my objections I looked up the subject of fossil reefs in Grabau's *Principles of Stratigraphy.* There, a drawing which I had merely glanced at before—a view of Triassic reef structures in the Tyrolean Alps—suddenly took on new meaning" (King, 1959, p. 37).

King continues, "nearly every geologist who was working in West Texas came independently to the same conclusion at about the same time; by 1929 the reef theory as an explanation of the Permian stratigraphy of the region was in full flower." In the Guadalupe Mountains a Shell Oil geologist, E. Russell Lloyd reached the same conclusion. His paper, "Capitan Limestone and Associated Formation of New Mexico and Texas," appeared in the AAPG *Bulletin* of 1929. In his paper Lloyd gave credit to W. van Holst Pellekan for pointing out the reef character of the Capitan limestone, acknowledging his debt (as did King) to European observation, but it was the local practical quest of the petroleum industry that had brought both Lloyd and King to consider the question originally. Finally and appropriately, it was through the interest of Wallace Pratt, owner of McKittrick Canyon, that Humble Oil later came to sponsor the classic Columbia University-American Museum of Natural History study of the Guadalupian reef facies (Newell and others, 1953).

CONCLUSION

This paper has endeavored briefly, through selected examples, to emphasize the debt geologic thinking owes to the stimulus of practical application or need, most often expressed in industrial activity. The examples chosen are a few among many. While the demands of war or national pride, pure intellectual curiosity, and a host of other factors are certainly involved in the making of geology what it is today, the role of utility as a prod to geologic thinking was very important in the first fifty years of the Geological Society of America.

ACKNOWLEDGMENTS

Thanks are extended to Michele Aldrich and Ellen Drake, who organized separate symposia for the Geological Society of America's History of Geology Division at which some of the above material was first presented.

REFERENCES CITED

Adams, F. D., 1938, The birth and development of the geological sciences: New York, Dover Publications, 506 p.

Bailey, S. W., 1981, The history of geology and geophysics at the University of Wisconsin-Madison 1848–1980: Madison, Wisconsin, Department of Geology and Geophysics, 174 p.

Barger, H. and Schurr, S. H., 1944, The mining industries 1899–1939: A study of output, employment and productivity: New York, National Bureau of Economic Research, Inc., 447 p. (reprinted 1972, New York, Arno Press).

Bates, C. C., Gaskell, T. F., and Rice, R. B., 1982, Geophysics in the affairs of man: A personalized history of exploration geophysics and its allied sciences of seismology and oceanography: Oxford, Pergamon Press, 492 p.

Chamberlin, T. C., 1924, Charles Richard Van Hise: Memoirs of the National Academy of Sciences, v. 17, p. 143–151.

Clark, F. R., 1928, Economic geology of the Castlegate, Wellington and Sunnyside quadrangles, Carbon county, Utah: U.S. Geological Survey Bulletin 793, 165 p.

Copithorne, W. L., 1982, From doodlebug to seismograph: The Lamp, v. 64, p. 42–47.

DeGolyer, E. L., 1947, Seventy-five years of progress in petroleum: *in* Parsons, A. B., ed., Seventy-five years of progress in the mineral industry, 1871–1946, New York, American Institute of Mining and Metallurgical Engineers, 817 p.

Fuller, J.G.C.M., 1969, The industrial basis of stratigraphy: John Strachey, 1671–1743, and William Smith, 1969–1839: American Association of Petroleum Geologist Bulletin, v. 53, p. 2256–2273.

Giebelhaus, A. W., 1983, Petroleum's age of energy and the thesis of American abundance: Materials and Society, v. 7, p. 279–293.

Gilbert, G. K., 1877, Report on the geology of the Henry Mountains: U.S. Geographic and Geological Survey of the Rocky Mountain Region: Washington, D.C., U.S. Government Printing Office, 160 p.

———— 1914, The transportation of the debris by running water: U.S. Geological Survey Professional Paper 86, 263 p.

———— 1917, Hydraulic-mining debris in the Sierra Nevada: U.S. Geological Survey Professional Paper 105, 153 p.

Gregory, C. E., 1980, A concise history of mining: New York, Pergamon Press, 258 p.

Guilbert, J. M., 1983, G.S.A.'s benefactor, R.A.F. Penrose, Jr.: Geological Society of America News and Information, v. 5, no. 9, p. 146–147.

Hagwood, J. J., Jr., 1981, The California debris commission: A history: U.S. Army Corps of Engineers, Sacramento District, p. 102.

Hewett, D. F., 1957, Charles Kenneth Leith, 1875–1956: National Academy of Sciences, Biographical Memoirs, v. 33, p. 180–204.

Jordan, W. M., 1979, Geology and the industrial-transportation revolution in early to mid-nineteenth century Pennsylvania, *in* Schneer, C. J., ed., Two hundred years of geology in America, Hanover, New Hampshire, University Press of New England, p. 385.

Kelley, R. L., 1959, Gold vs. grain, the hydraulic mining controversy in California's Sacramento Valley: A chapter in the decline of the concept of laissez faire: Glendale, California, Arthur H. Clark, p. 326.

King, P. B., 1948, Geology of the southern Guadalupe Mountains, Texas, U.S. Geological Survey Professional Paper 215, 183 p.

King, P. B., 1959, The evolution of North America, Princeton, New Jersey, University Press, p. 189.

Knowles, R. S., 1959, The greatest gamblers: The epic of American oil exploration: New York, McGraw-Hill, p. 347.

Lawson, A. C., 1885, Report on the geology of the Lake of the Woods region, Geological Survey of Canada, Annual Report, v. 1, pt. CC, p. 1–151.

———— 1888, Report on the geology of the Rainy Lake region, Geological Survey of Canada, Annual Report, v. 3, p. 1–196 F.

Leith, C. K., 1903, The Mesabi iron-bearing district of Minnesota: U.S. Geological Survey Monograph 43, p. 316.

———— 1920, Memorial of Charles Richard Van Hise: Bulletin Geological Society of America, v. 31, p. 100–110.

Leopold, L. B., 1980, Techniques and interpretation: The sediment studies of G. K. Gilbert, *in* Yochelson, E. L., ed., The scientific ideas of G. K. Gilbert, Geological Society of America Special Paper 183, 148 p.

Lewis, W. D., 1976, Iron and steel in America: Greenville, Delaware, The Hagley Museum, 64 p.

Lloyd, E. R., 1929, Capitan limestone and associated formations of New Mexico and Texas: American Association of Petroleum Geologists Bulletin, v. 13, p. 645–658.

Longwell, C. R., ed., 1949, Sedimentary facies in geologic history: Geological Society of America Memoir 39, 171 p.

Lund, R. J., 1956, Memorial to Charles Kenneth Leith: Proceedings volume of the Geological Society of America, Annual Report for 1956, p. 147–158.

McGrath, S. W., 1971, Charles Kenneth Leith: Scientific adviser: Madison, Wisconsin, University Press, 255 p.

Millbrooke, A., 1976, The Geological Society of Pennsylvania 1832–1836: Pennsylvania Geology, v. 7, no. 6, p. 7–11; v. 8, 1977, no. 1, p. 12–16.

Moore, R. C., 1949, Meaning of facies, *in* Longwell, C. R., ed., Sedimentary facies in geologic history, Geological Society of America Memoir 39, p. 1–34.

Morley, H. T., 1966, A history of the American Association of Petroleum Geologists: American Association of Petroleum Geologist Bulletin, v. 50, p. 669–820.

Nelson, C. F., 1985, Facies in stratigraphy: From "terrains" to "terranes": Journal of Geological Education, v. 33, p. 175–187.

Newell, N. D., and others, 1953, The Permian reef complex of the Guadalupe mountains region, Texas and New Mexico: A study in paleoecology: San Francisco, W. H. Freeman, 236 p. (reprinted 1972, New York, Hafner).

Owen, E. W., 1975, Trek of the oil finders: A history of exploration for petroleum: Memoir 6, American Association of Petroleum Geologist, 1,647 p.

Pierce, N. A. and Jordan, W. M., 1982, Economics, science, and politics: Establishment of the Second Geological Survey of Pennsylvania: Abstracts with programs: Northeastern and Southeastern combined section meetings, Geological Society of America, v. 14, nos. 1 and 2, 73 p.

Porter, R., 1977, The making of geology: Earth science in Britain, 1660–1815: New York, Cambridge University Press, 288 p.

Pyne, S. J., 1980, Grove Karl Gilbert: A great engine of research: Austin, Texas, University of Texas Press, 306 p.

Rabbitt, M. C., 1980, Minerals, lands, and geology for the common defense and general welfare: Volume 2, 1879–1904: Washington, D.C., U.S. Government Printing Office, 407 p.

Ritter, D. F., 1978, Process geomorphology: Dubuque, Iowa, William C. Brown, 603 p.

Russell, R. D. and Tener, R. C., 1981, SEPM—the first fifty years: Journal of Sedimentary Petrology, v. 51, p. 1401–1432.

Salvador, A., 1982, Memorial: Wallace Everette Pratt, 1885–1981: American Association of Petroleum Geologists Bulletin, v. 66, p. 1412–1416.

Tarbell, I. M., 1904, The history of the Standard Oil Company: New York, Macmillan, v. 1, 406 p.; v. 2, 409 p.

Turner, F. J., 1893, The frontier in American history: (Oral paper delivered in 1893, first printed in 1920; reprinted 1950, New York, Holt).

Van Hise, C. R. and Leith, C. K., 1909, Pre-Cambrian geology of North America: U.S. Geological Survey Bulletin 360, 939 p.

——— 1911, The geology of the Lake Superior region: U.S. Geological Survey Monograph 52, 641 p.

Vance, M. M., 1960, Charles Richard Van Hise, scientist progressive: Madison, Wisconsin, State Historical Society of Wisconsin, 246 p.

Williamson, H. F. and Daum, A. R. 1959, The American petroleum industry: The age of illumination, 1859–1899: Evanston, Illinois, Northwestern University Press, 846 p.

Williamson, H. F., Andreano, R. L., Daum, A. R., and Klose, G. C., 1965, The American petroleum industry: The age of energy, 1899–1959: Evanston, Illinois, Northwestern University Press, 928 p.

Wyman, M., 1979, Hard rock epic: Western miners and the industrial revolution, 1860–1910: Berkeley, University of California Press, 331 p.

Yochelson, E. L., ed., 1980, The scientific ideas of G. K. Gilbert: Geological Society of America Special Paper 183, 148 p.

Young, O. E., Jr., 1970, Western mining: An informal account of precious-metals prospecting, placering, lode mining, and milling on the American frontier from Spanish times to 1893: Norman, Oklahoma, University of Oklahoma Press, 342 p.

——— 1976, Black powder and hand steel: Miners and machines on the old western frontier: Norman, Oklahoma, University of Oklahoma Press, 196 p.

Zaslow, M., 1975, Reading the rocks: The story of the Geological Survey of Canada 1842–1972: Toronto, Macmillan of Canada, 599 p.

Geological Society of America
Centennial Special Volume 1
1985

Seismic exploration of the crust and upper mantle of the Basin and Range province

L. C. Pakiser
U.S. Geological Survey
MS 966, Box 25046
Denver Federal Center
Denver, Colorado 80225

ABSTRACT

The fundamental features of the seismic-velocity structure of the crust of the Basin and Range province have been known for 30 years. Tatel and Tuve (1955) expressed surprise in finding that the crust is only 28-34 km thick in the Basin and Range/Colorado Plateau transition zone of Arizona, New Mexico, and Utah. From isostatic considerations they had expected the crust of this elevated region to be about 70 km thick. Carder and Bailey (1958) obtained similar results in the Basin and Range province in the vicinity of the Nevada Test Site. Later, Press (1960), Berg and others (1960), and Diment and others (1961) discovered the characteristic low velocity of the upper mantle of the province, although they regarded the 7.6-7.8 km/sec velocities they found at depths of 24-28 km as possibly too low for the upper mantle. Nevertheless, the unexpectedly thin crust and low-velocity upper mantle have been accepted by most investigators as characteristic features of the Basin and Range province since U.S. Geological Survey seismic teams demonstrated that the discontinuity below which the velocity is about 7.8 km/sec is the most prominent one in the province and argued that it should be regarded as the Mohorovičić discontinuity (Pakiser, 1963).

Our knowledge of the gross features of the crustal and upper-mantle structure of the Basin and Range province has changed little in the past 20 years, but many important details have been added. Major problems, however, remain to be resolved. Reexamination of the evidence for extremely low upper-mantle velocities in the vicinity of the Wasatch front (Smith, 1978) suggests that velocities there may not be very different from those in other parts of the province. New, more nearly definitive seismic-refraction and -reflection experiments and reanalysis of existing data using new analytical techniques will be required to resolve such problems.

INTRODUCTION

Seismic investigations of the earth's crust and upper mantle are a relatively recent development in the history of geology in North America. Such investigations have been underway, more or less continuously, since Gutenberg and others (1932) began testing seismic methods for determining crustal structure in southern California, and Byerly and Dyk (1932) recorded a quarry blast at Richmond, California, to study layering of the earth in the vicinity of Berkeley. Thus, the history of seismic investigations of crustal and upper-mantle structure in North America spans more than half the history of the Geological Society of America as we approach the GSA's centennial year.

Tuve and his coworkers at the Carnegie Institution of Washington (Tatel and Tuve, 1955) made the first determinations of crustal thickness in the transition zone between the Basin and Range province and the Colorado Plateau in 1954. Thus, the history of seismic investigations of crustal and upper-mantle structure in the Basin and Range province spans a third of the first century of the history of the GSA. In their initial investigations in

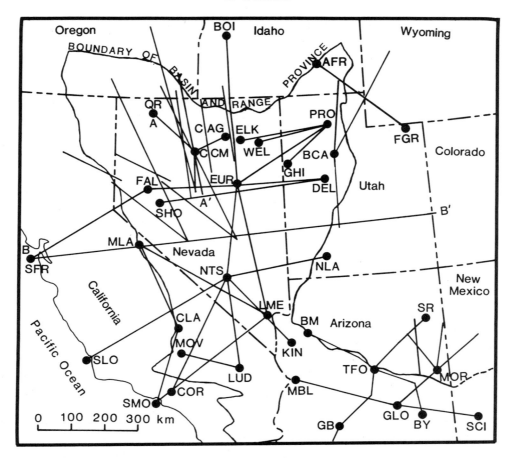

Figure 1. Map of the Basin and Range province showing locations of cross sections A-A' and B-B' and seismic profiles: SFR, San Francisco; SLO, San Luis Obispo; SMO, Santa Monica Bay; MLA, Mono Lake; CLA, China Lake; MOV, Mojave; COR, Corona; NTS, Nevada Test Site; LME, Lake Mead; LUD, Ludlow; KIN, Kingman; NLA, Navajo Lake; FAL, Fallon; SHO, Shoal; DEL, Delta; EUR, Eureka; QR, Quinn River; CCM, Copper Canyon Mine; CAG, Carlin gold mine; ELK, Elko; WEL, Wells; GHI, Gold Hill; PRO, Promontory; BCA, Bingham Canyon; BOI, Boise; BM, Blue Mountain; BY, Bylas; GB, Gila Bend; SR, Sunrise; MBL, Miser's Bluff; GLO, Globe; SCI, Silver City; MOR, Morenci; AFR, American Falls Reservoir; FGR, Flaming Gorge Reservoir.

Arizona, New Mexico, and Utah, Tuve and his coworkers discovered one of the characteristic features of the Basin and Range province—the unexpectedly thin (28-34 km thick) crust. Following these experiments, they wrote of the thin crust (Tatel and Tuve, 1955): "This is quite different from the expected value of about 70 km based on the previous experience in the East and the fact that the average surface elevation of the Arizona-New Mexico region is some three times that of its eastern counterpart." They then went on to confirm that discovery by an experiment in Utah. In the 30 years since the pioneering work of Tatel and Tuve, the most extensive network of seismic profiles in North America has been completed in the Basin and Range province (Fig. 1).

Why has the Basin and Range province been studied more thoroughly than others? The primary reason is scientific interest, spurred by the search for economic mineral deposits, oil, and gas. The province has drawn America's most talented geologists to its

study since the pioneering work of G. K. Gilbert (1875), as reviewed by Nolan (1943), and by Stewart (1978). Based on the tectonic framework developed in geologic studies, geophysicists have been drawn to the province to explore the deeper foundations of mapped geologic features (Thompson and Burke, 1974; Smith, 1978). Institutional interest has also played an important role. Four universities with significant geophysical capability (California Institute of Technology, Stanford University, University of Nevada, and University of Utah) are located in or near the margins of the Basin and Range province, and all have conducted seismic investigations there. The U.S. Geological Survey, also, has had a long-term interest in the province, and by the late 1950s had acquired the ability to conduct crustal studies there (Diment and others, 1961). Another important reason for the intensive exploration of the crust and upper mantle of the Basin and Range province is that the Nevada Test Site (NTS) is located there. The NTS provided strong and accurately timed sources and an opportunity to perform research on problems related to the location and identification of underground nuclear explosions funded by the VELA UNIFORM program of the Department of Defense (Pakiser, 1963). In the remainder of this paper, I will review the seismic exploration of the crust and upper mantle of the Basin and Range province, summarize the state of knowledge of the velocity structure and composition of the crust and upper mantle there, identify some significant unsolved problems, and suggest future research to help solve those problems.

HISTORICAL SKETCH

Following the pioneering work of Tuve and his coworkers, seismic exploration of the crust and upper mantle in the Basin and Range province was continued by many investigators. Carder and Bailey (1958) made the first use of seismic waves generated by nuclear explosions at the NTS to study crustal structure in the region. They determined that the earth's crust in Nevada is about 35 km thick. Press (1960) also recorded seismic waves generated at the NTS along a southwest-trending profile that was reversed by seismic waves generated by chemical explosions in southern California. Press determined a depth of 24 km to a layer of velocity 7.66 km/sec, a velocity he considered too low for the upper mantle. He concluded that this layer is probably made up of a mixture of gabbroic and ultramafic rocks, later referred to by Cook (1962) as a mantle-crust mix. At about the same time, Berg and others (1960) made a seismic investigation of crustal structure in the eastern part of the Basin and Range province; their results and conclusions were virtually the same as those of Press. Diment and others (1961) recorded seismic waves from the NTS along a line to Kingman, Arizona, from which they also obtained results similar to those of Press (1960) and Berg and others (1960) and reached similar conclusions. The basic features of the velocity structure of the crust and upper mantle had been discovered by these workers by 1961, although later studies added many details and new insights. Steinhart and Meyer (1961) summarized the results of seismic investigations of crustal structure in Europe and North America to 1961.

In 1961, the U.S. Geological Survey began a seismic reconnaissance of the crust and upper mantle in the western United States as a part of the VELA UNIFORM program. This work provided an extensive network of seismic profiles in the southern and north-central portions of the Basin and Range province (Fig. 1). The results of these investigations, summarized by Pakiser (1963), generally confirmed those of the earlier workers. However, Pakiser and Hill (1963) reached the conclusion that the velocity boundary below which velocities ranging from 7.6 to 7.8 km/sec had been reported is the most prominent one in the Basin and Range province and that it should be taken to represent the Mohorovičić discontinuity or the crust-mantle interface. They concluded that the low upper-mantle velocities reported by Press (1960) and Berg and others (1960) "may be apparent down-dip velocities or the result of scatter in the data they interpreted."

Pakiser (1963) summarized the crustal and upper-mantle properties of the province as follows:

1. The crust is relatively thin (about 30 km).
2. The lower part of the crust has velocities appropriate for gabbro, but such material, if it exists, is relatively thin. An exception is the northern part of the Basin and Range province in the transition zone to the Snake River Plain, where a thick intermediate crustal layer is probably present. . . .
3. The upper mantle has P-wave velocities less than 8.0 km/sec and relatively low density.

Herrin and Taggart (1962) compiled a P_n-velocity map of the continental United States based on seismic waves generated by earthquakes and explosions. This map revealed upper-mantle velocities as low as 7.5 km/sec in the Basin and Range province. Stepp and others (1963) reported P_n velocities as low as 7.4 km/sec in the Basin and Range province based on a study of seismic waves generated at the NTS. Pakiser and Steinhart (1964) compiled maps showing variations in crustal thickness, mean crustal velocity, and P_n velocity in the continental United States. Their map of crustal thickness revealed the characteristic thin crust of the Basin and Range province, and differs little from more recent maps of crustal thickness (Warren and Healy, 1974; Smith, 1978).

Prodehl (1970, 1979) reinterpreted the seismic profiles obtained by the U.S. Geological Survey in the western United States. He reached conclusions similar to those of Pakiser (1963), but his velocity models of the crust and upper mantle in the Basin and Range province differed significantly in detail from those of earlier workers. Spence (1974), using P-wave residuals, discovered that a mass of high-velocity rocks in the upper mantle extends downward beneath the Silent Canyon volcanic center at the NTS from just below the Mohorovičić discontinuity to depths as great as 170–190 km.

The University of Utah and cooperating universities conducted seismic studies of the crust and upper mantle in the Basin and Range/Colorado Plateau transition zone. The results were summarized by Smith (1978). They found evidence for a very thin crust (25 km or less) and exceptionally low upper-mantle velocities (less than 7.5 km/sec) in the vicinity of the Wasatch front, generally confirming the results of Berg and others (1960).

Studies by the University of Nevada filled in a previously unexplored area in the northwestern part of the Basin and Range province. Priestley and others (1982) concluded from this work that the crust is as thin as 20-22 km in this area and that the upper-mantle velocity is fairly uniform, averaging 7.8 km/sec. Stauber (1980, 1983) obtained similar results but found evidence locally for upper-mantle velocities as low as 7.4 km/sec in the vicinity of the Battle Mountain heat-flow high. Recently, the University of Arizona and the University of Texas, El Paso, cooperated in a study of crustal structure in the southern part of the Basin and Range province. Results obtained there are generally similar to those elsewhere in the province (Sinno and others, 1981; Gish and others, 1981). An exciting new chapter of

TABLE 1. VELOCITY MODELS OF THE BASIN AND RANGE/COLORADO PLATEAU TRANSITION
ZONE IN THE VICINITY OF THE WASATCH FRONT

A

Velocity (km/sec)	Depth to bottom of layer (km)
5	29
8.0	-

B

Velocity (km/sec)	Depth to bottom of layer (km)
2.83	0.2
5.73	9
6.33	25
7.59	72
7.97	-

C

Mean crustal velocity, 6.16 km/sec; depth to strongest gradient, 29.3 km; velocity at depth of strongest gradient, 7.24 km/sec (see text).

D

Velocity (km/sec)	Depth to interval (km)
3.0^1	0-1.5
$5.8-6.0^2$	1.5-9.5
5.3	9.5-11.5
$6.3-6.6^2$	11.5-21.5
6.3^1	21.5-25.5
$7.3-7.5^2$	25.5-35.0

[1]Constant
[2]Uniform gradient

E

Velocity (km/sec)	Depth to bottom of layer (km)
3.57	2.2
6.06	9.4
5.80	14.6
6.40	19.6
6.90	28.6
7.60	-

F

Velocity (km/sec)	Depth to bottom of layer (km)
3.4	1.7
6.0	8.4
5.5	14.7
6.5	24.7
7.4	-

A, Tatel and Tuve (1955); B, Berg and others (1960); C, Prodehl (1970, 1979); D, Muller and Landisman (1971); E, Braile and others (1974); F, Keller and others (1975).

seismic-reflection studies of the crustal structure of the Basin and Range province was initiated by the Consortium for Continental Reflection Profiling (COCORP). The initial results of this work were reported by Allmendinger and others (1983).

HISTORICAL EVOLUTION OF SEISMIC MODELS

Two areas have been selected for a comparative review of crustal velocity models in the Basin and Range province: (1) the Basin and Range/Colorado Plateau transition zone and (2) the Nevada Test Site. Velocity models from these areas include those from the earliest studies of Tatel and Tuve (1955) and Carder and Bailey (1958) to the most recent ones of Müller and Mueller (1979) and Taylor (1983). In a third area, northwestern Nevada in the vicinity of the Battle Mountain heat-flow high, crustal thicknesses from refraction and wide-angle reflection data are seemingly incompatible with depths to a prominent reflecting horizon in the lower-crust-to-upper-mantle transition zone discovered in recent COCORP profiling. This problem area will also be discussed in detail. Results from other areas not included in those above will be summarized briefly.

During the nearly 30-year-long period in which studies of crustal structure were made, seismic instruments evolved from the single-point pen-and-ink recording units used by Tatel and Tuve (1955) to multi-channel FM analog tape-recording seismographs employed by later investigators, and finally to the 100 newly developed, portable, automatically operated, cassette-tape-recording units now being used by the U.S. Geological Survey. Interpretative techniques progressed from straightforward calculation of depths to velocity discontinuities from first-arrival data in the earlier studies, to construction of synthetic record sections for theoretical velocity models that can be compared directly with record sections prepared from field recordings (Keller and others, 1975), and to ray-tracing techniques (Hoffman and Mooney, 1983). In general, amplitudes of seismic waves have been used only qualitatively, and more recently by comparing relative amplitudes on synthetic seismograms with observed amplitudes. Müller and Mueller (1979) used amplitudes of wide-angle reflections to infer the existence of a crustal low-velocity zone west of Delta, Utah. The details of seismic models have been enhanced by improvement in recording and interpretative techniques. Details of crustal layering have been particularly enhanced, but the basic features of the models have remained surprisingly consistent since the early work of Tatel and Tuve (1955). This consistency gives us confidence in the overall reliability of the models, although they may be questioned in detail.

Unfortunately, many seismic profiles in the Basin and Range province were not reversed and were interpreted on the assumption of horizontal layering. If the Moho is a dipping boundary, true upper-mantle velocities would differ from apparent ones by

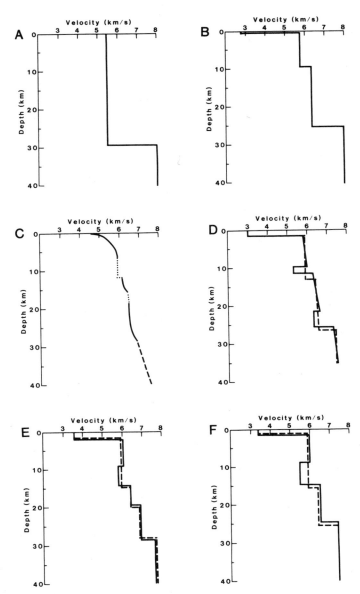

Figure 2. Seismic velocity models in the Basin and Range/Colorado Plateau transition zone: A, Tatel and Tuve (1955); B, Berg and others (1960); C, Prodehl (1970, 1979); D, Mueller and Landisman (1971); E, Braile and others (1974); F, Keller and others (1975). Solid lines, preferred models. Dashed lines, alternate models or extrapolated velocities. Dots, inferred velocities.

about 0.1 km/sec for each degree of slip. Computed depths to the Moho would be less severely affected by dip.

Basin and Range/Colorado Plateau transition zone

Study of the crustal structure of the Basin and Range/Colorado Plateau transition zone began with the early work of Tuve and his coworkers and has continued into the 1980s. Tatel and Tuve (1955) recorded seismic waves generated by the Kennecott Copper Company at Bingham Canyon, Utah (Fig. 1),

along a 240-km-long unreversed profile extending north from the shotpoint. In the velocity model calculated from first-arrival data (Fig. 2A, Table 1), a 29-km-thick crust overlies the upper mantle having a velocity of 8 km/sec. (Unless noted otherwise, crustal thicknesses in this paper are given in kilometers below the surface.) Tatel and Tuve recognized that the velocity of the crust is probably not uniform but increases with depth, and that the crust and upper mantle may be separated by a narrow transition zone rather than a first-order discontinuity. Depth to the Moho would be little affected by these modifications, however.

Berg and others (1960) recorded seismic waves generated by large quarry blasts at Promontory and Lakeside, Utah (Fig. 1). The traveltime data obtained along this unreversed profile were interpreted on the assumption that layers of uniform velocity are separated by plane, horizontal discontinuities. A two-layer crust overlying a layer of velocity 7.59 km/sec at a depth of 25 km was computed from the data (Fig. 2B, Table 1). The velocity of the layer below 25 km was regarded as too low for the upper mantle by Berg and others (1960). By plotting traveltime data at distances extending to 1,000 km from Promontory, they found evidence for a discontinuity at a depth of 72 km (not shown on Fig. 2B) below which the velocity was determined to be 7.97 km/sec.

Prodehl (1970, 1979) reinterpreted a profile recorded by the U.S. Geological Survey (Eaton and others, 1964) between a nuclear explosion near Fallon, Nevada (designated SHOAL), and a series of chemical explosions at a shotpoint near Delta, Utah (Fig. 1). He computed a crustal velocity model (Fig. 2C) from a record section prepared from recordings west of Delta using an approximation method developed by Giese (1966). Velocity models derived using this method contain segments in which velocity increases continuously with depth separated by segments of uniform velocity in low-velocity zones. The crust-mantle transition zone is represented by a velocity gradient rather than a discontinuity, and crustal thickness is represented by the depth to the strongest velocity gradient between the crust and upper mantle. Thus, crustal thicknesses from Prodehl are not directly comparable with those of other workers, in which the Moho is represented by a first-order discontinuity. The depth to this gradient west of the Delta shotpoint is 29.3 km, and the velocity there is about 7.3 km/sec. Velocity increases gradually in the upper mantle to 7.8 km/sec at a depth of about 50 km. The velocity within the crust increases gradually downward from 4.6 km/sec at the surface to 7.0 km/sec just above the strongest gradient (Fig. 2C). The average velocity of the crust is 6.16 km/sec. Prodehl's model is generally similar to other models in the Basin and Range/Colorado Plateau transition zone but differs from them markedly in the nature of the transition zone between the lower crust and upper mantle.

Mueller and Landisman (1971) made a second reinterpretation of the record section west of Delta. Their velocity model contains a 2-km-thick low-velocity zone at a depth of about 10 km and a 3-km-thick low-velocity zone just above the Moho at a depth of 25.5 km (Fig. 2D, Table 1). The Mueller and Landisman model without low-velocity zones, shown in dashed lines on Figure 2D, is similar to that of Berg and others (1960). Mueller and Landisman (1971) found that this simplified model does not adequately explain the observations.

Müller and Mueller (1979) conducted a test of the necessity for including a low-velocity zone in the upper crust beneath the Delta shotpoint. They found that peak amplitudes derived from the velocity models are displaced about 10 km toward larger distances than the peak amplitude observed for a wide-angle reflection from the mid-crust, unless a low-velocity zone is included. The low-velocity zone most compatible with the amplitude data extends from a depth of 7 km to 12 km, with velocity gradients at the top and bottom of the zone. The minimum compressional-wave velocity within the low-velocity zone in their preferred model (not shown in Fig. 2) is about 5.4 km/sec.

Braile and others (1974) interpreted seismic recordings from a 340-km-long unreversed profile, extending from Bingham Canyon northeast, from the Basin and Range province across the Middle Rocky Mountains of Wyoming (Fig. 1). Unfiltered and filtered record sections of excellent quality were prepared by digitizing the FM tapes obtained in the field. Braile and others (1974) derived two velocity models from the record sections (Fig. 2E, Table 1). Their preferred model features a three-layer crust beneath the near-surface layer and contains a crustal low-velocity zone at a depth of about 10 km (solid lines in Fig. 2E). The velocity of the upper mantle, 28.6 km deep, is 7.6 km/sec. The velocity model without the low-velocity zone (dashed lines in Fig. 2E) contains the same discontinuities, except for the upper boundary of the low-velocity zone, which is missing. The velocity just below the Moho in the simpler model is 7.7 km/sec.

Signals from radial and transverse horizontal seismometers were recorded along with those of each vertical seismometer, so Braile and others (1974) were able to prepare excellent record sections emphasizing shear waves. They found better evidence for a shear-wave low-velocity layer than for compressional waves. Shear-wave velocities (not shown on Fig. 2E) were anomalously low at intermediate crustal depths, resulting in a Poisson's ratio as high as 0.32 within the low-velocity layer. Braile and others concluded from the high Poisson's ratio that the low-velocity layer is a layer of low rigidity.

Braile and others (1974) prepared a crustal cross section from the Bingham Canyon mine in the Basin and Range province across the Middle Rocky Mountains and into the Green River Basin of Wyoming. The Moho in their model was assumed horizontal in the Basin and Range province. The crust thickens abruptly at a point about 40 km east of the Wasatch front, accompanied by an increase in the P_n velocity from 7.6 km/sec in the Basin and Range province to 7.7–7.9 km/sec in the Middle Rocky Mountains and 8.0 km/sec beneath the Green River Basin. The P_n velocity would be uniform and about 7.8 km/sec if the Moho dips uniformly along the profile. The crustal thickening was based on observed delays in P_n arrivals and reflections from the Moho on recordings made in the Middle Rocky Mountains and the Green River Basin, and on evidence for a thicker crust in

the Middle Rocky Mountains along a profile between American Falls Reservoir, Idaho, and Flaming Gorge Reservoir, Utah (Fig. 1), as interpreted by Willden (1965) and Prodehl (1970, 1979). Braile and others (1974) concluded that the crustal structure characteristic of the Basin and Range province extends beneath the Wasatch Mountains, and that the Wasatch Mountains have no crustal root.

The data interpreted by Braile and others (1974) are among the best obtained in the North American continent, but they assumed plane, horizontal discontinuities in computing their velocity models in the Basin and Range province. It would be equally plausible to assume uniform dip along the profile. Signals corresponding to reflections from the upper surface of the postulated low-velocity zone are clearly visible on their record sections, but those signals are not isolated events. Other prominent signals with the characteristic properties of reflections can also be identified on the record sections between the first (P_g) arrivals and the reflections from the Moho. Perhaps these reflections indicate merely that the crust is layered in the vicinity of the Wasatch front.

Keller and others (1975) continued the investigation of crustal structure along the Basin and Range/Colorado Plateau transition zone. They interpreted record sections along a south-trending, unreversed, 245-km-long profile extending from the Bingham Canyon copper mine to a point east of the Wasatch front (Fig. 1). Their record sections for both compressional and shear waves are similar to those obtained in the earlier study. The velocity models obtained by Keller and his coworkers are similar to those of Braile and others, but consist of only two crustal layers beneath the near-surface layer with a low-velocity zone in the preferred model (solid lines in Fig. 2F) which extends from a depth of about 8 km to about 15 km. The velocity of compressional waves just beneath the Moho at a depth of 24.7 km is only 7.4 km/sec. The simpler model without a low-velocity zone is shown in dashed lines in Figure 2F. Keller and others (1975) also determined shear-wave velocities for each crustal layer and computing a Poisson's ratio of 0.31 in the low-velocity zone. They suggested the existence of a mantle upwarp with a low P_n velocity in the transition zone between the Basin and Range province and the Colorado Plateau and centered beneath the Wasatch front.

The thin crust west of the Wasatch front is generally compatible with a prominent but discontinuous reflection with a two-way traveltime of about 11 sec beneath the Sevier Desert (Allmendinger and others, 1983). The estimated depth to the reflecting horizon, corrected for near-surface sediments, is somewhat deeper, about 30 km.

The limitations of the record sections interpreted by Keller and his coworkers are similar to those of Braile and others (1974). Keller and others (1975) acknowledged that the anomalously low P_n velocity of 7.4 km/sec "would represent an apparent velocity" if the Moho were dipping, and estimated that the true upper-mantle velocity could be as high as 7.7 km/sec if the crustal thickness increases from 25 km at the shotpoint to 40 km at the southern end of the profile. However, they found no evi-

TABLE 2. VELOCITY MODELS AT THE NEVADA TEST SITE

A			B	
Velocity (km/sec)	Depth to bottom of layer (km)		Velocity (km/sec)	Depth to bottom of layer (km)
3.8	2.5		3.0	1
6.10	3.6		6.11	24
8.1	-		7.66	50
			8.11	-

C			D	
Velocity (km/sec)	Depth to bottom of layer (km)		Velocity (km/sec)	Depth to bottom of layer (km)
5.2	1.7		3.0	0.7
6.15	18.4		6.03	27.7
7.81	-		7.84	-

E	F	
Mean crustal velocity, 6.30 km/sec; depth to strongest gradient, 30.7 km; velocity at depth of strongest gradient, 7.57 km/sec (see text).	Velocity (km/sec)	Depth to bottom of layer (km)
	1.8	0.6
	2.4	1.7
	5.3	10.0
	5.7	17.0
	6.4	32.0
	7.9	-

A, Carder and Bailey (1958); B, Press (1960); C, Diment and others (1961); D, Pakiser and Hill (1963); E, Prodehl (1970, 1979); F. Taylor (1983).

dence for significant dip along the profile and concluded that the upper-mantle velocity there is 7.5 ± 0.1 km/sec.

Tatel and Tuve (1955), Warren (1969), Sinno and others (1981), and Gish and others (1981) investigated crustal structure in the Colorado Plateau/Basin and Range transition zone in the vicinity of the Mogollon rim (Fig. 1). Warren's results are the most detailed. He found that the crust thickens from as little as 21 km near Gila Bend, Arizona, in the Basin and Range province to about 34 km under the Tonto Forest Seismological Observatory near the Mogollon rim in the transition zone and 40 km in the Colorado Plateau near Sunrise Springs, Arizona (Fig. 1). The upper-mantle velocity determined by Warren is 7.85 km/sec. Crustal thicknesses determined by Tatel and Tuve (1955), Sinno and others (1981), and Gish and others (1981) are generally compatible with those of Warren (1969). Upper-mantle velocities determined by them ranged from 7.6 to 8.1 km/sec along unreversed profiles.

Nevada Test Site

Study of seismic waves generated by nuclear explosions at the Nevada Test Site (Fig. 1) has provided a virtually continuous history of development of velocity models of the crust and upper mantle in the southern Great Basin since the work of Carder and Bailey (1958). Their initial NTS velocity model (Fig. 3A, Table 2) included a single-layer crystalline crust overlying the upper mantle (velocity 8.1 km/sec) at a depth of 36 km. Since their work, crustal velocity models at the NTS have been determined by Press (1960), Diment and others (1961), Pakiser and Hill (1963), Ryall and Stuart (1963), Prodehl (1970, 1979), Taylor

(1983), and Hoffman and Mooney (1983). The basic features of these models, spanning a period of 25 years, are surprisingly consistent (Fig. 3). Spence (1974), Monfort and Evans (1982), and Taylor (1983) also studied lateral velocity variations in the upper mantle beneath the NTS to depths as great as 170-190 km.

Press (1960) interpreted seismic waves generated by nuclear explosions at the NTS and quarry blasts at Victorville and Corona, California, along a 500-km-long profile (Fig. 1). The model computed from the traveltime data (Fig. 3B, Table 2) revealed a crystalline crustal layer extending to a depth of 24 km. The material below 24 km, with a velocity of 7.66 km/sec, was not interpreted as the upper mantle by Press (1960). He regarded this velocity as too low for the upper mantle and found evidence at recording distances beyond 330 km for first arrivals having a velocity of 8.11 km/sec. Press computed the depth to the layer of this velocity, which he regarded as ultramafic rock, to be 50 km (not shown on Fig. 3B). Press (1960) studied the average crust between the NTS and Corona, which crosses portions of the Basin and Range province, the Transverse Ranges, and the Peninsular Ranges. This is an area of complex geology in which significant lateral variations in crustal velocities are to be expected. These variations might result in traveltime variations that could account for significant changes in the apparent velocity of P_n waves along the profile.

Diment and others (1961) studied seismic waves from an unreversed profile extending southeast from the NTS to Kingman, Arizona (Fig. 1). From the resulting traveltime data, they calculated a velocity model which has properties midway between those of Carder and Bailey (1958) and Press (1960). In their velocity model (Fig. 3C, Table 2), a crystalline crustal layer

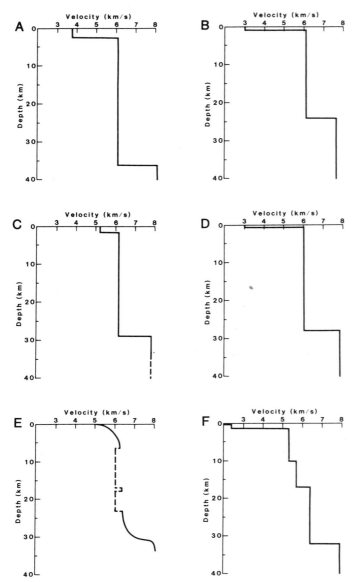

Figure 3. Seismic velocity models at the Nevada Test Site: A, Carder and Bailey (1958); B, Press (1960); C, Diment and others (1961); D, Pakiser and Hill (1963); E, Prodehl (1970, 1979); F, Taylor (1983). Dashed lines represent extrapolated or inferred velocities.

refraction traveltime segment. Later workers have made extensive use of such wide-angle reflections in determining crustal models.

Pakiser and Hill (1963) interpreted seismic waves from nuclear explosions at the NTS along a profile extending more than 500 km to the north. Their velocity model (Fig. 3D, Table 2) is nearly identical to that of Diment and others (1961). A single layer of crust extends to a depth of 27.7 km, below which the velocity is 7.84 km/sec. Pakiser and Hill (1963) identified this layer as the upper mantle.

Neither Diment and others (1961) nor Pakiser and Hill (1963) found evidence for a layer of intermediate velocity in the lower part of the crust. Strong first arrivals of seismic waves generated by nuclear explosions commonly visually obscure wide-angle reflections and other secondary arrivals from crustal layers. These phases are commonly observed on record sections from weaker sources, however. If such a layer, having a velocity of, say, 6.6 km/sec, is present in the lower one-third of the crust, it would require that the crustal thickness beneath the Nevada Test Site be increased from 28 to 31 km.

Prodehl (1970, 1979) reinterpreted the profile from the NTS to Kingman. His velocity model (Fig. 3E, Table 2) is similar to that of Diment and others (1961), but it differs in detail. The velocity increases downward from 5.16 km/sec at the surface to 7.57 km/sec at the depth of the strongest gradient, 30.7 km, which Prodehl defined as crustal thickness, and to 8.06 km/sec at a depth of 33.7 km. The crust contains a low-velocity zone (6.0 km/sec) between depths of 7 and 23 km, with a thin layer of higher velocity at a depth of about 18 km. The average crustal velocity is 6.3 km/sec. The crust-to-upper-mantle transition zone extends over a depth range of 5 to 10 km in Prodehl's model (Fig. 3E). Prodehl (1970, 1979) interpreted several profiles radiating outward from the NTS. Depths to the strongest gradients within thick crust-to-upper-mantle transition zones of these models range from about 31 to as much as 36 km.

Taylor (1983) derived a three-dimensional model of the crust and upper mantle beneath the NTS by combining teleseismic P-wave traveltime residuals with P_n traveltime variations. He computed velocity models beneath Pahute Mesa, Rainier Mesa, and Yucca Flat. The models are similar but contain significant horizontal variations in crustal and upper-mantle velocities. I have selected the velocity model at Yucca Flat for comparison with those of earlier workers (Fig. 3F, Table 2). Taylor's Yucca Flat model has three layers of the crystalline crust (Fig. 3F, Table 2) overlying the Moho at a depth of 32 km. The velocity of the upper mantle is 7.9 km/sec. (The significance of the horizontal velocity variations will be discussed later.)

Ryall and Stuart (1963) interpreted recordings of seismic waves generated by nuclear explosions at the NTS along a line extending east beyond Navajo Lake, Utah, to Ordway, Colorado. They found a crustal thickness of 25 km at the NTS and an upper-mantle velocity of 7.6 km/sec, which they interpreted as an apparent velocity (true velocity about 7.9 km/sec) suggesting crustal thickening to about 42 km in the western part of the Colorado Plateau.

extends to a depth of 28.4 km. The velocity is 7.81 km/sec below that depth. Diment and others (1961) did not identify the 7.81-km/sec layer as the upper mantle. They designated it the H_2 layer and concluded that crustal thickening may occur mainly in this layer. Diment and others (1961) plotted computed reflections from the top of the 7.81-km/sec layer and from the top of a layer 52.8 km deep on a carefully constructed record section. The computed reflections from the 7.81-km/sec layer are consistent with a very strong phase observed on the record section. The computed reflections from the deeper layer (Press' 8.11-km/sec layer) fall near a fairly strong phase following the 7.81-km/sec

Hoffman and Mooney (1983) conducted a seismic investigation of crustal structure at the Nevada Test Site from 1980 to 1982. In addition to recording seismic waves from nuclear explosions, they and their coworkers recorded seismic waves from a chemical explosion detonated near Beatty, Nevada. Their preliminary velocity model, derived from refraction and wide-angle reflection phases revealed on record sections of excellent quality, is similar to that of Taylor (1983). This model (not shown on Fig. 3) has three layers of crystalline crust, a crustal thickness of 35 km, and an upper-mantle velocity of 7.8 km sec.

Diment and others (1961) and Pakiser and Hill (1963) obtained nearly identical values for crustal thickness (28 km) and upper-mantle velocity (7.8 km/sec) in the area north and southeast of the NTS. The profiles they interpreted extend in nearly opposite directions from the NTS (Fig. 1). Therefore, if the Moho beneath the profiles they interpreted is a nearly plane surface, it is also nearly horizontal, and the true upper-mantle velocity is about 7.8 or 7.9 km/sec. This tentative conclusion is reinforced by the recent results of Taylor (1983) and Hoffman and Mooney (1983). The upper-mantle velocity beneath the NTS on a more local scale is not uniform, however, as Spence (1974), Monfort and Evans (1982), and Taylor (1983) have shown.

Using P-wave residuals, Spence (1974) detected a zone of high velocity (at least 8.1-8.2 km/sec) extending downward from a depth of about 50 km to a depth of 170-190 km beneath the Silent Canyon volcanic center. The mean radius of the high-velocity structure (which broadens with depth) was estimated to be 50-60 km by Spence. Spence interpreted the structure as representing a part of the former low-velocity zone of the upper mantle that has been depleted of a low-melting fraction by eruption of the partial melt through the Silent Canyon caldera. Similar results were obtained by Monfort and Evans (1982) using similar methods.

Taylor (1983), using P-wave residuals and lateral variations in the apparent velocity P_n waves, located a body of high-velocity material extending downward from the middle to lower crust to depths of at least 100 km beneath the Silent Canyon caldera. The velocities within this body are 0.1-0.2 km/sec greater than those in the surrounding rocks of the crust and upper mantle.

Northwestern Basin and Range province

Recently, Stauber (1980, 1983) and Priestley and others (1982) investigated the crustal and upper-mantle structure of the northwestern Basin and Range province. They independently defined an area of exceptionally thin crust, with a minimum thickness of only 20 km, in the vicinity of Battle Mountain, Nevada. The zone of thin crust extends roughly from Battle Mountain to Reno, Nevada, according to Stauber. The crust thickens to the northwest, east, and south to 30-35 km. Earlier, Eaton (1963) found seismic evidence that the crust is only 22-24 km thick in the vicinity of Fallon, Nevada. Stauber (1980, 1983) also found evidence for upper-mantle velocities as low as 7.4 km/sec beneath the Battle Mountain heat-flow high of Sass and others

(1971), although Priestley and others (1982) did not. More recently, the Consortium for Continental Reflection Profiling (COCORP) recorded a seismic-reflection profile across Nevada. This profile crosses the area of thin crust. They mapped a prominent reflection with a two-way reflection traveltime of about 10 sec, corresponding to a depth of about 30-32 km (R. W. Allmendinger, oral communication, 1984). The 10-sec reflecting horizon is flat over much of Nevada, and it seems to coincide with the Moho of Eaton (1963) and Hill and Pakiser (1966) in central Nevada.

Eaton (1963) interpreted a seismic profile recorded between shotpoints at Fallon and Eureka, Nevada (Fig. 1). The crust in Eaton's model thickens from 24.1 km beneath Fallon to 33.9 km beneath Eureka if a deep crustal layer with velocity of 6.6 km/sec is included in the lower one-third of the crust. Confidence in the existence of this deep layer is based on strong wide-angle reflections from an intracrustal boundary recorded on seismograms from explosions at both the Fallon and Eureka shotpoints. The velocity of the upper crust is 6.0 km/sec, and the velocity of the upper mantle is 7.82 km/sec in the Eaton model. Prodehl (1970, 1979) reinterpreted the Fallon-to-Eureka profile. The depth to the strongest velocity gradient in his model is 29.3 km at Fallon and 34.5 km at Eureka. The width of the transition zone between the lower crust and the upper mantle in Prodehl's models is 5 km or more, and his models do not include the discontinuity or narrow transition zone in the lower crust suggested by the prominent wide-angle reflections recorded along the profile.

Stauber (1980, 1983) recorded seismic data along three seismic profiles in the vicinity of the Battle Mountain heat-flow high in north-central Nevada (Fig. 1). A crustal cross section (A-A', Figs. 1 and 4) constructed by Stauber (1980, 1983) indicates that the crust ranges in thickness from 30.5 km near a shotpoint at Quinn River (QR) on the north to 23.5 km near the location of the blasts at Copper Canyon Mine (CCM) and to 34 km at the southern end of the profile. The profile south of CCM was partially reversed by seismic waves generated by nuclear explosions at the NTS.

A two-layer crust in the Stauber model (Fig. 4) overlies the upper mantle in which the velocity ranges from 7.8 km/sec beneath the northern and southern ends of the profile to a minimum of 7.4 km/sec between QR and CCM. The crustal model on Stauber's cross section and Eaton's is virtually identical where the two profiles intersect.

Stauber assumed plane, dipping interfaces between layers in his analysis, so the low P_n velocity of 7.4 km/sec can not be attributed to failure to take dip into account. The weak first arrivals on which this velocity was based were recorded over a distance range of only a few kilometers along the profile extending south from QR, but they are compatible with strong wide-angle reflections interpreted as coming from the Moho by Stauber. Stauber (1980) also recorded shear waves in his investigation. The velocities of shear waves (not shown in Fig. 4) indicate that Poisson's ratio is high in the lower crust. Stauber and Boore (1978) and Priestley and others (1982) did not find evi-

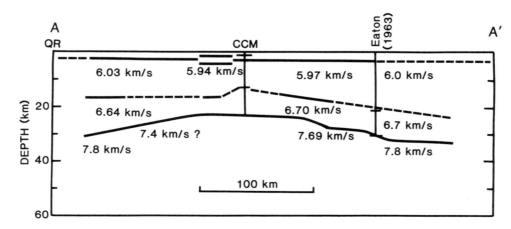

Figure 4. Crustal and upper-mantle structure along profile A-A' from Stauber (1980, 1983). Designations of shotpoints same as Figure 1.

dence for upper-mantle velocities as low as those suggested by Stauber (1980, 1983). Priestley and his coworkers concluded that the upper-mantle velocity is about 7.8 km/sec in this area and that observations of apparent velocities very different from this value are "principally related to variations in crustal thickness." Priestley and others (1982) also concluded that any anisotropy in P_n waves in this area is less than ±0.1 km/sec.

The prominent 30- to 32-km-deep reflecting horizon mapped in the COCORP profile, if taken to be the Moho, is seemingly incompatible with the shallow upper mantle reported by Eaton (1963), Stauber and Boore (1978), Stauber (1980, 1983), and Priestley and others (1982). It is more nearly compatible with the 29-km-deep Moho reported by Prodehl (1970, 1979). The Moho of Prodehl, however, is based on the strongest velocity gradient in a 5-km-wide transition zone, whereas the reflection data, including wide-angle reflections analyzed by Eaton and Stauber, suggest a narrower transition zone.

If the COCORP reflecting horizon and the Moho based on refraction and wide-angle reflection data in western Nevada are the same boundary, conflicting depths would be difficult to reconcile. However, if they are not the same boundary, there are two alternative interpretations. The COCORP reflection (near vertical incidence) may come from a velocity contrast a few km below the Moho, or the boundary based on refraction and wide-angle reflection data may be a different boundary representing the upper surface of a wedge of high-velocity material in the lower crust. Stauber's 7.4-km/sec velocity might indicate the velocity of such a wedge. Refraction, wide-angle reflection, and near-vertical-incidence reflection results need not necessarily agree precisely. Results from the different methods are critically dependent on different seismic wave paths and frequency spectra, and some differences should be expected. Results should be generally compatible, however.

Koizumi and others (1973) studied P-wave residuals at stations of the Nevada seismic network for teleseismic earthquakes distributed over a range of azimuths and epicentral distances. They found that for South American earthquakes, teleseismic

P-waves at northern Nevada stations arrived early by as much as 1.4 sec with respect to the Tonopah station to the south. Koizumi and his coworkers interpreted these early arrivals to indicate the presence of a thin high-velocity lithospheric plate (a paleosubduction zone) in the upper mantle, striking northeast and dipping southeast, under northern Nevada. More recent observations by Iyer and others (1977) indicate that the high-velocity zone has a greater northwest-to-southeast extent than was suggested by Koizumi and others (1982), and may broaden with increasing depth. The NTS-to-Boise, Idaho, profile interpreted by Hill and Pakiser (1966) crosses the high-velocity zone in the vicinity of Elko, Nevada. At three locations just north of Elko, weak but clear high-frequency forerunners of waves generated at the NTS preceded P_n by 0.3 to 0.5 sec. Hill and Pakiser suggested that the forerunners may have resulted from a zone of high-velocity material, perhaps a series of thin intrusions of mafic rock injected into the crust from a mantle source. These high-velocity features lie along the northeasterly trend of the zone of thin crust proposed by Stauber (1980, 1983) and Priestley and others (1982).

Other studies

Several other investigators, interpreting data obtained primarily by the U.S. Geological Survey, have contributed to our knowledge of the velocity structure of the Basin and Range province in areas other than those selected for detailed study. Roller and Healy (1963), Roller (1964), Johnson (1965), Gibbs and Roller (1966), and Prodehl (1970, 1979) determined crustal thicknesses ranging from 27 to 40 km and upper-mantle velocities of 7.8-7.9 km/sec in portions of the Basin and Range province (Table 3). The analysis of Priestley and others (1982) suggests that the crust is thinner in the vicinity of Mono Lake than the 40 km reported by Johnson. Hill and Pakiser (1966) found that the depth to the Moho increases from about 30 km in the Basin and Range province to more than 40 km in the Snake River Plain. The thickening occurs abruptly in the lower crust (velocity 6.7 km/sec) near the Nevada-Idaho border in the transi-

TABLE 3. RESULTS FROM OTHER STUDIES

Shotpoint[1]	Crustal thickness (km)	Upper-mantle velocity (km/sec)	Investigator(s)
LME	28	7.8	Roller and Healy (1963)
KIN	27	7.8	Roller (1964)
LME	30	7.8	Johnson (1965)
MLA	40	7.8	Johnson (1965)
LUD	27	7.8-7.9	Gibbs and Roller (1966)
MOV	31	7.8	Prodehl (1970, 1974)
CLA	34	7.8	Prodehl (1970, 1979)
NLA	36	7.8	Prodehl (1970, 1979)

[1]See Figure 1 for shotpoint names.

tion zone between the Basin and Range province and the Snake River Plain. The upper crust (velocity 5.2-6.0 km/sec) thins from about 20 km south of the transition zone to less than 10 km north of the zone. The well-determined upper-mantle velocity in the profile interpreted by Hill and Pakiser (1966) is 7.9 km/sec between the NTS and Boise, Idaho.

VELOCITY OF THE UPPER MANTLE

Various investigators have reported upper-mantle velocities in the Basin and Range province ranging from as low as 7.3 km/sec (Mueller and Landisman, 1971) to as high as 8.1 km/sec (Tatel and Tuve, 1955; Carder and Bailey, 1958). Herrin and Taggart (1962) analyzed seismic waves from earthquakes and explosions to prepare a map of regional variations in P_n velocity in the continental United States. These velocities range from 7.5 km/sec to more than 8.0 km/sec in the Basin and Range province. Later, Herrin (1972) revised the map to eliminate effects of variations in crustal thickness. The P_n velocity in the Basin and Range province on the revised map is generally in the 7.8-7.9 km/sec range and similar to that in the P_n-velocity map of Pakiser and Steinhart (1964).

Smith (1978) prepared maps of variations in crustal thickness and P_n velocity in the western United States from the compilation of Warren and Healy (1974) and, in the Wasatch front area, from Berg and others (1960), Braile and others (1974), Keller and others (1975), and Smith and others (1975). These maps (Fig. 5) reveal a bilateral symmetry in both crustal thickness and P_n velocity in the Basin and Range province. P_n velocities range from less than 7.5 km/sec in the Basin and Range/Colorado Plateau transition zone near the Wasatch front to 7.9 km/sec in southern Nevada. The low velocities in the Wasatch front area were based on unreversed profiles.

Sources of seismic waves for profiles used to determine the low P_n velocities were located in the area of a structural high on the Moho, and all profiles radiated outward from the high toward areas of thicker crust. Thus, the P_n velocities, based on the assumption that the Moho is horizontal, may be interpreted as apparent down-dip velocities. The velocity of the lower crust just

above the Moho in the transition zone reported by different investigators ranges from 6.3 km/sec (Berg and others, 1960; Mueller and Landisman, 1971) to 6.9 km/sec (Braile and others, 1974). These velocities are based on unreversed profiles, however; I have adopted a value of 6.5 km/sec as representative of the lower crust for the eastern Basin and Range province. If the true velocity of the upper mantle is 7.8 km/sec in this area, a dip of 2½ to 3 degrees would result in an apparent velocity of about 7.6 km/sec, and a dip of about 5 degrees would result in an apparent velocity of about 7.4 km/sec.

The P_n velocity of 7.59 km/sec reported by Berg and others (1960) is based on traveltimes between Gold Hill, Utah, and Eureka, Nevada (Fig. 5). The crustal thickness, as estimated from Smith's map, increases by about 7 km from Gold Hill to Eureka, indicating that the Moho dips down by about 2½ degrees between these two locations. Thus, the P_n velocity reported by Berg and others (1960) can be explained as an apparent down-dip velocity and that the true P_n velocity may be about 7.8 km/sec.

The average dip along the profile interpreted by Braile and others (1974) is about 2½ to 3 degrees (Fig. 5), so the P_n velocity of 7.6 km/sec reported by them can be interpreted as an apparent down-dip velocity and the true P_n velocity is also about 7.8 km/sec. The average dip as estimated from the contour map (Fig. 5) is about 2½ degrees along the profile interpreted by Keller and others (1975). This dip is not large enough to account completely for the 7.4 km/sec upper-mantle velocity reported by them. However, the dip along the segment of the profile beneath which the P_n waves propagated may be steeper, about 5 degrees as estimated from the contours, so the P_n velocity reported by Keller and others (1975) can also be explained as an apparent velocity and the true P_n velocity beneath the profile may be about 7.8 km/sec.

I cannot fully account for the very low P_n velocity of 7.3-7.4 km/sec reported by Mueller and Landisman (1971) west of the Delta, Utah, shotpoint as an apparent down-dip velocity. But, Eaton and others (1964) made a preliminary interpretation of the same profile between Delta and the SHOAL explosion in western Nevada. They concluded that variations in P_n velocity can be accounted for by variations in crustal thickness and that the true

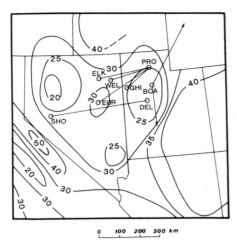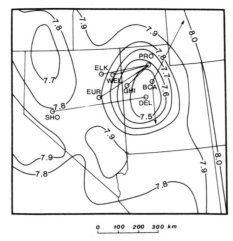

Figure 5. Contour maps of crustal thickness (left) and P_n velocity (right) modified from Figures 6-2 and 6-3 in Smith (1978). Contour interval for crustal-thickness map is 5 km and for P_n-velocity map is 1 km/s. Designations of shotpoints same as Figure 1. A 30-km contour surrounding EUR has been added. P_n-velocity map has been recontoured to make contour interval consistent throughout area.

upper-mantle velocity along this profile is about 7.8 km/sec. On the other hand, Pechmann and others (1984) have interpreted traveltimes from earthquakes recorded on the University of Utah seismic network to suggest the existence of a "double Moho" beneath the Wasatch front, one part 25-28 km deep, below which the velocity is 7.4-7.6 km/sec, and a second part about 40 km deep, below which the velocity is 7.9 km/sec. They postulate that the 7.9 km/sec layer is continuous with material of similar velocity beneath the thin crust of the central Basin and Range province to the west and the thicker crust of the Colorado Plateau to the east.

Although uncertainties may remain, it seems reasonable to suggest that the upper-mantle velocity is generally 7.8-7.9 km/sec in the Basin and Range province and that reported velocities much different from this, such as those shown on Figure 5, may reflect primarily variations in crustal thickness. It is possible that the Moho may be nearly flat in the area of the indicated structural high and that the variations in apparent velocity of P_n may result in part from severe lateral variations in the velocity of the lower crust, perhaps as a high-velocity wedge similar to that postulated for the Battle Mountain area. It must be remembered, also, that the seismic-refraction profiles from which these velocities were obtained tend to yield average P_n values over horizontal distances of 100 km or more. There may be local variations in upper-mantle velocities, as suggested by the results of Spence (1974), Koizumi and others (1973), and Taylor (1982), that may not be revealed on long refraction profiles. These local variations imply that the upper mantle is not uniform, but heterogeneous.

Many investigators (Niazi and Anderson, 1965; Johnson, 1967; Archambeau and others, 1969; Burdick and Helmberger, 1978; Priestley and Brune, 1978; Priestley and others, 1980; Stauber, 1980, 1983; and Mavko and Thompson, 1983) have determined velocity-depth models for the upper mantle of the Basin and Range province. Most models place the base of the

lithosphere (which corresponds to the base of the upper-mantle lid) at an average depth of about 65 km. But Archambeau and others (1969) suggested that the upper-mantle lid is very thin or missing. This interpretation is similar to a suggestion made by Press (1960). Although Press discounted the notion that the 7.66-km/sec layer he found at a depth of about 25 km is the upper mantle, he concluded that if "we accept the alternative (and to us the less likely) refraction interpretation of a shallow, low-velocity mantle then the interesting possibility occurs that the mantle low-velocity zone has migrated upwards to the Mohorovičić discontinuity in this tectonically active region."

Stauber (1980, 1983) suggested that the lithosphere might be 20-30 km thinner in the area of the Battle Mountain heat-flow high than the 65 km suggested by Priestley and Brune (1978) for central Nevada. Mavko and Thompson (1983) estimated that the lithosphere is about 60 km thick in the vicinity of Mono Lake. Zoback and Lachenbruch (1984) estimated that the lithosphere is 55-65 km thick in the northern Basin and Range province and 50-60 km thick in the southern part of the province.

BOUNDARIES OF THE BASIN AND RANGE PROVINCE

The Great Basin portion of the Basin and Range province is bounded by the Sierra Nevada and Cascade Ranges on the west, the Columbia Plateau and the Snake River Plain on the north, and the Middle Rocky Mountains and Colorado Plateau on the east. The province extends southeastward into the Sonoran Desert.

Eaton (1963, 1966) placed the eastern edge of the root of the Sierra Nevada about 30 km west of Fallon, Nevada, within the Basin and Range province. Carder and others (1970) and Carder (1973) interpreted seismic waves that propagated westward from the Nevada Test Site to arrive unexpectedly early in the Sierra Nevada as indicating that the crust thins from 35-40 km beneath the western part of the Basin and Range province to form an "antiroot" about 30 km or less deep beneath the highest mountains of the Sierra Nevada. Pakiser and Brune (1980) reinterpreted the data of Carder and his coworkers and concluded that the sharp, eastern edge of a thick Sierran root is located in the narrow Sierra Nevada/Basin and Range transition zone. Pakiser and Brune (1980) explained the early arrivals recorded in the Sierra Nevada by Carder and his coworkers, which they interpreted as evidence for crustal thinning, as waves propagating upward from the edge of the Sierran root through high-velocity crustal rocks. They proposed two possible models to explain the early arrivals, one involving diffraction at the edge of the root and a second involving propagation along a 5-km-thick high-velocity "wave guide" that rises to within a few km of the surface in the vicinity of the ophiolite belt in the western foothills of the Sierra Nevada.

The high-velocity layer in the wave-guide model proposed by Pakiser and Brune (1980) is slightly curved and convex upward. Bolt and Gutdeutsch (1982) tested a model similar to this one by ray-tracing. They found that body waves propagating in the curved wave guide would leak energy and die out with distance. Bolt and Gutdeutsch also tested the diffraction model of Pakiser and Brune and the thin-crust models of Carder and his coworkers and found that both are generally compatible with ray theory. The thin-crust models, however, are incompatible with convincing seismic evidence cited by Eaton (1963, 1966) and Pakiser and Brune (1980) that the crust is about 50 km thick or more beneath the highest mountains of the Sierra Nevada.

Nelson and others (in press) interpreted a COCORP reflection profile across the northern Sierra Nevada. They mapped a prominent reflection from beneath the Sierra that defines a planar zone that rises westward from mid-crustal depths. This zone projects upward into the Melones fault zone, which lies within the ophiolite belt of the western foothills of the Sierra Nevada. If the slightly curved wave guide of Pakiser and Brune (1980) is really bounded by planar surfaces, the objections to this model raised by Bolt and Gutdeutsch (1982), that it would leak energy and lose its wave-guide effect, might disappear.

Hill and Pakiser (1966) found that the crust thickens from about 30 km in the Basin and Range province to more than 40 km in the Snake River Plain in a horizontal distance as small as 25 km. The lower, high-velocity crust represents about one-third of the crust in the Basin and Range province and more than three-fourths of the crust in the Snake River Plain. Prodehl (1970, 1979) reinterpreted the data of Hill and Pakiser, confirming in general their results. His seismic models, however, suggest a more gradual thickening of the crust from the Basin and Range province to the Snake River Plain.

Braile and others (1974) suggested that the thin crust of the Basin and Range province extends about 40 km east into the Wastach Mountains. Keller and others (1975) supported this result and suggested that the mantle upwarp beneath the Wasatch front extends at least 50 km east of the front. Ryall and Stuart (1963) suggested that the Moho dips gently downward (about 2½

degrees) between the NTS and the western part of the Colorado Plateau.

In the southern Basin and Range province, Warren (1969) found that the transition zone between the Basin and Range province and the Colorado Plateau in the vicinity of the Mogollon rim is broad, and that the Moho dips gently to the northeast from about 22 km below the surface at Gila Bend, Arizona, to about 42 km in the Colorado Plateau. Roller (1965) found that the crust is 40-43 km thick in the Colorado Plateau.

DISCUSSION

Our knowledge of the general characteristics of the crustal structure of the Basin and Range province has changed little in the past 20 years, but our understanding of crustal and upper-mantle processes and their relations to structure has changed dramatically as a result of the revolutionary changes in geologic thinking brought about by plate tectonics. This change is illustrated by two cross sections of the crust and upper mantle, one published in 1963 (B-B', Fig. 1, upper part of Fig. 6) and the other is revised for this review (lower part of Fig. 6). The configuration of the Moho is virtually unchanged in the two cross sections, except that the eastern edge of the Sierran root has been made sharper, and the root has been deepened to about 55 km. The suggested density and velocity variations in the upper mantle have been changed significantly, however, and the crust has been separated into upper and lower units in the new cross section.

Pakiser (1963) concluded that "to maintain isostatic equilibrium . . . it is necessary that the material of P-wave velocity 7.8 km/sec and density 3.30 g/cm^3 in the Basin and Range province extend to a depth of about 80 km into the standard mantle of P-wave velocity 8.2 km/sec and density 3.45 g/cm^3" (upper part of Fig. 6). Although some part of isostatic compensation is undoubtedly accomplished by lateral variations in the density of the uppermost mantle (as well as variations in crustal thickness and density), a significant part of compensation must take place at the base of the lithosphere (lower part of Fig. 6). Crough and Thompson (1977) have proposed such an upper-mantle source for the uplift of the Sierra Nevada. Mavko and Thompson (1983) have estimated from teleseismic residuals and isostatic considerations that the depth to the bottom of the mantle lid may vary from about 110 km at the north end of the Sierra to about 60 km near Mono Lake.

Warren (1969) concluded that, in order to reconcile seismic results with gravity data and isostasy, higher-density material in the upper mantle must rise from a depth of about 100 km in the Basin and Range province to the Moho in the Colorado Plateau. A more likely interpretation, consistent with gravity, seismic, surface-elevation, and heat-flow data, is that the lithosphere thickens from 50-65 km beneath the Basin and Range province to 90-100 km beneath the Colorado Plateau, as proposed by Zoback and Lachenbruch (1984).

A summary of our present knowledge of the gross velocity structure of the crust and uppermost mantle of the Basin and

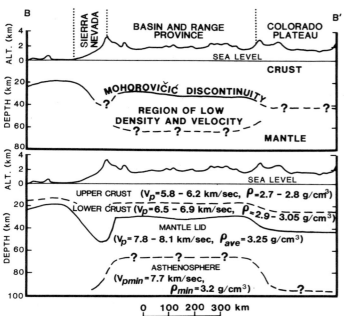

Figure 6. Crustal and upper-mantle structure along profile B-B'. Upper part is from Pakiser (1963). Lower part: Vp, compressional-wave velocity; V_{pmin}, minimum compressional-wave velocity in asthenosphere; ρ, density ρ_{min}, minimum density in asthenosphere.

Range province (except for the dramatic new discoveries of reflection seismology) differs little from our knowledge of the early 1960s. By 1961 researchers had discovered the thin crust (Tatel and Tuve, 1955; Carder and Bailey, 1958) and low-velocity upper mantle (Press, 1960; Berg and others, 1960; Diment and others, 1961) that have been recognized for more than 20 years as characteristic features of the crust and upper mantle of the Basin and Range province (Pakiser, 1963).

We can be more confident now that the crust is layered, and that much of the province is separated into the upper crust, with P-wave velocities averaging about 6.0 km/sec, and the lower crust, with P-wave velocities averaging about 6.6-6.7 km/sec. The discontinuity or, more likely, narrow transition zone that separates the two crustal layers has been recognized by many investigators (Eaton, 1963; Roller and Healy, 1963; Johnson, 1965; Hill and Pakiser, 1966; Mueller and Landisman, 1971; Keller and others, 1975), mainly from wide-angle reflections.

The existence of a low-velocity layer in the crust remains an open question. Stewart and Peselnick (1978) have clearly indicated in laboratory experiments that temperature gradients common in the Basin and Range province are sufficiently high to result in velocity decreases with depth in some crustal rock types. The analysis of Müller and Mueller (1979), based on the horizontal position of a prominent mid-crustal reflection west of Delta, Utah, and which appears to require a low-velocity zone, is persuasive. However, the lateral scatter in observed amplitudes is notoriously severe on many seismic profiles. Therefore, I would not accept the low-velocity crustal layer for extensive areas of the

Basin and Range province until studies such as those of Müller and Mueller are repeated by other investigators in different parts of the province. Nevertheless, the evidence for multiple layering of the crust in the Basin and Range/Colorado Plateau transition zone is compelling (Braile and others, 1974; Keller and others, 1975).

Evidence for large variations in the velocity of P_n should be discounted. Much of this evidence (Berg and others, 1960; Braile and others, 1974; Keller and others, 1975; Smith, 1978) is based on unreversed profiles that yield only apparent velocities that are strongly dependent on dip. The velocity of the uppermost mantle is probably in the range 7.8-7.9 km/sec in the Basin and Range province, except possibly for local variations of the type discovered by Koizumi and others (1973) and Spence (1974). The area of northwestern Nevada in the vicinity of the Battle Mountain heat-flow high, and the area of the suggested mantle upwarp in the vicinity of the Wasatch front remain problem areas, requiring additional study.

The geologic significance of geophysical investigations of the Basin and Range province has been summarized by Thompson and Burke (1974), Smith (1978), Thompson and Zoback (1979), and others. I will not attempt here a comprehensive geologic analysis of the results reported in this review, except to suggest that geologic theories based on unusually low upper-mantle velocities and crustal low-velocity zones should be reexamined.

The average composition of the upper crust in the Basin and Range province can reasonably be inferred to be silicic from its velocity and density and from the results of geologic mapping and drilling. From direct observations, however, we know that the upper crust is compositionally very heterogeneous. The composition of the lower crust is problematical. The velocity (averaging about 6.6-6.7 km/sec) suggests that the lower crust may be made up of granulite of intermediate composition (Christensen and Fountain, 1976), if it can be regarded as more-or-less homogeneous. This inference of homogeneity is most doubtful.

The lower crust has been extensively invaded by predominantly mafic magma from the asthenosphere as a result of the processes of an extending lithosphere discussed by Lachenbruch and Sass (1978). Some of the magma has fused portions of the crust and broken through to the surface. We can expect the lower crust to contain a complex system of intrusives, such as dikes and sills, primarily of gabbroic composition. If the intruded crust was initially more silicic, as the velocities suggest (Christensen and Fountain, 1976), a grossly heterogeneous lower crust composed of rocks ranging from silicic to mafic seems a reasonable model. The average composition of such a heterogeneous lower crust would, of course, be intermediate.

Earth scientists now agree that the upper mantle is ultramafic, composed predominantly of peridotite and a lower-melting fraction. The upper mantle of the extending lithosphere of the Basin and Range province has also been extensively invaded by magma from the asthenosphere. Therefore, it can be expected to be grossly heterogeneous, consisting primarily of peridotite but containing large intrusive masses of predominantly mafic composition. Such upper-mantle rocks would tend to have lower velocities and densities than those of stable areas of the continent. Higher temperatures would also tend to decrease upper-mantle velocities relative to stable areas. Plagioclase, a low-velocity mineral, might be stable in the mafic fraction of the shallowest portions of the upper mantle (Ringwood, 1975, p. 214–215).

SUGGESTIONS FOR FUTURE STUDY

Although the velocity structure of the crust and upper mantle in the Basin and Range province is better known than that of any other area of the continental United States, much remains to be learned. Important problems remain to be solved. High priority should be given to recording a west-to-east seismic-refraction profile between, say, Reno and Elko, Nevada, along the axis of the zone of thin crust found by Stauber (1980, 1983) and Priestley and others (1982). The explosive sources should be strong enough to produce strong P_n waves, and the geometry of the shotpoints and recording units should be such that true subsurface path reversal of P_n waves is obtained. Multiple shotpoints should be used along the profile to provide reliable information on the velocity structure of the crust.

Because our knowledge of crustal and upper-mantle structure in the vicinity of the Wasatch front is based primarily on unreversed profiles, a similar field experiment should be conducted there. A reversed refraction profile with multiple shotpoints in the Basin and Range province and generally parallel to the Wasatch front could provide data critical to resolving the problems associated with very low upper-mantle velocities reported in the Basin and Range/Colorado Plateau transition zone. Synthesis of the results of this profile, the earlier work, and recordings of both near and teleseismic earthquakes on the Wasatch seismic net operated by the University of Utah should lead to a much more reliable model of the structure of the crust and upper mantle in this critical area.

Our knowledge of the transition zones between the Basin and Range province and bordering provinces, especially the Sierra Nevada and the Colorado Plateau, is still inadequate. Additional detailed seismic profiling should be done in those areas. In particular, a great deal has been learned in recent years about the tectonic and magmatic evolution of the eastern and western province boundaries. The potential for integrating that new knowledge into interpretation of crustal structure and processes holds great promise. Recordings of seismic waves generated at the Nevada Test Site (and elsewhere) should be reexamined for evidence of layering within the upper mantle, as suggested by (probably reflected) phases following P_n.

Finally, reflection profiling, such as that of COCORP, should be closely coordinated with refraction profiling. In particular, the near-vertical reflections of COCORP and wide-angle reflections obtained from long profiles should be interpreted together. Do they, and refracted arrivals as well, come from the same, or different boundaries? The results of refraction and reflection profiling must be reconcilable, but at the time of this writing they have not been reconciled.

Studies such as those suggested above are within our present capability. They can be done now. If done, they will assure continuation of the productive and exciting history of seismic exploration of the crust and upper mantle of the Basin and Range province inaugurated by Tuve and his coworkers 30 years ago.

ACKNOWLEDGMENTS

I thank T. P. Barnhard, who prepared the figures, and Louise Hobbs, who prepared the mansucript text; R. E. Anderson, W. H. Diment, J. C. Pechmann, R. B. Smith, W. Spence, and M. L. Zoback, who reviewed the manuscript; and R. W. Allmendinger, W. H. Diment, W. D. Mooney, K. D. Nelson, J. E. Oliver, K. F. Priestley, R. B. Smith, D. A. Stauber, G. A. Thompson, and M. L. Zoback, with whom I held valuable discussions prior to writing this paper. I am especially indebted to my colleagues W. H. Jackson and J. H. Healy who coordinated the field operations and analysis of the U.S. Geological Survey's seismic exploration of the Basin and Range province in the 1960s, which work provided much of our knowledge of the crust and upper mantle of the Basin and Range province. Most of all, I thank the authors whose papers are cited in this review. Its scientific content is based mainly on their work.

REFERENCES CITED

Allmendinger, R. W., Sharp, J. W., Von Tish, Douglas, Serpa, Laura, Brown, Larry, Kaufman, Sidney, Oliver, Jack, and Smith, R. B., 1983, Cenozoic and Mesozoic structure of the eastern Basin and Range province, Utah, from COCORP seismic-reflection data: Geology, v. 11, p. 532–536.

Archambeau, C. B., Flinn, E. A., and Lambert, D. G., 1969, Fine structure of the upper mantle: Journal of Geophysical Research, v. 74, p. 5825–5865.

Berg, J. W., Jr., Cook, K. L., Narans, H. D., Jr., and Dolan, W. D., 1960, Seismic investigation of crustal structure in the eastern part of the Basin and Range province: Seismological Society of America Bulletin, v. 50, p. 511–535.

Bolt, B. A., and Gutdeutsch, Rolf, 1982, Reinterpretation by ray tracing of a transverse refraction seismic profile through the California Sierra Nevada, Part I: Seismological Society of America Bulletin, v. 72, p. 889–900.

Braile, L. W., Smith, R. B., Keller, G. R., and Welch, R. M., 1974, Crustal structure across the Wasatch front from detailed seismic refraction studies: Journal of Geophysical Research, v. 79, p. 2669–2677.

Burdick, L. J., and Helmberger, D. V., 1978, The upper mantle P velocity structure of the western United States: Journal of Geophysical Research, v. 83, p. 1699–1712.

Byerly, P., and Dyk, K., 1932, Richmond quarry blast of Sept. 12, 1931, and the surface layering of the earth in the region of Berkeley: Seismological Society of America Bulletin, v. 22, p. 50–55.

Carder, D. S., 1973, Trans-California seismic profile, Death Valley to Monterey Bay: Seismological Society of America Bulletin, v. 63, p. 571–586.

Carder, D. S., and Bailey, L. F., 1958, Seismic wave traveltimes from nuclear explosions: Seismological Society of America Bulletin, v. 48, p. 377–398.

Carder, D. S., Quamar, Anthony, and McEvilly, T. V., 1970, Trans-California seismic profile—Pahute Mesa to San Francisco Bay: Seismological Society of America Bulletin, v. 60, p. 1829–1846.

Christensen, N. I., and Fountain, D. M., 1975, Constitution of the lower continental crust based on experimental studies of seismic velocities in granulite: Geological Society of America Bulletin, v. 86, p. 227–236.

Cook, K. L., 1962, The problem of the mantle-crust mix—Lateral inhomogeneity in the uppermost part of the earth's mantle: Advances in Geophysics, v. 9, p. 295–360.

Crough, S. T., and Thompson, G. A., 1977, Upper mantle origin of Sierra Nevada uplift: Geology, v. 5, p. 396–399.

Diment, W. H., Stewart, S. W., and Roller, J. C., 1961, Crustal structure from the Nevada Test Site to Kingman, Arizona, from seismic and gravity observations: Journal of Geophysical Research, v. 66, p. 201–214.

Eaton, J. P., 1963, Crustal structure from San Francisco, California, to Eureka, Nevada, from seismic-refraction measurements: Journal of Geophysical Research, v. 68, p. 5789–5806.

—— 1966, Crustal structure in northern and central California from seismic evidence, *in* Bailey, E. H., ed., Geology of California: California Division of Mines and Geology Bulletin 190, p. 419–426.

Eaton, J. P., Healy, J. H., Jackson, W. H., and Pakiser, L. C., 1964, Upper-mantle velocity and crustal structure in the eastern Basin and Range province, determined from SHOAL and chemical explosions near Delta, Utah [abs.]: Seismological Society of America, 1964, Annual Meeting Program, p. 30–31.

Gibbs, J. F., and Roller, J. C., 1966, Crustal structure determined by seismic-refraction measurements between the Nevada Test Site and Ludlow, California: U.S. Geological Survey Professional Paper 550-D, p. D125–D131.

Giese, Peter, 1966, Versuch einer Gliederung der Erdkruste in nördlichen Alpenvorland, in den Ostalpen and in Teilen der Westalpen mit Hilfe charakteristischer Refraktions-Laufzeitkurven, sowie eine geologische Dentund, Habil. Schrift, Math. Naturwiss Fak. Freie Univ. Berlin, 143 p.

Gilbert, G. K., 1875, Report on the geology of portions of Nevada, Utah, California, and Arizona: U.S. Geographical Survey West of the 100th Meridian Reports, v. 3, p. 17–187.

Gish, D. M., Keller, G. R., and Sbar, M. L., 1981, A refraction study of deep crustal structure in the Basin and Range—Colorado Plateau of eastern Arizona: Journal of Geophysical Research, v. 86, p. 6029–6038.

Gutenberg, B., Wood, H., and Buwalda, J., 1932, Experiments testing seismographic methods for determining crustal structure: Seismological Society of America Bulletin, v. 22, p. 185–246.

Herrin, Eugene, 1969, Regional variations of P-wave velocity in the upper mantle beneath North America, *in* Hart, P. J., ed., The Earth's crust and upper mantle: American Geophysical Union Geophysical Monograph 13, p. 242–246.

Herrin, Eugene, and Taggart, James, 1962, Regional variations in Pn velocity and their effect on the location of epicenters: Seismological Society of America Bulletin, v. 52, p. 1037–1046.

Hill, D. P., and Pakiser, L. C., 1966, Crustal structure between the Nevada Test Site and Boise, Idaho, from seismic-refraction measurements, *in* Steinhart, J. S., and Smith, T. J., eds., The earth beneath the continents: American Geophysical Union Geophysical Monograph 10, p. 391–419.

Hoffman, L. R., and Mooney, W. D., 1983, A seismic study of Yucca Mountain and vicinity, southern Nevada; data report and preliminary results: U.S. Geological Survey Open-File Report 83-588, 50 p.

Iyer, H. M., Hitchcock, T., Roloff, J., and Coally, J., 1977, P-wave residual measurements over the Battle Mountain High, Nevada [abs.]: EOS (American Geophysical Union Transactions), v. 58, p. 1238.

Johnson, L. R., 1965, Crustal structure between Lake Mead, Nevada, and Mono Lake, California: Journal of Geophysical Research, v. 70, p. 2863–2872.

—— 1967, Array measurements of P velocities in the upper mantle: Journal of Geophysical Research, v. 72, p. 6309–6325.

Keller, G. R., Smith, R. B., and Braile, L. W., 1975, Crustal structure along the Great Basin-Colorado Plateau transition from seismic refraction studies: Journal of Geophysical Research, v. 80, p. 1093–1098.

Koizumi, C. J., Ryall, Alan, and Priestley, K. F., 1973, Evidence for a high-velocity lithospheric plate under northern Nevada: Seismological Society of America Bulletin, v. 63, p. 2135–2144.

Lachenbruch, A. H., and Sass, J. H., 1978, Models of an extending lithosphere and heat flow in the Basin and Range province, *in* Smith, R. B., and Eaton, G. P., eds., Cenozoic tectonics and regional geophysics of the western Cordil-

lera: Geological Society of America Memoir 152, p. 209–250.

Mavko, B. B., and Thompson, G. A., 1983, Crustal and upper mantle structure of the northern and central Sierra Nevada: Journal of Geophysical Research, v. 88, p. 5874–5892.

Monfort, M. E., and Evans, J. R., 1982, Three-dimensional modeling of the Nevada Test Site and vicinity from teleseismic P-wave residuals: U.S. Geological Survey Open-File Report 82-409, 66 p.

Mueller, S., and Landisman, M., 1971, An example of the unified method of interpretation for crustal seismic data: Geophysical Journal of the Royal Astronomical Society, v. 23, p. 365–371.

Müller, G., and Mueller, S., 19779, Travel-time and amplitude interpretation of crustal phases on the refraction profile Delta-W, Utah: Seismological Society of America Bulletin, v. 69, p. 1121–1132.

Nelson, K. D., Zhu, T. F., Gibbs, A., Harris, R., Oliver, J. E., Kaufman, S., Brown, L., and Schweikert, R. A., 1984, COCORP deep seismic reflection profiling in the northern Sierra Nevada Mountains, California: Tectonics (in press).

Niazi, Mansour, and Anderson, D. L., 1965, Upper mantle structure of western North America from apparent velocities of P waves: Journal of Geophysical Research, v. 70, p. 4633–4640.

Nolan, T. B., 1943, The Basin and Range province in Utah, Nevada, and California: U.S. Geological Survey Professional Paper 197-D, p. 141–196.

Pakiser, L. C., 1963, Structure of the crust and upper mantle in the western United States: Journal of Geophysical Research, v. 68, p. 5747–5756.

Pakiser, L. C., and Brune, J. N., 1980, Seismic models of the root of the Sierra Nevada: Science, v. 210, p. 1088–1094.

Pakiser, L. C., and Hill, D. P., 1963, Crustal structure in Nevada and southern Idaho from nuclear explosions: Journal of Geophysical Research, v. 68, p. 5757–5766.

Pakiser, L. C., and Steinhart, J. S., 1964, Explosion seismology in the Western Hemisphere, *in* Odishaw, H., ed., Research in geophysics, Volume 2—Solid earth and interface phenomena: Cambridge, Mass., M.I.T. Press, p. 123–147.

Perchmann, J. C., W. D. Richins, and R. B. Smith, 1984, Evidence for a "Double Moho" beneath the Wasatch front, Utah [abs.]: EOS, Transactions of the American Geophysical Union, v. 65, p. 988.

Press, Frank, 1960, Crustal structure in the California-Nevada region: Journal of Geophysical Research, v. 65, p. 1039–1051.

Priestley, Keith, and Brune, James, 1978, Surface waves and the structure of the Great Basin of Nevada and western Utah: Journal of Geophysical Research, v. 83, p. 2265–2272.

Priestley, Keith, Orcutt, J. A., and Brune, J. N., 1980, Higher-mode surface waves and structure of the Great Basin of Nevada and western Utah: Journal of Geophysical Research, v. 85, p. 7166–7174.

Priestley, K. F., Ryall, A. S., and Fezie, G. S., 1982, Crust and upper mantle structure in the northwest Basin and Range province: Seismological Society of America Bulletin, v. 72, p. 911–923.

Prodehl, Claus, 1970, Seismic refraction study of crustal structure in the western United States: Geological Society of America Bulletin, v. 81, p. 2629–2646.

—— 1979, Crustal structure of the western United States: U.S. Geological Survey Professional Paper 1034, 74 p., 3 pls.

Ringwood, A. E., 1975, Composition and petrology of the earth's mantle: New York, McGraw-Hill, 618 p.

Roller, J. C., 1964, Crustal structure in the vicinity of Las Vegas, Nevada, from seismic and gravity observations: U.S. Geological Survey Professional Paper 475-D, p. D108–D111.

—— 1965, Crustal structure in the eastern Colorado Plateaus province from seismic-refraction measurements: Seismological Society of America Bulletin, v. 55, p. 107–119.

Roller, J. C., and Healy, J. H., 1963, Seismic-refraction measurements of crustal structure between Santa Monica Bay and Lake Mead: Journal of Geophysical Research, v. 68, p. 5837–5849.

Ryall, Alan, and Stuart, D. J., 1963, Travel times and amplitudes from nuclear explosions, Nevada Test Site to Ordwa, Colorado: Journal of Geophysical Research, v. 68, p. 5821–5835.

Sass, J. H., Lachenbruch, A. H., Munroe, R. J., Green, G. W., and Moses, T. H.,

Jr., 1971, Heat flow in the western United States: Journal of Geophysical Research, v. 76, p. 6376–6413.

Sinno, Y. A., Keller, G. R., and Sbar, M. L., 1981, A crustal seismic refraction study in west-central Arizona: Journal of Geophysical Research, v. 86, p. 5023–5038.

Smith, R. B., 1978, Seismicity, crustal structure, and intraplate tectonics of the interior of the Western Cordillera, *in* Smith, R. B., and Eaton, G. P., eds., Cenozoic tectonics and regional geophysics of the Western Cordillera: Geological Society of America Memoir 152, p. 111–144.

Smith, R. B., Braile, L., and Keller, G. R., 1975, Crustal low velocity layers—possible implications of high temperatures at the Basin Range-Colorado Plateau transition: Earth and Planetary Science Letters, v. 28, p. 197–204.

Spence, William, 1974, P-wave residual differences and inferences on an upper mantle source for the Silent Canyon volcanic centre, Southern Great Basin, Nevada: Geophysical Journal of the Royal Astronomical Society, v. 38, p. 505–524.

Stauber, D. A., 1980, Crustal structure in the Battle Mountain heat-flow high in northern Nevada from seismic refraction profiles and Rayleigh wave phase velocities [Ph.D. thesis]: Stanford, Calif., Stanford University, 315 p.

——1983, Crustal structure in northern Nevada from seismic refraction data, *in* The role of heat in the development of energy and mineral resources in the northern Basin and Range province: Geothermal Resources Council Special Report 13, p. 319–332.

Stauber, D. A., and Boore, D. M., 1978, Crustal thickness in northern Nevada from seismic refraction profiles: Seismological Society of America Bulletin, v. 68, p. 1049–1058.

Steinhart, J. S., and Meyer, R. P., 1961, Explosion studies of continental structure: Carnegie Institution of Washington Publication 622, 409 p.

Stepp, J. C., Spence, W. J., Harding, S. T., Sherburne, R. W., and Algermissen, S. T., 1963, A study of P_n velocities, amplitudes and traveltime residuals for a portion of the western United States: Earthquake Notes, v. 34, p. 38–49.

Stewart, J. H., 1978, Basin-range structure in western North America—a review, *in* Smith, R. B., and Eaton, G. P., eds., Cenozoic tectonics and regional geophysics of the Western Cordillera: Geological Society of America Memoir 152, p. 1–31.

Stewart, R. M., and Peselnick, L., 1978, Systematic behaviour of compressional velocity in Franciscan rocks at high pressure and temperature: Journal of Geophysical Research, v. 83, p. 831–839.

Tatel, H. E., and Tuve, M. A., 1955, Seismic exploration of a continental crust, *in* Poldervaart, Arie, ed., Crust of the Earth: Geological Society of America Special Paper 62, p. 35–50.

Taylor, S. R., 1983, Three-dimensional crust and upper mantle structure at the Nevada Test Site: Journal of Geophysical Research, v. 88, p. 2220–2232.

Thompson, G. A., and Burke, D. B., 1974, Regional geophysics of the Basin and Range province: Annual Review of Earth and Planetary Sciences, v. 2, p. 213–238.

Thompson, G. A., and Zoback, M. L., 1979, Regional geophysics of the Colorado Plateau: Tectonophysics, v. 61, p. 149–181.

Warren, D. H., 1969, A seismic-refraction survey of crustal structure in central Arizona: Geological Society of America Bulletin, v. 80, p. 257–282.

Warren, D. H., and Healy, J. H., 1974, Structure of the crust in the conterminous United States, *in* Mueller, Stephan, ed., The Structure of the Earth's Crust: Elsevier Scientific Publishing Company, Developments in Geotectonics 8, p. 203–213.

Willden, Ronald, 1965, Seismic-refraction measurements of crustal structure between American Falls Reservoir, Idaho, and Flaming Gorge Reservoir, Utah: U.S. Geological Survey Professional Paper 525-C, p. 44–50.

Zoback, M. L., and Lachenbruch, A. H., 1984, Upper mantle structure beneath the western U.S.: Geological Society of America Abstracts with Programs, v. 16, no. 6, p. 705.).

Geological Society of America
Centennial Special Volume 1
1985

The development of earthquake seismology
in the western United States

Bruce A. Bolt
Department of Geology and Geophysics
University of California at Berkeley
Berkeley, California 94720

ABSTRACT

This paper traces the growth of seismology in the western United States, giving emphasis to geological connections. Pre-instrumental foundations of a science of earthquakes were laid late last century by geologists, especially G. K. Gilbert. These early studies were based largely on Quaternary tectonic structural features evident in the west. The 1906 San Francisco earthquake resulted in a tremendous increase in seismological research and knowledge, including H. F. Reid's "elastic rebound" fault theory of the seismic source, theses on seismic hazards, prediction, seismometry, and other related topics. Significant growth of seismographic networks in California under the auspices of the University of California, Berkeley, and the California Institute of Technology, led to many geologically valuable results on such issues as seismic activity, tectonic patterns, and focal mechanisms. This discussion covers the period up to about 1960.

INTRODUCTION

No thorough historical exegesis has yet been made of the role that seismology has played in the evolution of geological science in North America. The subject is not without difficulties involving priorities and interpretations. As an introduction, a far-from-thorough survey of the subject during the first critical formative years is presented here. The starting point is the end of the nineteenth century. While it is clear that study of earthquakes during the last century had geological as well as engineering importance, it was not until the beginning of the twentieth century that seismology made significant impacts on geological knowledge (Byerly, 1958; Richter, 1958). The impacts grew until the major seismological contributions could be made to the definition of plate boundaries, transform faults, and subduction zones in the sea-floor spreading paradigm of the 1960's.

The emphasis of this article, aspects of earthquake studies in the western United States that bear most closely on the growth of geological knowledge, should not detract from the wide implications of seismology to planetology, engineering, mineral prospecting, petroleum exploration, underground nuclear explosion discrimination, and environmental matters. There are so many themes in this history that even when one limits the coverage to

the geological core, the subject is still too large for a short account. I will focus on the tremendous impact that the 1906 San Francisco earthquake had on seismological science and trace the subsequent evolution of seismological observatories in California and in other western states. Principal earthquake research is covered up to about 1960, with only a hint of the greatly expanded activity of the last two decades. Finally, the crucial contributions of seismology to geology in the west are summarized.

BEGINNINGS

As Wallace (1980) points out, a number of fundamental ideas on seismology were conceived at an early date by the renowned geologist, G. K. Gilbert. Although for some reason his ideas have been nearly ignored in seismological literature, Gilbert appears to be the first member of the U.S. Geological Survey to have written a paper on earthquakes (Gilbert, 1884) and, indeed, is the only author listed in the *Bibliography of North American Geology* from 1785 to 1922 with a paper on earthquake prediction (Gilbert, 1909).

Gilbert's 1884 discussion is of particular importance to any historical study of the development of the modern view of earthquake genesis. The present explanation that tectonic earthquakes (i.e., seismic waves) are produced by the sudden release of elastic strain in the dislocation of a geological fault is in sharp contrast to the older belief that (surface) fault rupture is caused by earthquake forces. Gilbert's early paper is descriptive rather than physically quantitative, and the argument proceeds in generalities by the use of analogy, but his conclusions were remarkably modern. He maintained that his observations of "fault scarps . . . found at the bases of so many ranges of the Great Basin" strongly suggested a fault mechanism as the cause of earthquakes. Furthermore, Gilbert did not hesitate to apply his theory—slow accumulation of strain to the mountainous region followed by "instant yielding" along boundary faults—to future seismic hazards along the fault scarps of the Wasatch Mountains.

Gilbert's insight is in harmony with evolving geological thought on earthquakes worldwide. Indeed, various geological reports in the last decades of the 19th century described changes in ground elevation and vertical and horizontal fault offsets accompanying great contemporary earthquakes. Quantitative mechanical models of the seismic source began to emerge. For example, after quoting several cases, including fresh fault scarps seen after large earthquakes in Baluchistan in 1892 and Greece in 1894, Davison (1905) states "There can be no doubt that a fault scarp is formed in the first place with great rapidity." Nevertheless, the essential quantitative step to the elastic rebound theory of earthquakes had to wait until the 1906 San Francisco earthquake.

The history of observational seismology in North America began in 1887 with continuous recording at two seismographic stations set up at the University of California (Berkeley) and at Lick Observatory (Mt. Hamilton). This early history of the University of California stations has been documented by George D. Louderback (1942), former professor of geology at Berkeley. During the 1880s, astronomy had made important strides in California and, surprisingly, astronomers, not geologists, were responsible for bringing instrumental earthquake recording to the western United States. Because the California astronomical observatories were subject to earthquakes, the President of the University, astronomer E. S. Holden (1898), considered it necessary to "keep a register of all earthquake shocks in order to be able to control the positions of the astronomical instruments." A horizontal motion and a vertical motion seismograph, as well as a duplex pendulum seismograph, were installed at both University observatories in 1887.

At Lick Observatory, the first record of an earthquake was on the duplex pendulum on April 24, 1887. At Berkeley, the first record was also on the duplex pendulum and was due to an explosion at the powder works in North Berkeley, August 11, 1887, at 1:20 a.m. Only a few reproductions of the early seismograms remain because the recordings were made on smoked-glass plates.

One of Holden's ideas was to develop a coordinated group of stations in order to learn more profoundly the seismic characteristics of California. He obtained copies of the Ewing duplex pendulum and arranged for eight regional stations equipped with them to cooperate with Lick Observatory and report all earthquake action. All the stations were prepared to operate in 1887. Louderback noted that, if not the first, this "California System" was one of the earliest sets of closely coordinated seismographic stations in the world and certainly the first in America.

The system operated until 1898 when Holden left Lick Observatory. At that time he published the *Catalogue of Earthquakes on the Pacific Coast, 1769–1897.* This list was later to form a cornerstone of the centrally important catalog of historical western seismicity, the *Descriptive Catalog of Earthquakes of the Pacific Coast of the United States, 1769–1928,* by S. D. Townley and Maxwell W. Allen (1939).

When Holden left Lick Observatory, the stations continued to function independently, although there was no longer a central station to which they all reported. As Louderback pointed out, Holden's underlying idea was sound, and in later years the central and northern California network was developed more effectively with more refined instruments into the multipurpose telemetry network that exists today. Many of the early stations furnished crude records of the California earthquake of April 18, 1906, the most important being the strong ground motions recorded in a fragmentary way at Lick by the Ewing three-component seismograph. This seismograph provided key evidence for the recent relocation of the focus of the 1906 earthquake (near the Golden Gate) and for recent studies of the faulting mechanism involved on the San Andreas fault.

Holden initiated regular publication of the records of the California seismographic stations - the first one being the *List of Recorded Earthquakes in California, Lower California, Oregon and Washington Territory,* published by the Regents of the University of California in 1887. It included reports of the Berkeley and Lick observatories. Subsequently, these reports were printed by the U.S. Geological Survey in their bulletins, the first of which was Bulletin 68 (1890) by J. E. Keeler, *Earthquakes in California in 1889,* and in *U.S. Earthquakes,* a series begun in 1928 by the former U.S. Coast and Geodetic Survey (now the National Oceanic and Atmospheric Administration). It is now issued jointly by the U.S.G.S. and N.O.A.A.

IMMEDIATE EFFECTS OF THE 1906 SAN FRANCISCO EARTHQUAKE ON SEISMOLOGY

The great California earthquake of 1906 aroused widespread interest in earthquake studies and, without question, was a milestone in the history of seismology, not only in North America but worldwide. There was a realization that the science was particularly unorganized in America and that seismographic stations were too few and far between. In studies of the earthquake itself, a causal connection between geological faulting and seismic intensity was clearly demonstrated and geodetic data were used to establish a quantitative physical theory of earthquakes.

A number of leading geologists and scientists at Berkeley, Stanford, and the U.S. Geological Survey took some part in the investigation of the earthquake, the most notable being A. C.

Lawson, David Starr Jordan, G. K. Gilbert, and George D. Louderback. (The latter did important research later in his career into historical evidence on early earthquakes and pioneering studies of the properties of active fault zones.) Lawson was made Chairman of the State Earthquake Investigation Commission by the Governor of California three days after the earthquake, and was responsible for much of the first volume of *The California Earthquake of April 18, 1906 - Report of the State Earthquake Commission.* Four members of the eight-member Commission were geologists. This treatise remains a model of an effective study of a great earthquake, and it should be required reading for all those interested in what is likely to happen in the next great California earthquake. An almost immediate outgrowth of the earthquake was adoption of a constitution of the Seismological Society of America in November 1906.

The major theoretical advance on earthquake genesis was made by H. F. Reid (1910), Professor of Applied Mechanics at Johns Hopkins University. In his theory, an earthquake consists of radiated seismic waves caused by sudden slippage along a geological fault, thus reducing the elastic strain energy in the vicinity. According to Reid, "A rupture takes place and the strained rock rebounds under its own elastic stresses, until the strain is largely or wholly relieved. In the majority of cases, the elastic rebounds on opposite sides of the fault are in opposite directions."

This theory for the first time clearly stated that faulting is the cause of earthquakes and not a consequence of them. Other field workers had come close to this explanation and mechanism, particularly G. K. Gilbert in 1884 and B. Koto, a professor of geology in Japan, after the great Mino-Owari earthquake of October 28, 1891; but Reid had before him not only the clear expression of right-lateral strike-slip faulting along the San Andreas fault (Reid did not use such terminology), but also unequivocal geodetic measurements (Figure 1). Similar near-surface features were to occur in other western earthquakes in the ensuing decades, and careful studies of strain release repeatedly verified Reid's elastic rebound mechanism. It can be said with some confidence that even the most careful studies of earthquakes in the eastern United States and Europe, including the great 1811–1812 New Madrid earthquakes and the 1857 Italian Neapolitan earthquake, would have been unable to produce acceptance of an elastic rebound theory because of the lack of contemporaneous surface faulting and associated regional strain observations. C. E. Dutton (1888) offered no theory on the origin of the 1886 Charleston earthquake, for which no definitive surface faulting occurred. By 1905, Davison, from general seismological considerations, could only hypothesize, although in some detail, a thrust fault as the seismic source for the Charleston earthquake.

The 1906 San Francisco earthquake stimulated thinking on earthquake prediction, already a respectable subject in seismology, particularly in Japan. Reid (1910) specifically made a quantitative prediction of the length of the next cycle on the central San Andreas fault, based on the strain measurements. Later, popular and sometimes contradictory discussions of predic-

tion were given by Lawson, C. W. Hayes, Gilbert, and H. O. Wood, among others. The discussion by Gilbert (1909), who was in Berkeley at the time of the 1906 earthquake, was particularly perceptive. He argued that scientific forecasts as to places were practical, at least in tectonic regions like the Basin and Range and along San Andreas-type "rifts," if appropriate field geological evidence was available. He predicted that seismically hazardous regions would "eventually be subdivided with confidence by means of geologic criteria."

However, Gilbert did not have as much confidence in eventual forecasting of the time of earthquakes. He considered hidden periodicities in seismicity, the reliability of foreshock occurrence, and the repeatability of strain patterns. He examined the hypothesis now known as the "seismic gap theory" in the light of Omori's inferences on circum-Pacific earthquakes and his own, made 25 years earlier, concerning a large earthquake on the Wasatch fault. In California, notably Lawson and Hayes also engaged for a time in this imprecise type of general forecasting, now regarded as rather naive. Gilbert also examined critically the possible use of premonitory signs such as minor seismicity, anomalous animal behavior, and trigger forces such as Earth tides. He concluded that "there was little practical value in any quality of time precision (then) attainable" and, indeed, under the contemporary circumstances, "the cost in anticipatory terror would be great."

After the 1906 earthquake the two University of California stations were supplied with the latest seismographic equipment. Berkeley became the central station, and the University set up its own publication series, the *Bulletin of the Seismographic Stations.* The first issue was published on January 2, 1912. The *Bulletin* has continued to the present and has included the Berkeley and Lick Observatory records as well as those of the other cooperating or branch stations as they were established (Bolt and Miller, 1975).

Investigators of the 1906 earthquake were impressed by the need for a regional program of systematic earthquake studies. Lawson circulated a statement calling for such a plan on April 18, 1907. In 1916, H. O. Wood published in the *Bulletin of the Seismographic Stations* a comprehensive program, including station locations, for the study of local earthquakes and all related phenomena. The idea was pursued under the auspices of the Carnegie Institution of Washington, D.C.

Wood, a mineralogist and former professor of geology at Johns Hopkins University, became very interested in seismology while at Berkeley as a result of his investigations of the intensity of the 1906 earthquake. When the Berkeley station was placed under the administration of the Department of Geology and Mineralogy in 1910, Wood was put in charge of the instruments, and he prepared the first reports for the new *Bulletin of the Seismographic Stations.*

DEVELOPMENT OF EARTHQUAKE RECORDING IN THE WESTERN UNITED STATES (TO 1930)

In the period 1910–1930, the great developments of the seismological research program in California began, particularly

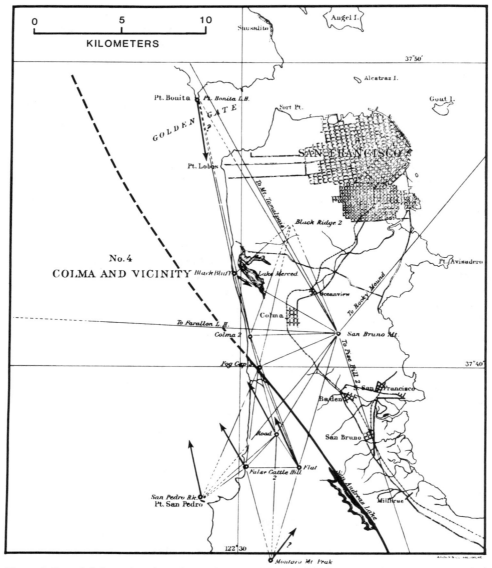

Figure 1. Crustal deformation along the newly mapped San Andreas fault as revealed by displacement of triangulation stations of the U.S. Coast and Geodetic Survey and published in the *Report of the State Earthquake Investigation Commission* (Lawson, 1908). The vectors, determined by the resurvey of 1906–1907, were used by H. F. Reid.

at the California Institute of Technology in Pasadena. From here much of central importance to the science of earthquakes was to emerge during the next six decades.

Southern California's seismological studies (Goodstein, 1983) began under the direction of H. O. Wood in 1921, who had come from Berkeley by way of the Hawaii Volcano Observatory. For the first six years, Wood ran the operations from the Mount Wilson Observatory. While at Berkeley, Wood had worked out his ideas on the contributions of the new science to geology. He placed southern California at center stage because he judged that the next major earthquake in California might well strike in the southern part, since the 1857 Fort Tejon earthquake had occurred there. Wood realized that there was urgent need in

southern California for seismographic stations, particularly with short-period instruments to locate local earthquakes so that accurate seismicity maps could be drawn. He also stressed the need for field studies of earthquakes, including geodetic work similar to that used by Reid in his analysis of the source of the 1906 earthquake.

Wood put considerable emphasis on the mapping of active faults and argued that if such faults could be identified using small earthquakes, geologists could then "produce the place where strong shocks originate, considerably in advance of their event." This line of reasoning led him to speculate that "generalized prediction" of large earthquakes might be possible; he argued that "weak shocks are indicative of growing strain and are preliminary to failure in faulting, with resulting strong shocks." Seismologists still follow many of the ideas set out by Wood, although his method based on the frequency of small earthquakes has not proved effective.

At Mount Wilson Observatory, Wood found a skilled collaborator in astronomer John A. Anderson and together they produced the short-period torsion seismograph with free period of 0.8 sec which became the basis of Charles Richter's earthquake magnitude scale. The Wood-Anderson torsion seismometer, placed into operation in the fall of 1922, proved to be an ideal instrument for recording horizontal ground shaking from local earthquakes and later became the basis of early reliable strong-motion seismographs. The oldest Wood-Anderson records now available date from mid-January 1923. By 1955, seismographic stations with these torsion instruments had been established by Cal Tech at Pasadena, Riverside, Santa Barbara, Tinemaha, Barrett, Woody, and Haiwee (Richter, 1958).

In 1925, Cal Tech established a Department of Geology, and in the following year President Robert A. Millikan invited the Carnegie Institution to conduct its earthquake research at a new seismological laboratory in the foothills of the San Rafael Mountains. The destructive earthquake at Santa Barbara on June 29, 1925, did much to stimulate further interest in regional seismology throughout the state. In 1927, Wood moved into the new building and a broad program of seismological research at Cal Tech began in earnest.

In 1924, back in Berkeley, Professor Lawson asked the Reverend James B. Macelwane, S. J., to join the Geology Department and make the Seismographic Stations his life work. In June 1925, Macelwane returned to Saint Louis University where he began his distinguished career in geophysics, and Perry Byerly took over at Berkeley as the seismology specialist. In 1925, the University of California still operated only two seismographic stations, at Berkeley and on Mt. Hamilton, both equipped with Wiechert and Bosch-Omori seismographs. The time was right to modernize the network and a special committee of the Seismological Society of America, chaired by Professor Bailey Willis of Stanford, raised the money. The first installation of a Wood-Anderson seismograph was completed on November 21, 1927, at Stanford University, under the supervision of S. D. Townley. Wood-Anderson seismographs were subsequently installed at Mt. Hamilton (1928), Berkeley (1930), the University of San Francisco (1935), Mineral (1939), and Arcata (1948). All stations reported to the Berkeley observatory, and amplitude data have been regularly published ever since in the *Bulletin of the Seismographic Stations*. Given the demands of modern studies of seismicity statistics and earthquake risk, the Wood-Anderson seismographs remain among the most valuable instruments operated at the stations.

DEVELOPMENT OF EARTHQUAKE RECORDING FROM ABOUT 1930 TO 1950

For a time, the acceptability of geological inferences based on the elastic rebound theory as the cause of earthquakes was in question (Bolt, 1979a), because verifiable consequences of the theory had not been checked against accessible surface faulting. As early as 1925, Macelwane and Byerly had examined the direction of first motions of both P and Love waves from seismograms of a large 1925 Chilean earthquake. Prior to this study, efforts to deduce the forces at the earthquake focus from the first motion on seismograms had been restricted to Japan, using only Japanese seismograms. Byerly (1938) broke free of these constraints and, based on a mechanism theory of Nakano and the elastic rebound theory of Reid, argued that sudden slippage along a fault would send out waves of compression and dilatation that would form quadrants separated by a fault plane and a plane perpendicular to it (the auxiliary plane). Wide coverage of points on the surface of the Earth by seismographic stations would lead to locations of these fault planes.

Appropriate projections were worked out by Byerly (1938) to enable the nodal lines from P-wave motions to be plotted so that strike and dip of the causative fault could be estimated (Figure 2). Byerly and his students continued to work on aspects of focal mechanisms in the succeeding decades.

At the beginning of these researches on remote sensing of tectonic stresses and displacements, there was some skepticism because of lack of standardization of seismographs and their poor calibration. These observational difficulties contributed to many incorrect reports of wave polarity. (The establishment of the World-Wide Standardized Network of Seismograph Stations [WWSSN] in the 1960s provided much more reliable recordings and seismologists then found that the fault-plane method could be used rapidly, with few exceptions, to obtain consistent focal patterns.) Gradually, verifications of the Byerly algorithm accumulated by correlations with actual surface faulting and the method became accepted as a powerful tool in determining faulting in all tectonically active regions including mid-oceanic rifts, transform faults, and subduction zones.

Stimulated by the 1906 earthquake and with the California stations as an example, earthquake observatories appeared in other western states. At the Tucson, Arizona, magnetic observatory, established in 1909, seismographs were added in 1926. This station has low microseismic background and therefore provided a much improved coverage of U.S. regional earthquakes and teleseisms. Readings from Tucson have also contributed considerably to the location of Pacific earthquakes used by the Honolulu-based Tsunami Early Warning System.

The western mountain states share a significant earthquake hazard, having experienced a number of destructive earthquakes in the last century. Correlation between geologic field studies and available seismological recordings has been of great importance in the extension of knowledge of earthquake mechanisms and the geodynamics of the Cordillera. The Wasatch fault along the Wasatch Mountains and passing through Salt Lake City has clear geological evidence for movement in Quaternary time. As Gilbert (1884) indicated, this fault is a matter of considerable concern in terms of future earthquakes. Montana has also been the center of

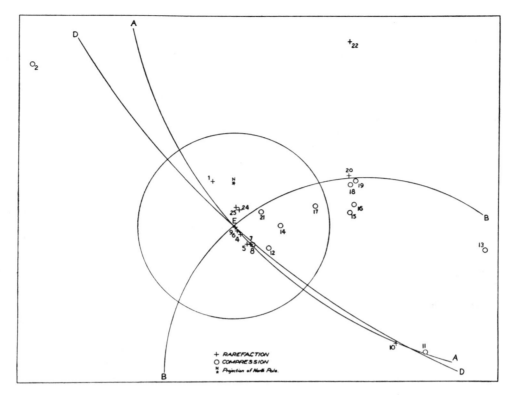

Figure 2. First known published fault-plane solution of P. Byerly (1938). The stereographic projection is of first P wave polarities at worldwide stations from the California earthquake of July 6, 1934.

major earthquakes, notably the magnitude 7.6 (revised in 1984) Hebgen Lake earthquake on August 17, 1959. This great earthquake killed 28 people, caused a landslide which formed a large lake, and did $11 million damage to roads and timber. Faulting was found extending to 30 km, with a maximum dip-slip offset of about 6 meters.

In Washington, damaging earthquakes have occurred around Puget Sound and smaller earthquakes have been associated with volcanic activity in the Cascades, such as those at Mount St. Helens. The University of Washington established stations in the 1950s, with the important Longmire station near Mt. Rainier beginning operation in 1958. This station monitors earthquakes in western Washington and rockfalls and glacier activity on Mt. Rainier.

In Alaska, early seismographs were operated at the Sitka Magnetic Observatory and at College, Alaska. It was not until after the Good Friday earthquake of March 1964 that a major expansion of seismographic observatories occurred. Currently, the main tasks of the Alaskan observatories are the mapping of seismicity and participation in the Tsunami Early Warning System. In the Alaskan coastal region, it is crucial to get early warnings within fifteen minutes of a major earthquake on or adjacent to the continental margin. Nowadays there are many automated instruments along the Aleutian chain as well as on the mainland, with a major central observatory operating at Palmer.

Mention should also be made of the role of the Society of

Jesus in the growth of seismology in the United States. The first Jesuit network was developed following a letter of February 2, 1909, by Father F. L. Odenback, S. J., of St. Ignatius College (now John Carroll University in Cleveland), addressed to all Jesuit colleges in the United States and Canada, suggesting that they buy standard seismographs, such as the new Wiechert instrument, and start a seismological service. In the western states, stations were established at Denver, Santa Clara and Spokane. Following World War II, Father Macelwane, who had by this time developed an important research and teaching center in seismology at St. Louis University, organized a meeting to resuscitate the Jesuit efforts and the Jesuit Seismological Association that had been formed in 1925. The Association published an important *Preliminary Bulletin of Epicenters* for the United States and abroad and collaborated closely with the U.S. Coast and Geodetic Survey on listing of earthquakes. The central Jesuit seismographic station was established at Saint Louis University.

DEVELOPMENT OF SEISMOLOGICAL RESEARCH FROM ABOUT 1930 TO 1950

In 1930, President Millikan of the California Institute of Technology invited Beno Gutenberg to come from Frankfurt to be Director of the Seismological Laboratory. This started an era of great productivity in seismology centered largely on the researches of Gutenberg, Richter, and Benioff (Figure 3).

Figure 3. Conference on the future of seismological studies at Cal Tech at Pasadena in 1929. Front row, left to right: assistant, Leason Adams, Hugo Benioff, Beno Gutenberg, Harold Jeffreys, Charles Richter, Arthur Day, Harry Wood, Ralph Arnold, John P. Buwalda. Back row: assistant, Perry Byerly, Harry Fielding Reid, John Anderson, Father J. B. Macelwane.

Gutenberg made fundamental contributions to geology based on seismological and other geophysical methods. Sometimes working alone, sometimes with Charles Richter, he covered the gamut of seismology. He had a remarkable ability in reading seismograms and made seminal studies on seismicity, travel-time curves, improvements in the magnitude scale, and energy release in earthquakes. He investigated the consequences of anomalous "low velocity" layers in the crust and upper mantle and made crucial contributions on the structure of the deep interior both before and after arriving at Cal Tech. The first definitive estimates of the thickness of the crustal layers in California were obtained by Gutenberg in 1932 for southern California (Gutenberg, 1932) and by Byerly in 1934 for central California (Byerly, 1934).

In the decades from 1930 to 1950, perhaps the basic seismological contributions in the west to the science of geology related to systematic mapping of seismicity. Precise seismicity maps laid the basis for the major advances of the 1960s and 70s in the prediction of strong ground motion and the delineation of active faults. One of the keys to the necessarily precise measurements of small earthquakes was the design of seismographs of high reliability and sensitivity by Hugo Benioff. Short-period vertical seismographs of Benioff type installed both at Cal Tech and at Berkeley about 1935 provided high resolution of local and regional seismicity as well as an ability to record, with high fidelity, seismic waves from earthquakes around the world. During this period, Benioff made many other innovations in seismometry, most notably on strainmeters and dilatometers. An associated development of central importance was the definition in 1935 of

the instrumental magnitude scale by Richter. The process of development is described briefly in Richter's book (1958).

By 1960, a more or less stable network of permanent seismographic stations with modern photographic equipment and reasonably reliable clocks had been established both in northern California by the University of California at Berkeley and in southern California by the Cal Tech Seismological Laboratory. There was also a concurrent slow growth of seismographic stations in other western states, several under the auspices of the U.S. Coast and Geodetic Survey. Some of these stations also had Wood-Anderson instruments which permitted the earthquake catalogs to begin listing not only epicentral locations but also magnitude estimates.

In the same period, fault plane solutions began to be estimated in more detail, and vital cross-checks were found with the geologic field evidence of fault displacement. Important field studies were made of the sources and effects of earthquakes, such as the 1940 Imperial Valley earthquake. This earthquake near El Centro provided one of the first strong-motion records written by an accelerometer and related to a surface fault rupture; and this recorded ground motion became the standard for earthquake-resistant design around the world for several decades. Such acceleration seismographs, capable of remaining on scale in even the strongest earthquakes, were gradually improved and installed throughout the west.

In the depression years, applied seismology developed rapidly and began to make major contributions to petroleum exploration (Bates and others, 1982). From about 1930, many oil company geologists returned to college or were otherwise re-trained in seismic exploration methods. Specialized methods of structural interpretation were developed that required close collaboration between geologists and seismologists.

Of immediate geological value, details of the crustal structure based on idealized models were worked out for central and southern California (Gutenberg, 1959). Depths to the Mohorovicić discontinuity were established and reasonable estimates for the velocity of seismic waves within the crust were determined by many studies of travel times of P and S waves from local earthquakes. It was clearly demonstrated that almost any earthquake which occurred within an area where there were seismographic stations was worthy of seismological study and that inferences well beyond the immediate characteristics of that earthquake could be drawn from analysis of the seismograms. Comprehensive and high resolution seismic studies of the western crust are, however, a rather recent endeavor, and much remains to be done.

In 1930, Byerly provided seismological evidence for a "root" under the southern Sierra Nevada Mountains. His interest (Byerly, 1958) had been stimulated by work of A. C. Lawson on isostasy. He conceived the idea of comparing travel times of seismic waves from earthquake foci near the Pacific coast recorded at the Fresno seismographic station to the west of the Sierra Nevada, with those from the same sources recorded at the Tinemaha station on the east side of the mountains. At Tinemaha, there was a lag of several seconds in the travel times, which Byerly demonstrated mathematically to be consistent with a region about 40 km wide beneath the mountains in which the velocity is smaller than crustal velocities on each side. Gutenberg and others followed up this inference (Gutenberg, 1959). Recent studies have shown that the structure is more complicated than the model originally put forward, but there has been general confirmation of an asymmetrical root of the Sierra Nevada; similar roots have been found by seismologists under several other large mountain ranges around the world.

MORE RECENT DEVELOPMENTS IN RESEARCH IN THE WEST

In the last two decades there has been a veritable explosion of seismological research in the U.S., and contributions in the western states have been numerous. No longer can the history be given in terms of two or three centers; many universities have strong departments in the earth sciences, special seismological programs have been formed by federal and state agencies, and industry has established geotechnical groups.

The pioneer seismological centers at U.C. Berkeley and Cal Tech have continued strongly, if no longer uniquely, with teaching and research in seismology. In 1957, Gutenberg retired as Director at the Seismological Laboratory and Frank Press took his place until 1965, when he became head of a new Earth Science Department at M.I.T. The Cal Tech position was then taken up by Don Anderson. In 1963, Perry Byerly retired as

Director at Berkeley after 38 years, and Bruce A. Bolt was appointed.

Although all the details of recent history cannot be included in a short treatment, many strong interactions and points of contact have been established in recent decades between geologists and seismologists as a result of federal efforts to establish a basis for international treaties controlling underground testing of nuclear explosive devices, the regulations on reduction of hazards from large critical structures such as nuclear power reactors and large dams, attempts to find methods of earthquake prediction, exploration for geothermal energy, and exploration for oil and gas.

The main opportunity for thorough upgrading of U.S. seismology and earthquake observatories came as a result of nuclear test ban discussions held during the late 1950s. A special Presidential panel released the Berkner Report which recommended the establishment of the World-Wide Standardized Seismograph Network (WWSSN) with up to 200 stations collaborating under the auspices of the U.S. Coast and Geodetic Survey in the United States. Between 1960 and 1967, 120 WWSSN systems were installed, most at existing seismographic stations. Six seismometers were installed at each station, three short-period seismometers of the Benioff type and three long-period seismometers of Press-Ewing design. In the western United States, the WWSSN stations were installed at Berkeley (Byerly), Goldstone, Tucson, Albuquerque, Dugway, Missoula, Rapid City, Corvallis, Longmire, and College (Alaska). There is also a WWSSN station in Kipapa on the Island of Hawaii.

Radical improvements in regional seismographic networks also occurred after 1960 in the western United States. The feasibility and advantages of a connected regional network of stations, forming a large-scale seismic array with telemetry over commercial telephone lines were demonstrated successfully at the University of California at Berkeley. By 1963, there was a reliable telemetered network of some 15 stations of short-period vertical instruments in central California and in the foothills of the Sierra Nevada. Such a network, with its common time base, provided much more precise and rapid location of earthquake hypocenters than ever before. Almost at once, two results of geologic importance from the Berkeley network became apparent: in central California, earthquake foci are shallower than about 15 km, with most from 5 to 10 km deep, and most foci are situated along the major active fault systems (Bolt and others, 1968). Similar networks have since become commonplace in many earthquake areas for monitoring regional earthquakes, studying microearthquakes, and monitoring seismicity around critical structures.

A leading role in seismology was taken recently by the U.S. Geological Survey, which, in the 1960s, established its Office of Earthquake Studies at Menlo Park, California, and the National Earthquake Information Service in Denver, Colorado (Rabbitt, 1979).

In 1966, the U.S. Geological Survey began to operate an extensive telemetry seismographic network in California, primarily to record microearthquakes (Eaton, 1983). The first cluster was eight stations along the San Andreas fault near the south end of San Francisco Bay. The network was extended gradually by about 15 stations per year to cover a strip 500 km long and 100 km wide to include about 200 stations. Other special purpose telemetry seismographic networks have also been installed in California and other western states by the U.S.G.S. or by universities with support by the U.S.G.S. A striking early result was the unequivocal confirmation of the concentration of microearthquakes along fault planes.

By 1974, five telemetry networks had been installed in southern California areas by the U.S. Geological Survey, some as part of a cooperative program with the California Institute of Technology. A 20-station network was installed in 1973 in the Imperial Valley, related at first to the association between microearthquake activity and development of geothermal resources. The network, however, also provided key information following the moderate Imperial Valley earthquake of 1979. A 22-station network was installed in the Mojave Desert in 1974, supported by the Atomic Energy Commission in relation to a proposed nuclear reactor site. And in 1974, a 32-station San Bernardino network was established for studies of local seismicity and developing methods of earthquake prediction. Also, there was a 7-station network in the Los Angeles Basin, initially installed to study seismicity associated with the injection of water in oil fields within the Basin. In addition, there was a 16-station network near Santa Barbara Channel and Point Mugu, designed to provide better resolution of earthquakes occurring within the Santa Barbara Channel. The varied purposes of these networks illustrate the changing and growing applications of seismology in the years since the 1906 San Francisco earthquake.

The California Division of Mines and Geology also became significantly involved in seismological problems at the time of the Arvin-Tehachapi earthquake of July 21, 1952. The Division produced the comprehensive *Bulletin 171*, with many contributions from seismologists, geologists, and structural engineers. It was not until 1970, however, that the Division hired the first State Civil Service seismologist and has since become specially involved with geologic hazards aspects and strong ground motion (Sherburne, 1981).

CONTRIBUTIONS OF WESTERN SEISMOLOGY TO NORTH AMERICAN GEOLOGY

Even a short survey demonstrates the wide scope of the contributions of seismology to geology. Seismological studies in the western United States were unique because of the evidence they provided on geodynamics and seismic hazards.

First, the precise location of earthquake foci and specific localization of source regions in the western United States provided essential statistics for the occurrence of earthquakes of all sizes and evaluation of the rate and distribution of energy and stress release (Bolt, 1979b).

For at least a century, seismological studies have been crucial in the understanding of Quaternary geology and tectonics of

North America. The definition of the modes of earth movements and mountain building has been much advanced by the application of the fault plane and focal mechanism algorithms. Syntheses of collections of these studies provided quantitative measures of current regional crustal-strain patterns (Smith and Lindh, 1978). Not only has mapping of the present stress field been of great importance along the North American plate margin in California and in the Cascade Mountains, but it has complemented geologic inferences on the deformation of the whole interplate zone. In particular, the relation between the focal depths of western earthquakes and the heat flow has provided important hypotheses concerning the nature of crustal deformation and faulting and the effect of temperature on the elastic and plastic states of rocks in the crust (Sibson, 1982).

There have been imaginative geomorphological studies, following Gilbert's early work, on land forms such as the uplifted features in the Basin and Range province and the ancient landslides in the Columbia River Basin. Striking results have been achieved on dating ancient fault movements by radiocarbon methods applied to sand and peat layers exposed in trenches dug across active faults (Sieh, 1978) and by measurement of the erosional geometry of fault scarps (Wallace, 1980). Often stimulated by seismological questions, geological field studies of Quaternary fault displacements have been outstanding (Sherard and others, 1974).

Originally, the volcanic history and associated orogenies in the western U.S. were studied using geological and geochemical methods, but after about 1920 it gradually became apparent that evidence from local earthquake patterns and seismologically-based structures produced more adequate physical models of the eruptive and deformational mechanisms. Quite recently, major geological surveys of Mount St. Helens and at Mammoth Lakes, have included seismological techniques as an essential part of the mapping of a volcanic system. Seismic exploration methods have also been used effectively in extensive mapping of geothermal fields such as The Geysers in central California.

Finally, seismological work in the western states led to fundamental results on seismic risk theory and to maps predicting the intensity of future earthquakes (Dewey and others, 1972). The precise mapping of seismicity (e.g., Hileman and others, 1973; Bolt and Miller, 1975; Smith, 1978) delineated the present activity of faults. Indications of this kind are now on special geological maps, for example, by the California Division of Mines and Geology under their responsibilities for the state earthquake hazard study zones. The development of correlations between earthquake magnitudes and geological studies of fault dimensions, such as fault offset and fault length, has provided the basis for the estimation of strong ground motion parameters needed for design of critical engineering structures such as nuclear power stations, dams and bridges.

ACKNOWLEDGMENTS

I am indebted to Professor C. Allen, Professor P. Dehlinger,

Dr. G. B. Oakeshott, Professor W. Stauder, and Dr. R. Wallace for their assistance and criticism in preparing this paper.

REFERENCES CITED

Bates, C. C., T. G. Gaskell and R. B. Rice, 1982, Geophysics in the affairs of man: Pergamon, Oxford, 492 p.

Berg, J. W., and C. D. Baker, 1963, Oregon earthquakes, 1841–1958: Seismological Society of America Bulletin, v. 53, p. 95–108.

Bolt, B. A., C. Lomnitz, and T. V. McEvilly, 1968, Seismological evidence on the tectonics of central and northern California and the Mendocino escarpment: Seismological Society of America Bulletin, v. 58, p. 1725–1767.

Bolt, B. A., and R. D. Miller, 1975, Catalogue of earthquakes in northern California and adjoining areas, 1910–1972: Berkeley, University of California, Seismographic Station, Special Publication, 576 pp.

Bolt, B. A., 1979a, Memorial, Perry Byerly: Seismological Society of America Bulletin, v. 69, p. 928–945.

Bolt, B. A., 1979b, Seismicity of the western United States: Reviews in Engineering Geology, Geological Society of America, v. 4, p. 95–107.

Byerly, P., 1938, The earthquake of July 6, 1934: Amplitudes and first motion: Seismological Society of America Bulletin, v. 28, p. 1–13.

Byerly, P., 1958, The beginnings of seismology in America: Symposium on Physical and Earth Sciences, University of California, p. 43–52.

Byerly, P., and J. T. Wilson, 1935, The Central California earthquakes of May 16, 1933 and June 7, 1934: Seismological Society of America Bulletin, v. 25, p. 223–246.

Coffman, J. L., and C. A. von Hake, 1973, Earthquake history of the United States: Washington, D.C., U.S. Department of Commerce Publication No. 41-4, revised edition (through 1970), 208 p.

Cook, K. L., and R. B. Smith, 1967, Seismicity in Utah, 1850 through June 1956: Seismological Society of America Bulletin, v. 57, p. 689–718.

Crosson, R. S., 1974, Compilation of earthquake hypocenters in western Washington: Washington Department of Natural Resources, Information Circular 56.

Davison, C., 1905, A study of recent earthquakes: Walter Scott Publishing, London, 355 p.

Dewey, J. W., W. H. Dillinger, J. Taggert, and S. T. Algermissen, 1972, A technique for seismic zoning: analysis of earthquake locations and mechanisms in northern Utah, Wyoming, Idaho and Montana, *in* Proceedings, Microzonation Conference: Seattle, University of Washington, v. 2, p. 879–894.

Dutton, C. E., 1888, The Charleston earthquake of August 31st, 1886; U.S. Geological Survey, Ninth Annual Report, p. 209–528.

Eaton, J. P., 1983, Temporal variation in the pattern of seismicity in central California: U.S. Geological Survey, Menlo Park (unpublished).

Gilbert, G. K., 1884, A theory of the earthquakes of the Great Basin, with a practical application: American Journal of Science, v. 27, p. 49–53.

Gilbert, G. K., 1909, Earthquake forecasts: Science, v. 29, p. 121–138.

Goodstein, J., 1983, Archiving early seismological records at the California Institute of Technology: Earthquake Information Bulletin, v. 15, p. 177–184.

Gutenberg, B., 1932, Travel time curves at small distances and wave velocities in Southern California, Geol. Beitr. z. Geophpysik, v. 35, p. 6–50.

Gutenberg, B., 1959, Physics of the Earth's interior: Academic Press, New York, 240 p.

Hileman, J. A., C. Allen, and J. M. Nordquist, Seismicity of the southern California region, 1 January 1932 to 31 December 1972: Pasadena, California Institute of Technology, Seismological Laboratory, 487 p.

Holden, E. S., 1898, A catalogue of earthquakes on the Pacific coast, 1769–1897: Smithsonian Institution, Miscellaneous Contribution 1087, 253 p.

Keeler, J. E., 1980, Earthquakes in California in 1889, U.S. Geological Survey Bulletin 68, p. 1–14.

Koto, B., 1893, On the cause of the great earthquake in central Japan, 1891: Journal College of Science, Imperial University of Japan, v. 5, p. 296–353.

Lawson, A. C., 1908, The California earthquake of April 18, 1906, Report of the State Earthquake Investigation Commission, v. 1, Carnegie Institution, Washington, D.C.,

Louderback, G. D., 1942, History of the University of California seismographic stations and related activities: Seismological Society of America Bulletin, v. 32, p. 205–229.

Rabbitt, M. C., 1979, The Survey's first venture into seismology: U.S. Geological Survey Earthquake Information Bulletin, v. 11, p. 50–52.

Rasmussen, N., 1967, Washington state earthquakes, 1840–1965: Seismological Society of America Bulletin, v. 57, p. 463–476.

Real, C. R., and T. Toppozada, 1978, Earthquake catalog of California, January 1, 1900 - December 31, 1974 (first edition): California Division of Mines and Geology Special Publication 52, 000 p

Richter, C. F., 1958, Elementary seismology: W. H. Freeman, San Francisco, 768 p.

Reid, H. F., 1910, The California Earthquake of April 18, 1906, The mechanics of the earthquake, Report of the State Earthquake Investigation Commission, v. 2, Carnegie Institution, Washington, D.C., p. 1–192.

Sherard, J. L., L. S. Cluff and C. R. Allen, 1974, Potentially active faults in dam foundations: Geotechnique, v. 24, p. 367–428.

Sherburne, R. W., 1981, Seismology program: California Division of Mines and Geology: Earthquake Information Bulletin, v. 13, p. 65–68.

Sibson, R. H., 1982, Fault zone models, heat flow and the depth distribution of earthquakes in the continental crust of the United States: Seismological Society of America Bulletin, v. 72, p. 151–163.

Sieh, K., 1978, Prehistoric large earthquakes produced by slip on the San Andreas fault at Pallett Creek, California: Journal of Geophysical Research, v. 83, p. 3907–3939.

Slemmons, D. B., A. E. Jones, and J. I. Gimlet, Catalog of Nevada earthquakes, 1852–1960: Seismological Society of America Bulletin, v. 55, p. 537–583.

Smith, R. B., 1978, Seismicity, crustal structure, and intraplate tectonics of the interior of the western Cordillera, in Smith, R. B., and Eaton, G. P., eds., Cenozoic tectonics and regional geophysics of the western Cordillera: Geological Society of America Memoir 152, p. 111–144.

Smith, R. B., and A. G. Lindh, 1978, Fault-plane solutions of the Western United States: A compilation, *in* Smith, R. B., and Eaton, G. P., eds., Cenozoic tectonics and regional geophysics of the western Cordillera: Geological Society of America Memoir 152, p. 107–110.

Townley, S., and M. W. Allen, 1939, Descriptive catalogue of earthquakes of the Pacific coast of the United States, 1769 to 1928: Seismological Society of America Bulletin, v. 29, p. 1–297.

von Hake, C. A., 1974a, Earthquake history of Montana: Earthquake Information Bulletin, v. 6, no. 4, p. 30–34.

von Hake, C. A., 1974b, Earthquake history of Nevada: Earthquake history of Nevada: Earthquake Information Bulletin, v. 6, no. 6, p. 26–29.

von Hake, C. A., 1975, Earthquake history of New Mexico: Earthquake Information Bulletin, v. 7, no. 3, p. 23–26.

Wallace, R. E., 1980, G. K. Gilbert's studies of faults, scarps, and earthquakes: Geological Society of America Special Paper 183, p. 35–44.

Wood, H. O., 1911, The observation of earthquakes: Seismological Society of America Bulletin, v. 1, p. 48–82.

Geological Society of America
Centennial Special Volume 1
1985

Geology from space: A brief history of orbital remote sensing

Paul D. Lowman, Jr.
Geophysics Branch (Code 622)
NASA/Goddard Space Flight Center
Greenbelt, Maryland 20771

ABSTRACT

This paper reviews the development of geologic remote sensing of the Earth and earth-like bodies from space in the 1946–1984 period. Topics include sounding rocket photography, hand-held photography from Mercury, Gemini, Apollo, Apollo-Soyuz Test Project, and Skylab missions, the Earth Resources Experiments Package carried on Skylab, Landsats 1 through 5, the Heat Capacity Mapping Mission, Seasat, and geologic remote sensing experiments carried on two Shuttle missions. Extraterrestrial bodies discussed include the Moon, Mercury, Mars, Venus, and the outer planets and satellites. Data from earth-orbital experiments have been used in oil, mineral, and ground-water exploration, for research, and for geologic education, the most promising applications being in structural geology. Visual range images from Landsat have been the most widely used, but infrared data have also been used successfully for compositional mapping and as indirect indicators of structure, composition, and ground-water distribution. Orbital radar images have proved valuable for structural studies. Extraterrestrial remote sensing has in effect opened up a new solar system by providing close-range data for 17 extraterrestrial bodies previously accessible only by astronomical techniques. Remote sensing from space has been one of the most valuable and unexpected results of space flight.

INTRODUCTION

By the early 1950s most now-familiar space achievements, such as communications satellites and expeditions to the Moon, had been predicted in considerable detail by Arthur Clarke, Willy Ley, and others. But remote sensing from space was almost totally *un*predicted, except for weather and military reconnaissance observations; nothing in even the most perceptive forecasts resembled Landsat, and there were only hints of the many automated missions to the Moon and planets carried out since 1959. Remote sensing from space is thus interesting not only by itself but as an example of the unexpected by-products of space flight. This paper summarizes the history of geologic remote sensing from space, stressing investigations of the surface of the Earth. The main techniques applied to other bodies are also covered. The paper is confined to American missions and data therefrom; the Soviet Union has carried out extensive remote sensing from space, but the subject is beyond the scope of this paper.

The term "remote sensing" will follow current usage to mean the long-distance non-contact study of objects or phenomena by electromagnetic radiation, independent of transmission by any medium other than space itself, thus excluding potential field surveys, exploration seismology, and sonar. Although, as Allum (1980) has pointed out, remote sensing is by no means synonomous with "remote sensing from space," this review concerns only space-borne instruments.

The literature on remote sensing in general is immense and can be cited only briefly here. A fundamental source for aerial and space remote sensing is the encyclopedic "Manual of Remote Sensing" (Colwell, 1983), which covers all aspects of data acquisition and handling with an extensive section on geological applications. Other useful publications include those by Holz (1973), Woll and Fischer (1977), Sabins (1978), Siegal and Gillespie (1980), Goetz and Rowan (1981), Slaney (1981), Short (1982), and Goetz and others (1983). Annotated collections of Landsat pictures have been published by Short and others (1976), Wil-

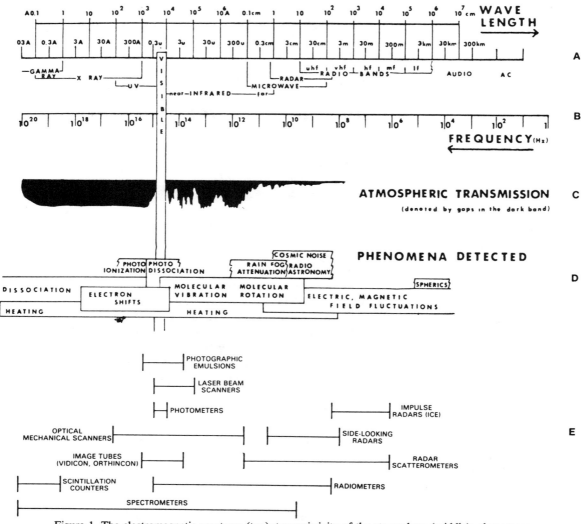

Figure 1. The electromagnetic spectrum (top); transmissivity of the atmosphere (middle); phenomena detected (usually, also the radiation source); and spectral sensitivity of sensors (bottom). From Short (1982); reproduced with permission.

liams and Carter (1976), and Sheffield (1983), and the Landsat mosaic of the United States at 1:1,000,000 scale by Ryder (1981). Detailed technical descriptions of sensors used from space have been compiled by Tanner (1982); see also the appropriate chapters in Colwell (1983). Summaries of the *results* of extraterrestrial remote sensing (as distinguished from techniques, covered here) have been published by Short (1975), King (1976), Cadogan (1981), Murray and others (1980), Glass (1982), Beatty and others (1982), and Taylor (1982).

The physical principles of remote sensing as defined here are covered by chapters in Holz (1973) and Colwell (1983), and by Hunt (1980). The electromagnetic spectrum is reproduced here (Fig. 1), together with associated parameters. An interesting relationship implied by "phenomena detected" but rarely pointed out is the dependence of wavelength on the size of radiation sources. As shown by the diagram, the larger the source, the longer the wavelength: gamma rays come from atomic nuclei, thermal in-

frared radiation from molecules, and so on, up to low frequency radio waves transmitted by antennas hundreds or thousands of meters long.

Some understanding of orbital mechanics is helpful in the use of data from artificial satellites and manned spacecraft, and a typical earth satellite ground track is therefore presented here (Fig. 2). Such ground tracks can be easily visualized by imagining the satellite to follow a great circle, fixed in space, while the Earth rotates under it. The inclination of the great circle to the equator is also the latitude limit, subject to in-flight plane changes (which are rarely done because of the high energy requirements).

REMOTE SENSING OF THE EARTH

Progressive stages in remote sensing of the Earth from space have largely overlapped in time. However, a rough chronological order is followed in this account. Many of the examples presented

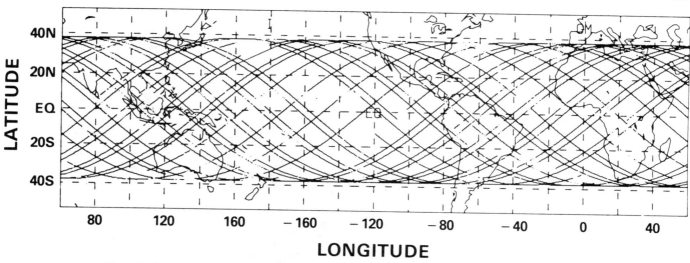

Figure 2. Ground track of shuttle on OSTA-1 mission; circular orbit with 38° inclination and 262 km average altitude.

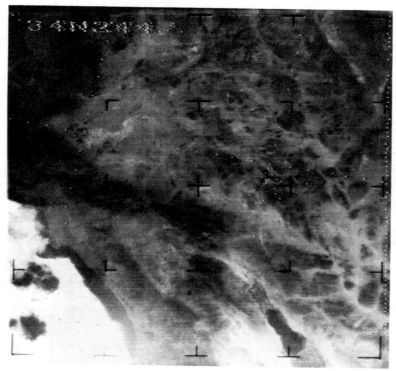

Figure 3. Nimbus 1 Advanced Vidicon Camera System showing S.W. United States; Salton Sea at lower right. AVCS used a 2.54 cm vidicon tube with 800 scan lines; estimated ground resolution of this scene 300 to 400 m (Sabatini and others, 1971).

cover the same area, the southwest U.S., to permit convenient comparison of different types of imagery.

Sounding Rockets and Weather Satellites

The first views of the Earth from space were obtained by small cameras at altitudes of 250 km or so by V-2, Aerobee, and Viking rockets launched from White Sands Proving Ground,

New Mexico. They showed large parts of the southwest U.S. and Mexico (Holliday, 1950). Although put to little direct geologic use, these pictures demonstrated the resolving power of even modest focal lengths, and led P. M. Merifield in 1962 to suggest that Mercury astronauts photograph geologically interesting areas. Weather satellites began returning low-resolution television pictures (Fig. 3) in 1960, and contributed to the increasing interest in orbital surveillance. Some major faults in snow-covered

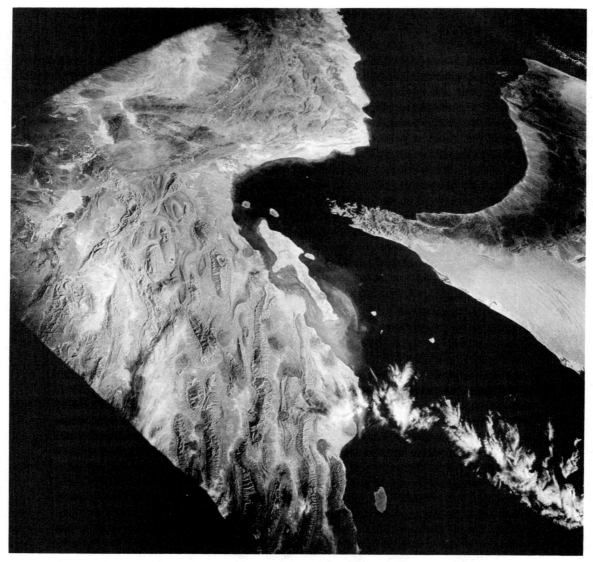

Figure 4. Gemini 12 photograph showing Zagros Mountains, Iran (foreground), Strait of Hormuz, and Oman Range (upper right). Original picture taken with 80mm lens on 70mm Ektachrome from about 200 km altitude. For detailed description see Lowman (1968, 1972). NASA photo number S-66-63082.

mountains are well-displayed on weather satellite pictures (Lowman, 1972), and thermal infrared imagery from Nimbus satellites has been used by Andre and Blodget (1984) to map major structures in the Arabian Peninsula.

Early Manned Missions

Beginning in 1962, astronauts on most Mercury, Gemini, and earth-orbital Apollo missions carried out an extensive program of terrain photography with hand-held 70mm cameras and color film (Lowman, 1968), whose objective was to obtain pictures of selected areas for geologic, geographic, and oceanographic research. The experiment was extremely successful, with several thousand usable pictures (Fig. 4) being obtained, many of

which were used in geologic publications. The Apollo 9 earth-orbital mission carried in addition to the single-camera experiment a multispectral terrain photography experiment suggested by A. P. Colvocoresses. Using four coaxially-mounted 70mm cameras with different film-filter combinations (Lowman, 1969a; Kaltenbach, 1970), the experiment's objective was to investigate the feasibility of multispectral orbital photography in preparation for the Earth Resources Technology Satellite (now Landsat). The four-band sets obtained proved useful for geological, agricultural, and forestry applications, and served as an analogue simulation of Landsat imagery. One of the sets, covering part of southern California (Figs. 5, 6) indicated that Cretaceous structures were not offset by the Elsinore fault, leading to a new interpretation of Peninsular Range tectonics (Lowman, 1980b).

The Apollo-Soyuz Test Project, a joint Soviet-American mission flown in 1975, returned a large number of 70mm terrain photographs similar to those from earlier missions. These have been catalogued and interpreted by El-Baz (1979).

The first American space station, Skylab, carried in 1973-4 a large array of remote sensing experiments (Fig. 7), the Earth Resources Experiment Package (EREP), hard-mounted in the Multiple Docking Adapter. The EREP experiments were highly successful as efforts in technique development, with immense amounts of data returned as digital tape and photographs. These data were successfully applied to a wide range of disciplines including geology, land-use management, geophysics, agriculture, and water resources (NASA, 1974). Several Skylab/EREP investigations were carried out in parallel with Landsat-1 investigations, providing valuable comparisons. In addition, the three successive Skylab crews took several thousand 35mm and 70mm photographs with hand-held cameras, supplementing them with verbal descriptions (Wilmarth and others, 1977). The S190B Earth Terrain Camera, with an 18-inch focal length, provided the highest resolution orbital photographs obtained from civilian satellites up to that time, having ground resolutions in the 20- to 40-meter range (Fig. 8).

The Skylab experience in remote sensing is of interest in relation to the proposed permanently-manned Space Station under study by NASA. As summarized by Garriott (1974) and Lowman (1980a), Skylab's main advantages as a remote sensing platform stemmed from the long mission length, its favorable orbit (50° inclination), and the presence of highly-trained scientist-astronauts. The mission itself was salvaged by the celebrated extravehicular repair of the damaged structure during the first occupation. Several months of succeeding occupations permitted the crews to acquire and record transient phenomena and appearances both on the Earth and on the Sun, demonstrating the inherent versatility of manned observing platforms. The return of photographic film, though possible with unmanned missions, permitted detailed comparisons between conventional photographs and Landsat digital data.

The only significant disadvantages of Skylab for remote sensing were, first, conflicting pointing requirements among various earth-looking and astronomical instruments, a problem re-encountered in recent Shuttle missions; and second, the use of remote sensing instruments with limited independent pointing capability required orientation of the entire cluster, involving fuel and power consumption. These problems, though generally circumvented by the efforts of flight controllers and the Skylab crews, suggest the desirability of flexible instrument mounts or co-orbiting free-flyers if remote sensing research is to be done from space stations.

Landsat

The Landsat Program, widely considered one of the most valuable space applications projects ever carried out, was the synergistic outgrowth of several independent developments in the

1960s. The NASA Natural Resources Program, first directed by P. C. Badgley, began planning remote sensing experiments from space stations in 1963, about the same time the U.S. Geological Survey's Earth Resources Observation Satellite (EROS) Program was organized by W. T. Pecora and W. A. Fischer. The sudden success of the Mercury and Gemini terrain photography aroused interest in geologic orbital photography, leading to the proposal by the U.S.G.S. in 1966 for a dedicated electronic observation satellite, also to be called EROS (Waldrop, 1982). The USGS concept eventually became the NASA Earth Resources Technology Satellite (ERTS), later re-named Landsat, the first of which was placed in a circular 917-km orbit on July 23, 1972.

The first three Landsats (Table 1) were based on the Nimbus meteorological satellite design (Short and others, 1976). The Multispectral Scanner (MSS) images are by far the most widely used at this writing, and have been seen by most geologists. Landsat D (Landsats 4 and 5 after launch) was a completely new design and will be discussed separately.

Landsat has more than met its original objectives, the imagery having been successfully applied to almost all sciences concerned with the Earth's surface. Geology-related applications (Williams and Carter, 1976) include cartography, water resources monitoring, glaciology, land-use mapping and planning, environmental monitoring, conservation, and oceanography. Specifically geologic applications fall into three main categories: structural investigations, lithologic mapping, and geologic education.

Structural investigations were identified at an early stage (Lowman, 1964, 1967) as the most promising geologic use for space photography. Orbital altitudes provide synoptic coverage (Fig. 9) that permits first-hand study of regional structure that is impractical with air photos or mosaics thereof (Gold, 1980), and the global nature of satellite orbits provides access to areas anywhere on Earth (subject to cloud cover, daylight, and latitude constraints). These advantages have been demonstrated by Molnar and Tapponnier (1975), Ni and York (1978), and others who have revolutionized knowledge of the tectonics of western China and adjacent regions that had been almost totally unknown. Using Landsat pictures, they have constructed a generally-accepted explanation for the intense tectonism of this area (Fig. 10) as resulting from the northward movement of peninsular India and the consequent lateral movement of the crust in China.

The study of lineaments has understandably been greatly assisted by Landsat. Lineament investigations, stimulated initially by Hobbs (1911), were eclipsed in the late 1960s by plate tectonics, but interest revived when Landsat imagery became available (Williams and others, 1983). Many papers on the subject have been published in proceedings of several Basement Tectonics Symposia and elsewhere (Hodgson, 1977; Kutina and Carter, 1977). Nur (1982) has concluded, on the basis of Landsat pictures and other evidence, that the pervasive fracturing of basement areas is the result of tensile failure rather than lateral shear. A study of fractures shown on 60 Landsat pictures of the Cana-

Figure 5. Apollo 9 photograph showing Peninsular Ranges, California. Original picture taken with 80mm lens on B-W infrared film sensitive to reflected IR (Lowman, 1969a), as part of S065 experiment. NASA photo number AS9-26C-3798.

dian Shield (Fig. 11a) by Lowman and others (in preparation) arrived at a similar conclusion for the majority of lineaments, finding in addition that there was no coherent pattern that could represent the long-sought "regmatic" shear system.

Lithologic mapping, or, more generally, determination of surface composition, from space proved to be much more difficult than structure mapping. Unlike stars, which radiate energetically on all wavelengths, planetary surfaces are relatively inert, and remote sensing is largely restricted to reflected radiation in the visible and near-visible range, including the thermal infrared (Lyon, 1977). However, Landsat data lend themselves well to digital processing techniques, and substantial progress has been made in identification of minerals, rocks, and rock alteration products. As summarized by Goetz and others (1983), lithologic remote sensing has been most successful using reflection spectra

in the 1 to 3 micrometer region. Specifically, it has been possible to detect OH-bearing minerals, typical of hydrothermal alteration, although absorption by water in the atmosphere (Fig. 1) is an inherent problem for orbital techniques. Rowan and Wetlaufer (1976) have used band-ratioing and contrast-stretching of MSS data to identify alteration zones in the Goldfield district of Nevada. Blodget and others (1978) used similar techniques to discriminate among several bedrock types and gossans overlying massive sulfide deposits in the Arabian highlands (Figs. 12, 13).

A major problem in composition mapping by remote sensing is vegetative cover; in large parts of the world, the ground is simply not visible from the air or from space. However, considerable progress has been made in geobotanical methods, with which compositional anomalies in soil or bedrock can be inferred indirectly through their effects on the spectral properties of vege-

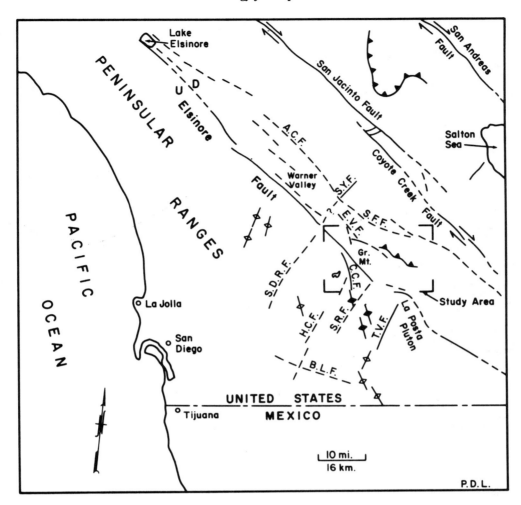

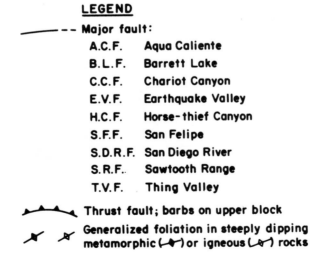

LEGEND

—— - - - **Major fault:**

A.C.F.	Aqua Caliente
B.L.F.	Barrett Lake
C.C.F.	Chariot Canyon
E.V.F.	Earthquake Valley
H.C.F.	Horse-thief Canyon
S.F.F.	San Felipe
S.D.R.F.	San Diego River
S.R.F.	Sawtooth Range
T.V.F.	Thing Valley

Thrust fault; barbs on upper block

Generalized foliation in steeply dipping metamorphic ( ) or igneous ( ) rocks

Figure 6. Geologic sketch map of Fig. 5 (Lowman, 1980).

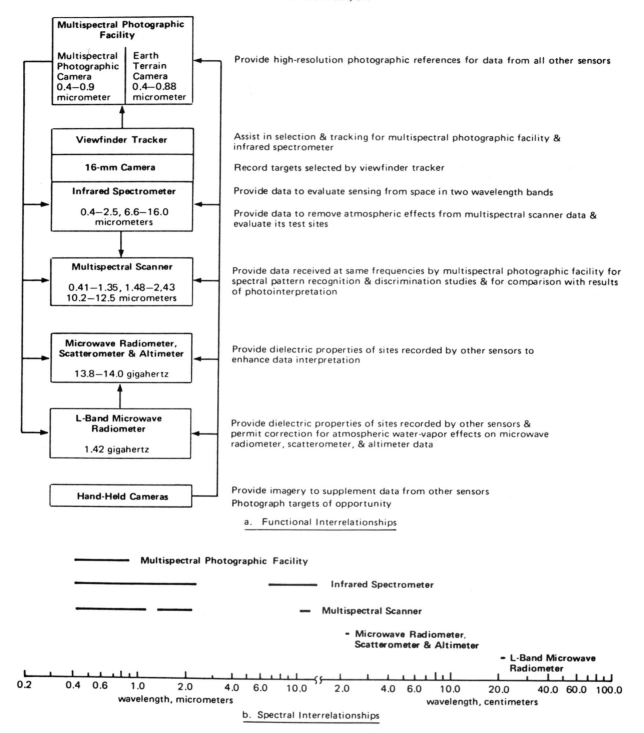

a. Functional Interrelationships

b. Spectral Interrelationships

Figure 7. Remote sensing instruments carried on Skylab (NASA, 1974).

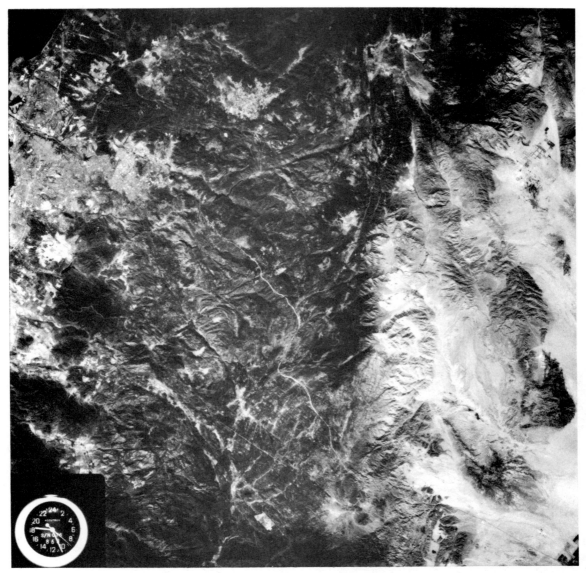

Figure 8. Skylab Earth Terrain Camera photograph showing Peninsular Ranges, California. Original picture taken with 18-inch lens on 5-inch SO-242 film using image motion compensation from 500 km altitude (NASA, 1974).

tation (Goetz and others, 1983; Lyon, 1975). In a field investigation designed to support Landsat D interpretation, Labovitz and others (1983) found that Bands 3 and 5 (Fig. 17) were sensitive to variations in the reflectance of oak leaves caused in turn by concentrations of Pb and Cu in the soil.

A third field of application for Landsat imagery lies in geologic education. Virtually all recent introductory texts use Landsat pictures at least for illustrating regional structures and landforms, and for teaching global tectonics. A global tectonic activity map (Fig. 14), prepared with the aid of orbital imagery, has proved useful both by itself and in conjunction with Landsat slides, presented as an "orbital field trip" (Lowman, 1982). For an undergraduate structural geology course, black-and-white offset prints of Landsat scenes and mosaics (Fig. 15) have been used by the writer at the University of Maryland as laboratory exercises and for examinations. When Landsat and similar orbital pictures are used for purposes beyond mere textbook decoration and lecture entertainment, geology students should develop a grasp of global geology previously acquired only on the verge of retirement.

The discussion up to this point has been based only on data from the first three Landsats, which were basically modified weather satellites. The first civilian satellite designed and launched specifically for visual-range remote sensing of land surfaces was Landsat D (Landsat 4 after its launch in July, 1982). Landsat 4, followed in March 1984 by the identical Landsat 5

TABLE 1. THE LANDSAT SYSTEM*

Coverage of nearly the entire land surface and selected ocean areas of the Earth.
Spacecraft altitude: 917 km (570 nautical miles).
14 revolutions/day; repetitive 18-day cycle; multiseasonal.
Sun-synchronous near polar orbit; fixed time (ca. 9:30 a.m. local Sun time at Equator) of pass
   over imaged scene.
185 km x 185 km (115 statute mile x 115 statute mile) ground scene in each image; orthographic;
   standard image scale in 9-in format: 1:1,000,000.
Stereo sidelap viewing: 85% near polar to 14% equatorial.
Effective ground resolution of image: 79 m (260 ft); 0.45 hectare or 1.1 acre.
Sensor Systems:

Multispectral Scanner (MSS)	Return Beam Vidicon (RBV)
Band 4: 0.5-0.6 m (green)	Band 1: 0.48-0.57 m (green)
5: 0.6-0.7 m (red)	2: 0.58-0.68 m (red)
6: 0.7-0.8 m (near IR)	3: 0.69-0.83 m (IR)
7: 0.8-1.1 m (near IR)	
8: 10.4-12.6 m (thermal IR)	(RBV on Landsat 3: 0.50-0.75 m)
(Landsat-3 only)	

Onboard digitization of data; recording on tape of data collected when the satellite is beyond
   the line-of-sight of any Landsat Ground Receiving Station.
Principal data products:
   Black-and white and color prints and transparencies.
   Computer-compatible tapes (CCT's); 7 and 9 track; 800 and 1600 bits per inch (bpi).
   Computer-processed data bases:
      -Statistical evaluation of radiometric parameters.
      -Density-sliced images or printouts.
      -Contrast-stretched images.
      -Edge-enhanced (band pass filtered) images.
      -Band ratio images or printouts.
      -Classification (thematic) printouts or images ("maps").

---

*Landsat-1 launched on July 23, 1972, and ceased operation on January 6, 1978; Landsat-2
launched on January 22, 1975, stopped on January 22, 1980, but resumed operation on May 27,
1980; Landsat-3 launched on March 5, 1978, MSS turned off on December 17, 1980, reactivated
April 13, 1981.

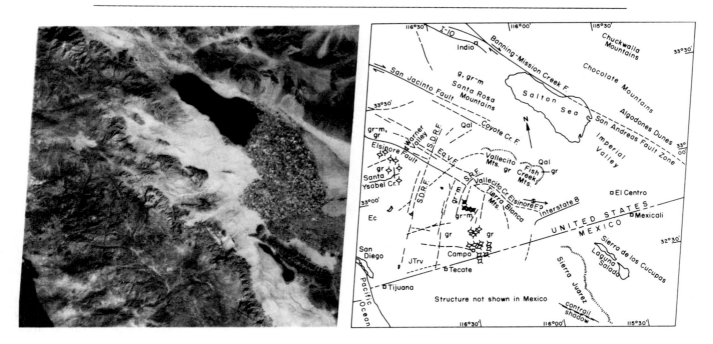

STRUCTURE SKETCH MAP
Peninsular Ranges, San Diego County, California
ERTS-1 Image 1106-17504 (6 Nov. 72)

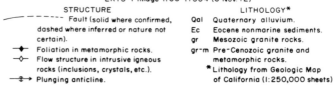

STRUCTURE	LITHOLOGY*
⎯ ⎯ ⎯ ⎯ Fault (solid where confirmed, dashed where inferred or nature not certain).	Qal Quaternary alluvium.
	Ec Eocene nonmarine sediments.
	gr Mesozoic granite rocks.
⟶ Foliation in metamorphic rocks.	gr-m Pre-Cenozoic granite and
⟶ Flow structure in intrusive igneous rocks (inclusions, crystals, etc.).	metamorphic rocks.
⟶ Plunging anticline.	*Lithology from Geologic Map of California (1:250,000 sheets)

Figure 9. Landsat Multispectral Scanner (MSS) picture of Peninsular Ranges, California, and index
map.

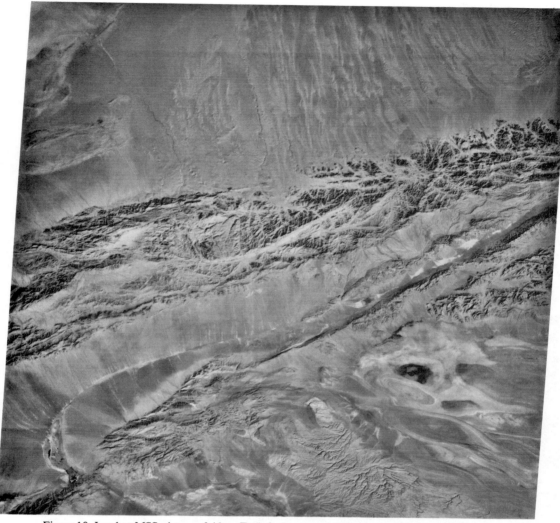

Figure 10. Landsat MSS picture of Altyn Tagh fault, western China: active left-lateral strike-slip fault with estimated 400 km offset (Molnar and Tapponnier, 1975).

(Fig. 16), was an enormous step forward in orbital remote sensing. Operationally, the Landsat spacecraft was but one element in a complex space network that included other satellite systems for navigation, tracking, and data relay, demonstrating the synergistic effect of progress in various areas of space applications. The spacecraft itself was a version of the Multimission Modular Spacecraft (MMS), the first designed for retrieval or repair in orbit by the Shuttle, as actually done in 1984 for the Solar Maximum Mission satellite. The sensors carried include the same Multispectral Scanner (Table 1 and Fig. 17) used on previous Landsats and the new Thematic Mapper (TM). The Thematic Mapper was a major improvement on the highly successful MSS, having 7 spectral bands with higher spectral resolution. The ground resolution (the instantaneous field of view corresponding to one pixel) was 30 meters, compared to 79 meters for the MSS.

Landsats 4 and 5 were also important as the beginning of operational orbital remote sensing of land areas, unlike earlier Landsats, which were labeled experimental. The new status was formalized by transfer of Landsat from NASA to the National Oceanic and Atmospheric Administration after initial checkout in orbit, with the USGS Eros Data Center retaining responsibility for dissemination of the data. After several months of successful operation, Landsat 4 developed problems with the power supply, and Landsat 5 was therefore launched on March 1, 1984. Planning was immediately begun to lower the orbit of Landsat 4 with onboard thrusters to permit retrieval or repair by the Shuttle.

Only preliminary results from Landsat 4 and 5 investigations are available at this writing. The high quality of the Landsat 4 pictures is shown by Figures 18 and 19. A preliminary study by Abrams and others (1983) using an airborne Landsat 4 TM simulator over porphyry copper deposits in Arizona clearly showed the value of the increased spectral coverage and ground resolution of Landsat 4 over the MSS for detecting both lineaments and surface composition. An early evaluation of actual

Figure 11a. Landsat MSS picture of Georgian Bay and adjacent Ontario segment of Canadian Shield.
Low sun elevation (28°) accentuates topography. See 11b for details.

Landsat 4 imagery has been made by Bernstein and others (1984), covering the New Almaden mercury district of California, and showing that the TM data permitted rendition of all mapped faults and several unmapped ones in the area. In addition, geobotanical anomalies possibly indicative of mercury mineralization were visible in principal components transformation images. This and other early studies indicated that Landsat 4 had double the capabilities of previous Landsats (Salomonson and Koffler, 1983).

What can be said about the impact of Landsat on geology so far? The question can be answered with respect to applied geology and to geologic research, education having been discussed previously. It is clearly in applied geology—oil, gas, and mineral exploration, engineering geology, and regional mapping—that Landsat has had its greatest effect. Because of the many stages involved between initial exploration and marketing, it is hard to attribute discovery of oil pools or mineral deposits specifically to Landsat. Like most regional exploration techniques, Landsat imagery serves to focus more precise surveys, themselves preliminary to evaluation and extraction. However, as such a focussing technique, Landsat has become a routinely-applied tool in oil, mineral, and ground-water exploration (Halbouty, 1976; Lyon, 1977; Salas, 1977), and recommendations of the mineral industries, represented by the Geosat Committee, have been taken into account in space mission planning since the mid-1970s (Henderson and Swann 1976). It is safe to say that Landsat has been extraordinarily successful in the general field of applied geology, to say nothing of its uses in agriculture, cartography, oceanography, and other areas.

In terrestrial geologic research, the situation is not so clear.

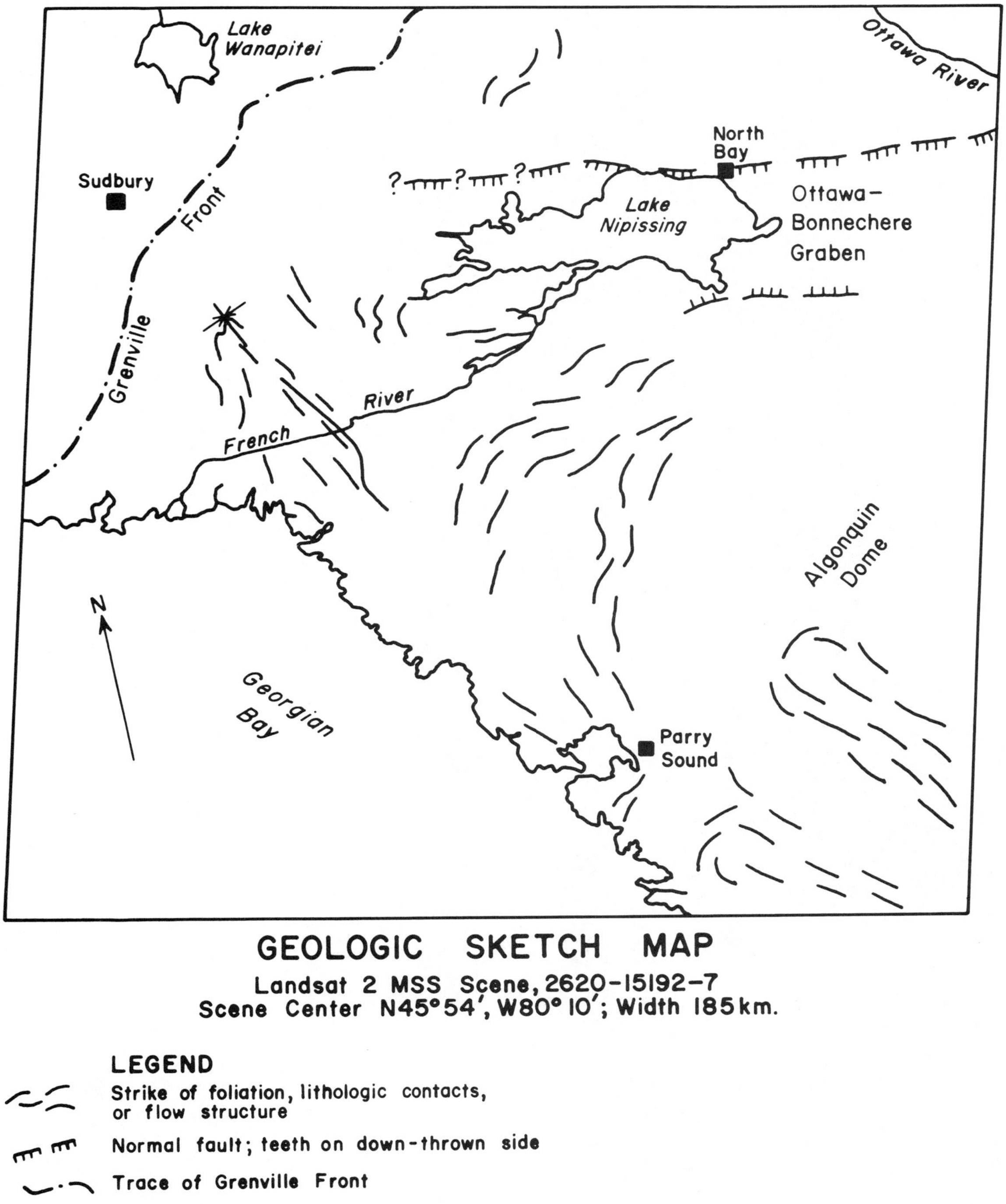
Lake Wanapitei
Ottawa River
North Bay
Sudbury
Grenville Front
Lake Nipissing
Ottawa–Bonnechere Graben
French River
Algonquin Dome
N
Georgian Bay
Parry Sound
GEOLOGIC SKETCH MAP
Landsat 2 MSS Scene, 2620-15192-7
Scene Center N45°54′, W80° 10′; Width 185 km.
LEGEND
Strike of foliation, lithologic contacts, or flow structure
Normal fault; teeth on down-thrown side
Trace of Grenville Front

Figure 11b. Sketch map of Figure 11a.

The example cited earlier, Landsat studies of tectonics in western China, is nearly the only case in which Landsat imagery has had a primary impact on knowledge of a given area or topic. Most other examples (e.g., Mohr, 1974; Abdel-Gawad, 1971; Lowman, 1980b) not only from Landsat but from other types of orbital photography are largely refinements or expansions of previous knowledge. A quick examination of current geologic journals will show only a handful of research papers inspired by or based largely on Landsat data. Even in areas such as California, ideally-suited for remote sensing and plagued by innumerable geologic problems, only a few geologists have published papers using Landsat or other orbital imagery. The reasons for this curious and unsatisfying situation are not at all clear, and seem worth studying.

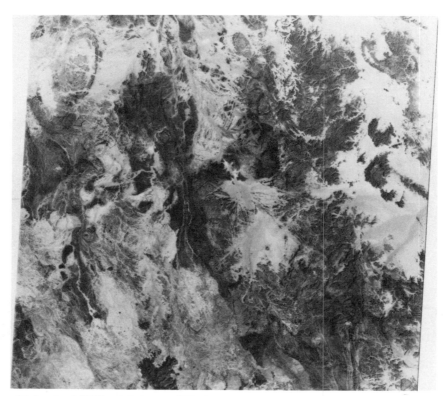

Figure 12. Landsat MSS Band 6 picture of Wadi Wassat area, Saudi Arabia (Blodget and others, 1978). Bedrock chiefly Precambrian igneous and metamorphic rock. Compare with Fig. 13.

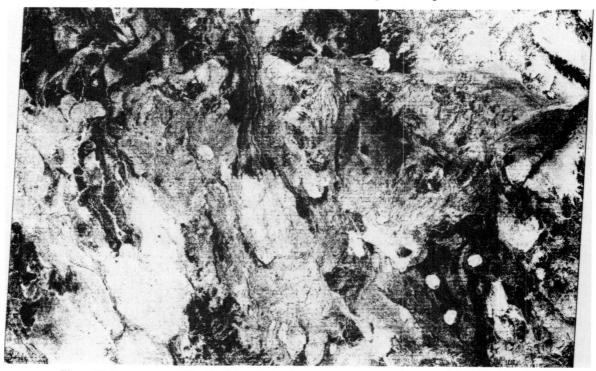

Figure 13. Lower part of Fig. 12: contrast-enhanced, ratio color composite version. Original in color. Picture is a composite of three band-ratio images, chosen for maximum rock discrimination. Three bright circular features at lower right are peralkalic granite intrusions. See Blodget and others (1978) for detailed description.

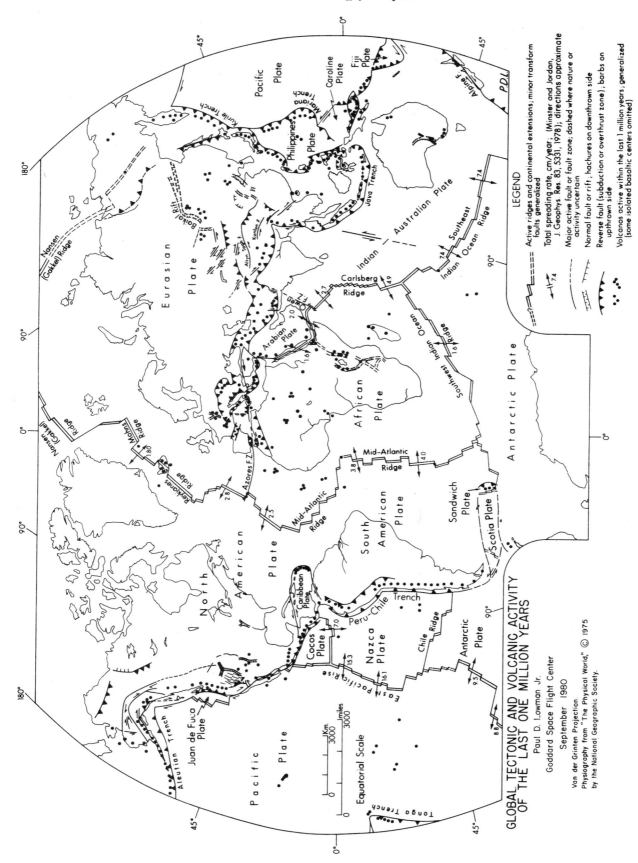

GLOBAL TECTONIC AND VOLCANIC ACTIVITY
OF THE LAST ONE MILLION YEARS

Paul D. Lowman Jr.

Goddard Space Flight Center
September 1980

Van der Grinten Projection.
Physiography from "The Physical World," © 1975
by the National Geographic Society.

LEGEND

Active ridges and continental extensions; minor transform
faults generalized

Total spreading rate, cm/year; (Minster and Jordan,
J. Geophys. Res. 83, 5331, 1978); directions approximate
or generalized

Major active fault or fault zone; dashed where nature or
activity uncertain

Normal fault or rift, hachures on downthrown side

Reverse fault (subduction or overthrust zone), barbs on
upthrown side

Volcanos active within the last 1 million years; generalized
(some isolated basaltic centers omitted)

Figure 14. Map showing tectonic and volcanic features active within the last one million years. Drawn
partly from satellite imagery (Lowman, 1981).

Figure 15. Landsat mosaic of S. W. United States (Band 5) with index map. Fault abbreviations: S.J.F., San Jacinto; S.G.F., San Gabriel; W.F., Whittier; S.N.F., Sierra Nevada; O.V.F.Z., Owens Valley; L.V.F.Z., Las Vegas; F.C.F., Furnace Creek; K.C.F., Kern Canyon; W.W.F., White Wolf; S.Y.F., Santa Ynez; A.B.F., Agua Blanca; S.M.F., San Miguel; H.F., Hayward.

LANDSAT D FLIGHT SEGMENT

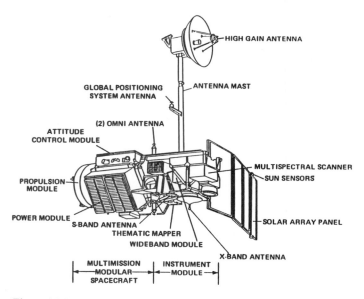

Figure 16. Landsat-D components; diagram applicable to both Landsat 4 and 5.

LANDSAT-D

Thematic Mapper Spectral and Radiometric Characteristics

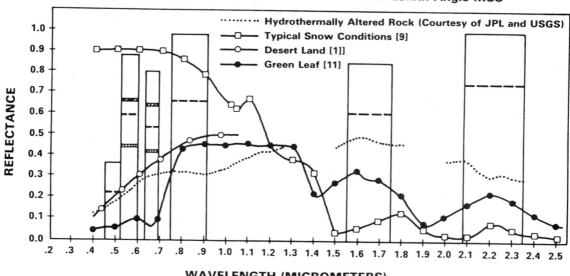

Figure 17. Comparison of Landsat-D Thematic Mapper spectral bands with various natural spectra. NEΔP is spectral sensitivity, literally noise equivalent reflectance difference: the reflectance difference between two objects producing a signal difference equal to the noise. From Short (1982); reproduced with permission.

Band	Wavelength (μm)	NEΔp	Basic Primary Rationale for Vegetation
TM 1	0.45–0.52	0.008	Sensitivity to chlorophyll, and carotinoid concentrations
TM 2	0.52–0.60	0.005	Slight sensitivity to chlorophyll plus green region characteristics
TM 3	0.62–0.69	0.005	Sensitivity to chlorophyll
TM 4	0.76–0.90	0.005	Sensitivity to vegetational density or biomass
TM 5	1.55–1.75	0.01	Sensitivity to water in plant leaves
TM 6	2.08–2.35	0.024	Sensitivity to water in plant leaves
TM 7	10.4–12.5	0.5 K	Thermal properties

Figure 18. Landsat 4 Thematic Mapper picture of San Francisco Bay area, Dec. 31, 1982. Spacecraft altitude about 710 km. Original in color.

Heat Capacity Mapping Mission

The thermal (or emitted) infrared region of the spectrum, 3 to 14 micrometers, has been used by weather satellites since the early 1960s, but the first civilian satellite intended for thermal sensing of land areas was the Heat Capacity Mapping Mission (HCMM), also called the Applications Explorer Mission (AEM-1). This relatively low-cost spacecraft, using a leftover Nimbus 5 radiometer and the solid-fuel Scout launch vehicle, was put in orbit on April 26, 1978. Operating until September 30, 1980, well past its design lifetime of one year, the HCMM satellite transmitted over 25,000 images. A comprehensive report on the mission has been compiled by Short and Stuart (1982), from which the following summary is taken. For background, the chapter on thermal IR remote sensing by Sabins (1978) will be helpful.

The HCMM sensor was a scanning radiometer (Table 2) sensitive to the reflected (0.5 to 1.1 micrometer) and thermal (10.5 to 12.5 micrometer) infrared. In this section, following usage in the HCMM report, reflected infrared imagery is labeled VIS because it covers part of the visible spectrum, and thermal infrared imagery as IR. Samples of Day-VIS, Day-IR, and Night-IR are given in Figures 20, 21, and 22. Investigations summarized by Short and Stuart also used computer-derived images showing temperature differences between image pairs (ΔT) and apparent thermal inertia (ATI). The ATI images express the property de-

Figure 19. Landsat 4 Thematic Mapper picture of Salton Sea and Peninsular Ranges. Black-and-white print of 2, 3, 4 color composite, shown for comparison of ground resolution.

TABLE 2. HEAT CAPACITY MAPPING RADIOMETER CHARACTERISTICS

Orbital Altitude	620 km
Instantaneous Field of View	Infrared: 0.6 x 0.6 km at nadir
	Visible: 0.5 x 0.5 km at nadir
Scan angle	60° (full angle)
Scan rate	.14 revolutions per second
Swath width	716 km
Information bandwidth	53 KHz per channel
Spectral range	Thermal channel: 10.5 to 12.5 micrometers
	Visible channel: 0.55 to 1.1 micrometers
Dynamic range	0 to 100% albedo
Telescope optics diameter	20 cm

From Short and Stuart, 1982.

scribed by

$$P = \sqrt{K\rho c}$$

where P is thermal inertia, K thermal conductivity, ρ density of the material, and c the specific heat of the material (Kahle, 1980).

Thermal inertia is a measure of the resistance of a substance to temperature change. A rock with high thermal inertia, a dense sandstone for example, will respond more slowly to external temperature changes than one with low thermal inertia, such as a pumice.

The HCMM spacecraft was designed primarily to measure

Figure 20. Heat Capacity Mapping Mission (HCMM) Day VIS image of S.W. United States and N.W. Mexico, taken Oct. 24, 1979.

the temperature of the Earth's surface. The temperature at any one point on the ground will be affected by many factors, such as weather, topography, time of day, moisture content, density of rock, and others. In addition, the thermal radiation received by the radiometer will be affected by the temperature of the atmosphere through which it passes (Fig. 21). Interpretation of HCMM imagery is thus complex, but the potential applications are correspondingly numerous, and the imagery was used in geology, agriculture, hydrology, and urban environment studies.

Geologic applications unique to HCMM can be placed in several interrelated categories. Perhaps the most important is rock discrimination: the separation, though not necessarily the identification, of rock units. Rock discrimination with thermal IR data (Kahle, 1980) is an indirect technique based on the apparent thermal inertia, rather than the direct use of spectral signatures. In her HCMM report (Short and Stuart, 1982), Kahle demonstrated discrimination among such diverse rock types as limestone, basalt, rhyolite, and alluvium. Other HCMM investigators, including K. Watson, J. Y. Scanvic, R.J.P. Lyon, T. W. Offield, and

M. M. Cole reported comparable results (all in Short and Stuart, 1982), and it appears that the basic feasibility of this technique with space data has been demonstrated. (For a comparable study using airborne IR data, see Rowan and others, 1970).

Several structural investigations were carried out with HCMM data, most of them indirect approaches based on the thermal properties of the terrain. As reported in Short and Stuart (1982), J. Y. Scanvic mapped lineaments in France using Day-Ir and Night-IR images, suggesting that most of them were large fracture zones that had concentrated ground water and hence had different thermal properties. R. Nuñez de las Cuevas and K. Watson similarly mapped lineaments in, respectively, Spain and Wyoming.

The nature of the HCMM sensors and its orbit provided an unusually large swath width (716 km, vs. 185 km for Landsats 1 through 3). Although such coverage is not inherently unique to HCMM, it produced pictures of unusual geologic interest because of the large physiographic regions shown with constant sun illumination. For example, Figure 20 covers a large area of great

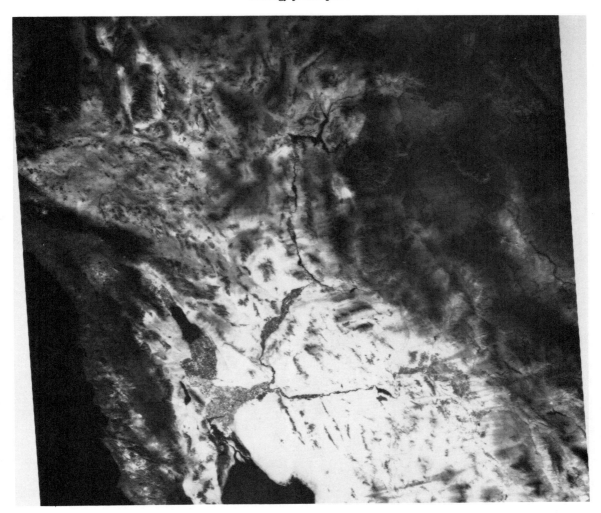

Figure 21. HCMM Day IR image taken at same time as Fig. 20. Nature of east-trending dark streaks (cooler) not known; presumably associated with wind patterns since they are not visible on the Day VIS (Fig. 20). Note apparent disappearance of Pinacate volcanic field, Sonora (bottom center of Fig. 20), presumably indicating lack of thermal contrast with surroundings.

tectonic importance as the junction of the San Andreas fault system and the Basin and Range Province, giving an objective view with perspective unmatched by individual Landsat scenes and imperfectly by Landsat mosaics.

Seasat

The geologic value of side-looking airborne radar has been well-known for some time (MacDonald, 1980) and there was considerable interest in the potential value of *orbital* radar during the 1970s, especially after the success of Landsat (Short and Lowman, 1973). In June 1978, the Jet Propulsion Laboratory launched Seasat, carrying the first American scientific synthetic aperture radar (SAR) to be put in earth orbit. Seasat was intended primarily for ocean surveillance: sea state, sea ice, and the marine geoid (by radar altimetry), using 23.5 cm (L-band) imaging radar as well as other sensors. During its 3.5-month operating

lifetime, cut short by a power failure, Seasat produced a large amount of valuable radar imagery over the oceans. However, the land imagery proved almost as interesting, and provided geologists with their first experience with orbital radar. The best terrain images were collected by Ford and others (1980); an example is presented in Figure 23 with a Landsat picture for comparison (Fig. 24).

The advantages and disadvantages of orbital radar in general will be discussed after the radar experiments carried on Shuttle missions have been discussed. The Seasat experience can be summarized as a valuable rehearsal for and stimulus to the later radar-carrying orbital missions.

Shuttle Remote Sensing Experiments

The second flight of the space shuttle Columbia, in November 1981, carried a series of remote-sensing experiments la-

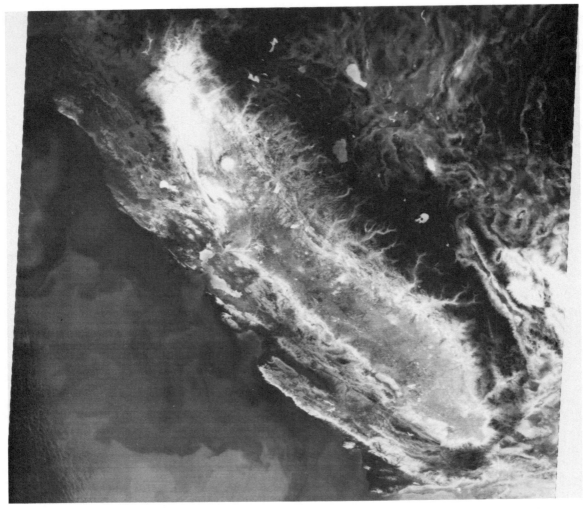

Figure 22. HCMM Night IR image of California, taken July 17, 1978. Note strong topographic control of temperature; high areas such as Sierra Nevada generally much cooler (darker).

beled the OSTA-1 payload, after the NASA Office of Space and Terrestrial Applications (Settle and Taranik, 1982). The spacecraft spent 54 hours in a 262 km circular orbit inclined 38° to the equator, before having to make an early landing because of electrical and weather problems. Despite the short duration of the mission, the OSTA payload produced much useful data, which will be summarized briefly here; more detailed accounts were published in *Science* for December 3, 1982 (v. 218, no. 4576). A second remote sensing payload, OSTA-3, was carried by the shuttle Challenger in October 1984, with a 57° orbital inclination and an 8-day mission length. The data from this mission are being analyzed at this writing.

In terms of quantity and novelty of results, the most interesting remote sensing experiment on OSTA-1 was the shuttle imaging radar, SIR-A, an L-band synthetic aperture imaging radar whose design was descended from the Seasat radar (Table 3). The SIR-A operated for 8 hours, covering about 10 million square kilometers that included all major terrain types found within the latitude limits of the orbit (Fig. 25).

As reported by Elachi and others (1982), the SIR-A imagery was of high quality and proved to be of great potential value in a number of geologic applications. Structure mapping appeared to be well-suited to the SIR-A characteristics, primarily because of its superior rendition of topography, especially in well-vegetated areas. However, the viewing geometry has a strong effect on structural interpretations. Imagery from the SIR-B acquired in 1984 over a test site in southern Ontario (Lowman and others, 1984), similar to the SIR-A data, showed that at least under some conditions linear features such as ridges or valleys parallel to the illumination direction are nearly invisible. Although compensated to some extent by correspondingly good rendition of features normal to the illumination, the biasing of radar imagery by viewing geometry is a serious problem that must be taken into account in experiment planning and data interpretation.

Gross lithology could be identified on SIR-A imagery to the extent that it was expressed topographically. Sabins (1983) demonstrated this capability for heavily-forested areas of Indonesia,

Figure 23. Seasat radar image of Los Angeles area. Note strong directional biasing; bright rectangular areas are cities of Burbank (top) and Anaheim (lower right center), highlighted because one set of streets is normal to illumination direction (from left to right). Note also strong layover in San Gabriel Mountains (upper right); compare with Fig. 24.

pointing out that such interpretation depends *not* on ability of the L-band radiation to penetrate vegetation but on the fact that the treetops conform roughly to the underlying topography.

One of the most striking results of the SIR-A experiment was the ability of the radar to penetrate several meters of dry sand in the eastern Sahara (Fig. 26), revealing buried Pleistocene to Oligocene drainage that was completely invisible on Landsat pictures (McCauley and others, 1982). This phenomenon, reminiscent of the deep penetration on 60 meter radar from Apollo 17 in the Moon (to be discussed), is due largely to the low electrical conductivity of dry desert sand. Field work performed to check interpretations uncovered innumerable Stone Age artifacts along

former watercourses, demonstrating an archaeological use for orbital remote sensing. In effect, a spaceship has been used to find stone axes some 200,000 years old—certainly one of the most curious by-products of the Space Age.

The SIR-A and Seasat results have done much to demonstrate the geologic value of orbital radar imagery. A major advantage is that common to all radar, namely its ability to operate regardless of cloud cover and daylight. The pioneering geologic uses of radar by MacDonald (1969) in Panama and more recently by Sabins (1983) in Indonesia involved areas for which coverage by aerial photography or Landsat was essentially impossible because of persistent cloud cover. However, imaging radar is

Figure 24. Landsat MSS picture of Los Angeles area for comparison with Fig. 23.

not simply cloud-penetrating photography. Radar reflections can in principle provide information on wavelength-scale roughness, electrical properties, and topography not available from visual range sensors. Sorting out the various factors affecting radar imaging will take much time and experience, but it seems clear that orbital radar, used with other techniques, will be a valuable geologic tool.

An improved imaging radar, the SIR-B, was flown in 1984, as mentioned previously. It differed from SIR-A (Table 3) in having an antenna 10.7 m long (compared to 9.4 m for SIR-A), and more important, in that the antenna could be adjusted in flight to produce different look angles without moving the entire spacecraft. The 1984 mission (41G) encountered problems with the Ku band antenna and the TDRS, which greatly reduced the amount of data collected. However, some imagery returned was of excellent quality (Fig. 27) and it is clear that the SIR-B system itself was basically successful.

A second geologic experiment carried as part of the OSTA-1 payload in 1981 was the shuttle multispectral infrared radiometer (SMIRR), designed to record reflected radiation in 10 spectral bands between 0.4 and 2.5 micrometers (Goetz and others, 1982). The data were analyzed with the aid of laboratory measurements of mineral reflectance, specifically calcite, kaolinite, and montmorillonite, as well as simultaneous 16mm photographs of the target areas. The initial interpretation of flight data was concentrated on a section of the Egyptian desert for which recent geologic maps and laboratory reflectance data were available. It was possible to identify specific areas of carbonate, kaolinite, and possibly montmorillonite, the first such direct identification (as distinguished from differentiation) of a mineral other than limonite with orbital data. The area studied was admittedly ideal, in being unvegetated (the reason it was chosen), but the results were nevertheless encouraging for the future of orbital compositional mapping. Another study with SMIRR data

Figure 25. SIR-A image of California coastline between Point Conception (lower right) and Ventura (top); illumination from right to left (south to north). Bright dots in ocean at right are oil rigs in Santa Barbara Channel. See Ford, Cimino, and Elachi (1983) for discussion.

TABLE 3. SEASAT SAR, SIR-A, AND SIR-B CHARACTERISTICS

	Seasat	SIR-A	SIR-B
Orbit			
Altitude, km	795	259	225
Inclination to equator	108°	38°	57°
Radar			
Frequency, GHz	1.28	1.28	1.28
Wavelength, cm	23.5	23.5	23.5
Bandwidth, MHz	19	6	12
Transmitted peak power, W	1000	1000	1000
Polarization	HH	HH	HH
Antenna			
Dimensions, m	10.7 x 2.2	9.4 x 2.2	10.7 x 2.2
Look angle (from vertical), deg.	20 ± 3	47 ± 3	Variable: 15 to 60
Swath width, km	100	50	30 to 60
Resolution, m	25 x 25	40 x 40	25, azimuth 57 to 17, range
Data recording	ground station, digital	onboard, optical	onboard, optical and digital; relay through tracking and data relay satellite, digital

was carried out by Baird (1984), who used data from the OSTA-1 mission collected over Baja California and the Eastern Desert of Egypt. Baird found it possible to infer iron contents of the target areas consistent with mapped lithologies, and pointed out that wide field of view (100 m wide) of the SMIRR from orbital altitudes was an advantage geologically in that it permitted averaging local composition variability of the bedrock.

The OSTA-3 payload included an instrument being used in space for the first time, the Large Format Camera (LFC), whose characteristics are summarized in Table 4. The LFC was the first camera designed and built in the U.S. specifically for topographic mapping from space by civilian agencies, although innumerable map corrections had been made earlier on the basis of previous space photography and Landsat imagery. Several hundred pictures were taken by the LFC on the 41G mission, one of which is reproduced in Figure 28. The expected geologic value of LFC pictures stems not only from their high resolution but also the stereo coverage provided.

Both OSTA-1 and OSTA-3 included a remote sensing experiment that should be a valuable supporting device for satellites such as Landsat: the Feature Identification and Location Experiment (FILE), carried on the OSTA pallet (Fig. 29) in the shuttle cargo bay. The FILE consists basically of two solid-state low-resolution television cameras, one sensitive to the 0.65 micrometer part of the spectrum and the other to the 0.85 micrometer part. The relative brightness in these two bands is reasonably characteristic of clouds, snow, ice, water, bare land, and vegetation (Sivertson and others, 1982), and can be combined with other data using a simple algorithm to decide if a given scene is largely cloud-covered or not. This technique, used with some

success on the OSTA-1 mission, should make it possible to cut down the amount of unusable cloud-covered imagery transmitted by remote sensing satellites intended primarily for terrain coverage.

EXTRATERRESTRIAL REMOTE SENSING

This review would be incomplete without a brief summary of techniques used from spacecraft on missions to other planets and satellites. Only those bodies approachable through geologic methods will be discussed; the giant planets such as Jupiter are the province of meteorology, chemistry, and physics.

The Moon

Being close, small, and airless, the Moon has been investigated with a wider variety of orbital remote sensing methods than any other body, including the Earth. Optical imaging techniques, involving television, telemetered film photographs (Fig. 30), and returned film (Table 5), have all been used successfully (Saunders and Mutch, 1980), but since they are similar in principle to those already discussed they will not be discussed here. Earth-based astronomical studies of the Moon have been greatly aided by the newer data and by the *ground truth* provided by samples returned on Apollo missions.

Compositional mapping has been uniquely successful from lunar orbiting spacecraft because the lack of an atmosphere permits methods impossible on bodies such as the Earth. Apollos 15 and 16 carried three complementary experiments, described in the Preliminary Science Reports for those missions (Fig. 31).

Figure 26. SIR-A image of Selima Sand Sheet, southern Egypt/northern Sudan, superimposed on Landsat scene 2307-08011 by USGS Flagstaff Image Processing Facility. SIR-A picture shows paleo-drainage now concealed under windblown sand up to a few meters thick, illustrating ability of L-band radiation to penetrate dry material. See McCauley and others (1982) for discussion.

TABLE 4. LARGE FORMAT CAMERA CHARACTERITSICS

Focal Length	30.5 cm (12 in.)
Field of view	73.7° along track, 41.1° across track
Format	22.9 x 45.7 cm (9 x 18 in.)
Ground resolution	27 m (67 ft.) (maximum, with black-and white film and minus-blue filter, 278 km altitude)
Film capacity	1220 m (4000 ft.), 2400 images
Total weight, incl. film and N_2 for film advance	429.6 kg (947.1 lb)
Optics	Eight element lens, between element filter Minus haze or minus blue filter Image motion compensation by platen movement Cast iron lens cone
Nominal scale of images, 278 km. altitude	1:912,000

Figure 27. SIR-B image of northern Peru, acquired on shuttle mission 41G, Oct. 7, 1984. Center of picture roughly 5°S, 78°W; Maranon River at right. Illumination from left to right; north to upper right (arrow). Area underlain by east-dipping sedimentary rocks. Note flatiron-like features produced by layover.

Figure 28. Large Format Camera picture taken from 370 km altitude during shuttle mission 41G on October 6, 1984, showing section of Himalayas centered on 29°N, 86°W, including Mt. Everest. Area covered measures 278 by 555 km. NASA photo number 84-H-716.

Figure 29. Shuttle pallet undergoing checkout; as shown, pallet is carrying shuttle imaging radar (wrapped in plastic), FILE, and SMIRR (out of view, behind radar antenna).

X-ray fluorescence (Adler and others, 1972), depending on fluorescent X-rays stimulated from the lunar surface by solar X-rays, permitted mapping the relative concentrations of Al, Si, and Mg with a spatial resolution of a few tens of kilometers (Fig. 32). The instrument consisted essentially of proportional counters to measure X-rays coming from the Moon's surface and, simultaneously, the incoming solar X-rays. An experiment of this sort carried out in earth orbit (an impossibility because of the atmosphere) would be relatively uninformative. However, aluminum behaves coherently in a geochemical sense on the Moon because plagioclase is the chief mineral involved in lunar differentiation, and is a major component of the lunar highlands. Thus, the X-ray fluorescence experiment provided the first good evidence for the existence of a

global differentiated crust. Gamma-ray spectroscopy (Arnold and others, 1972), using a scintillation-counting technique, was similarly successful. This experiment measured the lunar surface gamma radiation coming from both "natural" (potassium-40, uranium, and thorium) and cosmic-ray-induced radioactivity. Compositional maps showing relative concentrations of uranium, thorium, potassium-40, Fe, Mg, and Ti were produced along the orbital tracks of the Apollo 15 and 16 missions. Finally, an alpha-particle spectrometer (Gorenstein and Bjorkholm, 1972) detected excess radon over the general area of the crater Aristarchus. With a half-life of only 3.7 days, any radon detected must have come recently from the interior of the Moon. The alpha-particle experiment thus produced strong evidence of recent gas

Figure 30. Lunar Orbiter III wide-angle view of crater Kepler, looking north, taken with 80mm lens. Picture produced by Bimat process; consists of framelets made on 35mm film produced on earth from telemetered scans of Bimat film exposed and developed on spacecraft (Table 5). Each framelet has 17,000 scan lines across width of picture as shown here.

TABLE 5. LUNAR AND PLANETARY REMOTE SENSING TECHNIQUES*

Body	Mission	Camera Design	Camera Type	Lens Focal Length (mm)	Lens Angular Coverage (deg)	Lens Aperture	Image Format (mm)
Moon							
Primary Sources	Lunar Orbiter 4, 5	Eastman readout	70-mm film	80	37.º09x44.^{o}2	f/5.6	55x65
	Apollo 15, 16, 17	Fairchild	127-mm film return	76	74°x74^o	f/4.5	114x114
	Soviet Zond 6, 8		190-mm film return	400	18.º2x5.^{o}4	f/6.3	130x180
Also Important	Apollo 15, 16, 17	Itek pan	127-mm film return	610	10.º7x108.^{o}0	f/3.5	114x1148
	Apollo 8-17	Hasselblad	70-mm film return	various			55x55
	Apollo 14	Hycon	127-mm film return	457	14.º2x14.^{o}2	f/4	114x114
	Lunar Orbiter 1-5	Eastman	70-mm film readout	610	5.º2x20.^{o}4	f/5.6	55x219
	Lunar Orbiter 1-3	Eastman	70-mm film readout	80	37.º09x44.^{o}2	f/5.6	55x65
Mercury	Mariner 10	JPL	Vidicon	1500	0.º4x0.^{o}5	f/8.5	9.6x12.35
Mars	Mariner 9	JPL	Vidicon	52, 500	10.º5x13.^{o}5, 1.ºx1.^{o}4	f/4, f/2.4	9.6x12.5
	Viking 1, 2	JPL	Vidicon	475	1.º5x1.^{o}7	f/3.5	12.5x14.0
Jupiter Satellites	Voyager 1, 2	JPL	Vidicon	200, 1500	3.º2x2.^{o}2, 0.º4x0.^{o}4	f/3.5, f/8.5	11.14x11.14
Saturn	Voyager 1, 2	JPL	Vidicon	200, 1500	3.º2x3.^{o}2, 0.º4x0.^{o}4	f/3.5, f/8.5	11.14x11.14

*Extracted from "Planetary Cartography in the Next Decade (1984-1994), by The Planetary Cartography Working group, NASA SP-475: Washington, National Aeronautics and Space Administration, 77 p.

emission from the Aristarchus area, which had already been of interest because of well-confirmed visual observations of glowing red spots inside the crater (Greenacre, 1965).

Another remote sensing technique that produced valuable geologic data was the Apollo lunar sounder experiment (ALSE), a three-frequency radar experiment generating signals at 5, 15, and 150 MHz (Phillips and others, 1973). These frequencies, corresponding to wavelengths of 60, 20, and 2 meters, can penetrate to several hundred meters in materials of suitable electrical properties. The technique is sometimes called "electromagnetic sensing" or, in this case, radio sounding to distinguish it from imaging radar, operating at wavelengths of a few centimeters in the microwave region. The Apollo 17 experiment was very successful, producing subsurface and surface profiles, images, and information on electrical properties of lunar materials.

Fruitful as these remote sensing investigations were, they were limited in geographic coverage of the Moon by the low orbital inclinations (less than 30) of the Apollo missions. Accordingly, they should be considered only a preview of what might be accompanied by remote sensing from a polar-orbiting spacecraft.

Mercury, having no atmosphere and a Moon-like topography, could in principle be studied by the same techniques used for the Moon. However, it has to date been visited only by Mariner 10, carrying optical instrumentation, and hence will not be discussed further here.

Mars

Thanks to several years of data collected on the surface and from orbit by the Viking missions, we know more about Mars than any other extraterrestrial body except for the Moon. Several orbital remote sensing techniques have been used successfully on various Mars missions, beginning with the low-resolution slow-scan television cameras carried by the Mariner 4 spacecraft in 1965. Successive U.S. missions—Mariners 6, 7, and 9, and Viking Orbiters 1 and 2—have carried increasingly better optical sensors, in each case dual camera arrays for wide-angle and telephoto coverage, equipped with filters permitting reconstruction of color pictures (Hartmann and Raper, 1974; Carr, 1981). An important lesson learned from the first systematic orbital coverage by Mariner 9, is that earlier fly-by missions, especially Mariner 4, produced very misleading impressions of Martian geology because of their limited geographic coverage. The global coverage from later missions revealed a complex and highly-evolved planet (Fig. 33) quite unlike the primitive Moon-like body suggested by the first pictures. Even a brief summary of Martian geology is impossible here; good summaries have been published by Carr (1981), Beatty and others (1982), and Glass (1982).

Of the non-imaging techniques used near Mars, the most interesting from a geologic viewpoint is the infrared spectroscopy experiment (Hanel and others, 1972). The infrared interferometer

Figure 31. In-flight photograph of Apollo 15 Command and Service Module (CSM) Endeavour taken from Lunar Module Falcon by A.M. Worden, August 1971. Picture shows remote sensing instruments in Scientific Instrument Module (exposed sector of SM). CSM was not taking data at this time.

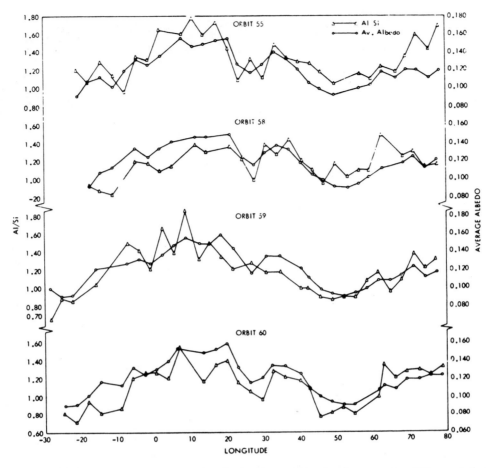

Figure 32. Apollo 16 X-ray fluorescence data for four lunar orbits, showing general positive correlation between Al/Si and optical albedo caused by concentration of plagioclase in lunar highlands. From Lowman (1976); see also Adler and others (1972).

spectrometer (IRIS) was designed to provide spectral measurements of the thermal radiation from Mars in the 6.2 to 50 micrometer region. The instrument was a Michelson interferometer using a beam-splitter to divide incoming radiation into two components, eventually recombined to produce an interferogram from which the radiation intensity as a function of wavelength, i.e. the spectrum, could be determined. Over 2000 spectra were produced during the Mariner 9 mission, providing information on atmospheric composition and temperature. Of particular geologic importance were spectra (Fig. 34) from which the chemical composition of the suspended dust could be roughly estimated. It appeared to have an SiO_2 content of between 50 and 70 percent, implying widespread differentiation of Mars if the dust was representative of surface composition. Later *in situ* analyses by X-ray fluorescence on the Viking Landers suggested that the IRIS estimate was on the high side; nevertheless, the conclusion that Mars is globally differentiated has been generally accepted because of other convergent lines of evidence.

Another infrared experiment, the infrared thermal mapper (IRTM) (Kieffer and others, 1977) carried on the Viking Orbi-

ters, produced immense amounts of information about the thermal properties and behavior of the surface and atmosphere of Mars. The IRTM was a 4-telescope radiometer that produced spectra in six bands between 0.3 and 24 micrometers. Among the geologic findings of the IRTM was the discovery of a negative correlation between albedo and thermal inertia; i.e., bright areas had low thermal inertias. This was interpreted as resulting from the distribution of wind-blown dust on the surface, which is quite consistent with the occurrence of dust storms visible from Earth. It was found possible to calculate characteristic thermal inertia values for various surface geologic units as mapped from visual range images. In addition, inferences as to surface temperature, atmospheric and polar cap composition, and other characteristics of Mars are of great indirect geologic importance.

Being restricted to geologic remote sensing, this summary gives only a hint of the great variety of investigations performed on or near Mars by various spacecraft, ranging from seismology to life detection. Interested readers should consult the *Journal of Geophysical Research* for September 30, 1977 (V. 82, No. 28), a special issue devoted to initial results of the Viking missions.

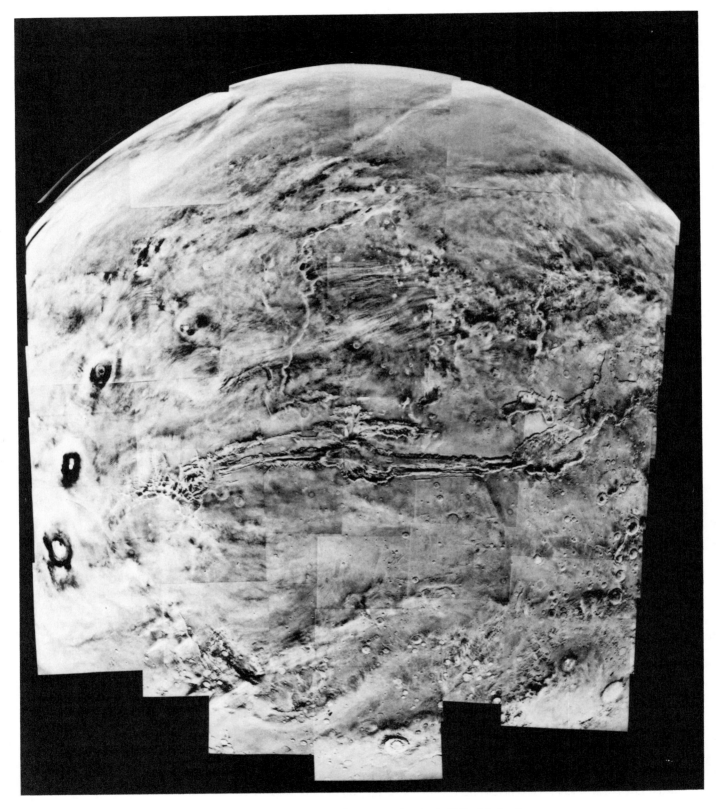

Figure 33. Mosaic of Viking Orbiter pictures of Mars (north at top). Valles Marineris, at center, are a system of interlocking canyons formed primarily by block-faulting. Dimensions are: over 4000 km long, 80 to 100 km wide, up to 4 km deep.

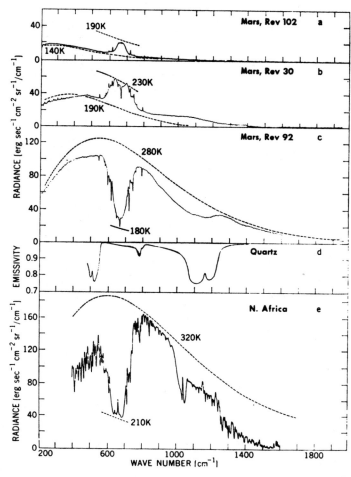

Figure 34. Mariner 9 IRIS thermal emission spectra (Hanel and others, 1972): (a) north polar hood of Mars; (b) south polar hood; (c) midlatitude; (d) SiO_2 laboratory emissivity (e) IRIS spectrum from Nimbus 4 showing suspended dust over Sahara. See Lowman (1976) for discussion.

Venus

Because of its dense cloud cover, Venus has always been the most difficult of the terrestrial planets to study. However, remote sensing from spacecraft, combined with data from earth-based radar and *in situ* images and analyses from Soviet Venera landers, has begun to produce a coherent though still vague picture of the geology of Venus. Mariner 10 and Pioneer Venus, launched in 1973 and 1978, respectively, have been the most fruitful American missions. Good summaries have been published by Fimmel and others (1983) and in special issues of the *Journal of Geophysical Research* (v. 85, no. A13, 1980) and *IEEE Transactions on Geoscience and Remote Sensing* (v. GE-18, no. 1, 1980).

Of the many investigations carried out by Soviet and American missions to Venus, the only one appropriate for this review is the Orbiter Radar Mapper (ORAD) (Pettengill and others, 1980). The ORAD is an S-band radar operating at 1757.0 MHz (17 cm wavelength). Although intended primarily as an altimeter,

it could also be used in an imaging mode when aimed to the side at altitudes under 550 km. When used in the altimetric mode, the ORAD could produce topographic data with a spatial resolution of up to 23 by 7 km at periapsis, the altitude above the surface being subtracted from the spacecraft's orbital radius as determined from tracking data. By combining the altimetric data from many months of orbital operation, a topographic map covering over 93 percent of the Venusian surface with a vertical accuracy of better than 200 m has been produced. When combined with imagery and other data (Masursky and others, 1980), some tentative but fundamental inferences about the geology of Venus have been drawn (Phillips and others, 1981; Head and Solomon, 1981).

First, it apparently does not exhibit the bimodal topography that, on the Earth, corresponds to continents and ocean basins. There are several continent-like elevated areas, but they make up only about 5 percent of the planet's area, and only 15 to 20 percent of the area is in distinct basins. Venus may be what has been called a "one-plate planet," (Head and Solomon, 1981) but there are indirect indications of continuing volcanic and tectonic activity. It is hoped the recently-approved Venus Radar Mapper, an imaging radar mission scheduled for the late 1980s, will settle these questions with its higher spatial resolution.

The Outer Planets and Satellites

Jupiter and Saturn and their satellites were studied by remote sensing techniques from the Pioneer 10 and 11 and Voyager 1 and 2 spacecraft during flyby missions in the 1970s, with spectacular success (Fig. 35). Jupiter and Saturn, having no detectable solid surfaces and consisting largely of hydrogen, will not be covered here. However, the many satellites have proved susceptible to a geologic approach. This brief review will cover only the remote sensing techniques used; the fascinating results are summarized by Beatty and others (1982), Glass (1982), and others. The instruments on Pioneer 10 and 11 were used only for Jupiter and Saturn and hence will also be omitted.

The remote sensing arrays on Voyager 1 and 2 were basically identical, consisting of coaxially-mounted sensors carried on a steerable scan platform; those of geologic importance were the imaging science and infrared experiments. The imaging science subsystem consisted of two slow-scan television cameras, one with a wide-angle 200 mm lens and the other with a 1500 mm lens, bore-sighted to provide nested coverage. Each camera had an eight-position filter wheel, covering a wide spectral range and permitting reconstruction of now-famous color pictures of the Jovian and Saturnian systems (Morrison, 1982). The slow-scan Voyager cameras were much more versatile than those used in commercial television, permitting a wide range of exposure times and transmission in digital form either in real time or more slowly after recording on tape. The theoretical resolution of the narrow-angle camera, i.e., the width covered by a scan-line pair, was about 19m at 1000 km.

The pictures produced by the Voyager television cameras

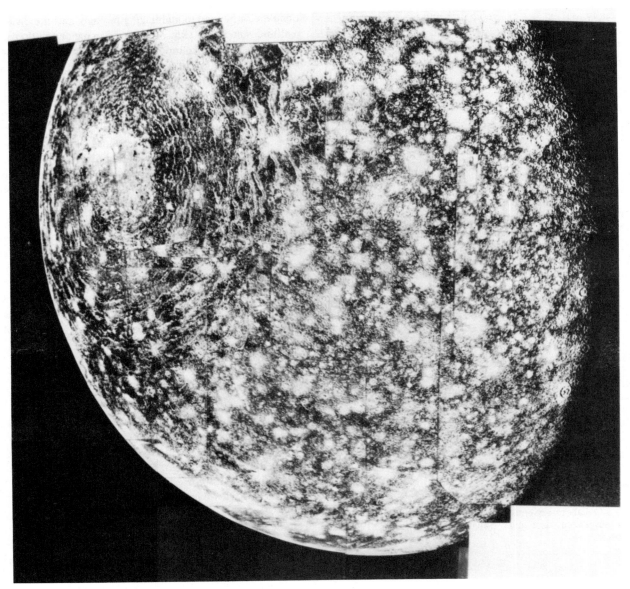

Figure 35. Voyager 1 television mosaic of Callisto taken from about 400,000 km distance; surface resolution about 10 km. Large concentric structure is Valhalla, thought to be a major impact feature (central diameter roughly 600 km) greatly modified by flowage in ice crust and mantle.

are well-known, but it is worth emphasizing that the television data proved usable for establishing photometric properties, control networks, and rotation rates for even small satellites. The dedicated issue of the *Journal of Geophysical Research* for November 1, 1983 (V. 88, No. All) included examples of such applications.

The Voyager spacecraft also carried an infrared spectroscopy and radiometry (IRIS) instrument basically similar to that previously described for Mariner 9. The IRIS experiment produced much valuable information on the thermal and compositional properties of the giant planets, and thermal data on the satellites (Hanel and others, 1979).

SUMMARY AND CONCLUSIONS

The impact of orbital remote sensing on geology is best discussed under two headings: extraterrestrial and terrestrial. It is obvious that remote sensing from lunar and planetary missions has created an entire new field of geology; George Gamow's (1948) whimsically-coined "comparative planetology" is now a thoroughly respectable and fruitful discipline, one touched only briefly in this review. With regard to terrestrial geology, the importance of orbital remote sensing is not so clear.

Landsat imagery, the only type of geologically-useful orbital data so far available on a wide basis, is by now familiar at least superficially to most western geologists, and has become a generally-accepted tool in many areas of applied geology. On the other hand, neither Landsat nor any other sort of orbital remote sensing data has had the impact on geological theory of, for example, the results of the Deep Sea Drilling Project. Will this situation continue? Will orbital surveys eventually have fundamental effects on geologic concepts and theories?

It is the writer's opinion that the answer is a qualified "yes." The qualification is that full appreciation and utilization of orbital remote sensing will have to wait for the next generation of geologists, those still in school in 1984. When today's geology students enter the profession, they will have been exposed to Landsat pictures, orbital radar, and similar ways of seeing the Earth the way today's geologists were exposed to the air photo and the petrographic microscope. This coming generation will have a grasp of world geology and the Earth's place in the solar system that was simply unattainable 20 years ago, and they will be able to use a far wider part of the electromagnetic spectrum than visible light. The "geology" of the 21st century will thus be a far broader and richer field than it is in 1984, and the view from space will have helped make it so.

ACKNOWLEDGMENTS

In a review paper of this sort, the reference list is an acknowledgment of the many people whose work is summarized here. In addition, I thank the following for direct help and encouragement in writing the paper: E. T. Drake, H. W. Blodget, A. P. Colvocoresses, L. Demas, F. Doyle, H. Fuhrmann, A. M. Lohmann, P. Masuoka, L. C. Rowan, V. V. Salmonson, M. Settle, N. M. Short, L. M. Stuart, P. T. Taylor, R. W. Underwood, and A. R. Washington.

REFERENCES CITED

Abdel-Gawad, M., 1971. Wrench movements in the Baluchistan arc and relation to Himalayan-Indian Ocean tectonics: Geological Society of America Bulletin, v. 82, p. 1235–1250.

Abrams, M. J., Brown, D., Lepley, L., and Sadowski, R., 1983, Remote sensing for porphyry copper deposits in southern Arizona: Economic Geology, v. 78, p. 591–604.

Adler, I., Trombka, J., Gerard, G., Lowman, P., Schmadebeck, R., Blodget, H., Eller, E., Yin, L., Lamothe, R., Osswald, G., Gorenstein, P., Bjorkholm, P., Gursky, H., and Harris, B., 1972, Apollo 16 geochemical X-ray fluorescence experiment: preliminary report: Science, v. 177, p. 256–259.

Allum, J.A.E., 1980, Photogeology—early days: Geoscience Canada, v. 7, p. 155–158.

Andre, C. G., and Blodget, H. W., 1984, Thermal infrared satellite data for the study of tectonic features: Geophysics Research Letters, v. 11, p. 983–986.

Arnold, J. R., Peterson, L. E., Wetzger, A. E., and Trombka, J. I., 1972, Gamma-ray spectrometer experiment, *in* Apollo 15 preliminary science report, NASA SP-289: Washington, U.S. Government Printing Office, p. 16–1—16–6.

Baird, A. K., 1984, Granitic terrains viewed remotely by shuttle infrared radiometry: some compositional predictions: Journal of Geophysical Research, v. 89, p. 9439–9447.

Beatty, J. K., O'Leary, B., and Chaikin, A., 1982, The new solar system: Cambridge, Cambridge University Press, 240 p.

Bernstein, R., Lotspeich, J. B., Myers, H. J., Kilsky, H. G., and Lees, R. D., 1984, Analysis and processing of LANDSAT-4 sensor data using advanced image processing techniques and technologies: IEEE Transactions on Geoscience and Remote Sensing, v. GE-22, p. 192–221.

Blodget, H. W., Gunther, F. J., and Podwysocki, M. H., 1978, Discrimination of rock classes and alteration products in southwestern Saudi Arabia with computer-enhanced Landsat data, NASA Technical Paper 1327: Washington, National Aeronautics and Space Administration, 34 p.

Cadogan, P., 1981, The Moon—our sister planet: Cambridge, Cambridge University Press, 391 p.

Carr, M. H., 1981, The surface of Mars: New Haven, Yale University Press, 232 p.

Colwell, R. N., 1983, Manual of remote sensing: Falls Church, Va., American Society of Photogrammetry, 2440 p.

Duck, K. I., and King, J. C., 1983, Orbital mechanics for remote sensing, *in* Colwell, R. S., ed. Manual of remote sensing: Falls Church, Va., American Society of Photogrammetry, 2440 p.

Elachi, C., Brown, W. E., Cimino, J. B., Dixon, T., Evans, D. L., Ford, J. P.,

Saunders, R. S., Breed, C., Masursky, H., McCauley, J. F., Schaber, G., Dellwig, L., England A. W., MacDonald, H., Martin-Kaye, P., and Sabins, F., 1982, Shuttle imaging radar experiment, Science, v. 218, p. 996–1003.

El-Baz, F., 1979, Catalog of earth photographs from the Apollo-Soyuz Test Project, NASA Technical Memorandum 58218: Washington, U.S. Government Printing Office, 292 p.

Fimmel, R. O., Colin, L., and Burgess, E., 1983, Pioneer Venus, NASA SP-461: Washington, U.S. Government Printing Office, 253 p.

Ford, J. P., Blom, R. G., Bryan, M. L., Daily, M. I., Dixon, T. H., Elachi, C., and Xenos, E. C., 1980, Seasat views North America, the Caribbean, and western Europe with imaging radar, JPL Publication 80-67: Pasadena, Jet Propulsion Laboratory, 160 p.

——, Cimino, J. B., and Elachi, C., 1983, Space shuttle Columbia views the world with imaging radar: the SIR-A experiment, JPL Publication 82-95: Pasadena, Jet Propulsion Laboratory, 179 p.

Gamow, G., 1948, Biography of the Earth: New York, New American Library, 194 p.

Garriott, O. K., 1974, Skylab report: man's role in space research: Science, v. 186, p. 219–226.

Glass, B. P., 1982, Introduction to planetary geology: Cambridge, Cambridge University Press, 469 p.

Goetz, A.F.H., and Rowan, L. C., 1981, Geologic remote sensing: Science, v. 211, p. 781–791.

——, Rock, B. N., and Rowan, L. C., 1983, Remote sensing for exploration: an overview: Economic Geology, v. 78, p. 573–590.

——, Rowan, L. C., and Kingston, M. J., 1982, Mineral identification from orbit: initial results from the shuttle multispectral infrared radiometer: Science, v. 218, p. 1020–1024.

Gold, D. P., 1980, Structural geology, *in* Siegal B. S., and Gillespie, A. R., eds., Remote sensing in geology: New York, John Wiley, p. 419–484.

Gorenstein, P., and Bjorkholm, P., 1972, Alpha-particle experiment, *in* Apollo 15 preliminary science report, NASA SP-289: Washington, U.S. Government Printing Office, p. 18–1—18–7.

Greenacre, J. C., 1965, The 1963 Aristarchus events: Annals of the New York Academy of Sciences, v. 123, p. 811–816.

Halbouty, M., 1976, Applications of Landsat imagery to petroleum and mineral exploration: American Association of Petroleum Geologists Bulletin, v. 60, p. 745–793.

Hanel, R., Conrath, B., Hovis, W., Kunde, V., Lowman, P., Maguire, W., Pearl, J., Pirraglia, J., Prabhakara, C., Schlachman, B., Levin, G., Straat, P., and Burke, T., 1972, Investigation of the Martian environment by infrared spectroscopy on Mariner 9: Icarus, v. 17, p. 423–442.

——, Conrath, B., Flasar, M., Kunde, V., Lowman, P., Maguire, W., Pearl, J., Pirraglia, J., Samuelson, R., Gautier, D., Giersch, P., Kumar, S., and Ponnamperuma, C., 1979, Infrared observations of the Jovian system from Voyager 1: Science, v. 204, p. 972–976.

Hartmann, W. K., and Raper, O., 1974, The new Mars, NASA SP-337: Washington, U.S. Government Printing Office, 179 p.

Head, J. W., and Solomon, S. C., 1981, Tectonic evolution of the terrestrial planets: Science, v. 213, p. 62–76.

Henderson, F. B., III, and Swann, G. A., 1976, Geological remote sensing from space: Berkeley, Lawrence Berkeley Laboratory, University of California.

Hobbs, W. H., 1911, Repeating patterns in the relief and in the structure of the land: Geological Society of America Bulletin, v. 22, p. 123–176.

Hodgson, R. A., 1977, Regional linear analysis as a guide to mineral resource exploration using Landsat (ERTS) data, *in* Geological Survey Professional Paper 1015: Washington, U.S. Government Printing Office, 370 p.

Holliday, C. T., 1950, Seeing the earth from 80 miles up: The National Geographic Magazine, v. XCVIII, p. 511–528.

Holz, R. K., 1973, The surveillant science: Boston, Houghton Mifflin, 390 p.

Hunt, G. R., 1980, Electromagnetic radiation, *in* Siegel, B. S., and Gillespie, A. R., eds: Remote sensing in geology: New York, John Wiley, p. 5–46.

Kahle, A. B., 1980, Surface thermal properties, *in* Siegal, B. S., and Gillespie, A. R., eds.: Remote sensing in geology: New York, John Wiley, p. 257–274.

Kaltenbach, J. L., 1970, Apollo 9 multispectral photographic information, NASA TM X-1957: Washington, National Aeronautics and Space Administration, 34 p.

Kieffer, H. H., Martin, T. Z., Peterfreund, A. R., Jakosky, B. M., Miner, E. D., and Palluconi, F. D., 1977, Thermal and albedo mapping of Mars during the Viking primary mission: Journal of Geophysical Research, v. 82, p. 4249–4291.

King, E. A., 1976, Space geology: New York, John Wiley, 349 p.

Kutina, J., and Carter, W. D., 1977, Landsat contributions to studies of plate tectonics, *in* Geological Survey Professional Paper 1015: Washington, U.S. Government Printing Office, 370 p.

Labovitz, M. L., Masuoka, E. J., Bell, R., Siegrist, A. W., and Nelson, R. F., 1983, The application of remote sensing to geobotanical exploration for metal sulfides—results from the 1980 field season at Mineral, Virginia: Economic Geology, v. 76, p. 750–760.

Lowman, P. D., Jr., 1964, A review of photography of the Earth from sounding rockets and satellites, NASA TN D-1868: Washington, National Aeronautics and Space Administration, 25 p.

——1967, Geologic applications of orbital photography, NASA TN D-4155: Washington National Aeronautics and Space Administration, 37 p.

——1968, Space panorama: Zurich, Weltflugbild Reinhold A. Muller, 141 p.

——1969a, Apollo 9 multispectral photography: geologic analysis, X-644-69-423: Greenbelt, Md., Goddard Space Flight Center, 53 p.

——1969b, Geologic orbital photography: experience from the Gemini Program: Photogrammetria, v. 24, p. 77–106.

——1972, The third planet: Zurich, Weltflugbild Reinhold A. Muller, 170 p.

——1976, Crustal evolution in silicate planets: implications for the origin of continents: Journal of Geology, v. 84, p. 1–26.

——1980a, The evolution of geological space photography, *in* Siegal, B. S., and Gillespie, A. R., eds.: Remote sensing in geology: New York, John Wiley, p. 91–116.

——1980b, Vertical displacement on the Elsinore fault of southern California: evidence from orbital photographs: Journal of Geology, v. 88, p. 415–432.

——1981, A global tectonic activity map: Bulletin of the International Association of Engineering Geology, v. 23, p. 37–49.

——1982, A more realistic view of global tectonism: Journal of Geologic Education, v. 30, p. 97–107.

——, Blodget, H. W., Webster, W. J., Jr., Pala, S., Singhroy, V. H., and Slaney, V. R., 1984, Structural investigations of the Canadian Shield by orbital radar and Landsat, *in* The SIR-B science investigations plan, JPL Publication 84-3: Pasadena, Jet Propulsion Laboratory, p. 4–99—4–101.

Lyon, R.J.P., 1975, Correlation between ground metal analysis, vegetation reflectance, and ERTS brightness over a molybdenum skarn deposit, Pine Nut Mountains, western Nevada: *in* Proceedings of the Tenth International Symposium on Remote Sensing of Environment, Environmental Research Institute of Michigan, Ann Arbor, Michigan, v. 2, p. 1031–1044.

——1977, Mineral exploration applications of digitally-processed Landsat imagery, *in* Geological Survey Professional Paper 1015: Washington, U.S. Government Printing Office, 370 p.

MacDonald, H. C., 1969, Geologic evaluation of radar imagery from Darien Province, Panama: Modern Geology, v. 1, p. 1–63.

——1980, Techniques and applications of imaging radars, *in* Siegel, B. S., and Gillespie, A. R., eds.: Remote sensing in geology: New York, John Wiley, p. 297–336.

Masursky, H., Eliason, E., Ford, P. G., McGill, G. E., Pettengill, G. H., Schaber, G. G., and Schubert, G., 1980, Pioneer Venus radar results: geology from images and altimetry: Journal of Geophysical Research, v. 85, p. 8232–8260.

McCauley, J. F., Schaber, G. G., Breed, C. S., Grolier, M. J., Haynes, C. V., Issawi, Elachi, C., and Blom, R., 1982, Subsurface valleys and geoarcheology of the eastern Sahara revealed by shuttle radar: Science, v. 218, p. 1004–1020.

Mohr, P. A., 1974, Mapping of the major structures of the African rift system, final report, Contract NAS 5-21748: Cambridge, Mass., Smithsonian Institution, 85 p.

Molnar, P., and Tapponnier, P., 1975, Cenozoic tectonics of Asia: effects of a continental collision: Science, v. 189, p. 419–426.

Morrison, D., 1982, Voyages to Saturn, NASA SP-451: Washington, U.S. Government Printing Office, 197 p.

Murray, B., Malin, M. C., and Greeley, R., 1980, Earthlike planets: San Francisco, W. H. Freeman, 387 p.

NASA, 1974, Skylab earth resources data catalog, JSC 09016: Houston, Lyndon B. Johnson Space Center, 359 p.

Ni, J., and York, J. E., 1978, Late Cenozoic tectonics of the Tibetan plateau: Journal of Geophysical Research, v. 83, p. 5377–5384.

Nur, A., 1982, The origin of tensile fracture lineaments: Journal of Structural Geology, v. 4, p. 31–40.

Pettengill, G. H., Horwood, D. F., and Keller, C. H., 1980, Pioneer Venus Orbiter radar mapper: design and operation: Geoscience and Remote Sensing, v. GE-18, p. 28–31.

Phillips, R. J., Adams, G. F., Brown, W. E., Jr., Eggleton, R. E., Jackson, P., Jordan R., Linlor, W. I., Peeples, W. J., Porcello, L. J., Ryu, J., Schaber, G., Sill, W. R., Thompson, T. W., Ward, S. H., and Zelenka, J. S., 1973, Apollo lunar sounder experiment, *in* Apollo 17 preliminary science report, NASA SP-330: Washington, U.S. Government Printing Office, p. 22–1—22–26.

——, Kaula, W. M., McGill, G. E., and Malin, M. C., 1981, Tectonics and evolution of Venus: Science, v. 212, p. 879–887.

Rowan, L. C., Offield, T. W., Watson, K., Cannon, P. J., and Watson, R. D., 1970, Thermal infrared investigations, Arbuckle Mountains, Oklahoma: Geological Society of America Bulletin, v. 81, p. 3549–3562.

——, and Wetlaufer, P. H., 1976, Discrimination of rock types and detection of hydrothermally-altered areas in south-central Nevada, *in* Geological Survey Professional Paper 929: Washington, U.S. Government Printing Office, p. 102–105.

Ryder, N. G., 1981, Ryder's standard geographic reference: Denver, Ryder Geosystems, 215 p.

Sabatini, R. R., Rabchevsky, G. A., and Sissala, J. E., 1971, Nimbus earth resources observations, Technical report No. 2, Contract NAS 5-21617: Concord, Mass., Allied Research Associates, Inc., 256 p.

Sabins, F. F., Jr., 1978, Remote sensing: San Francisco, W. H. Freeman, 426 p.

——1983, Geologic interpretation of space shuttle radar images of Indonesia: American Association of Petroleum Geologists Bulletin, v. 67, p. 2076–2099.

Salas, G. P., 1977, Relationship of mineral resources to linear features in Mexico as determined from Landsat data, *in* Geological Survey Professional Paper 1015: Washington, U.S. Government Printing Office, 350 p.

Salomonson, V. V., and Koffler, R., 1983, An overview of Landsat 4 status and results: Paper presented at the Seventeenth International Symposium on Remote Sensing of Environment, Ann Arbor, Michigan, May 9-13, 1983.

Saunders, R. S., and Mutch, T. A., 1980, Extraterrestrial geology, *in* Siegal, B. S., and Gillespie, A. R., eds: Remote sensing in geology: New York, John Wiley, p. 659–678.

Settle, J., and Taranik, J. V., 1982, Use of the space shuttle for remote sensing research: recent results and future prospects: Science, v. 218, p. 993–995.

Sheffield, C., 1983, Man on earth: New York: Macmillan, 160 p.

Short, N. M., 1975, Planetary geology, Englewood Cliffs, N.J., Prentice-Hall, 361 p.

—— 1982, The Landsat tutorial workbook, NASA Reference Publication 1078: Washington, U.S. Government Printing Office, 553 p.

——, and Lowman, P. D., Jr., 1973, Earth observations from space: outlook for the geological sciences, X-650-73-316: Greenbelt, Md., Goddard Space Flight Center, 115 p.

——, and Stuart, L. M., Jr., 1982, The heat capacity mapping mission (HCMM) anthology, NASA SP-465: Washington, U.S. Government Printing Office, 264 p.

——, Lowman, P. D., Jr., Freden, S. C., and Finch, W. A., Jr., 1976, Mission to earth: Landsat views the world, NASA SP-360: Washington, U.S. Government Printing Office, 459 p.

Siegel, B. S., and Gillespie, A. R., 1980, Remote sensing in geology: New York, John Wiley, 702 p.

Sivertson, W. E., Jr., Wilson, R. G., Bullock, G. F., and Schappell, R. T., 1982, Feature identification and location experiment: Science, v. 218, p. 1031–1033.

Slaney, V. R., 1981, Landsat images of Canada—a geological appraisal, Geological Survey of Canada Paper 80-15: Ottawa, Geological Survey of Canada, 102 p.

Tanner, S., 1982, Handbook of sensor technical characteristics, NASA Reference Publication 1087: Washington, U.S. Government Printing Office, 326 p.

Taylor, S. R., 1982, Planetary science: a lunar perspective: Houston, Lunar and Planetary Institute, 481 p.

Waldrop, M. M., 1982, Imaging the earth (I): the troubled first decade of Landsat: Science, v. 215, p. 1600–1603.

Williams, R. S., Jr., and Carter, W. D., 1976, ERTS-1, A new window on our planet, Geological Survey Professional Paper 929: Washington, U.S. Government Printing Office, 362 p.

Williams, R. S., Jr., and 51 others, 1983, Geological applications, *in* Colwell, R. S., ed.: Manual of remote sensing: Falls Church, Va., American Society of Photogrammetry, 2440 p.

Wilmarth, V. R., Kaltenbach, J. L., and Lenoir, W. B., 1977, Skylab explores the Earth, NASA Sp-380: Washington, U.S. Government Printing Office, 517 p.

Woll, P. W., and Fischer, W. A., 1977, Proceedings of the First Annual William T. Pecora Memorial Symposium, Geological Survey Professional paper 1015: Washington, U.S. Government Printing Office, 370 p.

Index

Typeset by WESType Publishing Services, Inc., Boulder, Colorado
Printed in U.S.A. by Malloy Lithographing, Inc., Ann Arbor, Michigan